RISC-V System-on-Chip Design

David Harris
James Stine
Rose Thompson
Sarah Harris

ELSEVIER

MK
MORGAN KAUFMANN PUBLISHERS
AN IMPRINT OF ELSEVIER

Morgan Kaufmann is an imprint of Elsevier
50 Hampshire Street, 5th Floor, Cambridge, MA 02139, United States

Front cover art is by Sarah Harris.

Back cover image is a die photograph of the FE310, released in 2016 as the world's first commercially-available RISC-V chip. Image © 2016 SiFive, reproduced with permission.

ISBN: 978-0-323-99498-9

For information on all Morgan Kaufmann publications visit our website at https://www.elsevier.com/books-and-journals

Publisher: Peter Linsley
Acquisitions Editor: Stephen Merken
Editorial Project Manager: Vicky Sharma
Production Project Manager: Prasanna Kalyanaraman
Cover Designer: Greg Harris

Typeset by TNQ Tech

Contents

Preface xviii

How to Use This Book xxi

Acknowledgements xxiii

About the Authors xxv

Foreword xxvii

Chapter 1 A Brief History of Computer Design 1

1.1 Moore's Law and Beyond 2
1.2 System-on-Chip 6
1.3 Birth of Computing 9
1.4 Mainframes and Minicomputers 13
1.5 Microprocessors 16
1.6 CISC and RISC 20
1.7 RISC-V 26
1.8 International Economic and Security Competition 26
1.9 Summary 29

Chapter 2 Introduction to RISC-V 31

2.1 RISC-V Assembly Language 32
2.1.1 *R-Type Instructions* 32
2.1.2 *I-Type Instructions* 34
2.1.3 *S-Type Instructions* 34
2.1.4 *B-Type Instructions* 35
2.1.5 *J-Type Instructions* 35
2.1.6 *U-Type Instructions* 35
2.1.7 *Other Instructions* 36
2.1.8 *RV64I Additional Instructions* 36
2.1.9 *Register Conventions* 37
2.1.10 *Endianness* 37
2.1.11 *Pseudoinstructions* 37
2.1.12 *C to Assembly Language Examples* 38
2.2 RISC-V Machine Language 43

2.3 Have a Hart ... 45
2.3.1 *Privilege Modes* ... 46
2.3.2 *Control and Status Registers* ... 46
2.3.3 *Processes and Threads* ... 46
2.3.4 *Cores and Harts* ... 47
2.3.5 *Exceptions, Interrupts, and Traps* ... 49
2.3.6 *Execution Environment* ... 49
2.4 Memory Map ... 51
2.5 RISC-V Extensions and Profiles ... 52
2.5.1 *RISC-V Profiles* ... 53
2.6 Comparison with Other Architectures ... 54
2.7 RISC-V Microarchitecture ... 56
2.7.1 *Single-Cycle Microarchitecture* ... 56
2.7.2 *Pipelined Microarchitecture* ... 65
2.7.3 *Verification* ... 67
2.8 Wally System-on-Chip ... 70
2.8.1 *Wally Block Diagram* ... 73
2.8.2 *Wally Configurations* ... 73
2.8.3 *Wally Code Base* ... 74
2.8.4 *Verification* ... 76
2.8.5 *Benchmarking* ... 76
2.8.6 *Implementation* ... 77
2.9 RISC-V Market ... 80
2.10 Summary and a Look Ahead ... 85

Chapter 3 RISC-V Software Tool Flow ... 89

3.1 Platform Requirements and Tool Overview ... 90
3.2 Getting Started ... 90
3.3 GCC Assembler and Spike Simulator ... 91
3.3.1 *A Quick Tutorial* ... 91
3.3.2 *Linking and ELF* ... 93
3.3.3 *Application Binary Interface* ... 95
3.3.4 *Supervisor Binary Interface* ... 99
3.3.5 *Machine Architecture* ... 99
3.3.6 *Driving GCC* ... 101
3.3.7 *Assembler Directives* ... 103
3.3.8 *Spike Simulation* ... 105
3.3.9 *Performance Measurement* ... 107
3.3.10 *Putting It All Together* ... 107
3.4 GCC GNU C Compiler ... 114
3.4.1 *A First C Program* ... 115
3.4.2 *Inline Assembly Language* ... 120
3.4.3 *Mixed C and Assembly Language* ... 121
3.4.4 *Standard Library and Startup Code* ... 122

3.4.5 *Compiler Explorer* 123
3.5 Sail Simulation 124
3.6 QEMU Simulation 127
3.6.1 *Simulating with QEMU* 127
3.6.2 *Debugging With GDB* 128
3.7 Test Suites 133
3.7.1 *riscv-arch-test* 133
3.7.2 *RISCOF* 137
3.7.3 *wally-riscv-arch-test* 138
3.8 Summary 140

Chapter 4 HDL Design Practices 141

4.1 HDL Design Methodology 142
4.1.1 *Design Strategies* 142
4.1.2 *Version Control* 144
4.1.3 *Reuse* 144
4.2 Hardware Design with SystemVerilog 145
4.2.1 *Modules* 145
4.2.2 *Combinational Logic* 148
4.2.3 *Sequential Logic* 153
4.2.4 *Verilog Syntax* 168
4.2.5 *Configurable Hardware* 173
4.3 Wally Style Guidelines 181
4.3.1 *Signals* 181
4.3.2 *Spacing and Indentation* 185
4.3.3 *Synchronous Design* 187
4.3.4 *Modules* 188
4.3.5 *Configurable Hardware* 189
4.3.6 *Comments* 189
4.4 Wally Configuration 190
4.5 Summary 191

Chapter 5 Design Verification 193

5.1 Verification Strategies 195
5.1.1 *Test Vector Generation* 196
5.1.2 *Reference Models* 198
5.1.3 *Unit vs. System Testing* 198
5.1.4 *Emulation* 198
5.1.5 *Technology Readiness Levels* 199
5.1.6 *Design and Verification Approaches* 200
5.2 Design Verification Plan 202
5.3 Lint 203
5.4 Testbenches 204

5.5 Logic Simulators 207
5.5.1 *Questa* 207
5.5.2 *Verilator* 210
5.5.3 *VCS* 212
5.6 Lock-Step-Compare 212
5.6.1 *RVVI-TRACE Interface* 214
5.7 Assertions 214
5.7.1 *Immediate Assertions* 215
5.7.2 *Concurrent Assertions* 216
5.8 Code Coverage 217
5.9 Functional Coverage 221
5.9.1 *riscvISACOV* 222
5.9.2 *Directed Tests* 222
5.9.3 *Constrained Random Tests* 226
5.10 Formal Verification 226
5.11 Wally Verification 227
5.11.1 *Design Verification Plan* 227
5.11.2 *Reference Model* 229
5.11.3 *Top-Level Testbench and Testvectors* 229
5.11.4 *Lint* 230
5.11.5 *Simulation* 231
5.11.6 *Regression* 233
5.11.7 *Debugging Hints* 235
5.11.8 *Floating-Point Unit Testing* 239
5.11.9 *Lock-Step-Compare* 240
5.11.10 *Assertions* 241
5.11.11 *Code Coverage* 241
5.11.12 *Functional Coverage* 245
5.12 Summary 247

Chapter 6 Logic Synthesis 249

6.1 Performance, Power, and Area (PPA) 250
6.1.1 *Performance* 250
6.1.2 *Power* 255
6.1.3 *Area* 257
6.2 Fabrication Technologies 257
6.2.1 *Inverter Characteristics* 259
6.2.2 *Datapath Units* 260
6.2.3 *Memories* 266
6.3 Running Synthesis 268
6.4 Reviewing Log Files 269
6.4.1 *synth.out* 269
6.4.2 *Mapped Output Files* 270
6.4.3 *Quality of Results (QoR) Reports* 275

6.4.4 *Performance Reports* 277
6.4.5 *Power Reports* 282
6.4.6 *Area Reports* 284
6.4.7 *Other Reports* 284
6.5 Wally Synthesis Results 287
6.5.1 *Synthesis Configurations* 287
6.5.2 *Area and Delay vs. Target Clock Period* 287
6.5.3 *Area and Delay vs. Configurations* 291
6.5.4 *Area and Delay vs. Features in rv64gc* 295
6.6 Summary 297

Chapter 7 Pipelined Core 299

7.1 Single-Cycle Core 301
7.1.1 *Architectural State* 302
7.1.2 *Immediate Extension* 303
7.1.3 *Arithmetic Circuitry* 303
7.1.4 *Controller Design* 311
7.1.5 *Instruction Fetch Unit* 314
7.1.6 *Load/Store Unit* 314
7.1.7 *Sample Program* 316
7.2 Pipelining the RV32I Core 317
7.2.1 *Synchronous Memories* 319
7.2.2 *Hazards* 321
7.3 SystemVerilog 328
7.4 Test Plan 329
7.5 Summary 333

Chapter 8 Privileged Operations 335

8.1 Principles 336
8.1.1 *Security and Protection* 336
8.1.2 *Traps* 337
8.1.3 *Control and Status Registers (CSRs)* 338
8.2 RISC-V Practices 339
8.2.1 *Privilege Modes* 340
8.2.2 *Privileged Instructions* 340
8.2.3 *Control and Status Registers (CSRs)* 342
8.2.4 *Reset Behavior* 349
8.2.5 *Traps* 350
8.2.6 *Trap Handler* 357
8.3 Test Plan 360
8.3.1 *Generating Exceptions* 362
8.3.2 *Generating Interrupts* 363

8.4 Wally Implementation 364
8.4.1 *Wally Privileged Unit* 364
8.4.2 *Privileged Instructions* 370
8.4.3 *Control and Status Registers* 370
8.4.4 *Traps* 371
8.4.5 *Pipeline Hazards* 373
8.4.6 *Instruction Fetch Unit (IFU) and Privileged Changes* 377
8.4.7 *Endian Swapper* 378
8.5 Summary 380

Chapter 9 Bus Interface 383

9.1 Principles 383
9.1.1 *Bidirectional vs. Unidirectional Buses* 385
9.1.2 *Handshaking* 386
9.1.3 *Multiple Peripherals* 386
9.1.4 *Multiple Controllers* 387
9.1.5 *Other Bus Features* 389
9.1.6 *High-Speed Off-Chip Links* 390
9.2 AMBA Buses 391
9.2.1 *APB* 392
9.2.2 *AHB* 395
9.2.3 *AXI* 404
9.3 Other SoC Buses 410
9.4 Test Plan 410
9.5 Wally Implementation 411
9.5.1 *LSU and IFU AHB Interface* 412
9.5.2 *External Bus Unit* 415
9.5.3 *System-on-Chip* 417
9.5.4 *Combined Bus and Tightly Integrated Memories* . 420
9.5.5 *Bus Hazards* 423
9.6 Summary 427

Chapter 10 Caches 429

10.1 Principles 430
10.1.1 *Cache Performance* 430
10.1.2 *Locality* 432
10.1.3 *Direct-Mapped Cache* 432
10.1.4 *Cache Writes* 434
10.1.5 *Associative Caches* 436
10.1.6 *Cache Design Tradeoffs* 438
10.1.7 *Replacement Policy* 439
10.1.8 *Cache Flush and Invalidation* 441

10.2 RISC-V Practices 442
10.3 Test Plan 442
10.4 Wally Implementation 443
10.4.1 *Interfaces* 445
10.4.2 *Cache Way Module* 446
10.4.3 *Cache FSM* 454
10.4.4 *pLRU Implementation* 456
10.4.5 *Top-Level Cache Logic* 459
10.4.6 *Pipeline Stalls* 466
10.4.7 *Exceptions and Interrupts* 466
10.5 Summary 467

Chapter 11 Memory Management Unit 469

11.1 Principles 469
11.1.1 *Virtual Memory* 469
11.1.2 *Physical Memory Checking* 477
11.2 RISC-V Practices 478
11.2.1 *RISC-V Virtual Memory* 478
11.2.2 *RISC-V Memory Protection* 489
11.2.3 *Memory Exceptions* 494
11.2.4 *Fences* 495
11.2.5 *HPTW* 495
11.3 Test Plan 495
11.4 Wally Implementation 495
11.4.1 *MMU* 497
11.4.2 *HPTW* 507
11.4.3 *MMU Faults* 511
11.4.4 *Fence Instruction* 511
11.5 Summary 512

Chapter 12 Load/Store Unit 515

12.1 Wally Load/Store Unit Overview 515
12.2 Data Cache Integration 518
12.2.1 *Cache Line Fetches* 520
12.2.2 *Cache Line Writeback* 523
12.2.3 *Burst EBU* 525
12.3 Memory Management and HPTW Integration 526
12.3.1 *PMA Checker* 527
12.3.2 *PMP Checker* 529
12.3.3 *TLB and Cache Integration* 529
12.3.4 *HPTW Integration* 531

12.4 Stalls, Flushes, and Traps 532
12.4.1 *LSU Stall Generation* 532
12.4.2 *LSU Trap Generation* 533
12.4.3 *Disabling Interrupts* 533
12.4.4 *LSU Response to Stalls and Flushes* 533
12.5 Examples 534
12.5.1 *Cache Load Hit* 534
12.5.2 *Cache Store Hit* 534
12.5.3 *Cache Miss* 536
12.5.4 *Uncached Access* 537
12.5.5 *TLB Miss* 540
12.5.6 *TLB Access Fault* 543
12.6 Test Plan 544
12.7 Summary 544

Chapter 13 Instruction Fetch Unit 545

13.1 Branch Prediction Principles 545
13.1.1 *Static Branch Prediction* 548
13.1.2 *Simple Dynamic Branch Prediction* 548
13.1.3 *Gshare* 551
13.1.4 *Branch Target Buffer* 554
13.1.5 *Return Address Stack* 556
13.1.6 *TAGE* 557
13.2 Test Plan 558
13.2.1 *Correctness* 558
13.2.2 *Performance* 558
13.3 Wally Implementation 559
13.3.1 *Instruction Cache* 560
13.3.2 *PC Logic* 560
13.3.3 *Branch Predictor* 562
13.3.4 *ITLB Misses* 574
13.3.5 *Reset, Stalls, Flushes, and Traps* 574
13.4 Summary 576

Chapter 14 Extensions: C (Compressed) 579

14.1 RISC-V Practices 579
14.1.1 *Zc* Extension* 580
14.2 Test Plan 582
14.3 Wally Implementation 582
14.3.1 *Instruction Decompression* 582
14.3.2 *PC Logic* 585
14.3.3 *Spills* 586
14.4 Summary 590

Chapter 15 Extensions: M (Multiply and Divide) 591

15.1 Principles 591
15.1.1 *Signed and Unsigned Arithmetic* 591
15.1.2 *Carry-Save Addition* 592
15.1.3 *Multiplication* 593
15.1.4 *Division* 602
15.2 RISC-V Practices 604
15.2.1 *Signed and Unsigned Operations* 604
15.2.2 *Division Special Cases* 604
15.2.3 *Software Arithmetic Implementations* 605
15.2.4 *Zmmul* 606
15.3 Test Plan 606
15.4 Wally Implementation 607
15.4.1 *Multiplication* 607
15.4.2 *Division and Remainder* 610
15.4.3 *Multiply and Divide Unit* 613
15.4.4 *Hazard Unit* 614
15.5 Summary 615

Chapter 16 Extensions: F/D/Q/Zfh/Zfa (Floating-Point) 617

16.1 Principles 617
16.1.1 *Normalized Floating-Point Formats* 618
16.1.2 *Special Cases* 620
16.1.3 *Rounding* 622
16.1.4 *Flags* 626
16.1.5 *Fixed-Point Notation* 628
16.2 Floating-Point Unit Design 630
16.2.1 *Unpack Unit* 631
16.2.2 *Fused Multiply-Add* 633
16.2.3 *Division and Square Root* 637
16.2.4 *Conversion* 639
16.2.5 *Comparison* 640
16.2.6 *Postprocessing Unit* 641
16.3 SoftFloat 641
16.3.1 *SoftFloat Demo* 643
16.3.2 *fpcalc* 643
16.4 RISC-V Practices 646
16.4.1 *Floating-Point Control and Status Register* 646
16.4.2 *Registers and NaN Boxing* 648
16.4.3 *NaNs in RISC-V* 648
16.4.4 *Floating-Point Move Instructions* 649
16.4.5 *Zfa Additional Floating-Point Instructions* 649
16.5 Test Plan 651
16.5.1 *riscv-arch-test* 651
16.5.2 *TestFloat* 652

16.6 Wally Implementation 655
16.6.1 *Floating-Point Unit* 656
16.6.2 *Unpacking Unit* 659
16.6.3 *Fused Multiply-Add* 660
16.6.4 *Divide and Square Root* 660
16.6.5 *Convert* 661
16.6.6 *Move* 662
16.6.7 *Compare* 662
16.6.8 *Sign Injection* 662
16.6.9 *Classify* 662
16.6.10 *Zfa Additional Floating-Point Instructions* 663
16.6.11 *Postprocessing* 666
16.6.12 *Privileged Unit Changes* 666
16.6.13 *Integer Execution Unit Changes* 666
16.6.14 *Hazard Unit Changes* 667
16.6.15 *LSU Changes* 669
16.7 Summary 669

Chapter 17 Extensions: A (Atomic) 673

17.1 Principles 673
17.1.1 *Concurrency* 674
17.1.2 *Coherence* 676
17.1.3 *Consistency* 677
17.1.4 *Synchronization* 682
17.1.5 *Deadlock, Livelock, and Starvation* 696
17.2 RISC-V Practices 698
17.2.1 *Consistency Models* 698
17.2.2 *Fences* 699
17.2.3 *Atomic Instructions* 700
17.2.4 *Atomic Subsets* 705
17.2.5 *Profile Extensions* 705
17.3 Test Plan 706
17.4 Wally Implementation 706
17.4.1 *Atomic Module* 708
17.4.2 *Timing Diagrams* 711
17.5 Summary 712

Chapter 18 Extensions: Zb* & Zk* (Bit Manipulation and Cryptography) 715

18.1 Cryptography Principles 716
18.1.1 *Finite Fields* 716
18.1.2 *AES: Advanced Encryption Standard* 722
18.1.3 *Hash Functions* 731

18.2 RISC-V Practices 734
18.2.1 *Zba: Address Generation* 734
18.2.2 *Zbb: Basic Bit Manipulation* 736
18.2.3 *Zbc: Carry-less Multiplication* 738
18.2.4 *Zbs: Single-Bit Operations* 739
18.2.5 *Zbkb: Bit Manipulation for Cryptography* 739
18.2.6 *Zbkx: Crossbar Permutations* 741
18.2.7 *Zknd and Zkne: AES Decryption and Encryption* 742
18.2.8 *Zknh: SHA-2 Instructions* 747
18.3 Test Plan 748
18.4 Wally Implementation 748
18.4.1 *ALU Modifications for Bit Manipulation* 748
18.4.2 *ALU Modifications for Cryptography* 757
18.4.3 *Controller Modifications* 764
18.5 Summary 765

Chapter 19 Other Extensions 767

19.1 Cache Management: Zicbom, Zicboz, Zicbop 767
19.1.1 *RISC-V CMO Instructions* 767
19.1.2 *Test Plan* 769
19.1.3 *Wally Implementation* 769
19.2 Misaligned Loads and Stores: Zicclsm 775
19.2.1 *Test Plan* 776
19.2.2 *Wally Implementation* 776
19.3 TLB Management: Svinval 777
19.4 Pause Hint: Zihintpause 779
19.5 Conditional Instructions: Zicond 779
19.5.1 *Test Plan* 781
19.5.2 *Wally Implementation* 782
19.6 Custom Extensions 783
19.7 Summary 784

Chapter 20 Peripherals 784.e1

20.1 GPIO 784.e2
20.1.1 *Memory Map* 784.e3
20.1.2 *Loopback Testing* 784.e5
20.1.3 *Test Plan* 784.e5
20.1.4 *Wally Implementation* 784.e6
20.2 CLINT 784.e8
20.2.1 *Memory Map* 784.e8
20.2.2 *Test Plan* 784.e9
20.2.3 *Wally Implementation* 784.e10

20.3 PLIC . . . 784.e10
20.3.1 *Memory Map* . . . 784.e11
20.3.2 *Test Plan* . . . 784.e12
20.3.3 *Wally Implementation* . . . 784.e13
20.4 UART . . . 784.e13
20.4.1 *Memory Map* . . . 784.e16
20.4.2 *Test Plan* . . . 784.e18
20.4.3 *Wally Implementation* . . . 784.e19
20.5 SPI . . . 784.e19
20.5.1 *Memory Map* . . . 784.e20
20.5.2 *Loopback Testing* . . . 784.e22
20.5.3 *Test Plan* . . . 784.e23
20.5.4 *Wally Implementation* . . . 784.e23
20.6 Summary . . . 784.e23

Chapter 21 Benchmarking . . . 784.e24

21.1 CoreMark . . . 784.e24
21.1.1 *Porting CoreMark* . . . 784.e26
21.1.2 *CoreMark Compiler Flags* . . . 784.e27
21.1.3 *Instruction Set Impact* . . . 784.e27
21.2 Embench . . . 784.e27
21.2.1 *Porting Embench* . . . 784.e34
21.2.2 *Embench Flags* . . . 784.e35
21.2.3 *Instruction Set Impact* . . . 784.e36
21.2.4 *Troubleshooting* . . . 784.e39
21.3 Performance Verification . . . 784.e39
21.3.1 *Cache Performance Verification* . . . 784.e40
21.3.2 *Branch Predictor Performance Verification* . . . 784.e41
21.4 Summary . . . 784.e47

Chapter 22 Linux . . . 784.e48

22.1 Principles . . . 784.e48
22.1.1 *Embedded Linux Hardware Requirements* . . . 784.e50
22.1.2 *Bootloader* . . . 784.e50
22.1.3 *Device Tree* . . . 784.e50
22.1.4 *Kernel* . . . 784.e51
22.1.5 *Root Filesystem* . . . 784.e51
22.1.6 *Init* . . . 784.e52
22.1.7 *Shell* . . . 784.e52
22.2 Buildroot Linux on Wally . . . 784.e52
22.2.1 *Building Buildroot Linux* . . . 784.e53
22.2.2 *Source, Binary, and Disassembled Files* . . . 784.e53
22.2.3 *Linux Memory Map* . . . 784.e56

22.2.4 *Linux Boot Overview* 784.e57
22.2.5 *Zero Stage Bootloader* 784.e57
22.2.6 *OpenSBI Firmware* 784.e60
22.2.7 *Kernel* 784.e63
22.2.8 *Root File System (`rootfs`)* 784.e65
22.2.9 *Device Tree* 784.e66
22.2.10 *Init Process* 784.e66
22.2.11 *Trap Handlers* 784.e68
22.3 Booting Linux 784.e70
22.3.1 *Booting Linux on QEMU* 784.e70
22.3.2 *Booting Linux on Wally using a Logic Simulator* 784.e73
22.3.3 *Modifying Buildroot* 784.e75
22.4 Summary 784.e76

Chapter 23 FPGA Implementation 784.e78

23.1 FPGA Principles 784.e79
23.1.1 *FPGAs* 784.e79
23.2 Wally FPGA Implementation 784.e80
23.2.1 *SD Card* 784.e83
23.2.2 *Arty A7 Setup* 784.e84
23.2.3 *Supporting Other Boards* 784.e96
23.2.4 *Wally Zero Stage Bootloader* 784.e99
23.3 FPGA Results 784.e99
23.3 Interesting Bugs 784.e100
23.3.1 *Bug 1: AHB Bursts* 784.e101
23.3.2 *Bug 2: Concurrent Exceptions* 784.e102
23.3.3 *Critical Path Analysis* 784.e102
23.4 Summary 784.e105

Appendix A Wally Synopsis 784.e106

A.1 Wally Repository 784.e106
A.2 Wally Configurations 784.e106
A.3 Wally Regression and Test Suites 784.c109
A.4 Key Wally Commands 784.e109
A.5 Wally Diagrams 784.e109

Appendix B Hitchhiker's Guide to Linux 784.e153

B.1 Connecting to a Linux Server 784.e153
B.1.1 *X2Go* 784.e153
B.1.2 *Changing Your Password* 784.e154
B.2 Working With Files 784.e154

B.3 More Handy Commands 784.e154
B.3.1 *Cast of Characters* 784.e157
B.3.2 *Standard Input and Output* 784.e157
B.3.3 *Common Commands* 784.e158
B.3.4 *Working With Processes* 784.e159
B.3.5 *Compressing, Bundling, and Transferring Files* 784.e161
B.4 Linux Productivity 784.e161
B.4.1 *Multiple Terminal Windows* 784.e161
B.4.2 *Text Editors* 784.e162
B.4.3 *The Linux File System* 784.e163
B.4.4 *Environment Variables* 784.e164
B.4.5 *Web Browsing* 784.e164
B.4.6 *Document Viewing and Printing* 784.e165
B.4.7 *Installing Packages* 784.e165
B.5 Scripting and Programming 784.e165
B.5.1 *Shell Scripts* 784.e165
B.5.2 *Python* 784.e166
B.5.3 *C* 784.e167
B.5.4 *Makefiles* 784.e167

Appendix C Version Control Using Git 784.e170

C.1 Gitting Started 784.e171
C.2 Setting Up a Repository 784.e172
C.3 Basic Git Flow 784.e172
C.4 Snapshots and HEAD 784.e175
C.5 Merge Conflicts 784.e175
C.6 Branching and Merging 784.e176
C.7 Tags 784.e178
C.8 Staging and Undo 784.e179
C.9 Submodules 784.e180
C.10 Other Git Capabilities 784.e181
C.11 Summary 784.e182

Appendix D Tcl Book of Armaments 784.e185

D.1 Tcl Usage 784.e185
D.1.1 *Arguments and Variables in Tcl* 784.e186
D.1.2 *Nested Commands* 784.e187
D.1.3 *Summary of the Basic Tcl Command Syntax* 784.e187
D.2 Additional Commands 784.e188
D.2.1 *Lists* 784.e188
D.2.2 *Outputting Data to the Screen With Tcl* 784.e189
D.2.3 *File, Glob, and Other File-Related Commands* 784.e189
D.2.4 *Control Flow* 784.e189

D.2.5 *Procedures* . . . 784.e191
D.2.6 *Running Tcl Scripts* . . . 784.e191
D.2.7 *Aliases* . . . 784.e192
D.2.8 *Exiting Tcl* . . . 784.e192

Appendix F Floating-Point Implementation . . . 784.e193

F.1 Fused Multiply-Add . . . 784.e193
F.1.1 *Multiply Significands* . . . 784.e193
F.1.2 *Product and Addend Signs* . . . 784.e193
F.1.3 *Add Exponents* . . . 784.e193
F.1.4 *Alignment Shift* . . . 784.e194
F.1.5 *Add Aligned Significands* . . . 784.e195
F.1.6 *LZA: Leading Zero Anticipator* . . . 784.e198
F.1.7 *Multipath FMA Optimization* . . . 784.e200
F.2 Floating-Point Division and Square Root . . . 784.e201
F.2.1 *Recurrence Division Algorithm* . . . 784.e202
F.2.2 *Radix-2 Recurrence Division* . . . 784.e207
F.2.3 *Radix-4 Recurrence Division* . . . 784.e213
F.2.4 *Divider Latency* . . . 784.e219
F.2.5 *Quotient Conversion* . . . 784.e220
F.2.6 *Early Termination* . . . 784.e222
F.2.7 *Subnormal Floating-Point Preprocessing* . . . 784.e223
F.2.8 *Floating-Point Postprocessing* . . . 784.e224
F.2.9 *Integer Division* . . . 784.e225
F.2.10 *Square Root* . . . 784.e234
F.2.11 *Unified Divide and Square Root* . . . 784.e249
F.2.12 *Multiplicative Algorithms for Division and Square Root* . . . 784.e257
F.3 Conversion . . . 784.e258
F.3.1 *Result Exponent* . . . 784.e258
F.3.2 *Shift Amount Selection* . . . 784.e261
F.3.3 *Examples* . . . 784.e262
F.4 Postprocessing . . . 784.e263
F.4.1 *Shift Calculation* . . . 784.e263
F.4.2 *Initial Result Mux* . . . 784.e265
F.4.3 *Normalization Shift* . . . 784.e265
F.4.4 *Correction Shift* . . . 784.e265
F.4.5 *Sign Logic* . . . 784.e267
F.4.6 *Rounding* . . . 784.e268
F.4.7 *Special Cases and Flags* . . . 784.e269

Bibliography . . . 785
Index . . . 799
RISC-V Instruction Set Summary . . . 815

Preface

We wrote this textbook to close the gap in computer engineering education between principles of computer architecture and practices to build real microprocessors and systems-on-chip (SoCs). Many excellent books address computer architecture, operating systems, and digital design, but none that we know of connect these ideas together to reach an SoC capable of running an operating system (OS). Moreover, the open RISC-V instructions set architecture (ISA), pioneered at the University of California, Berkeley, has energized the computer architecture field by offering a modern instruction set suitable for hardware implementation in commercial systems, avoiding the intellectual property barriers of ARM, the historical baggage of x86, and the warts of earlier RISC architectures such as MIPS and SPARC. This book shows you how to apply contemporary professional practices to design, verify, and implement a RISC-V SoC with a processor core, memories, and peripherals to support its booting Linux.

We named our RISC-V SoC Wally in homage to the Harvey Mudd College mascot Wally Wart, which personifies the flat, square protrusions that adorn many brutalist-style buildings at Harvey Mudd College, and also in honor of Pixar's adorable WALL-E robot.

This book is intended for a graduate or advanced undergraduate course or for self-study. It assumes that you have had introductory courses on digital design and computer architecture at the level of *Digital Design and Computer Architecture: RISC-V Edition* or *Computer Organization and Design: RISC-V Edition*. Specifically, you should be comfortable reading and writing C, assembly language, and Verilog and working in a Linux environment. You may get even more out of this book if you've taken a second course at the level of *Computer Architecture: A Quantitative Approach* or studied operating systems, compilers, field programmable gate arrays (FPGAs), or chip design.

While most textbooks offer simplified processor designs to illustrate general principles, this book takes the next step to explore optimized implementations of a complete processor. Our CORE-V Wally processor and SoC is part of the free and open-source OpenHW Foundation suite of RISC-V processors. Wally is described in SystemVerilog because it is an industry standard understood by most engineers and is supported by nearly all commercial simulation and synthesis tools. Wally is configurable in most regards, ranging from a tiny RV32E embedded core to a full-featured SoC, which is capable of booting Linux, with an RV64GC core, caches, virtual memory, and peripherals. This book delves into design issues such as bus interfaces, exceptions, privileged operations, division, and floating-point arithmetic that are glossed over in most texts. Although detailed implementation issues could become tedious, they are where most practicing engineers spend most of their time. We aim to give insight into good designs and the choices that lead to such designs.

The book applies professional tools and methodologies that students will see in industry. These include SystemVerilog for design, verification, and synthesis, Spike and QEMU for RISC-V functional simulation, Siemens Questa or Verilator for SystemVerilog simulation, Verilator for lint, Synopsys Design Compiler for chip implementation, and Xilinx Vivado for FPGA implementation. The system uses a Linux development environment, including GCC tools for programming and debugging in C and assembly language, Git for version control, and make-based scripts to drive compilation, simulation, and synthesis. The SoC itself also can boot Linux based on the embedded Buildroot flow, including the boot loader, kernel, device drivers, and command prompt. It has open-source implementations on an FPGA and on the Google Skywater 130-nm CMOS process design kit. All of the design files are open-source and available in a Git repository, although some commercial tools require licenses to run.

This book is divided into four parts:

- **Introduction:** The big picture, guided by an example of designing, verifying, and implementing a simplified single-cycle RISC-V processor
- **Tools:** C, Assembly, and SystemVerilog compilers, simulators, validation, synthesis
- **RISC-V Microarchitecture:** Pipelining, privileged operation, bus interface, caches, memory management, branch prediction, compressed instructions, multiplication and division, floating-point, atomic instructions, bit manipulation instructions, and other extensions
- **Implementation:** Peripherals, benchmarking, Linux boot, and FPGA implementation

This book took three years of development and another year of production. There are many more important topics in computer architecture and SoC design that were omitted from this edition for schedule reasons; we hope to address them in a future volume:

- **Multicore:** Multilevel cache coherency, synchronization, interprocessor communication
- **Microarchitecture:** Superscalar, out-of-order, deeper pipelines
- **RISC-V Extensions:** Vector, hypervisor, debug, trace
- **Verification:** Functional coverage metrics, hardware emulation

- **SoC Design:** Intellectual property block design and integration, AXI interfaces, accelerators, memory controllers, network-on-chip
- **SoC Implementation:** Timing and power optimization, clock gating, clock domain crossings, external interfaces, phase-locked loops (PLLs), verification, design-for-test, tapeout

Additional materials related to the book are available on the Companion Site: https://www.elsevier.com/books-and-journals/bookcompanion/9780323994989.

Please note that chapters 20–23 and Appendices A–D and F are available on the above Companion Site. These chapters and appendices are also in the electronic version of the book.

For qualified instructors, additional materials are available on the Instructor Site: https://educate.elsevier.com/9780323994989.

How to Use This Book

This book is most satisfying to use in conjunction with the CORE-V Wally processor source code, along with the compiler toolchain, simulation environment, and synthesis scripts. The complete code base can be cloned from github.com/openhwgroup/cvw. This repository also has documentation on how to install Wally and supporting tools.

The book comes with a set of lecture slides and labs that may be freely used or modified for non-commercial purposes as long as the source is credited. These supplements and an errata list are also available on the GitHub site.

OPEN SOURCE

All of the design files, including the Wally SoC, verification scripts, and implementation scripts, are provided open source under the Solderpad Hardware License version 2.1 (solderpad.org/licenses/SHL-2.1). We hope that others will enhance and extend the design and contribute to a vibrant open RISC-V ecosystem in collaboration with the OpenHW Foundation.

BUGS

Despite our best efforts, we are confident that this book and the Wally SoC still have bugs. Moreover, the processor relies on numerous external tools and repositories, all of which are likely to change and break things over time (a phenomenon called *bit rot*).

The textbook errata is on the GitHub site. We appreciate your corrections via a pull request to this page:

github.com/openhwgroup/cvw/blob/main/errata.md

Wally uses standard GitHub issues and pull requests for bug reports and fixes. We appreciate contributors.

FURTHER READING

This book has been influenced by many great works in computer architecture and systems. Hennessy and Patterson's *Computer Architecture: A Quantitative Approach* revolutionized the field when it was first introduced and remains the bible for principles of computer architecture. The *RISC-V Instruction Set Manual* by Waterman and Asanović is not just the authoritative definition of the RISC-V architecture but also an outstanding model of a specification that is succinct yet provides insight into the factors influencing design decisions. The Berkeley CS152 *Computer Architecture* course has an excellent web page that gave us many ideas. SiFive's datasheets for various RISC-V chips define de-facto standard peripherals. Bryant and O'Hallaron's *Computer Systems: A Programmer's Perspective* has excellent coverage of the interaction between software and hardware. *Operating Systems: Three Easy Pieces* has particularly good coverage of concurrency and atomic operations. Ercegovac and Lang's *Digital Arithmetic* is the leading general book on computer arithmetic, and Muller's *Handbook of Floating-Point Arithmetic* has the most comprehensive coverage of floating-point hardware and software.

We learned a great deal writing *Digital Design and Computer Architecture, RISC-V Edition*, and the processor in this book is an outgrowth of the simplified processor in that introductory book. That book provides sufficient background for readers to understand this book. If you are interested in chip design and hardware implementations of arithmetic and memories, *CMOS VLSI Design* by Weste and Harris covers these topics in great detail.

Acknowledgements

The Wally RISC-V processor development has been a team effort with contributions from many students at Harvey Mudd College, Oklahoma State University, the University of Nevada, Las Vegas, and the University of Engineering and Technology, Lahore. The authors thank our students who contributed to the Wally processor and the chapters and to reviewing this book. Katherine Parry was the main architect of the floating-point unit. Cedar Turek, Alessandro Maiuolo, and Kevin Kim contributed to the divide/square root unit. Jordan Carlin and Jessica Liu contributed to quad and Zfa additional floating-point. Kelvin Tran and Ryan Swann developed the crypo unit. Noah Boorstin, Kaveh Pezeshki, Ben Bracker, and Skylar Litz brought up Linux on Wally and contributed to the Linux chapter. Ben also implemented and tested peripherals. Thomas Fleming and Jessica Torrey developed the memory management unit. Kip Macsai-Goren developed and debugged privileged tests, especially for the memory management and physical memory protection units. Kip and Kevin Kim developed the bit manipulation unit. Nic Lucio developed peripheral tests. Madeleine Masser-Frye and Shreya Sanghai optimized critical paths in Design Compiler synthesis. Daniel Torres, Elizabeth Hendenberg, and Abraham Casteneda ported the benchmarks. Daniel also brought up RISCOF tests. Jacob Pease contributed to the SDC peripheral and Linux boot. Naiche Whyte-Aguayo developed the SPI peripheral. Limnanthes Serafini did cache performance verification and parameterization. Teo Ene contributed and mentored in many areas. The HMC and OSU SoC Design classes of 2023 contributed coverage tests for their class projects, along with Harshini Srinath. Other students contributing to the project have included Jarred Allen, Ethan Falicov, Thomas Kidd, Tony Lei, Domenico Ottolia, and Udeema Shakya.

Jordan Carlin has made extraordinary contributions to this work in countless dimensions, especially related to the flow and verification, improving the manuscript, and leading teams. He has become the OpenHW Foundation's most prolific contributor and also contributes to the Sail model and other teams at RISC-V International. As one of the CORE-V Wally committers, his eagle eyes have improved hundreds of pull requests.

Hamza Ali, Marina Bellido, Kaden Cassidy, Noah Fotenos, Corey Hickson, Isabella Hottenrott, Hamza Jamal Vikram Krishna, Muhammad

Abdullah, Roman De Santos, Georgia Tai, Daniel Torres, and Muhammad Zain at HMC and UET made notable contributions to functional verification.

The OpenHW Foundation has welcomed CORE-V Wally to its robust family of open RISC-V cores. We especially appreciate the advice and support from Mike Thompson, Duncan Bees, Rick O'Connor, and Flo Wohlrab.

We have benefited from advice and support from many leaders of the RISC-V community, including Andrew Waterman, Krste Asanović, Andrea Gallo, David Patterson, Megan Lehn, Pawan Kumar, and Jeremy Bennett.

We have particularly enjoyed collaborating on verification with several teams, including Simon Davidmann, Lee Moore, Aimee Sutton, and Duncan Graham at Synopsys; Umer Shahid, Bilal Zafar, and Huda Sajjad at 10x Engineers; Adnan Hamid and David Kelf at Breker Verification Systems; and Allen Baum, Derek Hower, and James Ball with the RISC-V Certification Steering Committee.

Our work in microarchitecture has been influenced heavily by our experience designing microprocessors at Intel, Hewlett-Packard, Broadcom, Air Force Research Laboratory, and elsewhere.

We are indebted to our reviewers: Andrea Bartolini, Duncan Bees, Javier Bruguera, Daniel Chaver, Simon Davidmann, Ashish Dixit, Michael Engel, Andrea Gallo, Firas Hassan, Brian "Redbeard" Herrington, Andreas Koch, Bill McSpadden, Lee Moore, Simon Moore, Robert Mullins, Torbjørn Ness, John Nestor, Xiaoning Qi, Alex Solomatnikov, Dwight Sunada, Mike Thompson, and Peiyi Zhao.

About the Authors

David Harris is the Harvey S. Mudd Professor of Engineering Design at Harvey Mudd College. He received his Ph.D. from Stanford University and S.B. and M.Eng. degrees from MIT. His interests are in the area of microprocessor design, computer arithmetic, and high speed and low power circuits. David is the author of *Digital Design and Computer Architecture*, *CMOS VLSI Design*, *Logical Effort*, and *Skew-Tolerant Circuit Design*, and he holds sixteen patents in the field. He has designed chips at Intel, Broadcom, Sun Microsystems, Hewlett-Packard, and elsewhere. David's other interests include writing Southern California hiking guidebooks, building experimental aircraft, and traveling with his family.

James Stine is the Edward Joullian Professor in Engineering in the School of Electrical and Computer Engineering at Oklahoma State University. He received his Ph.D. in electrical engineering from Lehigh University. He was previously with the Illinois Institute of Technology in Chicago, IL. He specializes in research and teaching in VLSI, digital arithmetic, computer system architecture, and digital design. He is also a leading developer of commercial and academic system on chip design flows for Google, Cadence Design Systems, Synopsys, Mentor Graphics and several public-domain tools. James's interests include spending time with his family.

Rose Thompson is a processor architect at Skyworks Solutions, Inc. She received her Ph.D. in Electrical Engineering from Oklahoma State University and two B.S. degrees in Electrical Engineering and Computer Engineering from the University of Washington. She has also designed chips at the Air Force Research Laboratory. Her professional interests include SoC design and verification, custom instruction set architectures, branch prediction, memory systems, and secure computing. Rose also enjoys biking, hiking, rock climbing, and playing the piano.

Sarah Harris is a Professor of Electrical and Computer Engineering at the University of Nevada, Las Vegas. She received her M.S. and Ph.D. at Stanford University. Before joining UNLV in 2014, she was a faculty member at Harvey Mudd College for ten years. She has also worked at Hewlett Packard, NVIDIA, and the Technical University of Darmstadt. She is the co-author of three other textbooks including *Digital Design and Computer Architecture: RISC-V Edition*. Her research interests include computer architecture and applications of embedded systems and machine learning to biomedical engineering and robotics. Sarah also enjoys playing music and spending time with her family.

Foreword

I'm honored and rather humbled to have been invited to write the foreword for this *RISC-V System-on-Chip Design* textbook. But, how did we get here? How is it possible to have a truly open-source system-on-chip (SoC) design? How does this textbook help readers achieve this outcome?

Open-source history traces back to the early days of computing when researchers and developers shared code to propel innovation. In the 1960s and 1970s, the concept began to solidify with the emergence of time-sharing systems and the advent of the Unix operating system, which was distributed along with its source code. In the early 1980s, as proprietary software proliferated, the open-source movement took on a more structured approach with the founding of the Free Software Foundation by Richard Stallman in 1985. Stallman's GNU Project was pivotal because it sought to develop a completely free Unix-like operating system. The 1990s marked a fundamental turning point for open-source software with the development of the Linux kernel by Linus Torvalds in 1991. Linux, combined with GNU tools, provided a complete open-source operating system that gained rapid popularity among developers and enterprises. GitHub, launched in 2008, has also revolutionized open-source development by providing a platform for collaborative coding and distribution. In contemporary engineering, open source plays an integral role in cloud computing, artificial intelligence, and the Internet of Things (IoT).

The success of open-source software has inspired a revolution in open-source hardware, exemplified by the RISC-V architecture. Developed at Berkeley in 2010 by Professor Krste Asanović, his graduate students, including Andrew Waterman and Yunsup Lee, and Professor Dave Patterson as a fifth-generation reduced instruction set computing architecture, RISC-V has become the fastest-growing computer architecture in the industry. To foster open community-based collaboration and manage overall development of the RISC-V instruction set architecture (ISA) specifications, the project was spun out of the University of California, Berkeley, and the RISC-V Foundation was formed in 2015. Working with Krste, Dave and executives from top technology companies, including Google, NVIDIA, Western Digital, and many more, I had the privilege of leading the effort to incorporate the RISC-V Foundation nonprofit organization and serve as executive director for close to 5 years. The open nature of the RISC-V ISA specifications has had a profound impact on the semiconductor industry, introducing a new level of flexibility and innovation by allowing companies to tailor their processors to specific applications without needing cumbersome licensing agreements. This textbook is an excellent example of the breadth of the RISC-V architecture's overall impact, with the majority of the CORE-V Wally SoC development flow and key building blocks being available via open source.

With the RISC-V ISA specifications having gained significant industry traction, the need for high-quality, well-written, SystemVerilog-based, open-source RISC-V cores became of interest to several key industry players. This need drove the creation of the OpenHW Foundation, which was established in 2019 as a not-for-profit global organization aimed at fostering the collaborative development of high-quality, production-worthy, open-source hardware designs, particularly RISC-V processors. After my tenure running the RISC-V Foundation, I was asked to spearhead the creation of the OpenHW Foundation, and together with industry leaders from companies such as Alibaba, Bluespec, Imperas, NXP, SiLabs, and Thales, the OpenHW Foundation was formed with initial CORE-V Family projects based on the excellent work of Professor Luca Benini, Frank Gürkaynak, and the rest of the PULP Platform team at ETH Zurich. The CORE-V Family has grown to include 10 different cores, including CORE-V Wally (cvw), of varying sizes and capabilities for use in application and embedded class systems. These CORE-V cores have been instrumental in demonstrating the capabilities and benefits of open-source hardware collaboration. Hundreds of millions of these cores have been sold commercially in SoCs across the industry. By leveraging the collective expertise of its community, OpenHW has contributed to the wider acceptance and implementation of RISC-V cores in various applications, from automotive to IoT and embedded systems.

This textbook is an excellent way for readers to become immersed in the vast area of open-source, the RISC-V ISA, and processor SoC design methodologies. The book is best used in conjunction with the CORE-V Wally processor source code and related tools, available in the OpenHW CORE-V Wally GitHub repository. To get started, readers are guided through an example of designing, verifying, and implementing a simplified single-cycle RISC-V processor before diving into the five-stage, 64-bit, Linux-capable Wally processor.

This textbook provides comprehensive coverage of both open-source and commercial compilers, simulators, and synthesis tools, including SystemVerilog, which is the de facto SoC design language used in the semiconductor industry. The authors have taken a practical approach to using both open-source tools such as GCC, Verilator, Spike, Sail, and Quick Emulator (QEMU) alongside tools from commercial electronic design automation companies, including Synopsys, Siemens, and Cadence. This open-source commercial tool pragmatic approach is required to achieve SoC designs suitable for deep submicron implementations and reflects the reality of the design environment graduating engineers will be working in after completing their studies.

The RISC-V microarchitecture section of the textbook provides fine-grain detail regarding processor design considerations, including pipelining, buses, caches, and memory as well as the various RISC-V instruction extensions. This is all brought together through a "soft-physical" implementation using a field programmable gate array, which demonstrates the use of SoC peripherals, benchmarking, and booting Linux. The deep dive the authors take into processor design topics, such as buses, privileged modes, virtual memory, and floating-point, which are often glossed over in other engineering textbooks, provides excellent learning for students because these are the topics with which most practicing engineers spend a majority of their time. Some 4 years in the making, this *RISC-V System-on-Chip Design* textbook is destined to become a "must-have" tome covering processor and SoC design for years to come.

Good luck with your open-source RISC-V development journey!

Rick O'Connor
Founding OpenHW Foundation CEO
Founding RISC-V Foundation Executive Director

A Brief History of Computer Design

1

1.1 Moore's Law and Beyond
1.2 System-on-Chip
1.3 Birth of Computing
1.4 Mainframes and Minicomputers
1.5 Microprocessors
1.6 CISC and RISC
1.7 RISC-V
1.8 International Economic and Security Competition
1.9 Summary

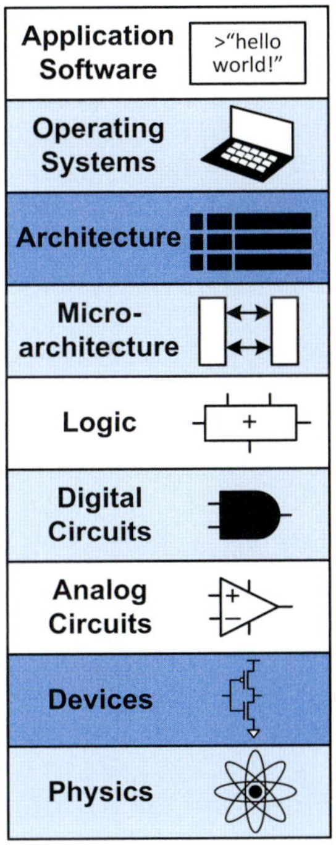

Computers are among the most significant inventions in the history of technology. Their cost, size, power consumption, and performance have improved exponentially, with greater impact and for a longer duration than any other human invention. A microprocessor, the brain of a computer, is smaller than a jellybean. A simple microprocessor can be manufactured on a chip for one-hundredth of a cent. Microprocessors are built from electrically controlled switches called transistors. Decades of miniaturization have shrunk transistors such that more than ten billion can be packed onto a square-centimeter chip. The transistors can switch on or off in a few picoseconds, consuming less than a femtojoule of energy each time. A *system-on-chip* (*SoC*) integrates multiple microprocessors and memories with various communication links, peripherals, and accelerators, enabling low-cost mobile and wearable devices. These astonishing hardware advances have transformed nearly every aspect of society in the past half century, including medicine, communication, transportation, consumer products, energy, warfare, education, and work. Computers have put much of humanity's collective knowledge at our fingertips while upending how we interact in ways with which society is still grappling.

Meanwhile, the free and open-source software movement has been transforming software by sharing source code with a license that allows others to freely use and improve it. For example, open-source software and standards such as Linux, TCP/IP, Apache HTTPD, and Python underlie much of the operation of the Internet. The movement has historical roots, such as the Automobile Manufacturers Association openly sharing automotive innovations in the first half of the 20th century and IBM's release of its operating system (OS) code in the 1950s and 1960s. Richard Stallman created the Free Software Foundation in 1985, and the Open Source Initiative began in 1998. However, computer architectures have historically been proprietary, protected by patents and licensing agreements that limit adoption and extract royalties from manufacturers. To advance a widely adopted open-source architecture, a team at Berkeley developed a fifth-generation reduced instruction set computer (RISC)

RISC-V, pronounced "risk five," is the fifth RISC architecture developed at Berkeley (the University of California, Berkeley), with their first RISC designs being developed in the 1980s.

architecture called *RISC-V* in 2010 and established the nonprofit RISC-V Foundation (now called RISC-V International) as a collaborative community. The nonproprietary RISC-V architecture facilitates innovation, and RISC-V has become the fastest-growing computer architecture, drawing enormous attention from both industry and academia.

Most computer engineering students take courses in computer architecture, and many have the opportunity to build toy processors, but very few get to build a real full-featured microprocessor. This book takes you step by step through the design of a complete open-source RISC-V processor and SoC using contemporary design tools and methods. By the end of this book, you should have a solid understanding of computer architecture theory and practice and be ready to make your own contributions to the vibrant field of processor and SoC design.

Advances in transistors, computer hardware, and computer software have become intertwined in a *virtuous cycle* in which improvements in one lead to improvements in the others. This chapter sets the context for the book with a brief history of transistors and computer design, with emphasis on how advances in implementation technology have driven architecture and microarchitecture. The next chapter introduces the RISC-V architecture and previews the rest of the book.

1.1 MOORE'S LAW AND BEYOND

The semiconductor revolution began in 1947 when John Bardeen, Walter Brattain, and William Shockley invented the first working point-contact transistor at Bell Lab. *Transistors* are semiconductor devices with a control terminal that influences the flow of current between two other terminals. They operate as amplifiers or electrically controlled switches and are a key building block of digital logic. In 1958, Jack Kilby and Robert Noyce coinvented the first *integrated circuits* by placing multiple transistors on a single substrate. Fig. 1.1 shows the first planar integrated circuit prototype built by Noyce and his team at Fairchild in 1960. Bardeen, Brattain, Shockley, and Kilby all received Nobel Prizes in physics for their world-changing contributions.

An integrated circuit is now manufactured on a large circular silicon wafer using a series of deposition, masking, and etching steps to create transistors and interconnecting wires. *Photolithography* (Greek for *light-stone-writing*) is a key manufacturing step in which a photoresist is exposed to light through a mask to define the shapes of transistors and wires. As manufacturing improvements allow for smaller transistors, they become faster (less distance for electrons to flow), lower power (fewer electrons flowing), and cheaper (more fit on a wafer). The wafer is cut up, or *diced*, into rectangles called *dice* or *chips*, which are tested and packaged.

In 1965, Gordon Moore observed that the number of transistors in an integrated circuit was growing exponentially. This observation, now

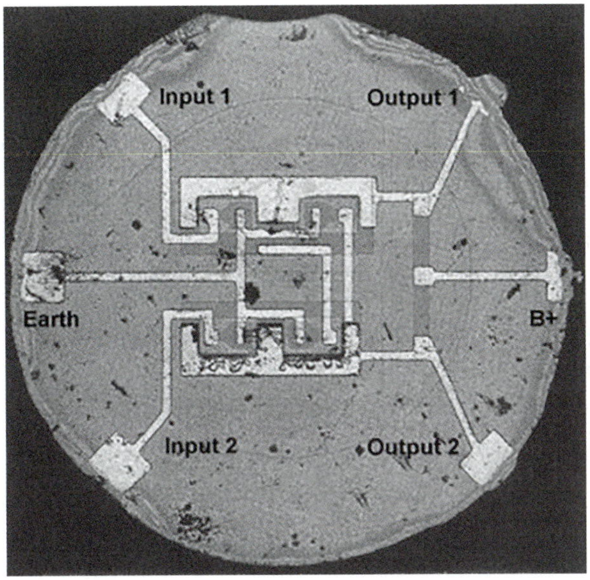

Fig. 1.1 First planar integrated circuit. (ON Semiconductor Corp., 1961 [onsemi61]. Reprinted with permission.)

known as *Moore's Law*, portended technical advances unrivaled by any other human achievement [Moore65, Moore03]. Since its inception, the semiconductor industry has increased the number of transistors on a chip by a ludicrous 11 orders of magnitude. Fig. 1.2 shows how microprocessor transistor counts have risen at a remarkably steady 40% compound annual growth rate for five decades, doubling almost every two years. In 2024 a cell phone chip such as the Apple A17 Pro had approximately 19 billion transistors on a 1 square centimeter chip for an estimated $40 manufacturing cost. In other words, the cost of a transistor is now about 2 nanodollars! However, no exponential curve can continue indefinitely. As shown in Table 1.1, Moore's Law has been slowing in recent years because of the rising cost and technical difficulty of manufacturing atomic-scale transistors far smaller than the wavelength of visible light. Only a handful of companies worldwide can invest the tens of billions of dollars necessary to develop cutting-edge manufacturing processes and factories, so the logic fabrication industry has largely consolidated to TSMC, Samsung, UMC, Global Foundries, SMIC, and Intel. Despite the small number of companies involved, global semiconductor sales rose from $33 billion in 1987 to approaching $700 billion in 2025, as shown in Fig. 1.3.

Gordon Moore (1929–2023) was a cofounder and CEO emeritus of Intel Corporation. He is best known for articulating Moore's Law, which drove the advancement of semiconductors for decades. Moore grew up in northern California and then attended San Jose State, Berkeley, and Caltech, receiving a PhD in chemistry. He cofounded Fairchild Semiconductor and then Intel. After retiring, Moore and his wife became philanthropists supporting environmental conservation and scientific research. Moore was a passionate fisherman, which inspired his conservation efforts.

(© 2006. Image courtesy of Intel Corporation. Reprinted with permission.)

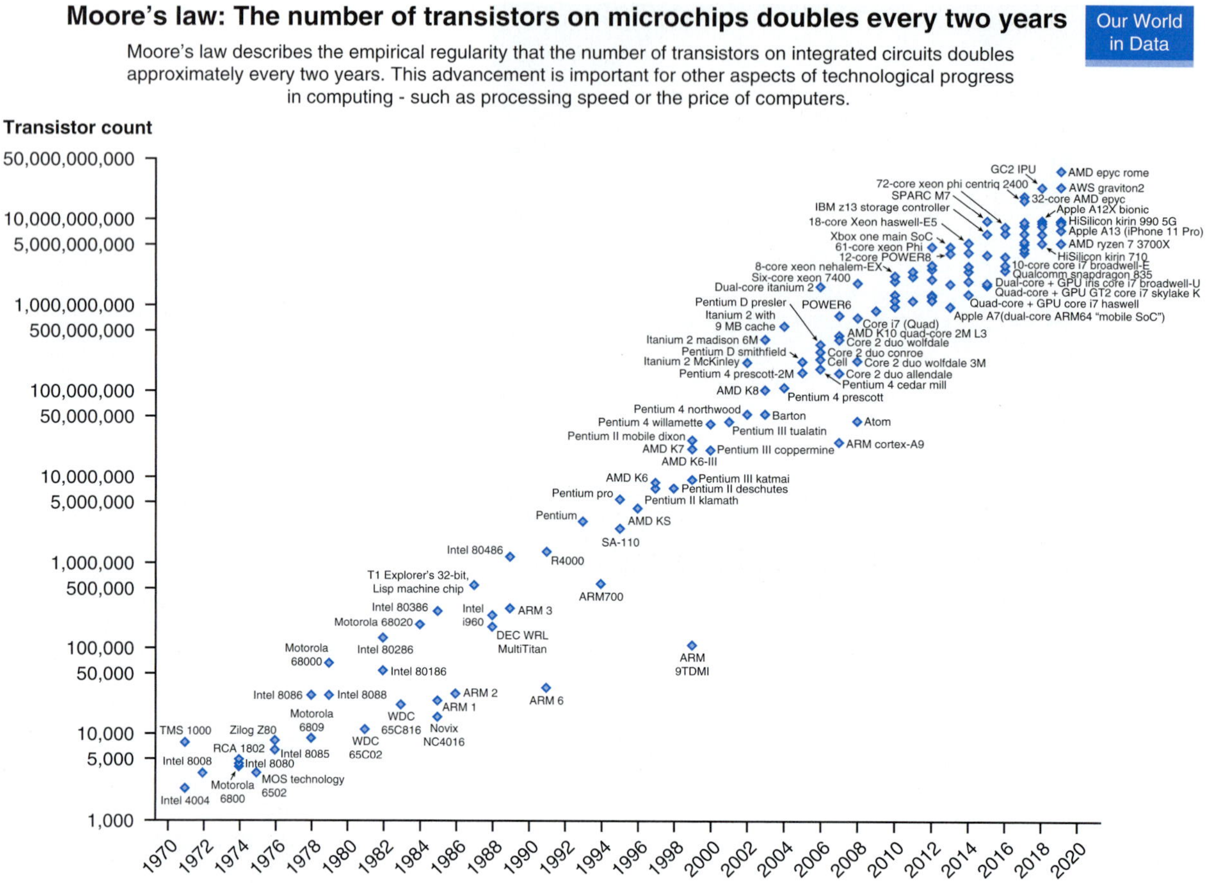

Fig. 1.2 Microprocessor transistor counts. (© 2020 Hannah Ritchie and Max Roser [Ritchie20]. This figure is openly licensed under CC BY 4.0.)

Table 1.1 Chip and transistor costs. [Zafar19]

Process Node	16 nm	10 nm	7 nm	5 nm	3 nm
Initial production	2014	2016	2018	2020	2022
Chip area (mm^2)	125	88	83	85	85
Billions of transistors	3.3	4.3	6.9	10.5	14.1
Gross dice / wafer	478	686	721	707	707
Net dice / wafer	360	512	546	530	510
Wafer cost ($)	5912	8389	9965	12500	15500
Die cost ($)	16.43	16.37	18.26	23.57	30.45
Transistor cost (nanobucks)	4.98	3.81	2.65	2.25	2.16

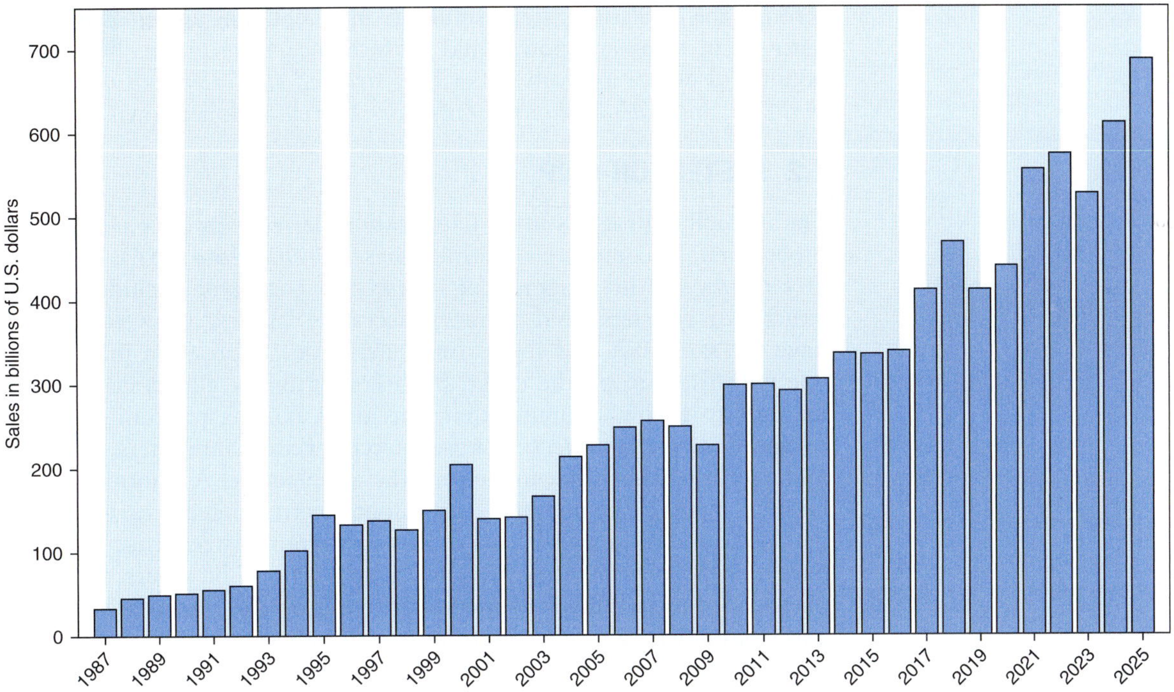

Fig. 1.3 Global semiconductor sales. [market25]

A corollary of Moore's Law is *Dennard's Law*, which describes how transistor speed and power consumption improve as transistors shrink [Dennard74, Bohr07]. Dennard scaling predicted that if a new process reduces transistor dimensions, supply voltage, and threshold voltage by 30%, then a chip of the same size can pack twice as many transistors that operate at 1.4 times the speed and half the power per transistor. Each such scaling is called a *process node*. Process nodes have historically been identified by their *feature size*, the gate length of a minimum-sized transistor. For example, a 180-nm process uses masks and photolithography to define a transistor with a channel length of 180 nm from source to drain. Process nodes from the mid-1990s to 2013 advanced from 180 nm to 130 nm, 90 nm, 65 nm, 45 nm, 32 nm, and 22 nm. To reduce leakage current at 22 nm or below, transistors became three-dimensional structures with gates wrapping around the channel. Now the process node no longer correlates with any physical feature, but manufacturers still have named their improved nodes 16 nm, 10 nm, 7 nm, 5 nm, 3 nm, 2 nm, and so on.

Dennard scaling tapered off around 2004 because continuing to reduce threshold voltage causes subthreshold leakage current to rise exponentially, and reducing gate oxide thickness below about 12 angstroms (4 atomic layers) allows large numbers of electrons to quantum-mechanically tunnel through the insulating gate, resulting in unacceptable gate leakage current. Since then, transistor power and performance have gradually improved because of device innovations such as *strained silicon*, *high-k metal gates*, *FinFETs*, and *gate-all-around transistors* as well as dimensional scaling.

1.2 SYSTEM-ON-CHIP

The incredible progress in miniaturization drives chip designs, especially processors. A computer's *instruction set architecture* (*ISA*), or simply *architecture*, is the programmer's view of a computer, including the instructions, registers, and memory accessing modes. If two computers have the same ISA and supporting capabilities, a program that runs on one will also run on the other. Such compatibility allows customers to upgrade to a new computer that runs their existing programs better. A computer's *microarchitecture* is the interconnection of registers, memories, and execution units that implement the architecture. As more transistors become available, computers can adopt microarchitectures with larger data sizes and more parallelism to run a program faster. Thus, processors have grown from operating on 8- to 16- to 32- to 64-bit values, from using microcoded to pipelined to superscalar to out-of-order microarchitectures, and from having a single core to using multiple cores.

Chips are sometimes categorized as *general purpose* (e.g., microprocessors, graphics processing units, or field-programmable gate arrays) versus *application-specific integrated circuits* (*ASICs*) (e.g., hard disk read channel controllers, Wi-Fi modems, Bitcoin mining engines). The distinction has blurred as ASICs incorporate more programmable components. Modern integrated circuits are often called *systems-on-chip* (*SoCs*) because they combine many processors, memories, accelerators, busses, and other components. For example, a contemporary cell phone chip such as the Apple A15 in Fig. 1.4 contains at least two high-performance and four energy-efficient application processors, tens of megabytes of level 2 and 3 caches, many graphics processing units and neural engine cores, video and image signal processing accelerators, and interfaces to a display and external memory. The components are clocked at several gigahertz (GHz) and collectively perform tens of trillions of math operations per second. Keeping all the components running simultaneously would overheat the chip, so only a portion of the chip operates at any given time; the remainder is known as *dark silicon*. The specialized *accelerators* are 2 to 3 orders of magnitude more power efficient than general-purpose processors [Taylor12], so tasks are offloaded to these accelerators when practical. The system is so complex that a small onboard microprocessor is used to manage power and clocks to all the other components.

The increasing number of transistors on a chip has led to rising development costs, as shown in Fig. 1.5. Chip design teams have grown from one or two people in the 1970s to hundreds of people spread across

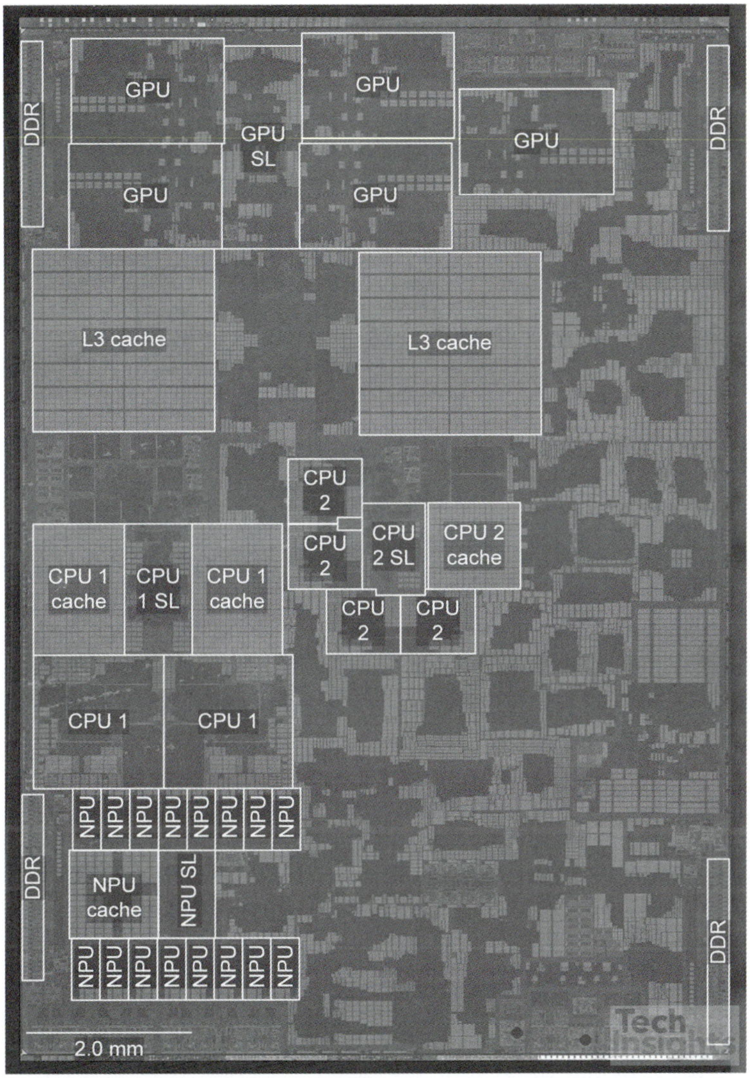

Fig. 1.4 Apple A15 Bionic die photo. (Image courtesy of TechInsights Inc. [iphone13]. Reprinted with permission.)

multiple divisions and companies for a modern SoC. Nevertheless, the number of transistors per designer has increased by five orders of magnitude, meaning engineers must be far more productive. In the 1990s, designers moved from transistor-level schematic design to *hardware description languages* (*HDLs*) such as Verilog or VHDL. An HDL allows a designer to specify logic functions using textual descriptions of its behavior or structure, thus increasing the level of abstraction and reducing the time to specify a function. Electronic design automation (EDA) tools

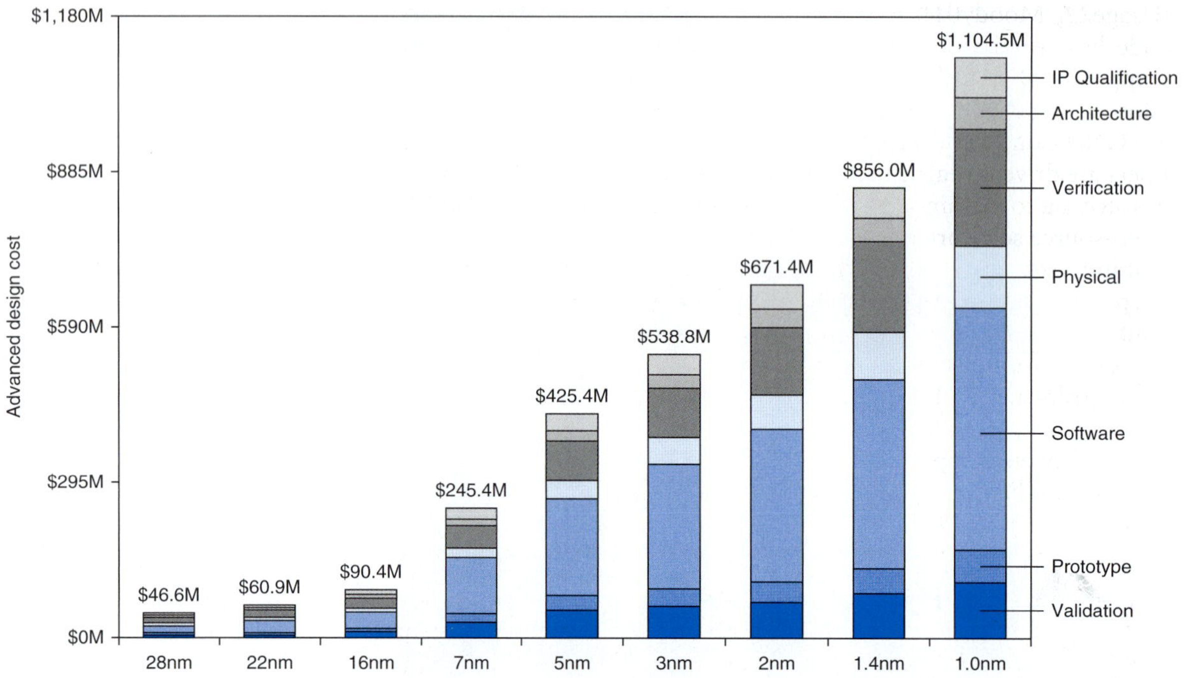

Fig. 1.5 SoC design cost. (Figure courtesy of International Business Strategies, Inc., September 2024. Reprinted with permission.)

such as Synopsys Design Compiler convert these HDL designs into gates and transistors, which is steadily improving productivity and quality of results. The industry has moved toward *design reuse*, in which *intellectual property* (*IP*) blocks such as processors, memories, accelerators, peripherals, and standard cell libraries are reused from a previous product or purchased from a third party. Designs have become so complex that verifying their correctness is more time-consuming and expensive than doing the design; IP blocks that come with robust test frameworks help reduce verification costs too. Despite all these advances, the cost of a new SoC design continues to rise exponentially. New SoCs in cutting-edge processes must target high-volume multi-billion-dollar markets, and the chip design industry has consolidated to a smaller number of companies (notably Intel, Samsung, Apple, Qualcomm, Broadcom, AMD, Texas Instruments, NVIDIA, MediaTek, STMicroelectronics, Infineon, NXP, Google, Meta, Amazon, Microsoft, and Renesas) capable of making these huge investments. The industry has learned that software is harder than hardware, and now the largest cost and source of errors in an SoC is the software running on it.

Especially because of these rising software development costs, the open-source movement has proven critical to the computer industry [opensource.org]. The Linux OS is the most prominent example; this robust kernel is the most widely used in the world, running on billions of processors from Android cell phones to warehouse-scale compute farms

[Usage22, Moody01]. More than 4,000 people contribute to the Linux code base each year, with the largest contributions coming from highly competitive for-profit companies such as Intel and RedHat [Larabel19]. Closely associated are many other open-source software tools such as the GNU Compiler Collection, Python, and PHP. Some open-source developers are driven by altruism, idealism, and curiosity, but others are capitalists striving to maximize profit. The computer industry has discovered that open-source software offers competitive advantages of flexibility, interoperability, security, and collaboration that often outweigh the benefits of keeping code proprietary. These open-source programs, such as the Android OS, run on many SoCs.

Open-source software has inspired an open-source hardware movement [oshwa.org]. Processor architectures, such as the Intel and ARM architectures, have historically been proprietary. However, RISC-V is a free and open computer architecture that has rapidly gained traction in the semiconductor industry. Chapter 2 describes the RISC-V architecture in detail.

1.3 BIRTH OF COMPUTING

Fig. 1.6 lays out some notable milestones in the development of computing.

Charles Babbage conceived the first digital computer in the 1830s. An English mathematician and Fellow of the Royal Society, Babbage was frustrated by the countless errors found in tables of mathematical functions in his day, finally exclaiming, "I wish to God these calculations had been executed by steam" [Eves69]. From 1823 until 1833, he designed the Difference Engine, an intricate array of 10-toothed gears that could rotate to perform additions, multiplications, and finite difference computations for tables of logarithms and trigonometric functions. The machine proved too difficult for Babbage and his machinists to build. Babbage eventually abandoned his efforts and moved on to the more general Analytical Engine with programmable instructions and a 1000-element memory that anticipated modern computers (Fig. 1.7). Running out of funding to complete the machine, Babbage died a broke and disappointed man in 1871. In 1991, engineers built two full-size operational Difference Engines from Babbage's plans. The 5-ton machines stand 11 feet long and 7 feet high. One of them can be seen at the Science Museum in London.

By World War II, militaries had urgent need for computation, such as ballistic firing tables needed to aim battleship turrets at targets 25 miles away. The first *computers* were people, usually young women, working in large teams to calculate and check these tables. Researchers strove to develop electromechanical computers using relays, electrically controlled switches, to automate such calculations. German scientist Konrad Zuse independently reinvented the notion of programmable computers and realized that binary numbers were most suitable because a switch can be in one

Charles Babbage (1791–1871) attended Cambridge University and married Georgiana Whitmore in 1814. He invented the Analytical Engine, the world's first mechanical computer. He also invented the cowcatcher and the universal postage rate and was interested in lock picking but abhorred street musicians. He was elected to the Royal Society in 1816 and served as the Lucasian chair of mathematics at Cambridge from 1828 to 1839.

(© 1833 Wellcome Collection, by R. Roffe [Babbage]. This image is openly licensed under CC BY 4.0.)

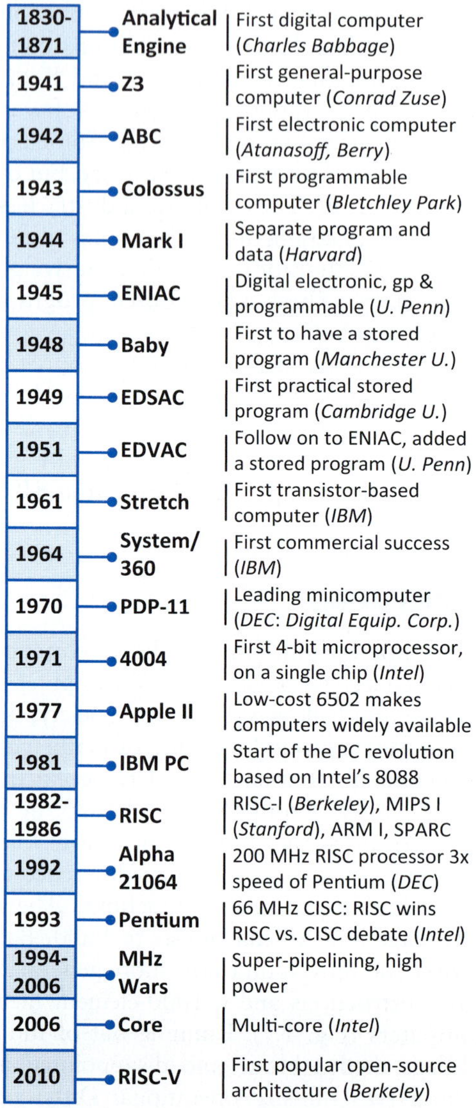

Fig. 1.6 Timeline of computer design.

of two states. His Z3 computer, built in Berlin in 1941, was the first functional general-purpose computer, but it was destroyed in a bombing raid. Meanwhile, Howard Aiken at Harvard, in collaboration with IBM, built the 5-ton Mark I (Fig. 1.8) in 1944 [Aiken46], and George Stibitz at Bell Telephone Labs built the nonprogrammable Complex Number Calculator in 1940.

Relays were both slow and unreliable. John Atanasoff at Iowa State College pioneered the application of vacuum tubes to perform digital calculations from 1937 to 1942. He and his student Clifford Berry built the

Fig. 1.7 Babbage's Analytical Engine, under construction at the time of his death in 1871. (This image is openly licensed under CC BY-SA 2.0.)

Fig. 1.8 Harvard Mark I. (Image courtesy of Harvard University [MarkI]. Reprinted with permission.)

Fig. 1.9 **Dorothy Du Boisson and Elsie Booker operating the Colossus in 1943.** (Image courtesy of the National Archives (UK), 1943 [colossus]. Public Domain.)

Vacuum tubes evolved from incandescent lightbulbs and look similar, with a vacuum-sealed glass container and electrodes. Like lightbulbs, vacuum tubes allowed current to flow in only one direction, which enabled the flow of current to be electrically controlled. The ENIAC used thousands of 6SN7 dual triode vacuum tubes for flip-flops, each 4 inches tall and dissipating up to 5 W. On average, one ENIAC tube burned out every two days.

6SN7 vacuum tube. (Vostok, 1980 [6SN7]. Public Domain: CC0 1.0 Universal.)

Grace Hopper found the first actual computer bug in the Harvard Mark II in 1947, a dead moth shorting one of the relays, as shown in Fig. 1.10.

first special-purpose digital electronic computer in 1942, the ABC (Atanasoff-Berry Computer) before getting called to serve in the war. In 1943, British codebreakers at Bletchley Park built the Colossus, the first programmable digital electronic computer, to decrypt German communication, arguably turning the tide of the war (Fig. 1.9). It used 1600 vacuum tubes and was programmed with plugs and switches rather than an internal stored program. Colossus was classified, so its significance was not widely recognized until much later. Another influential vacuum tube computer was the Electronic Numerical Integrator and Computer (ENIAC), built in 1945 by John Mauchly and J. Presper Eckert at the University of Pennsylvania. It ran at 5 kHz, weighed 30 tons, and consumed 174 kW. In collaboration with distinguished mathematician John von Neumann, they proposed a follow-on with a stored program called the Electronic Discrete Variable Automatic Computer (EDVAC) [vonNeumann45], but it did not begin operation until 1951. The Manchester University "Baby" computer in 1948 was the first to have a stored program. It was built by F.C. Williams and Tom Kilburn with influences from Alan Turing and Max Newman. The Electronic Delay Storage Automatic Calculator (EDSAC), developed by Maurice Wilkes at Cambridge University in 1949, is recognized as the first stored-program computer to go into regular use.

Like lightbulbs, vacuum tubes are prone to burning out over time, so keeping a computer with thousands of tubes running for an extended period was difficult. William Shockley, John Bardeen, and Walter Brattain invented the transistor in 1948 at Bell Labs (another Nobel Prize-winning discovery), and transistors were soon applied to digital computers. Gene Amdahl designed IBM's first transistorized computer, the 33-foot long 21-kW IBM 7030 "Stretch" behemoth, with 150,000 germanium bipolar transistors, that was the world's fastest computer in 1961. It was used by

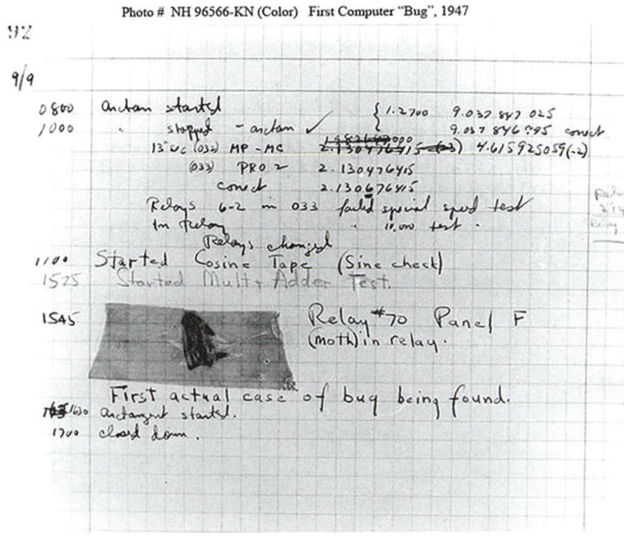

Fig. 1.10 First computer bug. (Image courtesy of the Naval Surface Warfare Center, 1945 [bug]. Public Domain.)

national labs in the United States for atomic weapons simulations. Although it was commercially disappointing, Stretch pioneered pipelining, multiprogramming, memory protection, and generalized interrupts that were later used on the System/360 computers.

1.4 MAINFRAMES AND MINICOMPUTERS

By the 1960s, large-scale *mainframe computers* became the focus of commercial computing. The name comes from the large cabinet, called the main frame, that holds the processor and memory. They initially operated in *batch* mode, operating on programs and data provided on punched cards or magnetic tape for applications such as census or payroll processing. Later, *timesharing* systems supported multiple users with interactive terminals, bringing applications such as airline booking and software development. Mainframes were so costly that they were mostly under centralized control of large organizations. Smaller *minicomputers* emerged in the 1960s and 1970s, typically costing tens of thousands of dollars, a price point that smaller organizations and universities could afford.

IBM bet its future on the System/360 series of computers in the 1960s (Fig. 1.11) with a common ISA shared by many microarchitectural implementations with different price/performance tradeoffs. IBM invested an astonishing $5 billion in development (half as much as the US spent on the Manhattan Project to develop the atomic bomb). Production delays nearly bankrupted the company, but the problems were resolved, and the wildly successful product line set the standard for *mainframe*

John von Neumann (1903–1957) was a mathematician and a founding figure in computing. Born in Budapest Hungary, von Neumann was a child prodigy. By 8 years of age, he was learning calculus and could speak six languages. He received his PhD in mathematics and soon joined the faculty at Princeton University's Institute for Advanced Study. Von Neumann made numerous fundamental contributions to pure and applied mathematics, established a mathematical framework for quantum mechanics, founded the field of game theory, and contributed to the development of the nuclear and hydrogen bombs and the notion of Mutual Assured Destruction. He is one of the founders of the field of computer architecture, publishing the idea of a stored program architecture. He died of cancer at 53 years of age, likely because of his work on the Manhattan Project.

(© 2008 Los Alamos National Laboratory [vonNeumann]. Reprinted with permission.)

Alan Turing (1912–1954) was a British mathematician and a founder of the fields of computer science and artificial intelligence (AI). Born in London, he attended the University of Cambridge and then received a PhD in mathematics from Princeton University in 1938. Turing developed the notion of a universal computing machine capable of carrying out any algorithm, now known as the Turing machine. He published a seminal proof that not all problems are computable. Turing returned to Britain to serve as a cryptanalyst during World War II and broke the German's Enigma Code. Historian Harry Hinsley estimated that the cryptanalysts may have shortened the war by two years and saved 14 million lives. Turing contributed to several early computer designs, wrote a program in 1950 to play chess, and coined the Turing Test for AI. Following the war, Turing was persecuted by the British government for being homosexual and had his security clearance stripped in 1952. He died of cyanide poisoning in 1954, which was ruled a suicide. The British government finally formally apologized for Turing's treatment in 2009.

(Alan Turing passport photo, age 16 years, 1928 or 1929 [Turing]. Public Domain.)

Fig. 1.11 IBM System/360 Model 85. (Image courtesy of IBM Corporation [IBM64]. Reprinted with permission.)

computers. Some models sold for up to $2 million. They spanned a wide range of performances, from the Model 30 with 8 KiB of magnetic core memory and 34.5 Kinstructions/second to the Model 91 with up to 6 MiB of memory and 16.6 Minstructions/second, filling a large room and consuming 94 KW of power.

As shown in Fig. 1.12, each bit of core memory was built from a ferrite donut hand woven into a mat of wires to flip orientation when a current was applied. Logic circuits were built with Solid Logic Technology (SLT), a precursor to integrated circuits [Davis64]. Fig. 1.13 shows an SLT module in which a transistor and several diodes and resistors are arranged on a shared substrate under a protective glass coat. Texas Instruments introduced the SN74xx-series of *small-scale integration* (*SSI*) chips in 1966, containing a handful of logic gates in a plastic package. These chips and their successors became widely used in both mainframes and minicomputers.

System/360 was succeeded by the compatible System/370 series in 1970, the System/390 series in 1990, and the z/Architecture in 2000 to provide software continuity over six decades. The z16 processor in 2022 has up to 200 cores, each running at up to 5.2 GHz, filling up to four refrigerator-sized frames, and it is still compatible with the original System/360 architecture. System/360 brought computing to the business mainstream and opened the ecosystem for many competitors, including Digital Equipment Corporation and Data General Corporation.

In the 1970s, Digital Equipment Corporation's 18-bit PDP-11 (Fig. 1.14) became the leading *minicomputer*, selling some 600,000 units at starting prices of $12,000. Early computers were built from racks of

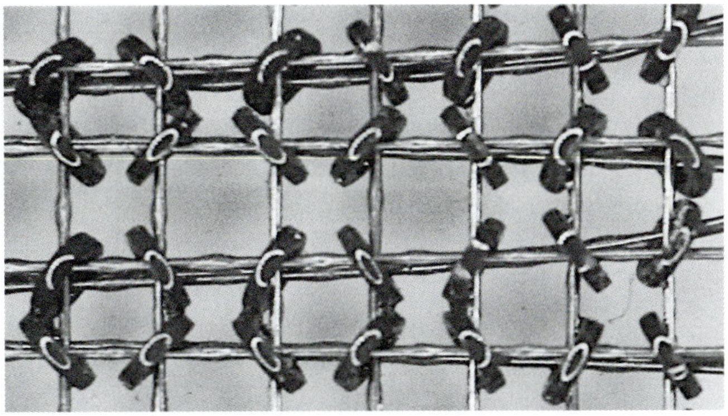

Fig. 1.12 Core memory. (Image courtesy of Ken Shirriff [memory]. Reprinted with permission.)

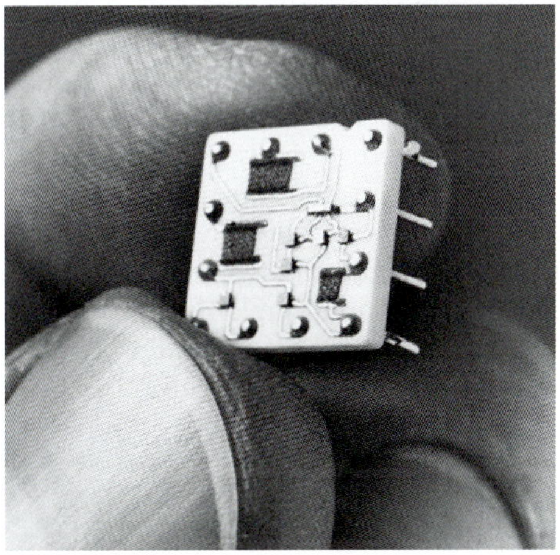

Fig. 1.13 IBM Solid Logic Technology. (Image courtesy of IBM Corporation [SLT]. Reprinted with permission.)

printed circuit boards containing hundreds of 74xx-series TTL logic gates. The PDP-11 soon found 18-bit addresses too constraining, so DEC followed up with the extremely successful 32-bit VAX *superminicomputer* that offered a large address space, virtual memory, floating-point operations, and vector processing.

The development of mainframes and minicomputers drove time-sharing OS advances. The Multics project, started in 1964 at MIT in collaboration with General Electric and Bell Labs, pioneered ideas in time-sharing, multiprocessing, virtual memory, security, and file systems. Multics ran behind schedule and required expensive hardware. When Bell Labs withdrew from the project, some of their developers took ideas from Multics to build a simpler OS capable of running on cheaper minicomputers such as a PDP-7 or PDP-11. These developers were Ken Thompson, Dennis Ritchie, Brian Kernighan, Douglas McIlroy, and Joe

Grace Hopper (1906–1992) was a pioneering computer scientist. She was born in New York City, received her PhD in mathematics from Yale in 1934, and joined the faculty at Vassar College. During World War II, she volunteered to join the Navy Reserve, where she programmed the Mark I computer to calculate the Navy's gunnery tables. In the 1950s, she proposed programming in high-level languages using English words and developed the COBOL language and compiler. She continued her Naval service and led initiatives to standardize the Navy's computer languages, retiring as a Rear Admiral (formerly called Commodore) at 79 years of age. She was awarded the National Medal of Technology and the Presidential Medal of Freedom for her service.

Rear Admiral Grace Hopper (Image courtesy of the United States Navy DN-SC-84-05971, 1984 [Hopper]. Public Domain.)

Fig. 1.14 DEC PDP-11/70. (Photo by Dave McGuire, 2009 [PDP-11]. Public Domain.)

Ossanna, who developed the Unix OS around 1970 [Ritchie84]. At the time, operating systems were usually written in assembly language. From 1972 to 1973, the team developed the C programming language and used it to rewrite Unix at a higher level. Unix became a broadly used OS for technical computing. However, it suffered from licensing issues and competing incompatible implementations. In 1991, Linus Torvalds began reimplementing Unix from scratch as open-source software. His Linux OS has become the leading Unix-like OS. Appendix B is a getting-started guide for chip designers using Linux, and Chapter 22 describes booting Linux in hardware.

Tracy Kidder's *The Soul of a New Machine* [Kidder81] gives a captivating account of the human side of developing the Data General Eclipse MV/8000 minicomputer, a 32-bit VAX competitor.

Ken Thompson (1943–) and Dennis Ritchie (1941–2011) received the Turing Award in 1983 for their contributions to operating systems, particularly their implementation of Unix at Bell Labs. Thompson received his bachelor of science and master's degrees in electrical engineering and computer science from Berkeley. Ritchie wrote his PhD thesis at Harvard University (but never bothered to pick up his degree). They both went directly to Bell Labs in the late 1960s.

Thompson *(left)* and Ritchie *(right)*. (Photo taken 1973 [ThompsonRitchie]. Public Domain.)

1.5 MICROPROCESSORS

Robert Noyce and Gordon Moore founded Intel in 1968. Their first product was the 3101 static RAM chip shown in Fig. 1.15. The chip had a whopping capacity of 64 bits, each clearly visible in the die photo, which was sufficient to begin the transition from magnetic core memories to integrated circuit RAM. In 1969 a Japanese calculator company called Busicom asked Intel to develop a 12-chip set for a desktop calculator. Ted Hoff was assigned to the design at Intel and soon realized that he could build the calculator more economically with a programmable processor. Too young to know it couldn't be done, Hoff and Federico Faggin designed Intel's 4-bit programmable 4004 in 1971. This chip is recognized as the world's first commercial microprocessor, in which all of the circuits were on a single chip. It used 2,300 transistors in a 10-μm silicon-gate pMOS process. Fig. 1.16 shows the 4004 die photo; some of the 4-bit datapaths are visible, such as the four repeating columns in the lower left. The layout was created from a red masking film called *rubylith*, cut by hand with a utility knife on a light table.

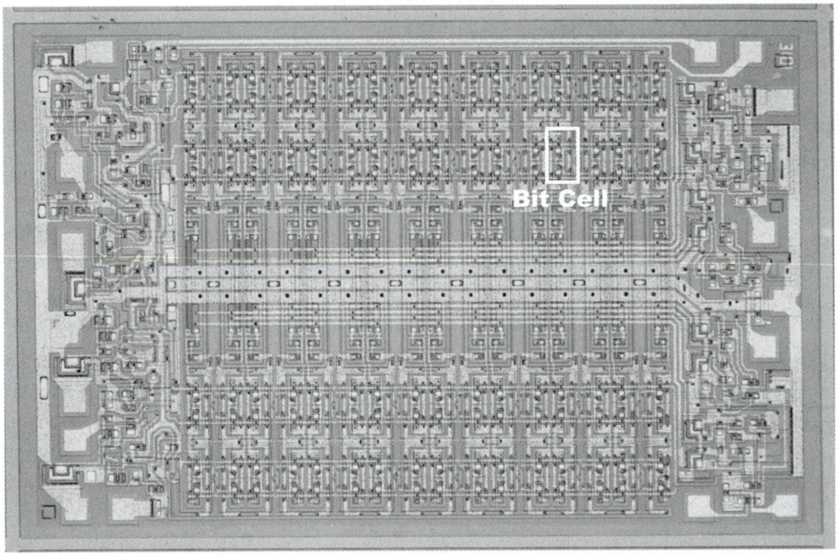

Fig. 1.15 3101 64-bit SRAM. (Image courtesy of Intel Corporation [Intel69]. Reprinted with permission.)

(a)

(b)

Fig. 1.16 Intel 4004 microprocessor: (a) die and (b) packaged chip. (Images courtesy of Intel Corporation [Intel71, Intel4004]. Reprinted with permission.)

The microprocessor revolution accelerated the progress of computing as integrated circuits rapidly improved transistor counts, speed, and cost. Soon processors became cheap enough that individuals could afford personal computers (PCs). Steve Wozniak and Steve Jobs founded the Apple Computer Company in 1976 to sell their Apple I computer, followed by the $1,300 Apple II in 1977, which was based on the 6502 microprocessor (Fig. 1.17). IBM introduced the PC in 1981 (Fig. 1.18). Because Intel's 8088 was available off the shelf in quantity at a good price, IBM's PC used that processor, which had a 16-bit instruction set but a low-cost 8-bit data bus. IBM was unwilling to gamble on young Intel as its sole supplier of processors, so it arranged a second-source agreement allowing another startup, AMD, to manufacture compatible chips. The PC was

Fig. 1.17 Apple II computer. (© 2016 Pratyeka [Applelle]. This image is openly licensed under CC BY-SA 2.0.)

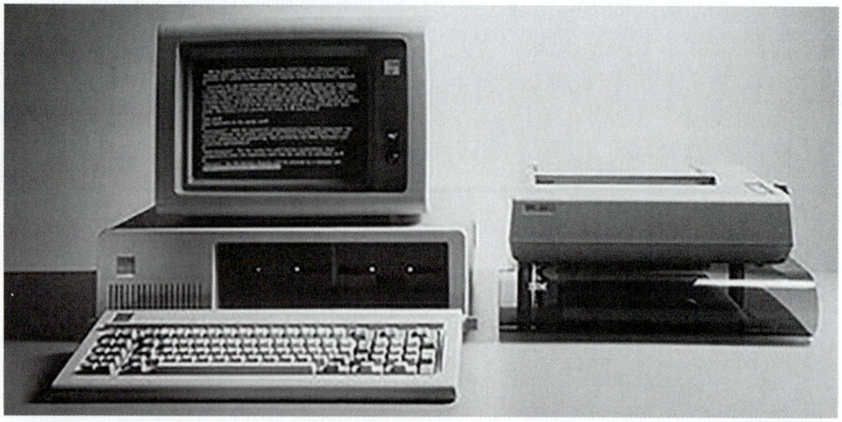

Fig. 1.18 IBM PC. (Figure courtesy of IBM Corporation [IBMPC]. Reprinted with permission.)

priced starting at $1,565. IBM hoped to sell 200,000 units a year but was soon selling that many per month, and the PC revolution took off. The 8088 was a cost-reduced version of Intel's 8086, which had a 16-bit data bus. The 80286 added 24-bit addresses and a memory management unit. The 80386 stepped up to a 32-bit instruction set. Fig. 1.19 shows the 80386 die photo. The 32-bit integer datapath is on the left. The external bus unit is on the upper right. A microcode programmable logic array (PLA) is in the lower right. The middle part of the chip has synthesized random logic. The 80486 added pipelining, a cache, and a floating-point unit. The Pentium introduced the execution (or *issuing*) of two instructions at once, called *2-way superscalar issue*, and the Pentium Pro/II/III/IV added wider out-of-order issue. A victim of its tremendous success, Intel's x86

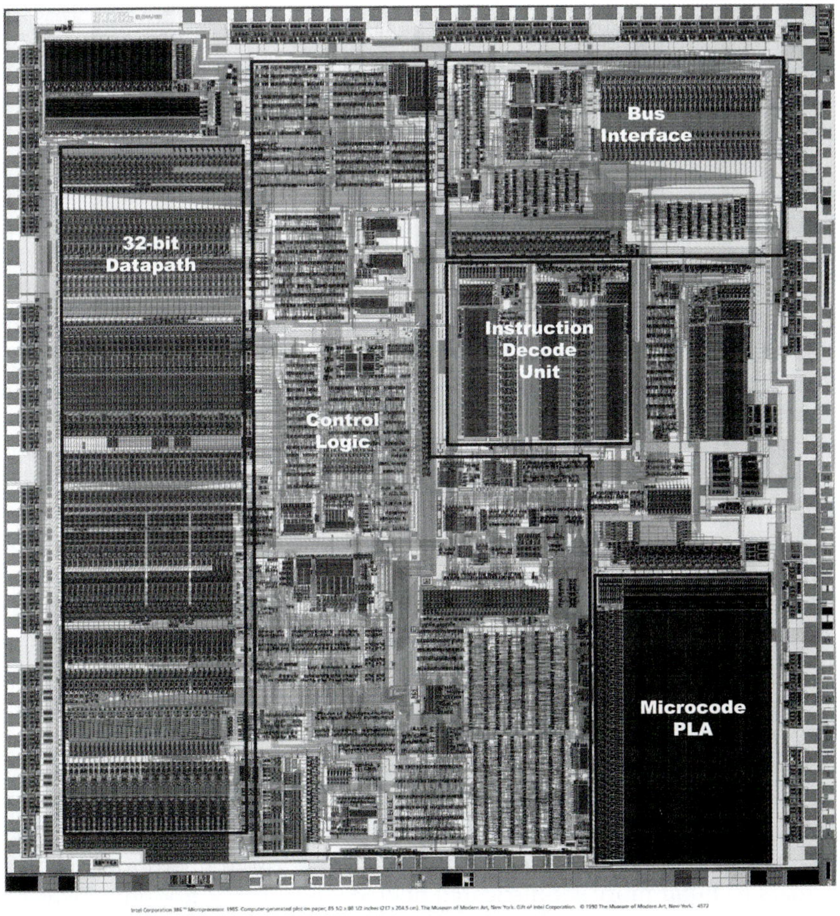

Fig. 1.19 Intel 80386 photo. (Image courtesy of Intel Corporation [Intel]. Reprinted with permission.)

architecture became a Frankenstein monster of backward-compatible variable-length instructions accumulated over the decades.

1.6 CISC AND RISC

In the 1970s, some computer architects sought to maximize the capability of assembly language instructions. For example, the x86 REP MOVS instruction moves an entire variable-length zero-terminated string from one address to another. Greenblatt [Greenblatt74] describes an entire machine optimized for running LISP. In contrast, David Patterson and David Ditzel laid out a 1980 manifesto [Patterson80] for *reduced instruction set computers (RISC)*. They argued that "the ultimate test for cost-effectiveness is the speed at which an implementation executes a given algorithm." Traditional *complex instruction set computers* (*CISC*), such as S/360, PDP11, and x86, had long design times, high rates of design errors, costly state to save during context switches, and long clock cycles. Only a quarter of the available instructions in some instruction sets account for 99% of the instructions used in real programs. Patterson and Ditzel proposed simplifying to a minimal set of instructions necessary to compile code and making these instructions run as fast as possible using a minimum number of clock cycles at a high clock rate. For example, RISC processors use a *load/store architecture* in which instructions operate on data in a small set of fast registers and use explicit load and store instructions to move data between memory and these registers.

The IBM 801, an experimental machine developed in the 1970s, is recognized as the first RISC processor. The architect, John Cocke, received the Turing Award for his contributions.

Many teams began developing RISC processors. Patterson's team at Berkeley developed the 32-bit RISC-I processor in 1982 [Sequin82]. The 32 rows of the datapath are visible in the die photo in Fig. 1.20. The large array on the left is the register file, with the address decoders at the top. The columns on the right are the integer execution unit datapath. The control circuits are in the upper right, implemented with some PLAs. John Hennessy's team at Stanford developed the MIPS-I processor in 1982–1984 [Hennessy82]. Steve Furber and Sophie Wilson at Acorn Computer developed the ARM1 processor in 1985. Hennessy and Patterson's *Computer Architecture, a Quantitative Approach* [Hennessy90] became the leading text on computer architecture, and the latest edition is a must-read for all serious students of computer architecture today.

David Patterson (1947–) has been a professor of computer science at Berkeley since 1976. He coinvented RISC with John Hennessy in 1984. It was later commercialized by Sun Microsystems as the SPARC architecture. Patterson helped develop the RISC-V architecture and continues to play an integral role in its development.

(Image courtesy of David Patterson. Reprinted with permission.)

Microprocessor design entered an intensely competitive period from 1986 to 2002, as shown in Fig. 1.21, when companies raced to claim market share by offering the best performance. The time to execute a program depends on the product of the number of instructions, the cycles per instruction, and the cycle time. Manufacturers chased Moore's Law aggressively to increase transistor speed and quantity, knowing their competitors were doing the same. Rising transistor counts allowed microprocessor designers to build more sophisticated microarchitectures, adopting features such as caches, branch prediction, superscalar, and out-of-order

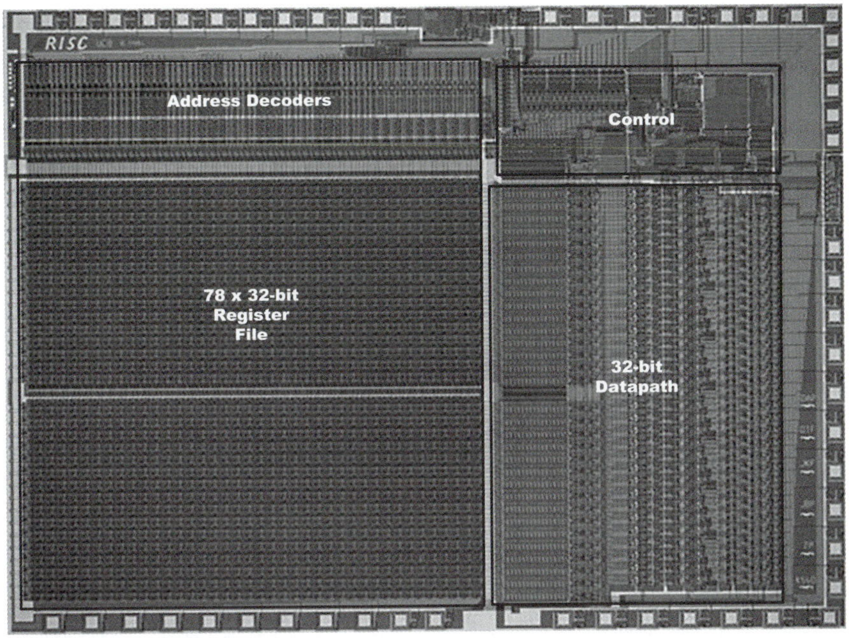

Fig. 1.20 RISC-I die photo. (Image courtesy of David Patterson. Reprinted with permission.)

execution pioneered by mainframes and minicomputers, driving down the cycles per instruction. Faster transistors and pipelining boosted clock speeds. Clever circuit designers innovated to reduce the number of gates between pipeline stages to improve clock speed even further. Digital Equipment Corporation introduced their Alpha 21064 RISC processor [Dobberpuhl92] running at 200 MHz in 1992, a year before Intel released their CISC (x86) Pentium chip running at 66 MHz. This stunning 3x clock rate advantage ended the debate of RISC vs. CISC. Intel became competitive again by translating its complex instructions into a sequence of one or more RISC-like microinstructions and then executing these microinstructions on a fast, efficient datapath, first seen in the Pentium Pro [Colwell05].

During the 1990s, processors were largely marketed based on clock rate. Intel, AMD, Alpha, MIPS, SPARC, and others competed for the highest clock rate during the dramatic "Megahertz Wars." For example, Intel's P6 microarchitecture came out at 200 MHz in the 1995 Pentium Pro processor in a 0.35-μm process and was aggressively pushed up to 1.13 GHz by the 1999 Pentium III in a 0.13-μm process. Intel then switched to their Netburst microarchitecture, reaching 3.4 GHz by 2002 for the Pentium 4 in the same 0.13-μm process. These clock frequencies were reached using superpipelining, with 14 pipeline stages in the Pentium III and 20 to 31 or more in the Pentium 4. The longer pipelines led to more pipeline hazards, so performance did not improve as much as clock rate

John Hennessy (1952–) is a professor of electrical engineering and computer science at Stanford University, where he served as president from 2000 to 2016. He coinvented RISC with David Patterson. He also developed the MIPS computer architecture and cofounded MIPS Computer Systems in 1984. The MIPS processor was used in many commercial systems, including products from Silicon Graphics, Nintendo, and Cisco. John Hennessy and David Patterson were given the Turing Award in 2017 for pioneering a quantitative approach to the design and evaluation of computer architectures.

(Image courtesy of John Hennessy. Reprinted with permission.)

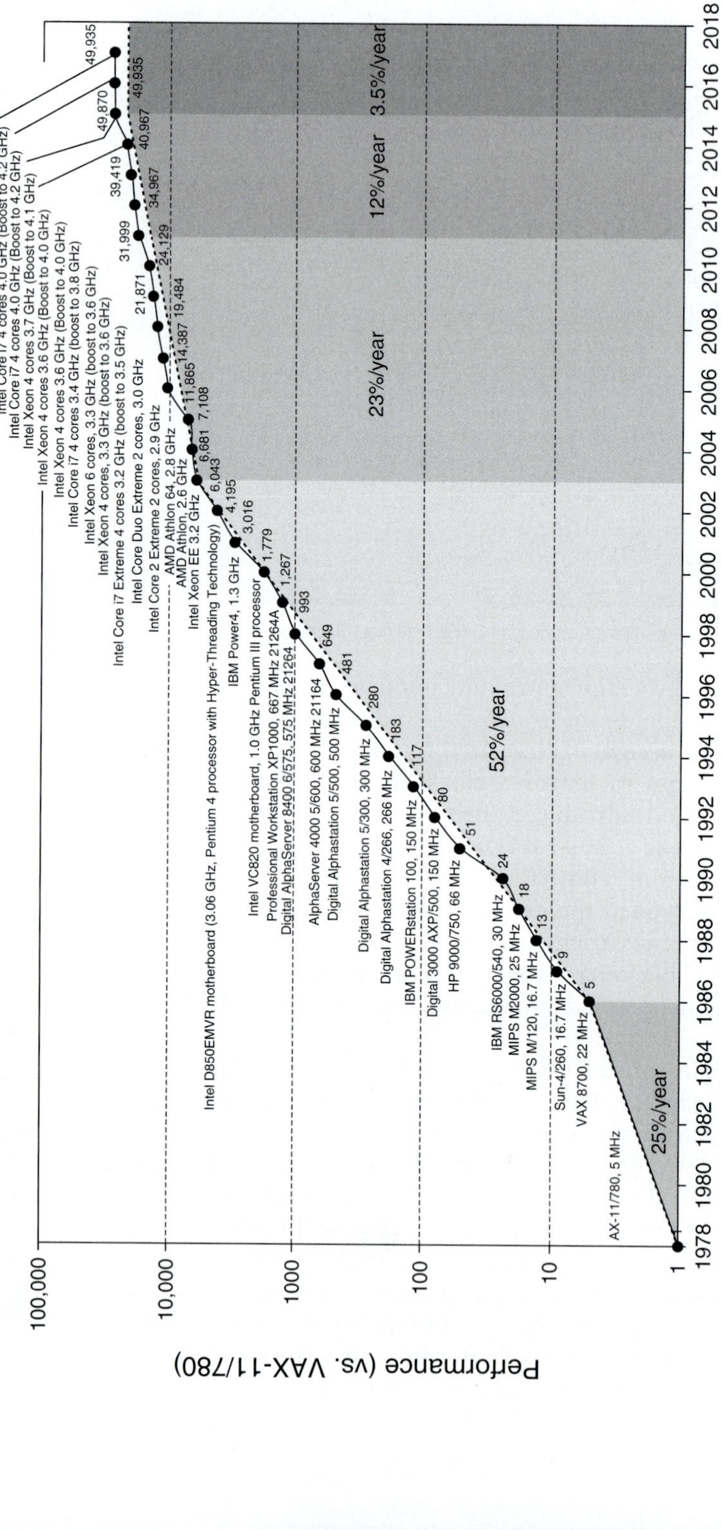

Fig. 1.21 **Microprocessor single-threaded performance.** (Figure courtesy of Elsevier [Hennessy19]. Reprinted with permission.)

increased, and in some cases, performance even decreased. Presently, pipeline depth ranges from 2 stages in low-cost microcontrollers to 5 to 8 stages in energy-efficient processors to 13 to 20 stages in high-performance processors.

Although long pipelines and high clock rates gave marginal performance benefits, they cost enormous amounts of power, often more than 100 W, making them expensive to cool and poorly suited to laptops. In a notorious keynote address at the International Solid-State Circuits Conference in 2001, Intel CTO Pat Gelsinger plotted the power density of processors over the years [Gelsinger01]. As shown in Fig. 1.22, he observed that "if scaling continues at present pace, by 2005, high-speed processors would have [the] power density of [a] nuclear reactor, by 2010, a rocket nozzle, and by 2015, [the] surface of sun." Investors may have taken this badly; Intel stock declined more than 30% in the subsequent months. But Gelsinger was arguing that scaling had to change. Processors had reached the *power wall*, in which clock frequency is constrained by power dissipation. In 2006, Intel abandoned the power-hungry Netburst microarchitecture and switched to the Core microarchitecture that attempts to get performance from parallelism (running multiple threads at once) using multiple power-efficient cores derived from the Pentium Pro. Fig. 1.23 shows that the rate of performance increase leveled off dramatically after microprocessors hit the power wall and the end of Dennard scaling around

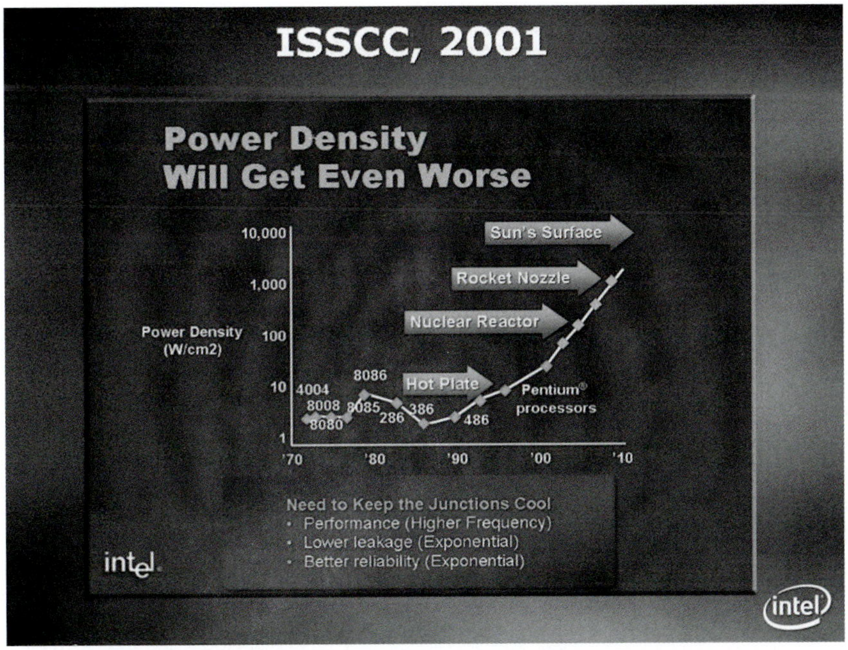

Fig. 1.22 Processor power density. (Figure courtesy of Intel Corporation [Gelsinger06]. Reprinted with permission.)

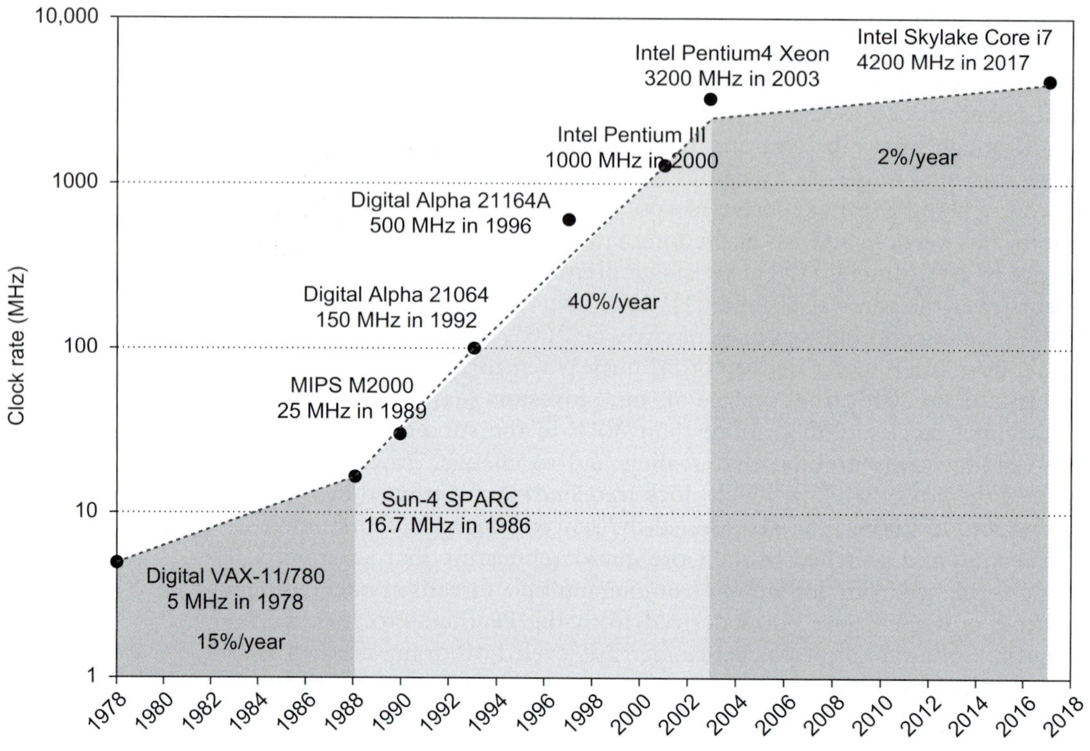

Fig. 1.23 Microprocessor clock speeds. (Figure courtesy of Elsevier [Hennessy19]. Reprinted with permission.)

2003, and Intel claimed most of the CPU market, reducing pressure for innovation. AMD took a sudden performance lead in 2020, and the renewed competition has been good for the industry and customers. Fig. 1.24 shows Intel's 13th-generation Core processor, Raptor Lake, with 24 total cores (16 energy-efficient cores and 8 high-performance cores running up to 5.7 GHz), 30 MiB of level 3 cache, graphics and video accelerators, and memory, display, and PCI expansion bus controllers, all fabricated on a 257-mm^2 die in the Intel7 process.

In recent years as innovation in PC processors has slowed, competition has shifted to the mobile device sector. Nearly all cell phones and tablets presently use ARM-based processors, which are also used in Apple computers and Raspberry Pi computers. ARM licenses its processor IP to companies such as Qualcomm, Samsung, MediaTek, and Apple, which design the rest of the SoC for mobile devices. Fig. 1.25 shows the logic board from the Apple iPhone 13 Pro, with an Apple A15 Bionic SoC. Recall the A15 die photo from Fig. 1.4.

Currently, more than 25 billion ARM processors ship each year, or more than 3 per human on earth, and most of these are embedded in systems where the customer may be unaware a computer exists. ARM makes

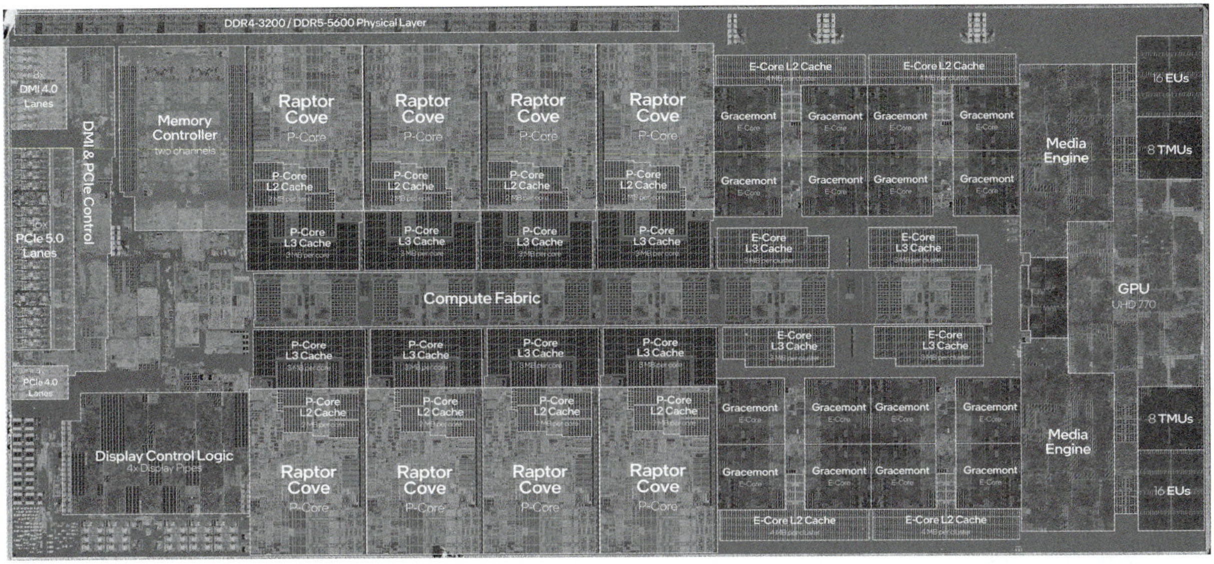

Fig. 1.24 **Intel Raptor Lake Core i9-13900K die, with 8 high performance and 16 energy-efficient cores.** (Image by JmsDoug and Fritzchens Fritz, 2023 [corei9]. Public Domain: CC0 1.0 Universal.)

Fig. 1.25 **Apple iPhone13 logic board with A15 processor.** (Image courtesy of TechInsights Inc. [iphone13]. Reprinted with permission.)

its revenue from licensing processors, so the company is known for jealously guarding its IP, to the point that it is difficult for universities to use the architecture in education, and ARM sued one of their largest customers, Qualcomm, in 2022 over a licensing dispute.

1.7 RISC-V

Krste Asanović started RISC-V as a summer project at Berkeley. He is a professor of computer science at Berkeley and served as chairman of the board for RISC-V International. He is also the cofounder and chief architect of SiFive, a company that develops RISC-V chips, boards, and supporting tools.

(Image courtesy of SiFive. Reprinted with permission.)

Since the RISC revolution, computer architects have gained more experience with architectural tradeoffs. The original ARM architecture made several choices that make high-performance implementations more difficult, including conditional execution, too few registers, flexible addressing modes using three source registers, and an optional shift before the arithmetic/logic unit (ALU). In 2010, Krste Asanović, Andrew Waterman, Yunsup Lee, David Patterson, and others at Berkeley developed a fifth-generation RISC architecture, dubbed RISC-V. It has many similarities to the MIPS instruction set but avoids the pitfalls known from MIPS and ARM (see Section 2.6). Most importantly, the RISC-V architecture is an open standard with no licensing fees, and it has a defined process to continue to evolve under the stewardship of the nonprofit organization RISC-V International. At the time of this writing, RISC-V has a tiny share of the overall processor market, but it has tremendous momentum. By 2024, more than 13 billion RISC-V cores shipped, including over a billion in NVIDIA GPUs. In 2024, the SHD Group forecast RISC-V revenue growing at a compound annual rate of 47% to $92 billion in 2030 [SHD24], as shown in Fig. 1.26. Based on current trends, it is likely that x86 processors will continue to run laptop and desktop computers, ARM will continue to run phones and tablets (perhaps with some competition from RISC-V), and RISC-V will displace most of the diverse proprietary architectures serving other markets such as embedded processors and data centers. According to Asanović at the 2022 RISC-V Summit, "RISC-V is inevitable." And according to RISC-V International's former CTO Mark Himelstein at the 2023 RISC-V Summit, "All your core are belong to us."

This quote refers to an early-2000s meme from the badly translated Zero Wing video game, "All your base are belong to us.")

1.8 INTERNATIONAL ECONOMIC AND SECURITY COMPETITION

In the 1970s, a handful of talented engineers could start a semiconductor company in a small building. A burst of startups in the 1970s and 1980s, centered in Silicon Valley but also active around the world, fed the highly innovative industry. Any company that wanted a fab could build its own. The exponentially increasing costs of developing advanced semiconductor technology have forced a shakeout, first eliminating smaller companies, then smaller countries, and now smaller continents. This drove manufacturing to foundries that produce chips for other companies. For example, in the United States, AMD and IBM, with large teams of brilliant technologists funded by formidable revenues from their microprocessor

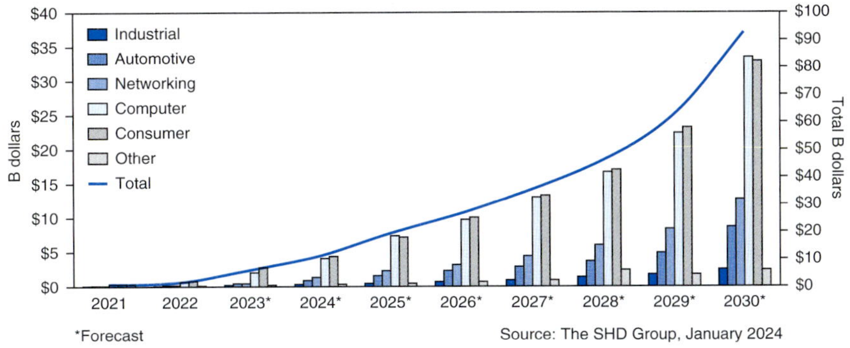

Fig. 1.26 Forecast RISC-V market growth. (Figure courtesy of The SHD Group [SHD24]. Reprinted with permission.)

Andrew Waterman is chief engineer at SiFive, a company he co-founded with Krste Asanović in 2015 to provide low-cost RISC-V cores and custom chips. He earned his PhD in computer science from Berkeley in 2016, where, weary of the vagaries of existing instruction-set architectures, he co-designed the RISC-V ISA and the first RISC-V cores.

(Image courtesy of Andrew Waterman. Reprinted with permission.)

and mainframe computer lines, ceased to be able to justify the cost of their fab development partnership. In 2008 and 2014, AMD and IBM sold their semiconductor business to the United Arab Emirates sovereign wealth fund which was seeking to diversify beyond oil. The resulting company, GlobalFoundries, manufactures chips in Singapore, Germany, and the United States. By 2015, Europe's largest semiconductor manufacturer, ST Microelectronics, was unable to keep up with the transition to a 14-nm FinFET process. They continue production, but not in the most advanced and profitable nodes. Japan, once home to many important semiconductor companies, fell to 2% of global foundry capacity by 2020. For decades, Intel led the industry, often a year or two ahead of the competition. Intel stumbled getting its 10-nm process into production and now trails behind the major foundries.

Fig. 1.27 shows the market share of the $102 billion global semiconductor foundry industry as of 2022 [IDC23]. Taiwan Semiconductor Manufacturing Company (TSMC) is by far the dominant foundry, producing chips for most of the biggest companies, including Apple, Qualcomm, AMD, NVIDIA, Broadcom, and MediaTek. Samsung (South Korea) is the second largest foundry, serving external customers as well as its enormous internal cell phone business. UMC (United Microelectronics Corporation in Taiwan) and GlobalFoundries (international) compete for the remaining business, largely by being lower-cost suppliers. Intel (United States) is the last manufacturer primarily building internal products (microprocessors and field programmable gate arrays), but it is starting to offer foundry services.

Countries that control semiconductor design and manufacturing have tremendous economic and security power. Innovations in technology have been a major driver of world economic growth, and countries such as the United States, South Korea, Taiwan, Israel, and India have benefited disproportionally. Artificial intelligence (AI) has the potential of being the next huge growth opportunity. NVIDIA's market value tripled to more than

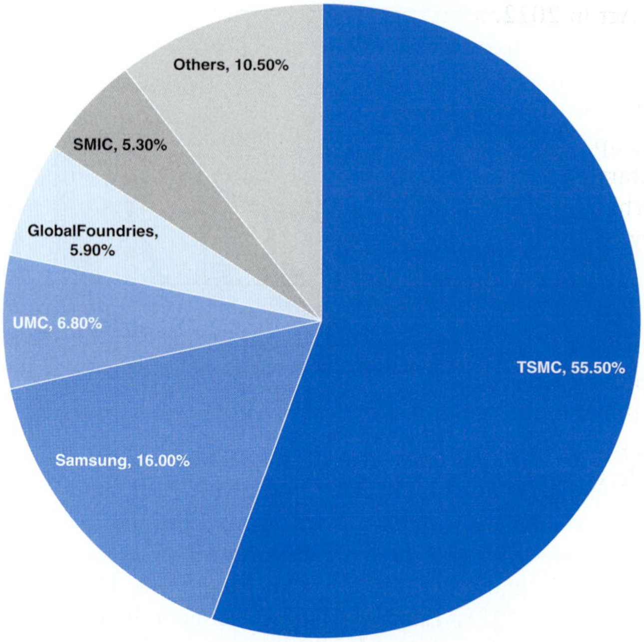

Fig. 1.27 **Global semiconductor foundry market share (2022).**

$3 trillion in 2024 because its chips are in great demand for machine learning. Vast amounts of the world's information pass through cell phone and fiberoptic networks. Governments have influenced the design of communication chips and systems to surreptitiously collect information. In 2018, the United States and other countries banned cellular providers from using Huawei and ZTE chips because of Chinese espionage concerns, and in 2023, China retaliated by banning government officials from using Apple iPhones. After Russia invaded Ukraine, Western countries attempted to enforce export controls to reduce Russia's access to chips used in their weapons systems. In a sign of growing economic competition, in 2023, the United States, Japan, and the Netherlands imposed export controls prohibiting the sale of advanced photolithography equipment to China, and China retaliated by cutting off exports of gallium, germanium, and other key materials used in chip manufacturing.

The two largest semiconductor producers are in areas of great geopolitical instability. South Korea is in range of conventional or nuclear attack by North Korea. China claims ownership of Taiwan and has been making heavy military investments, threatening invasion. Because many countries depend on TSMC for producing chips critical to both military and commercial activity, this has become a matter that threatens the national and economic security of most developed nations. The United States eventually recognized this risk, and a divided Congress passed the bipartisan

CHIPS Act in 2022, a rare act of industrial policy allocating $280 billion over a decade to spur chip research and production in the United States. The European Union, United Kingdom, and Japan have passed similar legislation to invest in this essential industry.

The open RISC-V architecture promises to change the power that the proprietary x86 and ARM architectures have held over international trade and perhaps the associated power the United States has held in the PC industry.

1.9 SUMMARY

Since the invention of the integrated circuit in 1958, chips have advanced to the point that we can build a chip with tens of billions of transistors for tens of dollars. The transistors switch in picoseconds using subfemtojoules of energy, allowing us to build mobile devices with a full day's battery life. Early electromechanical, vacuum tube, and transistor computers have given way to microprocessors, which have revolutionized nearly every aspect of society, from communication to medicine and from entertainment to warfare. The proprietary x86 and ARM architectures have long dominated personal computing and mobile devices. Since its inception in 2010, the open RISC-V architecture has grown at an unprecedented rate, now shipping more than 10 billion cores per year. RISC-V enables everything from simple microcontrollers that cost microcents to drop into an SoC, to application processors running Linux, to high-performance servers. It has also spurred development of a rich open-source software ecosystem.

The next chapter introduces the RISC-V architecture, including the base integer instruction set, privileged operations, the memory map, and standard extensions. It shows simplified single-cycle and pipelined implementations of a subset of the base instruction set, illustrating microarchitecture, logic design, and logic verification.

Introduction to RISC-V

2

2.1 **RISC-V Assembly Language**

2.2 **RISC-V Machine Language**

2.3 **Have a Hart**

2.4 **Memory Map**

2.5 **RISC-V Extensions and Profiles**

2.6 **Comparison with Other Architectures**

2.7 **RISC-V Microarchitecture**

2.8 **Wally System-on-Chip**

2.9 **RISC-V Market**

2.10 **Summary and a Look Ahead**

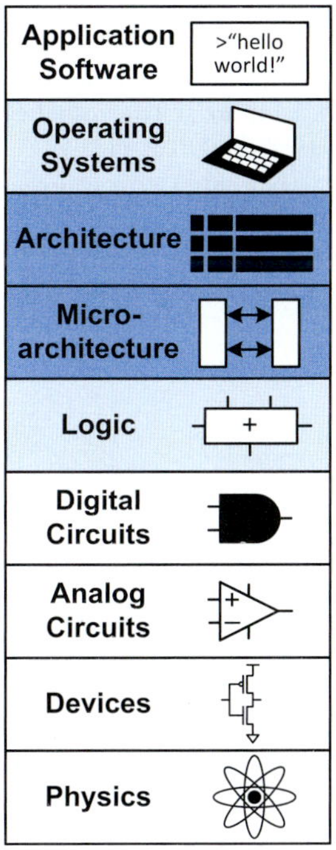

RISC-V is a fifth-generation *reduced instruction set computer* (*RISC*) architecture initially developed at Berkeley and now stewarded by the nonprofit RISC-V International. Its streamlined design reflects lessons from three decades of RISC processor sign, notably that clever optimizations in one era become implementation obstacles in a later era. Most importantly, RISC-V is a free and open standard that has led to widespread adoption and vibrant competition and innovation.

RISC-V is a load/store *instruction set architecture* (*ISA*) with 32-bit instructions, 32 general-purpose registers of width XLEN, and a program counter of width XLEN. The registers are named `x0` to `x31`, and `x0` is hardwired to 0. The base integer ISA comes in RV32I and RV64I flavors, corresponding to XLEN = 32 or 64. RV32I has 37 essential instructions summarized in Table I.1 (on the last pages of the book). RV64I adds 12 more instructions to handle 64-bit (double-word) loads and stores and to operate on only the bottom 32 bits of a 64-bit register.

The *architectural state* of a processor is the minimum set of information to determine what a program will do. For RISC-V, the architectural state consists of the memory, program counter, and registers, including optional *control and status registers* (*CSRs*). A processor can run many *threads*. Each thread has its own values of the program counter and registers, called the thread's *context*, but shares memory. The *operating system* (*OS*) gives the appearance of running all threads concurrently by switching from one to another every few milliseconds. This *context switch* involves saving the context of one thread to memory and reloading the program counter and registers with the context of the next thread.

RV32I also has `fence`, `ecall`, and `ebreak` instructions, bringing the official total to 40, but these can be ignored in a simple implementation.

A computer saves power by going into *sleep* mode. It saves the context of the current thread to memory and then powers off the processor while keeping the memory powered up. It can rapidly wake from sleep by powering up the processor and reloading the thread context from memory. A computer may reduce power consumption to zero by going into *hibernate* mode. It saves the entire architectural state, including the contents of memory, to disk, then powers off the entire system, including memory as well as the processor. It takes much longer to wake from hibernation because it has to reload the architectural state from disk.

Observe that the architectural state comprises the memory and the current thread's context.

Microprocessors are often classified as *application processors* or *embedded processors*. Application ("apps") processors generally run applications on a multitasking OS with a graphical user interface on a screen; these are the main processors in mobile devices, personal computers (PCs), and servers. Embedded processors are often invisible to the user; they may connect a switch to the window's motor controller in your car, make annoying noises in a child's toy, or manage voltage regulators and clock generators behind the scenes for the application processors in a cell phone system-on-chip (SoC). Because RV32 addresses are only 32 bits, the architecture has a 4-GiB address space. This is more than sufficient for embedded applications that often have kilobytes of code and data but is too small for many systems with a modern OS. Thus, application processors commonly use RV64.

The RISC-V architecture is officially defined at riscv.org/specifications/ratified, including the *Unprivileged Specification* that covers the main instruction set and extensions and the *Privileged Specification* that covers privilege modes, traps, and CSRs [RISCV-Unprivileged, RISCV-Privileged]. This section summarizes the RV32I and RV64I base instruction sets; see the Unprivileged Specification for more details. The Privileged Specification and standard extensions are covered in later chapters. The Unprivileged Specification is one of the best-written architecture specifications in existence, being succinct, clear, and annotated with commentary to explain the guiding principles behind the architectural choices. Those with serious interest in RISC-V should read these specifications rather than relying on the summaries in this book.

This chapter introduces the RISC-V architecture, including its assembly and machine language. It then illustrates the major design steps using toy single-cycle and pipelined RISC-V processors before introducing CORE-V Wally (cvw), a configurable RISC-V pipelined processor capable of running Linux. Wally supports the rich set of extensions in the RVA22S64 application processor profile, as well as virtual memory, caches, and branch prediction.

2.1 RISC-V ASSEMBLY LANGUAGE

The RV32I instruction set is summarized in Table I.1 (on the last pages of the book). The `op`, `funct3`, and `funct7` fields relate to the machine language encoding, which is discussed in Section 2.2. Instructions are classified into six types based on their operands and encoding, as shown in Table 2.1.

2.1.1 R-Type Instructions

R-type instructions have only register operands: two source registers and one destination register. For example, `add x3, x1, x2` computes the sum of

Table 2.1 Instruction types.

Type	Description	# Source Registers	# Destination Registers	Immediate
R	Register-type	2	1	N/A
I	Immediate-type	1	1	12-bit
S	Store-type	2	0	12-bit
B	Branch-type	2	0	13-bit branch offset
J	Jump-type	1	1	21-bit jump offset
U	Upper immediate-type	1	1	20-bit upper immediate

the values in source registers `x1` and `x2` and puts the result in destination register `x3`. R-type instructions are executed by the integer *arithmetic/logic unit (ALU)*, and they include:

- Arithmetic operations: `add`, `sub`
- Logical operations: `and`, `or`, `xor`
- Shift operations: `sll`, `srl`, `sra`
- Comparison operations: `slt`, `sltu`

Arithmetic and logical operations are the same, whether their operands are treated as unsigned or two's complement signed numbers. Recall that a *two's complement* signed binary number has a negative weight for the most significant bit. For example, the bits 1110_2 interpreted as a 4-bit *unsigned* number have a value of $1\times2^3 + 1\times2^2 + 1\times2^1 + 0\times2^0 = 14_{10}$. The same bits interpreted as a 4-bit signed two's complement number have a value of $1\times-(2^3) + 1\times2^2 + 1\times2^1 + 0\times2^0 = -2_{10}$. The bits $0001_2 = 1_{10}$ as either unsigned or two's complement. For example, `add` 1110, 0001 = 1111, which is 15_{10} as unsigned or -1_{10} as two's complement, and is correct in either interpretation.

Shift operations multiply or divide by powers of 2 by shifting left or right. Shift left logical (`sll`) fills the bottom bits with 0s. Shift right logical (`srl`) fills the upper bits with 0s. Shift right arithmetic (`sra`) fills the upper bits with the old sign bit. A left or right shift by k is equivalent to a multiply or divide by 2^k. When dividing, use `srl` if the number is unsigned or `sra` if it is signed. For example, `sll` 1110, 0001 = 1100. As a two's complement number, the result is $-2 << 1 = -2 \times 2^1 = -4$. As an unsigned number, the result is 14 << 1, which overflows to 12. `srl` 1110, 0001 = 0111. As an unsigned number, the result is $14 >> 1 = 14/2^1 = 7$. `sra` 1110, 0001 = 1111. As a two's complement number, the result is $-2 >> 1 = -2/2^1 = -1$.

Set less than (slt) comparison sets the result to 1 if the first operand is less than the second or 0 otherwise. Comparisons and branches (slt, blt, bge) ordinarily treat register values as signed two's complement integers, but the unsigned flavors (sltu, bltu, bgeu) treat them as unsigned integers instead. For example, slt 1110, 0001 = 1 because $-2 < 1$, but sltu 1110, 0001 = 0 because $14 \geq 1$.

2.1.2 I-Type Instructions

I-type instructions take a source register, a 12-bit signed *immediate* (i.e., a constant in the instruction that is immediately available without accessing memory), and a destination register. The immediate is sign-extended to XLEN before use. These instructions include:

- Loads: lw, lh, lhu, lb, lbu
- Jumps: jalr
- Arithmetic operations: addi
- Logical operations: andi, ori, xori
- Shift operations: slli, srli, srai
- Comparison operations: slti, sltiu

Note that subi is not needed because it can be performed by an addi with a negative immediate.

Beware that sltiu sign-extends the immediate to 32 bits before treating it as an unsigned number for comparison!

Loads read a value from memory and place it in the destination register. The memory address is computed by adding the sign-extended immediate to a base address in the source register. *Subword loads* come in normal and unsigned versions. The normal version (lb/lh) sign-extends the byte or half-word to the full XLEN length. The unsigned version (lbu/lhu) zero-extends.

Jump and link register (jalr) computes an address in the same way as loads: as the sum of the base address in the source register, and a sign-extended immediate. It jumps to this address. It also saves PC+4 in the destination register to remember where to return to later; this is called *linking*.

Other I-type arithmetic, logical, shift, and comparison instructions resemble R-type instructions but use the source register as the first operand and the immediate as the second operand. Immediate shift instructions (slli, srli, srai) shift by the unsigned number in the lower 5 bits of the immediate.

2.1.3 S-Type Instructions

S-type instructions are stores. They take two source registers and a 12-bit signed immediate. Like loads, they add the immediate to the first source register to compute an address. Then they store the contents of the second source register into that address. These instructions include store word/half/byte: sw/sh/sb.

2.1.4 B-Type Instructions

B-type instructions are branches. Like S-type instructions, they also take two source registers and a 13-bit signed immediate. They compare the two source registers. If the condition is satisfied, they take the branch by sign-extending the immediate (also called the *branch offset*) and adding it to the PC. Branches include branch if equal/not equal (`beq`/`bne`) and branch if less than/greater than or equal in both signed (`blt`/`bge`) and unsigned (`bltu`/`bgeu`) versions. Because the 13-bit branch offset will always be a multiple of 2, and thus its least significant bit will always be 0, only its upper 12 bits are encoded in the instruction's immediate field. The processor must append the implicit least significant 0 before sign-extending and adding it to the PC. So, the branch offset spans the range [−4096, 4094].

Ordinary RISC-V instructions are 4 bytes long, but compressed instructions are 2 bytes, so instructions may be on 2-byte boundaries. They are never at odd addresses, so the least significant bit (lsb) of the branch offset is always 0. RISC-V gains additional branching range by omitting the branch offset's lsb when encoding this immediate in the instruction.

2.1.5 J-Type Instructions

The only J-type instruction is *jump and link* (`jal`), which makes an unconditional PC-relative jump and saves PC+4 in the destination register as the return address. `jal` takes a destination register and a 21-bit two's complement immediate *(jump offset)*. Like branches, `jal` only encodes the upper 20 bits of the jump offset in the instruction because the lsb is always 0. When executing `jal`, the processor appends a 0 to the encoded 20 bits of the immediate before sign-extending it. Thus, the jump offset spans the range [−1048576, 1048574]. `jal` adds this jump offset to the PC and jumps to that address. For unconditional jumps (`j`), `x0` is used as the destination register to discard the return address.

2.1.6 U-Type Instructions

Like J-type instructions, U-type instructions also use a destination register and a 20-bit immediate. They shift the immediate left by 12 bits into the upper bits of a 32-bit word, filling the lower bits with 0s.

The load upper immediate (`lui`) instruction places the shifted immediate in the destination register. In combination with `addi`, this is useful for creating 32-bit constants.

Remember that 12-bit immediates are sign-extended, which could throw off the upper bits by 1. For example, to load the constant 0x00003800 into `x1`, use the following sequence. −2048 in 12-bit two's complement is 0x800, which is sign-extended to 0xFFFFF800.

The C language and RISC-V assemblers use `0x` and `0b` prefixes to indicate hexadecimal and binary, respectively.

```
lui  x1, 0x00004        # x1 = 0x00004000
addi x1, x1,-2048       # x1 = 0x00004000 + 0xFFFFF800 = 0x00003800
```

The more intuitive sequence of `lui x1, 0x00003; addi x1, x1, 0x800` would produce 0x00002800 instead of the intended result.

The *add upper immediate to PC* (`auipc`) instruction adds the shifted immediate to the program counter and places it in a destination register. `auipc rd, 0` is used to copy the PC into `rd`. `auipc` is also useful for computing arbitrary offsets relative to the program counter for either loads/stores or jumps. For either case, use `auipc` to add the upper 20 bits of the offset to the PC into a temporary register. Then, for a PC-relative load or store, use the 12-bit immediate offset to add the low bits to the temporary register to form the address. For a long PC-relative jump, use `jalr` to add the lower 12 bits. For example, if the PC is initially at 0x80001000, to jump ahead by 0x10002456 to 0x90003456, use the following code sequence:

Notice how `x1` is conveniently used as a temporary register because it will be overwritten with the return address.

```
auipc x1, 0x10002     # x1 = 0x80001000 + 0x10002000 = 0x90003000
jalr  x1, 0x456(x1)  # Jump to 0x90003000 + 0x00000456 = 0x90003456
                      # x1 = PC + 4 = 0x80001008
```

2.1.7 Other Instructions

The base instruction set defines `ecall` and `ebreak` for OS calls and debugger breakpoints, as discussed in Section 2.3.6. If there is no privileged `mtvec` trap vector control register, these instructions do nothing.

The base instruction set also defines the `fence` instruction for memory ordering. A simple single-core in-order processor such as Wally inherently executes memory accesses in program order and thus can ignore `fence`. Section 17.2.2 discusses fences for concurrent systems.

Consider adding the following two 32-bit integers on a 64-bit ALU. The 32-bit result should be 0:
0x8000_0000 + 0x8000_0000
= 0x0000_0000

These numbers are represented in 64-bit form by sign-extending from 32 to 64 bits: 0xFFFF_FFFF_8000_0000.

Computing the addition on a 64-bit ALU:
0xFFFF_FFFF_8000_0000+
0xFFFF_FFFF_8000_0000
= 0xFFFF_FFFF_0000_0000.

But when interpreted as a 32-bit result, the answer should be 0, so the upper bits are discarded and the 32-bit result is sign-extended, which instead results in 0x0000_0000_ 0000_0000, as desired.

2.1.8 RV64I Additional Instructions

Table I.2 (on the last pages of the book) summarizes the additional instructions in RV64I.

RV64I adds `ld` and `sd` to load and store 64-bit *double words*. It also adds `lwu` to load an unsigned word into the bottom 32 bits of a 64-bit register and fill the top 32 bits with 0s. Furthermore, shift amounts are extended to 6 bits.

In C-like languages, the 32-bit `int` data type is common even in 64-bit code. 32-bit `int`s are represented in 64-bit registers with bit 31 (the sign bit) extended into the upper bits (63:32). Ordinary 64-bit add, subtract, and shift instructions can produce incorrect results when the inputs and outputs are interpreted as 32 bits. These would require costly masking sequences to fix, so RV64 provides add, subtract, and shift instruction versions with a `w` suffix (which we call RW- and IW-type instructions) to support 32-bit operations. Any result bits overflowing the 32-bit word are discarded, and then the 32-bit result is sign-extended to fill the 64-bit register.

2.1.9 Register Conventions

The `x0` register is also called `zero` and is hardwired to 0 because it is such a commonly used constant. In principle, registers `x1`-`x31` could be used for any purpose. However, RISC-V establishes some conventions given in Table I.4 (on the last pages of the book) so that different programmers can write different blocks of code that play together nicely. `x1` is called `ra` and is used to hold the *return address* when making a function call with a `jal` instruction. `x2` is called `sp` and points to the *top of the stack*. The `s0`-`s11` *saved* registers are used for local variables that must preserve their value across function calls, while the `t0`-`t6` *temporary* registers hold intermediate results that can be overwritten during function calls. Function calls pass their inputs in the `a0`-`a7` *argument* registers and return their result in `a0`.

The strange ordering of register names was selected so that RV{32/64}E systems with only 16 registers (`x0`-`x15`) would still have some registers available for each purpose.

2.1.10 Endianness

A processor's *endianness* defines how it stores the bytes within a word, as shown in Fig. 2.1. In little-endian, the least significant byte in a 32-bit word is stored at an offset of 0 from the start of a word address, and the most significant byte is stored at an offset of 3. In big-endian, the order is reversed. For example, consider storing the word 0x1234F0E0 at address 0x8, as shown in Fig. 2.1. In a big-endian machine, address 0x8 contains byte 0x12, and address 0xB contains 0xE0. In a little-endian machine, address 0x8 contains 0xE0, and 0xB contains 0x12. A `lb t0, 8(x0)` instruction that reads a byte from address 8 will return 0x12 for big-endian and 0xE0 for little-endian machines.

RISC-V is little-endian by default. The `status` register can optionally switch the processor into big-endian mode for loads and stores, as described in Section 8.2.3.1. Instructions are always fetched in little-endian form.

2.1.11 Pseudoinstructions

RISC-V is missing many instructions that would be useful to a programmer, but these can be implemented using existing instructions.

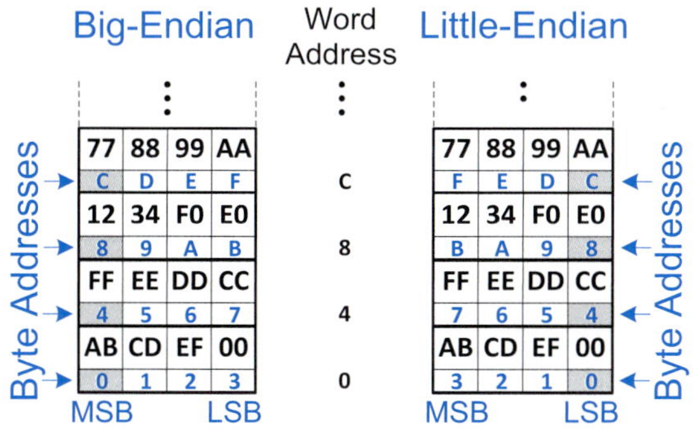

Fig. 2.1 Big and little-endian memory addressing.

For example, we might want a *load immediate* instruction `li rd, imm` that loads an immediate into register `rd`. For small immediates, this can be implemented using the real RISC-V instruction `addi rd, x0, imm`, which adds the immediate to 0 and places it in `rd`. Thus, `li` is called a *pseudoinstruction*. The programmer can use the pseudoinstruction, but the assembler replaces it with the real `add` instruction. Similarly, the *branch on less than or equal* pseudoinstruction `ble rs1, rs2, label` is replaced with the real instruction `bge rs2, rs1, label`; the *jump* pseudoinstruction `j label` is replaced by `jal x0, label`; and the *move* pseudoinstruction `mv rd, rs1` is replaced by `addi rd, rs1, 0`. Table I.7 (on the last pages of the book) summarizes common RISC-V pseudoinstructions, and is worth reviewing.

A full list of pseudoinstructions can be viewed at github.com/riscv-non-isa/riscv-asm-manual/blob/main/src/asm-manual.adoc#pseudoinstructions.

2.1.12 C to Assembly Language Examples

Programs are typically written in a high-level language and then compiled into assembly language. C is popular for system programming because it supports pointers, so programmers can directly access locations in memory. Other popular languages such as C++ and Java are closely related. This section reviews how C programs are translated to assembly language to help you think about both languages.

In each of the following programs, assume the following variables have been declared and that they are kept in specific registers. For example, `a` is in register `s0`, `b` is in register `s1`, and so on.

```
int a, b, c;       // a, b, c, in s0, s1, s2
int i, j;          // i, j in s3, s4
int grades[100];   // Base address of grades in s5
char name[80];     // Base address of name in s6
```

Code Example 2.1 shows how arithmetic in C translates into an R-type instruction in assembly language. Code Example 2.2 shows a logical instruction and how to use a constant expressed in hexadecimal. Code Example 2.3 shows how to break more complex arithmetic into a sequence of simpler steps; notice how it uses a temporary register `t0` to hold the intermediate result.

Code Example 2.1 SIMPLE ADDITION

C Code	Assembly Language
`a = b + c;`	`add s0, s1, s2        # a = b + c`

Code Example 2.2 ANDING WITH A CONSTANT

C Code

```
a = b & 0xFF;
```

Assembly Language

```
andi s0, s1, 0xFF            # 0xFF = 11111111
```

Code Example 2.3 TEMPORARY REGISTER (`t0`) HOLDING INTERMEDIATE CALCULATION

C Code

```
a = a + b − c;
```

Assembly Language

```
add t0, s0, s1         # t0 = a + b
sub s0, t0, s2         # a = t0 - c
```

Code Example 2.4 shows an `if` statement. The assembly language branches over the body if the opposite condition is true (`bne`, for `==`). Note the use of a label indicating where to resume. Code Example 2.5 shows an `if/else` statement. The assembly language branches to `else` if the opposite condition is true. Otherwise, it does the body of the `if` and then jumps past the `else`.

Code Example 2.4 `if` STATEMENT

C Code

```
if (i == j)
  a = 1;
```

Assembly Language

```
     bne s3, s4, done    # Skip if i != j
     li s0, 1
done:
```

Code Example 2.5 `if/else` STATEMENT

C Code

```
if (i == j)
  a = 1;

else
  a = 2;
```

Assembly Language

```
      bne s3, s4, else # i != j
      li s0, 1         # a = 1
      j done
else: li s0, 2         # a = 2
done:
```

Code Example 2.6 shows a `while` loop to add up even numbers from 0 to 10. The code initializes `a` and `i`, then loads the comparison value (10) into a temporary register. It exits the `while` loop if the opposite condition is satisfied: `bgt` (branch if greater than) for `<=` (less than or equal comparison). Otherwise, it executes the body of the loop and then jumps back to check the condition again.

Code Example 2.6 `while` LOOP TO ADD EVEN NUMBERS FROM 0 TO 10

C Code

```c
a = 0;
i = 0;

while (i <= 10) {
  a = a + i;
  i = i + 2;
}
```

Assembly Language

```
   li s0, 0              # a = 0
   li s3, 0              # i = 0
   li t0, 10             # Need 10 for compare
while:
   bgt s3, t0, done      # Done if i > 10
   add s0, s0, s3        # a = a + i
   addi s3, s3, 2        # i = i + 2
   j while               # Repeat while loop
done:
```

Code Example 2.7 shows a `for` loop to compute $a = 2^b$. A `for` loop has four parts: the initialization, comparison, postoperation, and body. The assembly language first does the initialization (`i = 1`). Then, at a label, it checks the opposite condition (>) to exit the loop. It performs the body (`a = a * 2`) and then the postoperation (`i++`) and then jumps back to the label to repeat the loop.

Code Example 2.7 `for` LOOP TO COMPUTE $A = 2^B$

C Code

```c
a = 1;
for (i = 1; i <= b; i++)
  a = a * 2;
```

Assembly Language

```
      li s0, 1           # a = 1
      li s3, 1           # i = 1
for:  bgt s3, s1, done   # Done if i > b
      slli s0, s0, 1     # a = a * 2
      addi s3, s3, 1     # i++
      j for              # Repeat loop
done:
```

In C, & is read as "address of." Hence, `&grades[i]` is the address of the *i*th element of the grades array.

Code Example 2.8 accesses fixed elements of an array using `lw` and `sw` with constant offsets. An `int` is 4 bytes, so `grades[1]` is found 4 bytes beyond the beginning of the array. Code Example 2.9 shows a loop accessing and incrementing elements of an array. The address of `grades[i]` is 4*`i` bytes past the beginning, so the loop uses a shift to multiply `i` by 4, then adds this offset to the base address to find the address of `grades[i]`.

Code Example 2.8 ACCESSING SINGLE ELEMENTS OF AN INTEGER ARRAY

C Code	Assembly Language

`grades[1] = grades[0] + 17;`	`lw t0, 0(s5)       # t0 = grades[0]`
	`addi t0, t0, 17     # t0 = t0 + 17`
	`sw t0, 4(s5)        # grades[1] = t0`

Code Example 2.9 `for` LOOP GRADE INFLATION

C Code

```
for (i = 0; i < 100; i++)
  grades[i] = grades[i] + 14;
```

Assembly Language

```
  li s3, 0              # i = 0
  li t0, 100            # t0 = 100 for <
for:
  bge s3, t0, done      # Done if t0 >= 100
  slli t1, s3, 2        # t1 = i * 4
  add t1, s5, t1        # t1 = &grades[i]
  lw t2, 0(t1)          # t2 = grades[i]
  addi t2, t2, 14       # t2 = grades[i] + 14
  sw t2, 0(t1)          # grades[i] = t2
  addi s3, s3, 1        # i++
  j for                 # Repeat
done:
```

Code Example 2.10 illustrates string handling by computing the length of a string. Recall that C strings are arrays of characters (bytes) terminated by a 0 (NULL) byte. Also recall that an expression in C is considered FALSE if 0 and TRUE if nonzero. Hence `while(name[i])` loops until reaching the NULL byte at the end of the string. The assembly language uses the `lb` instruction to access `name[i]`.

Code Example 2.10 `while` LOOP THAT COUNTS THE NUMBER OF LETTERS IN A NAME

C Code

```
i = 0;
while (name[i])
  i++;
```

Assembly Language

```
       li s3, 0               # i = 0
while: add t0, s6, s3         # t0 = &name[i]
       lb t1, 0(t0)           # t1 = name[i]
       beq  t1, zero, done    # (name[i]==0)?
       addi s3, s3, 1         # i++
       j while                # Repeat
done:
```

Code Example 2.11 shows a factorial function, which illustrates passing an argument, returning a result, and saving and restoring registers on the stack. The argument `n` arrives in `a0`. The function uses two local variables in `s0` and `s1`, so it saves the old values of these registers on the stack

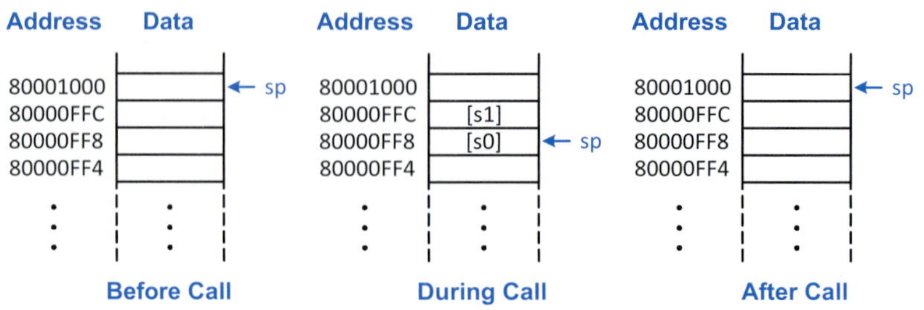

Fig. 2.2 Stack for factorial function from Code Example 2.11.

fact uses the mul (multiply) instruction from the M extension.

before changing them and restores these registers at the end. The stack pointer moves down and up by 8 bytes (two 4-byte words) to make room for these registers, as shown in Fig. 2.2. Like all functions, fact places the return value (in this case, result) in a0 before returning to the caller using the ret pseudoinstruction.

Code Example 2.11 FACTORIAL FUNCTION

C Code

```
int fact(int n) {
  int result = 1; // In s0
  int i;          // In s1

  for (i = 1; i <= n; i++)
    result = result * i;
  return result;
}
```

Assembly Language

```
fact: addi sp, sp, -8     # Room on stack
      sw s0, 0(sp)        # Save s0
      sw s1, 4(sp)        # Save s1
      li s0, 1            # result = 1
      li s1, 1            # i = 1
for:  bgt s1, a0, done    # i > n?
      mul s0, s0, s1      # result *= i
      addi s1, s1, 1      # i++
      j for               # Repeat
done: mv a0, s0           # Return result
      lw s1, 4(sp)        # Restore s1
      lw s0, 0(sp)        # Restore s0
      addi sp, sp, 8      # Restore stack
      ret                 # Return
```

An optimizing compiler will observe that there are no function calls within fact, so result and i can be kept in temporary registers rather than needing to save and restore s0 and s1, as shown in Code Example 2.12.

Code Example 2.12 OPTIMIZED FACTORIAL FUNCTION

C Code

```
int fact(int n) {
  int result = 1; // In t0
  int i;          // In t1

  for (i = 1; i <= n; i++)
    result = result * i;
  return result;
}
```

Assembly Language

```
fact: li t0, 1            # result = 1
      li t1, 1            # i = 1
for:  bgt t1, a0, done    # i > n?
      mul t0, t0, t1      # result *= i
      addi t1, t1, 1      # i++
      j for               # Repeat
done: mv a0, t0           # Return result
      ret                 # Return
```

Code Example 2.13 illustrates the factorial function again, this time written recursively. Before the function can call itself, it must save the previous return address and argument n on the stack.

Code Example 2.13 RECURSIVE FACTORIAL FUNCTION

C Code

```
int factr(int n) {
  if (n <= 1)
    return 1;

  else
    return n*factr(n-1);
}
```

Assembly Language

```
factr: li t0, 1              # For compare
       bgt a0, t0, else      # n > 1
       li a0, 1              # Return 1
       ret                   # Return
else:  addi sp, sp, -8       # Room on stack
       sw ra, 0(sp)          # Save return adr
       sw a0, 4(sp)          # Save n
       addi a0, a0, -1       # n-1
       jal factr             # Call factr(n-1)
       lw t0, 4(sp)          # Restore n
       lw ra, 0(sp)          # Restore ret adr
       addi sp, sp, 8        # Restore stack
       mul a0, t0, a0        # n * factr(n-1)
       ret
```

2.2 RISC-V Machine Language

Machine language is the binary representation of assembly language. Every assembly language instruction corresponds to a 32-bit machine language instruction stored in the computer's program memory. Fig. I.1 (on the last pages of the book) summarizes the six instruction formats used by RV{32/64}I and an R4 format used for floating-point multiply-accumulate instructions.

RV32I instructions are distinguished by their op, funct3, and funct7 fields, as given in Table I.1. The least significant 7 bits of an instruction always contain op, which defines the type of instruction so that an instruction decoder knows how to interpret the remaining bits. For example, all R-type instructions have op = 51 (0x33). The processor must examine the funct3 and funct7 fields to determine which R-type instruction is being executed.

The rs1, rs2, and rd fields are always found in the same bit positions of the instruction if they are needed. Thus, these fields can connect directly to address inputs of the register file without a need for multiplexing, saving time and delay compared with other ISAs.

The immediate bits are sometimes scrambled around the instruction in an odd way, as shown in Fig. 2.3, to keep the other fields in the same locations and to encode each immediate bit in the same instruction bit when possible. This bit swizzling does not impact performance because the immediate bits can be rearranged and extended while the register file is reading a source register.

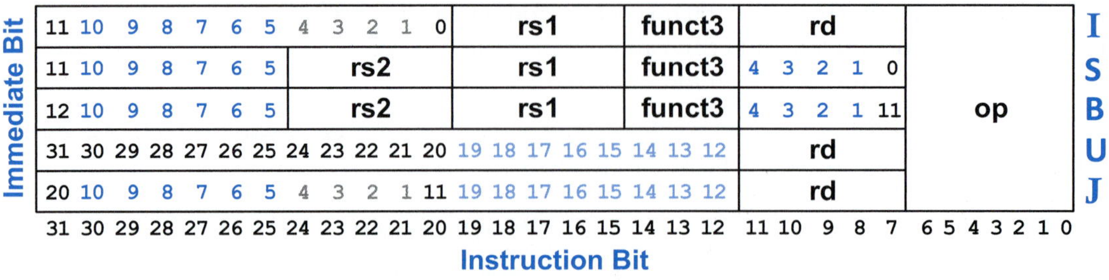

Fig. 2.3 Immediate bits in the instruction.

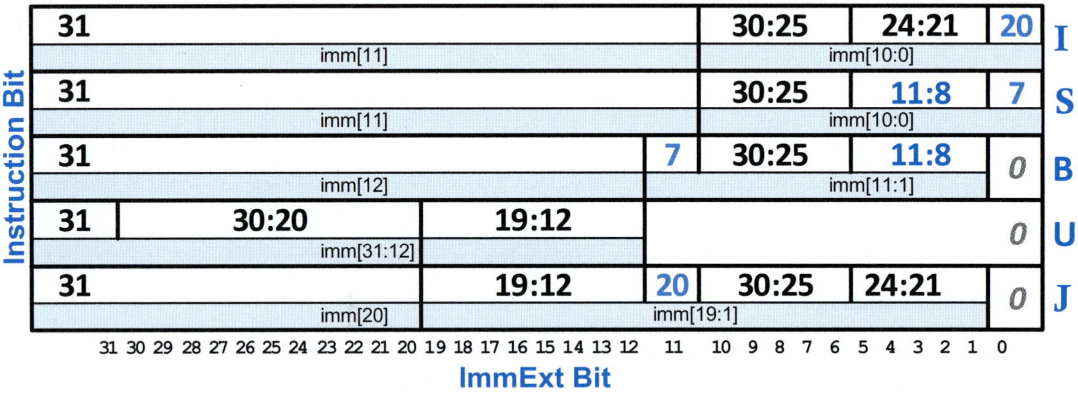

Fig. 2.4 Extension: formation of `ImmExt` from instruction (`Instr`) bits.

Fig. 2.4 shows how the extended immediate (`ImmExt`) is formed (bit-swizzled and extended) from the instruction bits (`Instr`) in each instruction format. The white and blue squares show how the instruction and immediate bits, respectively, form the 32-bit sign-extended immediate `ImmExt`. The blue italicized zeros show where `ImmExt` is hardwired to 0 instead of coming from `Instr`. Encoding a given immediate bit in the same instruction bit across formats, when possible, minimizes extension hardware; multiplexers can then select a given `ImmExt` bit from only a few options. For example, `ImmExt[31]` = `Instr[31]` for all instruction types. `ImmExt[20]` = `Instr[20]` for U-type instructions, or `Instr[31]` for other types. `ImmExt[0]` = `Instr[20]` for I-type, `Instr[7]` for S-type, or 0 for other types.

The most time-consuming part of extending the immediate is sign-extending its most significant bit (msb) into the upper bits because it must drive many loads (i.e., has a large *fanout*). Thus, RISC-V always places the immediate's msb in bit 31 of the instruction, so it is available without further decoding. RV64 also extends bit 31 into bits 63:32 of `ImmExt`.

The next two examples illustrate how converting between assembly and machine language is a methodical process.

Example 2.1 CONVERT ASSEMBLY TO MACHINE LANGUAGE

Convert `sub x3, x4, x5` to machine language.

Solution: According to Table I.1, `sub` is an R-type instruction with `op` = 0b0110011, `funct3` = 0b000, and `funct7` = 0b0100000. The source registers are `rs1` = 4 = 0b00100, `rs2` = 5 = 0b00101, and `rd` = 3 = 0b00011. Placing these fields in the order given in the R-type format of Fig. I.1, we obtain:

```
funct7  rs2   rs1   funct3 rd    op
0100000 00101 00100 000    00011 0110011 = 0x405201B3
```

Machine language instructions are generally written in hexadecimal for brevity.

Example 2.2 CONVERT MACHINE TO ASSEMBLY LANGUAGE

Convert RISC-V machine instruction 0x03E90523 to assembly language.

Solution: The instruction in binary is 0000 0011 1110 1001 0000 0101 0010 0011. The bottom 7 bits are 0100011, which, according to Table I.1, is an S-type instruction. Rewrite the spaces in the binary number to align with the S-type fields:

```
imm11:5 rs2   rs1   funct3 imm4:0 op
0000001 11110 10010 000    01010  0100011
```

The S-type instruction with `funct3` = 000 is `sb`. `rs1` = `x18` = `s2` and `rs2` = `x30` = `t5`. Cobble together the two halves of the immediate to get 000000101010 = 42. Hence, the instruction is:

```
sb t5, 42(s2)
```

You can find a convenient on-line RISC-V assembler at riscvasm.lucasteske.dev. Type in some assembly language code, optionally including pseudoinstructions, and the page will check your syntax and give you the machine language and disassembled code.

2.3 Have a Hart

A *hardware thread* has physical registers containing its context (values of the program counter and registers) and hardware to fetch and execute instructions independently of other hardware threads. A *single-threaded* core has one hardware thread, and a *multithreaded* or *multicore* processor has multiple hardware threads.

A RISC-V *hart* is an abstraction of a *har*dware *t*hread. A hart may be physical, virtual, or simulated. For example, a *physical hart* actually is a hardware thread. An OS may create the illusion of running several threads concurrently by time-multiplexing many *virtual harts* onto a smaller number of hardware threads. A simulator or emulator provides one or more *simulated harts*, possibly running on entirely different underlying hardware. From the program's perspective, all of these kinds of harts are indistinguishable from a true hardware thread except, perhaps, in performance.

Privilege modes provide security and prevent programs from interfering with each other. RISC-V harts may run in machine, supervisor, or user privilege modes. In addition to the ordinary registers, harts have CSRs, some of which are only accessible from higher privilege modes. When a special event such as a system call, illegal instruction, or interrupt occurs, the hart takes a trap to handle the event, often via an execution environment (EE) such as the OS or firmware running at a higher privilege level.

2.3.1 Privilege Modes

RISC-V processors boot in *machine mode*, which is the highest privilege mode and has full control over all processor capabilities. Processors may also optionally support *user* and *supervisor modes*, which provide security by restricting a program's access to certain portions of the memory and CSRs. Processor traps provide controlled access to higher privilege modes.

For example, whereas an OS such as Linux typically runs in supervisor mode, application programs run in user mode. A malicious program shouldn't be able to modify the OS or steal data from another program. Thus, the user-level program can only invoke the OS through a *system call* using the `ecall` instruction, which causes a trap, transferring control to the OS trap handler. The trap handler and system call validates the parameters to reject illegal requests, such as opening a file for which the program lacks permission.

2.3.2 Control and Status Registers

> The RISC-V Privileged Specification, which describes privilege modes, CSRs, instructions, and traps, is longer and more complex than the main instruction set specification [RISCV_Privileged]. Chapter 8 further describes privileged features.

Each privilege mode is associated with a set of CSRs that holds information such as flags from a floating-point operation, which RISC-V features are enabled, and where to go when a trap occurs. The CSRs also include performance counters to track things such as the number of cycles the processor has run, the number of instructions that have retired (i.e., finished successfully), and other implementation-dependent data.

2.3.3 Processes and Threads

A *process* is a running program. In an application processor with a multitasking OS, each user process generally has the illusion of owning the entire computer even though there are usually more processes than cores. The processor switches between processes every few milliseconds to give the appearance that all processes are running concurrently. One user process cannot directly interfere with the memory used by another (unless it requests a region of shared memory from the OS).

A process may be divided into one or more *software threads* that share a common memory and code space. Software threads appear to run concurrently. They are useful when a program needs to do several things

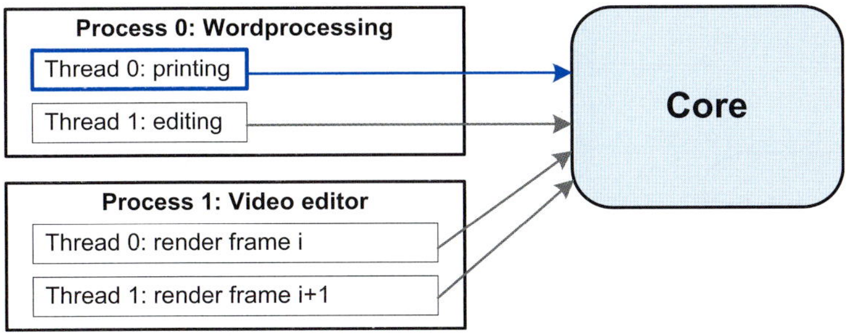

Fig. 2.5 Core running two processes, with two threads per process.

at once (e.g., a word processor spooling to a printer while you are still editing) and to improve performance (e.g., a video editor rendering many frames concurrently on many cores). Fig. 2.5 shows a core running these two processes. Only one thread (in this case, Thread 0, highlighted in blue) runs on the core at any given time. POSIX threads (*pthreads*) are a common type of software thread. Pthreads provide a programming interface to create and destroy threads and synchronize data between threads.

2.3.4 Cores and Harts

A *core* is an individual processor on a chip that fetches and executes instructions. At a minimum, it includes the context (PC, registers, and CSRs) and functional units needed to run a thread. A *multithreaded* core has multiple hardware copies of the PC and registers so that it can quickly switch between executing one thread and another by multiplexing between the contexts for each thread while sharing the common execution units. A *multicore* processor has two or more cores on the same chip, along with an interconnection network and usually an upper-level cache.

A microprocessor is often partitioned into core and *uncore* portions, with the uncore containing peripherals, shared memory or upper-level caches, and external interfaces.

Examples of harts, some of which are illustrated in Fig. 2.6, are:

- In a multicore processor, each RISC-V core is a *physical hart*.
- A *k*-way multithreaded core has *k* physical harts that operate in a time-interleaved fashion. A processor with 4 cores each capable of 2-way multithreading has 8 physical harts.
- An OS may time-multiplex multiple user-level *virtual harts* onto a smaller number of physical harts. A typical OS has one user-level virtual hart for each user-level software thread. The OS *scheduler* receives timer interrupts, causing it to *context switch* between threads every few milliseconds by saving the context of the current virtual hart to memory and loading the context of the next virtual hart onto the physical hart.

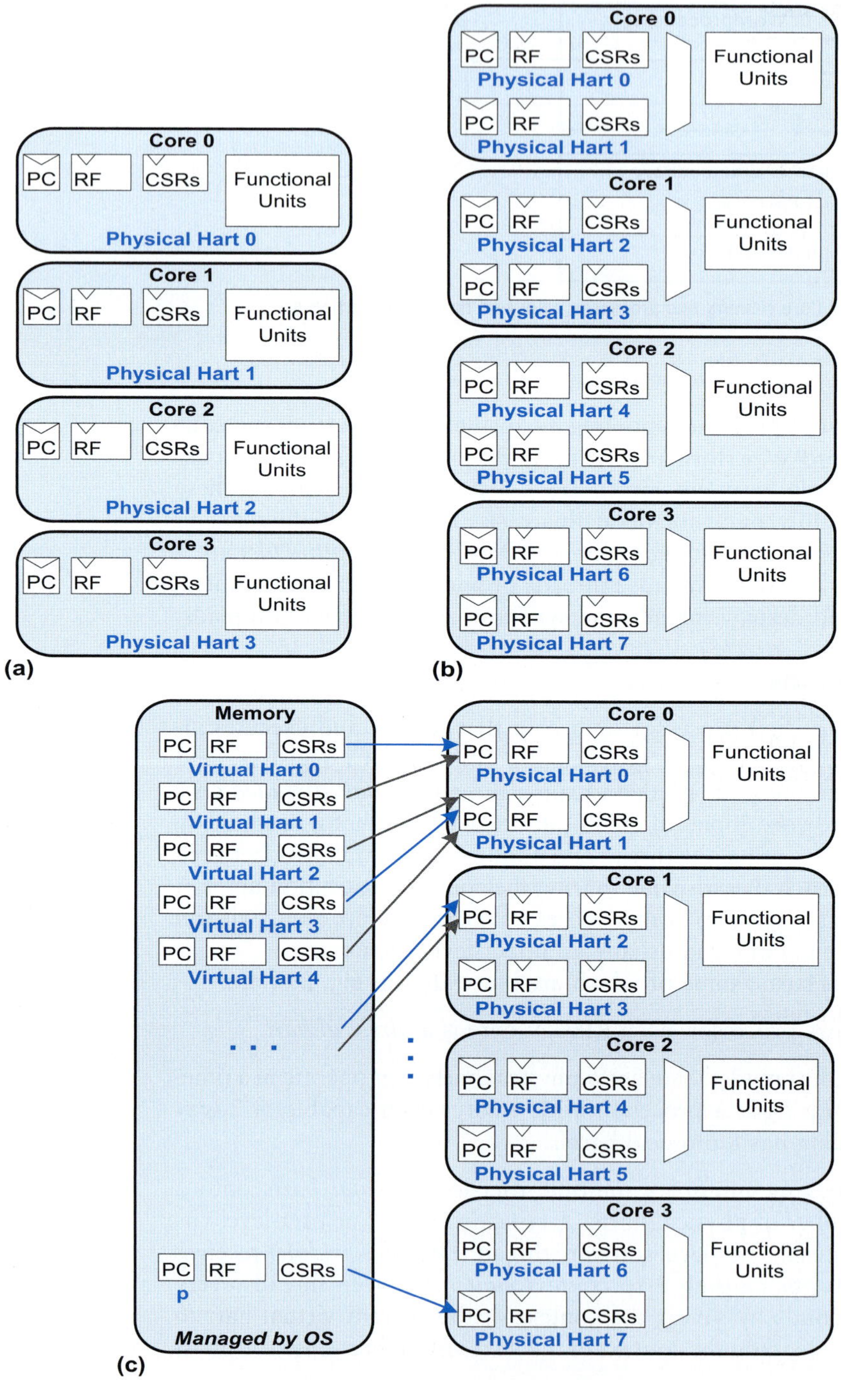

Fig. 2.6 Example harts: (a) multicore, (b) 2-way multithreaded cores, and (c) virtual harts.

- An emulator such as Spike or QEMU can run RISC-V programs using *simulated harts* on a computer with a different ISA.

This book does not cover hypervisors in detail. But, in general, a *hypervisor* runs more than one *guest OS*, time-multiplexing the physical harts to give each OS its own virtual harts, called *guest harts*. For example, VirtualBox is a hypervisor that allows Mac users to run Windows or Linux as a guest OS. Each OS in turn time-multiplexes multiple user-level virtual harts onto these guest harts.

2.3.5 Exceptions, Interrupts, and Traps

In RISC-V, an *exception* is an unusual condition caused by a program, such as an illegal instruction, an access to nonexistent memory, or an OS call. An *interrupt* is an asynchronous external event such as a timer reaching some trip point or a serial port indicating that a new piece of data has arrived. Both exceptions and interrupts cause a *trap*, which is a transfer of control from the current program to a *trap handler*, which is code that handles the event. Traps interact closely with the CSRs and also raise the privilege level, going to machine or supervisor mode. When a trap to machine mode occurs, the current program counter is stored in the *machine exception program counter* (`mepc`) CSR. A code indicating the trap's cause is stored in the *machine cause* (`mcause`) CSR. Supplementary information, such as the illegal instruction or illegal address being accessed, may be stored in the *machine trap value* (`mtval`) CSR. The processor changes to machine mode and jumps to the address specified by the *machine trap-vector base-address* (`mtvec`) CSR. Traps to supervisor mode are similar and use `sepc`, `scause`, `stval`, and `stvec`.

The trap handler is code that responds appropriately, for example by performing the OS call, placing the new serial data into a buffer, or terminating the program executing an illegal instruction. A machine- or supervisor-level trap handler returns to the original program using the `mret` or `sret` instruction, which returns to the address in `mepc` or `sepc` and restores the previous privilege level. Traps are discussed further in Chapter 8.

2.3.6 Execution Environment

A hart runs within an *execution environment* (*EE*) such as an OS. An *execution environment interface* (*EEI*) allows programmers to interact with its EE and defines the initial state of the program, the number and types of harts, including the privilege modes and instruction set they support, the memory map, and how traps are handled. An EE can be implemented in hardware, software, or a mix of the two.

The simplest EE is *bare metal*, in which the processor runs in machine mode (M-mode) with no OS. The programmer has direct access to all features and memory. Physical harts directly correspond to physical cores. There are no virtual harts and no protection between programs.

Fig. 2.7 shows a more sophisticated EE for a Linux-based OS on a processor that supports machine, supervisor, and user modes (M-, S-, and U-modes, respectively). The user can run multiple concurrent U-mode applications, such as a terminal, email client, compiler, and music player.

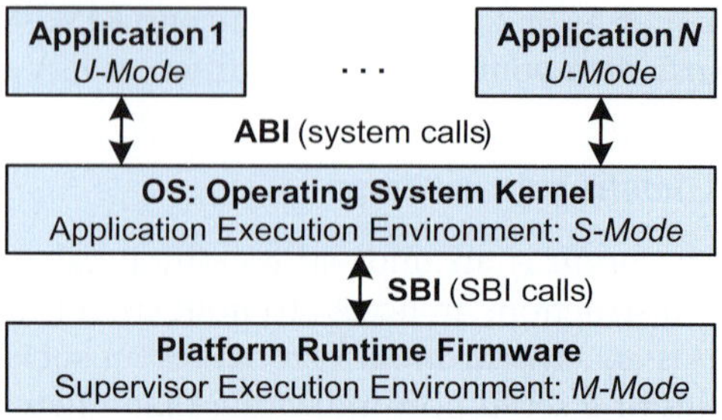

Fig. 2.7 Linux-based execution environments (EEs).

The OS, also called the *kernel*, runs in S-mode and provides each application with its own application EE that has its own virtual memory space so that one application can't interfere with another application's memory.

Fig. 2.7 shows two common EEIs, an *application binary interface* (*ABI*) and a *supervisor binary interface* (*SBI*). An ABI enables applications to interact with their EE, and an SBI enables the OS to interact with its EE, also called its runtime environment: peripherals, timers, and so on. As will be discussed further in Section 3.3.3, the ABI defines a set of *system calls* for applications to make requests to the OS for managing processes, memory, and files and synchronizing user-level threads. The OS kernel, in turn, uses SBI calls to access peripherals and services such as setting the system timer or starting, stopping, and synchronizing cores, as discussed in Section 3.3.4. The SBI allows the same OS to run on different RISC-V chips by only modifying the platform runtime firmware.

Software has no way to tell whether it is running on a physical, virtual, or simulated hart, except perhaps by performance. For example, consider a Linux OS running on a 4-core dual-threaded processor that has 8 physical harts, because it has 8 distinct contexts (PCs, register files, and CSRs) kept in hardware. The Linux OS runs on these physical harts in supervisor mode. Each user-level (software) thread is assigned to a virtual hart. The OS creates and destroys virtual harts as it spawns and kills software threads, and the OS scheduler time-multiplexes (periodically switches) which virtual hart is running on a given physical hart so that an unlimited number of threads can have the illusion of running on their own processor even though a finite number of physical harts exist. As long as the OS allocates time fairly, each U-mode virtual hart makes progress independent of the others.

As specified by the EEI, a hart typically makes system calls to its EE by placing arguments in the `a0...a7` registers and performing `ecall`. A hart may also transfer control to a debugger by performing `ebreak`. Each of these instructions causes an exception and jumps to the exception handler at `mtvec` or `stvec`.

A purely software EE is a RISC-V *simulator* or *emulator*, such as Spike, Sail, or QEMU, which runs RISC-V programs on underlying hardware such as an x86-based computer. In this case, there are no physical RISC-V cores, so the harts are simulated.

Spike, Sail, and QEMU are discussed further in Chapter 3.

Sometimes the term *simulator* is used for a tool that models only the harts and RAM, while an *emulator* also models *peripherals*, that is, input/output (I/O) devices such as keyboards, serial UART connections, and so forth. Spike is a RISC-V simulator that can run programs without peripherals, and QEMU and Sail are emulators that model the CLINT timer, PLIC interrupt controller, and UART serial port needed to boot Linux. The term *simulator* is used generically to describe all such tools.

The field of memory systems is inconsistent on whether address 0 is shown at the top or bottom of a memory diagram. This textbook uses the bottom.

Whereas fast simulators such as QEMU use *binary translation* to convert a program from the instruction set being simulated to the native instruction set, simpler simulators interpret programs one instruction at a time. QEMU runs approximately 1 billion instructions per second. Spike runs about 100 million instructions per second. Sail is focused on provable correctness rather than performance and runs only about 1 million instructions per second.

A mixed hardware/software EE can provide capabilities not directly available from the hardware. For example, consider a user-mode program running on hardware that lacks a floating-point unit (FPU). If the program tries to execute a floating-point instruction, an illegal-instruction exception occurs. The exception handler could perform the floating-point arithmetic in software, and then the user-mode program could continue.

2.4 MEMORY MAP

The *memory map*, a key part of the EE, defines the addresses at which memory and I/O peripherals are located. A representative memory map, used in the Wally processor, is shown in Fig. 2.8 (and in Table A.2 in Appendix A). This memory map contains a small boot ROM, a larger RAM, and various peripherals, including:

- CLINT: Core Local Interrupter, for handling software and timer interrupts
- PLIC: Platform-Level Interrupt Controller, for multiplexing external interrupts
- GPIO: General-purpose input/output, for reading and writing binary values from pins
- UART: Universal asynchronous receiver/transmitter, for serial communication, such as printing to a terminal
- SPI: Serial Peripheral Interface, for higher-speed clocked serial communication
- SDC: Secure Digital (SD) Card controller

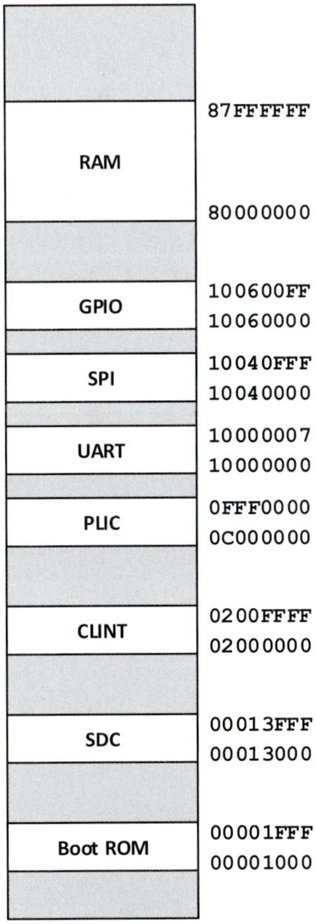

Fig. 2.8 Wally default memory map.

Wally's memory map is compatible with the SiFive FU540 processor and the QEMU *virt* machine (virt is a QEMU emulation model with peripherals suitable for booting Linux). An exception is that the UART is at 0x10010000 on the FU540 rather than 0x10000000 on Wally and virt. See Table 6 of [FU540]. Section 4.4 describes the Wally configuration files, which can be used to add or remove peripherals or change the memory map.

The RISC-V reset vector is platform dependent. In most Wally configurations, the reset vector is 0x80000000 because programs typically start at this address. The testbench automagically (automatically) loads a program into RAM at this address at reset. In the Buildroot Linux configuration and QEMU, the reset vector is 0x1000 to start a Linux boot loader preloaded in boot ROM. On the SiFive FU540 chip, the reset vector is 0x1004 for a bootloader.

Configurations without external RAM map the on-chip memories, data tightly-integrated memory (DTIM), and instruction read-only memory (IROM) to address 0x80000000 instead. Typically, the program begins at 0x80000000 and data follows.

On reset, the PC begins at an address called the *reset vector*, and some instructions must be available at that address. In a real system, the reset vector points to the boot ROM, which contains a *boot loader* that loads the next stage of an OS from the SD card (or flash memory or hard disk, if they are supported). In a simulator, the reset vector could point directly to RAM that was preinitialized on reset with the program of interest.

2.5 RISC-V EXTENSIONS AND PROFILES

The base instruction sets are known as RV32I or RV64I. These operate on 32- or 64-bit registers, respectively. The RISC-V Unprivileged Specification defines several other ratified extensions (see Table 2.2). The M extension supports multiply, divide, and remainder instructions. The F, D, and Q extensions support single-, double-, and quad-precision floating-point math, respectively. All of these could be emulated in software but are much faster with hardware support. The A extension introduces atomic instructions used for synchronization between multiple RISC-V harts. Most application processors support the M, F, D, and A extensions and CSRs and instruction fences, which is written as RV64IMAFD_Zicsr_Zifencei. A processor with all these extensions is abbreviated as RV32G or RV64G.

Table 2.2 Common RISC-V extensions.

Extension	Instructions
A	Atomic read-modify-write for synchronization
B	Bit manipulation (B = Zba + Zbb + Zbs)
C	Compressed (16-bit)
E	Embedded (reduce register set to x0-x15)
F, D, Q	Single-, double-, and quad-precision floating-point math
M	Multiply, divide, and remainder
Zicsr	CSRs and privileged instructions
Zifencei	`fence.i` instruction/data synchronization
V	Vectors data parallelism
U	User mode
S	Supervisor mode, for OS and virtual memory
H	Hypervisor for multiple concurrent operating systems
G	I, M, F, D, A, Zicsr, Zifencei

Nonvolatile program memory can be the most expensive component in an embedded processor, so these processors usually support a compressed instruction set (similar to the ARM Thumb instructions) that packs most frequently used instructions into only 16 bits. The RISC-V C extension provides these compressed instructions, reducing instruction bits fetched by up to 25% or 30%. RV64GC = RV64IMAFDC_Zicsr_Zifencei contains all of the standard extensions.

RV32E targets tiny embedded cores. The RV32I register file comprises about half the area of a minimal core, so RV32E eliminates registers x16-x31 to save a quarter of the core area.

The Zicsr extension introduces CSRs and instructions to read and write these registers; this is necessary to support traps, multiple privilege modes, and other configuration options. The Zifencei standard extension adds the `fence.i` instruction to synchronize writes and fetches for instruction memory. This is needed by operating systems that load programs into memory.

RISC-V also defines an unratified base instruction set RV128I supporting 128-bit memory addresses by using 128-bit registers, anticipating that the 64-bit addresses of RV64I (which support 16 exabytes of memory) will eventually become insufficient for large computers. RV128I adds quadword loads and stores and instructions to operate on the bottom 64 bits of a 128-bit register. No commercial need yet exists for 128-bit addresses, and the standard may evolve when these large systems are developed.

A large and rapidly expanding set of other extensions have been ratified, such as Zk* cryptographic instructions, Zicbo* cache management instructions, Sstc supervisor timer compare, and so forth.

RISC-V vendors are free to develop their own proprietary nonstandard extensions. Such extensions can provide a competitive advantage. However, vendors also have an incentive to publish and standardize their extensions to increase software adoption.

Some RISC-V systems provide *accelerators* for specialized tasks. An accelerator might be a fixed-function unit such as an H.265 video decompressor or a programmable unit such as a graphics processor. An accelerator typically communicates with the cores through shared memory.

All of the RISC-V extensions are optional. A program may read the *machine instruction set architecture* (`misa`) CSR (see Section 8.2.3.2) to determine which letter extensions are available and then either use an instruction from an extension or call a library function to emulate the missing instruction. RISC-V International is in the process of defining a standard way to determine whether other named extensions (e.g., Zifencei) are supported.

"There is only one mistake that can be made in computer design that is difficult to recover from—not providing enough address bits for memory addressing and memory management. The PDP-11 followed the unbroken tradition of nearly every computer." [Bell75]

2.5.1 RISC-V Profiles

RISC-V has a great deal of configurability and extensibility. Initially, it used ISA strings such as RV64IMAC to define the supported extensions, but these are becoming unwieldy as more extensions are added with longer names, resulting in strings such as RV64IMAFDC_Zicsr_Zifencei_Zicclsm_Zba_Zkt_Svnapot. RISC-V defines *profiles* to capture standard collections of extensions. Software developers can specify a profile with a minimum feature set needed to run the software. As of the time of this writing, the following profiles have been ratified [RISCV-Profiles]. Table A.6 details the extensions supported in each profile.

- RVI20U32: Basic unprivileged RV32I
- RVI20U64: Basic unprivileged RV64I
- RVA20U64: Application processor RV64GCU
- RVA20S64: Application processor RV64GCUS
- RVA22U64: Application processor RV64GCBU
- RVA22S64: Application processor RV64GCBUS with rich OS
- RVA23U64: Application processor RV64GCBVU with rich OS
- RVA23S64: Application processor RV64GCBVUSH with rich OS
- RVB23U64: Customized processor RV64GCBU
- RVB23S64: Customized processor RV64GCBUS

All profiles are little-endian. The profiles strongly recommend raising an illegal instruction exception on unimplemented opcodes.

Notice that the profiles standardize U- and S-mode features. Only a small amount of portable firmware runs in machine mode, so M-mode features are not standardized.

Table A.6 summarizes the various required and optional extensions for each profile. Of course, a core may have other extensions beyond the required and optional ones in the profile, but software written for that profile will not take advantage of those extensions. The table also indicates the chapter that describes the extension if it is presently supported by Wally.

RISC-V has many minor optional behaviors, such as whether it has hardware performance monitors (Zihpm), whether virtual memory page table entry dirty bits are set in hardware or via a software exception (Svade), and whether cache blocks are 64 bytes (Zic64b). The RISC-V profiles gave these behaviors names in March 2023 so that platforms can rely on consistent behavior.

2.6 COMPARISON WITH OTHER ARCHITECTURES

RISC-V is a fifth-generation RISC architecture, with a basic instruction set similar to other RISC processors. Its most important strength is that it is an open architecture that companies can implement without license fees and researchers can freely improve. A minor strength is that the machine language encodings keep all register fields and the sign bit of the immediate in the same position for all instructions, slightly reducing hardware and the decode critical path. RISC-V avoids various features that have not aged well in previous architectures, including those in ARM, MIPS, SPARC, and x86 architectures.

Such overly clever features in the 32-bit ARMv7 architecture include support of *conditional execution* of instructions, allowing a variable shift before ALU and memory instructions, and ALU and store instructions supporting three source registers. These features lead to more complex instruction encoding, leaving only enough remaining bits for encoding 16 registers, which has proved to be a performance limiter. The shifter is lightly used but impacts the critical path for all instructions. The conditional execution and third register file port increase the expense of building superscalar out-of-order processors. RISC-V eliminates all of these features. ARM recognized the same limitations of its instruction set and fixed some of them with its

64-bit ARMv8 ISA in 2013. ARMv8 has 32 64-bit registers and generally has many similarities to the RISC-V ISA. The 64-bit ARMv8 architecture has an irregular machine language encoding that is cruel for humans to manually translate; this is not a problem for assemblers, although it makes it difficult to teach in an introductory course. ARM processors are presently used in almost all mobile devices. ARM does not manufacture its own chips but rather licenses its synthesizable cores to other companies building SoCs. It also sells architecture licenses, so companies can design their own ARM cores. Apple has demonstrated with its M1 and subsequent chips that ARM processors can match or exceed the performance of x86 processors while consuming much less power. ARM's revenue is based on royalties from licensees, so it tightly controls access. ARM sued one of its biggest customers, Qualcomm, over a licensing dispute.

MIPS is similar to RISC-V. However, the MIPS architecture has a *branch delay slot*, meaning the instruction after the branch is always executed even if the branch is taken. This improved performance of the original pipelined implementation because it avoided flushing the next instruction when a branch was taken, but it proved difficult for microarchitects and compilers in higher-performance implementations [Hennessy19]. As ARM grew to dominate the market, MIPS market share has steadily declined and is now negligible.

SPARC has *register windows*, in which the hardware automatically rotates through a larger set of registers during function calls, so that the program does not have to save and restore registers on the stack. This worked well for benchmark programs on a pipelined microarchitecture with a poor compiler but is unnecessary with a better compiler, degrades performance on multitasking loads that have to save the registers during context switches, and is costly in high-performance implementations [Magnussen97]. SPARC is now primarily used by Oracle servers and has a tiny market share.

The Intel x86 complex instruction set computer (CISC) architecture evolved piecemeal from the 8088/8086 to today's modern processors. The complicated legacy instructions, variable-length encoding, and limited set of registers make high-performance x86 microarchitectures difficult. Since the Pentium Pro in the mid-1990s, Intel decoded the complex instructions into simpler micro-ops (RISC-like instructions) and then executed these micro-ops on superscalar out-of-order datapaths [Colwell02]. The variable-length decoder and micro-op cache add hundreds of thousands of transistors compared to a RISC processor but deliver comparable performance [Isen09]. Nearly all Windows PCs use x86 processors. Intel and AMD have a duopoly on the architecture and can, thus, charge high prices.

The greatest present drawbacks of RISC-V are its lack of binary compatibility with existing x86 and ARM software, its less-mature software and verification ecosystem, and its less-mature suite of implementations. However, software, verification, and implementations are advancing rapidly. The open architecture has potential to surpass its proprietary competitors in many markets. This is a new golden age to be a processor designer.

2.7 RISC-V MICROARCHITECTURE

Recall that microarchitecture is the interconnection of gates and registers that implement an architecture. Many microarchitectures may exist for a given architecture. Design teams write a *microarchitecture specification* describing the microarchitecture in enough detail that a logic designer can write hardware description language (HDL) code based on the spec. This section and the next explore microarchitectures for a simplified RV32I core that supports the following instructions:

- ALU Instructions: `add`, `sub`, `and`, `or`, `slt`, `addi`, `andi`, `ori`, `slti`
- Memory Instructions: `lw`, `sw`
- Branch Instructions: `beq`, `jal`

All microarchitectures must include the architectural state (memory, PC, and registers). Some microarchitectures may include other nonarchitectural state, such as pipeline registers, caches, or branch history, to make the processor run faster. However, this information could be flushed, and the processor will still restart correctly based on the architectural state. Different microarchitectures offer different cost/performance tradeoffs while running the same ISA.

Performance is measured as the *central processing unit's* (*CPU's*) time T_{CPU} to run a program of interest with a certain number of instructions. The total program execution time is given by Eq. (2.1) in terms of number of instructions, average number of *cycles per instruction* (*CPI*), and cycle time (T_c) as:

$$T_{CPU} = \#\boldsymbol{instructions} \times \boldsymbol{CPI} \times T_c \tag{2.1}$$

This section describes single-cycle and pipelined microarchitectures. The single-cycle microarchitecture executes one instruction in each clock cycle; that is, $CPI = 1$. It is simple but slow because the clock period is long. The pipelined microarchitecture uses a 5-stage pipeline so that five instructions can overlap their execution and the clock period T_c can be much shorter. This comes at the expense of pipeline registers and hardware to detect for dependencies and resolve them through forwarding and stalls; the stalls increase the CPI somewhat. The microarchitectures are described with block diagrams, truth tables, and HDL (SystemVerilog) code. We also describe a simple test program to verify basic functionality.

The microarchitectures are adapted from Chapter 7 of *Digital Design and Computer Architecture, RISC-V Edition*. The main differences are that the ALU and branch comparator have been reorganized to save hardware and that the hierarchical partitioning was changed to facilitate expanding to a full-featured processor.

2.7.1 Single-Cycle Microarchitecture

Fig. 2.9 shows a block diagram of a single-cycle microarchitecture. The core is partitioned into an *instruction fetch unit* (*IFU*), *integer execution unit* (*IEU*), and *load/store unit* (*LSU*). The IFU contains the PC, IROM, and the logic to increment the PC or handle branches and jumps. The IEU contains the *integer register file* (*IRF*), the datapath with an ALU, the branch comparator, and multiplexers to handle arithmetic instructions and memory/branch address generation, and a controller to decode instructions. The LSU contains the *data tightly-integrated memory* (*DTIM*) used for loads and stores. In these schematics, heavy lines indicate busses of size XLEN (32 for RV32 or 64 for RV64), medium lines indicate narrower multibit signals, and light lines indicate 1-bit signals. The processor fetches an instruction every cycle. Loads and stores each access memory in a single cycle.

The single-cycle processor has two memories and a register file. The register file has three ports: read ports 1 and 2 and write port 3. It receives a 5-bit address for each port (`A1`, `A2`, and `A3`) specifying one of the $2^5 = 32$ registers. The read ports read the specified registers onto the read data ports (`RD1` and `RD2`). The write port needs to know where to write (`A3`), what to write (`WD3`), whether to write (`WE3`, write enable), and when to write (on the rising edge of clock). The memories have a single port. The instruction memory only reads, so it is called the IROM. The DTIM may read or write depending on the write enable `WE` signal. The write enables for the register file and data memory come from the controller.

The comparator (`comp`) asserts `Eq` if `R1` == `R2`, for `beq` tests. The sign extender unit (`extend`) forms the 32-bit `ImmExt` from the immediate fields of the instruction (see Figs. 2.3 and 2.4 and Section 7.1.2). The ALU has an adder and various other logical operators. The adder directly drives `IEUAdr` = `PC` + `ImmExt` to the LSU and IFU for fast memory address calculation. `ALUResult` is selected from the various possible operations according to Table 2.3. The `ALUControl` signal comprises the {`Sub`, `ALUOp`} bits that indicate whether to subtract and whether to consider the `funct3` field to determine the operation.

A processor with a unified memory containing instructions and data is called a *von Neumann architecture*, based on John von Neumann's report on the EDVAC [vonNeumann45]. A processor with separate instruction and data memories is called a *Harvard architecture*, based on the Harvard Mark I that kept data in electronic storage and programs on a paper tape [Aiken46]. von Neumann architectures are more flexible because they allow for not only loading but also modifying programs, but they are slower because separate cycles are needed to access an instruction and then the data. Modern computers generally gain the advantages of both architectures by using a single unified main memory but separate instruction and data caches. The separate caches give single-cycle access to both instructions and data, while the unified main memory means that ordinary load and store instructions can be used to copy or modify instructions in the same way as data. Computers need a mechanism to update the instruction cache when changes are made to cached instructions. RISC-V uses the `fence.i` instruction to force this synchronization, as described in Section 10.4.2.6.

The DTIM is called *tightly integrated* because it connects directly to the IEU address and data busses without the overhead of an external bus protocol.

2.7.1.1 Sample Program

Consider running the program from Code Example 2.14 on the RISC-V single-cycle computer starting at address 0x1000. Each line contains the assembly language instruction, the instruction's address, and the machine language encoding.

Code Example 2.14 SAMPLE PROGRAM

Assembly Language	Address	Machine Language
`start: lw x3, 4(x1)`	`# 0x1000`	`0x0040A183`
`sub x5, x3, x2`	`# 0x1004`	`0x402182B3`
`beq x0, x0, 0x1000`	`# 0x1008`	`0xFE000CE3`

All numbers in Code Example 2.14 are 32-bit values and are expressed in hexadecimal; the 0x prefix is sometimes omitted for brevity. Binary control field values are indicated

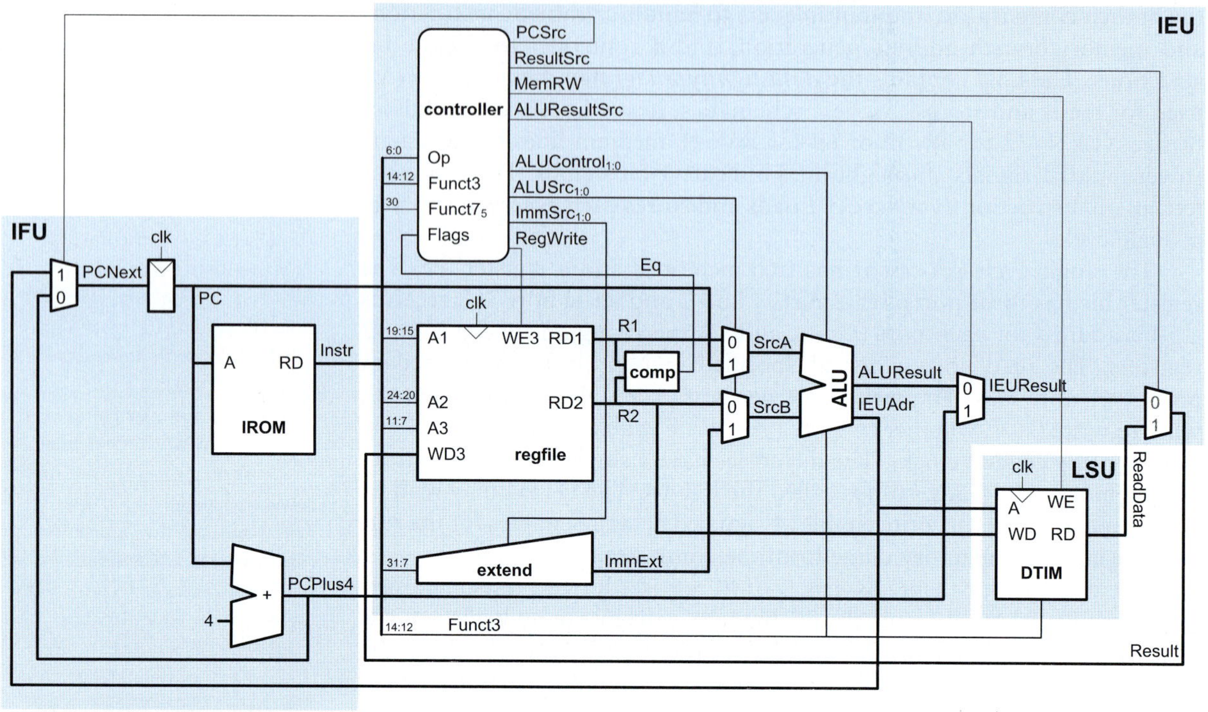

Fig. 2.9 Single-cycle processor block diagram.

Table 2.3 ALU operation.

`ALUControl[1]` (`Sub`)	`ALUControl[0]` (`ALUOp`)	`Funct3`	`ALUResult`
0	0	X	ADD
0	1	000	ADD
0	1	110	OR
0	1	111	AND
1	1	000	SUB
1	1	010	SLT

with a subscript 2. Suppose `x1` initially contains 0x2000, `x2` contains 7, and memory address 0x2004 contains the value 0x2A. The code executes as follows:

Cycle 1: `lw x3, 4(x1)`

- `PC` = 0x1000, so the instruction memory (the IROM) reads the `lw` instruction (0x0040A183) from address 0x1000.

- rs1 = 00001_2 = 1 (from instruction bits 19:15). The register file receives this number at its A1 port and reads the contents of x1 (0x2000) onto R1.
- At the same time, the extend unit extends the immediate field (4) to 32 bits: ImmExt = 0x000000004.
- The controller sets ALUSrc = 10_2 to choose SrcB = ImmExt and SrcA = R1. The ALU adder computes IEUAdr = 0x2000 + 4 = 0x2004.
- The data memory in the LSU (the DTIM) reads from this address and returns ReadData = 0x2A.
- The controller sets ResultSrc = 1 to choose Result = ReadData and sets RegWrite = 1 to write 0x2A into register 3 (obtained from rd in bits 11:7 of the instruction) on the rising edge of the clock at the end of the cycle.
- Meanwhile, PCPlus4 is computed as 0x1000 + 4 = 0x1004, and the PCSrc control signal is 0, so PCNext = PCPlus4 is written to PC at the end of the cycle.

Cycle 2: `sub x5, x3, x2`

- PC = 0x1004, so the instruction memory reads the sub instruction (0x402182B3) from address 0x1004.
- The register file receives A1 = 3 and A2 = 2. It thus reads SrcA = R1 = 0x2A and SrcB = R2 = 7.
- The ALU subtracts: ALUResult = 0x2A - 7 = 0x23.
- The Result mux chooses Result = ALUResult = 0x23 via IEUResult to send it to the register file. (Note that the data memory reads from this address, too, but the read data is not used.)
- The register file writes 0x23 to register 5 and increments PC to 0x1008.

Cycle 3: `beq x0, x0, 0x1000`

- PC = 0x1008, so the instruction memory reads the beq instruction (0xFE000CE3) from this address.
- The register file receives A1 = 0 and A2 = 0. Register 0 is hardwired to 0.
- Thus, the comparator compares 0 to 0 and asserts the Eq flag to indicate they are equal. Based on the instruction (beq) and this Eq flag, the controller sets PCSrc = 1.
- Meanwhile, the extend unit produces ImmExt = 0xFFFFFFF8 = −8. The ALU's adder computes IEUAdr = PC + ImmExt = 0x1008 + −8 = 0x1000.
- The PC multiplexer chooses PCNext = IEUAdr. RegWrite is 0, so no registers are written.
- At the end of the cycle, PC is updated to 0x1000. The processor repeats this loop indefinitely.

2.7.1.2 Controller Design

As we saw in the sample program, the controller must produce appropriate control signals to drive the multiplexers and write enable signals in the datapath to execute the current instruction. Fig. 2.10 shows a block diagram of the controller. The main decoder computes most of the control signals based on the op field of the instruction, as given in Table 2.4.

PC normally increments by 4, but it skips to the IEUAdr computed by the ALU for jumps and taken branches. The PCSrc logic determines whether this skip should occur.

ALUControl has an ALUOp bit and a Sub bit. Sub is asserted for subtracts and set less than instructions:

```
ALUControl[1] = Sub = ALUOp & ((Funct3 == 3'b000) & Funct7b5 & Op
[5] | (Funct3 == 3'b010))

ALUControl[0] = ALUOp
```

Distinguishing subtraction from add, addi, and other ALU instructions requires checking bits from funct3, funct7, and the opcode. This chapter uses a lazy decoder that only examines enough bits to distinguish the supported instructions; it produces unpredictable results for illegal instructions. The full Wally decoder uses a precise decoder that raises an illegal-instruction exception for any illegal instruction when exceptions (Zicsr) are supported.

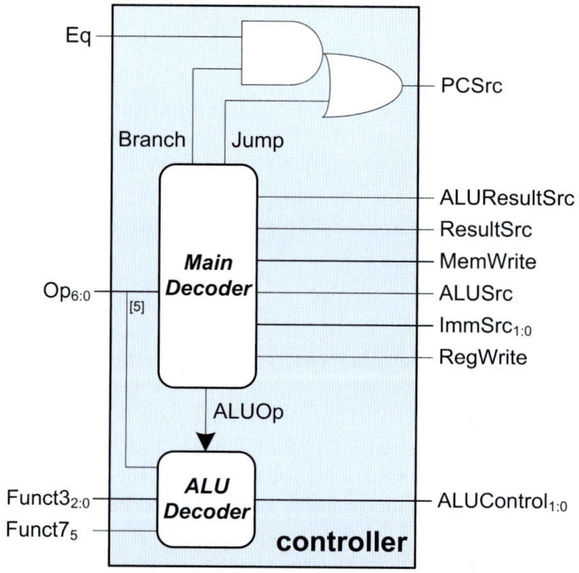

Fig. 2.10 **Single-cycle processor controller.**

Table 2.4 **Main decoder truth table.**

Instruction	Op	RegWrite	ImmSrc	ALUSrc	ALUOp	ALUResultSrc	MemWrite	ResultSrc	Branch	Jump
lw	0000011	1	00	01	0	0	0	1	0	0
sw	0100011	0	01	01	0	0	1	0	0	0
R-type ALU	0110011	1	xx	00	1	0	0	0	0	0
I-type ALU	0010011	1	00	01	1	0	0	0	0	0
beq	1100011	0	10	11	0	0	0	0	1	0
jal	1101111	1	11	11	0	1	0	0	0	1

2.7.1.3 SystemVerilog

Code Example 2.15 describes the single-cycle processor in SystemVerilog. SystemVerilog is covered in detail in Chapter 4, but even if you are unfamiliar with the details, you should be able to relate the modules and their inputs and outputs to the objects in the block diagrams.

This code is available in the CORE-V-Wally repository described in Section 3.2 in `cvw/examples/verilog/riscvsingle`.

Code Example 2.15 SIMPLIFIED SINGLE-CYCLE RISC-V PROCESSOR

```
// riscvsingle.sv
// RISC-V single-cycle processor
// David_Harris@hmc.edu
// Sarah.Harris@unlv.edu
// 2021

module riscvsingle(
  input  logic        clk, reset,
  output logic [31:0] WriteData, IEUAdr,
  output logic        MemWrite);

  logic [31:0] PC, PCPlus4, Instr, ReadData;
  logic        PCSrc;

  ifu ifu(.clk, .reset, .PCSrc, .IEUAdr, .Instr, .PC, .PCPlus4);
  ieu ieu(.clk, .reset, .Instr, .PC, .PCPlus4, .PCSrc, .MemWrite, .IEUAdr, .WriteData,
          .ReadData);
  lsu lsu(.clk, .MemWrite, .IEUAdr, .WriteData, .ReadData);
endmodule

module ifu(
  input  logic        clk, reset,
  input  logic        PCSrc,
  input  logic [31:0] IEUAdr,
  output logic [31:0] Instr, PC, PCPlus4);

  logic [31:0] PCNext;

  // Next PC logic
  flopr #(32) pcreg(clk, reset, PCNext, PC);
  adder       pcadd4(PC, 32'd4, PCPlus4);
  mux2 #(32)  pcmux(PCPlus4, IEUAdr, PCSrc, PCNext);
  irom        irom(PC, Instr);
endmodule
```

Continued

Code Example 2.15 SIMPLIFIED SINGLE-CYCLE RISC-V PROCESSOR—Cont'd

```
module irom(input  logic [31:0] a,
            output logic [31:0] rd);

  logic [31:0] ROM[63:0];

  initial
    $readmemh("riscvtest.memfile",ROM);

  assign rd = ROM[a[7:2]]; // Word aligned
endmodule

module ieu(
  input  logic        clk, reset,
  input  logic [31:0] Instr,
  input  logic [31:0] PC, PCPlus4,
  output logic        PCSrc, MemWrite,
  output logic [31:0] IEUAdr, WriteData,
  input  logic [31:0] ReadData);

  logic       RegWrite, Jump, Eq, ALUResultSrc, ResultSrc;
  logic [1:0] ALUSrc, ImmSrc;
  logic [1:0] ALUControl;

  controller c(.Op(Instr[6:0]), .Funct3(Instr[14:12]), .Funct7b5(Instr[30]), .Eq,
               .ALUResultSrc, .ResultSrc, .MemWrite, .PCSrc,
               .ALUSrc, .RegWrite, .ImmSrc, .ALUControl);
  datapath dp(.clk, .reset, .Funct3(Instr[14:12]),
              .ALUResultSrc, .ResultSrc, .ALUSrc, .RegWrite, .ImmSrc, .ALUControl, .Eq,
              .PC, .PCPlus4, .Instr, .IEUAdr, .WriteData, .ReadData);
endmodule

module controller(
  input  logic [6:0] Op,
  input  logic       Eq,
  input  logic [2:0] Funct3,
  input  logic       Funct7b5,
  output logic       ALUResultSrc,
  output logic       ResultSrc,
  output logic       MemWrite,
  output logic       PCSrc,
  output logic       RegWrite,
  output logic [1:0] ALUSrc, ImmSrc,
  output logic [1:0] ALUControl);

  logic        Branch, Jump;
  logic        Sub, ALUOp;

  logic [10:0] controls;

  // Main decoder
  always_comb
    case(Op)
    // RegWrite_ImmSrc_ALUSrc_ALUOp_ALUResultSrc_MemWrite_ResultSrc_Branch_Jump
      7'b0000011: controls = 11'b1_00_01_0_0_0_1_0_0; // lw
```

Code Example 2.15 SIMPLIFIED SINGLE-CYCLE RISC-V PROCESSOR—Cont'd

```
      7'b0100011: controls = 11'b0_01_01_0_0_1_0_0_0; // sw
      7'b0110011: controls = 11'b1_xx_00_1_0_0_0_0_0; // R-type
      7'b0010011: controls = 11'b1_00_01_1_0_0_0_0_0; // I-type ALU
      7'b1100011: controls = 11'b0_10_11_0_0_0_0_1_0; // beq
      7'b1101111: controls = 11'b1_11_11_0_1_0_0_0_1; // jal
      default:    controls = 11'bx_xx_xx_x_x_x_x_x_x; // Non-implemented instruction
    endcase

  assign {RegWrite, ImmSrc, ALUSrc, ALUOp, ALUResultSrc, MemWrite,
          ResultSrc, Branch, Jump} = controls;

  // ALU Control Logic
  assign Sub = ALUOp & ((Funct3 == 3'b000) & Funct7b5 & Op[5] | (Funct3 == 3'b010)); // Subtract or SLT
  assign ALUControl = {Sub, ALUOp};

  // PCSrc logic
  assign PCSrc = Branch & Eq | Jump;
endmodule

module datapath(
  input  logic        clk, reset,
  input  logic [2:0]  Funct3,
  input  logic        ALUResultSrc, ResultSrc,
  input  logic [1:0]  ALUSrc,
  input  logic        RegWrite,
  input  logic [1:0]  ImmSrc,
  input  logic [1:0]  ALUControl,
  output logic        Eq,
  input  logic [31:0] PC, PCPlus4,
  input  logic [31:0] Instr,
  output logic [31:0] IEUAdr, WriteData,
  input  logic [31:0] ReadData);

  logic [31:0] ImmExt;
  logic [31:0] R1, R2, SrcA, SrcB;
  logic [31:0] ALUResult, IEUResult, Result;

  // Register file logic
  regfile     rf(.clk, .WE3(RegWrite), .A1(Instr[19:15]), .A2(Instr[24:20]),
                 .A3(Instr[11:7]), .WD3(Result), .RD1(R1), .RD2(R2));
  extend      ext(.Instr(Instr[31:7]), .ImmSrc, .ImmExt);

  // ALU logic
  cmp         cmp(.R1, .R2, .Eq);
  mux2 #(32)  srcamux(R1, PC, ALUSrc[1], SrcA);
  mux2 #(32)  srcbmux(R2, ImmExt, ALUSrc[0], SrcB);
  alu         alu(.SrcA, .SrcB, .ALUControl, .Funct3, .ALUResult, .IEUAdr);
  mux2 #(32)  ieuresultmux(ALUResult, PCPlus4, ALUResultSrc, IEUResult);
  mux2 #(32)  resultmux(IEUResult, ReadData, ResultSrc, Result);
  assign WriteData = R2;
endmodule

module regfile(
  input  logic        clk,
  input  logic        WE3,
  input  logic [ 4:0] A1, A2, A3,
```

Continued

Code Example 2.15 SIMPLIFIED SINGLE-CYCLE RISC-V PROCESSOR—Cont'd

```
  input  logic [31:0] WD3,
  output logic [31:0] RD1, RD2);

  logic [31:0] rf[31:1];

  // 3-ported register file
  // Read 2 ports combinationally (A1/RD1, A2/RD2)
  // Write port 3 on rising edge of clock (A3/WD3/WE3)
  // Register 0 hardwired to 0

  always_ff @(posedge clk)
    if (WE3) rf[A3] <= WD3;

  assign RD1 = (A1 != 0) ? rf[A1] : 0;
  assign RD2 = (A2 != 0) ? rf[A2] : 0;
endmodule

module extend(
  input  logic [31:7] Instr,
  input  logic [1:0]  ImmSrc,
  output logic [31:0] ImmExt);

  always_comb
    case(ImmSrc)
                // I-type
      2'b00:    ImmExt = {{20{Instr[31]}}, Instr[31:20]};
                // S-type (stores)
      2'b01:    ImmExt = {{20{Instr[31]}}, Instr[31:25], Instr[11:7]};
                // B-type (branches)
      2'b10:    ImmExt = {{20{Instr[31]}}, Instr[7], Instr[30:25], Instr[11:8], 1'b0};
                // J-type (jal)
      2'b11:    ImmExt = {{12{Instr[31]}}, Instr[19:12], Instr[20], Instr[30:21], 1'b0};
      default: ImmExt = 32'bx; // Undefined
    endcase
endmodule

module cmp(
  input  logic [31:0] R1, R2,
  output logic        Eq
);

   assign Eq = (R1 == R2);
endmodule

module alu(
  input  logic [31:0] SrcA, SrcB,
  input  logic [1:0]  ALUControl,
  input  logic [2:0]  Funct3,
  output logic [31:0] ALUResult, IEUAdr);

  logic [31:0] CondInvb, Sum, SLT;
  logic        ALUOp, Sub, Overflow, Neg, LT;
  logic [2:0]  ALUFunct;

  assign {Sub, ALUOp} = ALUControl;

  // Add or subtract
  assign CondInvb = Sub ? ~SrcB : SrcB;
```

Code Example 2.15 SIMPLIFIED SINGLE-CYCLE RISC-V PROCESSOR—Cont'd

```
  assign Sum = SrcA + CondInvb + Sub;
  assign IEUAdr = Sum; // Send this out to IFU and LSU
  // Set less than based on subtraction result
  assign Overflow = (SrcA[31] ^ SrcB[31]) & (SrcA[31] ^ Sum[31]);
  assign Neg  = Sum[31];
  assign LT = Neg ^ Overflow;
  assign SLT = {31'b0, LT};

  assign ALUFunct = Funct3 & {3{ALUOp}}; // Force ALUFunct to 0 to Add when ALUOp = 0
  always_comb
    case (ALUFunct)
      3'b000:  ALUResult = Sum;          // add or sub
      3'b010:  ALUResult = SLT;          // slt
      3'b110:  ALUResult = SrcA | SrcB;  // or
      3'b111:  ALUResult = SrcA & SrcB;  // and
      default: ALUResult = 'x;
    endcase
endmodule

module lsu(
  input  logic        clk, MemWrite,
  input  logic [31:0] IEUAdr, WriteData,
  output logic [31:0] ReadData);

  logic [31:0] RAM[63:0];

  assign ReadData = RAM[IEUAdr[7:2]]; // Word aligned
  always_ff @(posedge clk)
    if (MemWrite) RAM[IEUAdr[7:2]] <= WriteData;
endmodule
```

2.7.2 Pipelined Microarchitecture

Fig. 2.11 shows how to convert the single-cycle processor into a classic five-stage pipeline by adding registers between stages. The processor executes an instruction across these stages as follows:

- **Fetch:** Reads the instruction from the instruction memory (IROM)
- **Decode:** Decodes the control signals and reads the source registers from the register file
- **Execute:** Performs the ALU operation
- **Memory:** Accesses the data memory to read or write data
- **Writeback:** Writes the result to the register file

Each signal is now annotated with a suffix corresponding to its pipeline stage. Notice how the control signals are also pipelined so they

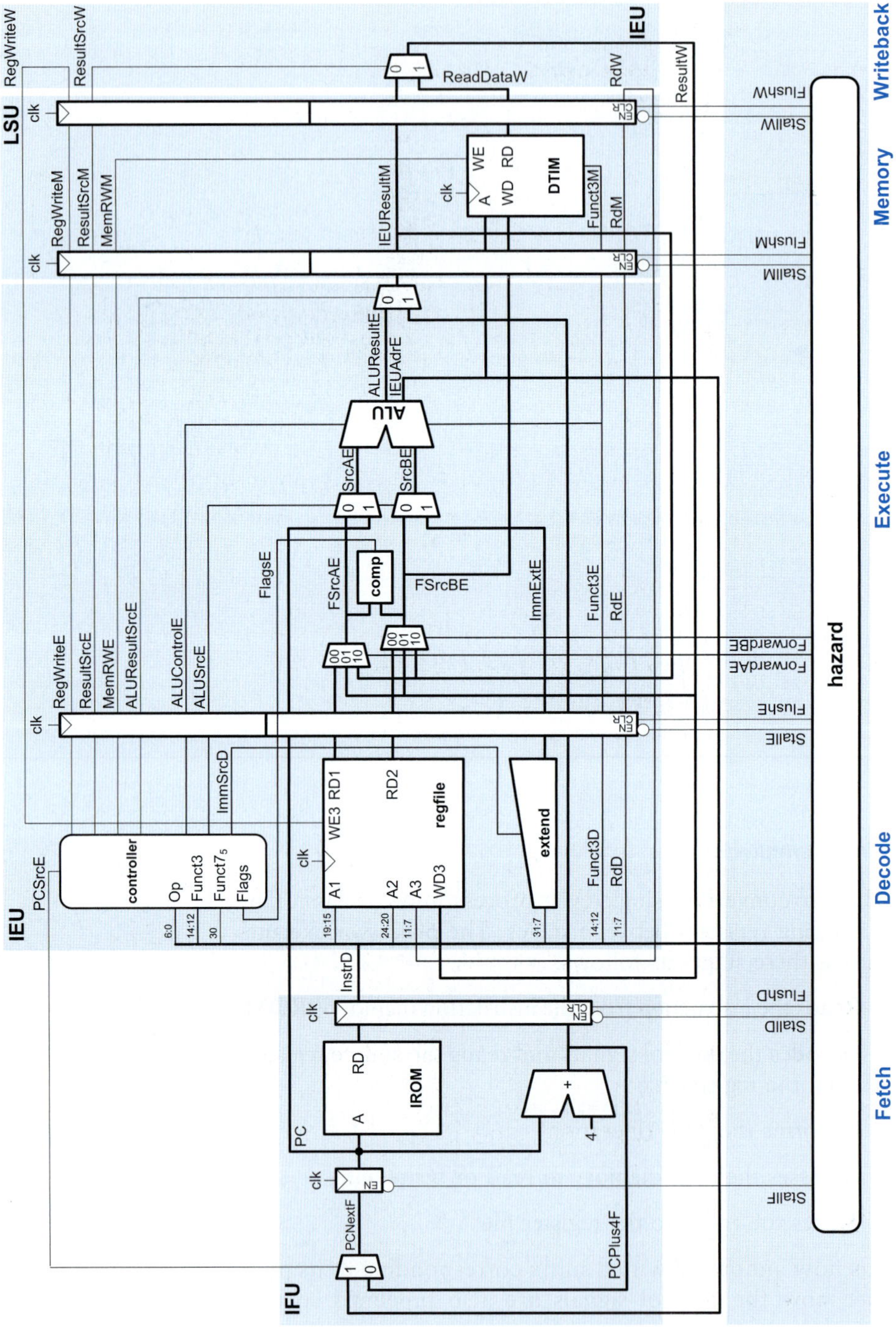

Fig. 2.11 Pipelined processor.

advance along with the instructions they control. Also note how the destination register `rd` must be pipelined so that the result is written to the proper register in the Writeback stage. The register file writes in the first half of a cycle and reads in the second half, so a value can be written and read back out in the same cycle. Only one major operation (ALU, memory, or register file access) occurs in each stage, so the cycle time is substantially faster—ideally, 5x if not for the sequencing overhead of the extra registers and the inability to perfectly balance each stage of the pipeline.

The IEU (integer execution unit) operates in the Decode, Execute, and Writeback stages of the pipelined processor.

The pipelined processor design is complicated by *pipeline hazards* that arise when one instruction depends on another that is still in the pipeline. The two main types of hazards are data hazards and control hazards. *Data hazards* are caused by pipeline delays in reading and writing registers. Some data hazards can be solved by *forwarding* results from a previous instruction, and others require *stalling* the pipeline. *Control hazards* occur when the decision of what instruction to fetch next has not been made by the time the fetch takes place. A branch predictor addresses control hazards by guessing which instruction should be fetched next; if the guess is wrong, the processor must *flush* the incorrect instructions and restart with the correct one. Fig. 2.11 shows the forwarding multiplexers (muxes) and a hazard unit to control the muxes and issue stalls and flushes. Hazard logic is discussed further in Section 7.2.2.

2.7.3 Verification

All hardware requires logic verification to catch the bugs and gain some confidence that the design is correct. Chapter 5 describes logic verification in more depth. This section uses a simple ad-hoc test program listed in Code Example 2.16 that exercises all types of instructions, including taken and not-taken branches. If the program runs correctly, it should store the value 7 into address 96 and then 25 into address 100. A store with any other address or value indicates failure.

A mark of an inexperienced engineer is to expect one's code is correct on the first try.

Code Example 2.16 TEST PROGRAM

```
# riscvtest.s
# Sarah.Harris@unlv.edu
# David_Harris@hmc.edu
# 2025
#
# Test the RISC-V processor.
#   add, sub, and, or, slt, addi, lw, sw, beq, jal
# If successful, it should write the value 25 to address 100

#         RISC-V Assembly           Description                  Address  Machine Code
main:     addi x2, x0, 5            # x2 = 5                     0        00500113
          addi x3, x0, 12           # x3 = 12                    4        00C00193
          addi x7, x3, -9           # x7 = (12 - 9) = 3          8        FF718393
          or   x4, x7, x2           # x4 = (3 OR 5) = 7          C        0023E233
          and  x5, x3, x4           # x5 = (12 AND 7) = 4        10       0041F2B3
          add  x5, x5, x4           # x5 = (4 + 7) = 11          14       004282B3
          beq  x5, x7, end          # Shouldn't be taken         18       02728863
          slt  x4, x3, x4           # x4 = (12 < 7) = 0          1C       0041A233
          beq  x4, x0, around       # Should be taken            20       00020463
          addi x5, x0, 0            # Shouldn't happen           24       00000293
around:   slt  x4, x7, x2           # x4 = (3 < 5)  = 1          28       0023A233
          add  x7, x4, x5           # x7 = (1 + 11) = 12         2C       005203B3
          sub  x7, x7, x2           # x7 = (12 - 5) = 7          30       402383B3
          sw   x7, 84(x3)           # [96] = 7                   34       0471AA23
          lw   x2, 96(x0)           # x2 = [96] = 7              38       06002103
          add  x9, x2, x5           # x9 = (7 + 11) = 18         3C       005104B3
          jal  x3, end              # Jump to end, x3 = 0x44     40       008001EF
          addi x2, x0, 1            # Shouldn't happen           44       00100113
end:      add  x2, x2, x9           # x2 = (7 + 18)  = 25        48       00910133
          sw   x2, 0x20(x3)         # Mem[100] = 25              4C       0221A023
done:     beq  x2, x2, done         # Infinite loop              50       00210063
```

The hexadecimal machine code is placed in a text file called `riscvtest.memfile`. The IROM loads this file at startup (see the `irom` module in Code Example 2.15).

Code Example 2.17 lists the SystemVerilog testbench. The testbench instantiates the top-level module, which could be either the single-cycle or pipelined processor. The testbench generates a clock repeating every 10 time units and applies a reset for the first two clock cycles. The processor runs the test program from the IROM. If all goes well, the testbench detects the write of 25 to address 100 and prints success. Similarly, it prints failure if an incorrect write occurs. The testbench will hang if no writes occur.

Code Example 2.17 SYSTEMVERILOG TESTBENCH

```
module testbench();
  logic        clk;
  logic        reset;

  logic [31:0] WriteData, DataAdr;
  logic        MemWrite;

  // Instantiate device to be tested
  top dut(clk, reset, WriteData, DataAdr, MemWrite);

  // Initialize test
  initial
    begin
      reset <= 1; # 22; reset <= 0;
    end

  // Generate clock to sequence tests
  always
    begin
      clk <= 1; # 5; clk <= 0; # 5;
    end

  // Check results
  always @(negedge clk)
    begin
      if (MemWrite) begin
        if (DataAdr === 100 & WriteData === 25) begin
          $display("Simulation succeeded");
          $stop;
        end else if (DataAdr !== 96) begin
          $display("Simulation failed");
          $stop;
        end
      end
    end
endmodule
```

Fig. 2.12 shows waveforms from the simulation. The 32-bit busses are displayed in hexadecimal. The testbench toggles `clk` with a 10-ns period and asserts `reset` for the first two cycles, initializing `PC` to 0. You can watch `PC` and `Instr` step through the instructions in accordance with the program from Code Example 2.16. Observe at time 110 ns how `PC` jumps from instruction address 0x20 to 0x28 because of the `beq around`. Also observe how `SrcA`, `SrcB`, and `Result` line up with the test program. For example, at 45 ns when `PC` = 8, the program adds 12 + (−9) = 3. `SrcA` is 0x0000000C = 12_{10}; `SrcB` is 0xFFFFFFF7, which is −9 as a two's complement number; `Result` = 3. At time 195 ns, `MemWrite` rises, and the program correctly writes 0x00000019 = 25_{10} to address 0x00000064 = 100_{10} and prints "`Simulation succeeded`".

To simulate with Questa or ModelSim, type:

```
$ cd cvw/examples/verilog/riscvsingle
$ vsim -do riscvsingle.do
```

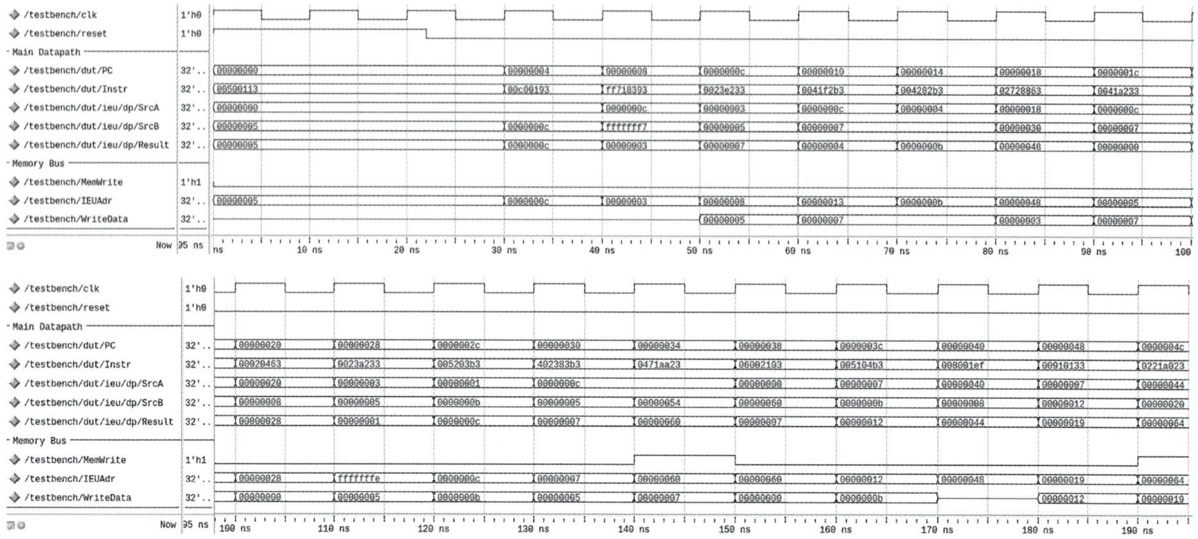

Fig. 2.12 `riscvsingle` **simulation (first and second half).**

2.8 WALLY SYSTEM-ON-CHIP

Wally is named in honor of the Harvey Mudd College mascot, Wally Wart, which personifies the flat, square protrusions that adorn the brutalist-style buildings at Harvey Mudd College.

Art by Sarah Harris (adapted from Jeff on Pinterest).

This book explores RISC-V SoC design using an open-source configurable pipelined processor affectionately known as Wally. Wally is described in SystemVerilog. The features are parameterized, and the values are set in a configuration file. Configurations range from a simple RV32E embedded core to a full-featured RV64GC application processor with caches, branch prediction, and virtual memory. Configurable features include those in Table 2.5.

Wally presently does not support:

- Advanced microarchitecture: superscalar, out-of-order, multithreading, multicore
- Clock gating and other power reduction techniques
- Certain extensions
  - V: Vector instructions
  - Zvk*: Vector cryptographic instructions
  - H: Hypervisor
  - Debug and trace interfaces
  - Various recent smaller extensions

Wally (formally CORE-V Wally: cvw) is a project of the OpenHW (Open Hardware) Foundation, which curates and supports robust open-source

Table 2.5 Wally configurable features.

Configurable Feature	Description
XLEN	32 or 64
Primary extensions	
A	Atomic
B	Bit manipulation
C	Compressed
D	Double-precision floating-point
E	Embedded 16 registers
F	Single-precision floating-point
Zk*	Cryptography
M	Multiplication and division
Q	Quad-precision floating-point
Additional extensions	
Zifencei	`fence.i` synchronization
Zfh	Half-precision floating-point
Zfa	Additional floating-point
Zmmul	Integer multiplication without division
Zbc	Carry-free multiplication
Zca, Zcb, Zcf, Zcd	Additional compressed instructions
Zicbom, Zicboz, Zicbop	Cache Management Operations
Zicclsm	Misaligned Loads and Stores
Zihintpause	Pause Hint
Zicond	Conditional zero
Privileged modes and features	
Zicsr	CSRs
Zicntr, Zihpm	Performance counters
Sm1p12	Machine mode (M) version 1.12
Ss1p12	Supervisor mode (S) version 1.12
U	User mode (U)

(Continued)

Table 2.5 Wally configurable features. Cont'd

Configurable Feature	Description
Sstc	Supervisor timer compare
Reset vector address	Typically 0x1000 or 0x80000000
PMP	Physical memory protection
Interrupts	Vectored or nonvectored
Sv32, Sv39, Sv48	Virtual memory
TLB sizes	Instruction and data TLB (ITLB and DTLB) sizes
Svnapot	Huge pages
Svpbmt	Page-based memory types
Svadu	A/D hardware updates
Svinval	TLB Management
Caches	I$ & D$ capacity, associativity
Branch Prediction	Gshare or predict not taken
Peripherals	
CLINT	Core Local Interruptor with timers
PLIC	Platform Level Interrupt Controller
UART	Universal asynchronous receiver/transmitter serial port
GPIO	General-purpose input/output pins
SPI	Serial Peripheral Interface
SDC	SD Card
DTIM / IROM	Data tightly-integrated memory, instruction ROM

cores. Many of the cores have evolved from academic projects to industry-supported open products. Other OpenHW cores include:

- CV32A60AX: RV32IMCA application class 6-stage pipeline based on Ariane
- CV32E40PV2: RV32IMFC microcontroller 4-stage pipeline based on RI5CY
- CV32E20: RV32[E|I][M]C simple, low-power microcontroller based on lowRISC Ibex

The Wally repository and how to access it are discussed in Section 3.2.

2.8.1 Wally Block Diagram

Fig. 2.13 shows a high-level block diagram of Wally. It uses the same 5-stage pipeline described in Section 2.7.2. The SoC core connects via a bus interface to the uncore block containing peripherals and memory.

The major blocks in the Wally core (module name: `wallypipelinedcore`) are given in Table 2.6. Main modules have a light blue background and submodules are shown within them.

These units are discussed in detail in the remainder of the book. Integer instructions have single-cycle latency and throughput. Loads, multiplies, and most floating-point instructions have single-cycle throughput but 2-cycle latency. A load immediately after a store also must stall for a cycle. Division and cache misses have long latency, and they stall the pipeline. Mispredicted branches are discovered in the Execute stage and thus must flush the next two instructions. Writes to the CSRs flush the pipeline to ensure consistency.

The large memory arrays in the caches and branch predictor are implemented as single-ported edge-triggered RAMs. The smaller multiported register files and TLB content-addressable memories (CAMs) are implemented using flip-flops or latches because many libraries for field programmable gate arrays (FPGAs) and application-specific integrated circuits (ASICs) lack specialized cells for these structures.

Popular open-source licenses include Apache, BSD, GNU, and MIT [opensource.org]. The GNU license is a copyleft license that requires all derivative works be distributed under the same terms, so proprietary derivatives are forbidden. The other licenses are essentially "anything goes," requiring only attribution and a lack of warranty. Apache is more complex than MIT or BSD but explicitly addresses patent rights. The SolderPad license is an open hardware flavor of the Apache license stewarded by the Free and Open Source Silicon Foundation. Wally uses the SolderPad license because it permissively supports hardware and software development while protecting from patent abuse and because it is the standard for the OpenHW Foundation.

2.8.2 Wally Configurations

Table A.1 lists standard configurations for which Wally has been tested, namely: rv32e, rv32i, rv32imc, rv32gc, rv64i, and rv64gc. The embedded configuration rv32e showcases the smallest size and highest frequency achievable in the TSMC 28-nm technology, with no on-chip memories. The simple CPU configurations, rv32i and rv64i, run only the integer instruction set and have small on-chip memories; for example, the rv32i configuration could run simple programs to control other portions of an SoC. The microcontroller configuration rv32imc adds a bus, peripherals, privileged modes, and compressed instructions; this configuration is common as a low-cost stand-alone chip to control appliances, toys, and vehicle systems. The application processor configurations, rv32gc and rv64gc, add the full set of instruction extensions, caches, branch prediction, supervisor mode, and virtual memory needed to boot Linux or similar operating systems. In particular, the rv64gc configuration supports the RVA22S64 profile. The more complex designs have a much lower frequency because the trap and floating-point logic is on the critical path. The area increases substantially as memories, floating-point, and branch prediction are added.

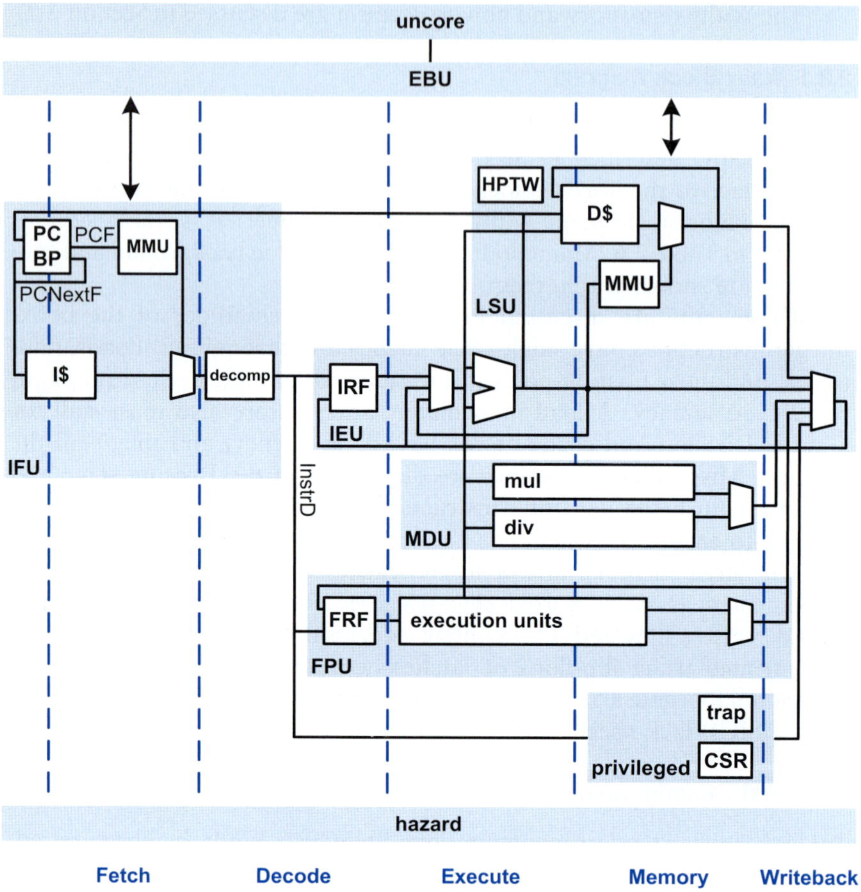

Fig. 2.13 Wally block diagram.

2.8.3 Wally Code Base

The Wally code is maintained in a Git repository at `github.com/openhwgroup/cvw`. Section 3.2 describes how to clone the repository and compile test cases. Fig. A.1 shows the organization of the repository.

The version described in this textbook is Wally 1.0. We anticipate Wally will continue to improve over time, and changes from Wally 1.0 will be described in the repository. Check out version 1.0 to see the processor exactly as it is in this book.

Table 2.6 Major blocks in the Wally core.

Module Name	Description
ifu	Instruction fetch unit (IFU)
cache or irom	Instruction cache (I$) or instruction ROM (IROM)
mmu	Memory management unit (MMU)
bpred	Branch predictor
decompress	Decompressor for compressed instructions
in multiple modules	PC logic
ieu	Integer execution unit (IEU)
regfile	Integer register file (IRF)
ieuresultmux	Result multiplexer (mux)
controller	Controller, including forwarding and mux selects
alu	Arithmetic logic unit (ALU)
mdu	Multiply/divide unit (MDU)
mul	Multiplier
div	Divider
fpu	Floating-point unit (FPU)
fpreg	Floating-point register file (FRF)
fma	Floating-point multiply/accumulate (FMA)
fdivsqrt	Floating-point divide/square root
fcvt	Floating-point convert
fcmp	Floating-point compare
lsu	Load/store unit (LSU)
cache or dtim	Data cache (D$) or DTIM
mmu	Memory management unit (MMU)
hptw	Hardware page table walker (HPTW)
ebu	External bus unit (EBU)
hazard	Hazard unit
privileged	Privileged unit
csr	Control and status registers (CSRs)
trap	Trap unit

2.8.4 Verification

Experience teaches us that untested features usually have bugs. The verification suites are not comprehensive enough to cover all corner cases, especially of the privileged architecture, so we are confident that bugs remain.

Wally is verified with a variety of test cases, as listed in Table 2.7. These tests are discussed further in Sections 3.7, 5.7, and A.3.

2.8.5 Benchmarking

Given that the logic is correct (or at least passes all test cases), the next step is to optimize for performance. Performance depends on a processor's cycle time and a program's number of instructions and execution cycles, as described in Eq. (2.1). We use benchmarking to measure the number of cycles and, in the next section, explore logic synthesis timing analysis to obtain cycle time. Compiler optimization flags affect the number of instructions. Benchmarks consist of a group of programs that are used to compare the performance of different processors. This section summarizes scores for the classic CoreMark and newer Embench benchmarks. Chapter 21 describes these benchmarks in more depth.

Table 2.7 Test Suites used to verify Wally.

Test Suite	Description
riscv-arch-test	Shows that each instruction in isolation performs its basic functions; provided by RISC-V International's Architecture Test Group
wally-riscv-arch-test	Tests functions not covered by riscv-arch-test, particularly for the MMU and privileged operations
Coverage tests	Exercises code coverage by showing that the lines of Verilog have been exercised by the tests. Functional coverage tests are in progress.
TestFloat	Shows that FPU passes a comprehensive set of tests; from Berkeley
Linux boot	Boots Linux in simulation and on an FPGA, confirming that Wally correctly executes 550 million instructions, including using privileged modes, CSRs, interrupts, peripherals, and virtual memory

Table 2.8 compares CoreMark scores for Wally and other embedded processors. One CoreMark is one iteration per second of the benchmark. Scores are usually reported as CoreMark/MHz to normalize for clock frequency. Wally has similar performance to ARM's Cortex M4 and OpenHW Foundation's RI5CY. Cadence's VeeR-EH2 is a dual-issue superscalar processor and has nearly twice the CoreMark performance of Wally.

Table 2.9 compares Embench 1.0 scores for Wally and other embedded processors, optimized for code size (using compiler flag -Os) and speed (using compiler flag -O2). Size and speed are normalized to the best achieved by an ARM Cortex M4, showing that Wally has similar code density and is about 9% faster at the same clock rate. Even though VeeR-EH2 is a superscalar processor with a tremendous CoreMark score, it is only 22% faster than Wally in Embench, showing that Embench is a more realistic benchmark that is harder to game. RI5CY speed appears suspicious, being equal for -Os and -O2.

2.8.6 Implementation

Wally is designed to balance performance, power, and area rather than favor one at great expense to the others. The SystemVerilog has been synthesized in 130-, 90-, and 28-nm complementary metal-oxide-semiconductor (CMOS) processes and on three FPGAs. The results were reviewed to detect and eliminate grossly inefficient structures caused by poor coding. The design was also tuned to shorten the cycle time by balancing the amount of logic in each pipeline stage and by optimizing cycle-limiting paths. Implementation is discussed further in Chapters 6 and 23.

We hope Wally will continue to evolve. It could be a good platform for evaluating implementation tradeoffs in RISC-V functional units. For example, we hope it can serve as a baseline against which improved implementations can be compared. Most of the SystemVerilog was written by five people working part time over 3 years, and we are confident that many important optimizations still should be made.

Table 2.8 CoreMark scores.

Processor	CoreMark/MHz
Wally (rv32imb)	3.36
Cortex M4	3.42[a]
RI5CY	3.19[b]
VeeR-EH2	6.3[c]
VeeR-EL2	3.9

[a]www.eembc.org/viewer/?benchmark_seq=1701
[b]www.pulp-platform.org/docs/RossiGdanskOrconf09-18.pdf
[c]riscv.org/wp-content/uploads/2019/12/12.11-14.20a3-Bandic-WD_SweRV_Cores_Roadmap_v4SCR.pdf

Table 2.9 Embench scores.[a]

Processor	Architecture	Compiler Directives	Size Optimized `-Os`		Speed Optimized `-O2`	
			Speed/MHz	Size	Speed/MHz	Size
Wally	RV32IMBC	GCC 13.2.0 `-ffunction-sections -fdata-sections`	1.01	1.01	1.09	1.18
Cortex M4	ARMV7M	GCC 9.2 `-ffunction-sections`	0.90	1.00	1.00	1.15
RI5CY	RV32IMC	GCC 9.2	0.93	1.05	0.93	1.27
VeeR-EH2	RV32IMACZb	GCC 10.2 `-ffunction-sections`	1.18	1.11	1.33	1.26

[a]github.com/embench/embench-iot-results

2.8.6.1 FPGA

Chapter 23 describes Wally implementations on a Xilinx FPGA on the Arty A7, VCU108, and VCU118 boards. Fig. 23.3 shows the Arty A7 board annotated with I/O and memory components. The FPGA boots Linux and prints boot messages over the USB cable to a terminal.

2.8.6.2 SoC

Chapter 6 describes logic synthesis with Design Compiler to convert the SystemVerilog description of the processor into a set of logic gates and memories placed on a chip. Fig. 2.14 shows a placement of the rv32gc configuration for a chip in the TSMC 28hpc+ 28-nm process. The components are placed during physical synthesis, but routing has not been performed. The chip area is 0.7 mm on a side. The memory arrays account for the largest part of the area. Each 16-KiB cache has four ways, each comprising four 1-KiB 64 × 128 data arrays and one 64 × 22 tag array. The logic blocks are color coded by their unit. Physical synthesis places cells to reduce delay on the critical path, so we observe interesting tentacles where the IFU, LSU, and privileged unit closely interact with the 16 physical memory protection (PMP) checkers. The FPU has the least communication with other blocks, so it is in the corner, near the IEU and LSU. This floorplan is automatically generated during synthesis; each run is different.

Table A.1 summarizes the maximum frequency for each configuration in TSMC28, ranging from 2.99 GHz for rv32e to 0.8 GHz for rv64gc. In this process, a fanout-of-4 inverter (FO4) gate delay is 11.6 ps for typical silicon at nominal (0.9 V) voltage and 25°C, which is a handy metric for normalizing frequency in a process-independent fashion. Hence, the clock

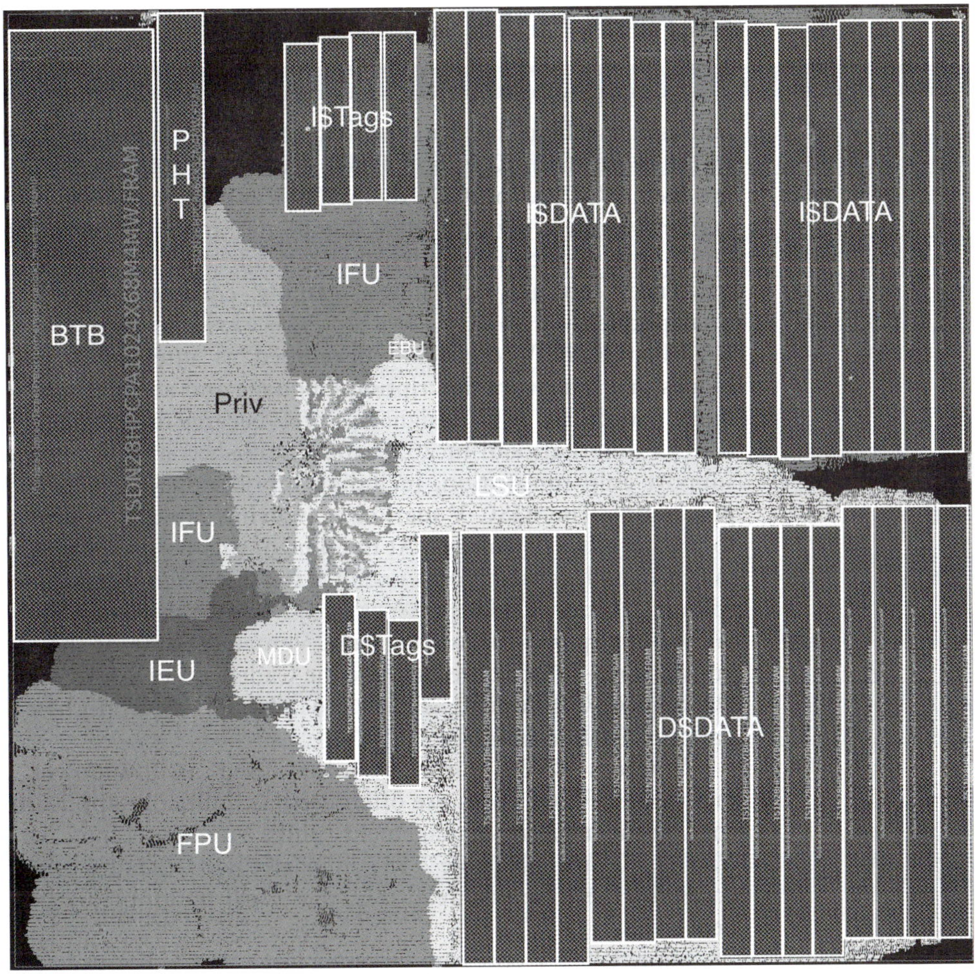

Fig. 2.14 rv32gc wallypipelinedcore layout in TSMC 28 nm.

period (reciprocal of frequency) ranges from 29 to 108 FO4 delays for the various configurations. The 5-stage pipelines are not as aggressive as contemporary high-performance processors with more pipeline stages that target closer to 50 FO4 delays per cycle.

A complete SoC needs bond pads or solder bumps to connect signals to the external world. These signals include clock, reset, peripheral pins, and the AHB bus, as well as power and ground. The core is so small in 28 nm that it is dwarfed by the pads.

Table 2.11 **Commercial RV32 single-core embedded processors.**

Vendor	SiFive	SiFive	SiFive	Andes	Andes	Synopsys	Synopsys
Product	E76[a]	E31[b]	E20[c]	N25F[d]	N22	RHX-100[e]	RMX-100[f]
RV32 ISA	IMAFC	IMAC	IMAC	GCB	{I/E}MAC	GCV	IC<MFD>
μArchitecture	2 issue Superscalar	1 issue Pipelined	1 issue Pipelined	1 issue Pipelined	1 issue Pipelined	2 issue Superscalar	1 issue Pipelined
Pipeline Stages	8	5	2	5	2	10	3
L1 I$/D$ (KiB)	32 /32 64B lines 2/4 way + TIM	16 KiB I$/ITIM 64 B line 2 way 64 KiB DTIM	None	8–64 1–4 way + local mem	1–32 1–2 way +local mem	4–128 + DTIM	2–64 I$ + DTIM
Branch predictor	1.3 KiB BHT 4-entry BTB 2-entry IJTP 3-entry RAS	512-entry BHT 28-entry BTB 6-entry RAS	None	256-entry BHT 32–256 entry BTB 4-entry RAS	optional	Yes	
Multiplier latency	3 cycles	2 cycles	5 cycles	2 cycles			1–9 cycles
Divider latency	3–32 cycles	2–33 cycles	Up to 32 cycles			16 cycles	
MMU	PMP8	PMP8	None	PMP	PMP	SV32/PMP	PMP
Privilege levels	MU	MU	M	MU	M, optional U	MSU	MSU
Bus	AHB/AXI	AHB/AXI	AHB/AXI	AHB/AXI	AHB/AXI	AHB/AXI	AHB/AXI
Frequency (GHz) (28hpc)	1.6 (28hpc)	0.9 (28hpc)	0.725 (28hpc)				
Area (mm^2) (28hpc)	0.14	0.187	0.046				
CoreMark/MHz	5.7	3.0	2.5				
Country	USA	USA	USA	Taiwan	Taiwan	USA	USA

[a]usermanual.wiki/m/51f036f7010a69b4bccfd2838ef299e126f18cbb65b682eb9076b35dedbc0e79
[b]usermanual.wiki/Document/SiFiveE31Manualv2p0.20928298.pdf
[c]static.dev.sifive.com/SiFive-E20-Manual-v1p0.pdf
[d]www.andestech.com/tw/產品與解決方案/andescore-processors/riscv-n25f/
[e]www.synopsys.com/dw/ipdir.php?ds¼arc-v-rhx-100
[f]www.synopsys.com/dw/ipdir.php?ds¼arc-v-rmx-100

Table 2.12 **Open-source RV64 application processors.**

Vendor	OpenHW	T-Head	Berkeley	Madras	lowRISC	BlackParrot (U. Washington+ Boston U.)	This Book (OpenHW)
Product	CVA6[a] (PULP Ariane)	C910[b]	BOOMv3[c]	Shatki C[d]	Muntjac[e]	BlackParrot[f]	Wally (64gc)
RV64 ISA	GC	GC	GC	GC	IMAC	IMAFD	GC
μArchitecture	1 issue Pipelined	3 issue Out-of-order	4 issue Out-of-order	1 issue Pipelined	1 issue Pipelined	1 issue Pipelined	1 issue Pipelined
Pipeline stage	6	12	~14	5	5	8	5
L1 I$/D$ (KiB)	Up to 32/32	32–64	32/32	Configurable	16/16	32/32	Configurable
L2$ (KiB)		128–8192	512	N/A	64/core	yes	N/A
Branch predictor	BHT BTB RAS	BHT BTB RAS	TAGE	Gshare	BHT BTB RAS	BHT BTB RAS	Gshare
Multiplier latency	2 cycles			Configurable	2–5 cycles	3 cycles	2 cycles
Divider latency	Up to 64 cycles				34–66 cycles	Configurable	Configurable
MMU	SV39 PMP	SV39 512–1024 entry TLB PMP16		SV39/48 PMP	SV39	SV39	SV39/48 PMP16/64
Privilege levels	MSU	MSU	MSU	MSU	MSU	MSU	MSU
Bus	AXI	AXI		TileLink/AXI	TileLink/AXI	Wishbone	AHB-Lite
Multicore		Up to 4			Yes	Yes	
Frequency (GHz)	1.5 (22 FDX)	2–2.5 (TSMC12nm)	1 (FinFET)		0.08 (Kintex 7 FPGA)		0.9 (TSMC 28hpc)
Area (mm^2)	0.23 (175 kgate	0.398					0.5
CoreMark/ MHz	2.5	7.1	6.2	2.84	3.01	3.04	3.06
Language	SystemVerilog	Verilog	Chisel3	BlueSpec	SystemVerilog	SystemVerilog	SystemVerilog
Nationality	Switzerland	China	USA	India	UK	USA	USA

[a]github.com/openhwgroup/cva6
[b]github.com/T-head-Semi/openc910
[c]github.com/riscv-boom/riscv-boom
[d]gitlab.com/shaktiproject/cores/c-class/-/tree/master
[e]github.com/lowRISC/muntjac
[f]github.com/black-parrot/black-parrot

Table 2.13 Open-source RV32 embedded processors.

Vendor	OpenHW	lowRISC	Chips Alliance	Chips Alliance	Madras	Papon	This Book (OpenHW)
Product	CV32E40P[a] (RI5CY)	Ibex[b] (Zero-riscy/ OpenHW CV32E20)	VeeR EH2[c] (Western Digital SweRV EH2)	VeeR EL2[d] (Western Digital SweRV EL2)	Shatki E[e]	VexRiscv[f]	Wally (32imc)
RV32 ISA	IMFC	{I/E}MCB	IMAC	IMC	IMAC	GC	IMC
µArchitecture	1 issue Pipelined	1 issue Pipelined	2 issue Superscalar dual-threaded	1 issue Pipelined	1 issue Pipelined	1 issue Pipelined	1 issue Pipelined
Pipeline Stage	4	2 or 3	9	4	3	5	5
L1 I$/D$ (KiB) and/or DTIM/IROM	N/A	Optional 4 KiB I$	Optional I$ + local mem	Optional I$ 32 or 64 B lines 2 or 4 ways + local mem		1–32 Up to 8 ways + local mem	DTIM + IROM
Branch predictor (entries)	N/A	N/A	32–4096 BHT 32–512 BTB 2–8 RAS	32–4096 BHT 32–512 BTB 2–8 RAS		BHT BTB	N/A
Multiplier latency	1– 5	1	3	1	4	3	2
Divider latency	3– 35	37	34	34	32	34	Configurable
MMU	N/A	PMP16	N/A	N/A	PMP16	SV39 PMP	
Privilegelevels	M	MU	M	M	MU	MSU	M
Bus	OBI	Custom OBI	AXI4	AXI4 or AHB-Lite	AXI4-Lite	AXI	AHB
Frequency (GHz)		0.5 (65 nm)					1.9
Area (mm^2)	70k gate	15–61k gate	0.067 (16 nm)	0.023 (16 nm)			0.043 (28nm)
CoreMark/MHz	3.19	0.9–3.1	6.3	4.3			3.34
Language	SystemVerilog	SystemVerilog	SystemVerilog	SystemVerilog	BlueSpec/ Verilog	SpinalHDL	SystemVerilog
Nationality	International	International	USA	USA	India	Switzerland	USA

[a]github.com/openhwgroup/cv32e40p
[b]github.com/lowRISC/ibex
[c]github.com/chipsalliance/Cores-VeeR-EH2
[d]github.com/chipsalliance/Cores-VeeR-EL2
[e]gitlab.com/shaktiproject/cores/e-class/-/tree/master
[f]github.com/SpinalHDL/VexRiscv

2.10 SUMMARY AND A LOOK AHEAD

This book is unusual in that it teaches you microprocessor and SoC design in the context of a real processor implementation. Each principle covered in the book is illustrated in the context of the Wally RISC-V processor. Logic verification is the most difficult and time-consuming part of processor design, so each feature begins with a test plan.

This chapter introduced the RISC-V instruction-set architecture. As a fifth-generation RISC architecture, RISC-V defines the RV32I instruction set with 40 simple instructions and a 32-entry register file of 32-bit registers. Register `x0` is tied to 0. The other register uses are flexible in principle, but the RISC-V ABI convention consistently uses certain registers for certain purposes, such as local variables (`s0`-`s11`), function arguments and return result (`a0`-`a7`), the stack pointer (`sp`), function call return address (`ra`), and temporary results (`t0`-`t6`). All instructions are encoded in a handful of 32-bit formats. The opcode and source/destination registers are in the same location for every format, facilitating a simple and fast decoder. The instructions are summarized below. Most operate on registers or immediates. R-type arithmetic and logical instructions operate on two source registers and place the result in a destination register. I-type arithmetic and logical operations are similar but use one source register and one 12-bit immediate. RISC-V is a load/store architecture, so explicit load and store instructions are used to move data between memory and registers. Conditional branch and unconditional jump instructions control program flow.

- **R-type arithmetic/logical:** `add`, `sub`, `and`, `or`, `xor`, `sll`, `srl`, `sra`, `slt`, `sltu`
- **I-type arithmetic/logical:** `addi`, `andi`, `ori`, `xori`, `slli`, `srli`, `srai`, `slti`, `sltui`
- **Loads and stores:** `lw`, `lh`, `lhu`, `lb`, `lbu`, `sw`, `sh`, `sb`
- **Branches:** `beq`, `bne`, `blt`, `bge`, `bltu`, `bgeu`
- **Jumps:** `jr`, `jalr`
- **Upper immediates:** `lui`, `auipc`
- **System instructions:** `ecall`, `ebreak`, `fence`

RV64I extends the register file to 64 bits and adds a handful of additional instructions to load/store double words and to operate on the bottom 32 bits of a register. Instructions remain 32 bits long. RISC-V avoids ideas that once seemed clever in other instruction sets but later proved problematic for high-performance implementations.

RISC-V also defines a privileged architecture with three privilege levels, traps, and CSRs. The memory map is system dependent, but a typical system has main memory starting at address 0x80000000 and perhaps a boot memory at 0x1000, along with various memory-mapped peripherals such as a timer, interrupt controller, and serial port. RISC-V has many standard extensions such as floating-point, which allows the architecture to scale from a simple embedded microcontroller core to a full-featured application processor. RISC-V profiles define a collection of extensions so that a programmer has a standardized target for development.

This chapter illustrated microarchitecture, logic design, SystemVerilog, testbenches, and verification with simple single-cycle and pipelined implementations of a toy processor supporting a subset of RV32I. It introduced Wally, a configurable open-source RISC-V processor and SoC capable of scaling from a tiny RV32I core to a full-featured SoC capable of booting Linux. The remainder of this book illustrates RISC-V SoC design by building up Wally's capabilities.

The next four chapters develop the tool flows for programming and hardware description. Chapter 3 begins with installing the RISC-V compiler and simulator tool chains and the Wally repository. It then covers low-level RISC-V programming in assembly language and C, including simulating with Spike, QEMU, and Sail. Modern digital systems are specified and synthesized using hardware description languages (HDLs), with SystemVerilog leading in industry. Chapter 4 addresses HDL practices, starting with a SystemVerilog tutorial and then covering design methodology and style guidelines. Chapter 5 describes logic verification and debugging using the Wally code base. Chapter 6 describes logic synthesis, mapping Wally configurations onto 130-, 90-, and 28-nm processes to quantify the speed and size of the SoC.

The largest part of this book covers the design of full-featured pipelined processors, including all the critical details usually glossed over in architecture texts. Most chapters begin with principles applicable across computer architectures. They then show how the principles are applied in practice in the RISC-V architecture. They present a test plan showing how the features are verified and then illustrate a real implementation with Wally. Chapter 7 implements the full RV{32/64}I instruction set, first on a single-cycle core and then with a 5-stage pipeline demonstrating hazards, forwarding, stalls, and flushes. Both designs assume ideal local instruction and data memories. Chapter 8 adds the privileged architecture, including privilege modes, traps, and CSRs.

The next five chapters focus on the memory system. Chapter 9 adds a bus interface to fetch instructions and data from real external memories. External memory is slow, so Chapter 10 describes caches to speed up access to frequently used memory. Chapter 11 describes memory management, including virtual memory and memory protection, essential for modern operating systems. Chapters 12 and 13 put this all together to describe the load/store unit (LSU) and instruction fetch unit (IFU) that can access data from caches or internal memory or over the bus while supporting virtual memory and protection. The IFU also contains a branch predictor, which must be highly accurate to give good performance in a pipelined processor.

The next six chapters describe the optional RISC-V extensions. Chapter 14 covers the compressed 16-bit instruction set, with a decoder that decompresses into standard RISC-V instructions executing on the regular datapath. Chapter 15 covers the integer multiply and divide unit (MDU). Chapter 16 and Appendix F describe the floating-point unit (FPU), including rarely addressed details of designing and verifying floating-point multiply/add, divide/square root, and other units. Chapter 17 covers atomic instructions, which are important for synchronizing multicore systems. Chapter 18 describes the bit manipulation and cryptography units. Chapter 19 addresses a variety of other smaller extensions.

The final chapters address system design. Chapter 20 describes common peripherals, including timers, interrupt controllers, and serial ports needed to boot Linux. Chapter 21 runs benchmarks to measure the processor performance. Chapter 22 examines how Linux boots on a full-featured processor, and Chapter 23 synthesizes Wally onto a field-programmable gate array to boot Linux on physical hardware.

The appendices cover practicalities of tools and development environments for the SoC designer. Appendix A provides a synopsys of the Wally design. Appendix B introduces Linux, especially regarding features helpful for the SoC workflow. Appendix C covers version control with Git. Appendix D covers TCL, used to control tools such as logic synthesis. The RISC-V instruction set is summarized for quick reference on the last pages of this book.

This project has kept us and our students awake late at night and early in the morning, wrestling with fascinating puzzles and challenging debug. It has been a thrill to draw together computer architecture, mathematics, operating systems, and digital design to build and optimize a complex system. We hope you share our excitement about demystifying the implementation of a real processor and SoC. The best way to learn this material is to make your own changes and improvements to the system.

Despite the hard work of a good team over three years, we are confident that bugs remain undetected. We appreciate your input pointing out and fixing these errors. GitHub facilitates an open-source development flow in which many people around the world can contribute to a shared code base.

We also see this book as a stake in the ground documenting a reasonable baseline for essential components of a pipelined processor. We look forward to reading your discoveries of better approaches.

RISC-V Software Tool Flow

3

3.1 Platform Requirements and Tool Overview
3.2 Getting Started
3.3 GCC Assembler and Spike Simulator
3.4 GCC GNU C Compiler
3.5 Sail Simulation
3.6 QEMU Simulation
3.7 Test Suites
3.8 Summary

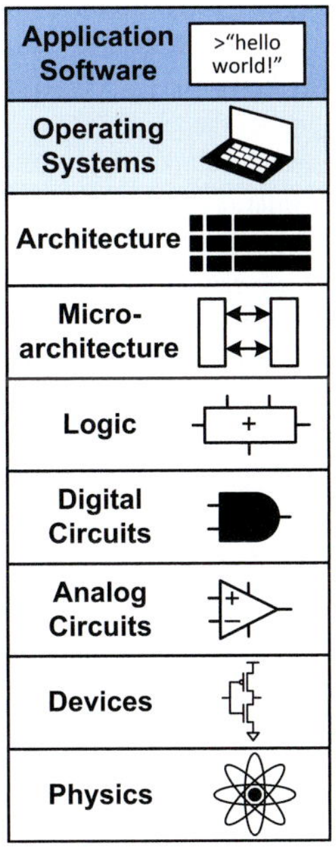

This chapter lays out the tool flow to work with the RISC-V *toolchain*, which consists of compilers, simulators, and other software development tools run successively to develop and test programs. Fig. 3.1 shows components of the RISC-V toolchain. The GCC (GNU Compiler Collection) assembler and compiler convert assembly language and C source code files into object files in Executable and Linkable Format (ELF). The GCC linker combines object files into an executable file, also in ELF. The executable file can be disassembled back into assembly language. A RISC-V simulator such as Spike or Sail runs the ELF and can log the results.

The chapter begins by describing platform requirements in Section 3.1 and how to clone the Wally git repository in Section 3.2. Section 3.3 describes how to write assembly language code for a bare metal platform and then assemble, link, and disassemble it with GCC driven by a Makefile. It also shows how to simulate the program using Spike. Section 3.4 then describes how to write, run, and disassemble C programs and how to mix assembly and C code using the same tools. Sections 3.5 and 3.6 introduce the formal Sail RISC-V model and simulator and the speedy QEMU

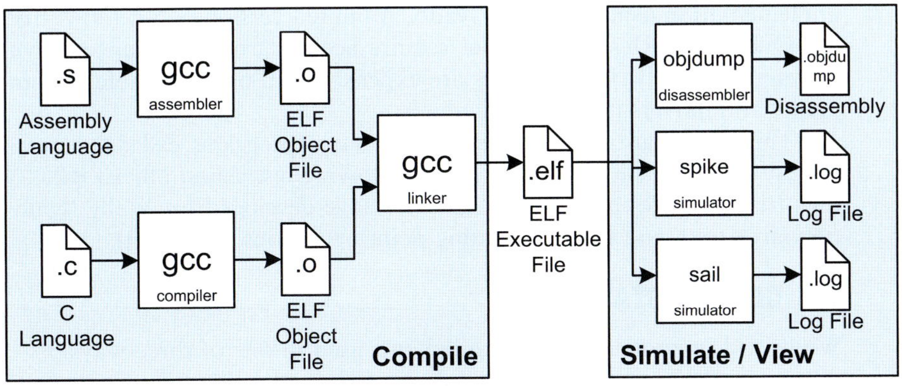

Fig. 3.1 **RISC-V toolchain.**

The term *compiler* generally refers to a tool that converts a program from a high-level language, such as C, into an executable. The compiler can also refer to just the first step of creating an executable, which is converting the program into assembly; the *assembler* converts the assembly into machine code, and the *linker* then groups programs and libraries into a single executable.

An *instruction set simulator (ISS)*, such as Spike, simulates how instructions running on a core change the architectural state and memory. It does not model peripherals, nor does it attempt to predict the number of cycles the program takes. In contrast, an *emulator*, such as QEMU, also models the behavior of peripherals but still does not accurately predict the number of execution cycles. An emulator is needed to boot Linux because this operating system (OS) depends on peripherals such as a timer, as well as terminal input/output. A *logic simulator*, also called a *cycle-accurate simulator*, such as Verilator or Questa, simulates the actual hardware as defined by the SystemVerilog code; it captures exactly how many cycles are required, accounting for microarchitectural details such as pipeline hazards, cache misses, or branch mispredictions. Logic simulators are orders of magnitude slower than Spike or QEMU, but they are critical in verifying the correct implementation. It is useful to compare the results of an ISS or emulator with a logic simulator to verify the SystemVerilog.

(Quick Emulator). Section 3.7 concludes by describing how to use these tools to build test suites for RISC-V implementations.

3.1 PLATFORM REQUIREMENTS AND TOOL OVERVIEW

Wally uses a Linux-based development flow. Although the tools may work under a variety of shells, the directions are tested on the Bash shell, which is the default on most Linux distributions. The main tools, which are freely available, are:

- **GCC:** The *software toolchain* for compiling programs to run on Wally
- **Spike:** RISC-V *instruction set simulator (ISS)*
- **Sail:** RISC-V ISS, formal architecture model
- **QEMU:** RISC-V *emulator* for simulating a core with peripherals
- **Verilator:** Logic simulator and lint

Optional tools are:

- **Questa:** Logic simulator (Siemens)
- **VCS:** Logic simulator (Synopsys)
- **Design Compiler:** Logic synthesis tool for custom chips (Synopsys)
- **ImperasDV:** RISC-V simulator for lockstep simulation (Synopsys)
- **Vivado:** Logic synthesis tool for FPGAs (AMD-Xilinx)

This chapter describes the tools for compiling and simulating RISC-V programs using GCC, Spike, Sail, and QEMU. Chapters 4 and 5 introduce hardware design and verification using SystemVerilog and logic simulation and show how to compare the results to the Sail reference. Chapter 6 discusses synthesis targeting a custom chip (application-specific integrated circuit [ASIC]) using Design Compiler, and Chapter 23 describes FPGA implementation using AMD's Vivado. A limited version of Vivado is free, but Questa and Design Compiler require licenses from Siemens and Synopsys, respectively. Both licenses are expensive and only available to companies and universities, not individuals.

This chapter also shows how to use a group of programs, called *test suites*, to verify functionality. Wally uses the following test suites: official RISC-V tests (riscv-arch-test), additional tests we have developed for Wally (wally-riscv-arch-test), and tests for floating-point operations (TestFloat).

3.2 GETTING STARTED

Wally and its tools may be installed and used in one of three ways:

- On a personal Linux computer, having administrative access to install tools. Clone the repository into a home directory.

- On a shared Linux server, where an administrator installs shared tools and users clone the Wally repository into their home directories. We expect this to be the most common case for universities and companies.
- Users install a Docker container that includes Ubuntu and some preinstalled tools.

The directions are likely to evolve over time, so they are part of the README in the CORE-V-Wally (cvw) repository at github.com/openhwgroup/cvw.

The directions include *New User Setup* and *Toolchain Installation and Configuration*. On a shared Linux server, each user performs the New User Setup, and the administrator performs Toolchain Installation. On a personal Linux computer, the user does both.

Wally uses the environment variables `$RISCV` to point to the shared RISC-V tools (typically in `/opt/riscv` or `~/riscv`) and `$WALLY` to point to the cvw repository in a home directory (typically `~/cvw`). The `setup.sh` script sets these variables and others to run RISC-V tools.

The New User Setup directions involve running `make` to build the tests, which takes a while. Then run `regression-wally`. If all goes well, you'll see it print "Success" on each test and then "SUCCESS! All tests ran without failures."

If installing the toolchain on your system is problematic (because of lack of administrator privileges, unsupported OS, general Linux software woes, and so on), the cvw repository `docs/docker.md` describes how to install a Docker container with many of the tools.

3.3 GCC ASSEMBLER AND SPIKE SIMULATOR

GCC is a high-quality and free open-source toolchain that includes cross-compilers to build RISC-V programs from another platform such as an Intel-based Linux computer. The toolchain and its associated software include an assembler and C compiler (`gcc`); the core C library (`newlib` or `glibc`); a debugger (`gdb`); and binary utilities, including a disassembler (`objdump`). Spike is a free and open-source RISC-V ISS originally developed at Berkeley. This section shows how to write, assemble, link, and simulate assembly language programs.

3.3.1 A Quick Tutorial

This section shows how to compile a program using GCC, disassemble it with objdump, and then simulate it on Spike. Code Example 3.1 shows a simple RISC-V assembly language program running on bare metal.

Spike is specific to RISC-V. All other tools can be used with other instruction sets or hardware.

The Wally RISC-V flow described in this chapter is tested on currently supported versions of Ubuntu LTS, Red Hat Enterprise Linux, AlmaLinux and Rocky Linux (free clones of Red Hat with no support), Debian Stable, and OpenSUSE Leap.

As of mid-2025, Synopsys and Cadence support their integrated circuit design CAD (computer-aided design) tools on Red Hat Enterprise Linux 8.4+ and 9.1+. Thus, Red Hat is suitable for organizations hosting CAD tools on the same server. Red Hat is expensive for universities, but AlmaLinux and Rocky are free and compatible Red Hat clones. We use AlmaLinux 9.5 in our development. Finding a Linux platform supported by all the necessary and rapidly evolving tools is an ongoing hassle for hardware developers.

Code Example 3.1 `example.S`

```
.section .text.init
.globl rvtest_entry_point   # Label rvtest_entry_point viewable by other files
rvtest_entry_point:         # Label expected by linker that indicates start of program
    li a0, 42               # a0 = 42

self_loop:
    j self_loop             # Infinite loop
```

In this book's command line examples, $ indicates the Linux prompt. Just type the commands in blue; do not type $.

The program will load 42 into register `a0`, then enter an infinite loop. The assembler directives starting with a dot define the initial text section (where the program begins) and a global label `rvtest_entry_point` that the linker expects at the start of the program.

Assemble the program by typing the following at a command line, which is fully explained in Section 3.3.6:

```
$ cd $WALLY/examples/asm/example
$ riscv64-unknown-elf-gcc -o example -march=rv32i -mabi=ilp32 -nostartfiles -mcmodel=medany
  -T../../link/link.ld example.S
```

C source code, assembly, object, and executable files end in `.c`, `.S`, `.o`, and `.elf` (or nothing), respectively.

Now use `objdump` to *disassemble the object file*, that is, view the compiled machine code in its assembly equivalent. For example, at address 0x80000000, the machine language instruction 0x02a00513 disassembles to `li a0, 42`.

All GCC tools require the `riscv64-unknown-elf-` prefix.

The compiler name is a hassle to type. It follows a GNU Target Triplet standard of `cpu-vendor-abi-` before the tool name. The cpu is riscv64, although GCC with multilib supports any of the RV32 and RV64 flavors using the `-march` flag. For RISC-V, the GCC vendor is normally called *unknown* because GCC is not specific to a particular flavor of Linux. The application binary interface (ABI) is defined by the ELF file. `riscv64-unknown-elf-gcc` uses a lightweight standard C library called *newlib*.

```
$ riscv64-unknown-elf-objdump -D example

example:     file format elf32-littleriscv


Disassembly of section .text:

80000000 <rvtest_entry_point>:
80000000:       02a00513        li      a0,42

80000004 <self_loop>:
80000004:       0000006f        j       80000004 <self_loop>

Disassembly of section .riscv.attributes:

00000000 <.riscv.attributes>:
   0:   1941            addi    s2,s2,-16
   2:   0000            unimp
...
```

Simulate the program with Spike using the `-d` flag for interactive debug mode. Expect a warning about undefined symbols. Press enter to advance through the program one line at a time. Spike inserts some startup code at address 0x1000 to initialize `a0` and `a1` to some values we don't care about and then jumps to the program at 0x80000000. Typing `reg 0 a0` will print the value of `a0` in hart 0[1], which should be 0x2a (42_{10}) after the `li` instruction. Press `q` to quit when done.

LLVM Clang is the major open-source alternative to GCC. Both are mature mainstream compiler tool chains that produce excellent code. LLVM is more modular. GCC was the first to support RISC-V, and it continues to be a default choice for RISC-V development.

```
$ spike --isa=rv32i -d example
warning: tohost and fromhost symbols not in ELF; can't communicate with target
:
core   0: 0x00001000 (0x00000297) auipc   t0, 0x0
:
core   0: 0x00001004 (0x02028593) addi    a1, t0, 32
:
core   0: 0x00001008 (0xf1402573) csrr    a0, mhartid
:
core   0: 0x0000100c (0x0182a283) lw      t0, 24(t0)
:
core   0: 0x00001010 (0x00028067) jr      t0
:
core   0: 0x80000000 (0x02a00513) li      a0, 42
: reg 0 a0
0x0000002a
:
core   0: 0x80000004 (0x0000006f) j       pc + 0x0
:
: q
```

3.3.2 Linking and ELF

GCC compiles source code files into *object* files and then automatically *links* one or more object files, libraries, and startup code together to form an *executable* file. The object files contain machine language code and also *unresolved external references* to other object files or libraries. Libraries contain standard functions such as `printf` or `abs`. To save space, the linker only links in the required parts of the library. Startup code may initialize the registers, including the stack and global pointers `sp` and `gp`, as well as set up an exception handler. The linker *relocates* the object files into a file with instructions at consecutive addresses and resolves the references between files to create the executable file.

`riscv64-unknown-elf-gcc` and most Linux-based systems represent both object and executable files as ELF files, which contain a 52- or 64-byte header followed by program and section header tables and the code and data that these headers point to. The file header begins with the magic number 0x7F 45 4C 46, indicating that it is an ELF file, followed by bytes specifying {32/64}-bit format, little or big endianness, the code ISA (instruction set architecture), the entry point address where the program should begin executing, and the size and location of the program and section header tables. Sections are marked by assembler directives, which begin with a

"GNU, a recursive acronym meaning GNU's not Unix, is the Unix-like OS Richard Stallman set out in 1983 to develop. The aim was to make GNU entirely free (freedom-respecting) software, so it would enable people to shun non-free software and recover lost freedom. GNU became self-hosting after its combination in 1992 with the kernel Linux. See gnu.org/gnu/the-gnu-project.html."
—Richard Stallman, 2022

[1]Spike supports simulating multihart processors, but this simulation only has hart 0.

period, and they include information needed to link object files together into executables, including code (`.text`), initialized variables (`.data`), the symbol table (`.symtab`) that lists the program's labels, and other named sections such as an attributes section (`.riscv.attributes`). Section 3.3.7 describes assembler directives further.

The `readelf` command prints the *metadata* from the ELF file. The first 16 *magic bytes* identify the type of file. The 45 4c 46 bytes are ASCII for "ELF". Subsequent metadata bytes describe the format and structure of the ELF file. The program uses 32-bit code (indicated by `ELF32`) in little-endian format, starting at 0x80000000 (indicated by `Entry point address`). The section headers list various sections and their locations in the ELF file. The `.text` section holds the program code. The `.symtab` section (the symbol table) lists the addresses of symbols (labels) in the program. The `.riscv.attributes` section (shown at the bottom `Attribute Section`) indicates that the code requires at least RV32I version 2.0 (indicated by `Tag_RISCV_arch: "rv32i2p1"`).

```
$ riscv64-unknown-elf-readelf -a example

ELF Header:
  Magic:   7f 45 4c 46 01 01 01 00 00 00 00 00 00 00 00 00
  Class:                             ELF32
  Data:                              2's complement, little endian
  Version:                           1 (current)
  OS/ABI:                            UNIX - System V
  ABI Version:                       0
  Type:                              EXEC (Executable file)
  Machine:                           RISC-V
  Version:                           0x1
  Entry point address:               0x80000000
  Start of program headers:          52 (bytes into file)
  Start of section headers:          4344 (bytes into file)
  Flags:                             0x0
  Size of this header:               52 (bytes)
  Size of program headers:           32 (bytes)
  Number of program headers:         1
  Size of section headers:           40 (bytes)
  Number of section headers:         6
  Section header string table index: 5

Section Headers:
  [Nr] Name              Type            Addr     Off    Size   ES Flg Lk Inf Al
  [ 0]                   NULL            00000000 000000 000000 00      0   0  0
  [ 1] .text             PROGBITS        80000000 001000 000008 00  AX  0   0  4
  [ 2] .riscv.attributes RISCV_ATTRIBUTE 00000000 001008 00001a 00      0   0  1
  [ 3] .symtab           SYMTAB          00000000 001024 000070 10      4   5  4
  [ 4] .strtab           STRTAB          00000000 001094 00002e 00      0   0  1
  [ 5] .shstrtab         STRTAB          00000000 0010c2 000033 00      0   0  1
Key to Flags:
  W (write), A (alloc), X (execute), M (merge), S (strings), I (info),
  L (link order), O (extra OS processing required), G (group), T (TLS),
  C (compressed), x (unknown), o (OS specific), E (exclude),
  D (mbind), p (processor specific)

There are no section groups in this file.
```

```
Program Headers:
  Type           Offset   VirtAddr   PhysAddr   FileSiz MemSiz  Flg Align
  LOAD           0x001000 0x80000000 0x00008 R E 0x1000

 Section to Segment mapping:
  Segment Sections...
   00     .text

There is no dynamic section in this file.

There are no relocations in this file.

The decoding of unwind sections for machine type RISC-V is not currently supported.

Symbol table '.symtab' contains 7 entries:
   Num:    Value  Size Type    Bind   Vis      Ndx Name
     0: 00000000     0 NOTYPE  LOCAL  DEFAULT  UND
     1: 80000000     0 SECTION LOCAL  DEFAULT    1 .text
     2: 00000000     0 SECTION LOCAL  DEFAULT    2 .riscv.attributes
     3: 00000000     0 FILE    LOCAL  DEFAULT  ABS cc21kp4v.o
     4: 80000004     0 NOTYPE  LOCAL  DEFAULT    1 self_loop
     5: 80000000     0 NOTYPE  GLOBAL DEFAULT    1 rvtest_entry_point
     6: 80001000     0 NOTYPE  GLOBAL DEFAULT    1 _end

No version information found in this file.
Attribute Section: riscv
File Attributes
  Tag_RISCV_arch: "rv32i2p1"
```

3.3.3 Application Binary Interface

As introduced in Section 2.3.6, an *application binary interface* (ABI) is the standard that defines how programs communicate during function calls and system calls, including:

- The binary format of executables and object and library files
- Which instruction set (ISA) is supported
- The size, layout, and alignment of data types
- The calling convention for how arguments are passed to functions in registers or on the stack
- OS calling conventions
- Code models of how instructions access a label in the program

An ABI is similar to an API (*application programming interface*). An API defines functions and variables exposed by library code to other programs. For example, the C library API provides functions to open and read files, print to the screen, and perform trigonometric math. An ABI defines conventions used by the compiler when building a program, such as which registers are used to pass inputs to a function.

The RISC-V ABI [RISCV-ABI] supports ELF for executable and object files. It defines a variety of 32- and 64-bit RISC-V ABI calling conventions summarized in Table 3.1. The name of these data type ABIs use the following convention: i = integer, l = long, and p = pointer. So, ilp32 means that integers, pointers, and longs are each 32 bits; lp64 means that longs and pointers are 64 bits (but ints are still 32 bits). Data types are aligned to address boundaries that are a multiple of their size; for example, address 0x3002 is legal for `char` and `short` but not for `int` or `long`. GCC needs to know the ABI to generate proper code, such as using `lw` vs `ld` to load a `long`. GCC takes a command argument such as `-mabi ilp32` to specify the ABI.

The RISC-V base integer calling convention (ilp32 or lp64) defines how functions pass arguments and use registers. In these ABIs, a function passes up to eight arguments in registers `a0`–`a7`. These registers need not be preserved. Functions can return results in `a0` (and `a1` if the result is 2XLEN wide). If a function needs more arguments than can be passed in `a0`–`a7`, the remainder are passed on the stack. The *saved registers*, `s0`–`s11`, are preserved across function calls and are suitable for local variables; the callee must save and restore any of these registers that it wants to use. `s0` may also be used as a frame pointer to access variables on the stack frame. Registers `t0`–`t6` are not preserved across a function call and are suitable for temporary (intermediate) results. Register `sp` (`x2`) is the stack pointer, and `ra` (`x1`) holds the return address during a function call. Register `gp` (`x3`) is the *global pointer*, used to access global variables. Table I.4 (on the last pages of the book) summarizes these register uses. The floating-point ABIs (ilp32f, ilp32d, and lp64d) pass `float` or `double` arguments in the floating-point registers `fa0`–`fa7`.

When running on a platform with an OS and supervisor mode, user-mode programs call the OS by placing the syscall number in `a7` and any other arguments in `a0`–`a5` and then use `ecall` (environment call instruction) to cause a

Table 3.1 **RISC-V ABI data type sizes for various calling conventions (in bits).**

<table>
<tr><th>Data Type</th><th>ilp32</th><th>ilp32f</th><th>ilp32d</th><th>lp64</th><th>lp64d</th></tr>
<tr><td>int</td><td colspan="5">32</td></tr>
<tr><td>long</td><td colspan="3">32</td><td colspan="2">64</td></tr>
<tr><td>pointers</td><td colspan="3">32</td><td colspan="2">64</td></tr>
<tr><td>char</td><td colspan="5">8</td></tr>
<tr><td>short</td><td colspan="5">16</td></tr>
<tr><td>float</td><td>n/a</td><td colspan="2">32</td><td>n/a</td><td>32</td></tr>
<tr><td>double</td><td colspan="2">n/a</td><td>64</td><td>n/a</td><td>64</td></tr>
</table>

Richard Stallman (1953–) is a pioneer of the free software movement and the founding developer of the GNU OS and Free Software Foundation (FSF). Stallman attended Harvard University and became a programmer at the MIT Artificial Intelligence Laboratory. After denouncing proprietary software as vicious and unjust, he defined the *four essential freedoms of free software* [Stallman86].

0. The freedom to run the program as you wish, for any purpose.
1. The freedom to study how the program works and change it so it does your computing as you wish.
2. The freedom to redistribute copies so you can help your neighbor.
3. The freedom to distribute copies of your modified versions to others. By doing this you can give the whole community a chance to benefit from your changes.

Stallman is the founding developer of GCC, the GDB debugger, `make`, and the emacs text editor.

(Photograph courtesy Richard Stallman. Reprinted with permission.)

The global pointer `gp` points to the static data segment where global variables are held.

trap and elevate privilege to supervisor mode. The trap handler in the OS (typically, Linux) kernel receives this `ecall` and looks at `a7` and the arguments to perform the appropriate system call, then places any return result in `a0`, and performs `sret` to return to the user-level program. For example, the `open` system call takes two arguments (in `a0` and `a1`) that specify a pointer to the file path name and flags conveying information such as whether to create a new file if it doesn't already exist. The call returns a *file descriptor*, which is an index into the OS's table of open files; the program can then pass this descriptor to other system calls such as `read`, `write`, or `close`.

```
int open(const char *pathname, int flags);
```

The RISC-V ABI prefers to access global variables using an immediate offset from the global pointer because this is tight and fast code. The two's complement immediate is only 12 bits, so if there are too many global variables, they cannot all be accessed with this short offset (−2048 to +2047 bytes). Instead, the code must explicitly calculate the address. RISC-V presently defines two code models: *small* (medlow) and *medium* (medany). The small model uses 32-bit *absolute addresses* and thus is suitable for RV32. It is not position independent, so the linker must change addresses when relocating code. The medium model, suitable for RV64, generates *PC-relative addresses*. It uses the `auipc` instruction to add a signed 32-bit offset to the program counter, and thus can access variables and code within ±2 GiB of the current instruction. It is position-independent, so the code does not have to be changed if it is relocated.

Code Example 3.2 and Fig. 3.2 illustrate the code generated for a simple C program under these two code models. The compiler placed `main` at address 0x10104, the global variable `a` at 0x129c0, and the global variable

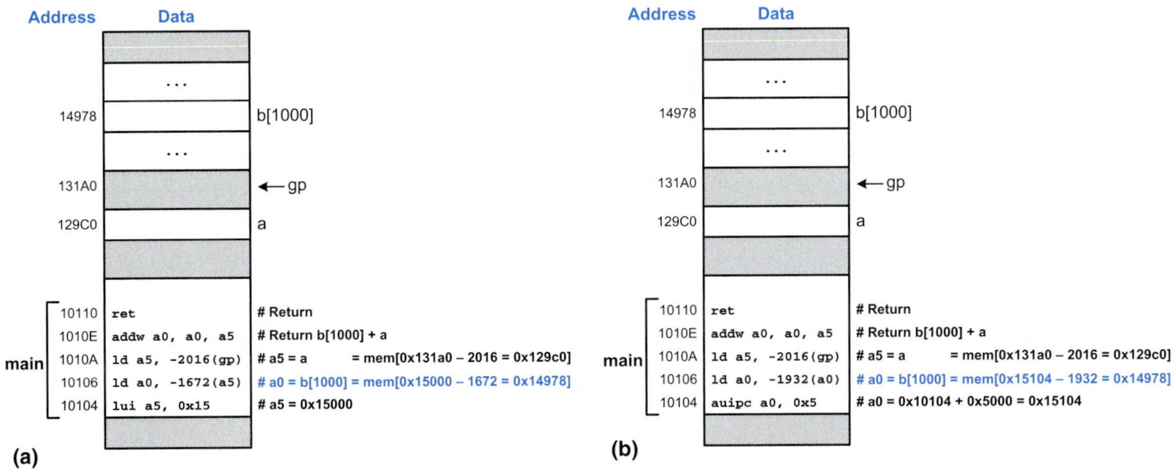

Fig. 3.2 Accessing global variables in (a) medlow and (b) medany code models.

`b[1000]` at 0x14978. The global pointer is initialized to 0x131a0. `a` is at 2016 bytes before `gp` (0x131a0 − 0x129c0 = 0x7E0 = 2016), so `a` is accessed in both code models with an immediate offset from the global pointer: `ld a5, -2016(gp)`. The `b` array has 2000 8-byte entries and thus is too large to access with an immediate offset.

In medlow, the absolute address of `b[1000]` is obtained by `lui` placing 0x15000 in `a5`, followed by `ld` with an offset of −1672 = −0x688, giving the absolute address 0x15000 − 0x688 = 0x14978. In medany, the relative address of `b[1000]` is 0x14978 − 0x10104 = 0x4874 bytes beyond the program counter (PC). This relative address is computed by adding 0x5000 to the PC and then subtracting 0x78c (0x5000 − 0x78c = 0x4874). First, `auipc` adds 0x5000 to the PC (`a0` = 0x10104 + 0x5000 = 0x15104). Then, `ld` accesses `b[1000]` with an offset of −1932 = −0x78C, again giving the address of `b[1000]` (0x15104 − 0x78c = 0x14978).

Code Example 3.2 CODE GENERATION WITH MEMORY MODELS

```
// mcmodel.c
long a;
long b[2000];

int main(void)
{
    return a + b[1000];
}
```

```
// Makefile
all:
    riscv64-unknown-elf-gcc -O3 -march=rv64gc -mcmodel=medlow mcmodel.c -o mcmodel_medlow
    riscv64-unknown-elf-objdump -d mcmodel_medlow > mcmodel_medlow.objdump
    riscv64-unknown-elf-gcc -O3 -march=rv64gc -mcmodel=medany mcmodel.c -o mcmodel_medany
    riscv64-unknown-elf-objdump -d mcmodel_medany > mcmodel_medany.objdump
```

```
// mcmodel_medlow.objdump
0000000000010104 <main>:
   10104: 67d5              lui    a5,0x15
   10106: 9787b503          ld     a0,-1672(a5) # 14978 <__global_pointer$+0x17d8>
   1010a: 8201b783          ld     a5,-2016(gp) # 129c0 <a>
   1010e: 9d3d              addw   a0,a0,a5
   10110: 8082              ret

// mcmodel_medany.objdump
0000000000010104 <main>:
   10104: 00005517          auipc  a0,0x5
   10108: 87453503          ld     a0,-1932(a0) # 14978 <__global_pointer$+0x17d8>
   1010c: 8201b783          ld     a5,-2016(gp) # 129c0 <a>
   10110: 9d3d              addw   a0,a0,a5
   10112: 8082              ret
```

Although programs normally follow the ABI, the ISA does not require this. For example, a hardware implementation cannot assume that the return address for a function call is always in `x1`.

3.3.4 Supervisor Binary Interface

The OS kernel can invoke platform runtime firmware via the supervisor binary interface (SBI). This firmware abstracts platform-dependent operations so that the OS does not need to change from one platform to another. Like ABI system calls, each SBI call is identified by a call number. Examples of SBI calls include the following:

- Set timer
- Send inter-processor interrupt (IPI, also called software interrupt) to another hart
- Remote fences to synchronize with other harts
- Start or stop harts
- Reboot system
- Read performance counters
- Check versions of the hardware and SBI implementation

To make an SBI call, the OS places the SBI call number in `a7` and `a6`, arguments in other `a` registers, and uses the `ecall` instruction, which causes an exception and elevates the privilege to machine mode. The trap handler in the SBI platform runtime firmware receives this `ecall`; does the necessary machine-mode work; places the error code, if any, in `a0` and the return result in `a1`; and performs `mret` to return to the OS. For example, the `sbi_set_timer` function sets the current value of the timer. It is called with an EID (extension ID) of 0x54494D45 in `a7`, an FID (function ID) of 0 in `a6` and a new 64-bit time value in `a0` (and `a1` on RV32). The timer is part of a memory-mapped peripheral such as the Core-Local Interruptor (CLINT) which can only be accessed in machine mode and whose address may be system dependent. For example, the SBI call to set the timer is:

```
struct sbiret sbi_set_timer(uint64_t stime_value);
```

See the SBI documentation [RISCV-SBI] for a list of SBI calls and their encodings. SBI calls are primarily made by the OS, so programmers rarely use them.

3.3.5 Machine Architecture

RISC-V has many ISA flavors, so GCC needs to be told which one to target. RISC-V defines the ISA with a base instruction set and letters or

.bss (also simply, bss) stands for *block starting symbol.* Historically, it meant a block of uninitialized data that was indicated (started) by a symbol (label). Most systems these days initialize the .bss section to 0 to avoid nondeterministic behavior caused by random values in uninitialized variables. Nevertheless, a prudent programmer always initializes variables before using them.

Assembler directives indicate section boundaries and specify variables and constants. Table 3.4 lists some of the common directives [RISCV-ASM]. Section assembler directives mark the beginning of that section. Text sections contain program code. Data sections contain variables. By convention, initialized global variables such as `int salary = 110000;` go in an ordinary data section while uninitialized global variables such as `double matrix[3][3];` go in a bss data section. Some directives have

Table 3.4 **RISC-V assembler directives.**

Assembler Directive	Description
	Sections
`.text`	Text (program code)
`.data`	Data: initialized variables
`.rodata`	Read-only data: initialized constants
`.bss`	Data: uninitialized variables
`.symtab`	Symbol table, including addresses of labels
`.section` *name*	User-defined section
`.end`	Optional end of file
	Data
`.byte` *x1, x2, ...*	8-bit comma-separated values
`.half/.short/.2byte` *x1, x2, ...*	16-bit comma-separated values
`.word/.long/.4byte` *x1, x2, ...*	32-bit comma-separated values
`.dword/.8byte` *x1, x2, ...*	64-bit comma-separated values
`.string` *"str"*	Null terminated string
`.space` *size*	Leave room for *size* bytes
`.fill` *num, size, value*	Fill memory with *num* entries of *size* bytes with *value*
	Other
`.align` *size*	Align next code or data address to a power of 2 *size*
`.global/.globl`	Define a global symbol visible to the linker
`.equ` *const, value*	Define a constant *const* to be *value*
`.macro ... .endm`	Define a macro

multiple names, shown separated by / in the table; for example, `.word` or `.4byte` both generate 32-bit numbers.

3.3.8 Spike Simulation

Spike was the original *golden reference model* of the RISC-V instruction set architecture: it served both as a simulator and as a reference against which other RISC-V hardware or software implementations can be compared. Sail has replaced Spike for this purpose, but Spike is more convenient to use. Both are instruction set simulators, so they only simulate the function of the instructions (not the actual hardware) and, thus, are not cycle accurate. They ignore delays such as cache misses, memory transactions, and so on. Spike shockingly lacks a user manual, but brief instructions can be found at [Spike].

Spike is named for the golden spike that completed the transcontinental railroad. The term *golden* is contentious because it implies perfection, and even the best reference model never perfectly matches a nontrivial written specification.

To simulate in interactive debug mode, run Spike with `-d`.

```
$ spike --isa=rv32gc -d example
```

All values in Spike are in hexadecimal. To see the contents of integer register `t0` on hart 0, type:

```
:reg 0 t0
```

This also works for control and status registers (CSRs):

```
:reg 0 mstatus
```

To see the contents of a single- or double-precision floating-point register, replace `reg` with `fregs` or `fregd`. To read the contents of physical memory at address 0x80000004, type:

```
:mem 80000004
```

Table 3.5 summarizes common Spike commands, and the listing below it illustrates more Spike capabilities (but some would hang on this example because it doesn't write to memory).

```
: until pc 0 80000004
: until reg 0 a0 2a
: until mem 0 80000020 1
: while mem 0 80000020 deadbeefdeadbeef
: run
: <Ctrl-c>
: q
```

Spike supports other commands such as system calls through the HTIF, but they are poorly documented. Spike uses another program called proxy kernel (`pk`) to handle system calls using the host OS. Spike also interfaces with `gdb` through `OpenOCD` for debugging, but this has been unreliable in our environment. If your program is complicated enough to need these features, you may be better off running with QEMU, described in Section 3.6.

Table 3.5 **Spike commands.**

Command	Description
Enter	Press enter key to step through instructions one at a time
`reg 0 t0`	Print the contents of register `t0` on hart 0
`reg 0 mstatus`	Print the contents of the mstatus CSR on hart 0
`mem 80000004`	Print the contents of memory at physical address 0x80000004
`mem 0 80000008`	Print the contents of memory at Core 0's virtual address 0x80000008
`until <condition>`	Run until a condition is satisfied
`while <condition>`	Run while a condition is satisfied
`run`	Run indefinitely, printing PC and instruction as it goes
`rs`	Run silently, without printing
Ctrl-c	Pause the simulation and reenters interactive debug mode
`q`	End the simulation
`help`	List available commands

The address of `tohost` doesn't matter; Spike detects a write to an address with a label of this name in the symbol table.

A RISC-V program used for simulation or testing typically ends with an infinite loop, so it doesn't wander into undefined instructions. Therefore, a Spike simulation will also end by looping indefinitely. Spike provides a basic *host target interface* (HTIF) which, among other features, allows a program to signal completion. To do so, a program must declare an 8-byte `.dword` at a label called `tohost`. In Spike, a write of 1 to this address causes the program to terminate successfully. A write of 3 causes it to print `*** FAILED ***` and terminate. A write of any even number to the HTIF triggers a system call, with the number interpreted as a pointer to an array in memory containing the system call number and arguments. For example, system call number 64 is sys_write, which takes three arguments: a file descriptor, a pointer to a string, and the length of the string (see github.com/riscv-software-src/riscv-isa-sim/blob/master/fesvr/syscall.cc). File descriptors of 0, 1, and 2 correspond to stdin, stdout, and stderr, respectively.

Spike also has a *signature* mechanism to write the contents of a block of memory to a file upon completion. Typically, the signature block should have a series of words that are only correct if the program ran properly. The signature block in a `.data` section is delimited by `begin_signature:` and `end_signature:` labels. Run Spike with the command below to write the values in the signature block to a file named `example.signature.output`, with 4 bytes per line in hexadecimal. The user can compare this file with an expected output to determine whether the program ran correctly.

In this case, the signature file would be omitted because `example` doesn't have a signature block. Code Example 3.5 in Section 3.3.10 is a more interesting program that writes a signature.

```
$ spike +signature=example.signature.output +signature-granularity=4 example
```

3.3.9 Performance Measurement

RISC-V has several CSRs used for performance measurement. `instret` indicates how many instructions have been retired (i.e., completed). `cycle` indicates how many clock cycles have passed. The `csrr reg, CSR` pseudoinstruction reads a `CSR` into `reg`, an integer register. The example in the next section illustrates using `instret` to count the number of instructions executed by a function call.

In Spike, `cycle` and `instret` both advance by one each cycle and have the same value because one instruction is executed in each cycle. In a real processor with a CPI other than 1.0, the two will differ.

3.3.10 Putting It All Together

The program `$WALLY/examples/asm/sumtest` illustrates an RV64 program linking multiple files with multiple sections, using assembler directives, writing to the host interface, and generating a signature with Spike. The example involves the two source files given in Code Examples 3.5 and 3.6.

RISC-V has a configurable and implementation-specific set of performance counters to measure microarchitectural events, such as cache misses or branch mispredictions. These are discussed further in Section 8.4.3.1.

`sumtest.S` initializes the stack pointer `sp`. Note that `la` is a pseudoinstruction that loads an address into a register with an `auipc` and `addi` sequence to compute the address relative to the PC. Also note the double-word (8-byte) `ld` and `sd` loads and stores in the RV64 program. The program reads the value of `N` (4) from the data segment and then calls the function `sum(N)` defined in the `sum.S` file, which adds up the numbers from 1 to `N`. `sumtest` also counts the number of instructions retired before and after the call and takes their difference to measure the call performance. It writes the return result and instruction count to the signature memory and then writes 1 to `tohost` to indicate completion before entering an infinite loop at `self_loop`.

Code Example 3.5 `sumtest.S`

```
// sumtest.S
// David_Harris@hmc.edu 24 December 2021

.global rvtest_entry_point
rvtest_entry_point:
    la sp, topofstack       # Initialize stack pointer
    la t0, N                # Get address of N in data
    ld a0, 0(t0)            # Load N
    csrr s8, instret        # Count # instructions completed before call made
    jal sum                 # Call sum(N)
    csrr s9, instret        # Count # instructions completed after call made
    sub s9, s9, s8          # Calculate # instructions completed during call
    la t0, begin_signature  # Address of signature
    sd a0, 0(t0)            # Store sum(N) in signature
    sd s9, 8(t0)            # Record performance (# instructions executed during sum)

write_tohost:
    la t1, tohost           # Load address of tohost label into t1
    li t0, 1                # 1 for success, 3 for failure
    sd t0, 0(t1)            # Send success code

self_loop:
    j self_loop             # Wait

.section .tohost
tohost:                     # Write to HTIF
    .dword 0
fromhost:
    .dword 0

.data
N:
    .dword 4                # N = 4

.EQU XLEN,64
begin_signature:
    .fill 2*(XLEN/32),4,0xdeadbeef    # Initialize signature memory with 0xdeadbeef
end_signature:

# Initialize stack with room for 512 bytes
.bss
    .space 512
topofstack:
```

Each 32-bit value in the signature is initialized to the hexadecimal number 0xdeadbeef, which should be different than any sensible signature later written by the program.

`sum(N)` adds the numbers from 1 to `N` and returns the result. The `sum` label is declared global so that the other file `sumtest.S` can reference it. The `sum` function uses `s0` and `s1`, so it begins by saving them on the stack and then restores them at the end. According to the RISC-V ABI, `sum` looks for `N` in `a0` during the function and returns the result in `a0` at the end of the function.

Code Example 3.6 **sum.S**

```
// sum.S
// David_Harris@hmc.edu 24 December 2021
// Add numbers from 1 to N.

// Result in s0, i in s1, N in a0, return answer in a0
//
// long sum(long N) {
//   long result, i;
//   result = 0;
//   for (i=1; i<=N; i++) result = result + i;
//   return result;
// }
.global sum
sum:
     addi sp, sp, -16     # Make room to save s0 and s1 on the stack
     sd s0, 0(sp)
     sd s1, 8(sp)

     li s0, 0             # result = 0
     li s1, 1             # i = 1
for: bgt s1, a0, done     # Exit loop if i > N
     add s0, s0, s1       # result = result + i
     addi s1, s1, 1       # i++
     j for                # Repeat

done:
     mv a0, s0            # Put result in a0 to return
     ld s0, 0(sp)         # Restore s0 and s1 from stack
     ld s1, 8(sp)
     addi sp, sp, 16
     ret                  # Return from function
```

Code Example 3.7 lists the expected results written to the signature memory. Counting instructions, we see the call to `sum` takes 13+4`N` instructions, including the `jal` and one of the `csrr` instructions. Remember that `bgt` executes one last time when the for loop exits. So, for `N` = 4, the expected return value is `sum(4)` = 1+2+3+4 = 10 = 0xA, and the number of instructions executed is 29 = 0x1D.

Code Example 3.7 **sumtest.reference_output**

```
000000000000000A
000000000000001D
```

The Makefile in Code Example 3.8 compiles `sumtest` for rv64gc. `make sim` runs Spike to generate `sumtest.signature.output`. It then compares against `sumtest.reference_output`. If the files are not identical, it quits with an exit code; otherwise, it prints `Signature matches! Success!` Build and test using the following commands:

> Notice how the Makefile in Code Example 3.8 declares a variable `TARGET` with the name of the program so we could rename the program without retyping `sumtest` many times. In contract, Code Example 3.4 had the file name `example` in many places, each of which would need to be changed.

```
Disassembly of section .data:

0000000080002000 <N>:
    80002000:       0004                    0x4
    80002002:       0000                    unimp
    80002004:       0000                    unimp
        ...

0000000080002008 <begin_signature>:
    80002008:       deadbeef                jal     t4,7ffdd5f2 <XLEN+0x7ffdd5b2>
    8000200c:       deadbeef                jal     t4,7ffdd5f6 <XLEN+0x7ffdd5b6>
    80002010:       deadbeef                jal     t4,7ffdd5fa <XLEN+0x7ffdd5ba>
    80002014:       deadbeef                jal     t4,7ffdd5fe <XLEN+0x7ffdd5be>

Disassembly of section .bss:

0000000080003000 <_end-0x200>:
        ...
```

Observe how `objdump` disassembles the data section into gibberish. Using `-d` instead of `-D` only shows disassembled sections expected to contain instructions and does not show the values in the data section.

Code Example 3.10 lists the ELF metadata of the `sumtest` file. The file has `.text`, `.tohost`, `.data`, and `.bss` sections specified in the assembly language files, as well as RISC-V attributes, debug symbols, and the symbol table generated by the linker. The symbol table contains the addresses of all the labels from the text and data segments. The compiler targeted version 2.0 of the RV64GC instruction set and 1.11 of the privileged specification, as indicated by the attributes section at the end of the file.

Code Example 3.10 **`sumtest` ELF METADATA**

```
$ riscv64-unknown-elf-readelf -a sumtest

ELF Header:
  Magic:   7f 45 4c 46 02 01 01 00 00 00 00 00 00 00 00 00
  Class:                             ELF64
  Data:                              2's complement, little endian
  Version:                           1 (current)
  OS/ABI:                            UNIX - System V
  ABI Version:                       0
  Type:                              EXEC (Executable file)
  Machine:                           RISC-V
  Version:                           0x1
  Entry point address:               0x80000000
  Start of program headers:          64 (bytes into file)
  Start of section headers:          9904 (bytes into file)
  Flags:                             0x5, RVC, double-float ABI
  Size of this header:               64 (bytes)
  Size of program headers:           56 (bytes)
  Number of program headers:         2
  Size of section headers:           64 (bytes)
  Number of section headers:         15
  Section header string table index: 14
```

```
Section Headers:
  [Nr] Name              Type             Address           Offset
       Size              EntSize          Flags  Link  Info  Align
  [ 0]                   NULL             0000000000000000  00000000
       0000000000000000  0000000000000000           0     0     0
  [ 1] .text             PROGBITS         0000000080000000  00001000
       0000000000000060  0000000000000000  AX       0     0     2
  [ 2] .tohost           PROGBITS         0000000080001000  00002018
       0000000000000010  0000000000000000           0     0     1
  [ 3] .data             PROGBITS         0000000080002000  00002000
       0000000000000018  0000000000000000  WA       0     0     1
  [ 4] .bss              NOBITS           0000000080003000  00002018
       0000000000000200  0000000000000000  WA       0     0     1
  [ 5] .riscv.attributes RISCV_ATTRIBUTE  0000000000000000  00002028
       0000000000000037  0000000000000000           0     0     1
  [ 6] .debug_line       PROGBITS         0000000000000000  0000205f
       0000000000000128  0000000000000000           0     0     1
  [ 7] .debug_line_str   PROGBITS         0000000000000000  00002187
       000000000000003e  0000000000000001  MS       0     0     1
  [ 8] .debug_info       PROGBITS         0000000000000000  000021c5
       0000000000000050  0000000000000000           0     0     1
  [ 9] .debug_abbrev     PROGBITS         0000000000000000  00002215
       0000000000000028  0000000000000000           0     0     1
  [10] .debug_aranges    PROGBITS         0000000000000000  00002240
       0000000000000060  0000000000000000           0     0     16
  [11] .debug_str        PROGBITS         0000000000000000  000022a0
       000000000000004a  0000000000000001  MS       0     0     1
  [12] .symtab           SYMTAB           0000000000000000  000022f0
       00000000000002a0  0000000000000018          13    25     8
  [13] .strtab           STRTAB           0000000000000000  00002590
       000000000000008a  0000000000000000           0     0     1
  [14] .shstrtab         STRTAB           0000000000000000  0000261a
       0000000000000096  0000000000000000           0     0     1
Key to Flags:
  W (write), A (alloc), X (execute), M (merge), S (strings), I (info),
  L (link order), O (extra OS processing required), G (group), T (TLS),
  C (compressed), x (unknown), o (OS specific), E (exclude),
  D (mbind), p (processor specific)

There are no section groups in this file.

Program Headers:
  Type           Offset             VirtAddr           PhysAddr
                 FileSiz            MemSiz              Flags  Align
  LOAD           0x0000000000001000 0x0000000080000000
                 0x0000000000000060  R E    0x1000
  LOAD           0x0000000000002000 0x0000000080002000
                 0x0000000000000018 0x0000000000001200  RW     0x1000
```

```
 Section to Segment mapping:
  Segment Sections...
   00     .text
   01     .data .bss

There is no dynamic section in this file.

There are no relocations in this file.

The decoding of unwind sections for machine type RISC-V is not currently supported.

Symbol table '.symtab' contains 28 entries:
   Num:    Value          Size Type    Bind   Vis      Ndx Name
     0: 0000000000000000     0 NOTYPE  LOCAL  DEFAULT  UND
     1: 0000000080000000     0 SECTION LOCAL  DEFAULT    1 .text
     2: 0000000080001000     0 SECTION LOCAL  DEFAULT    2 .tohost
     3: 0000000080002000     0 SECTION LOCAL  DEFAULT    3 .data
     4: 0000000080003000     0 SECTION LOCAL  DEFAULT    4 .bss
     5: 0000000000000000     0 SECTION LOCAL  DEFAULT    5 .riscv.attributes
     6: 0000000000000000     0 SECTION LOCAL  DEFAULT    6 .debug_line
     7: 0000000000000000     0 SECTION LOCAL  DEFAULT    7 .debug_line_str
     8: 0000000000000000     0 SECTION LOCAL  DEFAULT    8 .debug_info
     9: 0000000000000000     0 SECTION LOCAL  DEFAULT    9 .debug_abbrev
    10: 0000000000000000     0 SECTION LOCAL  DEFAULT   10 .debug_aranges
    11: 0000000000000000     0 SECTION LOCAL  DEFAULT   11 .debug_str
    12: 0000000000000000     0 FILE    LOCAL  DEFAULT  ABS ccOflQM8.o
    13: 0000000080003200     0 NOTYPE  LOCAL  DEFAULT    4 topofstack
    14: 0000000080002000     0 NOTYPE  LOCAL  DEFAULT    3 N
    15: 0000000080002008     0 NOTYPE  LOCAL  DEFAULT    3 begin_signature
    16: 0000000080000034     0 NOTYPE  LOCAL  DEFAULT    1 write_tohost
    17: 0000000080001000     0 NOTYPE  LOCAL  DEFAULT    2 tohost
    18: 0000000080000042     0 NOTYPE  LOCAL  DEFAULT    1 self_loop
    19: 0000000080001008     0 NOTYPE  LOCAL  DEFAULT    2 fromhost
    20: 0000000000000040     0 NOTYPE  LOCAL  DEFAULT  ABS XLEN
    21: 0000000080002018     0 NOTYPE  LOCAL  DEFAULT    3 end_signature
    22: 0000000000000000     0 FILE    LOCAL  DEFAULT  ABS ccZtnxq7.o
    23: 000000008000004e     0 NOTYPE  LOCAL  DEFAULT    1 for
    24: 0000000080000058     0 NOTYPE  LOCAL  DEFAULT    1 done
    25: 0000000080000044     0 NOTYPE  GLOBAL DEFAULT    1 sum
    26: 0000000080000000     0 NOTYPE  GLOBAL DEFAULT    1 rvtest_entry_point
    27: 0000000080003200     0 NOTYPE  GLOBAL DEFAULT    4 _end

No version information found in this file.
Attribute Section: riscv
File Attributes
  Tag_RISCV_arch: "rv64i2p1_m2p0_a2p1_f2p2_d2p2_c2p0"
  Tag_RISCV_priv_spec: 1
  Tag_RISCV_priv_spec_minor: 11
```

3.4 GCC GNU C COMPILER

The GCC tools are known for their excellent C compiler. This section shows how to *cross-compile* C programs for RISC-V. Cross-compile means to generate RISC-V assembly on a processor with a different architecture, such as x86. We disassemble the program and examine its memory map and the effects of optimization flags. Then we look at how to mix C and assembly by using inline assembly statements and by linking functions written in the different languages.

3.4.1 A First C Program

Code Example 3.11 shows a simple program, `sum.c`, that illustrates a C program to add the numbers from 1 to `N`. It also illustrates printing to the console, via the HTIF, and verifying results. The program is built using GCC and run with Spike. The program adds the numbers from 1 to 4 and successfully prints the sum of 10. It prints the number of cycles and instructions retired during the `sum` function call by sandwiching that call between the `setStats(1)` and `setStats(0)` calls. These calls have some latency, so `mcycle` and `minstret` are not sampled at exactly the same time and are conservative. Run the `sum.c` program on Spike as follows:

```
$ cd $WALLY/examples/C/sum
$ make
riscv64-unknown-elf-gcc -o sum -g -O \
  -march=rv64gc -mabi=lp64d -mcmodel=medany \
  -nostartfiles -T../common/test.ld -I../common \
  sum.c ../common/crt.S ../common/syscalls.c
riscv64-unknown-elf-objdump -S sum > sum.objdump
$ spike sum
s = 10
mcycle = 31
minstret = 38
```

Code Example 3.11 **sum.c**

```
// sum.c
// David_Harris@hmc.edu 24 December 2021
// Simple illustration of compiling C code

#include <stdio.h>  // Supports printf
#include "util.h"   // Supports verify

long sum(long N) {  // Adds up numbers from 1 to N
  long result, i;
  result = 0;
  for (i=1; i<=N; i++) {
    result = result + i;
  }
  return result;
}

int main(void) {
  int s[1], expected[1];
  setStats(1);             // Record initial mcycle and minstret
  s[0] = sum(4);           // Call assembly language sum function to compute 1 + 2 + 3 + 4 = 10
  setStats(0);             // Record elapsed mcycle and minstret
  printf("s = %d\n", s[0]);// Print sum of numbers, which should be 10
  expected[0] = 10;        // Set expected value
  return verify(1, s, expected); // Check the 1 element of s matches expected. 0 means success
}
```

GCC normally produces only the executable output file. Adding the `--save-temps` flag to the compile statement saves the preprocessor, compiler, and assembler intermediate outputs as `.i`, `.s`, and `.o` files, respectively, which may give you more insight into the whole process.

The optimization flag is the capital letter "O": `-O` (*not* the number zero: –0).

Code Example 3.12 shows the Makefile with the compiler flags. The backslash `\` is a continuation character allowing a command to exceed one line. `-g` generates debug symbols, and `-O` performs basic optimization. The compiler needs the `-march`, `-mabi`, and `-mcmodel` flags to generate proper code for rv64gc with the `lp64d` ABI. The next line of flags allows the program to run on bare metal without an OS. Specifically, `-nostdlib` disables linking both the C standard library `libc.a` and the startup code; it is a superset of `-nostartfiles`. `-static` prevents dynamic linking with shared libraries, so we can see the entire program in the ELF object dump. `-lm` links the `libm.a` math library if necessary. `-fno-tree-loop-distribute-patterns` disables an optimization that would replace loops with `memcpy` or `memset` from the disabled standard library. The `-T` and `-I` flags specify the linker script and include the file search path. Because we requested `-nostdlib`, we also must compile and link the `crt.S` minimal startup file and `syscalls.c` library (borrowed from the `riscv-software-src/riscv-tests` repository) to do `printf` and `exit` with the HTIF.

Optimization flags include:

`-O` Perform simple optimizations to reduce code size and execution time.

`-O2` Perform all optimizations that don't bloat the code (compiling takes longer).

`-O3` Perform all optimizations, including those that increase code size, to decrease execution time.

Code Example 3.12 MAKEFILE FOR C PROGRAMS

```
TARGET = sum

$(TARGET).objdump: $(TARGET)
    riscv64-unknown-elf-objdump -S $(TARGET) > $(TARGET).objdump

$(TARGET): $(TARGET).c Makefile
    riscv64-unknown-elf-gcc -o $(TARGET) -g -O\
      -march=rv64gc -mabi=lp64d -mcmodel=medany \
      -nostdlib -static -lm -fno-tree-loop-distribute-patterns \
      -T../common/test.ld -I../common \
      $(TARGET).c ../common/crt.S ../common/syscalls.c
clean:
    rm -f $(TARGET) $(TARGET).objdump
```

The disassembled program is too long to show here in full because of the startup and library code, but you can view it in `sum.objdump`. The code begins at 0x80000000 in the `.text.init` section at the `_start` label defined in `crt.S`. This startup code initializes all registers to 0 and turns on the floating-point unit (FPU). It sets the trap vector `mtvec` to be the address of the `trap_entry` function. If a trap (interrupt or exception) occurs, the `trap_entry` function causes the program to exit with an error code. The startup code initializes the global pointer, thread pointer, and stack pointer registers, `gp`, `tp`, and `sp`, to RAM above the program section. It then calls `_init` in `syscalls.c`, which calls `main()` from hart 0. Fig. 3.3 shows a memory map of the program.

`main()` calls `sum(4)`. Code Example 3.13 shows the disassembled code for `sum` with optimizations turned off (no `-O` flag) and Fig. 3.4 shows the stack usage. We added the comments to help explain the code. The program allocates 48 bytes (6 double words) on the stack. It saves `s0` on the stack and then uses `s0` as a frame pointer that points to the bottom of `sum`'s stack frame (i.e., the original top of stack before the function

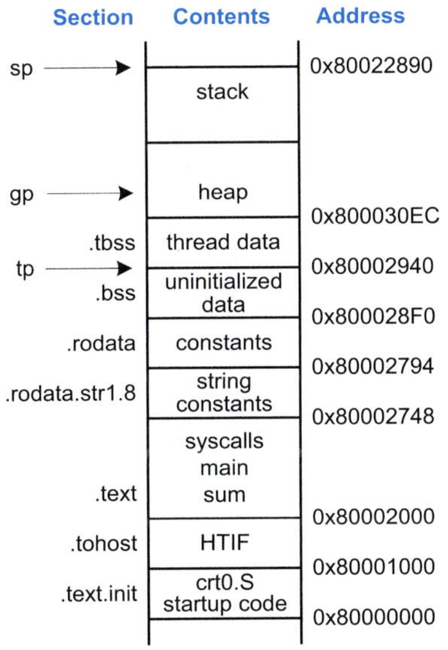

Fig. 3.3 Program memory map.

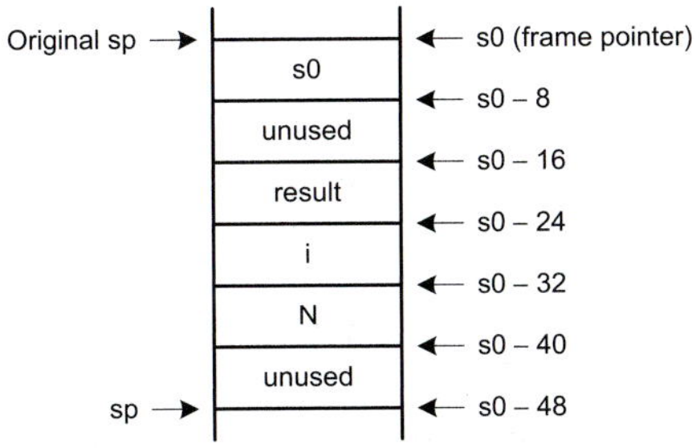

Fig. 3.4 `sum` stack usage (optimization turned off).

was called). It keeps local variables on the stack instead of in registers and accesses them relative to the frame pointer s0. The code is grossly inefficient.

Code Example 3.13 sum DISASSEMBLY (OPTIMIZATION TURNED OFF)

```
long sum(long N) {
    8000231a:  7179        addi  sp,sp,-48     # Room for 6 double-words on stack
    8000231c:  f422        sd    s0,40(sp)     # Save s0 on stack
    8000231e:  1800        addi  s0,sp,48      # s0 = frame ptr (old top of stack)
    80002320:  fca43c23    sd    a0,-40(s0)    # Save N on stack
  long result, i;
   result = 0;
    80002324:  fe043423    sd    zero,-24(s0) # result = 0
   for (i=1; i<=N; i++) {
    80002328:  4785        li    a5,1
    8000232a:  fef43023    sd    a5,-32(s0)    # i = 1
    8000232e:  a829        j     80002348 <sum+0x2e> # Check if done
       result = result + i;
    80002330:  fe843703    ld    a4,-24(s0)    # a4 = result
    80002334:  fe043783    ld    a5,-32(s0)    # a5 = i
    80002338:  97ba        add   a5,a5,a4      # result = result + i
    8000233a:  fef43423    sd    a5,-24(s0)    # Put result back in memory
   for (i=1; i<=N; i++) {
    8000233e:  fe043783    ld    a5,-32(s0)    # a5 = i
    80002342:  0785        addi  a5,a5,1       # i = i + 1
    80002344:  fef43023    sd    a5,-32(s0)    # Put i back in memory
    80002348:  fe043703    ld    a4,-32(s0)    # a4 = i
    8000234c:  fd843783    ld    a5,-40(s0)    # a5 = N
    80002350:  fee7d0e3    bge   a5,a4,80002330 <sum+0x16> # Loop if N >= i
   }
   return result;
    80002354:  fe843783    ld    a5,-24(s0)    # a5 = result
}
    80002358:  853e        mv    a0,a5         # Return value = result
    8000235a:  7422        ld    s0,40(sp)     # Restore s0 from stack
    8000235c:  6145        addi  sp,sp,48      # Clean up stack pointer
    8000235e:  8082        ret                 # Return
```

Code Example 3.14 shows the disassembled code for sum with basic optimizations (-O) turned on. Now the program doesn't use the stack at all. It receives N in a0 and stores N+1 in a4 as a completion condition. It keeps i in a5 and uses a0 for result, which saves copying a different register to a0 to return result at the end. It loops until i equals N+1 and then returns. The code is much faster. Observe that the disassembler has some difficulty associating the assembly language with the original C code because the optimizer reordered some of the code for performance.

Code Example 3.14 `sum` DISASSEMBLY (OPTIMIZATION TURNED ON WITH `-O`)

```
0000000080002000 <sum>:
#include "util.h"    // Supports verify

long sum(long N) {
  long result, i;
   result = 0;
   for (i=1; i<=N; i++) {
    80002000:   00a05b63              blez   a0,80002016 <sum+0x16>    # Return 0 if N <= 0
    80002004:   00150713              addi   a4,a0,1                   # a4 = N+1
    80002008:   4785                  li     a5,1                      # a5 = i = 1
   result = 0;
    8000200a:   4501                  li     a0,0                      # result = 0
       result = result + i;
    8000200c:   953e                  add    a0,a0,a5                  # result= result + i
   for (i=1; i<=N; i++) {
    8000200e:   0785                  addi   a5,a5,1                   # i++
    80002010:   fee79ee3              bne    a5,a4,8000200c <sum+0xc>  # Loop if i != (N+1)
    80002014:   8082                  ret                              # Return result
   result = 0;
    80002016:   4501                  li     a0,0                      # result = 0
   }
   return result;
}
    80002018:   8082                  ret                              # Return result
```

When `-O2` or `-O3` optimization is enabled, the compiler precomputes the result of calling `sum(4)` as 10 and replaces the sum function entirely with assigning `a1` to 10 as shown in Code Example 3.15. The call to verify is also eliminated because the compiler precomputes that verify will return 0 (indicating success).

Code Example 3.15 `sum(4)` DISASSEMBLY (OPTIMIZATION TURNED ON WITH `-O2`)

```
8000271a:      45a9                li          a1,10
```

`main()` uses the verify function to compare an array of integer results to expected values; in this case, there is only one array element, which is 1+2+3+4 = 10. It then returns with a code of 0 for success or 1 for failure. After main completes, `init` calls `exit` with the code for success or failure. Spike terminates silently if the code is 0 or prints `*** FAILED *** (tohost = 1)` on failure.

Table 3.6 Cycles to execute `sum(4)`.

Optimization Level	`mcycle`	`minstret`	True Cycles = True # Instructions
None	115	132	62
`-O`	31	38	21
`-O2` or `-O3`	11	16	1

The `asm` statement needs to determine which register contains a variable in the C program. %0 substitutes the argument, which is `"=r"(cycles)`, meaning the register containing the cycles variable. Therefore, the `csrr` instruction specifies the destination register as the register holding the variable `cycles`, which the compiler selected to be `a1`. If the program used more registers, `cycles` might be allocated to a different register, and the `asm` statement would pick the correct register. The `volatile` keyword prevents the compiler from optimizing away the assembly language instruction.

Table 3.6 shows the number of cycles and instructions retired to run `sum`, as reported by the program at various optimization levels. Spike simulates one instruction per cycle, so the two columns ideally would be the same, but the program has some overhead to call `setStats` and read each performance counter, and the overhead differs slightly between counters because of the order in which the counters are read. The table also shows the true number of instructions used by `sum`, measured by manually counting during the Spike simulation. These performance counters are more useful for longer programs where the overhead is negligible.

3.4.2 Inline Assembly Language

Code Example 3.16 illustrates *inline assembly language* used to add a single assembly language instruction that can't be implied by standard C code. It demonstrates the `asm` statement, which reads the `mcycle` CSR (0xB00). The result goes in the register holding the variable `cycles`. Code Example 3.17 shows the disassembly. The `asm` statement produces `csrr a1, mcycle`, where `a1` is the register holding the variable `cycles`.

Code Example 3.16 `inline.c`: SHOWS INLINE ASSEMBLY LANGUAGE

```
// inline.c
// David_Harris@hmc.edu 12 January 2022
// Illustrates inline assembly language

#include <stdio.h>

int main(void) {
  long cycles;
  asm volatile("csrr %0, 0xB00" : "=r"(cycles)); // read mcycle register
  printf ("mcycle = %ld\n", cycles);
}
```

Code Example 3.17 DISASSEMBLY OF **inline.c**

```
int main(void) {
    80002000:   1141                    addi    sp,sp,-16
    80002002:   e406                    sd      ra,8(sp)
    long cycles;
    asm volatile("csrr %0, 0xB00" : "=r"(cycles)); // Read mcycle register
    80002004:   b00025f3                csrr    a1,mcycle
    printf ("mcycle = %ld\n", cycles);
    80002008:   00000517                auipc   a0,0x0
    8000200c:   70850513                addi    a0,a0,1800 # 80002710 <atol+0x6a>
    80002010:   540000ef                jal     ra,80002550 <printf>
}
    80002014:   4501                    li      a0,0
    80002016:   60a2                    ld      ra,8(sp)
    80002018:   0141                    addi    sp,sp,16
    8000201a:   8082                    ret
```

3.4.3 Mixed C and Assembly Language

When a C program needs to integrate many lines of assembly language, the easiest approach is to put the C and assembly language in different files and have them communicate through function calls via the ABI's calling convention. Code Example 3.18 shows `sum_mixed.c`, which calls the `sum` function written in assembly language from Code Example 2.5. The code is located in `$WALLY/examples/C/sum_mixed`. The C file declares `sum` as extern, so the linker knows to look for it elsewhere. The `sum.S` file declared the label `sum` as global, so the linker finds it. According to the ABI, the C program passes `N` to `sum` in `a0` and gets the result back in `a0`. Recall that `sum.S` saved and restored `s0` and `s1` on the stack so that they are preserved across the function call. The optimizer does not modify the assembly language code.

Code Example 3.18 **sum_mixed.c**

```
// sum_mixed.c
// David_Harris@hmc.edu 12 January 2022
// Call assembly language from C

#include <stdio.h>          // Defines printf
#include "util.h"           // Defines setStats and verify
extern int sum(int);        // Prototype for the sum function written an external assembly file

int main(void) {
  int s[1], expected[1];

  setStats(1);              // Record initial mcycle and minstret
  s[0] = sum(4);            // Call assembly language sum function to compute 1 + 2 + 3 + 4 = 10
  setStats(0);              // Record elapsed mcycle and minstret
  printf("s = %d\n", s[0]);// Print sum of numbers, which should be 10
  expected[0] = 10;         // Set expected value
  return verify(1, s, expected); // Check the 1 element of s matches expected. 0 means success
}
```

3.4.4 Standard Library and Startup Code

C programmers use a variety of standard libraries, such as `stdio.h`, `string.h`, `time.h`, and `math.h` that include basic functions such as `printf()`, `strlen()`, `time()`, `sqrt()`, or `cos()`. GCC needs to know how to implement these functions. The implementation is platform specific; for example, printing is different to a Linux console versus a serial COM port, reading the time depends on the system timer, and math is different for processors with or without floating-point instructions.

Whereas the `libc` standard library defines most of these functions, `libm` defines math functions. Together, these are called the *C standard library*. Historically, `libc` and `libm` were separated because the math libraries could be large and not all programs needed them. `libc` is automatically linked unless `-nostdlib` is specified when compiling. `libm` is only linked if the `-lm` flag is provided. The libraries can be *statically linked*, such that a copy is incorporated into the program, or *dynamically linked*, such that only one copy of the standard library is in memory and many different programs can call this library. Dynamic linking saves memory but requires some OS support. Hence, platforms such as Linux and Windows tend to use dynamic linking, while simple embedded processors use static linking. `libc.so` and `libm.so` are the *shared object* dynamically linked libraries, while `libc.a` and `libm.a` are the statically linked *archive* libraries.

Many implementations of the C standard library exist. Some are proprietary and vendor specific, such as the Microsoft C runtime library for Windows on x86 processors. The leading open-source implementations are glibc and newlib. *glibc* is often used for Linux systems that already have it available as a shared library. *newlib* has more permissive licensing terms and thus is often preferred for embedded software. When installing *riscv-gnu-toolchain*, the `$WALLY/bin/wally-tool-chain-install.sh` script compiled the RISC-V GCC cross-compiler with *newlib*.

The newlib standard library is precompiled so that GCC doesn't have to recompile all of the library functions every time it compiles a program. Instead, GCC simply links the program to the precompiled libraries in `libc.a` and `libm.a`. The best implementation of a library depends on the machine architecture. For example, `sqrt()` could be implemented with `fsqrt.d` in rv64gc but would require a lengthy series of integer operations in rv32i. GCC solves this with the *multilib* feature, which precompiles many flavors of newlib targeting different machine architectures. For example, the installation script compiled all the flavors listed in Table 3.2. GCC automatically links the best library flavor consistent with the `-march` of the program being compiled.

In Section 3.4, the C Makefiles used the `-nostdlib` argument for GCC to avoid the startup code standard libraries, and instead have included `crt.S` and `syscalls.c` from `$WALLY/examples/C/common`. It is informative to read through each of these files. In brief, `crt.S` is an assembly language startup routine that initializes all registers to 0, enables floating-point, initializes a trap vector, sets up a global pointer and stack pointer, and defines the `.tohost` label to terminate a program in simulation. `syscalls.c` defines the `printf` function, which invokes `putchar`, which makes a system call to write a character to the console via stdout. `syscalls.c` defines a few other essential library functions such as `memcpy()`, `memset()`, `strlen()`, `strcmp()`, `strcpy()`, and `exit()`.

3.4.5 Compiler Explorer

The Compiler Explorer at godbolt.org is a handy tool to quickly understand how code compiles without having to set up a Makefile and type a series of commands. Choose the architecture and compiler (e.g., RISC-V rv32gc gcc 10.2.0), and optionally specify compiler flags such as `-O` or `-O3`. Type your C code in the left pane. The disassembled output will appear in the right pane. Fig. 3.5(a) shows how a recursive factorial function is compiled with `-O` optimization. Fig. 3.5(b) shows how the compiler is clever enough to eliminate the recursive function call with `-O2` optimization.

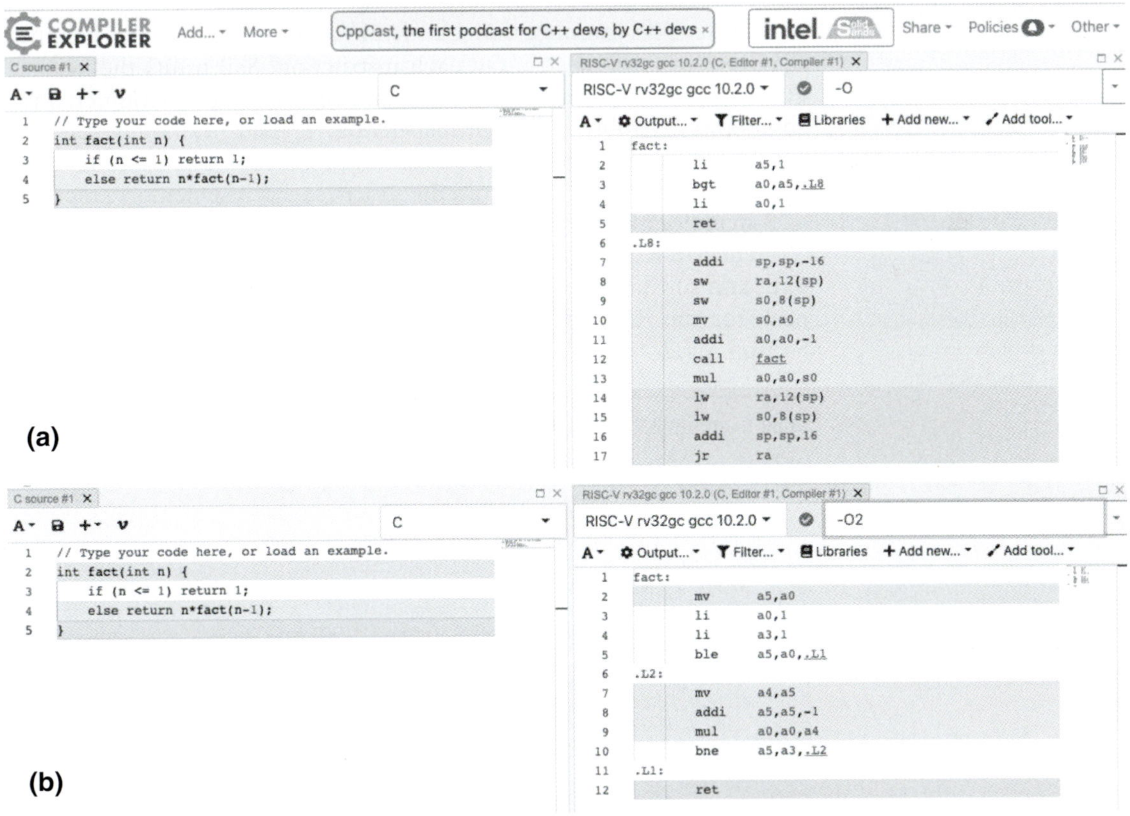

Fig. 3.5 Compiler Explorer factorial output: (a) -O, and (b) -O2. (© 2024 Matt Godbolt. Reprinted with permission.)

3.5 SAIL SIMULATION

Sail is written in an object-oriented functional programming language called OCaml, which is a derivative of a language called ML used in theorem provers. The Sail compiler can be downloaded from its GitHub page, github.com/rems-project/sail.

Sail descriptions of MIPS, ARM, and IBM Power ISAs also exist. The holy grail of formal language specification and theorem proving would be to prove that an HDL implementation of the ISA is logically correct. Much research remains before this can be automated for complex cores.

`sail_riscv_sim` uses a JSON configuration file to control supported RISC-V extensions and other implementation specific parameters. The default configuration can be viewed with the `--print-default-config` flag. The simulator defaults to RV64 with most extensions enabled and matches Spike's implementation choices. To specify the configuration, for example the Wally rv64gc configuration, run: `sail_riscv_sim --config $WALLY/tests/riscof/sail_cSim/rv64gc.json $WALLY/examples/asm/sumtest/sumtest`

The official formal and executable specification of the RISC-V architecture is written in a language called Sail [RISCV-SAIL]. Indeed, if the English specification is ever ambiguous or deviates from the Sail specification, Sail takes precedence. Code Example 3.19 shows the Sail description of RISC-V I-type instructions. Sail is a pseudocode-like language for describing ISAs. Sail serves several purposes:

- Documents the behavior of each instruction in an ISA in a precise and readable way.
- Automatically creates a simulator that exactly matches the specification.
- Generates definitions for theorem provers aiming to prove properties about an ISA or implementation.

Sail produces a RISC-V simulator called `sail_riscv_sim`. The simulator is compiled in C++ based on the Sail specification, so it is also known as sail_cSim. We use the term "Sail" colloquially for this simulator that is automatically produced from the Sail language description of RISC-V.

The following example (see the next two pages) shows an excerpt of `sail_riscv_sim` running `sumtest` from Section 3.3.10 and logging the signature to `sumtest.sig`. On each instruction, Sail prints the instruction number, processor mode (e.g., `[M]` for machine), program counter, and machine and assembly language instruction. It then prints any registers or memory that are modified. The signature `testsum.sig` contains the sum (0A) and cycle count (1D) from the `begin_signature` memory, represented as 8-byte double-words printed 4 bytes per line.

The Sail RISC-V simulator runs at about 300,000 instructions per second, and substantially slower for floating-point instructions. It is a bottleneck for the RISCOF test signature generation, as will be described in Section 3.7.2.

Code Example 3.19 SAIL DESCRIPTION OF RISC-V I-TYPE INSTRUCTIONS

```
// Excerpted from:
// github.com/riscv/sail-riscv/blob/master/model/riscv_insts_base.sail
function clause execute (ITYPE (imm, rs1, rd, op)) = {
  let immext : xlenbits = sign_extend(imm);
    X(rd) = match op {
     RISCV_ADDI  => X(rs1) + immext,
     RISCV_SLTI  => zero_extend(bool_to_bits(X(rs1) <_s immext)),
     RISCV_SLTIU => zero_extend(bool_to_bits(X(rs1) <_u immext)),
     RISCV_ANDI  => X(rs1) & immext,
     RISCV_ORI   => X(rs1) | immext,
     RISCV_XORI  => X(rs1) ^ immext
  };
  RETIRE_SUCCESS
}
```

```
$ cd $WALLY/examples/asm/sumtest
$ sail_riscv_sim sumtest --test-signature sumtest.sig --trace-all
using sumtest.sig for test-signature output.
ELF Entry @ 0x80000000
HTIF located at 0x80001000
begin_signature: 0x80004008
end_signature: 0x80004018
CSR mstatus (0x300) <- 0x0000000A00000000
CSR misa (0x301) <- 0x800000000034112F
CSR mcause (0x342) <- 0x0000000000000000

[0] [M]: 0x0000000000001000 (0x00000297) auipc x5, 0x0
x5 <- 0x0000000000001000

. . . More bootloader code, jumping to 0x80000000

[5] [M]: 0x0000000080000000 (0x00005117) auipc x2, 0x5
x2 <- 0x0000000080005000

[6] [M]: 0x0000000080000004 (0x20010113) addi x2, x2, 0x200
x2 <- 0x0000000080005200

[7] [M]: 0x0000000080000008 (0x00004297) auipc x5, 0x4
x5 <- 0x0000000080004008

[8] [M]: 0x000000008000000C (0xFF828293) addi x5, x5, -0x8
x5 <- 0x0000000080004000

[9] [M]: 0x0000000080000010 (0x0002B503) ld x10, 0x0(x5)
mem[R,0x0000000080004000] -> 0x0000000000000004
x10 <- 0x0000000000000004

[10] [M]: 0x0000000080000014 (0xC0202C73) csrrs x24, instret, x0
CSR instret (0xC02) -> 0x000000000000000A
x24 <- 0x000000000000000A

[11] [M]: 0x0000000080000018 (0x02C000EF) jal x1, 0x2c
x1 <- 0x000000008000001C

[12] [M]: 0x0000000080000044 (0x1141) c.addi x2, -0x10
x2 <- 0x00000000800051F0

[13] [M]: 0x0000000080000046 (0xE022) c.sdsp x8, 0x0(x2)
mem[W,0x00000000800051F0] <- 0x0000000000000000

[14] [M]: 0x0000000080000048 (0xE426) c.sdsp x9, 0x8(x2)
mem[W,0x00000000800051F8] <- 0x0000000000000000

[15] [M]: 0x000000008000004A (0x4401) c.li x8, 0x0
x8 <- 0x0000000000000000

[16] [M]: 0x000000008000004C (0x4485) c.li x9, 0x1
x9 <- 0x0000000000000001

[17] [M]: 0x000000008000004E (0x00954563) blt x10, x9, 0xa

[18] [M]: 0x0000000080000052 (0x9426) c.add x8, x9
x8 <- 0x0000000000000001

[19] [M]: 0x0000000080000054 (0x0485) c.addi x9, 0x1
x9 <- 0x0000000000000002

[20] [M]: 0x0000000080000056 (0xBFE5) c.j -0x8
```

```
. . . More loops of sum.S

[34] [M]: 0x0000000080000058 (0x8522) c.mv x10, x8
x10 <- 0x000000000000000A

[35] [M]: 0x000000008000005A (0x6402) c.ldsp x8, 0x0(x2)
mem[R,0x00000000800051F0] -> 0x0000000000000000
x8 <- 0x0000000000000000

[36] [M]: 0x000000008000005C (0x64A2) c.ldsp x9, 0x8(x2)
mem[R,0x00000000800051F8] -> 0x0000000000000000
x9 <- 0x0000000000000000

[37] [M]: 0x000000008000005E (0x0141) c.addi x2, 0x10
x2 <- 0x0000000080005200

[38] [M]: 0x0000000080000060 (0x8082) c.jr x1

[39] [M]: 0x000000008000001C (0xC0202CF3) csrrs x25, instret, x0
CSR instret (0xC02) -> 0x0000000000000027
x25 <- 0x0000000000000027

[40] [M]: 0x0000000080000020 (0x418C8CB3) sub x25, x25, x24
x25 <- 0x000000000000001D

[41] [M]: 0x0000000080000024 (0x00004297) auipc x5, 0x4
x5 <- 0x0000000080004024

[42] [M]: 0x0000000080000028 (0xFE428293) addi x5, x5, -0x1c
x5 <- 0x0000000080004008

[43] [M]: 0x000000008000002C (0x00A2B023) sd x10, 0x0(x5)
mem[W,0x0000000080004008] <- 0x000000000000000A

[44] [M]: 0x0000000080000030 (0x0192B423) sd x25, 0x8(x5)
mem[W,0x0000000080004010] <- 0x000000000000001D

[45] [M]: 0x0000000080000034 (0x00001317) auipc x6, 0x1
x6 <- 0x0000000080001034

[46] [M]: 0x0000000080000038 (0xFCC30313) addi x6, x6, -0x34
x6 <- 0x0000000080001000

[47] [M]: 0x000000008000003C (0x4285) c.li x5, 0x1
x5 <- 0x0000000000000001

[48] [M]: 0x000000008000003E (0x00533023) sd x5, 0x0(x6)
htif[0x0000000080001000] <- 0x0000000000000001
htif-syscall-proxy cmd: 0x000000000001
mem[W,0x0000000080001000] <- 0x0000000000000001

SUCCESS

$ more sumtest.sig
0000000a
00000000
0000001d
00000000
```

3.6 QEMU SIMULATION

QEMU is a powerful open-source emulator that handles many different computer architectures including RISC-V. It performs just-in-time translation from the architecture being simulated to the architecture of the host computer running QEMU (e.g., from RISC-V to x86). Thus, it can run on the order of 1 billion instructions per second, which is fast enough to comfortably boot an OS and an order of magnitude faster than Spike. It also interfaces with GDB to facilitate debugging, and emulates peripherals (e.g., a timer) that are important for booting an OS. QEMU is not a golden model for RISC-V, but no known discrepancies exist.

QEMU is variously pronounced kyoo-EM-yoo, kee-moo, kwemu, or Q.E.M.U. It is an acronym for Quick EMUlator.

3.6.1 Simulating with QEMU

Table 3.7 lists the options used in this chapter for running QEMU. To simulate `sumtest.c` using QEMU, do the following:

```
$ cd $WALLY/examples/asm/sumtest
$ qemu-system-riscv64 -S -M virt -nographic -bios none -kernel sumtest
```

The following excerpt (below) shows common QEMU commands, as summarized in Table 3.8. Just after invoking QEMU, it will appear to hang. Pressing Ctrl-a and then c enters the monitor. `info registers` prints the contents of all registers, showing that the PC is at 0x1000, where it runs a small bootloader as Spike does. `x` displays memory. Recall from Code Example 3.9 that the signature is two double words stored at address 0x80002008. Using `x /2xg 0x80002008`, we see the signature is initialized to 0xdeadbeefdeadbeef. Press `c` to continue. QEMU will automatically halt at the infinite self_loop, as you can confirm with `info registers` showing the PC at 0x80000042. Now the signature shows a sum of 0x0a (10) and an `instret` value of 0x48470! The sum is correct, but the number of instructions retired is wildly incorrect. QEMU by default relies on the host processor's performance counters, which run at a much different rate. Finally, `q` terminates the simulation.

As you are learning to use QEMU, it will often seem to lock up when you press the wrong keys. Kill it with `killall` in another terminal window:
`$ killall qemu-system-riscv64`

```
Ctrl-a c
QEMU 6.0.91 monitor - type 'help' for more information
(qemu) info registers
 pc       0000000000001000
… <other registers>
(qemu) x /2xg 0x80002008
0000000080002008: 0xdeadbeefdeadbeef
(qemu) c
(qemu) info registers
 pc       0000000080000042
… <other registers>
(qemu) x /2xg 0x80002008
0000000080002008: 0x000000000000000a 0x0000000000048470
(qemu) q
```

Table 3.7 QEMU run options.

Run Option	Description
-S	Start QEMU with the PC paused at the beginning of the program
-M virt	Use a generic virtual RISC-V platform instead of a specific board
-nographic	Disable graphical window support and redirect the virtual serial port to display on the console
-bios none	Skip loading the OpenSBI BIOS and run in bare metal mode
-kernel sumtest	Load sumtest.c program into memory so it's ready to execute
-icount 0	Enable the virtual instruction counter to get accurate cycle counts
-gdb tcp::1234	Interface QEMU with GDB to single step through the program

Table 3.8 QEMU commands.

Command	Description
Ctrl-a, then c	Enter monitor (right after QEMU starts)
info registers	Print all general-purpose registers
x <address>	Display memory at <address>
x /2xg <address>	Display two double words of memory in hexadecimal format
c	Continue running program
q	Terminate simulation

Learn more from QEMU's documentation at www.qemu.org/docs/master. The Invocation and QEMU Monitor sections are of particular relevance.

We could use the -icount 0 option to enable the virtual instruction counter to get accurate cycle counts, at a cost in simulation speed. Omit the -S flag to start immediately. After these modifications, shown below, see that instret is 0x1d, in agreement with Spike.

```
$ qemu-system-riscv64 -M virt -nographic -bios none -kernel
  sumtest -icount 0
Ctrl-a c
QEMU 6.0.91 monitor - type 'help' for more information
(qemu) x /2xg 0x80002008
0000000080002008: 0x000000000000000a 0x000000000000001d
(qemu) q
```

3.6.2 Debugging With GDB

The authoritative book on GDB is freely available [Stallman23].

QEMU can interface with the GDB debugger to single step through a program or use breakpoints. The interface is made with a TCP network connection to the QEMU virtual machine.

Invoke QEMU with a `-gdb tcp::<port>` option. Usually port 1234 is fine, but if somebody else is using it, pick a different port in the range 1024–49151.

```
$ qemu-system-riscv64 -S -M virt -nographic -bios none -kernel sumtest -gdb tcp::1234
```

Then invoke GDB in another terminal, connect it over the same port, and load the symbols from `sumtest` using this command:

```
$ riscv64-unknown-elf-gdb -ex "target remote :1234" -iex "file sumtest"
```

The following transcript (see the next two pages) shows an example debugging session, and Table 3.9 summarizes common GDB commands. With the symbol table, we can refer to an address by name, such as `&begin_signature`. In the transcript, we see the simple bootloader jump to the main program at 0x80000000, and GDB prints the associated source code. At the end of the transcript, we see the signature contains the expected sum of 0x0a and the registers contain expected values.

Table 3.9 GDB commands.

Command	Description
`info all-reg`	Print contents of all integer registers
`x <address>`	Print memory at address
`x <name>`	Print memory starting at name in symbol table
`stepi`	Single step through program
Enter	Press enter key to repeat previous command
`b <function>`	Set a breakpoint at a function, specified by its name
`b <address>`	Set a breakpoint at instruction address
`c`	Continues running
Ctrl-c	Pause program
`set $pc=<address>`	Set PC to `<address>` (for example, to 0x1000 to restart program)
`q`	End debug session

```
GNU gdb (GDB) 10.1
…
This GDB was configured as "--host=x86_64-pc-linux-gnu --target=riscv64-unknown-elf".
…
Reading symbols from sumtest...
Remote debugging using :1234
0x0000000000001000 in ?? ()
(gdb) info all-reg
zero           0x0     0
ra             0x0     0x0
sp             0x0     0x0
gp             0x0     0x0
tp             0x0     0x0
t0             0x0     0
…
--Type <RET> for more, q to quit, c to continue without paging--q
Quit
(gdb) display /i $pc
1: x/i $pc
=> 0x1000:        auipc    t0,0x0
(gdb) x/2xg &begin_signature
0x80002008:      0xdeadbeefdeadbeef        0xdeadbeefdeadbeef
(gdb) step
Cannot find bounds of current function
(gdb) stepi
0x0000000000001004 in ?? ()
1: x/i $pc
=> 0x1004:        addi     a2,t0,40
(gdb)
0x0000000000001008 in ?? ()
1: x/i $pc
=> 0x1008:        csrr     a0,mhartid
(gdb)
0x000000000000100c in ?? ()
1: x/i $pc
=> 0x100c:        ld       a1,32(t0)
(gdb)
0x0000000000001010 in ?? ()
1: x/i $pc
=> 0x1010:        ld       t0,24(t0)
(gdb)
0x0000000000001014 in ?? ()
1: x/i $pc
=> 0x1014:        jr       t0
(gdb)
rvtest_entry_point () at sumtest.S:6
6                 la sp, topofstack   # Initialize stack pointer
1: x/i $pc
=> 0x80000000 <rvtest_entry_point>:           auipc    sp,0x3
(gdb)
0x0000000080000004 in rvtest_entry_point () at sumtest.S:6
6                 la sp, topofstack   # Initialize stack pointer
1: x/i $pc
=> 0x80000004 <rvtest_entry_point+4>:         addi     sp,sp,512
(gdb)
7                 la t0, N            # Get address of N in data
1: x/i $pc
=> 0x80000008 <rvtest_entry_point+8>:         auipc    t0,0x2
(gdb)
0x000000008000000c      7                     la t0, N            # Get address of N in data
1: x/i $pc
=> 0x8000000c <rvtest_entry_point+12>:        addi     t0,t0,-8
(gdb)
```

```
8          ld a0, 0(t0)        # Load N
1: x/i $pc
=> 0x80000010 <rvtest_entry_point+16>:    ld      a0,0(t0)
(gdb)
9          csrr s8, instret    # Count instructions before call
1: x/i $pc
=> 0x80000014 <rvtest_entry_point+20>:    rdinstret  s8
(gdb)
10         jal sum             # Call sum(N)
1: x/i $pc
=> 0x80000018 <rvtest_entry_point+24>:    jal      ra,0x80000044 <sum>
(gdb)
sum () at sum.S:16
16         addi sp, sp, -16    # Make room to save s0 and s1 on the stack
1: x/i $pc
=> 0x80000044 <sum>:                      addi     sp,sp,-16
(gdb) b self_loop
Breakpoint 1 at 0x80000042: file sumtest.S, line 23.
(gdb) c
Continuing.

Breakpoint 1, self_loop () at sumtest.S:23
23         j self_loop         # Wait
1: x/i $pc
=> 0x80000042 <self_loop>:  j       0x80000042 <self_loop>
(gdb) x/2xg &begin_signature
0x80002008:    0x000000000000000a      0x00000017658e47a8
(gdb) info all-reg
zero           0x0                     0
ra             0x8000001c              0x8000001c <rvtest_entry_point+28>
sp             0x80003200              0x80003200
gp             0x0                     0x0
tp             0x0                     0x0
t0             0x1                     1
t1             0x80001000              2147487744
t2             0x0                     0
fp             0x0                     0x0
s1             0x0                     0
a0             0xa                     10
a1             0x87000000              2264924160
a2             0x1028                  4136
(gdb) set $pc=0x1000
(gdb) b sum
Breakpoint 1 at 0x80000044: file sum.S, line 16.
(gdb) c
Continuing.

Breakpoint 1, sum () at sum.S:16
16         addi sp, sp, -16    # make room to save s0 and s1 on the stack
(gdb) stepi
17         sd s0, 0(sp)
(gdb) q
A debugging session is active.

        Inferior 1 [process 1] will be detached.

Quit anyway? (y or n) y
Detaching from program: /home/ben/cvw/examples/asm/sumtest/sumtest, process 1
Ending remote debugging.
[Inferior 1 (process 1) detached]
```

To simulate a C program, change `-g` to `-gdwarf-2` in the Makefile to generate appropriate debugging symbols for C variables. Then type the following commands:

```
$ cd $WALLY/examples/C/sum
$ qemu-system-riscv64 -M virt -nographic -bios none -kernel sum -S -gdb tcp::1234
```

In another terminal, type:

```
$ riscv64-unknown-elf-gdb -iex "file sum" -ex "target remote:1234"
```

See the following transcript. QEMU hangs on the `printf` statement.

```
GNU gdb (GDB) 10.1
…
Reading symbols from sum...
Remote debugging using :1234
0x0000000000001000 in ?? ()
(gdb) b main
Breakpoint 1 at 0x8000201a: main. (2 locations)
(gdb) c
Continuing.

Breakpoint 1, main () at sum.c:17
17 int main(void) {
(gdb) s
19      setStats(1);
(gdb) n
20      s[0] = sum(4);
(gdb) s
sum (N=N@entry=4) at sum.c:11
11     for (i=1; i<=N; i++) {
(gdb) p N
$1 = 4
(gdb) s
12         result = result + i;
(gdb) p i
$2 = 1
(gdb) s
11     for (i=1; i<=N; i++) {
(gdb) p result
$3 = 1
(gdb) s
12         result = result + i;
(gdb)
11     for (i=1; i<=N; i++) {
(gdb)
12         result = result + i;
(gdb)
11     for (i=1; i<=N; i++) {
(gdb)
12         result = result + i;
(gdb)
11     for (i=1; i<=N; i++) {
(gdb)
main () at sum.c:21
21      setStats(0);
```

```
(gdb) p s[0]
$4 = <optimized out>
(gdb)
$5 = <optimized out>
(gdb) n
22      printf("s = %d\n", s[0]);
(gdb) n

^C
Program received signal SIGINT, Interrupt.
0x00000000800020c2 in syscall (arg2=7, arg1=<optimized out>, arg0=1, which=64)
    at ../common/syscalls.c:28
28   while (fromhost == 0)
(gdb)
```

3.7 TEST SUITES

RISC-V International's Architecture Test Special Interest Group (SIG) provides the RISCOF RISC-V Compatibility Framework, which includes the riscv-arch-test suite of tests that exercise every unprivileged RISC-V instruction. Passing these tests is a requirement to use the RISC-V trademark, but the tests are not sufficient to prove an implementation is completely correct. RISCOF compiles the riscv-arch-test tests and simulates them with Sail to produce signatures for each test. Section 5.6 describes verifying the SystemVerilog model by running the same tests and comparing the signatures.

RISCOF is written in Python and enables testing of RISC-V processors and comparing their results to a reference model such as Sail or Spike.

At the time of this writing, further test suites and formal verification methods are rapidly evolving. In the interim, we have created the wally-riscv-arch-test suite with additional tests, especially of privileged instructions, the memory management unit, and peripherals, described in Chapters 8, 11, and 20.

Passing the RISCOF tests and having the results approved by RISC-V International is a prerequisite to licensing the RISC-V trademarks in connection with the design.

Other more specialized Wally tests include floating-point tests, benchmarks, and booting Linux, described in Chapters 16, 21, and 22 and Appendix F.

Most of the test suites come as source code, which requires compilation. To compile all the test cases, go to the top-level `cvw` directory and type:

```
$ make --jobs
```

`make --jobs` launches jobs in parallel on a multicore computer, which is much faster than just `make`.

Section 5.11.1 describes how to run a regression to apply these tests to the Wally SystemVerilog model.

3.7.1 riscv-arch-test

riscv-arch-test provides test cases for the RV32I and RV64I instruction sets and various ratified extensions including C, M, F, D, Zb*, and Zk*, along with some basic privileged instruction tests [riscv-arch-test].

Each instruction has an assembly language test file that exercises the instruction and generates a signature containing the expected results. RISCOF compiles the test programs with GCC, runs them using Spike and Sail, and verifies that the signatures match.

riscv-arch-test is incorporated into the cvw repository as a git submodule so that improvements in the suite can easily be pulled into cvw.

3.7.1.1 Architectural Test Pool

`$WALLY/addins/riscv-arch-test/riscv-test-suite` contains the *architectural test pool* of tests for all the base RISC-V architectures including RV32I and RV64I. A tree of the most important files is given in Fig. 3.6. The current set of tests is designed to run in machine mode, so the top-level architectural test suite directories include `rv32i_m` and `rv64i_m`. Each test suite contains subdirectories such as `I`, `C`, `M`, and `F` for the core integer instructions and each of the optional extensions. It also contains a `privilege` directory that tests `ebreak`, `ecall`, and misaligned stores, loads, branches, and jumps. Each of these directories contains a `src` directory with assembly language test files for each test objective, such as `I/src/add-01.S`, that tests `add` instructions.

> Linux lists a folder's hierarchical contents using the `tree` command, as shown in Fig. 3.6 when running `tree` in the `$WALLY/addins/riscv-arch-test/riscv-test-suite` directory. By the time you read this, more extensions will be in the test pool.

```
├── env
│   ├── arch_test.h
│   ├── encoding.h
│   └── test_macros.h
├── Makefile.include
├── README.md
├── rv32e_m
│   ├── B
│   ├── C
│   ├── E
│   ├── M
│   ├── privilege
│   └── Zifencei
├── rv32i_m
│   ├── A
│   ├── B
│   ├── C
│   │   └── src
│   │       ├── cadd-01.S
│   │       ├── caddi-01.S
│   │       ├── ...
│   ├── CMO
│   ├── D
│   ├── D_Zfa
│   ├── F
│   │   └── src
│   │       ├── fadd_b10-01.S
│   │       ├── fadd_b1-01.S
│   │       ├── fadd_b11-01.S
│   │       ├── ...
│   ├── F_Zfa
│   ├── I
│   │   └── src
│   │       ├── add-01.S
│   │       ├── addi-01.S
│   │       ├── ...
│   ├── K
│   ├── M
│   ├── privilege
│   ├── P_unratified
│   ├── Zfh
│   ├── Zicond
│   └── Zifencei
└── rv64i_m
    ├── A
    ├── B
    ├── C
    ├── CMO
    ├── D
    ├── D_Zfa
    ├── F
    ├── F_Zfa
    ├── I
    ├── K
    ├── M
    ├── privilege
    ├── P_unratified
    ├── Zfh
    ├── Zicond
    └── Zifencei
```

Fig. 3.6 riscv-test-suite organization.

3.7.1.2 Test File Structure

If you need to write additional tests, the structure of the test case source files is important. Code Example 3.20 shows an excerpt of the `rv32i_m/I/src/add-01.S` file containing various macros. The code example has explanatory comments and disassembled code added in blue showing what the macros produce. The file tests two additions (0x7fffffff + 1 and 0x20000 + 0x20000), writes the results to signature memory, and calls the HTIF to halt.

> Users interested in testing instructions can gloss over most macros and just focus on the test cases, but if you are curious, you can learn about the macros at github.com/riscv-non-isa/riscv-arch-test/blob/main/spec/TestFormatSpec.adoc.

Code Example 3.20 **`rv32i_m/src/add-01.S`** EXCERPT

```
#include "model_test.h"
#include "arch_test.h"
RVTEST_ISA("RV32I")

.section .text.init
.globl rvtest_entry_point
rvtest_entry_point:                        # Initialize registers to predictable values
# 80000000:        feedc0b7                lui      ra,0xfeedc
# 80000004:        ead08093                addi     ra,ra,-339  # feedbead <_end+0x7eed63a9>
# 80000008:        ff76e137                lui      sp,0xff76e
# 8000000c:        f5610113                addi     sp,sp,-170  # ff76df56 <_end+0x7f768452>
...
RVMODEL_BOOT                               # Does nothing
RVTEST_CODE_BEGIN
RVTEST_CASE(0,"//check ISA:=regex(.*32.*);check ISA:=regex(.*I.*);def TEST_CASE_1=True;",add)

RVTEST_SIGBASE( x3,signature_x3_1)         # Set x3 (gp) to point to signature memory
800000f8:         00000013                 nop
800000fc:         00000013                 nop
80000100:         00006197                 auipc    gp,0x6
80000104:         01418193                 addi     gp,gp,20     # 80006114 <signature_x3_0>
80000108:         00000013                 nop
8000010c:         00000013                 nop
80000110:         00000013                 nop
80000114:         00000013                 nop
80000118:         00000013                 nop
8000011c:         00000013                 nop
inst_0: # Test add x24, x4, x24 with x4 = 0x7fffffff, x24 = 0x00000001
// rs2 == rd != rs1, rs1==x4, rs2==x24, rd==x24, rs1_val > 0 and rs2_val > 0, rs2_val == 1,
// rs1_val == (2**(xlen-1)-1), rs1_val != rs2_val, rs1_val == 2147483647
// opcode: add ; op1:x4; op2:x24; dest:x24; op1val:0x7fffffff;  op2val:0x1
TEST_RR_OP(add, x24, x4, x24, 0x80000000, 0x7fffffff, 0x1, x3, 0, x18)
80000120:         80000237                 lui      tp,0x80000
80000124:         fff20213                 addi     tp,tp,-1     # 7fffffff <absimm+0x3ad0dad0>
80000128:         00100c13                 li       s8,1
8000012c:         01820c33                 add      s8,tp,s8
80000130:         0181a023                 sw       s8,0(gp)
```

```
inst_1: # Test add x28, x10, x10 with x10 = 0x00020000
// rs1 == rs2 != rd, rs1==x10, rs2==x10, rd==x28, rs1_val > 0 and rs2_val < 0, rs2_val == -257,
// rs1_val == 131072
// opcode: add ; op1:x10; op2:x10; dest:x28; op1val:0x20000;  op2val:0x20000
TEST_RR_OP(add, x28, x10, x10, 0x40000, 0x20000, 0x20000, x3, 4, x18)
80000134:        00020537                lui      a0,0x20
80000138:        00020537                lui      a0,0x20
8000013c:        00a50e33                add      t3,a0,a0
80000140:        01c1a223                sw       t3,4(gp)
...
RVTEST_CODE_END
RVMODEL_HALT                             # Writes successful completion to HTIF
# 800032bc:        00100093              li       ra,1
# 800032c0 <write_tohost>:
800032c0:        00001297                auipc    t0,0x1
800032c4:        d412a023                sw       ra,-704(t0) # 80004000 <tohost>
800032c8:        ff9ff06f                j        800032c0 <write_tohost>
RVTEST_DATA_BEGIN                        # Room for initialized data values (for loads and stores)
.align 4
rvtest_data:
.word 0xbabecafe
RVTEST_DATA_END

RVMODEL_DATA_BEGIN                       # Start of signature
rvtest_sig_begin:
sig_begin_canary:
CANARY;

signature_x3_0:
    .fill 0*(XLEN/32),4,0xdeadbeef       # Unused

signature_x3_1:
    .fill 17*(XLEN/32),4,0xdeadbeef      # Room for 17 32-bit results

RVMODEL_DATA_END                         # End of signature
```

Code Example 3.21 shows the signatures produced by Sail and Spike, found in `tests/riscof/work/riscv-arch-test/rv32i_m/I/src/add-01.S/{ref/Reference-sail_c_simulator.signature` and `dut/DUT-spike.signature}`. Thus, the expected signature is 0x80000000 and 0x00040000. To change the tests, replace the code between `RVTEST_SIGBASE` and `RVTEST_CODE_END` and increase the amount of signature storage at the `.fill` statement as necessary.

Code Example 3.21 **tests/riscof/work/riscv-arch-test/rv32i_m/I/src/add-01.S/ref/reference-sail_c_simulator.signature**

```
80000000
00040000
```

3.7.2 RISCOF

RISCOF (github.com/riscv-software-src/riscof) is a set of scripts that compile test cases and run them on Sail and Spike to generate signatures, which can then be used to compare signatures produced by an HDL processor model. RISCOF was developed by the Indian Institute of Technology (IIT) Madras and Incore and has become part of the RISC-V International riscv-arch-test flow. The Wally toolchain includes RISCOF. RISCOF depends on configuration files in `$WALLY/tests/riscof/spike`. It places the compiled tests and signatures in `$WALLY/tests/riscof/work`.

RISCOF is invoked by running `make` from `$WALLY/tests/riscof`. This Makefile is also indirectly invoked when running `make` from `$WALLY` or `$WALLY/sim`.

The Makefile compiles suites of tests listed in Table 3.10. Each test references a configuration in the `riscof` directory, which in turn points to a YAML file in the `riscof/spike` subdirectory containing the instruction set architecture (RV{32/64}), extensions, and so on. Code Example 3.22 shows part of the rv32gc YAML configuration. The configuration has one core that supports RV32GC (RV32IMAFDCZicsr) and various other extensions such as bit manipulation. The YAML file configures the value of the machine instruction set architecture register (`misa`) to indicate standard extensions supported (see Section 8.2.3.2): A, C, D, F, I, M, and XLEN = 32. RISCOF examines all the tests in the specified directory, making a list in `riscof/riscof_work/{CONFIG}/test_list.yaml` of those with an `RVTEST_ISA` macro compatible with the ISA configuration.

YAML (*YAML Ain't Markup Language*) is a generic language for describing hierarchical configuration information. Various RISC-V tools are configured with YAML files. YAML has similarities to JSON, but hierarchies are expressed by indenting rather than nested braces.

Code Example 3.22 **tests/riscof/spike/spike_rv32gc_isa.yaml** CONFIGURATION (FOR **config32.ini**)

```
hart_ids: [0]
hart0:
  ISA: RV32IMAFDCZicsr_Zicond_Zifencei_Zfa_Zfh_Zca_Zcb_Zba_Zbb_Zbc_Zbkb_Zbkc_Zbkx_Zbs_Zknd_Zkne_Zknh
  physical_addr_sz: 32
  User_Spec_Version: '2.3'
  supported_xlen: [32]
  misa:
   reset-val: 0x4000112D
   rv32:
     accessible: true
     mxl:
       implemented: true
       type:
           warl:
              dependency_fields: []
              legal:
                - mxl[1:0] in [0x1]
              wr_illegal:
                - Unchanged
```

Table 3.10 Wally RISCOF test suites.

Suite	Configuration	Directory	Description
arch32	`config32.ini`	`riscv-arch-test`	Standard 32-bit compatibility tests
arch64	`config64.ini`	`riscv-arch-test`	Standard 64-bit compatibility tests
arch32e	`config32e.ini`	`riscv-arch-test`	Standard 32-bit compatibility tests for rv32e
wally32	`config32.ini`	`wally-riscv-arch-test`	Custom 32-bit tests: privileged, peripherals
wally64	`config64.ini`	`wally-riscv-arch-test`	Custom 64-bit tests: privileged, peripherals

Running RISCOF can take more than an hour on a slow machine. The largest part of the time comes from the slow Sail simulator producing signatures. Unfortunately, RISCOF is not yet smart enough to only process test cases that have changed.

Fig. 3.7 shows arch32 excerpts of the report generated by RISCOF. RISCOF compiles and runs the tests on both a "Device under test" (Spike) and a reference model (Sail_CSim) and checks that their signatures (`DUT-spike.signature` and `Reference-sail_c_simulator.signature`) match. It also invokes `elf2hex` to convert the ELF files into memfiles suitable to load into a Verilog testbench.

The Makefile copies the compiled tests and signatures into the `work` directory structured as shown in Fig. 3.8. The `dut` and `ref` directories each have an ELF, memfile, and signature. The `ref` directory also has an `.objdump`, which is convenient when debugging, and `.addr` and `.lab` files to associate labels with addresses in the Verilog testbench.

3.7.3 wally-riscv-arch-test

The `$WALLY/tests/wally-riscv-arch-test` directory is derived from riscv-arch-test but contains test cases written for this book that are not part of the riscv-arch-test suite. They include:

- Privileged tests, including traps, CSRs, and the MMU
- Peripheral tests

Sail does not presently model the peripherals used to generate interrupts, so it cannot be used to generate the expected values. Instead, scripts copy a `reference_output` file from wally-riscv-arch-test to the corresponding `riscof/work` directory. This means that the developer must predict the expected values and that there is no reference model to catch errors in these predictions.

```
make -C ../../tests/riscof/
make[1]: Entering directory '/home/harris/cvw/tests/riscof'
mkdir -p ./riscof_work
mkdir -p ./work
mkdir -p ./work/riscv-arch-test
mkdir -p ./work/wally-riscv-arch-test
sed 's,{0},/home/harris/cvw/tests/riscof,g;s,{1},32gc,g' config.ini > config32.ini
sed 's,{0},/home/harris/cvw/tests/riscof,g;s,{1},64gc,g' config.ini > config64.ini
sed 's,{0},/home/harris/cvw/tests/riscof,g;s,{1},32e,g' config.ini > config32e.ini
riscof run --work-dir=./riscof_work --config=config32.ini --suite=../../addins/riscv-arch-test/riscv-test-suite/ --env=../../addins/riscv-
arch-test/riscv-test-suite/env --no-browser
  INFO | ****** RISCOF: RISC-V Architectural Test Framework 1.25.2 *******
  INFO | using riscv_isac version : 0.11.0
  INFO | using riscv_config version : 3.5.0
  INFO | Reading configuration from: /home/harris/cvw/tests/riscof/config32.ini
  INFO | Preparing Models
  INFO | Input-ISA file
  INFO | Loading input file: /home/harris/cvw/tests/riscof/spike/spike_rv32gc_isa.yaml
  INFO | Load Schema /home/harris/.local/lib/python3.8/site-packages/riscv_config/schemas/schema_isa.yaml
  INFO | Processing Hart: hart0
  INFO | Initiating Validation
  INFO | No errors for Hart: 0 :)
  INFO |  Updating fields node for each CSR
  INFO | Initiating WARL legality checks.
  INFO | Initiating post processing and reset value checks.
  INFO | Initiating validation checks for indexed csrs
  INFO | Initiating validation checks for shadow fields
  INFO | Performing Checks on PMP CSRs
  INFO | Dumping out Normalized Checked YAML: /home/harris/cvw/tests/riscof/riscof_work/spike_rv32gc_isa_checked.yaml
  INFO | Input-Platform file
  INFO | Loading input file: /home/harris/cvw/tests/riscof/spike/spike_platform.yaml
  INFO | Load Schema /home/harris/.local/lib/python3.8/site-packages/riscv_config/schemas/schema_platform.yaml
  INFO | Initiating Validation
  INFO | No Syntax errors in Input Platform Yaml. :)
  INFO | Dumping out Normalized Checked YAML: /home/harris/cvw/tests/riscof/riscof_work/spike_platform_checked.yaml
  INFO | Generating database for suite: /home/harris/cvw/addins/riscv-arch-test/riscv-test-suite
  INFO | Database File Generated: /home/harris/cvw/tests/riscof/riscof_work/database.yaml
  INFO | Env path set to/home/harris/cvw/addins/riscv-arch-test/riscv-test-suite/env
  INFO | Running Build for DUT
  INFO | Running Build for Reference
  INFO | Selecting Tests.
  INFO | Running Tests on DUT.
  INFO | Running Tests on Reference Model.
  INFO | Initiating signature checking.
  INFO | Following 387 tests have been run :
  INFO | TEST NAME                                                : COMMIT ID                        : STATUS
  INFO | /home/harris/cvw/addins/riscv-arch-test/riscv-test-suite/rv32i_m/I/src/add-01.S : -: Passed
  INFO | /home/harris/cvw/addins/riscv-arch-test/riscv-test-suite/rv32i_m/I/src/addi-01.S : -: Passed
…
  INFO | Test report generated at /home/harris/cvw/tests/riscof/riscof_work/report.html.
rsync -a ./riscof_work/rv32i_m/ ./work/riscv-arch-test/rv32i_m/ || echo "error suppressed"
...
riscv64-unknown-elf-elf2hex --bit-width 32 --input ../../tests/riscof/work/riscv-arch-test/rv32i_m/I/src/add-01.S/ref/ref.elf --output
../../tests/riscof/work/riscv-arch-test/rv32i_m/I/src/add-01.S/ref/ref.elf.memfile
...
```

Fig. 3.7 RISCOF log for arch32 suite.

```
cvw/tests/riscof/work/riscv-arch-test/rv32i_m
├── I                                                Similar directories for C, M, F, D, etc.
│   └── src
│       ├── add-01.S                                 Similar directories for each instruction
│       │   ├── dut                                  Spike device under tests files
│       │   │   ├── DUT-spike.signature              Signature from Spike
│       │   │   └── my.elf                           ELF for Spike
│       │   └── ref                                  Sail reference files
│       │       ├── add-01.log                       Log file from Sail simulation
│       │       ├── ref.elf                          ELF for Sail
│       │       ├── ref.elf.memfile                  Machine language code from ELF for Verilog sim
│       │       ├── ref.elf.objdump                  Disassembled ELF
│       │       ├── ref.elf.objdump.addr             Addresses of labels in ELF for Verilog sim
│       │       ├── ref.elf.objdump.lab              Names of labels in ELF for Verilog sim
│       │       └── Reference-sail_c_simulator.signature   Signature from Sail
│       │
```

Fig. 3.8 RISCOF work output.

To add new tests to regression in `tests/wally-riscv-arch-test`:

- Create the test assembly language `.S` file following the form of other tests.
  - Preserve the existing riscv-arch-test macros.
  - Edit the `RVTEST_ISA` and `RVTEST_CASE` lines to match your required extensions (e.g., RV64I_Zicsr).
- Add the tests to a test suite in `tests.vh`.
- Add the test suite to the testbench if it is new.
- Add the test suite to regression.
- If Sail can simulate the test correctly (e.g., doesn't depend on interrupts or peripherals):
  - Run `make` in the `sim` directory, invoking RISCOF to compile the tests and generate `Reference-sail_c_simulator.signature`.
- Otherwise:

  Include "`NO_SAIL=True`" in the RV_TEST_CASE (see privileged tests for an example).
  - Write the expected values in `riscv-test-suite/rv{32/64}i_m/<testtype>/references/testname.reference_output`.
  - Run `make` in the `sim` directory, invoking RISCOF to compile the tests without running Sail.
- Run `regression-wally` to run the tests.
- Inspect the object dump and signature in `tests/riscof/work/wally-riscv-arch-test`.

3.8 SUMMARY

This chapter has described the RISC-V toolchain, simulators, and test suites used in an open-source Linux-based development environment widely used by professional developers. GNU's `gcc` provides an assembler, linker, and C compiler that support many architectures including RISC-V. It outputs RISC-V programs as ELF files that also contain headers, debug information, and the symbol table. `objdump` disassembles ELF files to assembly language.

Spike, Sail, and QEMU are three standard RISC-V simulators. Spike was the original golden RISC-V simulator and remains the easiest to use. Sail is a newer simulator that also serves as RISC-V's formal executable architecture specification. QEMU is an extremely fast open-source emulator, and it interfaces nicely with GDB for debugging. This chapter also showed how to use the toolchain and simulators to compile RISC-V verification test suites and generate expected results.

HDL Design Practices

4

4.1 HDL Design Methodology
4.2 Hardware Design with SystemVerilog
4.3 Wally Style Guidelines
4.4 Wally Configuration
4.5 Summary

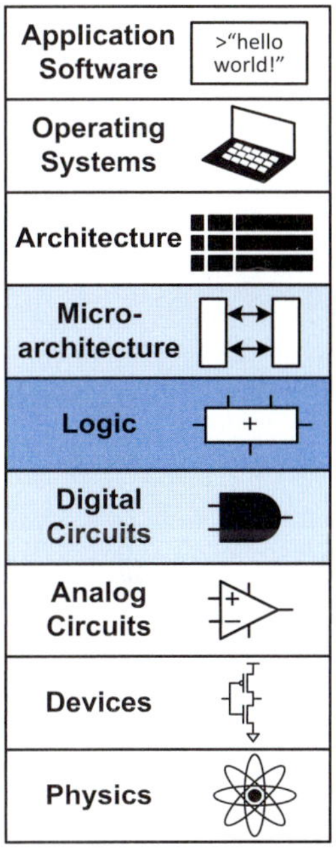

Digital designers need high-productivity design methods to develop complex systems-on-chip (SoCs) in a reasonable amount of time. Currently, most designs are specified using *hardware description languages* (*HDLs*), such as SystemVerilog and VHDL, that offer a textual representation of hardware. The HDL code can be simulated in a *logic simulator*, such as Siemens Questa, Synopsys VCS, Cadence Xcelium, or open-source Verilator, to *verify* that it performs the intended function and then synthesized with a *logic synthesis* tool, such as Synopsys Design Compiler, AMD Vivado, or Intel Quartus, to map the design onto a library of standard cells for a custom chip, also called an ASIC (application-specific integrated circuit), or onto configurable logic in an FPGA (field programmable gate array).

An advantage of HDLs over schematics is that the logic can be specified at a higher level of abstraction called *behavioral* or *register transfer level* (*RTL*). An RTL description focuses on the registers and the datapath elements, such as adders, shifters, and multiplexers, rather than individual logic gates or transistors. For example, the designer can write "+" for addition instead of drawing a chain of logic gates that perform addition, or they could describe an FSM based on its state transitions rather than in terms of Boolean equations. Moreover, the logic synthesis tool can automatically pick the most appropriate implementation for given constraints. For example, adders can be built using ripple carry, carry lookahead, or prefix tree architectures, with increasing speed but also increasing cost. Given timing constraints, a logic synthesis tool will try to pick the lowest-cost implementation that meets timing.

The Wally processor uses SystemVerilog [IEEE1800-2023], a modern flavor of Verilog. Verilog is the most widely used HDL in the industry. SystemVerilog fixes some syntactical annoyances of Verilog (e.g., the `reg` data type, which might or might not be a register) and adds powerful capabilities for verification that are beyond the scope of this book. SystemVerilog is nearly interchangeable with Verilog and is well supported in modern tools, so there is little reason to write new code in classic Verilog anymore. SystemVerilog has many nonsynthesizable constructs for writing testbenches, and the SystemVerilog standard is not explicit about what is synthesizable or not. This chapter emphasizes the standard idioms to describe hardware. At times, this book uses the names *Verilog* and *SystemVerilog* interchangeably.

The leading competitor to Verilog is VHDL [IEEE1076-2019], which once was dominant in Europe and at US military contractors but is steadily declining in market share [Golson16]. The synthesizable subset of VHDL is essentially interchangeable with SystemVerilog, but idiosyncrasies of the language lead to about 50% more code to describe the same hardware. SystemC [IEEE1666-2011] is another widely supported language suitable for simulation and synthesis. It is a C++ class library, offering a higher level of abstraction that can reduce design time for sophisticated users. Leading simulation and synthesis tools support a mix of all three languages. Other HDLs such as Chisel, SpinalHDL, and Bluespec promise to offer higher productivity but have not yet gained large market share.

This chapter begins with an overview of methodologies for designing with HDLs. It provides a tutorial about designing hardware using SystemVerilog, including combinational and sequential logic and memories. It examines how to use SystemVerilog to describe configurable systems such as the Wally processor. It then discusses coding style guidelines used on Wally. Chapters 5 and 6 address logic verification and synthesis with SystemVerilog.

4.1 HDL DESIGN METHODOLOGY

An SoC is too big for a single person to design in a competitive timeframe and too complex to keep in one's head in its entirety. Designers have developed several techniques for managing this complexity.

4.1.1 Design Strategies

Sophisticated design teams write specifications before writing code. The specifications define the system functions and design hierarchy. A good specification helps all members of a team understand how the system functions and where different features are implemented. Getting the hierarchy right the first time reduces effort repartitioning the system later. However, the ultimate goal of the project is working code, not polished documentation, so managers must remember that the specifications are merely a means toward an end. Less experienced designers often lack a frame of reference to write a proper specification and have to write some code first to understand the system before documenting it later. Startup companies also commonly omit specifications, rushing to get a minimally viable product before funding runs out. Startups often have to bet on the skill of a small team of designers who keep the design in their heads, hoping that none of the designers get hit by a truck before the product is ready.

After the requirements are defined, start by thinking of the design with block diagrams. Begin with a top-level block diagram and draw the inputs and outputs. Early in the design phase, it may be most practical to focus on the major input and output busses and leave their associated control signals for later. Then, sketch the sub-blocks and how they connect to the primary inputs and outputs and to internal signals between sub-blocks. Repeat this process hierarchically for each sub-block until it is simple enough to understand. Often, different blocks are owned by different designers and communicate via interfaces specified early in the design process. Later chapters of this book show hierarchical block diagrams of Wally's units and subsystems.

After you get down to simple blocks, design them in terms of hardware components such as registers, multiplexers, ALUs, memories, and combinational logic blocks. The most common beginner mistake is to write HDL like a program, and the more programming experience one has, the easier it is to make this mistake.

To obtain good power, performance, and area, and to minimize design errors, designers use only a synthesizable subset of SystemVerilog where they can readily understand the hardware implied by their code. For each hardware component, write the corresponding idiom from Section 4.2. If you aren't sure what hardware you intend to imply with some code, you are unlikely to get good results from that code. Section 4.3 addresses style guidelines that avoid common Verilog pitfalls and produce clear code.

Two of the most common design methodologies are waterfall and agile. *Waterfall* project management is a traditional approach drawn from mature industries such as construction. The project is divided into tasks, such as conception, initiation, design, building, testing, and deployment. Each step is reviewed and then frozen before moving on to the next step. It works well when the time and cost of each step can be predicted well from experience and when risk is low. Software projects were historically based on the waterfall model; for example, the US Defense Department DOD-STD-2167A standard of 1985 mandated this approach for defense contracts. Because testing of the overall design does not occur until much later in the project when using this method, major flaws may not be revealed until late in the process when they are most expensive to fix, and sometimes the problems in the underlying design are insurmountable so late in the process [Keating02]. In contrast, *agile* project management is an iterative design process that emphasizes delivering frequent working prototypes, getting rapid customer feedback, and embracing changing requirements. Key features are implemented first, problems are discovered early, and the design adapts to work around the problems. Except when an SoC is a minor update to an existing product, agile methods tend to be most appropriate for SoC design. Wally was developed using an agile methodology.

4.1.2 Version Control

Version control is the practice of tracking and managing changes to code. Version control systems maintain a database, called a *repository*, of all of the code files, tracking every modification. The main branch of the code is intended to be stable and robust. Developers create other branches as they write and test new features, then merge them into the main branch when ready. Merging is smart enough that different developers can modify different parts of a file concurrently without stomping on each other. If a bug is introduced, version control makes it easy to roll back to the last working version of the system and to identify exactly what changes have been made since that working version. Open-source development is driven by public repositories managed by a team of *committers* who review and accept contributions.

Version control is essential for complex projects, and Git has become the version control system of choice for most organizations that aren't deeply entrenched in some other tool. Wally uses Git for version control, as discussed in Appendix C. GitHub is a free and widely used cloud service for hosting Git repositories.

A student with a rich set of code in their public GitHub repositories and a history of making contributions to other projects has a substantial edge in an interview!

Even when you are working on a small project alone, version control provides file backups, an easy way to roll back to the last working design, and a repository to store and share your work. When developers get accustomed to version control, they rarely go back.

4.1.3 Reuse

The most effective strategy for controlling design cost is to reuse proven components. The SoC design team typically licenses cell, memory, and input/output (I/O) libraries from the chip foundry or reuses libraries developed internally. The team then adds components such as processor cores, peripherals, and accelerators. ARM and RISC-V are leading SoC processors. In 2023, SoCs shipped more than 30 billion ARM cores [Statista25]. As of 2022 an estimated 10 billion RISC-V cores had shipped [RISCV23], and RISC-V is increasing exponentially. The cores range in complexity from tens of thousands to tens of millions of transistors and vary in performance over two orders of magnitude. Peripherals include double-data-rate (DDR) and Flash memory controllers, radios for Wi-Fi, Bluetooth, or cellular communication, wired Ethernet and serial links, timers, and analog/digital and digital/analog data converters. Accelerators include graphics processing units (GPUs); video codecs; and specialized blocks for artificial intelligence, cryptography, and networking. Accelerators operate in parallel with processors and typically offer two orders of magnitude better performance and energy efficiency than general-purpose processors programmed for the same task. Most of these components are licensed from third parties or reused from a company's proprietary library. The design team may add some "special sauce" blocks customized to the target application and then integrates all the parts onto an SoC, verifies the system, and develops software to run on the chip.

4.2 HARDWARE DESIGN WITH SYSTEMVERILOG

A good hardware designer visualizes the desired hardware (e.g., registers, memories, FSMs, and combinational logic blocks) and then writes HDL idioms that imply such hardware. This section reviews these idioms. See also [Harris21] for a side-by-side introduction to SystemVerilog and VHDL. The most common novice error when using hardware description languages is to think of them as software programming and to simply write code rather than visualize hardware.

We design logic *hierarchically* or *structurally* by decomposing a complex system into modules with well-defined inputs and outputs. Each of the modules is recursively decomposed into smaller modules until the modules are simple enough that they can be described *behaviorally*.

Behavioral HDL (RTL) code defines the function performed by the module. The designer must first consider whether a piece of behavioral logic is combinational or sequential. Whereas *combinational logic* outputs depend only on current input values, *sequential logic* outputs depend on current and previous input values and thus have memory. For ease of design, we rely on the *synchronous sequential discipline*, in which the only sequential elements are edge-triggered flip-flops or memories.

SystemVerilog has many features not covered in this tutorial. Some, such as `$display`, are intended for writing testbenches and are not synthesizable. Others, such as `reg` and `wire`, are historical baggage and shouldn't be used in current designs. Except in testbenches, we recommend using the synthesizable subset of the language described in this chapter. A substantial fraction of the Verilog code you will discover in an Internet search is bad or has subtle errors, so it is much better to think of the hardware you want and then write the appropriate idiom. If you aren't sure of what hardware you want, you are likely to write bad code too.

4.2.1 Modules

Code Example 4.1 shows the SystemVerilog for a full adder. It begins with the keyword *`module`* and then the name of the module (`fulladder`). It declares the three inputs and two outputs. Note that signals in synthesizable SystemVerilog are declared of type `logic`, which can represent 0, 1, z (floating), or x (uninitialized or invalid). The outputs are described with `assign` statements specifying Boolean equations using `&` (and), `|` (or), `^` (xor). Fig. 4.1 shows various possible implementations. The synthesis tool chooses the lowest-cost solution that meets timing requirements in a target cell library.

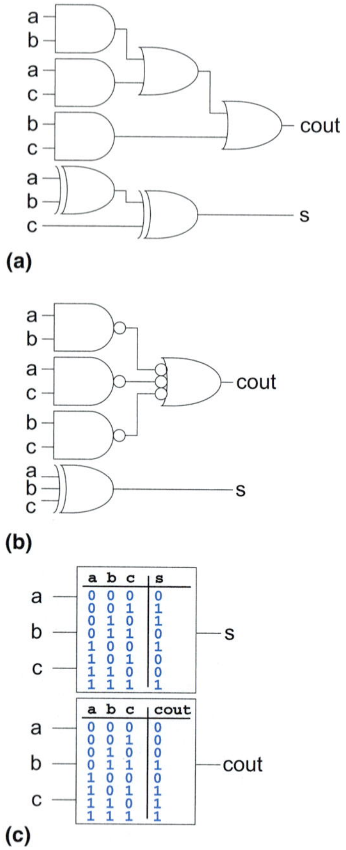

Fig. 4.1 **fulladder** **implementations: (a) basic logic gates, (b) optimized logic gates, and (c) FPGA lookup tables.**

Code Example 4.1 **fulladder**

```
module fulladder(input  logic a, b, c,
                 output logic s, cout);

  assign s = a ^ b ^ c;
  assign cout = (a & b) | (a & c) | (b & c);
endmodule
```

Code Example 4.2 shows the SystemVerilog for a 64-bit 2:1 multiplexer module. Observe that some of the inputs are 64 bits wide. It uses the *ternary operator* `? :` to choose the output `y` to be either `d1` or `d0` based on the value of the select line, `s`. The comments show a logically equivalent but less readable description with Boolean equations, illustrating the ~ (not) operator and replicating `s` into a 64-bit bus. Fig. 4.2 shows the circuit implied. It could be implemented in many ways, such as 64 1-bit 2:1 muxes, 128 2-input AND gates and 64 2-input OR gates, 192 2-input NAND gates, or 64 3-input FPGA lookup tables.

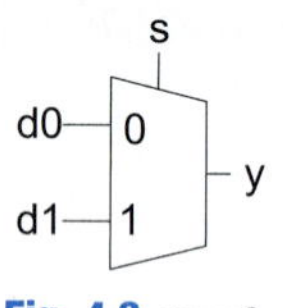

Fig. 4.2 **mux2.**

Code Example 4.2 **mux2**

```
module mux2(input  logic [63:0] d0, d1,
            input  logic        s,
            output logic [63:0] y);

  assign y = s ? d1 : d0;

  /* Alternative implementation using gates
  logic [63:0] s64;

  assign s64 = {64{s}}; // Replicate s 64 times
  assign y = ~s64 & d0 | s64 & d1; */
endmodule
```

Often it is helpful to create a *parameterized* module with adjustable bus width. Code Example 4.3 shows a parameterized 2:1 mux with a default width of 8 bits. Parameterized modules can save time and code and enable code reuse; `mux2` can be used for any 2:1 mux, regardless of bus width.

Code Example 4.3 PARAMETERIZED **mux2**

```
module mux2 #(parameter WIDTH = 8) (
  input  logic [WIDTH-1:0] d0, d1,
  input  logic             s,
  output logic [WIDTH-1:0] y);

  assign y = s ? d1 : d0;
endmodule
```

Code Example 4.4 shows how to *instantiate* the default (8-bit bus) version of mux2 and a 12-bit version. Each mux is instantiated by writing the *module name* (mux2) followed by the *instance name* (in this case, muxa or muxb) followed by the inputs and outputs. An instantiation is like grabbing parts (in this case, two muxes) from the hardware bin and then connecting their inputs and outputs as desired.

Code Example 4.4 INSTANTIATING PARAMETERIZED MODULES

```
mux2 muxa(d0, d1, s, y);                     // 2:1 mux with default 8-bit busses
mux2 #(12) muxb(data0, data1, sel, yout); // 2:1 mux with 12-bit busses
```

Code Example 4.5 shows four SystemVerilog representations of a 4:1 multiplexer with a default width of 64 bits. The first is a structural representation, in which the 4:1 mux is built from three 2:1 muxes, as shown in Fig. 4.3. The width is passed as a parameter to each mux2 to override the

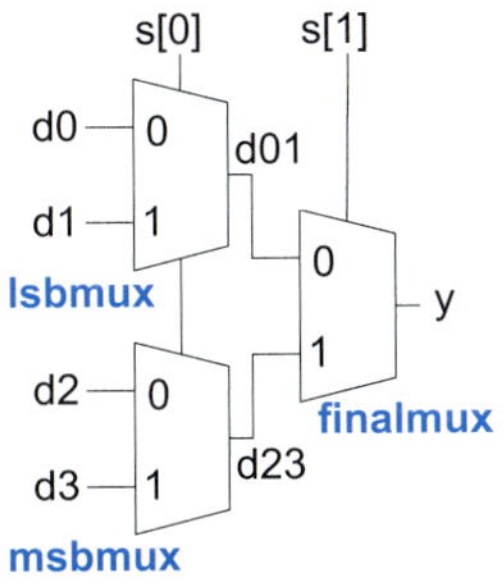

Fig. 4.3 mux4 **built from 2:1 multiplexers.**

Code Example 4.5 mux4

```
module mux4 #(parameter WIDTH = 64) (
  input  logic [WIDTH-1:0] d0, d1, d2, d3,
  input  logic [1:0]       s,
  output logic [WIDTH-1:0] y);

  logic [WIDTH-1:0] d01, d23;

  // Structural implementation using 2:1 muxes
  mux2 #(WIDTH) lsbmux(d0, d1, s[0], d01);
  mux2 #(WIDTH) msbmux(d2, d3, s[0], d23);
  mux2 #(WIDTH) finalmux(d01, d23, s[1], y);

  /* Alternative behavioral implementation using assign statement
  assign y = s[1] ? (s[0] ? d3 : d2) : (s[0] ? d1 : d0); */


  /* Alternative behavioral implementation using case statement
  always_comb
    case(s)
      0: y = d0;
      1: y = d1;
      2: y = d2;
      3: y = d3;
    endcase */


  /* Alternative behavioral implementation using if statements
  always_comb
    if      (s == 2'b00) y = d0;
    else if (s == 2'b01) y = d1;
    else if (s == 2'b10) y = d2;
    else                 y = d3; */
endmodule
```

Wally Structural Coding Style
Wally favors a structural coding style that decomposes datapath modules into simple blocks, such as registers, multiplexers, memories, and Boolean equations, to emphasize the hardware implied by the code. These lower-level modules are instantiated within a higher-level or top-level module in the hierarchy. Experienced designers who readily visualize the hardware implied by Verilog statements often use more of a behavioral style for brevity. Beginning designers who don't want to think about the hardware implied by Verilog statements also often use a behavioral style and end up writing code that implies wasteful hardware or that won't synthesize.

Instantiating vs. Calling
In a typical programming language, a function *calls* another function. The second function puts information into a stack frame, performs its task, and then cleans up the stack frame and vanishes. In contrast, a hardware description language *instantiates* modules, that is, creates an instance of the module that permanently exists within the design, as opposed to function calls, which only exist while they are being called. The `mux4` in Code Example 4.5 can be built from three `mux2` modules, which exist on a chip whether or not the inputs are changing.

default value. The internal signals `d01` and `d23` must be declared in the `mux4` module as well. The alternate representations are commented out. The second 4:1 multiplexer representation uses an `assign` statement, the third uses an `always` block with a `case` statement, and the fourth uses an `always` block with `if`/`else` statements. These approaches are discussed in Section 4.2.2. This example illustrates comments, which can proceed from `//` to the end of the line or from `/*` until the next `*/`. It also illustrates using `[]` to select bits from within a larger bus.

4.2.2 Combinational Logic

Recall that the output of *combinational logic* depends only on the current inputs. SystemVerilog describes combinational logic with `assign` statements or `always_comb` blocks.

4.2.2.1 Combinational Logic Using `assign` Statements

Boolean equations are described using an `assign` statement. Table 4.1 lists the various Verilog operators and their order of precedence. The precedence follows the usual standard of AND before OR and * before +. Use parentheses to change this order or if you don't want your reader to have to remember the order of more obscure operations.

Multiple `assign` statements all operate in parallel, just like the hardware they imply. Hence, the order of `assign` statements in a module does not affect operation. This is a major difference between HDLs and conventional sequential programming languages.

Boolean operators `&`, `^`, and `|` can work in either bitwise or reduction form. *Bitwise* form takes two *N*-bit inputs and produces an *N*-bit output. For example, `y = a & b;` performs the bitwise AND of `a` and `b` and puts the result in `y`, where `a`, `b`, and `y` are all *N*-bit signals. If `a` = 1010 and `b` = 0111, then `y` = 0010. The *reduction* form takes an *N*-bit input and produces a single-bit output; for example, `w = &a;` performs the AND of all bits in the *N*-bit signal `a` and puts the result in the 1-bit signal `w`. If `a` = 1010, `w` = 0 because not all bits of `a` are TRUE. The equality comparator in Code Example 4.6 checks if `a` is equal to `b` by bitwise XNORing each of the four columns of bits to see if each column is equal, then using a reduction AND to see if all four columns are equal, as shown in Fig. 4.4.

Code Example 4.6 `eqcmp`

```
module eqcmp(input  logic [3:0] a, b,
             output logic       eq);

  assign eq = &(a ~^ b); // Equivalent to assign eq = (a == b)
endmodule
```

Table 4.1 SystemVerilog operators.

Precedence	Operator	Meaning
Highest	`~`	NOT
	`* / %`	Multiply, divide, modulo
	`+ -`	Add, subtract
	`<< >>`	Logical left/right shift
	`<<< >>>`	Arithmetic left/right shift
	`< <= > >=`	Relative comparison
	`== !=`	Equality comparison
	`& ~&`	AND, NAND
	`^ ~^`	XOR, XNOR
	`| ~|`	OR, NOR
Lowest	`?:`	Conditional (ternary)

Verilog also has &&, || (logical AND, OR) operators that take 2 multibit inputs and return a 1-bit output. As in C, they interpret a multibit input as FALSE if zero and TRUE if nonzero. Hence, 011 && 111 = 1 because 011 and 111 are both nonzero. Some Verilog coding styles use logical operators for control logic and bitwise operators for datapaths. Our style uses bitwise operators exclusively and never attempts to interpret a multibit input as TRUE or FALSE.

Arithmetic operations can be performed with various power, performance, and area (PPA) trade-offs. The hardware for an *N*-bit add or subtract (`+` or `-`) operation depends on the speed requirements but involves at least *N* full adders (see [Weste12] for an extensive discussion of alternatives). Relative comparison ($<$, $\geq$, and so on) is almost as expensive as subtraction. As seen earlier, equality comparison is cheaper, requiring only *N* XNOR gates and one *N*-input AND. Constant shifts are free (just wires), but variable shifts require many multiplexers and are fairly costly (see Section 7.1.3.3). Multiplication by a power of two is just a shift, but general multiplication is quite costly, involving an array of partial product generators and an array of adders to sum the partial products as discussed in Section 15.1.3. Division and modulus by powers of two are equivalent to shifts or truncation and are best expressed that way. Division and modulo by other amounts are expensive and are avoided in a digital system when feasible; see Section 15.1.4 for CMOS division circuits.

Cost is proportional to chip area.

4.2.2.2 Combinational Logic Using `always_comb` Statements

Combinational logic can also be described with `if` statements (convenient for next state logic in FSMs) and `case` statements (convenient for things resembling truth tables). These statements can only be placed inside an `always` block. In classic Verilog, `always` blocks require a sensitivity list that is notoriously prone to accidentally imply sequential logic instead of combinational logic or cause mismatches between simulation and

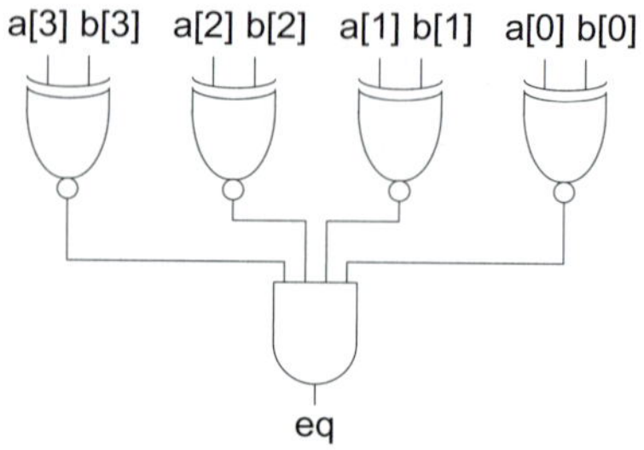

Fig. 4.4 Equality comparator.

synthesized logic, so SystemVerilog introduced the `always_comb` statement that will warn you if you make a mistake.

Code Example 4.5 showed a multiplexer optionally implemented with `case` and `if` statements. These are more verbose than `assign` statements or a structural design. Code Example 4.16 shows a better application of `if` statements to FSM next state logic.

A seven-segment display is shown in Fig. 4.5. For example, to display the digit 3, the display should light up segments a, b, c, d, and g, as shown. Code Example 4.7 shows a seven-segment-display decoder that receives a digit and determines which segments to light up. It is implemented with a `case` statement, which is far easier than grinding out seven Boolean equations by hand and coding `assign` statements for each output. Note that the `default` statement is necessary so that the outputs are well-defined even when an illegal input such as 1111 is applied; otherwise, the logic would remember the old output, which implies a latch (called *latch inference*) and is not combinational logic or synchronous sequential logic.

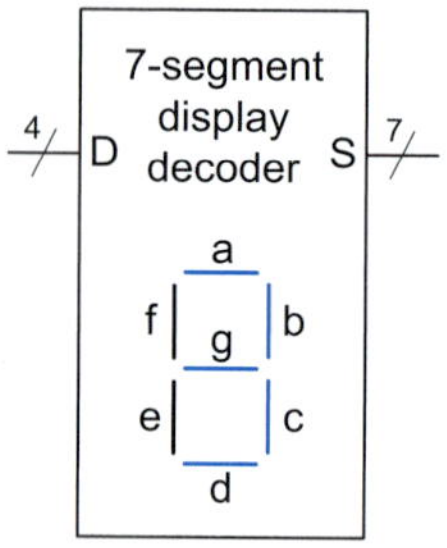

Fig. 4.5 Seven-segment display.

Placing underscores (_) in constants helps readability; they are ignored by simulation and synthesis tools.

Code Example 4.7 SEVEN-SEGMENT DISPLAY DECODER

```
module sevenseg(input  logic [3:0] digit,
                output logic [6:0] segments);

  always_comb
    case(digit)
                                // abc_defg
      0:       segments = 7'b111_1110;
      1:       segments = 7'b011_0000;
      2:       segments = 7'b110_1101;
      3:       segments = 7'b111_1001;
      4:       segments = 7'b011_0011;
      5:       segments = 7'b101_1011;
      6:       segments = 7'b101_1111;
      7:       segments = 7'b111_0000;
      8:       segments = 7'b111_1111;
      9:       segments = 7'b111_0011;
      default: segments = 7'b000_0000;
    endcase
endmodule
```

Code Example 4.8 implements a priority circuit using a casez statement, which is like case but allows don't cares (?) in the inputs. The priority circuit grants priority (outputs 1) to the most-significant input that requests priority (inputs 1). The comments show a second equivalent implementation with if/else statements. Both styles are comparably easy to read and write and imply the same hardware. Fig. 4.6 shows the priority circuit schematic.

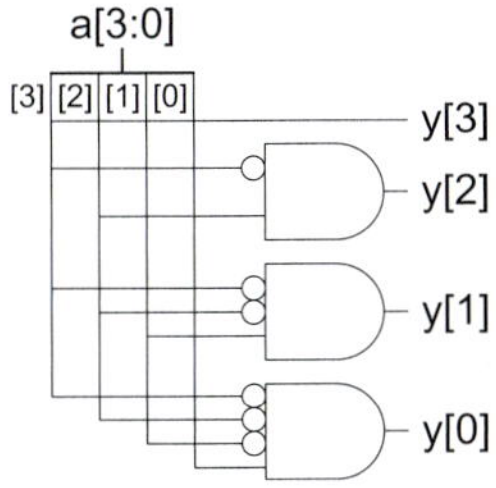

Fig. 4.6 Priority circuit.

Code Example 4.8 PRIORITY CIRCUIT

```
module priorityckt(input  logic [3:0] a,
                   output logic [3:0] y);

  // casez implementation
  always_comb
    casez(a)
      4'b1???: y = 4'b1000;
      4'b01??: y = 4'b0100;
      4'b001?: y = 4'b0010;
      4'b0001: y = 4'b0001;
      default: y = 4'b0000;
    endcase

  /* if implementation:
  always_comb
    if      (a[3]) y = 4'b1000;
    else if (a[2]) y = 4'b0100;
    else if (a[1]) y = 4'b0010;
    else if (a[0]) y = 4'b0001;
    else           y = 4'b0000; */
endmodule
```

4.2.2.3 Xs and Zs, Oh My

SystemVerilog logic signals have four states: 0, 1, x, and z. x indicates unknown, and z indicates *floating* (high impedance).

Unknown (x) values have several possible causes. A logic signal with no driver or multiple conflicting drivers has an unknown value. The output of a sequential element such as a flip-flop or memory cell is unknown until a value is written. If a gate receives an x or z input, it will produce an x output when it can't determine the correct output value. For example, 1'b1 & 1'bz = x. But 1'b0 & 1'bz = 0 because anything AND 0 is 0.

The presence of x in simulation is generally an indication of a logic problem in a design and should be tracked down and fixed. An exception is that memories are generally initialized to x. However, this is a problem when uninitialized memory is read, and this x percolates into logic that affects the state of the system. For example, if Wally reads an uninitialized

The Questa simulator, which is discussed further in Chapter 5, displays x as a red line and z as a blue line.

memory address, the x values can propagate through the hardware page table walker and hang the simulation.

SystemVerilog has another data type called bit that only represents 0 and 1. When logic is converted to bit, x and z map to 0. As shown in Section 4.2.3.3, Wally uses bit instead of logic in random access memory (RAM) to prevent uninitialized RAM from corrupting the system and to cut simulation memory usage (because bit takes half as much space as logic).

Some designers have tried to cleverly use x as a mechanism to indicate problems, such as an illegal instruction. Unfortunately, simulations may be optimistic or pessimistic in a way that hides real problems or indicates phantom problems, so it's best not to rely on x to detect design errors [Sutherland13].

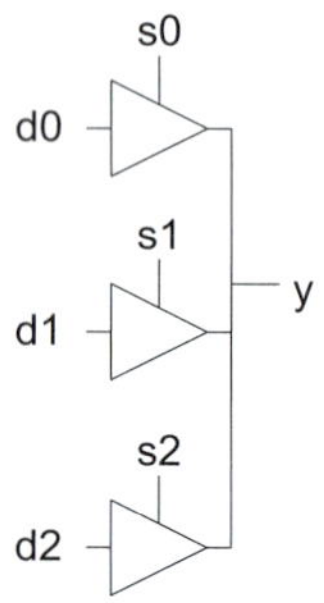

Fig. 4.7 Tristate distributed multiplexer.

A *tristate bus* has multiple drivers controlled such that at most one is active. If the bus has one active driver, it takes the value from that driver. If no driver is active, the bus floats. Tristate drivers are modeled with z. For example, Code Example 4.9 describes a distributed multiplexer. It defines a tristate buffer and then allows a bus y to be driven by d0, d1, or d2 based on s0, s1, or s2, as shown in Fig. 4.7. Each tristate output is floating (z) when not enabled. If exactly one of the tristates is enabled, it will overpower the other floating tristates and drive its value onto the bus. If multiple tristates attempt to drive the bus to different values, the bus will get x to indicate contention. If none of the tristates are enabled, the bus will be floating.

Note that y is declared tri rather than logic. logic signals can only have a single driver. Tristate busses have multiple drivers, so they should be declared as a tri net. Typically, exactly one driver on a net is active at a time, and the net takes on that value. Otherwise, the net floats (z). Also note that a tri output from a module can be used as a logic input to another module.

Note that 'z makes all the bits of y float, as discussed in Section 4.2.5.2.

Code Example 4.9 TRISTATE DISTRIBUTED MULTIPLEXER

```
module tristate #(parameter WIDTH=32) (
  input  logic [WIDTH-1:0] a,
  input  logic             en,
  output tri   [WIDTH-1:0] y);

  assign y = en ? a : 'z;
endmodule

module distributedmux(
  input  logic [3:0] d0, d1, d2,
  input  logic       s0, s1, s2,
  output tri   [3:0] y);

  tristate #(4) t0(d0, s0, y);
  tristate #(4) t1(d1, s1, y);
  tristate #(4) t2(d2, s2, y);
endmodule
```

Fig. 4.8 shows the waveforms from the distributed mux. d0 is held at 0000. d1 is held at 0101. d2 is unknown. Following the waveform from left to right: when none of the s signals are asserted, the output y floats;

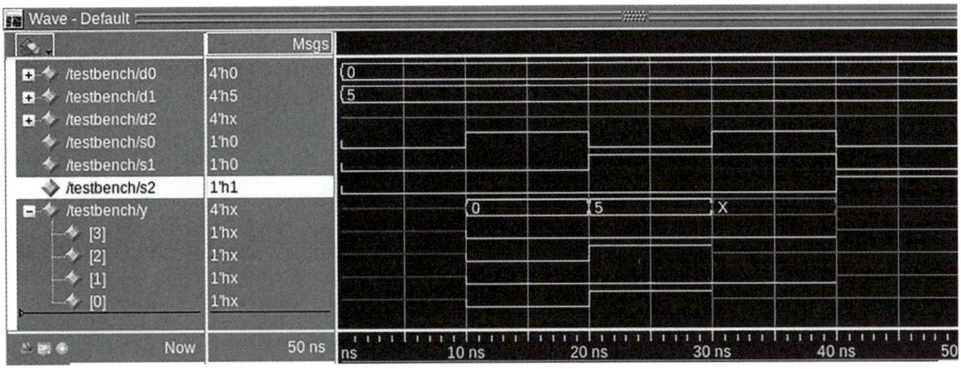

Fig. 4.8 Distributed multiplexer waveforms.

when `s0` is asserted, `y` is 0000; when `s1` is asserted, `y` is 0101; when `s0` and `s1` are both asserted, `y` is 0x0x because bits 0 and 2 are simultaneously driven high and low by tristates in contention; when `s2` is asserted, `y` is unknown because `d2` is unknown.

This coding style is not recommended within a chip. Tristate busses are slow and can have contention if one driver turns ON before another turns OFF. Many cell libraries don't even have tristate cells. Instead, tristate drivers are more appropriate for I/O pins that must access an off-chip shared bus or for shared bidirectional pins. Code Example 4.10 shows how to use tristate drivers for bidirectional general-purpose I/O, and Fig. 4.9 shows the synthesized circuit. Observe that the bidirectional pin is declared `inout` and must be a `tri` net because it could have multiple drivers.

Code Example 4.10 BIDIRECTIONAL I/O

```
module gpio #(parameter WIDTH=16) (
  input  logic [WIDTH-1:0] GPIOOutVal, GPIOEn,
  output logic [WIDTH-1:0] GPIOInVal,
  inout  tri   [WIDTH-1:0] GPIOPin);

  assign GPIOInVal = GPIOPin;
  tristate #(1) ts[WIDTH-1:0](GPIOOutVal, GPIOEn, GPIOPin);
endmodule
```

The tristates are defined as an array of `WIDTH` 1-bit modules because each bit needs an independent enable. Arrays of modules are described further in Section 4.2.5.5.

4.2.3 Sequential Logic

The output of *sequential logic* depends on previous as well as current inputs. Common forms of sequential logic include flip-flops, finite state machines, counters, shift registers, and memories. We are primarily interested

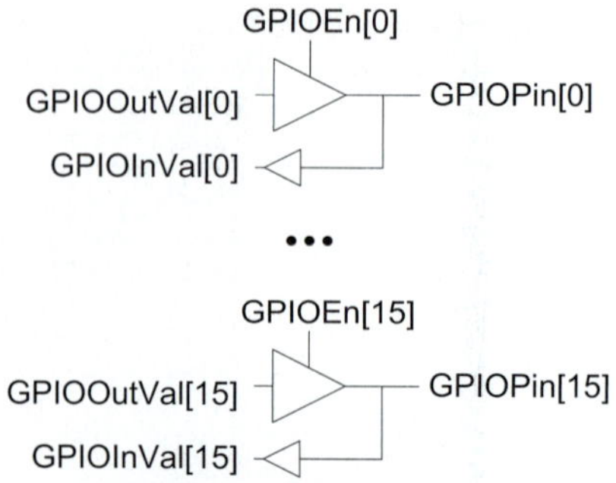

Fig. 4.9 **Bidirectional I/O.**

in *synchronous sequential logic* in which all outputs change on the edge of a common clock. SystemVerilog uses an `always_ff` block to describe flip-flops.

4.2.3.1 Flip-Flops

The SystemVerilog idiom for an ordinary *flip-flop* (also called a *D flip-flop*, *flop*, or *positive edge-triggered flip-flop*) is given in Code Example 4.11. On the rising edge of the clock, a flip-flop copies the input to the output. The `always_ff` construct will perform the operation in the body when the condition inside the sensitivity list (in this case, `posedge clk`) is true. So `q` gets the value of `d` at the rising edge of the clock. The <= operator is called a *nonblocking assignment* and is often pronounced "gets." All nonblocking assignments happen in parallel. A group of one or more flip-flops sharing a common clock and a bus of inputs and outputs is called a *register* as shown in Fig. 4.10. This flop module is parameterized so that the number of flip-flops is configurable.

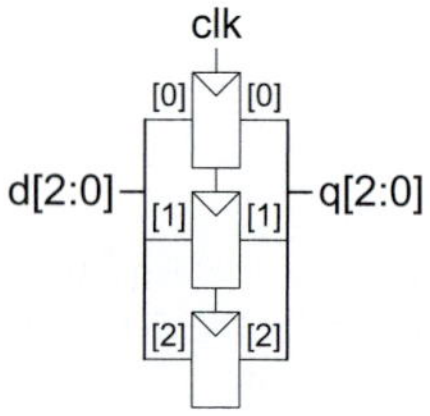

Fig. 4.10 **3-bit register.**

Code Example 4.11 `flop`

```
module flop #(parameter WIDTH = 3) (
  input  logic             clk,
  input  logic [WIDTH-1:0] d,
  output logic [WIDTH-1:0] q);

  always_ff @(posedge clk)
    q <= d;
endmodule
```

Combinational logic in an `always_comb` block is best described with blocking assignments (=) when one line depends on the previous. Nonblocking assignments also produce the correct result, but the simulation will be slower because the `always` block may need to be evaluated multiple times.

Sequential logic is best described with nonblocking assignments (<=) because <= captures the way real hardware operates in parallel rather than consecutively. For example, in the following code, on the rising edge of the clock, a is copied to b, and then the new value of b is copied to c, so both b and c receive the value of a, and the circuit implies only one flip-flop. If each = is replaced with <=, c will get the old value of b concurrently with b getting the old value of a, so the code implies two flip-flops. Fig. 4.11 shows the hardware implied by the blocking and nonblocking code.

(a) **Bad synchronizer using blocking assignments**

```
always_ff @(posedge clk) begin
  b = a;  // Blocking
  c = b;  // Blocking
end
```

(b) **Good synchronizer using nonblocking assignments**

```
always_ff @(posedge clk) begin
  b <= a;  // Non-blocking
  c <= b;  // Non-blocking
end
```

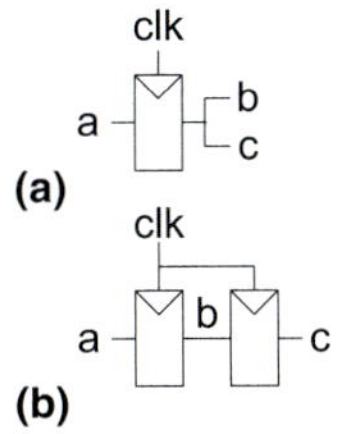

Fig. 4.11 (a) Blocking: one flip-flop and (b) Nonblocking: two flip-flops.

4.2.3.1.1 Resettable Flops

At startup, the output of a flip-flop is unpredictable. SystemVerilog represents this unknown state with an x. Most systems require that some or all flip-flops receive a reset at startup to place them in a known state. Code Example 4.12 shows the idioms for asynchronously and synchronously resettable flip-flops. Whereas *asynchronously resettable flip-flops* are reset as soon as the reset signal rises, *synchronously resettable flip-flops* only respond when the clock rises. Synchronous reset is also known as *clear*.

Asynchronous resets must be glitch-free to prevent a spurious reset.

Synchronous reset can be tricky if a system has gated clocks that might not all fire during reset. This could be solved by a more complex clock gater that always fires when reset is applied. Asynchronous reset avoids this issue and works even if some clocks come from external sources that aren't yet running. However, asynchronous reset requires substantially more transistors [Weste12], along with a complex distribution tree to get the reset to all flops at nearly the same time. Most FPGAs have asynchronous reset inputs built in to all flip-flops, so asynchronous

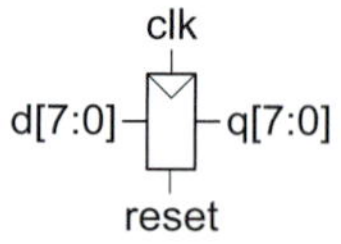

Fig. 4.12 **Resettable flip-flop.**

reset is preferred on FPGAs. ASIC designers reduce cost by using synchronous reset or eliminating reset entirely on some flip-flops. Synchronous clear is also used to flush a stage in a pipelined processor by deasserting the write control signals. Fig. 4.12 shows a flip-flop with a reset terminal.

Reset can be eliminated for registers that always get a value before they are used. For example, the register file does not need to be reset because a program should not depend on the values of registers before they are initialized. In simulation, registers without a reset start with an unknown value (x) until given a value. Occasionally this x propagates through other logic, leading to subtle simulation problems. x also can be a clue of a bug in simulation, and nonreset registers can obscure this clue. Wally resets all flip-flops to avoid these difficulties.

Code Example 4.12 ASYNCHRONOUSLY AND SYNCHRONOUSLY RESETTABLE FLIP-FLOPS

```
module flopr #(parameter WIDTH = 8) (
  input  logic             clk, reset,
  input  logic [WIDTH-1:0] d,
  output logic [WIDTH-1:0] q);

  always_ff @(posedge clk, posedge reset) // Asynchronous reset
    if (reset) q <= 0;
    else       q <= d;
endmodule

module flopc #(parameter WIDTH = 8) (
  input  logic             clk, reset,
  input  logic [WIDTH-1:0] d,
  output logic [WIDTH-1:0] q);

  always_ff @(posedge clk) // Synchronous reset
    if (reset) q <= 0;
    else       q <= d;
endmodule
```

4.2.3.1.2 Enabled Flops

Enabled flip-flops only copy the input to the output on the rising edge of the clock when the enable is true. The enable can be used to stall a pipeline stage in a processor or to save power. Code Example 4.13 shows the idiom for an enabled flip-flop. Code Example 4.14 puts everything together for a flip-flop with enable, reset, and clear. Fig. 4.13 shows flip-flops with these terminals.

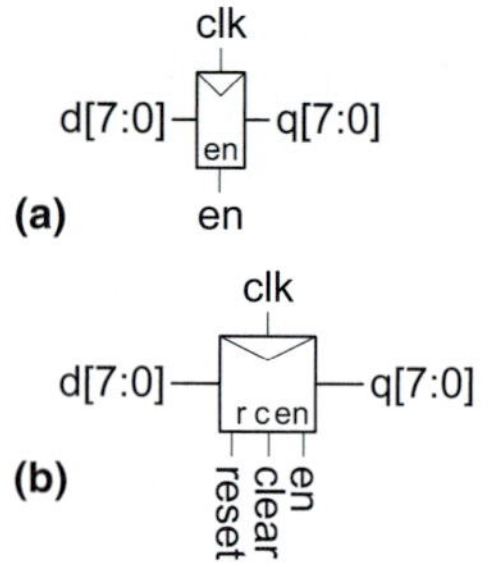

Fig. 4.13 **(a) Enabled flip-flop and (b) Enabled flip-flop with reset and clear.**

Code Example 4.13 ENABLED FLIP-FLOP

```
module flopen #(parameter WIDTH = 8) (
  input  logic             clk, en,
  input  logic [WIDTH-1:0] d,
  output logic [WIDTH-1:0] q);

  always_ff @(posedge clk)
    if (en) q <= d;
endmodule
```

Code Example 4.14 FLIP-FLOP WITH ENABLE, ASYNCHRONOUS RESET, SYNCHRONOUS CLEAR

```
module flopenrc #(parameter WIDTH = 8) (
  input  logic             clk, reset, clear, en,
  input  logic [WIDTH-1:0] d,
  output logic [WIDTH-1:0] q);

  always_ff @(posedge clk, posedge reset)
    if (reset)   q <= 0;
    else if (en)
      if (clear) q <= 0;
      else       q <= d;
endmodule
```

4.2.3.1.3 Latches

A *transparent* or *D latch* passes D through to Q as long as clock is asserted and retains Q otherwise. Code Example 4.15 shows the SystemVerilog idiom for a latch using the `always_latch` statement. Transparent latches are not edge triggered and thus generally should be avoided in a synchronous sequential design methodology, but they are needed in special-purpose circuits such as clock gaters. Fig. 4.14 shows the schematic symbol for a latch, which omits the triangle representing the edge-triggered clock.

Code Example 4.15 TRANSPARENT LATCH

```
module latch #(parameter WIDTH = 8) (
  input  logic             clk,
  input  logic [WIDTH-1:0] d,
  output logic [WIDTH-1:0] q);

  always_latch
    if (clk) q <= d;
endmodule
```

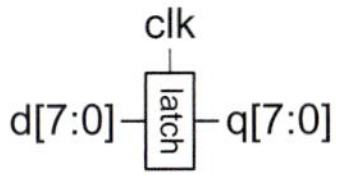

Fig. 4.14 Transparent latch.

4.2.3.2 Finite State Machines

Finite state machines (FSMs) are often used as controllers. They keep track of the state of a system or subsystem and calculate the next state depending on the current state and the inputs. FSMs are categorized as Moore or Mealy machines. In Moore machines, the outputs depend only on the current state. In Mealy machines, the outputs depend on the current state and inputs.

Fig. 4.15 shows generic block diagrams for both a Moore and Mealy FSM, each with M bits of input, N bits of output, and k bits of state. Both FSMs have a k-bit register to hold the current state, and the combinational next state logic computes the k bits of next state (inputs to the register)

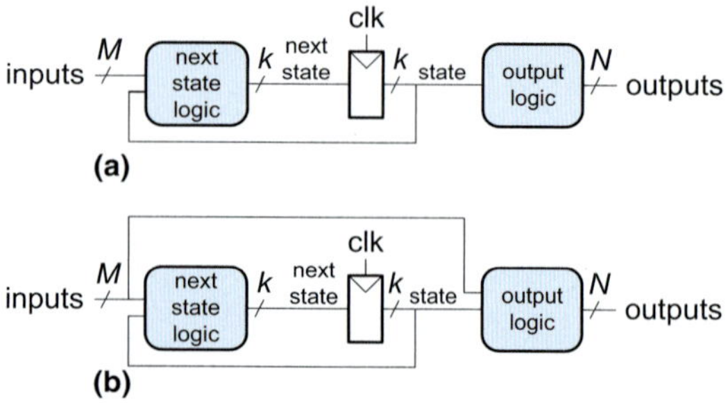

Fig. 4.15 FSM block diagrams: (a) Moore and (b) Mealy.

based on the inputs and current state. The output logic is combinational logic that computes the outputs based on the current state (for a Moore FSM) or the current state and inputs (for a Mealy FSM). SystemVerilog describes the FSM using the same format, with a register and two blocks of combinational logic. The next state logic is often described using a `case` statement that depends on the current state. Verilog's `enum` keyword allows us to give names instead of numbers to the states, making the FSM easier to read during design and simulation.

For example, Fig. 4.16 shows a conceptual Moore state transition diagram for a two-bit branch predictor FSM. It has an output `p` (predict taken), which is 1 when the branch should be predicted as taken, and an input `t` (taken), which indicates when the branch actually was taken. Reset and clock are not considered inputs for the next state logic. The two bits of state give four states: Strongly Not Taken (SNT), Weakly Not Taken (WNT), Weakly Taken (WT), and Strongly Taken (ST). The FSM resets into the Weakly Taken state and increases or decreases its prediction based on whether the branch was actually taken. Code Example 4.16 shows the SystemVerilog code for the FSM. The four states are enumerated. Because the enumeration encoding is not stated explicitly, it is in natural order, with SNT = 00 and ST being 11. The resettable state register is described with an `always_ff @(posedge clk)` statement. The next state logic is described with a `case` statement depending on the current state, containing `if` statements depending on the input(s). The output logic depends only on the current state and is conveniently described with an `assign` statement.

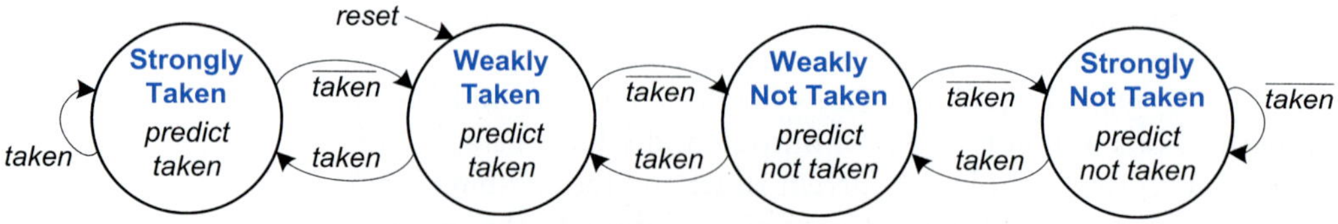

Fig. 4.16 Two-bit branch predictor state transition diagram.

Fig. 4.17 shows a schematic of the FSM. Logic synthesis automatically derives the Boolean expressions and maps them onto appropriate gates, saving the engineer from that time-consuming and error-prone task.

Code Example 4.16 TWO-BIT BRANCH PREDICTOR FSM

```
module twobitbranchpredictor(input  logic clk, reset,
                             input  logic t,
                             output logic p);

  typedef enum logic [1:0] {SNT, WNT, WT, ST} statetype;

  statetype state, nextstate;

  // State register
  always_ff @(posedge clk, posedge reset)
    if (reset) state <= WT;
    else       state <= nextstate;

  // Next state logic
  always_comb
    case (state)
      SNT: if (t) nextstate = WNT;
           else   nextstate = SNT;
      WNT: if (t) nextstate = WT;
           else   nextstate = SNT;
      WT:  if (t) nextstate = ST;
           else   nextstate = WNT;
      ST:  if (t) nextstate = ST;
           else   nextstate = WT;
      default:    nextstate = WT;
    endcase

  // Output logic
  assign p = (state == WT) | (state == ST);
endmodule
```

In Code Example 4.16, `typedef enum` defines a new *enumerated type* called `statetype`, which has an underlying type of `logic [1:0]`. This new type enables the user to use enumerated values, in this case `SNT`, `WNT`, `WT`, and `ST`, instead of binary encodings for the states (i.e., 00, 01, 10, and 11). This increases readability and ease of coding.

Code Example 4.17 shows a more complex FSM illustrating how to merge the state register with the next state logic, how to incorporate a counter into the FSM, how to use `begin`/`end` to take multiple actions, and how Verilog keeps the old value of signals when they aren't changed

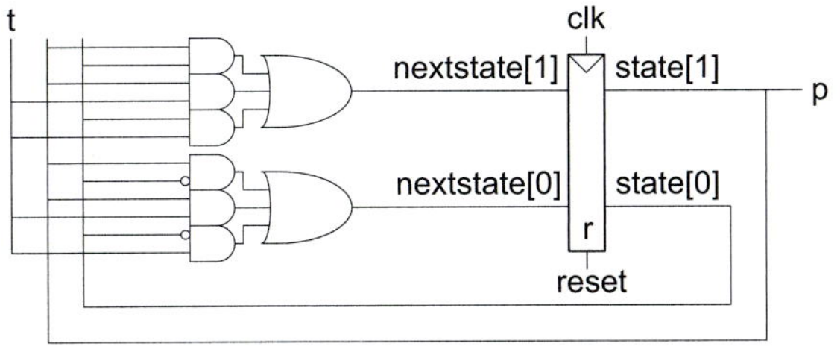

Fig. 4.17 Two-bit branch predictor logic.

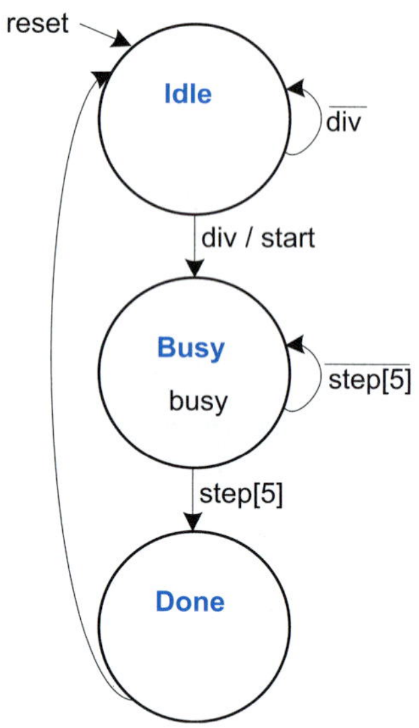

Fig. 4.18 Divider FSM.

in sequential logic. As shown in Fig. 4.18, the FSM produces `start` and `busy` signals for a divider that takes 32 clock cycles. When the divider is Idle and the `div` input is asserted, the FSM asserts `start` for one cycle and enters the Busy state. The `busy` signal is held high for 32 cycles, and then the FSM enters the Done state for one cycle before returning to Idle. The next state logic is implicit inside the `always_ff`. If the state is not changed, it keeps its old value. This is only legal in an `always_ff` because keeping an old value implies sequential logic rather than combinational logic. In some conditions, both the `state` and `step` signals are updated. Verilog provides the `begin/end` construct to contain multiple statements within the body of a block. The divider controller could be viewed as two interacting FSMs, one to track the state and the other to count steps. However, the two are merged here for brevity. Fig. 4.19 shows the hardware implied by this SystemVerilog. Avoid coding FSMs like this until you are experienced enough to be able to visualize the hardware you are implying.

Code Example 4.17 DIVIDER CONTROL

```
module dividercont(input  logic clk, reset,
                   input  logic div,
                   output logic start, busy);

  typedef enum logic [1:0] {Idle, Busy, Done} statetype;
  statetype state;
  logic [5:0] step;

  always_ff @(posedge clk)
    if (reset) state <= Idle;
    else case (state)
      Idle: if (div) begin
              state <= Busy;
              step  <= 1;
            end
      Busy: begin
              if (step[5]) state <= Done;
              step <= step + 1;
            end
      Done: state <= Idle;
    endcase

  assign start = div & (state == Idle);
  assign busy = (state == Busy);
endmodule
```

A group of statements within `begin/end` delimiters is known as a compound statement. Using `begin/end` is analogous to using curly brackets { } in C and Java.

In Fig. 4.18, checking `step[5]` in the Busy state is the same as checking (`step == 6'd32`), but it may produce less logic because of its 1-bit comparison instead of the latter's 6-bit comparison.

4.2.3.3 Memories

Memories are organized as arrays of 2^M-words × N-bits. The memory receives an address indicating which word to read or write. Digital systems

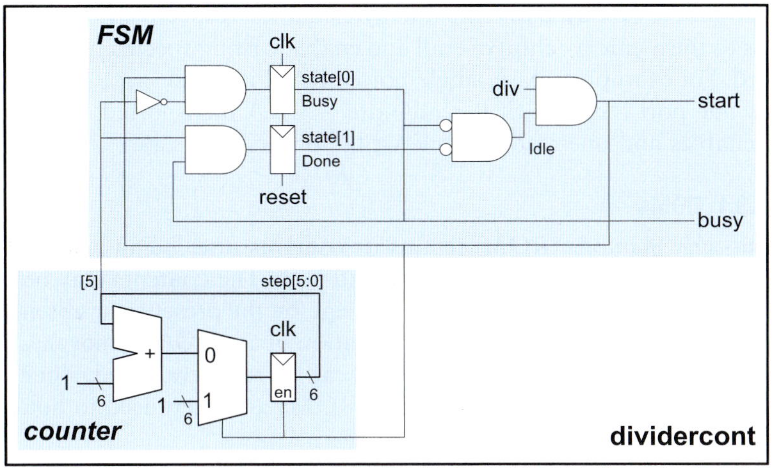

Fig. 4.19 Divider control schematic.

have many types of memories, and the designer must select an appropriate memory with the desired number of read and write ports, capacity ($2^M \times N$), and timing. Designing a custom memory is tricky, takes time, and introduces risk of error, so the designer is often constrained to pick memories from a limited library of options.

A *read port* consists of an address input `a` and a read data output `rd`. Memory reads can be either *asynchronous* or *synchronous*, as shown in Fig. 4.20. With asynchronous (also called *combinational*) read, the address stabilizes at some point, and the read data becomes valid after a propagation delay. With synchronous (also called *edge-triggered*) read, the address stabilizes at least some setup time before the rising edge of the clock. The read data arrives some clock-to-Q propagation delay after the clock edge. Memory writes are always synchronous because they need a timing edge indicating when to write (sometime after the address and write data are stable). We use synchronous memories in this textbook (starting in Section 7.2.1) because, for all but very small memories, they are typically smaller and faster than memories with asynchronous reads.

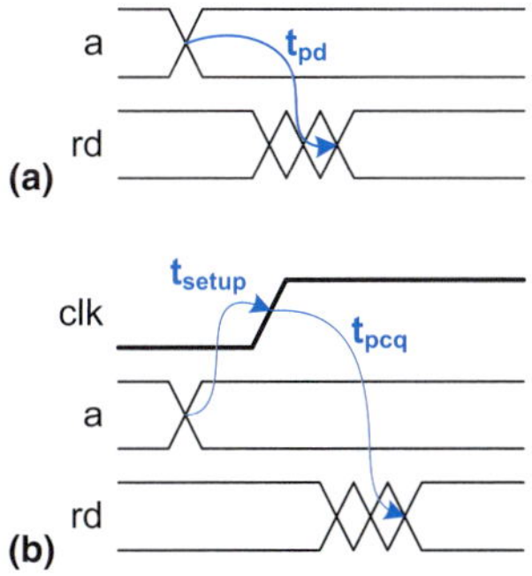

Fig. 4.20 Memory read timing: (a) asynchronous and (b) synchronous.

A *write port* consists of an address input `a` specifying where to write, a write data input `wd` specifying what to write, a write enable input `we` specifying whether to write, and a clock `clk` specifying when to write. Thus, write ports are inherently synchronous.

A *read/write port* has the same connections as a write port but also has a read data output. It writes when `we` is asserted and always reads.

Two types of memory exist, read-only memory (ROM) and RAM, which are discussed next. This book names memories based on the number of ports and how many can be used for read, write, or read/write.

e and b suffixes are used for chip and byte enables, respectively, that enable access to the memory (chip) overall and enable writing specific bytes within a word. For example, a basic single-ported RAM is called ram1p1rw because it has one port (1p), which is read/write (1rw). A dual-ported RAM with byte enables and one read and one write port is called ram2p1r1wb.

4.2.3.3.1 ROMs

A *read-only memory* (*ROM*) retains its contents after being powered off and back on. It typically has a single read port. The contents may be programmed into the ROM at design time (e.g., by the presence or absence of transistors in an ASIC or by the configuration of an FPGA memory). Other ROMs are built from flash memory and can be reprogrammed, albeit in a slow and cumbersome way. Such flash ROMs are often used to hold upgradable firmware.

Code Example 4.18 shows the idiom for a 2^M-word × N-bit ROM with one asynchronous read port, and Fig. 4.21 shows the symbol for the ROM. This module has two parameters: M and N. mem is declared as a 2-dimensional array of 2^M entries, each of size N. As discussed in Section 4.2.2.3, mem is declared as bit rather than logic, so uninitialized entries default to 0 rather than x. This reduces simulator memory requirements and avoids subtle bugs with uninitialized memory values corrupting a simulation. The initial $readmemh statement populates the memory array from a file. For example, Code Example 4.19 shows the romfile.dat for an 8-word × 16-bit ROM. It has only 6 lines, so the last two entries of the ROM default to 0000. Each line has 16 bits specified with four hexadecimal nibbles. For example, entry 2 is 0xA5A5 (binary 1010_0101_1010_0101). If the system lacks a ROM generator, logic synthesis might map this module onto a network of combinational logic producing the specified outputs.

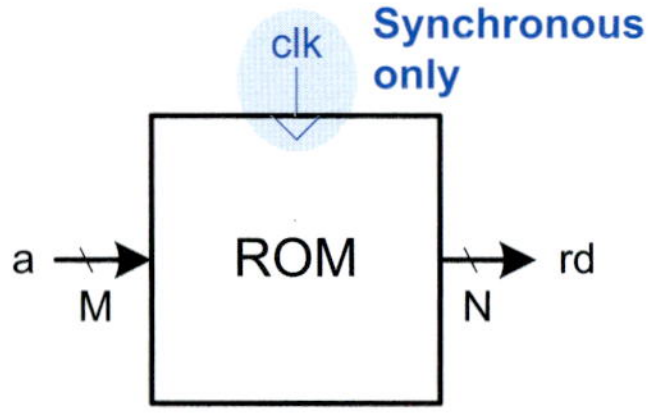

Fig. 4.21 ROM (rom1p1r) with 1 read port.

Note the exponentiation operator ** in the ROM example: 2**M = 2^M (=8 if M = 3). SystemVerilog also has a handy $clog2 (ceiling of $\log_2$) operator that is handy for calculating the number of bits needed to address a memory. For example, a memory containing 32 words needs $clog2(32) = 5 address bits.

Code Example 4.18 ASYNCHRONOUS 2^M X *N*-BIT ROM

```
module rom1p1r #(parameter M=3, N=16) (
  input  logic [M-1:0] a,
  output logic [N-1:0] rd);

  bit [N-1:0] mem[2**M];

  initial $readmemh("romfile.dat", mem);

  assign rd = mem[a]; // Asynchronous read
endmodule
```

Code Example 4.19 SAMPLE **romfile.dat**

```
0000
FFFF
A5A5
0001
DEAD
BEEF
```

FPGAs often build larger ROMs from synchronous block RAMs and thus expect synchronous read. Synchronous read is similar to the combinational read shown in Code Example 4.18 but would be coded as follows:

```
always_ff @(posedge clk) rd <= mem[a];
```

4.2.3.3.2 RAMs

A *random access memory* (*RAM*) is a historic and confusing term for a memory that can be read and written, generally with a delay that is short and independent of the address being accessed. Midsized memories are generally implemented with *static RAM* (*SRAM*), which retains its value as long as power is applied. SRAM is built from 6-transistor memory cells shown in Fig. 4.22. To write the cell, the bitline and its complement are driven to opposite values, and the wordline is pulsed to overpower the cross-coupled inverters. To read the cell, bitline and its complement are precharged high, and then wordline is pulsed to pull down one of the bitlines. The cell requires periphery circuits to precharge and sense bitlines, as well as a decoder. Thus, SRAM arrays typically come as a library component with a specified capacity, shape, and timing characteristics. They use a clock to sequence the precharge and sensing, so they read synchronously. Very small memories can be implemented with flip-flops or latches when the overhead of the periphery circuits is excessive or no suitable library component is available. Such memories generally read combinationally through a multiplexer tree.

Code Example 4.20 shows the SystemVerilog idiom for a 2^M-word × N-bit single-ported asynchronous RAM with one port capable of reading

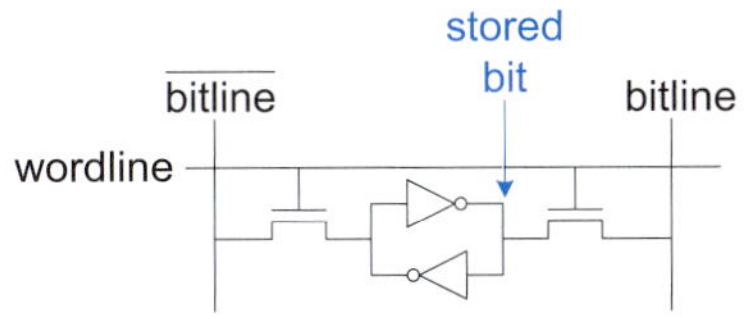

Fig. 4.22 **6T SRAM bit cell.**

and writing (`ram1p1rw_async`). The RAM reads asynchronously onto the read data port (`rd`) some propagation delay after the address (`a`) stabilizes, as shown in Fig. 4.20(a). If the write enable (`we`) is asserted, it also synchronously writes the write data (`wd`) to the same address (`a`) on the rising edge of the clock.

Code Example 4.20 ASYNCHRONOUS RAM

```
module ram1p1rw_async #(parameter M = 6, N = 32) (
  input  logic        clk,
  input  logic        we,
  input  logic [M-1:0] a,
  input  logic [N-1:0] wd,
  output logic [N-1:0] rd);

  bit [N-1:0] mem[2**M];

  assign rd = mem[a];
  always_ff @(posedge clk)
    if (we) mem[a] <= wd;
endmodule
```

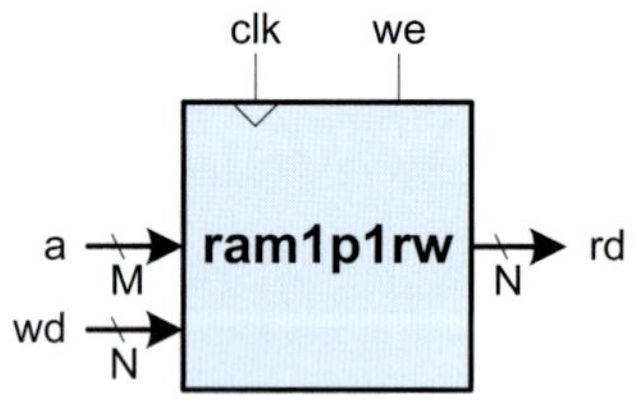

Fig. 4.23 **RAM with 1 read/write port.**

Synchronous read is again coded:

```
always_ff @(posedge clk) rd <= mem[a];
```

Fig. 4.23 shows a symbol for a RAM. The same symbol is used for both asynchronously and synchronously readable RAM so the designer must pay close attention to which form is used.

4.2.3.3.3 Byte Write Enables

Systems often need to write only some of the bytes within a memory word. For example, a cache memory array might have 128 words of 256 bits each. A write to the cache might need to update 32, 16, or 8 bits within the word. A brute force solution is to read the entire word, modify certain bytes using multiplexers, and write the word back. This may take an extra clock cycle. If a memory with a per-byte write enable is available, a more efficient solution is to only write the necessary bytes. Code Example 4.21 shows how to describe a RAM with one write enable signal per byte rather than a single one for the entire word. The design assumes `N` is a multiple of 8. For example, if `N` = 32, the memory has 32/8 = 4 byte enables `we[3:0]` controlling which bytes (bits [31:24], [23:16], [15:8], or [7:0]) are written of the 32-bit word at address a.

Code Example 4.21 RAM WITH BYTE ENABLES

```
module ram1p1rwb #(parameter M = 6, N = 32) (
  input  logic              clk,
  input  logic [N/8-1:0] we,
  input  logic [M-1:0]     a,
  input  logic [N-1:0]     wd,
  output logic [N-1:0]     rd);

  bit [N-1:0] mem[2**M];
  genvar i;

  always_ff @(posedge clk)
    rd <= mem[a];

  for (i = 0; i < N/8; i++) // Byte writes
    always_ff @(posedge clk)
      if (we[i]) mem[a][i*8 +: 8] <= wd[i*8 +: 8];
endmodule
```

Code Example 4.21 introduces two new pieces of Verilog syntax. The for loop is called a *generate loop* and implies multiple pieces of hardware as discussed in Section 4.2.5.4. In this example, it creates the byte-enabled columns of the RAM. The SystemVerilog +: operator is called an *indexed part-select*. It selects a specified number of bits starting at a specified offset. For example, wd[i*8 +: 8] selects 8 bits of wd starting at column i*8. When i = 2, it selects wd[23:16]. It is equivalent to wd[i*8 + 7 : i*8], but SystemVerilog synthesis prefers the indexed part-select notation.

4.2.3.3.4 Chip Enable

RAMs often provide a chip enable (ce) to save energy when the RAM is inactive. When enabled, the RAM behaves normally. When disabled, the RAM ignores all other inputs and retains the last read value rd. Code Example 4.22 shows SystemVerilog code for a RAM with a single read/write port with a chip enable and byte enables for writes.

Code Example 4.22 SYNCHRONOUS RAM WITH BYTE AND CHIP ENABLES

```
module ram1p1rwbe #(parameter M = 6, N = 32) (
  input  logic           clk,
  input  logic           ce,
  input  logic [N/8-1:0] we,
  input  logic [M-1:0]   a,
  input  logic [N-1:0]   wd,
  output logic [N-1:0]   rd);

  bit [N-1:0] mem[2**M];
  genvar i;

  always_ff @(posedge clk)
    if (ce) rd <= mem[a];

  for(i = 0; i < N/8; i++) // Byte writes
    always_ff @(posedge clk)
      if (ce & we[i]) mem[a][i*8 +: 8] <= wd[i*8 +: 8];
endmodule
```

4.2.3.3.5 Multiported RAMs

A multiported RAM has more than one read or write port. Code Example 4.23 shows the SystemVerilog idiom for a synchronous two-ported RAM with one read and one write port (`ram2p1r1w`). The port number is appended to each signal associated with it. So, port 1 is the read port (`a1`, `rd1`), and port 2 is the write port (`a2`, `wd2`, `we2`). Code Example 4.24 shows the idiom for a synchronous two-ported RAM, with each port capable of reading and writing (`ram2p2rw`). Code Example 4.25 shows the idiom for a three-ported *register file*, which is a small memory with two asynchronous read ports and a synchronous write port (`rf3p2r1w`). This example uses `logic` rather than `bit`, so a simulation can detect bad programs that use uninitialized registers. Fig. 4.24 shows symbol representations of each.

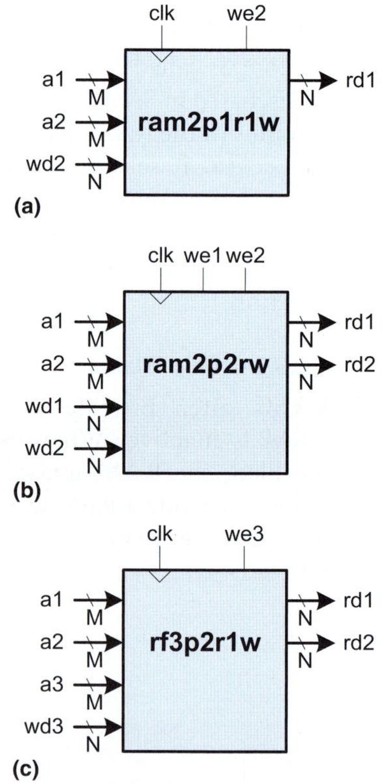

Fig. 4.24 (a) 2-port 1r1w RAM, (b) 2-port 2rw RAM, and (c) 3-port 2r1w register file.

Code Example 4.23 DUAL-PORT RAM WITH ONE READ AND ONE WRITE PORT

```
module ram2p1r1w #(parameter M = 6, N = 32) (
  input  logic         clk,
  input  logic         we2,
  input  logic [M-1:0] a1, a2,
  input  logic [N-1:0] wd2,
  output logic [N-1:0] rd1);

  bit [N-1:0] mem[2**M];

  always_ff @(posedge clk)  begin
    rd1 <= mem[a1];
    if (we2) mem[a2] <= wd2;
  end
endmodule
```

Code Example 4.24 DUAL-PORT RAM WITH TWO READ/WRITE PORTS

```
module ram2p2rw #(parameter M = 6, N = 32) (
  input  logic         clk,
  input  logic         we1, we2,
  input  logic [M-1:0] a1, a2,
  input  logic [N-1:0] wd1, wd2,
  output logic [N-1:0] rd1, rd2);

  bit [N-1:0] mem[2**M];

  always_ff @(posedge clk)  begin
    rd1 <= mem[a1];
    rd2 <= mem[a2];
    if (we1) mem[a1] <= wd1;
    if (we2) mem[a2] <= wd2;
  end
endmodule
```

Code Example 4.25 TRIPLE-PORTED REGISTER FILE WITH TWO READ AND ONE WRITE PORTS

```
module rf3p2r1w #(parameter M = 5, N = 32) (
  input  logic         clk,
  input  logic         we3,
  input  logic [M-1:0] a1, a2, a3,
  input  logic [N-1:0] wd3,
  output logic [N-1:0] rd1, rd2);

  logic [N-1:0] mem[2**M];  // Logic (not bit) to model uninitialized registers

  // Asynchronous read on ports 1, 2
  assign rd1 = mem[a1];
  assign rd2 = mem[a2];

  // Synchronous write on port 3
  always_ff @(posedge clk)
    if (we3) mem[a3] <= wd3;
endmodule
```

4.2.3.3.6 Library Considerations

ASIC vendor memory libraries vary widely in richness and configurability. Whereas a basic memory library has only single-ported RAMs, a better library may have RAMs with one read and one write port. For high speed, RAM arrays generally can't be larger than about 128 rows × 256 bits. However, larger RAMs can be constructed from multiple subarrays.

In an ASIC, the prior RAM idioms imply flip-flop-based memories. To target SRAM, an ASIC requires an explicit instantiation of a particular SRAM (see Section 6.5.3). When Wally is targeted to an FPGA (see Chapter 23), the RAM idioms are mapped to *block RAM* when possible.

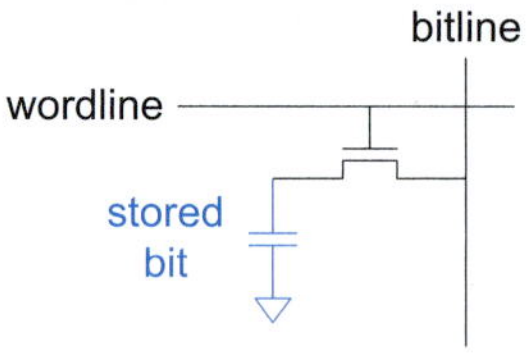

Fig. 4.25 1T DRAM cell.

Smaller memories are less area efficient because the decoder and column circuitry overhead is a greater proportion of the total area.

Custom designers can create a wide range of memories including many-ported register files and *content-addressable memories* (*CAMs*), but ASIC designers often have to synthesize these structures from flip-flops. Note that SRAM is about 10 times denser per bit than flop-based memory, and dynamic RAM (DRAM) is about 10 times denser than SRAM.

Very large memories generally use DRAM because of its density advantages. DRAM is built from a cell with one transistor and one capacitor as shown in Fig. 4.25. It requires complex periphery circuits and must be periodically read and refreshed before the charge on the capacitor leaks away. Commodity DRAM chips are more than an order of magnitude cheaper per bit than SRAM, but DRAM has longer latency, has much less bandwidth, and requires a controller to manage both refreshes and the high-speed off-chip interface. A double data rate (DDR) DRAM controller is tricky to design and would be the subject of its own textbook; most design teams purchase memory controllers as prevalidated IP blocks (intellectual property blocks). Some processes offer embedded DRAM on an SoC; this is less expensive per bit than SRAM but not as cheap as commodity DRAM.

4.2.4 Verilog Syntax

This section summarizes various other commonly used Verilog constructs, including numbers, bit swizzling, x and z values, delays, port connections, the preprocessor, and assertions.

4.2.4.1 Numbers

The format for declaring *integer literal constants* (numbers) is `N'Bvalue`, where `N` is the size in bits and `B` is a letter indicating the base, with `b` for binary, `d` for decimal, `h` for hexadecimal, and, less commonly used, `o` for octal. If the base is omitted, it defaults to decimal. Omitting the size leads to subtle gotchas and should generally be avoided [Sutherland06]. The number is padded with leading zeros to bring it to the specified size. Underscores are ignored and can be helpful to break long numbers into more readable chunks. Table 4.2 gives examples of representing constants.

An exception to the rule of always including the bit size for constants is when filling a signal with a repeated value of 0, 1, x, or z. For example, to fill the 5-bit `data` signal with all 1s (11111_2), use `assign data = '1;` To fill it with all 0s (00000), zs (zzzzz), or xs (xxxxx), use `assign data = '0`, `assign data = 'z`, or `assign data = 'x`, respectively. This syntax is particularly convenient when the bit size is parameterized.

Table 4.2 SystemVerilog numbers.

Number	Bits	Base	Decimal Value	Stored
3'b101	3	2	5	101
8'b1010_1011	8	2	171	10101011
8'b11	8	2	3	00000011
3'd6	3	10	6	110
8'hAB	8	16	171	10101011

4.2.4.2 Bit Swizzling

Bit swizzling is the process of selecting, replicating, and concatenating (joining together) signals and constants to form new signals. The `{}` operator concatenates bits. To replicate bits, use curly brackets around the number of times to replicate the bit; for example, `{5{a[2]}}` replicates bit 2 of signal `a`, `a[2]`, five times. Code Example 4.26 shows more examples of swizzling.

Code Example 4.26 BIT SWIZZLING

```
// Create 64-bit bus of all 0 or all 1 depending on the value of s
logic [63:0] s64;
assign s64 = {64{s}};

// Create 8-bit bus of all x or all 1s
logic [7:0] data0, data1;
assign data0 = 'x;      // data0 = xxxxxxxx
assign data1 = '1;      // data1 = 11111111

// Concatenate two 32-bit values to form a 64-bit value
logic [31:0] upper, lower;
logic [63:0] entire;
assign entire = {upper, lower};

// Form bus containing constants and parts of other busses
// y = a[2] a[2] b[2] b[1] b[0] a[1] a[0] 1 0 1
logic [2:0] a, b;
logic [9:0] y;
assign y = {{2{a[2]}}, b, a[1:0], 3'b101};

// Sign- and zero-extend signal c
// csignext = c[3] c[3] c[3] c[3] c[3:0]
// czeroext = 0    0    0    0    c[3:0]
logic [3:0] c;
logic [7:0] csignext, czeroext;
assign csignext = {{4{c[3]}}, c};
assign czeroext = {4'b0, c};

// Add two 32-bit numbers to produce a 32-bit sum and a 1-bit carry out
logic [31:0] a, b, sum;
logic        cout;
assign {cout, sum} = a + b;
```

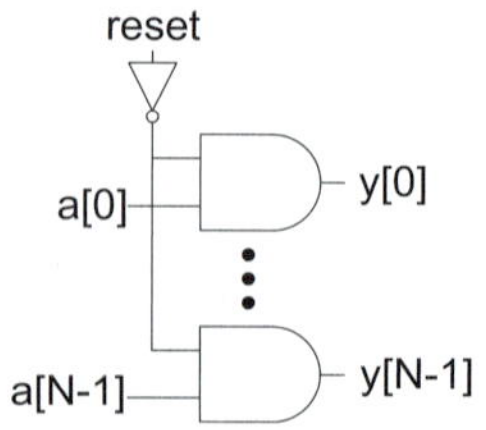

Fig. 4.26 Gating circuit.

4.2.4.3 One-Bit with Many-Bit Gating

Often many bits need to be *gated* by one bit. For example, we might want to force an N-bit bus `a` to all 0s during a reset condition. Verilog does not directly support combining one and many bits with a bitwise operator, but we can use either a conditional operator or swizzling as shown in Code Example 4.27. We prefer the conditional operator for brevity, although it is important to remember that it reduces to N AND gates rather than an N-bit 2:1 multiplexer. Both idioms produce the logic shown in Fig. 4.26.

Code Example 4.27 GATING

```
logic         reset;
logic [N-1:0] a, y1, y2;
assign y1 = reset ? '0 : a;   // Conditional operator: y1 = 000... upon reset
assign y2 = {N{~reset}} & a;  // Bit swizzling:        y2 = 000... upon reset
```

4.2.4.4 Delays

Verilog supports delays in simulation, indicated by #. Code Example 4.28 shows the single-ported RAM from Code Example 4.20 annotated with delays of 2 time units propagation delay on the `rd` output and 1 time unit of clock-to-q delay for writing the memory.

Code Example 4.28 RAM WITH DELAYS

```
module ram1p1rw_async_delay #(parameter M = 6, N = 32) (
  input  logic         clk,
  input  logic         we,
  input  logic [M-1:0] a,
  input  logic [N-1:0] wd,
  output logic [N-1:0] rd);

  bit [N-1:0] mem[2**M];

  assign #2 rd = mem[a];
  always_ff @(posedge clk)
    if (we) mem[a] <= #1 wd;
endmodule
```

Delays are necessary in testbenches to toggle signals, such as clocks, in the desired sequence, as will be illustrated in Code Example 5.1. Delays can be helpful (but not essential) for debugging to visualize the register outputs changing after the clock edge. Although Verilog has a `timescale` directive to indicate the value of each time unit, it's best to think of delays as just being relative and having arbitrary units. Delays are ignored during synthesis because the delay of a circuit depends on its physical implementation and varies with voltage, temperature, and manufacturing variations.

Delays are not necessary outside of testbenches. Some logic simulators, including Verilator, are significantly faster when all signals toggle on the same clock edge. The industry has mostly converged on a coding style with no # delays outside of testbenches, and Wally follows this style.

4.2.4.5 Port Connections

So far, we have connected ports by using the same order in the module instantiation as in the module declaration. This approach is succinct but can be difficult to read because you may have to look at both the instantiation and declaration to understand which signals are going to which ports. Moreover, if another designer reorders the ports in the module declaration, the instantiation might have subtle bugs because signals go to the wrong ports.

SystemVerilog has several other methods of specifying port connections illustrated in Code Example 4.29. The module describes a 4-bit register built from four 1-bit resettable flip-flops. `f0` is *connected by ordered list*. `f1` is *connected by name*. For example, the `clk` port of this flip-flop is connected to the `clk` input of `reg4`. The `d` port of this flip-flop is connected to the `x[1]` input of `reg4`. This solves the above issues with ordered lists but is substantially more typing. In this example, the signals are written one per line, just to be even less succinct. `f2` addresses this with *implicit port connections*. When the signal name is the same in the declaration and instantiation, it can just be listed once. For example, `.clk` connects the `clk` port of `f2` to the `clk` input of `reg4` without having to type it twice. The data terminal is still connected by name because it has a different name in the declaration (`d`) and the instantiation (`x[2]`). The ports could have been listed in any order and would still connect properly, although it is good practice to keep them in the same order. Finally, `f3` illustrates implicit port connections with `.*`, which automatically connects all of the ports that have the same name in the declaration and instantiation. In this case, `reset` and `clk` are connected using `.*`.

For brevity, we prefer ordered lists for simple modules such as flip-flops and muxes. We connect by name with implicit port connections for more complex modules. We find `.*` makes code difficult to read because it is harder to trace how signals propagate through the module hierarchy when they aren't listed in the instantiation, so we generally avoid it. However, `.*` can be helpful in a testbench or top-level module where signals are all connected by name, but the large number of names are tedious to type.

Code Example 4.29 PORT CONNECTIONS

```
module reg4(input  logic       clk,
            input  logic       reset,
            input  logic [3:0] x,
            output logic [3:0] y);

  flopr #(1) f0(clk, reset, x[0], y[0]);
  flopr #(1) f1(.clk(clk),
                .reset(reset),
                .d(x[1]),
                .q(y[1]));
  flopr #(1) f2(.clk, .reset, .d(x[2]), .q(y[2]));
  flopr #(1) f3(.d(x[3]), .q(y[3]), .*);
endmodule
```

Unused outputs can be discarded by omission, that is, by leaving that signal blank when connected by name, for example, `.outsigname()`.

4.2.4.6 Preprocessor

Like the C language, Verilog has a *preprocessor* that reads and modifies files before regular processing begins. Some of the common *compiler directives* it handles are `` `define `` and `` `include ``.

4.2.4.6.1 `` `define ``

`` `define `` is a *text macro substitution* directive. It is commonly used to define global constants and improve code readability, similar to `#define` in C and Java. For example, `` `define XLEN 64 `` replaces `` `XLEN `` with 64 everywhere it is found in the Verilog code. `XLEN` can then be easily changed to 32 by changing it only in the `` `define ``.

Code Example 4.30 shows an excerpt from a header file that uses `` `define `` to set `XLEN` to 32 and `PPN_BITS` to an expression that depends on `XLEN`, which evaluates to 22 in this case. Macros with a tick (`` ` ``) before their name are substituted into code verbatim during preprocessing. For example, everywhere `` `PPN_BITS `` appears in other code is replaced by `` (`XLEN==32 ? 22 : 42) ``, which in turn is replaced by `(32==32 ? 22 : 42)`. Macro expressions should have parentheses around them to avoid order-of-operations issues when substituted into code [Wilson10].

Code Example 4.30 `config.vh`

```
`define XLEN 32
`define PPN_BITS (`XLEN==32 ? 22 : 44)
...
```

The order of preprocessing and compiling files is not defined by SystemVerilog. If a macro is `` `define ``d in more than one place with different values, its behavior can be nondeterministic.

4.2.4.6.2 `` `include ``

`` `include `` inserts the entire contents of a file, similar to `#include` in C and Java. The included file is called a *header* file and is conventionally suffixed with `.vh` or `.svh`. A SystemVerilog file would incorporate `config.vh` above with the following:

```
`include "config.vh"
```

4.2.4.7 Assertions

SystemVerilog provides assertions to check properties during simulation. Assertions are of the form `assert (condition) else $error("message");` If the condition is not satisfied, the message is printed. Code Example 4.31 checks that `` `S_SUPPORTED `` is only 1 if `` `U_SUPPORTED `` is 1, too.

Substitute `$warning` for `$error` if the problem is less severe.

Code Example 4.31 ASSERTION

```
always_comb
  assert (`U_SUPPORTED | (`S_SUPPORTED == 0)) else $error ("S supported only if U is too");
```

4.2.5 Configurable Hardware

A complex SoC is usually configurable to support different needs. For example, the size of a branch predictor buffer should be configurable so that it can easily be changed without rewriting code. Wally takes such configuration to an extreme, making almost everything configurable, including XLEN (RV32 or RV64) and the supported extensions. This way, one code base can be used to support many configurations, from RV32E to RV64GC.

SystemVerilog has flexible ways to configure hardware based on compile-time constants, including disabling unused signals, setting the width of signals, and conditionally instantiating or generating multiple copies of a piece of hardware. We initially illustrate configurations with `` `define ``. This method does not work for SoCs with multiple cores that have different configurations because the order of defining the macros is nondeterministic. Therefore, we then introduce a more general method of passing parameters hierarchically.

4.2.5.1 Disabling Signals

Code Example 4.32 shows a macro used to disable a feature in a certain configuration. The file includes `config.vh`, which contains the configuration macros. When XLEN = 32, `WriteMSTATUSHM = CSRMWriteM & (CSRAdrM == MSTATUSH)`. If XLEN is 64, `WriteMSTATUSHM` is tied to 0 (and likely optimized away, along with any affected logic) because the `mstatush` register does not exist in this configuration.

Code Example 4.32 DISABLE FEATURE WITH A MACRO

```
`include "config.vh"
assign WriteMSTATUSHM = CSRMWriteM & (CSRAdrM == MSTATUSH) & (`XLEN == 32);
```

4.2.5.2 Values of Configurable Width

When a signal width is parameterized, the width of its value must be adjusted to match. Unfortunately, SystemVerilog does not accept expressions in bit widths; (`XLEN-1)'b0 is not recognized. The examples that follow show some ways to express values of configurable width using bit swizzling.

```
// Create XLEN-bit bus of all 0 or all 1 depending on s
logic [`XLEN-1:0] sx;
assign sx = {`XLEN{s}}; // Or: assign sx = s ? '1 : '0;

// Sign-extend Instr[31:20] to `XLEN bits
logic [`XLEN-1:0] extend;
assign extend = {{(`XLEN-12){Instr[31]}}, Instr[31:20]};

// Place LT in the bottom bit and WIDTH-1 0s in the upper bits
logic [`XLEN-1:0] SLT;
assign SLT = {{(`XLEN-1){1'b0}}, LT};
```

The number 0 always fills all bits and can be used safely in place of '0.

SystemVerilog also defines '0, '1, 'x, and 'z to place that value in each bit of the destination, thus filling the destination with all 0s, 1s, Xs, or Zs. Code Example 4.9 uses 'z to specify the floating output of a tristate of configurable width.

4.2.5.3 Conditional Generates

Verilog has an `ifdef compiler directive to conditionally compile code. Unused signals could be conditionally eliminated with `ifdef rather than just set to 0. However, we find `ifdef clutters the code excessively and complicates simulation because signals will exist in certain configurations and vanish in others. We prefer driving unused signals to 0 and allowing logic synthesis to optimize them away.

Conditional generate statements allow hardware to be created based on a condition. Code Example 4.33 shows how to instantiate a multiply/divide (muldiv) module only when the 'M_SUPPORTED configuration value is 1. Otherwise, the MulDivResultW and DivBusyE signals are tied to 0, so they aren't floating, and no hardware is created during synthesis.

Conditional generates may contain almost any Verilog statements, including signal declarations, module instantiations, continuous assignments, and always blocks. The condition must be known at compile time because hardware is either generated or not generated; it can't wink in and out of existence based on the value of a changing signal. Generates produce another level of hierarchy in simulation if they contain any modules or internal signals; it is good practice to give this level a name with a colon (:) after the begin statement. For example, if this module is called top, then the mdu module is top.mdugen.mdu. If the generate loop is not named, the simulator will name it automatically, such as top.genblk0.mdu.

Code Example 4.33 CONDITIONALLY INSTANTIATED MODULE (WITHIN **wallypipelinedhart**)

```
...
logic [`XLEN-1:0] MulDivResultW;
logic             DivBusyE;

/* Optional generate/endgenerate statement: generate */

if (`M_SUPPORTED) begin: mdugen
  mdu mdu(.clk, .reset, .ForwardedSrcAE, .ForwardedSrcBE,
    .Funct3E, .Funct3M, .MulDivE, .W64E,
    .MulDivResultW, .DivBusyE, .StallM, .StallW, .FlushM, .FlushW);
end else begin // No M instructions supported
  assign MulDivResultW = 0;
  assign DivBusyE = 0;
end
/* Optional generate/endgenerate statement: endgenerate */
```

Conditional generates (and generate loops described in Section 4.2.5.4) may optionally be surrounded by the `generate` and `endgenerate` statements as shown in comments in Code Example 4.33. We omit them because brevity is the soul of wit.

4.2.5.4 Generate Loops

A *generate loop* creates multiple pieces of hardware based on a loop variable. Code Example 4.34 shows a *priority circuit* of configurable size `N` that generates a *thermometer code* using a generate loop. It detects the least significant 1 digit in the input and puts 1s in all the less significant output bits and 0s in the upper output bits. For example, if `a` = 01010000, `y` = 00001111 because the four least significant input bits are 0 and the fifth input bit is 1. The generate loop implies a chain of *N*-1 AND gates and *N* inverters, as shown in Fig. 4.27(a). The loop index `i` is declared as a `genvar`, which is not a physical signal in the circuit but rather used to count the hardware being generated. As with Code Example 4.33, the generate loop is given an optional name: `therm`. Most synthesis tools, particularly Synopsys Design Compiler, are smart enough to recognize this priority circuit and optimize it to a tree structure rather than a ripple structure when speed matters, as shown in Fig. 4.27(b), thus reducing the delay from *N*-1 to $\log_2 N$ AND gates along the critical path.

Code Example 4.34 PRIORITY THERMOMETER CIRCUIT

```
module prioritythermometer #(parameter N = 8) (
  input  logic  [N-1:0] a,
  output logic  [N-1:0] y);

  // Create thermometer code mask
  genvar i;
  assign y[0] = ~a[0];
  for (i = 1; i < N; i++) begin: therm
    assign y[i] = y[i-1] & ~a[i];
  end
endmodule
```

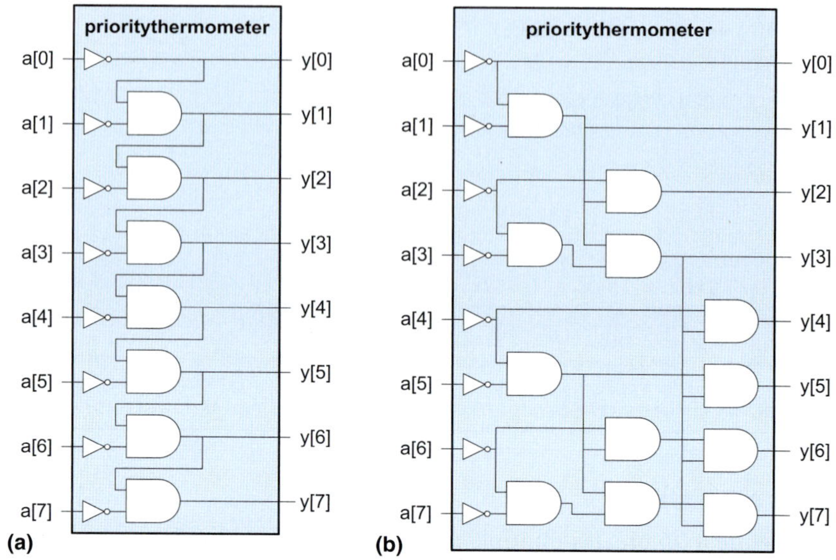

Fig. 4.27 Priority circuit: (a) Direct ripple implementation and (b) Faster tree implementation.

4.2.5.5 Arrays of Modules

To instantiate identical modules, SystemVerilog's array of instances notation is more succinct than a generate loop. Code Example 4.35 illustrates the `tlbram` instantiating an array of `TLB_ENTRIES` identical `tlbramline` modules. SystemVerilog connects signals to arrays as you would hope; for example, when the `Matches` signal (of width `TLB_ENTRIES`) is connected to the one-bit `tlbramline` array `re` port, each instance of the array receives the corresponding bit of `Matches`. The `q` outputs of the `tlbramline` array are `` `PPN_BITS ``+10 bits wide and are bundled together into the `RamRead` array which has `TLB_ENTRIES` entries of width `` `PPN_BITS ``+10 bits. Fig. 11.21 shows all these connections. The example also illustrates using module parameters (`TLB_ENTRIES`) and global configuration parameters (`` `PPN_BITS `` and `` `XLEN ``).

This example also illustrates a *distributed AND-OR mux* read circuit, which is used in many other components such as the cache reads in Section 10.4.5.1. Unlike the `ram1p1rw` from Code Example 4.20 that takes a binary address and uses array notation `rd <=mem[a]` to read from that address, the `tlbram` reads all entries and zeroes out the entries that don't match. Specifically, tlbram takes a one-hot `Matches` input that indicates which entry to read. Each `tlbramline` ANDs its `Match` signal with the line to produce `RamRead`, which is 0 when its `Match` signal is 0. The `or_rows` module in `tlbram` then performs a bit-wise OR on the `RamRead` signals to produce a `` `PPN_BITS ``+10-bit `PageTableEntry` signal.

Code Example 4.35 `tlbram` WITH `tlbramline` AND `or_rows`

```
`include "config.vh"

module tlbram #(parameter TLB_ENTRIES = 8) (
  input  logic                      clk, reset,
  input  logic [`XLEN-1:0]          PTE,
  input  logic [TLB_ENTRIES-1:0]    Matches, WriteEnables,
  output logic [`PPN_BITS-1:0]      PPN,
  output logic [7:0]                PTEAccessBits,
  output logic [TLB_ENTRIES-1:0]    PTE_Gs);

  logic [`PPN_BITS+9:0] RamRead[TLB_ENTRIES];
  logic [`PPN_BITS+9:0] PageTableEntry;

  // RAM implemented with array of flops and AND/OR read logic
  tlbramline #(`PPN_BITS+10) tlbramline[TLB_ENTRIES-1:0]
     (.clk, .reset, .re(Matches), .we(WriteEnables),
      .d(PTE[`PPN_BITS+9:0]), .q(RamRead), .PTE_G(PTE_Gs));
  or_rows #(TLB_ENTRIES, `PPN_BITS+10) PTEOr(RamRead, PageTableEntry);

  // Rename the bits read from the TLB RAM
  assign PTEAccessBits = PageTableEntry[7:0];
  assign PPN = PageTableEntry[`PPN_BITS+9:10];
endmodule

module tlbramline #(parameter WIDTH = 22) (
  input  logic             clk, reset,
  input  logic             re, we,
  input  logic [WIDTH-1:0] d,
  output logic [WIDTH-1:0] q,
  output logic             PTE_G);

  logic [WIDTH-1:0] line;

  flopenr #(WIDTH) pteflop(clk, reset, we, d, line);
  assign q = re ? line : 0;
  assign PTE_G = line[5]; // Send global bit to CAM as part of ASID matching
endmodule

module or_rows #(parameter ROWS = 8, COLS=2) (
  input  var logic [COLS-1:0] a[ROWS],
  output logic [COLS-1:0] y);

  genvar row;
  if(ROWS == 1)
    assign y = a[0];
  else begin
    logic [COLS-1:0] mid[ROWS-1:1];
    assign mid[1] = a[0] | a[1];
    for (row = 2; row < ROWS; row++)
      assign mid[row] = mid[row-1] | a[row];
    assign y = mid[ROWS-1];
  end
endmodule
```

4.2.5.6 Configuration With Parameter Structure

`define is the simplest way to configure code. However, it does not scale well when a piece of code is reused in different configurations. For example, consider building a chip with five Wally cores, four of which are rv64gc application processors and one that is an rv32e microcontroller for power and clock management. The different cores have identical Verilog but use different config.vh files. When the second config file is read, it overwrites macros from the first one, leading to unpredictable behavior.

The scalable solution for SoCs is to use Verilog's parameter capabilities, passing different parameters to the different cores. The top-level module in turn passes parameters to the submodules. A core may have dozens or hundreds of parameters, so typing out each parameter in every module is too cumbersome. Instead, we use Verilog's structure syntax to pass a single structure containing many parameters.

The following examples illustrate defining and using parameter structures in Wally. For brevity, they only show three of the parameters: XLEN, MISA, and M_SUPPORTED. The examples use a fair amount of new SystemVerilog syntax.

In Code Example 4.36, cvw.sv creates a *package* called cvw that defines the cvw_t struct with all of the parameters. Packages hold definitions that can be imported into other modules. The struct contains many elements of various data types. It is declared as packed, meaning that it comprises consecutive bitfields. The int fields are each 32 bits and logic is 1 bit, so the illustrated portion of cvw_t is 65 bits.

Code Example 4.36 **src/cvw.sv** PARAMETER STRUCTURE DEFINITION PACKAGE

```
package cvw;

typedef struct packed {
  int           XLEN;         // Machine width (32 or 64)
  int           MISA;         // Machine Instruction Set Architecture
  ...
  logic         M_SUPPORTED; // Mul/div extension supported
} cvw_t;

endpackage
```

In Code Example 4.37, the rv64gc config.vh defines configuration parameters that are specific to rv64gc, such as the XLEN and misa. The localparam statement declares named constant values, possibly dependent on other parameters. It also includes a config-shared.vh file (Code Example 4.38) that defines constants and parameters that are derived from the config.vh-specific parameters. For example, rv64gc = RV64I-MAFDC (with supervisor and user modes) has XLEN = 64 and a misa setting A, C, D, F, I, M, U, and S. The 12th bit of MISA indicates whether the M (multiply/divide) extension is supported. rv32i/config.vh in Code Example 4.39 contains a different configuration with XLEN = 32 and a MISA setting of only I. M_SUPPORTED is 1 for rv64gc and 0 for rv32i. The tedious parameter-defs.vh file in Code Example 4.40 creates a cvt_t structure named P (short for *Parameters*) with all of its fields initialized by the localparam values from the configuration file.

Code Example 4.37 **config/rv64gc/config.vh** CONFIGURATION DEFINITIONS

```
localparam XLEN = 32'd64;
localparam MISA = (32'h00000104 | 1<<5 | 1<<3 | 1<<18 | 1<<20 | 1<<12 | 1<<0);
...
`include "config-shared.vh"
```

Code Example 4.38 **config/shared/config-shared.vh** CONFIGURATION DEFINITIONS

```
localparam M_SUPPORTED = ((MISA >> 12) % 2 == 1);
```

Code Example 4.39 **config/rv32i/config.vh** CONFIGURATION DEFINITIONS

```
localparam XLEN = 32'd32;
localparam MISA = 32'h00000100;
...
`include "config-shared.vh"
```

Code Example 4.40 **config/shared/parameter-defs.vh** PARAMETER INITIALIZATION

```
localparam cvw_t P = '{
  XLEN :          XLEN,
  MISA :          MISA,
  ...
  M_SUPPORTED :   M_SUPPORTED
};
```

Next, we parameterize the SystemVerilog files using cvw_t P. Each file adds import cvw::*; #(parameter cvw_t P) to the module declaration line so that it reads the cvw package into its local namespace (to define cvw_t)

and accepts the parameter structure P. In Code Example 4.41, wallypipelinedsoc.sv is the top-level module. It instantiates wallypipelinedcore, passing the parameter P down to this module. Note that it also defines a reset synchronizer that does not require any parameters. In Code Example 4.42, wallypipelinedcore is the processor core with the various functional units. Observe how it defines a bus of width XLEN using the P.XLEN syntax instead of `XLEN to access this element of the parameter structure. It also instantiates the integer execution unit (IEU) and conditionally instantiates the multiply/divide unit (MDU) when the M extension is supported, passing the parameters down to each of these units.

Code Example 4.41 **src/wally/wallypipelinedsoc.sv** TOP-LEVEL MODULE

```
module wallypipelinedsoc import cvw::*; #(parameter cvw_t P)  (
  input  logic                clk,
  input  logic                reset_ext,
  output logic                reset,
  ...

  synchronizer resetsync(.clk, .d(reset_ext), .q(reset));
  wallypipelinedcore #(P) core(.clk, .reset, ...);
  ...

endmodule
```

Code Example 4.42 **src/wally/wallypipelinedcore.sv** TOP-LEVEL MODULE

```
module wallypipelinedcore import cvw::*; #(parameter cvw_t P)  (
  input  logic                clk,
  input  logic                reset,
  ...
  logic [P.XLEN-1:0] MDUResultW;

  ieu #(P) ieu(.clk, .reset, ...);
  if (P.M_SUPPORTED) begin:mdu
    mdu #(P) mdu(.clk, .reset, ..., MDUResultW);
  end
  ...

endmodule
```

Simulation requires a testbench to initialize the configuration structure P based on the desired configuration and to instantiate the top-level module with these parameters, as shown in Code Example 4.43. Section 6.3 shows how synthesis uses a *wrapper* to load the desired configuration.

In an SoC containing multiple different cores, replace `parameter_defs.vh` with files defining different parameters for each core.

Code Example 4.43 TESTBENCH IN **`testbench/testbench.sv`**

```
`include "config.vh"

import cvw::*;

module testbench;

 `include "parameter-defs.vh"

  wallypipelinedsoc  #(P) dut(.clk, .reset, ...);
  ...

endmodule
```

4.3 WALLY STYLE GUIDELINES

"A foolish consistency is the hobgoblin of little minds."

—**Ralph Waldo Emerson**

Most organizations and experienced designers establish a set of HDL style guidelines for large projects with multiple contributors to speed development, maintain consistency, and reduce errors. Many reasonable choices exist, but consistency is valuable for readability across the design, and the designer must be empowered to break these guidelines when the need arises. This chapter summarizes the guidelines used in this book for signal names, white space, modules, and configuration.

These style guidelines apply to synthesizable code. Testbenches are an entirely different matter and generally look more like software.

Two other extensive Verilog style guides are available from the following organizations:

- **lowRISC:** github.com/lowRISC/style-guides/blob/master/VerilogCodingStyle.md
- **Freescale:** people.ece.cornell.edu/land/courses/ece5760/Verilog/FreescaleVerilog.pdf

4.3.1 Signals

Signal naming conventions are important in all programming, and particularly in HDL. SystemVerilog is case sensitive; for example, `RegWrite`, `regwrite`, and `regWrite` are not the same signal. If capitalization is inconsistent, the developer may have to search across files to see how a particular signal is capitalized or may introduce bugs where the same signal is capitalized differently in different places and doesn't connect. Pipelined systems

Unlike Verilog, VHDL is case insensitive, which can cause problems with EDA (electronic design automation) tools such as Synopsys Design Compiler and Xilinx Vivado. Most EDA tools use *Tool Command Language* (*Tcl*) to process a design's RTL, such as defining a clock or specifying timing constraints. Tcl is case sensitive, so it is entirely possible for the EDA tool to not find the VHDL signal name if capitalization is not consistent throughout the VHDL files.

use suffixes to keep track of a signal's pipeline stage and prefixes to distinguish related signals from different blocks. In SystemVerilog, nearly all signals are declared of type `logic`. Ports are connected by order for simple generic modules and by name for more complex modules, preferably keeping the name the same across the hierarchy.

4.3.1.1 Capitalization

Table 4.3 lists various conventions for capitalizing signal names [NameConv]. *flatcase* works well for short simple names, such as in the C standard library, but is difficult to read for more complex signal names common in HDL design. *snake_case* involves reaching for the underscore key with the right pinky finger, which we find wearisome when typing. *camelCase* has the advantage that it requires no capitals for single-word variable names yet remains highly readable for multiword names, so this is a personal favorite of the authors when programming. However, *UpperCamelCase* is common in HDL design, looks natural in block diagrams, and was the choice of Patterson and Hennessy in the diagrams for *Computer Organization and Design*, from which Wally is indirectly descended [Hennessy20]. *MACRO_CASE* is the most difficult to type but makes constants and macros stand out from variables. In an empirical study, Binkley found that camelCase leads to higher accuracy and faster recognition than snake_case [Binkley09].

Guideline 4.1 Use flatcase for `clk` and `reset`.

These signals appear in almost every module and should have easy names.

Guideline 4.2 Use UpperCamelCase for signals.

Other conventions are reasonable, but it's important to pick one and stick with it. Wally inherits this convention choice from prior books.

Guideline 4.3 Use MACRO_CASE for constants and parameters.

Most developers use this convention.

Table 4.3 Capitalization conventions.

Example	Convention Name
`regwrite`	flatcase
`regWrite`	camelCase / dromedaryCase
`RegWrite`	UpperCamelCase / PascalCase
`reg_write`	snake_case
`REG_WRITE`	MACRO_CASE

4.3.1.2 Suffixes and Prefixes

In pipelined systems, the same signal often exists in multiple pipeline stages and must be named in a way to distinguish the stage. HDL designers normally do this with a suffix. For example, the Wally pipeline has Fetch, Decode, Execute, Memory, and Writeback stages. The suffix may be one letter (F, D, E, M, or W) or an abbreviation (FET, DEC, EXE, MEM, WRB). Single-letter abbreviations are more succinct, but complex systems with many pipeline stages may need longer stage suffixes for clarity. Signals outside the pipeline (e.g., in the bus interface or hardware page table walker) have no suffix.

While pipeline registers accept inputs from one pipeline stage and produce outputs for the next stage and, normally, combinational logic accepts inputs and produces outputs within the same pipeline stage, some components operate across multiple pipeline stages. Such components are a common source of bugs, and care must be taken in signal naming. Some components that necessarily break these rules include the register file (reading in the Decode stage and writing in the Writeback stage), the external bus interface (accepting requests from the Fetch and Memory stages), and the hazard unit (combining hazards from many stages and sending stalls and flushes to each stage). For such components, signals generated within the pipeline use the suffix of the pipeline stage in which they were produced. Signals generated outside the pipeline (i.e., in the hazard unit) use the suffix of the stage in which they will be used.

Guideline 4.4 Use suffixes on all signals associated with pipeline stages.

For the pipelined Wally processor, use F, D, E, M, and W suffixes.

Guideline 4.5 Use prefixes to distinguish related signals from different sources.

Sometimes signals are prefixed with their source unit to distinguish similar signals from different places. For example, the datapath uses `IEUResultM`, `MDUResultW`, and `SCResultW` to distinguish results from the IEU, MDU, and store conditional units. Source units are typically abbreviated in all caps (e.g., `IEU`, `IFU`, `LSU`, `FPU`, `SC`, and `MDU`).

4.3.1.3 Declaring Signals

Guideline 4.6 Declare all internal signals, including 1-bit signals.

Undeclared signals are a source of bugs.

Guideline 4.7 Only use (bidirectional) inout pins at the top-level chip module. Define them as `inout tri`.

Bidirectional pins should only be used when they are an essential part of an external I/O standard, such as general-purpose I/O (GPIO).

Internally, use unidirectional buses and multiplexers rather than bidirectional busses and tristates.

Guideline 4.8 Declare all other signals as `logic`.

The `logic` datatype can be used anywhere except on tristate nets, supports x and z, and avoids the ugly rules of `wire` and `reg` in old-school Verilog.

Guideline 4.9 Declare busses as little-endian.

For consistency with prevailing custom, declare a bus as [31:0] rather than [0:31].

Guideline 4.10 Avoid passing arrays through ports if possible but declare the port as `var logic` if necessary.

Arrays are usually used to represent memories and are kept internal to a module. In unusual cases when arrays have to be passed through ports, SystemVerilog wants the `var logic` syntax shown in Code Example 4.44, which is otherwise outside the scope of this chapter.

Code Example 4.44 illustrates many of these guidelines.

Code Example 4.44 IFU SIGNAL DECLARATIONS (PARTIAL)

```
module ifu import cvw::*;  #(parameter cvw_t P) (
  input     logic                clk, reset,                                  // Lower case
  input     logic                StallF, StallD, StallE, StallM, StallW,      // Suffix
  input     logic                FlushF, FlushD, FlushE, FlushM, FlushW,      // Suffix
  input     logic [P.XLEN-1:0]   SATP_REGW,                                   // Little-endian
  input     logic [1:0]          STATUS_MPP,
  input var logic [7:0]          PMPCFG_ARRAY_REGW[P.PMP_ENTRIES-1:0],        // Passing array
  output    logic                InstrAccessFaultF);

  logic [P.XLEN-1:0]             PCNextF;
  logic                          BranchMisalignedFaultE;
  ...
```

4.3.1.4 Port Connections

Guideline 4.11 Connect ports by order for simple generic modules such as flip-flops and multiplexers.

The connections are usually obvious, and this convention is most succinct, as shown for `mux2` in Code Example 4.45.

Guideline 4.12 Connect ports by name for other modules.

Use `.sig` to connect a port named `sig` to an external signal of the same name. Use `.sig(name)` to connect a port named `sig` to an external signal called `name`, as shown for ports such as `PAdr` connecting to the `dmmu` in Code Example 4.45.

Guideline 4.13 For non-generic modules used only once, use the same name for ports and signals connecting to those ports.

Tracing logic is easiest when signals don't change names unnecessarily at module boundaries. It also allows `.sig` notation in the port connections, which is most succinct. For example, the `dmmu` module in Code Example 4.45 does this for many ports such as `clk` and `PrivilegeModeM`.

Guideline 4.14 Avoid `.*` notation except in wrapper modules.

When a module uses `.*`, the reader needs to open up the module declaration to identify the inputs and outputs, which makes tracing logic more cumbersome. An exception is a top-level or wrapper module where nearly all signals have the same name, and humans find them tedious to read.

The *top-level* module is the highest module in the design hierarchy and is typically a structural module that instantiates modules. A *wrapper* module maps signals to desired interfaces, for example, pins on an FPGA that are connected to the system clock, switches, and so on.

Guideline 4.15 Pack multiple port connections on one line of the instantiation.

Brevity improves the ability to see the big picture. For example, in Code Example 4.45, all of the port connections (`ALUResultE,..., IEUResultE`) are on one line.

Code Example 4.45 PORT CONNECTIONS

```
mux2 #(P.XLEN) ieuresultmux(ALUResultE, AltResultE, ALUResultSrcE, IEUResultE);
mmu #(.TLB_ENTRIES(P.DTLB_ENTRIES), .IMMU(0)) dmmu(
  .clk, .reset, .SATP_REGW, .STATUS_MXR, .STATUS_SUM,
  .STATUS_MPRV, .STATUS_MPP,
  .PrivilegeModeW, .DisableTranslation(SelHPTW),
  .PAdr(PreLSUPAdrM), .VAdr(IEUAdrM),
  ...
  .PMPCFG_ARRAY_REGW, .PMPADDR_ARRAY_REGW);
```

4.3.2 Spacing and Indentation

These guidelines aim to use whitespace in a consistent way to improve readability without bloating code.

Guideline 4.16 Indent with spaces instead of tabs.

Different editors interpret tabs differently.

Guideline 4.17 Indent by two spaces for each level of nesting.

Two and four spaces are common conventions. We prefer two because space is at a premium in a textbook layout.

Guideline 4.18 Align signal declarations within a block.

Signals are more readable when they are aligned.

Guideline 4.19 Leave a space around operators and after `if`/`case`/`for` statements.

This space is conventional and improves readability.

Guideline 4.20 Avoid space before semicolons, commas, closing parentheses, and open parentheses in module instantiations, and inside bus width declarations.

Extra space here doesn't help readability.

Guideline 4.21 Only use `begin` and `end` when more than one statement is required in a block.

Avoid unnecessary syntax.

Guideline 4.22 Pack the `begin` statements on the same line where feasible.

Avoid code length bloat.

Guideline 4.23 Limit lines to a maximum of 95 characters.

This is the most that fits on a line in this textbook layout and is also a comfortable number to fit in a text editor. The historic line length of 80 characters is also reasonable but no longer necessary with wider screens.

Guideline 4.24 Use blank lines sparingly to logically separate portions of code.

Place blank lines after internal signal declarations, between module declarations, and to divide unrelated portions of code.

Guideline 4.25 Use banner comments sparingly to separate major portions of code in a complex module when using hierarchy isn't a good alternative.

Banners take up space but could help the reader recognize different portions of code.

Code Example 4.46 shows an example that follows these guidelines.

Code Example 4.46 `redundantmul`

```
module redundantmul #(parameter WIDTH = 8)(
  input  logic [WIDTH-1:0]   a,b,
  output logic [2*WIDTH-1:0] out0, out1);

  if (P.DESIGN_COMPILER == 1) begin: mul
    logic [2*WIDTH-1+2:0] tmp_out0;
    logic [2*WIDTH-1+2:0] tmp_out1;
    DW02_multp #(WIDTH, WIDTH, 2*WIDTH+2) mul(
      .a, .b, .tc(1'b0), .out0(tmp_out0), .out1(tmp_out1));
    assign out0 = tmp_out0[2*WIDTH-1:0];
    assign out1 = tmp_out1[2*WIDTH-1:0];
  end else begin: mul
    assign out0 = a * b;
    assign out1 = 0;
  end
endmodule
```

4.3.3 Synchronous Design

Guideline 4.26 Use synchronous sequential design.

Every component should be a positive edge-triggered flip-flop, a standard memory, or combinational logic. All components should share a common clock named `clk`. Avoid latches, negative edge-triggered flip-flops, and asynchronous components. Exceptions to synchronous design should only occur when essential (e.g., at an interface with an asynchronous peripheral) and should be carefully documented with comments and reviewed.

Large systems with multiple subsystems inevitably have more than one clock domain. For example, a RISC-V core might run at a different frequency than a Wi-Fi packet processor or GPU or DRAM controller. Use synchronous sequential design within a clock domain. Clock domain crossings, sending signals between clock domains, are beyond the scope of this book.

Guideline 4.27 Instantiate memories and flip-flops using modules.

This makes it easier to identify the number of bits of state in the system and to swap out synchronous or asynchronous reset, scan, or low-power features.

An exception is that FSM state registers often use enumerated state types that are incompatible with the logic datatype of a flip-flop module. These state registers are defined with `always_ff` in the FSM (see Code Example 4.16).

Guideline 4.28 Do not use # delays except in the testbench.

Some simulators are significantly faster when all signals toggle on the cycle boundary.

Comments at the beginning of a module that explain the principles of operation are most valuable. Good variable name choices eliminate the need for many line-by-line comments.

Bad Example:

```
assign ab = ~a;          // Flip the bits
assign minusa = ab + 1; // Add one
```

Good example:

```
assign minusa = ~a + 1; // Take the 2's complement to negate a
```

4.4 WALLY CONFIGURATION

The Wally (cvw) repository organization is listed in Fig. A.1.

As described in Section 2.8.2, Wally is configurable to scale from a tiny rv32e core to simple rv{32/64}i cores to an rv32imc microcontroller with peripherals to rv{32/64}gc application processors with floating-point, branch prediction, caches, and virtual memory, all using the same code base. Each configuration is defined by:

```
$WALLY/config/<configname>/config.vh
```

In addition, Wally has many derived configurations, which are perturbations of the base configurations. They are used to test optional features, such as whether caches are enabled, how branch predictors perform with different table sizes, and whether various combinations of floating-point data sizes can be supported. These derived configurations are summarized in Table A.5 and defined in:

```
$WALLY/config/deriv/<configname>/config.vh
```

The derived configurations are generated with the `bin/derivgen.pl` script using the `config/derivlist.txt` table. They are regenerated by running `sim/make deriv` or `sim/make`.

Two notable derived configurations are buildroot and fpga, intended to boot Linux in simulation and on an FPGA. These are variants of rv64gc, with the reset vector pointing to a boot ROM at 0x1000 with a zero-stage boot loader.

Simulation and synthesis scripts point to the desired configuration. The `config.vh` file defines many parameters used in SystemVerilog generate blocks. Table A.3 lists key configuration parameters for each configuration, and Table A.4 lists additional configuration parameters.

The source code is in `$WALLY/src`, with subdirectories for each unit.

`testbench.sv` has a module called `riscvassertions` that checks legal values for configuration parameters. Examples of checks include:

- Optional supported modes are self-consistent (e.g., D requires F).

- Parameters have a limited set of legal values (e.g., `PMP_ENTRIES` = 0, 16, or 64).
- Cache organization is legal (e.g., way size doesn't exceed VM page size).

4.5 SUMMARY

Hardware description languages (HDLs) are used to specify the design of an SoC in a form that is both human and machine readable. The HDL code is simulated for logic verification and synthesized onto a chip or FPGA. Wally is written in SystemVerilog, the most widely used HDL.

SystemVerilog is a complex language, but only a tiny subset is used to describe synthesizable hardware. A good designer visualizes the desired type of logic, then writes SystemVerilog idioms that imply such logic. The designer should first determine whether a component is combinational or sequential logic. In combinational logic, the output depends on the current value of inputs. In sequential logic, the output depends on current and previous inputs; in other words, it has state.

SystemVerilog has several idioms for combinational logic. `assign` statements define Boolean or arithmetic equations. `always_comb` blocks can contain `case` statements or `if/else` statements. `case` statements define combinational logic with truth tables. `if/else` can be helpful for complex control logic. Synthesis excels at optimizing logic onto gates, so the designer can focus on writing a clear and concise description of what the logic should do and let the tool perform the Boolean optimization.

Synchronous sequential logic is built from edge-triggered flip-flops and combinational logic. The `always_ff @(posedge clk)` idiom describes edge-triggered flip-flops. FSMs can be described with a bank of flip-flops to hold the state, and combinational logic blocks to compute the next state and output. The next state code may be merged into the `always_ff` block for brevity. Memories, such as RAMs, ROMs, and register files, are a special case of sequential logic that contain an array of flip-flops or bit cells indexed by an address.

Larger modules are built from smaller modules, taking advantage of hierarchy and reuse. Parameterized modules can adjust bus widths based on the parameter value. Conditional generates, described by an `if` statement whose condition depends on a configuration parameter, instantiate modules only under certain conditions (e.g., when a feature is supported). Generate loops, described by a `for` statement, instantiate multiple pieces of hardware. SystemVerilog also has an array notation that is more succinct when the hardware is identical.

Compared to Verilog, SystemVerilog has a much larger set of features for writing testbenches to test hardware. These features include reading test vectors from files, checking results against expectations, and reporting discrepancies. Indeed, SystemVerilog has an entire object-oriented feature set for modeling and verification comparable to C++. However, this book focuses on widely used constructs and the basic features needed for testbenches.

Good style is important for readable and maintainable HDL code. Wally uses coding guidelines similar to many other large SystemVerilog systems. It adopts UpperCamelCase signal names, with single-letter suffixes to indicate the pipeline stage. Except in generic or multiply instantiated modules, the same name is used for a signal throughout the hierarchy of modules. Ports are connected by order for simple modules (flip-flops and muxes), with the `.sig` notation for more complex modules, or with `.sig(name)` when a signal must change names where it enters a generic module. We find that a dense coding style improves readability by letting the developer see more of the code base on a screen. Wally is heavily configurable, allowing it to scale from a tiny rv32e core to a full-featured rv64gc application processor. The configurations avoid unnecessary hardware, keeping the size and critical path as tight as possible.

5 Design Verification

5.1 Verification Strategies
5.2 Design Verification Plan
5.3 Lint
5.4 Testbenches
5.5 Logic Simulators
5.6 Lock-Step-Compare
5.7 Assertions
5.8 Code Coverage
5.9 Functional Coverage
5.10 Formal Verification
5.11 Wally Verification
5.12 Summary

"Everyone knows debugging is twice as hard as writing a program in the first place."

—**Brian Kernighan, inventor of the C language (1974)**

"If you didn't test it, it doesn't work."

—**Bob Colwell, chief architect on Pentium Pro through Pentium 4 [Colwell02]**

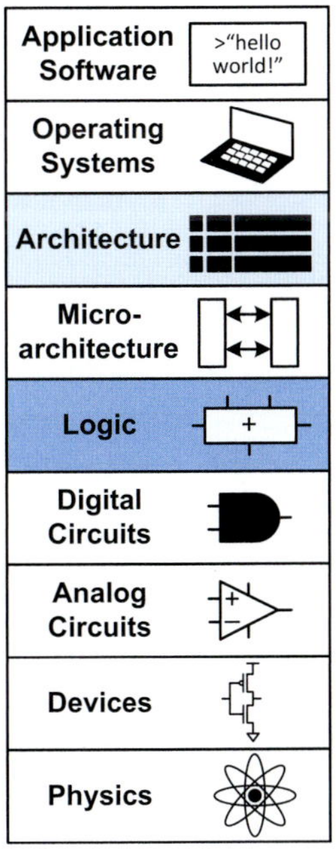

Logic verification is the process of checking that logic conforms to the specification. Verification is generally more time-consuming and expensive than design. Verification engineers make up about half of an SoC development team. Verification engineers need a broad understanding of the system to define tests that exercise the full range of features and because errors they discover might come from anywhere. The mantra of SoC design is "If it is not verified, it will not work."

Logic verification involves simulating the design with a *testbench* that applies a set of *test vectors* to the *device under test (DUT)* and checks that results match a reference model. Verification challenges include the following:

- Defining what features need to be tested
- Determining what test vectors to use
- Creating a reliable reference model
- Locating and fixing bugs when they are detected
- Coping with the enormous runtimes of large simulations

The Universal Verification Methodology (UVM) [IEEE 1800.2-2017] is a standardized method of logic verification with SystemVerilog that is widely accepted in industry. It takes an object-oriented approach involving drivers, monitors, stimulus generators, and scoreboards. The methodology is sufficiently complex that it is primarily used by verification specialists and is beyond what can be covered in this chapter. *Getting Started with UVM: A Beginner's Guide* [Cooper13] is a reasonable place to start learning UVM.

SystemVerilog has become the standard language for verification. The language is complex, and engineers use different subsets of the language for different purposes, as summarized in Table 5.1.

Section 5.1 begins by discussing verification strategies, from exhaustive and ad-hoc strategies that might be suitable for a class project or initial prototype to coverage-driven verification that systematically identifies what features must be covered, then generates directed and random tests to cover these features. Basic tests may run for minutes to hours, and comprehensive testing can take years, so tests are often run in hardware on a field programmable gate array (FPGA) emulator that is orders of magnitude faster than a software simulator. Technology Readiness Levels (TRLs), originally developed by the National Aeronautical and Space Administration (NASA) to manage multibillion dollar space programs, are a way of characterizing the maturity of design verification. Design and verification go hand in hand. Good logic designers test as they go and keep basic tests always passing rather than throwing untested RTL over the wall to a verification engineer.

A *design verification plan* (*DV plan*) is a structured document defining the coverage requirements, as outlined in Section 5.2. It must cover both the trees and the forest: it must have enough detail that a verification engineer can translate each requirement into a machine-readable test but also be organized in a way that the reader can be confident that all the system's functions are covered.

Designers commonly run a *lint* tool, described in Section 5.3, to detect static code problems such as mismatched port sizes. It is much faster to fix these problems immediately than to wait for one to cause a subtle bug and then track down the root cause by studying waveforms. Section 5.4

Table 5.1 SystemVerilog language subsets.

Subset	Section	Purpose
Synthesizable	4.2	Register transfer level (RTL) design
Testbench	5.4	Initialize memory and read tests from files, check results, log discrepancies
Assertions (SVA)	5.7	Specify properties a design must satisfy
Functional coverage (FC)	5.9	Coverpoints and covergroups to measure how thoroughly RTL is verified
Object-oriented	n/a	Generic reusable testbenches, particularly with the Unified Verification Methodology (UVM) library; also used in functional coverage

introduces a simple testbench, and Section 5.5 shows how to simulate the testbench using the open-source Verilator simulator as well as the commercial Questa and VCS simulators.

Some verification methodologies generate a *memory signature* where key results are written to memory and checked against expectations. This approach can miss bugs that do not corrupt the signature. A more robust approach, described in Section 5.6, is *lock-step-compare*, in which the entire architectural state of the system is compared against an independent reference model after each step.

Section 5.7 describes *assertions*, which check properties, such as whether a configuration is legal or whether a bus protocol is violated. Assertions can be a fast way to identify problems where they occur in the RTL rather than waiting for the problem to corrupt the architectural state.

Code coverage, discussed in Section 5.8, measures whether the simulations fully exercise every part of every line of code in the RTL. This is necessary but not sufficient. For example, code coverage would not detect that a feature is entirely absent from the code. *Functional coverage*, discussed in Section 5.9, tests that all features are exercised. The DV plan is translated into a large number of *coverpoints*, such as exercising each instruction with each possible register and corner case values. The coverpoints are organized into *covergroups*, such as one group for each instruction. When the DV plan has been carefully reviewed and converted to coverpoints and enough test vectors have been simulated to reach 100% code and functional coverage while completely agreeing with the reference model, it is reasonable to declare that the system is adequately verified.

It would be even better if one could mathematically prove that the RTL correctly implemented a specification, such as the Sail architectural model. Such *formal verification (FV)* has long been a subject of research and is gradually becoming practical for an increasing set of use cases, as briefly described in Section 5.10.

Section 5.11 concludes with details of Wally's verification. In the context of RISC-V processor verification, the DUT is the SystemVerilog RTL model of the processor, the test vectors are RISC-V programs exercising various features, and the reference model is Sail, Spike, or ImperasDV. The architectural state of the RTL model is compared to the lock-step-compare reference model or the functional coverage monitors through a standard protocol called the *RISC-V Verification Interface* (RVVI). Test vectors include the riscv-arch-test compatibility test suite, custom directed tests, and constrained random tests.

5.1 VERIFICATION STRATEGIES

A testbench loads and applies *test vectors* to a DUT and checks the results. Creating good test vectors and a reference model of expected results is a key challenge of logic verification. Verification strategies, discussed in this chapter, must address the following questions: When is it best to test each unit individually and then assemble components and test the whole system, and when is the overhead of the unit testbench high enough that it is more expedient to skip unit testing? Simulations run slowly but give complete visibility into the design to locate bugs. Hardware emulation runs orders of magnitude faster but may have restricted visibility. Design and verification are an iterative process, steadily improving over the lifetime of the system. How does one characterize the maturity of the system at each phase? How do design and verification engineers collaborate to minimize the overall development effort?

5.1.1 Test Vector Generation

If students were asked to verify a simple module with only a few inputs and outputs on a homework assignment, they would likely create test vectors listing every possible input and the expected output. For example, a full adder has three inputs and two outputs, so it could be fully tested with $2^3 = 8$ vectors. Unfortunately, such *exhaustive testing* becomes intractable even for moderately interesting blocks such as an adder with two 64-bit inputs, which requires 2^{128} test vectors.

The next level of sophistication is ad-hoc testing using a modest set of test vectors informed by an understanding of the design. The RISC-V test program from Code Example 2.16 is another example of *ad-hoc* or *exploratory* test vectors. The program was crafted by hand to test each instruction in the implemented instruction set, to verify that both taken and untaken branches work, and to check that a value could be stored to memory and correctly loaded back. Ad-hoc testing is a good way to quickly get basic confidence that a system performs its key functions. However, it does not give high confidence that the system works for the complete range of inputs.

Booting Linux is another readily available ad-hoc test that exercises some of the complicated interactions between instructions, exceptions, caches, and virtual memory. However, booting Linux is no substitute for rigorously testing each feature individually [Thompson23].

Directed testing attempts to systematically cover *corner cases*. In this context, corner cases are inputs that would be most likely to trigger unusual behavior. For example, for a 64-bit adder, corner cases for each input could include {0, 1, 2, 3, 2^{32}, $2^{63}-1$, 2^{63}, $2^{63}+1$, $2^{64}-2$, $2^{64}-1$, and two random values in the middle of the range}. The adder can be tested with the *cross-product* of these 12 corner cases, that is, where every combination of the corners is used as the two inputs. If the adder works correctly for all these $12 \times 12 = 144$ corner cases, it is likely to work for intermediate values as well unless the system has unusually subtle design bugs.

Corner cases are not always enough. Imagine a carry lookahead adder coded by hand. It would be easy to introduce a typo in one of the columns that works for these corner cases but does not work for all cases. Similarly, a manufacturing defect in one of the transistors that causes column 12 of the adder to depend only on input *A* but not input *B* would go undetected by these cases.

Intel's infamous Pentium floating-point divide bug (FDIV) in 1994 was caused by a bad tool that optimized out 5 of 1,066 minterms from a quotient digit selection programmable logic array (PLA), resulting in an error that occurred around the 9th or 10th decimal digit in about 1 division of every 9 billion for randomly selected inputs [Halfhill95]. Directed tests would be unlikely to catch such a subtle error. Moreover, testing was performed on the quotient selection equations, which were actually correct, rather than the transistor-level output of the PLA tool, which had the missing minterms. Intel initially dismissed the bug as unimportant, affecting an average spreadsheet user once every 27,000 years. However, Intel's leading customer, IBM, concluded the bug could occur once every 24 days in more demanding financial spreadsheets, which caused Intel to stop shipping Pentium-based computers. Intel recalled the defective processors at a cost of $475 million and replaced them with corrected chips.

Directed tests work well for some modules, but the cross-product can still be too large for modules with many inputs, and determining sensible corners can be difficult for other modules. *Coverage-driven testing* is a technique that defines *coverage points* (also called *coverpoints*) that need to be tested. For example, a processor might define basic coverage points to be every source and destination register, every instruction opcode, and extreme values (0, 1, −1, max positive, max negative, and so on). The riscv-arch-test suite satisfies coverage points for each instruction. It writes a signature to show that the test worked.

A *constrained random test generator* produces test vectors, consisting of random instructions. Constrained random test generation is particularly good at hitting hard-to-find cases, which are also called *corner cases* in a somewhat different usage of the term. For example, it is difficult to write a directed test that causes an interrupt while the instruction fetch unit is stalled waiting for the load/store unit to service a translation lookaside buffer miss. But a constrained random test generator that produces millions or billions of code sequences exercising interrupts and virtual memory may stumble across an instruction sequence that hits this corner case. Moreover, experience shows that the constrained random test generator often unearths bugs in areas unforeseen by the DV plan. How do we know if we have generated enough code sequences? The DV plan specifies code and functional coverage requirements, and we generate random tests until coverage reaches 100% or it becomes easier to write some directed tests for coverpoints that the random generator is not reaching.

In practice, verification engineers use a mix of all these techniques. For example, Wally was tested with simple ad-hoc programs early in development to get basic functionality. Then it was tested with riscv-arch-test that involves constrained random tests for a set of coverage points. At the time of this writing, riscv-arch-test doesn't cover much of the privileged specification, so we also write ad-hoc and directed tests for these features.

5.1.2 Reference Models

Creating a reference model for a new system can be an enormous task. Although it does not need to be cycle accurate and handle pipelining or out-of-order execution, it still must faithfully implement every feature of the architecture. Moreover, an architecture sometimes has unspecified behavior; for example, RISC-V highly recommends but does not require an illegal-instruction fault on an undefined opcode. The reference model and the DUT must produce the same results in these situations. The reference model and RTL should be written by different teams to reduce the chance that both models contain the same conceptual or implementation errors.

Other RISC-V simulators include QEMU, Whisper, and IDL.

RISC-V is fortunate to be mature enough to have multiple reference models, notably Sail, Spike, and ImperasDV. Sail is the official executable reference model and now takes priority over the written architecture specification in the event of disagreement. However, Sail does not yet have sufficient support for interrupts and related peripherals to be an adequate reference for certain privileged behavior. Spike is a more mature simulator with interrupt support and is the reference for some OpenHW Group cores. ImperasDV is a commercial reference model, widely used in industry.

5.1.3 Unit vs. System Testing

Development teams must weigh the pros and cons of unit versus system testing. In a hierarchical development approach, different teams could develop different units, such as the integer execution unit (IEU), floating-point unit (FPU), instruction fetch unit (IFU), and load/store unit (LSU). Each unit could be verified in isolation with unit-level testing. Then the units could be integrated and verified with system-level testing. Ideally, the individual units would already be correct, so system-level testing would only catch integration bugs. Moreover, testing a unit is faster than testing the system because it involves fewer lines of code and fewer tests irrelevant to the unit.

Unfortunately, testbench development is quite time-consuming. Creating test vectors tailored to unit interfaces can also be difficult. For example, it is much easier to load a program into memory and run a system-level test on an entire processor than to determine all of the control, data, and stall signals going into and coming out of an LSU.

Consequently, teams tend to emphasize testing relatively large blocks with simple interfaces. In an SoC, tests may focus on each different core or accelerator and on overall SoC integration, rather than on units within a core.

5.1.4 Emulation

An SoC with thousands to millions of lines of RTL generally simulates with a system clock of several kilohertz or even single hertz. Running large simulations is painful, especially if these simulations are repeated after each design change. Simulations are accelerated by running many jobs in parallel on different cores, but this takes an expensive compute farm and even more expensive simulator licenses. In contrast, if the RTL is synthesized onto one or more FPGAs, it can run at megahertz rates. Such *hardware emulation* can accelerate verification by orders of magnitude.

Moderately complicated systems fit on one large FPGA that costs thousands or tens of thousands of dollars. FPGA-based emulation involves a moderate amount of infrastructure development to apply and check test vectors via the chip interface. The design may take minutes to hours to synthesize, place and route, and download onto the FPGA. Tests that require days in simulation may finish in seconds on the FPGA. However, if the outputs do not match expectation, it may take hours to isolate the bug using a clunky on-chip logic analyzer interface.

Commercial SoCs are usually too big to emulate on a single FPGA. They often use commercial emulation systems, such as Cadence Palladium or Synopsys ZeBu, containing racks of FPGAs, RAM, and input/output (I/O) to handle hundreds of billions of logic gates. Associated software automatically partitions the RTL onto many FPGAs and streamlines the debug process. These systems cost millions of dollars and are a precious resource for large-scale verification.

5.1.5 Technology Readiness Levels

NASA developed a TRL scale in the 1970s to discuss the level of technical maturity for space systems. The scale was adopted by the US Defense Department and European Space Agency and is now used more broadly in research and development outside space systems. Whereas academia usually develops prototypes in TRL 1–4, industry commercializes products in TRL 7–9. TRLs 4–7 are sometimes called the *Valley of Death*, where neither researchers nor industry prioritizes funding.

The OpenHW Group uses the TRL scale to characterize microprocessors, as shown in Table 5.2. Most cores aim for TRL5, comparable to new commercially licensed processor IP. The OpenHW Group aims for collaborations between industry and academia to bridge the Valley of Death.

OpenHW hosts many open-source cores, including Wally, officially called CORE-V Wally (cvw). Wally reached TRL3 by passing the riscv-arch-tests and booting Linux in simulation. It then reached TRL4 with synthesis results and an FPGA implementation booting Linux in the lab. To reach TRL5 (eventually), Wally will need to meet the following guidelines:

> CORE-V is a family of open-source RISC-V cores hosted by the OpenHW Foundation, including cva6, cvw, cve2, and cve4.

- Design validation plans for all features of the design under test
- 100% functional coverage of all items in the DV plan
- 100% code coverage (line, condition, branch, FSM)
- All tests passing in regression
- No open issues

Table 5.2 Technology Readiness Levels (TRLs) for OpenHW cores. (Adapted and reprinted with permission from The OpenHW Foundation.)

TRL	OpenHW Utilization
1. Basic Principles Observed	• OpenHW research projects may target TRL-1 as project output, e.g., to develop novel approaches to core or accelerator architecture.
2. Concept Formulation	• Core IP or accelerator development projects are typically initiated as TRL-2 concepts, identifying principles and applications of the IP. • The OpenHW Project Concept Gate output includes a TRL-2 description of the Core IP.
3. Proof of Concept	• Core IP or accelerator development projects will pass through TRL-3 as the (RTL) design completes. • Proof of concept is shown by core compilation and demonstration of basic operations (e.g., Linux booted, CoreMark results, hello-world).
4. Component Protype	• Core IP or accelerator projects will pass through TRL-4 as they produce preliminary PPA results (via synthesis scripts for FPGA or ASIC) and/or run preliminary application code, such as an accelerator running machine learning code.
5. Subsystem Designed and Tested	• Core IP projects reach TRL-5 as they complete full verification. The OpenHW RTL Freeze checklist process verifies that the design is fully ready for industrial adoption.
6. Functional (Alpha) Prototype	• OpenHW IP that is integrated into an MCU system or other device reaches TRL-6 as prototype silicon is fabricated and demonstrated on a development board or other platform.
7. Field Demonstration Prototype	• OpenHW IP that is integrated into an MCU system or other device reaches TRL-7 as prototype silicon is fabricated, deployed, and demonstrated in the field.

5.1.6 Design and Verification Approaches

Back in the bad old days of the 1990s, design and verification were often rigidly separated. Designers wrote RTL, perhaps ran basic unit testing, and then threw the code "over the wall" to a verification team, often in a different country. The verification team would locate bugs and throw them back to the designer to fix. This process was slow because of the loose coupling between teams and tended to increase verification cost because the designer made decisions without awareness of how they would affect the verification effort. As verification cost came to exceed design cost, teams adopted more collaborative approaches. Successful SoC projects now begin with the goal of verification in mind.

One verification strategy is to develop the test cases before or concurrently with the hardware description language (HDL) so that the HDL can be tested as soon as it is written. Designers use a suite of *regression* tests to make sure that new changes don't break previously working features. These tests may be a group of known inputs and expected outputs or, for a processor, a specified program with expected results. Designers may run an abbreviated regression test every time they change code and a longer regression nightly.

A pitfall strategy for inexperienced designers is to get HDL to pass test cases by trial and error rather than by designing logically from the specifications. Test cases are rarely completely comprehensive, and it is likely that undiscovered bugs remain when the HDL is just tweaked until the tests pass.

Verification strategies can be classified as *always-broken* or *always-working* [Keating07]. In the always-broken strategy, a team of designers builds individual units and debugs them using unit tests. Then they integrate the units into an entire system and run regression on the system. Inevitably, upon integration, difficult bugs surface because of interdependencies and take a long time to debug. During this time, the design is broken, and it is difficult for others to continue development because they lack a functional prototype to improve. During integration, while the system is broken, the unit owners individually add new features to units and run unit testing but not system testing. Eventually, when the system works briefly, the painful integration process repeats. Overall regression is always broken except during brief intervals after successful integration steps.

In the always-working strategy, the team goes through the difficult integration just once, preferably on as small a set of features as possible, and gets those features working. Then team members incrementally add new features and perform system-level regression before checking the features back into the design to ensure the changes don't break existing features. Overall, regression is always working after the initial prototype is developed. New features and tests are gradually added after the enhanced system is debugged by individual contributors. When a proposed feature breaks regression, it is relatively easy to fix because the designer knows exactly what change caused the problem. In the real world, this is sometimes impractical, due to adding a very large new feature, restructuring the design or test cases, or because of license, hardware, or software problems that temporarily stall regression. However, when these problems arise, the team should prioritize getting back to a working code base as quickly as possible.

Test-driven development [Beck03] is a development process in which the test cases are written before the code. It can be viewed as a disciplined form of always-working verification. The general process is:

1. Add a new test for added capability.
2. Run all the tests and confirm the new one fails.
3. Make a code change to implement the added capability.
4. Run all the tests and confirm the new one succeeds.
5. Refactor the code to remove duplication.

Tracking down a complex bug in a large RTL system through simulation is so time consuming and painful that developers strive to minimize doing this. Wally has gravitated toward an always-working test-driven methodology. Whenever possible, small changes are made and immediately tested to confirm that they work and don't break other features. If a change introduces a bug, it may be faster to discard the change and re-create it carefully in smaller steps than to diagnose the bug in a waveform viewer. The developer makes the changes in a development (dev) branch of their fork of the repository, then issues a pull request to incorporate the changes into the main branch when the feature is complete and tested.

5.2 DESIGN VERIFICATION PLAN

A design verification plan (DV plan or test plan) needs to be organized in a way that human reviewers can determine if all features of the system are tested and in a way that the verification engineer can write tests to cover each feature and measure whether each feature was covered. This generally involves a high-level statement enumerating the major areas (e.g., base instruction set, unprivileged extensions, privileged instructions, privileged extensions), pointing to spreadsheets that list each feature and subfeature. For example, a spreadsheet for the M extension could contain a feature for each instruction. Each feature would contain subfeatures, such as the following:

> Division instructions have additional subfeatures for specific corner cases, such as division by 0 and overflow.

- Using all registers for each source and destination
- Using all combinations where source and destination register match
- Corner cases of inputs: (0, 1, 2, random number, max positive, max negative, −1)

The spreadsheet should contain the following columns for each feature and subfeature.

- A reference to a specification

- A description of the feature to be verified
- A description of the required stimulus to exercise the feature
- A method of checking the correct operation of the feature
- The coverage specification that indicates that the feature was actually verified
- A reference to the coverage specified in the previous point

See the Wally DV plan in Section 5.11.1 for more examples.

5.3 LINT

Debugging a large design is time-consuming. Engineers save a lot of time by detecting problems during static code analysis (compilation) rather than through simulation. HDL simulators such as Questa indicate some errors and warnings that should always be corrected, but dedicated code analysis tools called *linters* are more comprehensive and faster. Some of the types of warnings flagged by lint tools include:

- Unsynthesizable logic
- Unintended latches and combinational loops
- Undeclared, unused, and undriven signals
- Mismatched port sizes
- Always and case statement gotchas such as missing defaults
- Poor coding style
- Deprecated Verilog constructs
- Mismatched vector sizes in assignment

Many lint tools exist, with Ascent Lint from Real Intent and Synopsys SpyGlass Lint being current commercial market leaders. Design teams often produce customized lint rule sets reflecting their organization's coding standards. Wally uses Verilator lint because it is free and good enough to catch many bugs.

Lint tools produce spurious warnings on certain types of code. For example, Verilator warns about incomplete `case` statements without a `default` even if the value is set elsewhere in an `always_comb` block. To turn off a specific lint warning, place the following comment in the SystemVerilog code. `CASEINCOMPLETE` is the name of the warning being waived.

```
// verilator lint_off CASEINCOMPLETE
```

It is wise to turn the warning back on after the code that would cause the warning:

```
// verilator lint_on CASEINCOMPLETE
```

A good designer keeps the design lint clean every time it is checked in to reduce bugs and prevent warnings that confuse other designers.

5.4 TESTBENCHES

Testbenches apply stimulus to a DUT, and preferably check that the results match expectations. SystemVerilog offers a vast set of nonsynthesizable features for writing testbenches, including generating clocks, reading files, and printing messages to the screen. An entire object-oriented portion of the language caters to software engineers writing large verification testbenches [IEEE 1800-2017], but this is beyond the scope of this chapter.

A simple self-checking SystemVerilog testbench performs the following functions, as illustrated in Fig. 5.1:

- Instantiate DUT.
- Generate clock and reset inputs.
- Read a set of *test vectors* (inputs and expected outputs) from a file.
- Apply the inputs on successive clock cycles and check the outputs.
- Log any mismatches.

Code Example 5.1 shows a testbench for the full adder from Code Example 4.1. The DUT has three inputs and two outputs. The testbench declares internal signals, including the clock, reset, inputs, outputs, and expected output values. It also has two variables to keep track of the current vector number and number of errors seen so far and an array of test

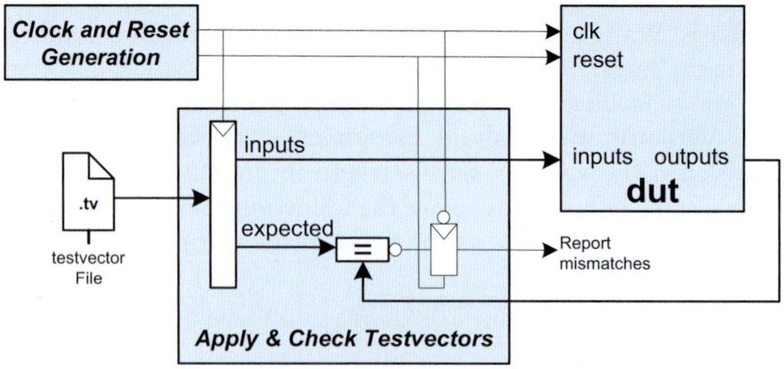

Fig. 5.1 Self-checking testbench.

Code Example 5.1 **fulladder** TESTBENCH

```
module testbench();
  logic        clk, reset;
  logic        a, b, c, s, cout, sexpected, coutexpected;
  logic [31:0] vectornum, errors;
  logic [4:0]  testvectors[10000:0];

  // Instantiate device under test
  fulladder dut(a, b, c, s, cout);

  // Generate clock
  always
    begin
      clk = 1; #5; clk = 0; #5;
    end

  // At start of test, load vectors and pulse reset
  initial
    begin
      $dumpfile("fulladder.vcd");
      $dumpvars;
      $readmemb("fulladder.tv", testvectors);
      vectornum = 0; errors = 0;
      reset = 1; #22; reset = 0;
    end

  // Apply test vectors on rising edge of clk
  always @(posedge clk)
    begin
      #1; {a, b, c, coutexpected, sexpected} = testvectors[vectornum];
    end

  // Check results on falling edge of clk
  always @(negedge clk)
    if (~reset) begin // Skip during reset
      if (s !== sexpected | cout !== coutexpected) begin  // Check result
        $display("Error: inputs = %b", {a, b, c});
        $display("  outputs cout s = %b%b (%b%b expected)",cout, s, coutexpected, sexpected);
        errors = errors + 1;
      end
      vectornum = vectornum + 1;
      if (testvectors[vectornum] === 'bx) begin
        $display("%d tests completed with %d errors",
                 vectornum, errors);
        $stop;
      end
    end
endmodule
```

vectors. The array is 5 bits wide to accommodate the inputs and outputs and 10,001 entries deep so it has plenty of space even if one modifies the file to use many more vectors. The `always` block uses delays to produce a clock continuously oscillating every 10 ticks. At the start of simulation, the testbench reads the test vectors from `fulladder.tv`, interpreting them as binary numbers. It asserts `reset` for slightly more than 2 cycles (22 simulation ticks). It then applies test vectors just after each rising clock edge and checks the outputs on the falling clock edge when reset is not asserted. If either output disagrees with expectations, it reports the failing vector and increments the error count. In any case, it then advances to the next test vector. If that vector is all x, then the testbench has completed all of the vectors in the file, and the simulation should stop.

Observe that the simulation uses several new commands. The `$dumpfile` and `$dumpvars` write a Verilog Change Dump (.vcd) file used in Section 5.5.2 to view waveforms after a Verilator simulation; they are unnecessary for other simulators. `$readmemb` reads a text file in binary format into a memory array. `$readmemh` does the same in hexadecimal format. `$display` writes to the console. `$stop` halts a simulation. `initial` statements are run once at the start of a simulation. `always` statements without a sensitivity list are repeated forever. Testbenches use === and !== instead of == and != for equality and inequality comparison to work properly on signals that might include `x` or `z`.

Code Example 5.2 shows the test vector file with an exhaustive set of vectors. The file lists the three inputs and two outputs. The optional underscores are included for readability but are ignored by the testbench. The file can have comments starting with `//`.

Code Example 5.2 **fulladder.tv**

```
// abc_cout s
000_00
001_01
010_01
011_10
100_01
101_10
110_10
111_11
```

Questa prefers `$stop` when the testbench is done applying vectors; `$finish` terminates without an opportunity to generate coverage statistics or inspect waveforms. However, Verilator prefers to halt the testbench with `$finish`.

A testbench may halt with `$stop` or `$finish`. `$stop` suspends the simulation. This is useful in a graphic user interface environment where the user may want to inspect the waveforms for troubleshooting. `$finish(errcode)` terminates the simulation and prints a diagnostic message according to Table 5.3. If `errcode` is omitted, it defaults to 1.

Table 5.3 **$finish codes.**

errcode	Diagnostic Message
0	None
1	Simulation time and location of `$finish`
2	Simulation time, location of `$finish`, and memory statistics

5.5 LOGIC SIMULATORS

A logic simulator applies test vectors to a model of a system and computes the outputs. We use SystemVerilog to describe both the system model and input generation. Our system model is coded with synthesizable Verilog because it describes hardware, whereas the test vectors are generated by a nonsynthesizable testbench. Traditional *event-driven logic simulation* maintains a list of signal changes called *events* that happen at a particular time. Each event causes the logic that depends on the signal to be reevaluated. For example, when the code on the right-hand side of an `assign` statement or a <= or = in an `always_comb` block changes, the signal on the left-hand side is recomputed. If the left-hand signal changes, this creates a new event that gets added to the list. If the assignment has a delay, the new event will be added after that delay. Otherwise, the new event will be added after an infinitesimal *delta* delay that appears to be zero in simulation yet keeps events ordered to reflect cause and effect. Older simulators are interpreted, but most modern simulators compile the Verilog into a faster executable form.

Leading commercial HDL simulators include Questa from Siemens (formerly called ModelSim, from Mentor Graphics), Cadence Xcelium, and Synopsys VCS. All are mature products that support SystemVerilog, VHDL, and SystemC. Verilator is a leading free open-source Verilog/SystemVerilog simulator.

5.5.1 Questa

Simulate the `fulladder` testbench in the Questa GUI using:

```
$ cd $WALLY/examples/verilog/fulladder
$ vsim fulladder.sv
```

Choose Compile → Compile... and select `fulladder.sv` to compile. Then choose Simulate → Start Simulation..., expand the `work` library, and select `testbench`. Click on Optimization Options and change to "Apply full visibility to all modules (full debug mode)" so that Questa doesn't optimize away the internal signals; then click OK.

In the Objects pane, select all of the signals. Choose Add → To Wave or drag them to the waveform window to add them.

Type `run 100` or `run all` to run the simulation. `run 100` runs the simulation for 100 simulation ticks (10 clock cycles), which is enough to run all 8 test vectors for this example. The testbench from Code Example 5.1 generates a clock to sequence the inputs, and pulses `reset` for two cycles at the start. It then applies all 8 possible combinations of `a`, `b`, and `c`, and checks

> If you are a heavy Questa user, be sure to download the manuals, which describe many features beyond what there is space to cover here. The Questa SIM User's Manual is a good place to start.

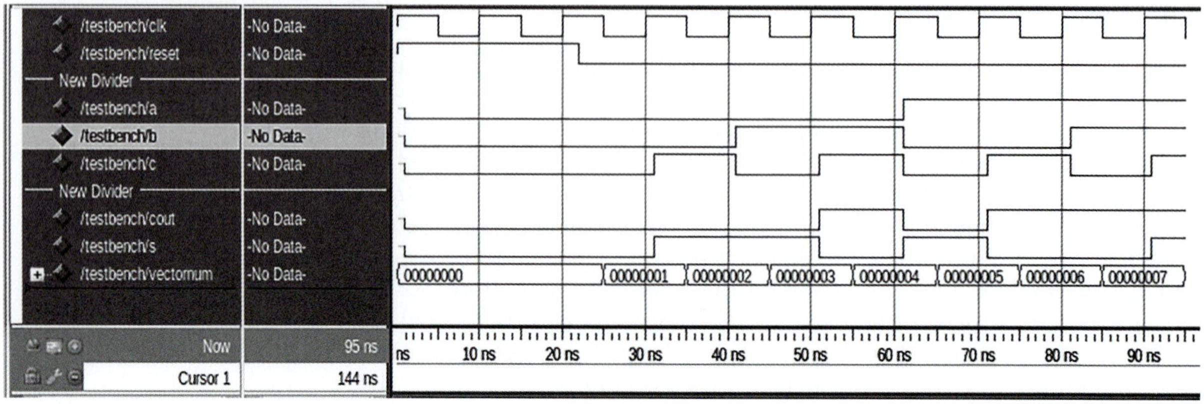

Fig. 5.2 `fulladder` **simulation waveforms in Questa.**

that `s` and `cout` are correct. You will see the following report and waveforms resembling Fig. 5.2. You can type `quit` to terminate Questa.

```
VSIM > run 100
#          8 tests completed with          0 errors
# ** Note: $stop    : /home/harris/cvw/examples/verilog/fulladder/fulladder.sv(42)
#    Time: 95 ns  Iteration: 1  Instance: /testbench
# Break at /home/harris/cvw/examples/verilog/fulladder/fulladder.sv line 42
VSIM2 > quit
```

If you need to fix a bug, reopening Questa and repeating all these clicks becomes tedious. You can recompile with the compile menu. Then repeat to simulation with:

```
VSIM > restart -f   # Restarts the simulation without losing your waveforms
VSIM > run 100
```

Better yet, you can create a script called a *do file* that compiles and simulates for you. Code Example 5.3 shows a do file for the full adder. First, it uses the `vlog`, `vopt`, and `vsim` commands to compile, optimize, and start the simulation. By default, the compiled design is placed in a directory called `work`. The do file adds all the interface signals to the waveform window and then runs until the simulation completes.

```
$ vsim -do fulladder.do
```

Code Example 5.3 **fulladder.do**

```
# fulladder.do

# Compile, optimize, and start the simulation
vlog fulladder.sv
vopt +acc work.testbench -o workopt
vsim workopt

# Add waveforms and run the simulation
add wave *
run -all
view wave
```

If you don't need the GUI to debug, you can run in batch mode, which is faster. Code Example 5.4 shows a batch mode script that doesn't add waveforms and quits when the simulation is complete. This is handy for regressions, which run many simulations and report whether they passed. Note the `-c` flag to `vsim` to invoke a batch mode simulation.

```
$ vsim -c -do fulladder-batch.do
```

If you don't have an X2Go or VNC connection to your server, you can be productive with command line tools using an `ssh -Y` connection. Batch simulations work great from the command line. The Questa GUI will open over `ssh -Y` but is painfully slow to use.

Code Example 5.4 **fulladder-batch.do**

```
# fulladder-batch.do
vlog fulladder.sv
vopt +acc work.testbench -o workopt
vsim workopt
run -all
quit
```

To avoid having to remember the `vsim` command and its arguments, we can write a 1-line script containing the command. Name it `sim-fulladder-batch`, change its permissions to be executable (`chmod 755`), and invoke the script with:

```
$ ./sim-fulladder-batch
```

If the simulation is simple, you can use the Questa `qrun` command to automate all of this batch simulation instead of writing a do file. `qrun` identifies the top-level testbench based on the module hierarchy and compiles, optimizes, and simulates the file.

```
$ qrun fulladder.sv
```

5.5.2 Verilator

Verilator is an open-source SystemVerilog simulator. Verilator converts the SystemVerilog code into a C program that simulates its behavior, then compiles and runs the C program. Because it is open source, Verilator is a key component of open-source hardware design. Verilator is gaining traction at companies as well because simulator licenses are expensive and because Verilator can be as fast or faster than commercial simulators.

Verilator was conceived by Paul Wasson at Digital Equipment Corporation in 1994 and was released as open source in 1998. Wilson Snyder has been the lead developer since 2001. Initially, it had many limitations, such as handling only a subset of Verilog and only supporting testbenches written in SystemC. During the time this book was in development, Verilator became capable of simulating SystemVerilog testbenches and all of Wally's features, and many of Verilator's functional and performance bugs were fixed. Verilator can run all of the regression tests on the Wally testbench. Verilator does not yet support UVM (see Table 5.1), but work is in progress.

Verilator involves three steps: compiling the Verilog into an executable file, running the executable to simulate the testbench, and optionally viewing waveforms. Invoke these steps with the following:

```
$ cd $WALLY/examples/verilog/fulladder
$ verilator --binary --top-module testbench --trace fulladder.sv
$ obj_dir/Vtestbench
```

Verilator takes a list of Verilog files to compile (`fulladder.sv`). The `--top-module` argument usually indicates the testbench. The `--binary` argument asks Verilog to generate C code to model the behavior of Verilog and compile it into a binary executable file. The optional `--trace` argument enables generating a `.vcd` (Verilog Change Dump) file for a waveform viewer when the testbench invokes `$dumpvars`. Verilator creates `obj_dir` containing the executable (`Vtestbench`) along with the C code and supporting files.

Alternatively, a script invokes both of these commands, so you don't need to remember the arguments. See the output printed below, showing Verilator compile the files and run the simulation. The testbench prints the vectors with the inputs and expected outputs. The simulation also records the signals to `fulladder.vcd`.

```
$ ./verilate
make: Entering directory '/home/harris/cvw/examples/verilog/fulladder/obj_dir'
ccache g++ -Os  -I.  -MMD -I/usr/local/share/verilator/include -
I/usr/local/share/verilator/include/vltstd -DVM_COVERAGE=0 -DVM_SC=0 -DVM_TRACE=1 -
DVM_TRACE_FST=0 -DVM_TRACE_VCD=1 -faligned-new -fcf-protection=none -Wno-bool-operation -Wno-
shadow -Wno-sign-compare -Wno-tautological-compare -Wno-uninitialized -Wno-unused-but-set-
parameter -Wno-unused-but-set-variable -Wno-unused-parameter -Wno-unused-variable    -
DVL_TIME_CONTEXT   -fcoroutines -c -o verilated.o
/usr/local/share/verilator/include/verilated.cpp
...
ccache g++ -Os  -I.  -MMD -I/usr/local/share/verilator/include -
I/usr/local/share/verilator/include/vltstd -DVM_COVERAGE=0 -DVM_SC=0 -DVM_TRACE=1 -
DVM_TRACE_FST=0 -DVM_TRACE_VCD=1 -faligned-new -fcf-protection=none -Wno-bool-operation -Wno-
shadow -Wno-sign-compare -Wno-tautological-compare -Wno-uninitialized -Wno-unused-but-set-
parameter -Wno-unused-but-set-variable -Wno-unused-parameter -Wno-unused-variable    -
DVL_TIME_CONTEXT   -fcoroutines -c -o Vtestbench__ALL.o Vtestbench__ALL.cpp
echo "" > Vtestbench__ALL.verilator_deplist.tmp
g++ -fuse-ld=mold   verilated.o verilated_vcd_c.o verilated_timing.o verilated_threads.o
Vtestbench__ALL.a    -pthread -lpthread -latomic   -o Vtestbench
rm Vtestbench__ALL.verilator_deplist.tmp
make: Leaving directory '/home/harris/cvw/examples/verilog/fulladder/obj_dir'



- V e r i l a t i o n   R e p o r t: Verilator 5.025 devel rev v5.024-12-ga9664b8d1
- Verilator: Built from 0.013 MB sources in 3 modules, into 0.049 MB in 11 C++ files needing
0.000 MB
- Verilator: Walltime 4.949 s (elab=0.000, cvt=0.212, bld=4.734); cpu 0.000 s on 1 threads;
alloced 19.750 MB
cycle: 00000001 vectornum 00000000 testvectors[vectornum]: 00000
cycle: 00000002 vectornum 00000000 testvectors[vectornum]: 00000
cycle: 00000003 vectornum 00000001 testvectors[vectornum]: 00101
cycle: 00000004 vectornum 00000002 testvectors[vectornum]: 01001
cycle: 00000005 vectornum 00000003 testvectors[vectornum]: 01110
cycle: 00000006 vectornum 00000004 testvectors[vectornum]: 10001
cycle: 00000007 vectornum 00000005 testvectors[vectornum]: 10110
cycle: 00000008 vectornum 00000006 testvectors[vectornum]: 11010
cycle: 00000009 vectornum 00000007 testvectors[vectornum]: 11111
cycle: 0000000a vectornum 00000008 testvectors[vectornum]: 00000
cycle: 0000000b vectornum 00000009 testvectors[vectornum]: 00000
        10 tests completed with          0 errors
- fulladder.sv:49: Verilog $finish
- S i m u l a t i o n   R e p o r t: Verilator 5.025 devel
- Verilator: $finish at 120ps; walltime 0.001 s; speed 0.000 s/s
- Verilator: cpu 0.000 s on 1 threads; alloced 249 MB
```

The waveforms recorded in the VCD file can be displayed with commercial or open-source viewers. For example, invoke the open-source GTKWave viewer uaing:

```
$ gtkwave -f fulladder.vcd
```

Click into the TOP -> testbench hierarchy to see a list of internal signals; then click to add them to the waveform viewer, as shown in Fig. 5.3.

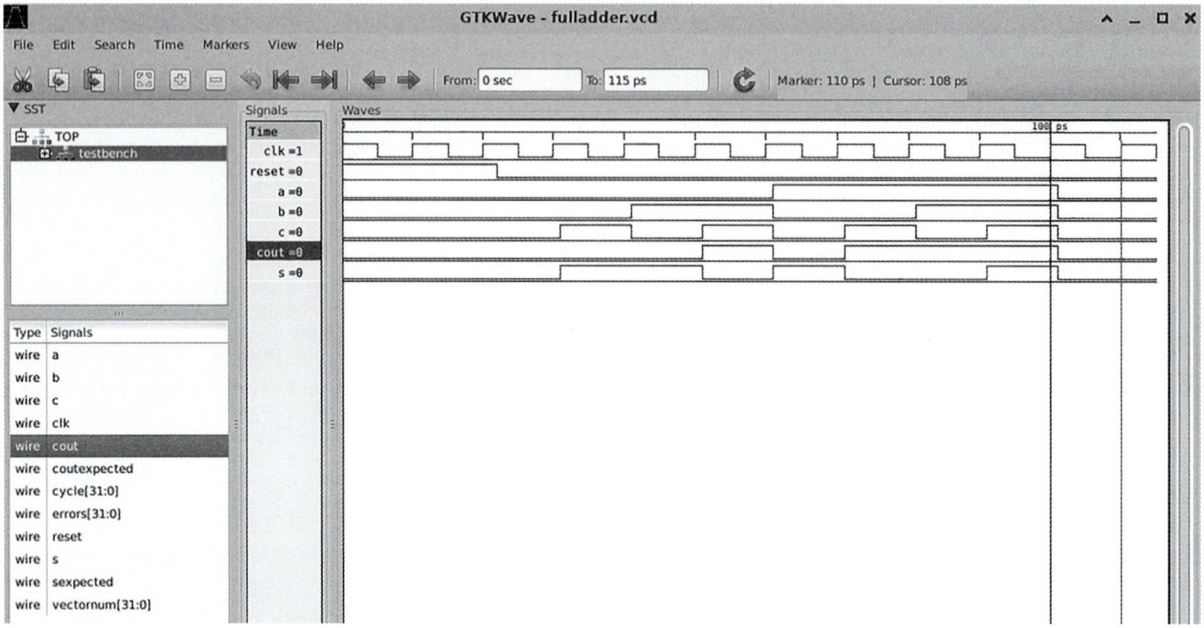

Fig. 5.3 `fulladder` **simulation waveforms in GTKWave.**

5.5.3 VCS

Synopsys VCS is another commercial HDL simulator. Like Verilator, it compiles HDL code into an executable, which is then run. By default, the executable is named `simv`. Invoke VCS with the following:

```
$ cd $WALLY/examples/verilog/fulladder
$ vcs -sverilog -full64 fulladder.sv
$ ./simv
```

5.6 LOCK-STEP-COMPARE

Signature checking has several limitations in verification. A program typically produces one signature word after a sequence of instructions, and that might not capture all design errors. For example, an early version of riscv-arch-test floating-point tests created a signature with the result of the instruction but not the floating-point flags, thus missing bugs in the flag logic. Moreover, a system test such as running a benchmark or booting Linux doesn't produce a signature or only indicates success (e.g., a login prompt) after millions of instructions. A more robust verification methodology is *lock-step-compare*, in which a testbench logs all the architectural state that changes when an instruction completes and compares against a reference model, reporting every discrepancy. The state includes the program counter (PC), integer and floating-point registers, control and status registers (CSRs), and memory. In this way, deviations are discovered immediately rather than possibly being hidden or causing a subtle failure much later. A key requirement of operation is to inject the same interrupts into the reference model as the DUT, so they remain in *lockstep*, following the same program counter sequence and producing exactly the same architectural state after each instruction. Another challenge is that the RISC-V specification is both configurable and ambiguous in many places, so the reference model must be configured to exactly match the behavior of the DUT when multiple behaviors are legal. The reference model must not only agree on XLEN and supported extensions but also on subtle behavior such as whether the `UXL` bits of the `mstatus` register are writable, such that user-mode `XLEN` can be changed at runtime.

ImperasDV is a widely used commercial lock-step-compare tool. It requires a tracer in the testbench to log the changing state using the RVVI-TRACE standard interface [RVVI-TRACE]. The DUT runs programs in the usual fashion, while the Imperas reference model runs the same programs, takes the same interrupts, reads the same values from memory and peripherals, and reports tracer discrepancies, as shown in Fig. 5.4 (github.com/riscv-verification/riscvISACOV/blob/v20231026/source/coverage/RV32I_coverage.svh). Synopsys acquired Imperas in 2024.

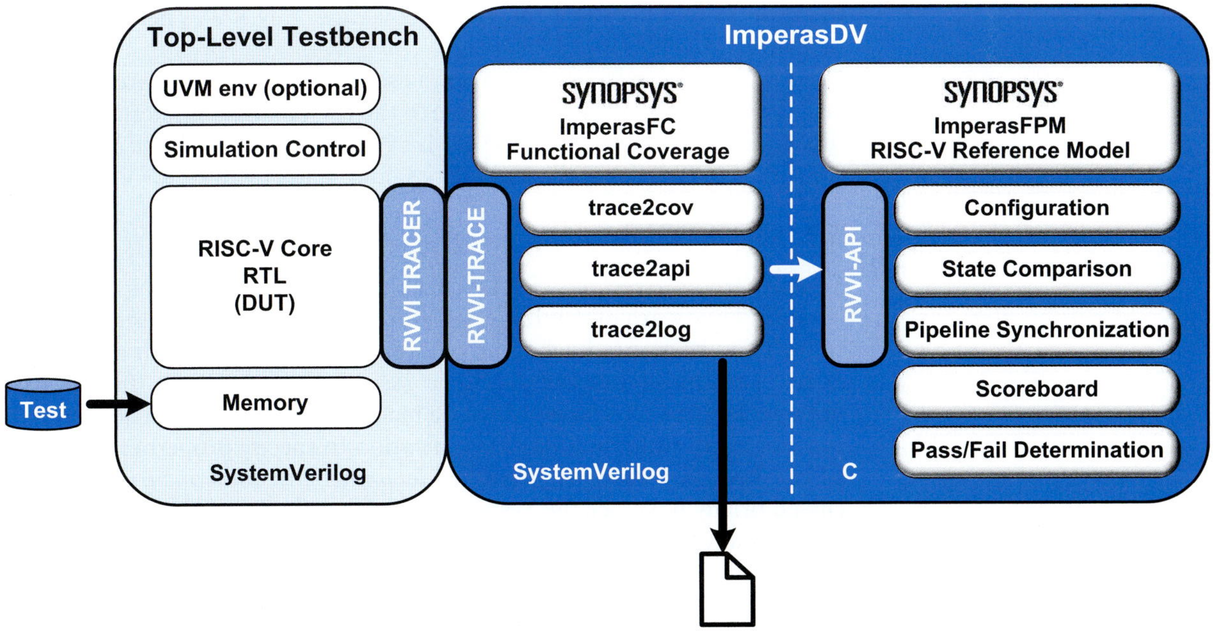

Fig. 5.4 ImperasDV lock-step-compare system. (Figure adapted and reprinted with permission, courtesy of Imperas.)

Table 5.4 RVVI-TRACE signals.

Signal	Size	Meaning
clk	1	System clock updating state
insn	32	Current instruction
trap	1	Trap rather than retiring instruction
pc_rdata	XLEN	PC of instruction
x_wdata	XLEN × 32	New value(s) being written to integer register file
x_wb	32	Which integer registers are being written
f_wdata	FLEN × 32	New value(s) being written to floating-point register file
f_wb	32	Which floating-point registers are being written
csr	XLEN × 4,096	New value(s) being written to CSRs
csr_wb	4,096	Which CSRs are being written

5.6.1 RVVI-TRACE Interface

RVVI-TRACE defines a SystemVerilog interface, capturing the state changes that occur during each instruction. Wally uses RVVI both to communicate with ImperasDV during lockstep simulation and to provide the state to riscvISACOV functional coverage. Key RVVI fields are listed in Table 5.4.

`testbench/common/wallyTracer.sv` sends register changes (register files and CSRs) from the Writeback stage to the RVVI-TRACE interface. It also pipelines other signals (such as `PCF`, `InstrF`, and `TrapM`) into the Writeback stage, accounting for stalls and flushes, and sends them to RVVI-TRACE.

5.7 ASSERTIONS

SystemVerilog Assertions (SVAs) check that the RTL conforms to a specification. They warn of trouble long before it propagates to the architectural state. Thus, they speed debug by pinpointing the root cause of a problem. Some of the most common places for assertions are in FIFOs (first in, first out queues), bus interfaces, and state machines. Assertions are also used in functional coverage and formal verification to specify properties that must be checked. Assertions are classified as *immediate* and *concurrent*, as discussed briefly next. See the SystemVerilog specification for syntax nuances or [Mehta20] for an in-depth tutorial about using SVA for verification.

5.7.1 Immediate Assertions

Immediate assertions test an expression at the time they are executed. They are placed in `initial` or `always` (`always_comb`/`always_ff`) blocks, which determines when they execute. Assertions are written so that the expression is TRUE unless there is a defect in the design. The syntax of an assertion is as follows:

```
assert (expression) <statement1> <else statement2>
```

If `expression` evaluates to TRUE (1), `statement1` is executed; otherwise, `expression` must be 0, x, or z, and `statement2` is executed. If `statement1` is omitted, the assertion is silent when the `expression` is TRUE. If `else statement2` is omitted, the assertion calls `$error` when `expression` is not TRUE.

The most common statements print messages associated with levels of severity.

- `$fatal(errcode, message)`
- `$error(message)`
- `$warning(message)`
- `$info(message)`

Each of these statements prints the `message`. `$fatal` also terminates the simulation by calling `$finish(errcode)`, with the codes described in Table 5.3. Assertions typically use `$error` to report a violation.

Code Example 5.5 gives examples of immediate assertions in the `riscvassertions` module, which is part of the Wally testbench. The module contains many assertions to ensure the configuration parameters describe legal architectural and microarchitectural choices. The assertion statements are in an `initial` block, so they are checked just once at the beginning of simulation. For example, the RISC-V architecture specification allows 0, 16, or 64 entries in the Physical Memory Protection (PMP) checker. Virtual memory is only available on systems with supervisor mode because it depends on the supervisor-mode `satp` CSR, discussed further in Chapter 11. The Wally load/store unit (LSU) microarchitecture reads a maximum of XLEN bits from the bus or tightly integrated memory but can read longer lines from a data cache. Thus, for example, Wally RV32 only supports FLEN = 64-bit double-precision floating-point if the cache is enabled. If any of these configuration parameters are unsupported, the simulation terminates with a fatal error and an explanation of the problem. Hence, the assertions enforce legal configurations as part of the code base rather than part of the documentation that a user might lose or not read. For example, if the config file sets PMP_ENTRIES to 1, the assertion will terminate the simulation with:

```
# ** Fatal: Illegal number of PMP entries: PMP_ENTRIES must be 0, 16, or 64
```

Code Example 5.5 **riscvassertions.sv**

```
module riscvassertions import cvw::*; #(parameter cvw_t P);
  initial begin
    assert (P.PMP_ENTRIES == 0 || P.PMP_ENTRIES==16 || P.PMP_ENTRIES==64)
      else $fatal(1, "Illegal number of PMP entries: PMP_ENTRIES must be 0, 16, or 64");
    assert (P.S_SUPPORTED || P.VIRTMEM_SUPPORTED == 0)
      else $fatal(1, "Virtual memory requires S mode support");
    assert (P.DCACHE_SUPPORTED || ~P.F_SUPPORTED || P.FLEN <= P.XLEN)
      else $fatal(1, "Data cache required to support FLEN > XLEN because
AHB/DTIM bus width is XLEN");
    ...
end
```

Sometimes the assertion may fire while expression is glitching, producing a faulty error during a transient invalid state that resolves to a good state. *Deferred immediate assertions* add the final keyword, which causes the assertion to wait until expression settles before evaluating.

```
assert final (expression) <statement1> <else statement2>
```

5.7.2 Concurrent Assertions

Concurrent assertions describe behavior that spans clock cycles. They are called concurrent because they execute in parallel with the rest of the design. They use a succinct and powerful notation suited to describe multicycle events in synchronous sequential systems. The SystemVerilog 1800-2017 specification devotes more than a hundred pages to concurrent assertions, but this section just touches on a simple example.

Concurrent assertions are expressed using *properties*, which in turn may depend on *sequences*. Properties use the *implication operator* |=> to imply a relationship between an *antecedent* and a *consequent*. If the antecedent is satisfied, the concurrent assertion checks the consequent and reports if it is violated.

For example, consider checking a property of a bus protocol. According to the AHB standard ([AHB21], Section 3.7.1), when the HREADY signal is low, indicating that a peripheral needs a wait state, the controller must not change the transfer type HTRANS from one bus cycle to the next, except:

- The controller can change HTRANS from IDLE to NONSEQ.

- If the controller was busy in the previous cycle of a burst transfer, HTRANS can go from BUSY to SEQ.
- During reset, the protocol does not apply.

Code Example 5.6 gives an example of a concurrent assertion to check this property. The assertion has the name `transaction_type_stability` used in reporting. The concurrent assertion is evaluated on the rising edge of the HCLK bus clock. It is disabled during bus reset. Otherwise, if HREADY is low (antecedent), one of three consequents must be true: HTRANS is stable (has the same value as the previous cycle), HTRANS was IDLE on the previous cycle, or HTRANS was BUSY on the previous cycle during a burst.

Code Example 5.6 **fulladder-batch-coverage.do**

```
// Example courtesy Umar Yaqoob and Bilal Zafar of 10xEngineers.
  transaction_type_stability: assert property (
    @(posedge HCLK) disable iff (!HRESETn)
      !HREADY |=> $stable(HTRANS)            |
                  ($past(HTRANS) == AHB_IDLE) |
                  (($past(HTRANS) == AHB_BUSY) & (HBURST != 3'b000))
  );
```

5.8 CODE COVERAGE

A logic simulator can collect and report coverage metrics for HDL code. A minimum level of coverage is that the test vectors should activate every line of code, called *line coverage*. A next level of coverage is to ensure every expression in a logic equation and every condition in a case or branch has been exercised, called *statement*, *expression*, *condition*, *FSM*, and *branch coverage*.

For example, Code Example 5.7 shows how to log code coverage into a Unified Coverage Database (UCDB) file using Questa. Compared to Code Example 5.4, the do file adds a `+cover=sbecf` to the `vopt` command to measure statement, branch, expression, condition, and FSM coverage. It runs `vsim` with the `-coverage` flag. Finally, it saves all the coverage information from `/testbench/dut` into `fulladder.ucdb`.

Code Example 5.7 **fulladder-batch-coverage.do**

```
vlog fulladder.sv
vopt +acc work.testbench -o workopt +cover=sbecf
vsim workopt -coverage
run -all
coverage save -instance /testbench/dut fulladder.ucdb
quit
```

`/testbench/dut` in Code Example 5.7 refers to the fulladder module. Lines 50-51, containing the fulladder body, are listed in Code Example 5.8.

Code Example 5.8 **fulladder.sv** BODY

```
50:  assign s = a ^ b ^ c;
51:  assign cout = (a & b) | (a & c) | (b & c);
```

Statement coverage counts whether each statement in the code has been executed. It is similar to *line coverage* but more specific because statements could span more than one line, and more than one statement could be placed on a line. *FSM coverage* checks that all states have been reached and all transitions between states have been exercised. *Branch coverage* checks if the Boolean expression in control structures (e.g., `if`, `case`) have evaluated to both TRUE and FALSE. Condition and expression coverage check that each input of a Boolean expression is exercised TRUE and FALSE while the other inputs are in a state that make the output depend on the input being exercised. *Condition coverage* applies to the condition in an if statement, and *expression coverage* applies to the right-hand side of assignment statements. For example, in the statement `y = a & b | c`, `y` only depends on `a` when `b` = 1 and `c` = 0. Hence, the full expression coverage requires exercising six cases:

- `a` = 0 / 1 when `b` = 1, `c` = 0
- `b` = 0 / 1 when `a` = 1, `c` = 0
- `c` = 0 / 1 when `a` and `b` are not both 1

Invoke the simulation with the following `vsim` command, then use `vcover` to produce a detailed human-readable report from the UCDB:

```
$ vsim -c -do fulladder-batch-coverage.do
$ vcover report -details fulladder.ucdb > fulladder_coverage_details.rpt
```

The report in Code Example 5.9 shows that the full adder has six expressions (e.g., `(a & b)`, `(a ^ b)`, etc.) and two statements (lines 50–51 assigning `s` and `cout`), as highlighted in blue. The simulation applied the `fulladder.tv` vectors from Code Example 5.2 to exhaustively check all 8 combinations of `a`, `b`, and `c`, so the full adder's expressions and statements are fully covered.

Code Example 5.9 `fulladder-coverage_details.rpt`

```
QuestaSim-64 vcover 2022.4_2 Coverage Utility 2022.12 Dec  9 2022
Start time: 10:58:48 on May 23,2023
vcover report -details fulladder.ucdb
Coverage Report by instance with details

=== Instance: /dut
=== Design Unit: work.fulladder
Expression Coverage:
    Enabled Coverage              Bins    Covered    Misses  Coverage
    ----------------              ----       ----    ------  --------
    Expressions                      6          6         0   100.00%

================================Expression Details================================

Expression Coverage for instance /dut -

  File fulladder.sv
-----------Focused Expression View (Bimodal)------------
Line       50 Item    1  ((a ^ b) ^ c)
Expression totals: 3 of 3 input terms covered = 100.00%

----------------Focused Expression View-----------------
Line       51 Item    1  (((a & b) | (a & c)) | (b & c))
Expression totals: 3 of 3 input terms covered = 100.00%


Statement Coverage:
    Enabled Coverage              Bins       Hits    Misses  Coverage
    ----------------              ----       ----    ------  --------
    Statements                       2          2         0   100.00%

===============================Statement Details===============================

Statement Coverage for instance /dut -

    Line         Item                      Count     Source
    ----         ----                      -----     ------
  File fulladder.sv
    50              1                          9
    51              1                          9

Total Coverage By Instance (filtered view): 100.00%
```

To see an example of incomplete coverage, comment out the sixth and seventh lines of `fulladder.tv`, as shown in Code Example 5.10. Now the tests no longer contain a vector that makes `cout` = 1 due to `a` being 1 (the last row has `cout` = 1, but `b` and `c` are also 1 so `cout` doesn't depend on `a`). Rerun `vsim` and regenerate the coverage report, as shown in Code Example 5.11. Now the `cout` logic at line 51 is incompletely covered. Specifically, the vectors no longer are sufficient to show that setting input `a` = 1 affects `cout`. Row 2 of the Focused Expression View indicates that `a` is never 1 when `b` and c are in a state that affects the output. The *nonmasking condition* is tricky to interpret in this report but essentially means when b ^ c is 1 such that cout is influenced by a. The fulladder shows 91.66% coverage overall.

Code Example 5.10 MODIFIED **fulladder.tv**

```
// abc_cout s
000_00
001_01
010_01
011_10
100_01
//101_10
//110_10
111_11
```

Code Example 5.11 **fulladder-coverage_details.rpt** (MISSING 101 AND 110 TEST CASES)

```
Coverage Report by instance with details

=== Instance: /dut
=== Design Unit: work.fulladder
Expression Coverage:
    Enabled Coverage              Bins    Covered    Misses  Coverage
    ----------------              ----    ----       ------  --------
    Expressions                      6       5            1    83.33%

================================Expression Details================================

Expression Coverage for instance /dut -

  File fulladder.sv
-----------Focused Expression View (Bimodal)-----------
Line       50 Item    1  ((a ^ b) ^ c)
Expression totals: 3 of 3 input terms covered = 100.00%

----------------Focused Expression View----------------
Line       51 Item    1  (((a & b) | (a & c)) | (b & c))
Expression totals: 2 of 3 input terms covered = 66.66%

  Input Term   Covered  Reason for no coverage   Hint
  -----------  --------  ----------------------  --------------
           a         N  '_1' not hit             Hit '_1'
           b         Y
           c         Y

     Rows:        Hits  FEC Target            Non-masking condition(s)
  ---------  ---------  --------------------  -------------------------
  Row   1:           1  a_0                   (~(b & c) && b), (~(b & c) && c)
  Row   2:     ***0***  a_1                   (~(b & c) && ~(a & c) && b), (~(b & c) && ~(a & b) && c)
  Row   3:           1  b_0                   (~(a & c) && a), (~((a & b) | (a & c)) && c)
  Row   4:           1  b_1                   (~(b & c) && ~(a & c) && a), (~((a & b) | (a & c)) && c)
  Row   5:           1  c_0                   (~(a & b) && a), (~((a & b) | (a & c)) && b)
  Row   6:           1  c_1                   (~(b & c) && ~(a & b) && a), (~((a & b) | (a & c)) && b)
```

```
Statement Coverage:
    Enabled Coverage              Bins      Hits    Misses  Coverage
    ----------------              ----      ----    ------  --------
    Statements                       2         2         0   100.00%

================================Statement Details================================

Statement Coverage for instance /dut -
    Line          Item                      Count       Source
    ----          ----                      -----       ------
  File fulladder.sv
    50               1                          6
    51               1                          6

Total Coverage By Instance (filtered view): 91.66%
```

In a larger design, multiple independent tests are run, often in parallel. To combine each of their UCDB files to get overall coverage, use:

```
$ vcover merge -out combined.ucdb test1.ucdb test2.ucdb
```

As discussed in Section 5.1.5, Technology Readiness Level 5 (TRL5) requires 100% code coverage to ensure the entire code has been exercised. This typically involves writing a set of directed tests that cover the obvious cases (e.g., tests each instruction), using a random test generator to stress unobvious cases, and then writing more custom directed tests to check stranger cases (e.g., illegal instructions).

Nevertheless, it is often impossible to actually exercise all of the statements because certain parts of the code may be unreachable, particularly if the code is configurable. For example, a 32-bit configuration cannot exercise 64-bit instructions. The designer adds *coverage exclusions* to tell the coverage tool to ignore certain uncovered cases that the designer is satisfied are unreachable. Section 5.11.11.2 gives an example of Wally coverage exclusions. Adding exclusions requires caution because poorly written or outdated exclusions can hide legitimate bugs.

5.9 FUNCTIONAL COVERAGE

Code coverage measures whether all code is exercised, while functional coverage measures whether all features have been exercised [Mehta20]. Functional coverage includes *architectural* coverage, which applies to any core supporting a particular extension, and *microarchitectural* coverage, which is specific to an implementation, stressing features such as pipeline hazards, interrupts, cache and bus systems, and branch prediction.

Functional coverage begins with the DV plan defining what features need to be covered. For example, in RISC-V, every instruction should be covered, using every register and corner cases appropriate to that instruction. Experience shows that coverage that is easy to describe doesn't catch many bugs, and coverage that catches hard bugs takes a lot of thought to describe.

SystemVerilog `covergroup` and `coverpoint` statements define how the coverage is measured. A covergroup consists of related coverpoints sampled at a particular time, such as coverpoints for a particular instruction. The coverpoints define the specific coverage, such as using every destination register. Each coverpoint has one or more bins. For example, the destination register coverpoint has 32 bins, one for each possible register. Wally uses a modified version of the Imperas riscvISACOV coverpoints, discussed below.

Functional coverage measurement involves running tests while the simulator monitors which coverpoint bins and assertions were hit. By also running the tests in lockstep mode against a reference model, you can ensure that the RTL produces the correct results for each of these coverpoints.

Functional coverage tests can be directed or random. Many features are easy to test with directed tests, but some corner cases are hard to reach and are easier to cover with constrained random tests. When functional coverage reaches 100%, the verification engineer knows enough tests have been run and all features in the DV plan have been exercised.

5.9.1 riscvISACOV

riscvISACOV is a set of SystemVerilog architectural functional coverpoints developed by Imperas (now Synopsys). They cover most of the common RISC-V extensions. A set of RV32I coverpoints are on a public repository [github.com/riscv-verification/riscvISACOV], and a larger set are available under a commercial license.

riscvISACOV has coverage definition files for every extension it supports (e.g., RV32I, RV64I, RV64M, etc.). Each file contains a covergroup for each instruction in that extension. Each covergroup contains various coverpoints. Table 5.5 summarizes coverpoints used for RV{32/64}I instructions. For example, each instruction is tested with every source and destination register and with various corner cases of signs and values. Not all instructions use all coverpoints; for example, R-type instructions don't use immediates, I-type instructions don't use `rs2`, shifts don't cover negative shift amounts, and only branches use branch offsets.

riscvISACOV uses the RVVI interface (see Section 5.6.1) to access the architectural state. The `ins` variable contains the current instruction sampled on the rising edge of the clock when it completes, including information such as the name, source and destination register identifiers and values, and whether a trap occurs. Fig. 5.5 lists an excerpt of the riscvISACOV RV32I add covergroup and its coverpoints. Only a few lines of repetitive definitions are shown.

5.9.2 Directed Tests

Test generators produce directed tests to cover the easy functional coverage cases systematically and efficiently.

Table 5.5 riscvISACOV RV{32/64}I coverpoints.

Name	Description	Number of Bins
cp_asm_count	Is the instruction used?	1
cp_rd	Each register is exercised	32
cp_rs1		32
cp_rs2		32
cp_rd_sign	Positive, negative	2
cp_rs1_sign		2
cp_rs2_sign		2
cp_offset		2
cp_imm_sign	0, positive, negative	3
cr_rs1_rs2	Cross-product of source signs	4
cr_rs1_imm		6
cmp_rd_rs1_eq	Destination and source uses same regs	32
cmp_rd_rs2_eq		32
cmp_rd_rs1_rs2	All three registers are the same	32
cmp_rd_rs1_eqval	Contents of registers match	2
cmp_rd_rs2_eqval		2
cp_rd_toggle	Each bit is 0 or 1	2XLEN
cp_rs1_toggle		2XLEN
cp_rs2_toggle		2XLEN
cp_rd_maxvals	0, 1, 2XLEN-1, 2^{XLEN-1}-1, 2^{XLEN-1}+1, 2^{XLEN}-1	6
cp_rs1_maxvals		6
cp_rs2_maxvals		6
cp_gpr_hazard	None, RAW, WAR, WAW	4
cp_mem_hazard	None, RAW	2
cp_mem_unaligned	Yes, no	2

```
covergroup add_cg with function sample(ins_rv32i_t ins);
    option.per_instance = 1;
    option.comment = "Add";

    cp_asm_count : coverpoint ins.ins_str == "add"  iff (ins.trap == 0)  {
        option.comment = "Number of times instruction is executed";
        bins count[]  = {1};
    }
    cp_rd : coverpoint ins.get_gpr_reg(ins.current.rd)  iff (ins.trap == 0) {

        option.comment = "RD (GPR) register assignment";
    }
    cp_rd_sign : coverpoint int'(ins.current.rd_val)  iff (ins.trap == 0)  {
        option.comment = "RD (GPR) sign of value";
        bins neg  = {[$:-1]};
        bins pos  = {[1:$]};
    }
    cp_rs1 : coverpoint ins.get_gpr_reg(ins.current.rs1)  iff (ins.trap == 0)  {

        option.comment = "RS1 (GPR) register assignment";
    }
    cp_rs1_sign : coverpoint int'(ins.current.rs1_val)  iff (ins.trap == 0) {

        option.comment = "RS1 (GPR) sign of value";
        bins neg  = {[$:-1]};
        bins pos  = {[1:$]};
    }
    cp_rs2 : coverpoint ins.get_gpr_reg(ins.current.rs2)  iff (ins.trap == 0)  {

        option.comment = "RS2 (GPR) register assignment";
    }
    cp_rs2_sign : coverpoint int'(ins.current.rs2_val)  iff (ins.trap == 0) {

        option.comment = "RS2 (GPR) sign of value";
        bins neg  = {[$:-1]};
        bins pos  = {[1:$]};
    }
    cr_rs1_rs2 : cross cp_rs1_sign,cp_rs2_sign  iff (ins.trap == 0)  {
        option.comment = "Cross coverage of RS1 sign and RS2 sign";
    }

`ifdef COVER_LEVEL_COMPL_EXT
    cmp_rs1_rs2_eq : coverpoint ins.current.rs1 == ins.current.rs2  iff (ins.trap == 0)  {
        option.comment = "Compare RS1 and RS2 register assignment";
        bins x0  = {1} iff (ins.current.rs1 == "x0");
        bins x1  = {1} iff (ins.current.rs1 == "x1");
        bins x2  = {1} iff (ins.current.rs1 == "x2");
        ...
        bins x31  = {1} iff (ins.current.rs1 == "x31");
`endif
    }
```

Fig. 5.5 riscvISACOV RV32I add covergroup.

```
cmp_rs1_rs2_eqval : coverpoint ins.current.rs1_val == ins.current.rs2_val
iff (ins.trap == 0)  {
    option.comment = "Compare RS1 and RS2 register values";
    bins rs1_eqval_rs2  = {1};
    bins rs1_neval_rs2  = {0};
    }
cp_rd_toggle : coverpoint unsigned'(ins.current.rd_val)  iff (ins.trap ==0)  {
     option.comment = "RD Toggle bits";
     wildcard bins bit_0_0  = {32'b???????????????????????????????0};
     wildcard bins bit_1_0  = {32'b??????????????????????????????0?};
     wildcard bins bit_2_0  = {32'b?????????????????????????????0??};
     ...
     wildcard bins bit_31_0  = {32'b0???????????????????????????????};
     wildcard bins bit_0_1  = {32'b???????????????????????????????1};
     wildcard bins bit_1_1  = {32'b??????????????????????????????1?};
     wildcard bins bit_2_1  = {32'b?????????????????????????????1??};
     ...
     wildcard bins bit_31_1  = {32'b1???????????????????????????????};
}
cp_rs1_toggle : coverpoint unsigned'(ins.current.rs1_val)  iff (ins.trap == 0) {
     ...
     }
cp_rs2_toggle : coverpoint unsigned'(ins.current.rs2_val)  iff (ins.trap == 0)  {
     ...
cp_rd_maxvals : coverpoint unsigned'(ins.current.rd_val)  iff (ins.trap == 0)  {
     option.comment = "RD Max values";
     wildcard bins zeros  = {0};
     wildcard bins min  = {32'b10000000000000000000000000000000};
     wildcard bins max  = {32'b01111111111111111111111111111111};
     wildcard bins ones  = {32'b11111111111111111111111111111111};
}
cp_rs1_maxvals : coverpoint unsigned'(ins.current.rs1_val)  iff (ins.trap == 0)  {
     ...
     }
cp_rs2_maxvals : coverpoint unsigned'(ins.current.rs2_val)  iff (ins.trap == 0)  {
     ...
     }
`endif


`ifdef COVER_LEVEL_DV_UP_BAS
cmp_rd_rs1_eq : coverpoint ins.current.rd == ins.current.rs1  iff (ins.trap == 0)  {
        option.comment = "RD and RS1 register (assignment) WAR Hazard";
        bins x0  = {1} iff (ins.current.rd == "x0");
        bins x1  = {1} iff (ins.current.rd == "x1");
        ...
    }
```

Fig. 5.5 **riscvISACOV RV32I add covergroup. Cont'd**

```
cmp_rd_rs1_eqval : coverpoint ins.current.rd_val == ins.current.rs1_val iff (ins.trap == 0) {
    option.comment = "Compare RD and RS1 register values";
    bins rd_eqval_rs1  = {1};
    bins rd_neval_rs1  = {0};
}
cmp_rd_rs2_eq : coverpoint ins.current.rd == ins.current.rs1  iff (ins.trap == 0)  {
    ...
`endif
}
cmp_rd_rs2_eqval : coverpoint ins.current.rd_val == ins.current.rs2_val iff (ins.trap == 0)  {
    ...
}
cp_gpr_hazard : coverpoint check_gpr_hazards(ins.hart, ins.issue)  iff (ins.trap == 0)  {
    option.comment = "GPR Hazard";
    bins hazards[]  = {NO_HAZARD, RAW_HAZARD, WAR_HAZARD, WAW_HAZARD};
}
`endif
endgroup
```

Fig. 5.5 riscvISACOV RV32I add covergroup. Cont'd

These paths may change as testing evolves.

The `addins/cvw-arch-verif/bin/testgen.py` script produces directed tests in assembly language. They are placed in `addins/cvw-arch-verif/tests`. `testgen.py` reads the coverage files from the `fcov` directory to determine which directed tests to create for which instructions. See Section 5.11.12 for how to run the directed test generator.

5.9.3 Constrained Random Tests

At the time of this writing, Wally is not regularly using riscv-dv, and scripts have broken.

riscv-dv is a widely used open-source constrained random test generator for RISC-V [github.com/chipsalliance/riscv-dv]. It was originally developed at Google and is now maintained by Chips Alliance. It consists of Python scripts that invoke SystemVerilog UVM modules that emit assembly language test files. riscv-dv is included as a submodule in Wally.

5.10 FORMAL VERIFICATION

A holy grail for logic verification would be to mathematically prove that HDL code faithfully implements a machine-readable specification. Such theorem provers are known as *formal verification* (FV) and are an active area of research. Some commercial offerings exist, but they presently require guidance from an experienced application engineer and are difficult to apply to the entirety of a complex design. Seligman provides an introduction to FV [Seligman23].

The general concept of FV is to annotate a DUT with SVA directives (see Section 5.7). Inputs and certain internal nodes are given `assume` directives to constrain their values. Outputs and other internal nodes are given `assert` directives to check their values. The chain of logic from assertions back to assumptions is called the *cone of influence*. The FV tool analyzes the logical relationships in the RTL to attempt to find a counterexample (consistent with the input assumptions) that violates the output assertions. If no counterexample exists, the assertion is proved. The number of cases to analyze increases exponentially with the size of the cone of influence, so FV becomes computationally intractable for blocks with large cones.

Algorithmic improvements have shrunk the cones requiring analysis, and methodology advancements have improved guidance about what blocks are suitable for FV and how to write the SVA directives. FV is likely to be practically relevant to an increasing set of designs in the future. For now, it is beyond the scope of this book.

For example, the OpenHW Group ran FV on the CV32E40P v2.0.0 core to compare it against the Sail specification, which was translated into a set of assertions using a proprietary tool. The process discovered many interesting bugs. However, the relatively simple core (4-stage RV32IMC) would have taken an impractical amount of compute time to completely prove that the RTL matched the Sail specification.

5.11 WALLY VERIFICATION

Wally targets TRL3 verification of all configurations and TRL5 verification of the rv64gc configuration. Section 5.11.1 describes the overall DV plan to achieve these readiness levels. Section 5.11.2 discusses how Sail and ImperasDV are used as reference models. Sections 5.11.3 to 5.11.6 discuss the testbench, test vectors, lint, simulation, and regression testing used to achieve TRL3 and to test every time the code changes, as well as for larger nightly regression. Code changes usually introduce bugs, so Section 5.11.7 describes debugging techniques to quickly locate the failure and get to the root cause. Section 5.11.8 addresses floating-point unit-level testing. Next, we examine deeper testing for the rv64gc configuration. Section 5.11.9 describes lockstep simulation with ImperasDV. Sections 5.11.10 to 5.11.12 explore assertions, code coverage, and functional coverage for rv64gc.

5.11.1 Design Verification Plan

Table 5.6 summarizes the elements of the Wally DV plan. The table lists each type of test that is run and the section that describes that test. For TRL3, it lists the applicable configurations. The primary configurations are summarized in Tables A.1, A.3, and A.4, and the derivative configurations are summarized in Table A.5 (see Appendix A). TRL5 verification

Table 5.6 **Wally DV plan.**

Tests	Section	TRL3	TRL5
Verilator Lint	5.3, 5.11.4	All configs	rv64gc
riscv-arch-test	3.7	All configs	rv64gc
wally-riscv-arch-test	3.7	All configs	rv64gc
TestFloat	5.11.8, 16.5.2	rv{32/64}gc + derived	rv64gc
CoreMark	21.1	Many configs	rv64gc
Embench	21.2	rv32*	n/a
Cache PV	21.3.1	rv{32/64}gc	rv64gc
BranchPred PV	21.3.2	rv{32/64}gc	rv64gc
Linux Boot	22.3.2	rv64gc	rv64gc
ImperasDV Lock-Step	5.6, 5.11.9		rv64gc
FPGA Linux Boot	23.2		rv64gc
Code Coverage	5.11.11		rv64gc
Functional Coverage	5.11.12		rv64gc

runs on the rv64gc configuration. Benchmarks and performance verification (PV) tests are also done to detect performance bugs.

The DV plan will continue to evolve with the code. The most up-to-date DV plan is in the cvw repository under `docs/testplans`.

Recall that the tests must be compiled before running simulations for the first time:

```
$ cd $WALLY
$ make --jobs
```

Invoke Wally's full set of verification tests (except FPGA Linux Boot) using:

```
$ regression-wally --nightly
```

5.11.2 Reference Model

RISCOF uses Sail as the reference model to generate reference signatures. Recall from Section 3.7.2 that RISCOF crosschecks Sail signatures against Spike. The wally-riscv-arch-test tests that use CSRs, interrupts, and peripherals that are not supported by Sail set the `NO_SAIL=True` directive and use a manually written reference signature. The flow of GCC, RISCOF, Sail, and Verilator compile and simulates regression tests entirely with open-source tools.

Wally also uses the commercial ImperasDV as a reference model for rv64gc lock-step-compare. ImperasDV confirms that the manually written signatures are correct, catches subtle discrepancies that might not appear in a signature, and detects bugs in Sail. For example, ImperasDV detected an incorrect implementation of the Zfa `fcvtmod.w.d` instruction in Sail.

5.11.3 Top-Level Testbench and Testvectors

The regression tests include Verilator lint, riscv-arch-test, and wally-riscv-arch-test. They are run concurrently to complete as fast as possible (usually several minutes). The riscv-arch-test *pool* of programs is divided into suites (e.g., RV32 integer instructions, or RV64 multiply/divide instructions), each of which contain files (e.g., `rv32i_m/I/src/add-01.S`), each of which contain tens to thousands of tests (e.g., corner cases for operands and source and destination registers). A key issue is that most of the files are small and run in less than a second, but the startup time to launch the simulator and compile Wally takes many seconds. Therefore, the testbench is designed to simulate all of the files in a suite, one after another, to reduce the startup overhead.

Testbenches are in `$WALLY/testbench`. The main testbench is called `testbench.sv`. It uses various helper modules in the `common` subdirectory. The testbench takes a `TEST` argument from the command line, which specifies the test suite. `testbench` performs the following functions:

- Selects a suite of tests from `tests.vh` based on the `TEST` argument
- Instantiates `wallypipelinedsoc` as the DUT
- Generates a clock and pulses reset
- Verifies that the configuration is self-consistent using the `riscvassertions` module
- Provides instrumentation to decode the names of instructions and labels
- For each program in the selected suite
  - Loads the test program at 0x80000000 in hexadecimal format from `test.elf.memfile`

- Runs tests until the program reaches an infinite loop or `write_tohost`.
- Flushes the data cache back to memory
- Compares a specified range of memory to `test.signature.output`
- Reports success or failure

The test suites defined in `testbench/tests.vh` and summarized in Table A.7 include test cases from the riscv-arch-test and wally-riscv-arch-test suites described in Section 3.7. Note that one RISCOF test suite contains many testbench suites. For example, the RISCOF arch32 suite contains arch32i, arch32c, arch32m, and so on. Details about the tests in each suite are described in subsequent chapters.

Interrupts are generated from software with test programs that configure the CLINT, PLIC, and GPIO as described in Section 8.3.2.

The testbench can also load and run Coremark and Embench, as described in Chapter 21. It also runs the coverage test suite and is used to collect code coverage measurements.

The buildroot test case is different than the others. It simply runs the first 1.6 million instructions of the Buildroot Linux boot and verifies that it does not hang. Section 22.3.2 describes running the Linux boot in lockstep with ImperasDV to verify that it behaves correctly; this is a multiday simulation.

5.11.4 Lint

After making significant edits, start by running lint and fixing any warnings, as described in Sections 5.1 and 5.3. The `lint-wally` script runs lint on rv32e, rv64gc, rv32gc, rv32imc, rv32i, rv64i, and fdqh_rv64gc configurations to detect configuration-dependent problems. It uses the `-I` flag to point to the desired configuration and the `testbench/wallywrapper.sv` *wrapper module* to load the configuration parameters and instantiate the top-level `wallypipelinedsoc`.

```
$ lint-wally
```

5.11.5 Simulation

Wally uses the `wsim` wrapper script to invoke simulation using Verilator, Questa, or VCS. Usage is as follows:

```
usage: wsim [-h] [--elf ELF] [--sim {questa,verilator,vcs}] [--tb
{testbench,testbench_fp}] [--gui] [--ccov] [--fcov] [--args ARGS] [--params PARAMS]
[--define DEFINE] [--vcd] [--lockstep] [--locksteverbose] [--rvvi] config [testsuite]

positional arguments:
  config                    Configuration file
  testsuite                 Test suite or path to .elf file

options:
  -h, --help                Show this help message and exit
  --elf ELF, -e ELF          ELF File name; use if name does not end in .elf
  --sim {questa,verilator,vcs}, -s {questa,verilator,vcs}
                            Simulator
  --tb {testbench,testbench_fp}, -t {testbench,testbench_fp}
                            Testbench
  --gui, -g                 Simulate with GUI
  --ccov, -c                Code Coverage
  --fcov, -f                Functional Coverage using cvw-arch-verif
  --args ARGS, -a ARGS      Optional arguments passed to simulator via $value$plusargs
  --params PARAMS, -p PARAMS Optional top-level parameter overrides of the form param=value
  --define DEFINE, -d DEFINE Optional define macros passed to simulator
  --vcd, -v                 Generate testbench.vcd
  --lockstep, -l            Run ImperasDV lock, step, and compare
  --lockstepverbose, -lv    Run ImperasDV lock, step, and compare with tracing enabled
  --rvvi, -r                Simulate RVVI hardware interface and ethernet
```

Example 5.1 USING **wsim** TO RUN SIMULATIONS

Simulate integer tests on the rv64gc configuration in Questa:

```
$ wsim rv64gc arch64i
<messages starting up Questa and compiling Verilog>
# rv64i_m/I/src/add-01.S succeeded.  Brilliant!!!
# rv64i_m/I/src/addi-01.S succeeded.  Brilliant!!!
...
# rv64i_m/I/src/xori-01.S succeeded.  Brilliant!!!
# SUCCESS! All tests ran without failures.
# ** Note: $stop : /home/harris/cvw/testbench/testbench.sv(424)
#    Time: 4449160 ns  Iteration: 1  Instance: /testbench
# Break at /home/harris/cvw/testbench/testbench.sv line 424
# Stopped at /home/harris/cvw/testbench/testbench.sv line 424
# End time: 04:08:23 on May 06,2024, Elapsed time: xxx
# Errors: 0, Warnings: 0
```

Run integer tests in Questa GUI for debugging:

```
$ wsim rv64gc arch64i --gui
```

Run floating-point testbench for addition tests. Many other derived configurations are available.

```
$ wsim f_rv64gc add --tb testbench_fp
<messages starting up Questa and compiling Verilog>
# Running f32_add_rne.tv vectors
# Running f32_add_rz.tv vectors
# Running f32_add_ru.tv vectors
# Running f32_add_rd.tv vectors
# Running f32_add_rnm.tv vectors
#
# All Tests completed with          0 errors
```

Verilator requires the user to increase the stack size before the first `wsim` run using `ulimit`. This is done in `setup.sh`.

Simulate multiply/divide instructions on rv32i configuration in Verilator, generating VCD file to view waveforms later. The VCD file can occupy gigabytes, so delete it when you are done.

```
$ wsim -s verilator rv32imc arch32m --vcd
<messages about compiling files>
g++ -fuse-ld=mold   wrapper.o verilated.o verilated_dpi.o verilated_vcd_c.o
verilated_timing.o verilated_threads.o Vtestbench__ALL.a    -pthread -
lpthread -latomic   -o Vtestbench
rm Vtestbench__ALL.verilator_deplist.tmp
make[1]: Leaving directory
'/home/harris/cvw/sim/verilator/wkdir/rv32imc_arch32m'
- V e r i l a t i o n   R e p o r t: Verilator 5.025 devel rev v5.024-12-
ga9664b8d1
- Verilator: Built from 308.865 MB sources in 244 modules, into 9.008 MB in
33 C++ files needing 76.297 MB
- Verilator: Walltime 9.125 s (elab=0.068, cvt=0.201, bld=8.733); cpu 0.000 s
on 1 threads; alloced 73.996 MB
mkdir -p /home/harris/cvw/sim/verilator/logs
wkdir/rv32imc_arch32m/Vtestbench +TEST=arch32m
rv32i_m/M/src/div-01.S succeeded.  Brilliant!!!
rv32i_m/M/src/divu-01.S succeeded.  Brilliant!!!
...
rv32i_m/M/src/mulhu-01.S succeeded.  Brilliant!!!
SUCCESS! All tests ran without failures.
- /home/harris/cvw/testbench/testbench.sv:422: Verilog $finish
- S i m u l a t i o n   R e p o r t: Verilator 5.025 devel
- Verilator: $finish at 799ns; walltime 0.126 s; speed 0.000 s/s
- Verilator: cpu 0.000 s on 1 threads; alloced 4914 MB
make: Leaving directory '/home/harris/cvw/sim/verilator'
```

View waveforms from Verilator simulation:

```
$ gtkwave -f $WALLY/sim/verilator/testbench.vcd
```

Simulate integer tests on the rv64gc configuration in VCS:

```
$ wsim -s vcs rv64gc arch64i
Read memfile ../../tests/riscof/work/riscv-arch-test/rv32i_m/I/src/add-01.S/ref/ref.elf.memfile
rv32i_m/I/src/add-01.S succeeded.  Brilliant!!!
...
Read memfile ../../tests/riscof/work/riscv-arch-test/rv32i_m/I/src/xori-01.S/ref/ref.elf.memfile
rv32i_m/I/src/xori-01.S succeeded.  Brilliant!!!
SUCCESS! All tests ran without failures.
$finish called from file "/home/harris/cvw/testbench/testbench.sv", line 427.
$finish at simulation time                 1034140
Info-[HSE-SIMPROF]  Simulating for the full-feature CPU time profiler
           V C S   S i m u l a t i o n   R e p o r t
Time: 1034140
CPU Time:       xxx seconds;        Data structure size: 264.1Mb
Tue May  7 12:59:36 2024
```

wsim launches the selected simulator in a simulator-dependent fashion. For example, to launch Questa:

```
$ wsim rv32gc arch32i --sim questa
Config=rv32gc tests=arch32i sim=questa gui=False args="
params=" define="
Running Questa with command: cd $WALLY/sim/questa;
MTI_VCO_MODE=64 vsim -c -do "do wally.do rv32gc arch32i
testbench --args \" \" --params \" \" --define \" \" "
```

wally.do, a *do file*, is a TCL file instructing Questa to compile, optimize, and simulate the code. It uses the +TEST flag to override a testbench variable called TEST with the name of the test suite. The code is placed in a library that is named based on the configuration and test suite so that many simulations can be performed concurrently during regression. wally.do also loads up waveforms using wave.do, which organizes commonly used signals into convenient groups. These scripts pass the configuration with the +incrdir argument so the compiler includes the appropriate config.vh file.

5.11.6 Regression

The regression-wally script launches lint-wally and all the test suite simulations in parallel and reports any errors. It uses Verilator because this simulator is free and fast. Table A.7 lists the various test suites run on various configurations. Some suites are repeated; for example, the arch32i integer tests are run on rv32i, rv32imc, and rv32gc to check all these configurations. If a test suite fails, go into the `logs` subdirectory and look at the `<config>_<suite>.log` file. Fix all compiler warnings and errors. If there is a real logic problem in the Verilog, you may need to simulate it using `wsim` to track it down. To keep the repository in always-working state, it is essential to run `regression-wally` before committing any changes.

```
$ regression-wally
```

5.11.6.1 Nightly Regression

Regression testing confirms that a recent change has not broken the system. The basic Wally regression tests (`regression-wally`) run architecture compatibility tests on the main Wally configurations. Experience has shown that it is important to run a larger suite of tests on a regular basis. These take longer to run, so they are often run automatically while the designer is away and are called *nightly regression*.

Wally's nightly regression includes:

- `regression-wally` expanded to run applicable tests on all derived configurations (Section 4.4)
- `lint-wally` on all derived configurations
- Normal `regression-wally` on Verilator, Questa, and VCS
- ImperasDV lock-step tests for the rv64gc configuration on riscv-arch-test, wally-riscv-arch-test, coverage, and Linux boot
- TestFloat on all floating-point configurations
- Functional coverage tests
- Cache performance validation tests
- CoreMark and Embench benchmarks

Eventually it should also include branch predictor performance validation tests.

Invoke nightly regression using:

```
$ regression-wally --nightly
```

5.11.6.2 Nightly Build

The `nightly_build.py` script clones a fresh copy of the cvw repository, makes all of the test cases, and runs these regression tests using `regression-wally --nightly`. It sends an email to a user or mailing list with a summary of the results, highlighting any failures. Invoke using:

```
$ nightly_build.py --path ~/nightly --target all --tests all
  --send_email ben@hmc.edu, alyssa@hmc.edu
```

`--path` specifies where to check out the repository and build the tests. `--target all` specifies building both Linux and RISCOF tests. `--tests all` runs `regression-wally --nightly --buildroot`. `--send_email` is followed by a comma-separated list of email addresses. The results are saved in `nightly/<date>/results`. If particular tests fail, consult the logs in `nightly/<date>/logs`.

The local copies of the repository and old logs can take a significant amount of disk space. `nightly_build.py` deletes old files, keeping only the most recent copy of the repository and the last 30 days of logs.

To automatically run `nightly_build.py` at 9 PM every evening, edit the cron table using:

```
$ crontab -e
```

Paste the following line into the crontab editor:

```
0 21 * * * curl -L https://raw.githubusercontent.com/openhwgroup/cvw/
refs/heads/main/bin/nightly_build.py | /opt/riscv/riscv-python/bin/
python - --path PATH/TO/NIGHTLY --target all --tests all --send_email
ben@hmc.edu, alyssa@hmc.edu
```

5.11.7 Debugging Hints

This section introduces a few bugs into the design and then shows how to systematically find them. The bugs will include:

- A syntax error caught by lint
- A compilation error caught by Questa
- A logic error in the design caught by a regression failure

Make the following edits in `$WALLY/src/`:

- In `ieu/ieu.sv`, misspell one of the `input` ports as `inputt`.
- In `ieu/datapath.sv`, change the output of the `fbemux` from `ForwardedSrcBE` to `ForwardedSrcAE` (which is also the output of `faemux`, resulting in contention).
- In `ieu/alu.sv`, change the line for the OR gate from `A | CondMaskInvB` to `A + CondMaskInvB`.

The error messages, line numbers, `wave.do` file, and code for this example may change as tool releases and Wally continue to evolve. You may need to use your judgement to debug.

Now try to find these bugs. First run lint:

```
$ lint-wally
Error: ../src/ieu/ieu.sv:31: syntax error, unexpected logic, expecting ')' or ','
  inputt logic      clk, reset,
         ^~~~~
%Error: ../src/ieu/ieu.sv:33: syntax error, unexpected input
  input logic [31:0]     InstrD,
  ^~~~~
```

Focus on the first error reported and don't get distracted by the many other errors. Carefully reading the message, we see an error right after `inputt` that points us to fix the misspelling.

Fix and run lint again, and it should pass. Now try running regression.

```
$ regression-wally
...
wsim --sim questa  rv32gc arch32m >
/home/harris/cvw/sim/questa/logs/rv32gc_arch32m.log: Failures detected in output
  Check /home/harris/cvw/sim/questa/logs/rv32gc_arch32m.log
...
Regression failed with 72 failed configurations
```

Given that almost every test failed, there's clearly something majorly wrong with the code. Look at any one of the failing log files and scroll through until you find the error reported:

```
$ tail $WALLY/sim/questa/logs/rv32gc_arch32m.log
...
# ** Error (suppressible): ../src/ieu/datapath.sv(108): (vopt-7045)
'ForwardedSrcAE' is driven by more than one continuous assignment. See
../src/ieu/datapath.sv(109).
```

Go to `src/ieu/datapath.sv` and look or search for components driving `ForwardedSrcAE`. The error message points to lines 108 and 109. You'll see that it is the output of two multiplexers. Fix this by changing the output of the `fbemux` back to `ForwardedSrcBE`.

Run regression again to find the next issue. Some of the tests pass and some fail, indicating an error affecting certain tests. There's no need to wait for everything to run; press Ctrl-c to halt when you are satisfied there is a problem.

```
$ regression-wally
rv32e passed lint
rv64gc passed lint
wsim --sim questa  rv32i arch32i >
/home/harris/cvw/sim/questa/logs/rv32i_arch32i.log: Failures detected in output
  Check /home/harris/cvw/sim/questa/logs/rv32i_arch32i.log
wsim --sim questa  rv32imc arch32c >
/home/harris/cvw/sim/questa/logs/rv32imc_arch32c.log: Failures detected in output
  Check /home/harris/cvw/sim/questa/logs/rv32imc_arch32c.log
wsim --sim questa  rv32imc wally32periph >
/home/harris/cvw/sim/questa/logs/rv32imc_wally32periph.log: Success
wsim --sim questa  rv32imc arch32m >
/home/harris/cvw/sim/questa/logs/rv32imc_arch32m.log: Success
...
```

Pick either the first failure or a failure in a test suite that you are most familiar with. The arch32i tests are a good choice because they relate to simple integer instructions that we understand well. Look at the log file for the problem:

```
$ less $WALLY/sim/questa/logs/rv32imc_arch32i.log
...
# Read memfile ../../addins/riscv-arch-test/work/rv32i_m/I/lw-align-01.elf.memfile
# rv32i_m/I/lw-align-01 succeeded.  Brilliant!!!
# Read memfile ../../addins/riscv-arch-test/work/rv32i_m/I/or-01.elf.memfile
# Error on test rv32i_m/I/src/or-01.S result 2: adr = 80006118 sim (D$) 00000004
  signature = 00000002
# ** Note: $stop    : /home/harris/cvw/testbench/testbench.sv(884)
#    Time: 758300 ns  Iteration: 1  Region: /testbench/genblk3
# Break at /home/harris/cvw/testbench/testbench.sv line 884
# Stopped at /home/harris/cvw/testbench/testbench.sv line 884
...
```

This shows that the problem is in the `rv32i_m/I/or-01` test suite during test number 2 (indicated by the "`2:`"). The expected signature is 00000002, but the simulation computes 00000004. Start by studying the test suite to understand the failing test. Look in `tests/riscof/work/riscv-arch-test/rv32i_m/I/src/or-01.S/ref/ref.elf.objdump` to see the disassembled file with corresponding source code lines. You'll see lines of code initializing registers and then setting `gp` to `signature_x3_0` (0x80006114). Then you'll see the test cases. The failing test is reproduced below. The key instruction at 0x80000118 computes 2 OR 2. The answer should be 2 (consistent with the signature reported in the error message above).

```
80000130 <inst_1>:
80000130:      00200313           li     t1,2
80000134:      00200313           li     t1,2
80000138:      006368b3           or     a7,t1,t1
8000013c:      0111a223           sw     a7,4(gp)
```

We will need to look at the simulation waveforms to troubleshoot. To make the failure occur early in the simulation, edit `tests.vh` and place the failing simulation at the beginning of the suite of tests:

```
string arch32i[] = '{
  `RISCVARCHTEST,
  "rv32i_m/I/or-01",      # Added this test at start
  "rv32i_m/I/add-01",
  ...
```

Now manually invoke the failing simulation with a GUI so we can inspect the waves. Note that the do file requires the configuration (rv32imc) and the name of the test suite (arch32i).

```
$ wsim --gui rv32imc arch32i
```

The simulation will run for a little while. You may see harmless errors about no object matching certain signal names because some configurations do not have all signals. It should fail with the same error as above. Explore the different groups of waves. Scroll to the start and click to put the yellow cursor at the start. Expand the Memory Stage wave group and click on `IEUAdrM`, the load/store address in the Memory stage. In the search box in the toolbar, enter the address of the bad store (80006118) and click the adjacent Search Next icon. Zoom in until the waves are legible. Expand the "alu execution stage" wave group and look back two cycles to see `SrcAE` and `SrcBE` are 00000002 and `ALUResultE` is 00000004. Now we need to figure out why `ALUResultE` is incorrect. Click on `ALUResultE` and use the Add → To Wave → Contributing Signals menu to add all the signals causing `ALUResultE`. `PreALUResult` should show up. Add contributing signals for `PreALUResult`, and `FullResult` should show up. Add the contributing signals for `FullResult`. Fig. 5.6 shows all of these signals, including `A` = 2, `CondMaskInvB` = 2, `ALUSelect` = 6, and `FullResult` = 4.

Browse or search in a text editor to find the portion of code that computes this result, shown below. We see for `ALUSelect` = 6 that the system selects `FullResult = A + CondMaskInvB`. This is the error. Fix it by replacing the blue + with |.

```
always_comb
  case (ALUSelect)
    3'b000: FullResult = Sum;                        // add or sub
    3'b001: FullResult = Shift;                      // sll, sra, or srl
    3'b010: FullResult = {{(WIDTH-1){1'b0}}, LT};   // slt
    3'b011: FullResult = {{(WIDTH-1){1'b0}}, LTU};  // sltu
    3'b100: FullResult = A ^ CondMaskInvB;           // xor, xnor, binv
    3'b101: FullResult = ... : Shift;                // bext (or IEU shift)
    3'b110: FullResult = A + CondMaskInvB;           // or, orn, bset
    3'b111: FullResult = A & CondMaskInvB;           // and, bclr
  endcase
```

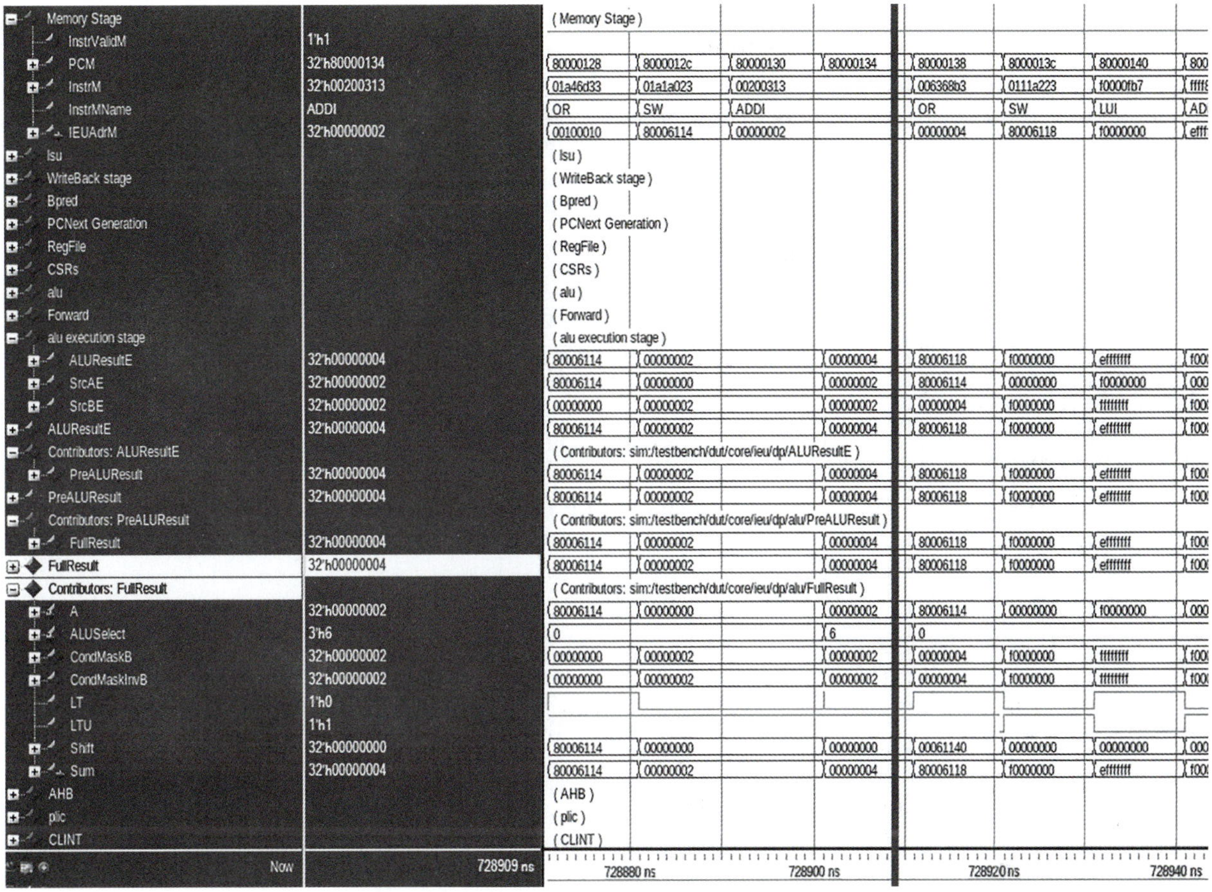

Fig. 5.6 rv32imc `arch32i` simulation waveforms.

Rerun `regression-wally` to confirm that this has fixed the problem.

5.11.8 Floating-Point Unit Testing

The riscv-arch-test vectors exercise the floating-point unit (FPU). The Berkeley TestFloat vectors, described in Section 16.5.2, exercise the FPU in different ways that have detected bugs not found by riscv-arch-test. They also test quad precision, which is not part of riscv-arch-test at the time of this writing.

TestFloat has a large number of vectors, so it is well-suited to unit-level testing. Such testing is faster because it only runs the FPU execution units, rather than the entire core, and because it applies vectors directly to the execution units rather than running supporting code to move values into registers and write signatures to memory.

FPU unit testing uses the unit-level `testbench/testbench-fp.sv` and vectors from `tests/fp/vectors`. See Section 16.5.2 for directions on how to invoke the tests. `regression-wally --nightly` runs TestFloat for all supported precisions (half-, single-, double-, and quad-precision) on both primary and derived configurations.

5.11.9 Lock-Step-Compare

ImperasDV is a commercial tool that requires a license. It is not part of the Synopsys academic bundle. This section assumes that you have a license and have installed the Imperas tools in `$RISCV/ImperasDV-OpenHW`.

As discussed in Section 5.6, ImperasDV performs lock-step-compare of running an Executable and Linkable Format (ELF) file on the Verilog model and on an Imperas reference model. It captures any discrepancies between the state of the system, including registers, the PC, CSRs, and memory. The reference model watches memory/peripheral reads from the Verilog to return the same values. It also watches the Verilog to take interrupts at the same time. This technique has located at least 20 bugs in Wally [Thompson23].

To run a single ELF file, use `wsim` with the `--lockstep` option:

```
$ wsim rv64gc $WALLY/tests/riscof/work/riscv-arch-test/rv64i_m/I/src/add- 01.S/ref/ref.elf
  --lockstep
...
# Info (IDV) -----------------------------------------------
# Info (IDV) ImperasDV INITIALISED
  (IDV) -----------------------------------------------
...
# Info (IDV) ImperasDV VERIFICATION REPORT
...
# Info (IDV)   Total compares    : 24,371
# Info (IDV)   Mismatches        : 0
```

Use `--lockstepverbose` instead to record each instruction that is committed. The output is large, so pipe it to a log file:

```
$ wsim rv64gc $WALLY/tests/riscof/work/riscv-arch-test/rv64i_m/I/src/add-01.S/ref/ref.elf
  --lockstepverbose > log
```

The following excerpt from the log file shows execution of the first instruction: `lui x16, 0x7d5c0`, highlighted in blue. It is represented in machine language as 7d5c0837 at address 80000000. After the instruction runs, `x16` changes to 000000007d5c0000.

```
# Info (RISCV_PMP) CPU 'refRoot/cpu': PMP PRIV=rwx 0x00000000: 0xfffffffffffff (mode Machine)
# Info 1: 'refRoot/cpu', 0x0000000080000000(rvtest_init): Machine 7d5c0837 lui x16,0x7d5c0
# Info   MEMX 0x80000000 0x80000000 2 0837
# Info   MEMX 0x80000002 0x80000002 2 7d5c
# Info   x16 0000000000000000 -> 000000007d5c0000
```

Section 22.3.2 describes how to run the Linux boot in lockstep with ImperasDV.

`regression-wally --nightly` runs lockstep simulations on all of the wally-riscv-arch-test and coverage tests, as well as the Linux boot.

5.11.10 Assertions

At the time of this writing, the only Wally assertions are in `riscvassertions.sv`, as described in Section 5.7.1. They check that the configuration options in the selected `config.vh` are self-consistent.

5.11.11 Code Coverage

Wally measures code coverage on the rv64gc configuration. It runs the functional coverage tests (described in Section 5.11.12) along with some extra hand-written directed tests for code coverage. The directed tests exercise features such as illegal instructions and each entry of the TLB and PMP. These extra tests must be compiled first:

```
$ cd $WALLY/tests/coverage
$ make --jobs
```

Then run regression with the `--ccov` flag.

```
$ regression-wally --ccov
```

Regression runs all the usual rv64gc tests, plus the extra coverage64gc tests just compiled. It also applies exclusions from the `$WALLY/sim/coverage-exlusions-rv64gc.do` file. It puts the results in `$WALLY/sim/questa/cov` and `$WALLY/sim/covhtmlreport/index.html`.

The HTML report is a good place to see the coverage percentages for the top-level modules. `rv64gc_coverage_hierarchical.rpt` lists each type of coverage for each instance in the design. Most modules are fully covered, so the `rv64gc_uncovered_hierarchical.rpt` is more useful, showing only the instances with incomplete coverage. For each top-level module, the `rv64gc_uncovered_<module>.rpt` contains a detailed report sufficient to identify the uncovered code.

Hint: in Microsoft's Visual Studio Code, install the HTML Preview extension. Then when you open `index.html`, click on the Show Preview button in the upper right to view the HTML web page.

The `coverage-exclusions-rv64gc.do` file waives coverage in a variety of areas. In particular, the load/store unit (LSU) and instruction fetch unit (IFU) share common memory management unit (MMU) and cache modules, which are configured to support reads and writes for the LSU and instruction fetches for the IFU. Unsupported access types cannot occur, so they must be excluded from coverage. Questa excludes coverage based on ranges of lines. This is unstable in that code changes that change line numbering break the exclusions in ways that are difficult to maintain. Wally defines a `GetLineNum` function that looks for a particular string in the code and substitutes the corresponding line number. For example, cachefsm.sv has a comment //exclusion-tag: cache AnyMiss. The coverage exclusion file searches for that line and excludes certain coverage related to cache misses.

5.11.11.1 Coverage Example: Illegal Instruction Test Case

As an example, consider test coverage for the controller decoding branch instructions. According to Table I.1 (on the last pages of the book), branches have Op = 1100011 and Funct3 of anything except 010 and 011. The relevant section of the controller is shown in Code Example 5.12. Line 211 is bolded.

Code Example 5.12 **controller.sv** BRANCH DECODING

```
assign BFunctD = (Funct3D[2:1] != 2'b01); // Legal branches
always_comb begin
  ControlsD = `CTRLW'b0_000_00_00_000_0_0_0_0_0_0_0_0_0_00_1; // Default: Illegal instr.
  case(OpD)
    ...
    7'b1100011: if (BFunctD) // Branches
                   ControlsD = `CTRLW'b0_010_11_00_000_1_0_0_0_0_0_0_0_0_00_0;
```

Branches with an illegal Funct3D should throw an Illegal Instruction exception, but these ill-formed branches are not part of the riscv-arch-test suite. Therefore, we add a line to tests/coverage/ieu.S with Op = 1100011 and Funct3 = 010 to check that illegal branches are detected. Zeroing all the other bits of the instruction that the decoder ignores gives an illegal instruction of 0x00002063. This line is shown in Code Example 5.13. Note that the .word assembler directive is used because there is no assembly language opcode for an illegal branch.

Code Example 5.13 **tests/coverage/ieu.S** ILLEGAL BRANCH

```
.word 0x00002063  // Illegal branch instruction
```

To see that this test is important, comment out the line and invoke make in the tests/coverage directory to recompile. Then run regression-wally --ccov. Code Example 5.14 shows an excerpt of the rv64gc_uncovered_ieu.rpt produced, showing that one branch is not covered. The uncovered branch is at line 211, with an All False Count of 0. This means that if (BFunctD) is never false when OpD is 1100011. Restoring the commented out line from ieu.S creates a branch test where BFunctD = 0, exercising a malformed illegal branch.

Code Example 5.14 **sim/cov/rv64gc_uncovered_ieu.rpt**

```
================================================================================
=== Instance: /core/ieu/c
=== Design Unit: work.controller
================================================================================
Branch Coverage:
    Enabled Coverage              Bins      Hits    Misses  Coverage
    ----------------              ----      ----    ------  --------
    Branches                        44        43         1    97.72%
...
-----------------------------------IF Branch-----------------------------------
    211                                    15647     Count coming in to IF
    211            2                       15647
                                         ***0***     All False Count
Branch totals: 1 hit of 2 branches = 50.00%
```

5.11.11.2 Coverage Example: Unreachable Logic Exclusion

As another example, consider adding an exclusion for unreachable logic. The instruction and data caches share the same code. As shown in bolded line 112 of Code Example 5.15, `cachefsm` receives an `InvalidateCache` signal used to invalidate the instruction cache; this signal is never asserted for data caches. Therefore, it is impossible for tests to exercise this `cachefsm` code for the data cache.

Code Example 5.15 **cachefsm.sv EXCERPT**

```
always_comb begin
  NextState = STATE_ACCESS;
  case (CurrState)
    STATE_ACCESS: if(InvalidateCache)
                    NextState = STATE_ACCESS;
                  else if (FlushCache & ~READ_ONLY_CACHE)
                    NextState = STATE_FLUSH;
                  else if (AnyMiss & (READ_ONLY_CACHE | ~LineDirty))
                    NextState = STATE_FETCH;
                  else if (AnyMiss & ~READ_ONLY_CACHE)
                    NextState = STATE_WRITEBACK;
                  else
                    NextState = STATE_ACCESS;
    STATE_FETCH:  if (CacheBusAck)
                    NextState = STATE_WRITE_LINE;
                  else
                    NextState = STATE_FETCH;
```

Code Example 5.16 shows an excerpt of `rv64gc_lsu_uncovered.rpt` illustrating this problem. The report indicates that most of the `if` conditions are activated hundreds or thousands of times, but item 2 of line 112 is never exercised, as highlighted in blue.

Code Example 5.16 **sim/cov/rv64gc_lsu_uncovered.rpt**

```
================================================================================
=== Instance: /core/lsu/bus/dcache/dcache/cachefsm
=== Design Unit: work.cachefsm
================================================================================
Branch Coverage:
    Enabled Coverage              Bins      Hits    Misses  Coverage
    ----------------              ----      ----    ------  --------
    Branches                        27        26         1    96.29%

===============================Branch Details===============================

Branch Coverage for instance /core/lsu/bus/dcache/dcache/cachefsm

    Line         Item                      Count     Source
    ----         ----                      -----     ------

-----------------------------------CASE Branch------------------------------------
    111                                   460817     Count coming in to CASE
    112             1                     405537
    117             1                      21035
    119             1                      10397
    120             1                      21468
    123             1                         32
    126             1                       1183
    129             1                       1137
    133             1                         28
Branch totals: 8 hits of 8 branches = 100.00%

-----------------------------------IF Branch------------------------------------
    112                                   405537     Count coming in to IF
    112             2                    ***0***
    113             1                        556
    114             1                      21033
    115             1                         49
    116             1                     383899
Branch totals: 4 hits of 5 branches = 80.00%
```

To waive covering this line in the data cache but not instruction cache, we add a unique comment to the end of line 112:

```
// exclusion-tag: dcache InvalidateCheck
```

Next, we add a line to the `coverage-exclusions-rv64gc.do` file, as shown in Code Example 5.17 highlighted in blue. It uses a custom function called `GetLineNum` to search for the line of `cachefsm.sv` containing this exclusion tag comment. It applies an exclusion with the `-scope` option to only waive coverage on the data cache, not the instruction cache. The `-item b 2` option excludes the uncovered branch from the coverage report.

This line is actually already in the do file, so the unreachable logic is waived and not reported as uncovered.

Code Example 5.17 **sim/coverage-exclusions-rv64gc.do**

```
coverage exclude -scope /dut/core/lsu/bus/dcache/dcache/cachefsm -linerange
[GetLineNum ../src/cache/cachefsm.sv "exclusion-tag: dcache
InvalidateCheck"] -item b 2
```

5.11.12 Functional Coverage

To collect functional coverage, Wally needs the covergroup files and test cases. This is a work in progress at the time of this writing.

Wally presently focuses on architectural functional coverage. It uses cvw-arch-verif, which is an extension of the nonproprietary riscvISACOV functional covergroups. Wally runs Questa simulations of test programs in lockstep with ImperasDV to detect logic bugs while measuring coverage. The goal is to have adequate covergroups for every RISC-V feature, test programs that give 100% coverage of these covergroups, and no logic bugs.

To create the covergroups and test programs and then run them on Wally:

```
$ cd $WALLY/addins/cvw-arch-verif
$ make --jobs
$ regression-wally --fcov
```

`make` first invokes `covergroupgen.py`, which reads the `.csv` files from the `testplans` directory specifying which coverpoints apply to which unprivileged instructions in each extension for RV32 and RV64. It generates functional coverage groups in the `fcov` subdirectory. `make` then runs `testgen.py`, which reads these covergroups and produces assembly language tests in the `tests` subdirectory to hit each coverpoint.

`regression-wally --fcov` runs these programs from the `tests` directory in lockstep while collecting coverage on a list of covergroups from the `fcov` directory specified in `$WALLY/config/<configuration>/coverage.svh`. It produces `.ucdb` files for each test, merges them, and puts reports in the `work` directory.

An excerpt of the coverage reports is shown in Code Examples 5.18 and 5.19. These reports show some RV32I instructions with 100% coverage and some with partial coverage. For example, the `beq` hits 100% of the bins in each coverpoint except that it does not yet test backward branches.

Code Example 5.18 `$WALLY/addins/cvw-arch-verif/work/summary_rv32gc.txt`

```
Covergroup Coverage:
    Covergroups                        79        na         na       69.70%
            Covergroup Bins         30958     20426      10532       65.97%
Covergroup                          Metric     Goal       Bins       Status
 Covergroup instance obj_add        100.00%     100          -       Covered
 Covergroup instance obj_addi       100.00%     100          -       Covered
 Covergroup instance obj_and        100.00%     100          -       Covered
 Covergroup instance obj_andi       100.00%     100          -       Covered
 Covergroup instance obj_auipc      88.57%      100          -       Uncovered
 Covergroup instance obj_beq        96.42%      100          -       Uncovered
```

Code Example 5.19 `$WALLY/addins/cvw-arch-verif/work/report_rv32gc.txt`

```
TYPE /RISCV_coverage_pkg/RISCV_coverage__1/beq_cg      96.42%      100      -     Uncovered
  covered/total bins:                                     404      405      -
  missing/total bins:                                       1      405      -
  % Hit:                                               99.75%      100      -
  Coverpoint cp_asm_count                             100.00%      100      -     Covered
  Coverpoint cp_rs1                                   100.00%      100      -     Covered
  Coverpoint cp_rs2                                   100.00%      100      -     Covered
  Coverpoint cmp_rd_rs1_eqval                         100.00%      100      -     Covered
  Coverpoint cp_rs1_corners                           100.00%      100      -     Covered
  Coverpoint cp_rs2_corners                           100.00%      100      -     Covered
  Coverpoint cmp_rs1_rs2                              100.00%      100      -     Covered
  Coverpoint cp_rs1_sign                              100.00%      100      -     Covered
  Coverpoint cp_rs2_sign                              100.00%      100      -     Covered
  Coverpoint cp_offset                                 50.00%      100      -     Uncovered
      covered/total bins:                                   1        2      -
      missing/total bins:                                   1        2      -
      % Hit:                                           50.00%      100      -
      bin neg                                               0        1      -     ZERO
      bin pos                                             539        1      -     Covered
  Coverpoint cp_rs1_toggle                            100.00%      100      -     Covered
  Coverpoint cp_rs2_toggle                            100.00%      100      -     Covered
  Cross cr_rs1_rs2_corners                            100.00%      100      -     Covered
  Cross cr_rs1_rs2                                    100.00%      100      -     Covered
```

By the time you read this, the unprivileged instructions are likely to be fully covered. Functional coverage will eventually include covergroups and tests for privileged features, including virtual memory, PMP, and CSRs.

5.12 SUMMARY

Logic verification now accounts for nearly twice as much effort as all the other elements of chip design combined. Although customers expect software to contain bugs and are accustomed to frequent updates, they need hardware to work reliably. Serious hardware bugs that require creating a new mask set and fabricating new chips cost millions of dollars in engineering costs and even more in opportunity costs when a chip is late to market. Defining the set of features that must be tested, the test vectors that exercise these features, and the expected results is harder than coding the features themselves.

Lint tools save a great deal of debugging time by warning the designer of bad code, such as undriven signals, combinational loops, and mismatched port sizes. The next verification step is to simulate the design in a testbench that applies test vectors and logs any discrepancies. Verification engineers also measure code coverage to ensure that the test vectors exercise every part of the code. Regression testing runs the full suite of tests after each code change to ensure that a new feature doesn't break the existing system; this is important to know right away because it is much easier for a developer to fix bugs in code as it is being written than for somebody else unfamiliar with the buggy code to isolate and fix the bug later.

The RISC-V Architecture Test Special Interest Group provides the riscv-arch-test suite of programs that exercise each unprivileged instruction and produce a signature for each test. Wally's testbench loads each program into memory and runs it on the Verilog, then checks that the signature in memory matches expectations. Wally also uses its own wally-riscv-arch-test suite to exercise privileged instructions, CSRs, exceptions, and the virtual memory system.

Wally also uses the Imperas lock-step-compare to run programs and check that the architectural state (memory, PC, registers, and CSRs) in the Verilog model matches the Imperas reference model after each instruction. This methodology has caught many subtle errors that are difficult to detect with signature-based verification.

Despite best efforts at verification, any system complicated enough to be interesting almost certainly has bugs lurking. An experienced design team has the humility to recognize human imperfection. Complex systems often have firmware-programmable configuration (known as "chicken bits") to disable or work around components that turn out to be defective.

6 Logic Synthesis

6.1 Performance, Power, and Area (PPA)
6.2 Fabrication Technologies
6.3 Running Synthesis
6.4 Reviewing Log Files
6.5 Wally Synthesis Results
6.6 Summary

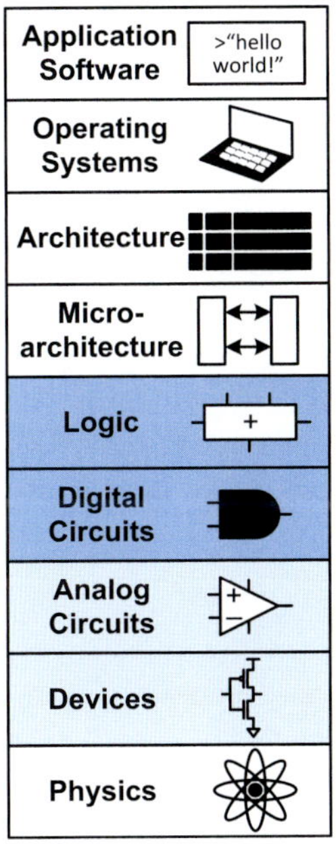

Logic synthesis converts register transfer level (RTL) to hardware by mapping the RTL (Verilog or VHDL) onto a library of cells. Synthesis attempts to minimize area and power while meeting timing constraints. Thus, the key metrics are performance, power, and area (PPA). More advanced (smaller feature size) fabrication processes offer better PPA, fueling the semiconductor industry's virtuous cycle of better products at lower prices.

Logic synthesis begins by reading RTL files, constraints, standard cell libraries, and a technology file. Constraints include the floorplan (including chip area and blockages), switching activity factors, target clock frequency, clock skew, and characteristics of input and output pins (including delays, drivers, or loads). The *standard cell library* contains gates such as inverters, NAND2-NAND4 (2- to 4-input NAND gates), NOR2-NOR4 (2- to 4-input NOR gates), compound AND-OR-Invert gates, and flip-flops. A robust cell library has multiple drive strengths for each cell to efficiently drive different loads while optimizing PPA. An SoC also needs a library of memories and a library of input/output pad cells. Each cell in the library is characterized with delay and power for various loads under different voltage and temperature operating conditions. The cell also contains physical views suitable for placement and routing to build a chip. The technology file contains design rules such as the number of metal layers and the width and spacing of the metal and vias between layers.

Next, the synthesis tool *analyzes* the design to find syntax errors and translates it into an HDL-independent intermediate format. Then it *elaborates* the design to map it onto a technology-independent generic library including registers and arithmetic operations. It *compiles* the design to map it onto the actual cell library and optimizes the choice of cells attempting to meet timing while minimizing chip area. The compiler may *flatten* the design to optimize better across module boundaries, but this produces a giant blob of gates that is difficult for the designer to understand or improve. A *physical synthesis* tool also places cells and estimates wire lengths because wires significantly affect timing. The synthesis tool produces a mapped *netlist* that specifies the cells and their connections; timing, area, and power reports; and possibly placement information. After synthesis, a *place and route* (*P&R*) tool performs detailed placement and routing to produce a manufacturable chip layout.

As with most other CAD (computer-aided design) tools, logic synthesis is controlled by a Tcl (Tool Command Language) script that specifies files, constraints, and steps to perform.

See Fig. 7.13 for a general block diagram of the rv32e configuration. However, note that rv32e does not include a DTIM or IROM; instead it goes to the AHB bus to access memory.

This chapter begins with an overview of fabrication technologies, including three example CMOS (complementary metal oxide semiconductor) processes (Skywater 130 nm, Skywater 90 nm, and TSMC 28 nm: SKY130, SKY90, and TSMC28) and the characteristics of inverters, datapath components, and memories. It describes how to run the industry-standard Synopsys Design Compiler synthesis tool and interpret the results. Finally, it describes synthesizing the rv32e configuration of Wally onto SKY130 and TSMC28 and synthesizing the larger configurations onto TSMC28 with RAM (random access memory) libraries.

6.1 PERFORMANCE, POWER, AND AREA (PPA)

The primary figures of merit for synthesis are performance, power, and area (PPA). *Performance* is indicated by the clock period at which the system operates. Increasing performance improves the responsiveness and capabilities of a system. Products such as personal computer (PC) microprocessors are marketed largely on clock speed, and modest improvements can lead to billions of dollars of revenue. *Power* is the number of watts drawn; it includes both *active power* while the system is running and *leakage power* consumed when the clock is stopped. Power determines the battery size and life of mobile systems. Power is also important for systems plugged into the wall because it determines the cooling cost and environmental footprint. *Area* is the chip space used by the design, which also indicates the design cost. It is expressed in square microns for ASICs (application-specific integrated circuits) or numbers of logic elements and memories for field programmable gate arrays (FPGAs). Area dictates how many features (cores, memories, accelerators, and so on) can be placed on a profitable chip. These figures of merit are influenced by the RTL design, the fabrication technology and cell library (see Section 6.2), and the logic synthesis tool. The performance generally trades off against power and area.

6.1.1 Performance

Most synthesized designs are synchronous sequential circuits composed of combinational logic and clocked memory elements, in which all the clocked elements in a block share a common clock with a specified target frequency. The clocked elements are either flip-flops or other edge-triggered memories. The *clock period* is the time from one rising clock edge to the next. The *clock frequency* is the reciprocal of the clock period.

For example, if the clock period is 0.5 ns (0.5×10^{-9} s), the clock frequency is 2 GHz (2×10^9 Hz).

On the rising edge of the clock, each flip-flop's outputs may change. These new values propagate through cones of logic to reach the inputs to the next flip-flops. The new values must settle sometime before the next rising edge of the clock. This is called the *setup time constraint*. If new values reach the next flip-flop too soon after the clock edge, they might slip through and corrupt the result. This is called the *hold time constraint*. The rest of this section reviews these constraints and describes how logic synthesis timing analysis checks the constraints.

6.1.1.1 Timing Constraints

Fig. 6.1 shows the timing characteristics of synchronous sequential building blocks. The horizontal axis indicates time. Fig. 6.1(a) illustrates combinational logic (CL) with input(s) A and output(s) Y. The timing diagram shows that A has some initial value and changes to some new value. Y starts at an initial value and takes some time to respond. After a contamination delay t_{cd}, Y might start to change. For some time, the value of Y is unknown, indicated by the crosshatch pattern. After a propagation delay t_{pd}, Y has definitely settled to the new value. The contamination and propagation delays are the minimum and maximum values, indicated by gray and blue, respectively. They may differ because of input patterns, rising or falling edges, operating conditions such as voltage and temperature, manufacturing variation that affects transistor currents or capacitances, and aging that degrades transistors over time. Fig. 6.1(b) illustrates a flip-flop with a clock *clk*, input D, and output Q. Just as a camera needs the subject to be still when the shutter is pressed to prevent blurring, the flip-flop needs the input to be stable around the rising edge of the clock to be sampled reliably. Specifically, the input must stabilize at least t_{setup} before the rising clock edge and remain stable for at least t_{hold} after the rising clock edge. Like combinational logic, the output will not start changing until some contamination delay t_{ccq} and will settle to a final value after some propagation delay t_{pcq} from clock to Q.

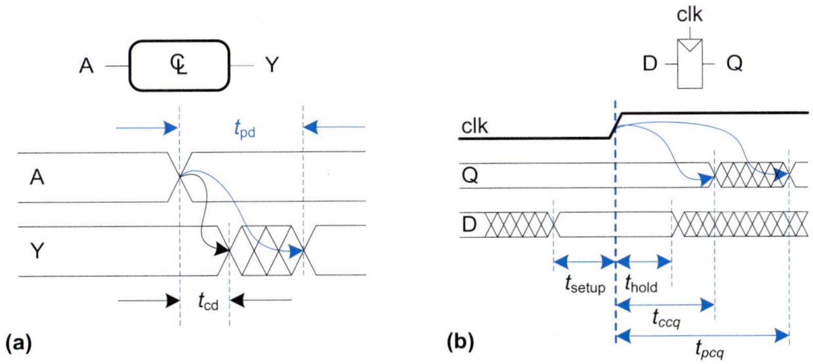

Fig. 6.1 Timing characteristics: (a) combinational logic and (b) flip-flops.

The *setup time constraint* determines the minimum reliable clock period, as illustrated in Fig. 6.2. A path in a synchronous sequential system starts at a register R1, goes through combinational logic, and ends at a register R2. The registers are made of one or more flip-flops sharing a common clock. When the clock rises, the first register's output *Q1* will stabilize after t_{pcq}. The combinational logic then takes t_{pd} to reach the second register's input *D2*. This input must stabilize t_{setup} before the next clock edge. The clock period T_c must be long enough for all of this:

$$T_c \geq t_{pcq} + t_{pd} + t_{\text{setup}} \tag{6.1}$$

The *hold time constraint* determines whether an input to a flip-flop might change too soon after the clock edge, causing the system to malfunction, as illustrated in Fig. 6.3. On the rising edge of the clock, the first register output *Q1* might change after t_{ccq}. The combinational logic might change after t_{cd}, reaching the input *D2* to the second register. This input must not change for at least a hold time t_{hold} after the rising edge of the clock lest R2 sample the wrong value, become slower, or go metastable.

$$t_{ccq} + t_{cd} > t_{\text{hold}} \tag{6.2}$$

In a real system, clocks do not all switch at exactly the same time. The switching time might vary because of systematic errors (e.g., different length clock wires going to two registers), random manufacturing variation (e.g., a faster clock buffer in one path than another), and environmental variation (e.g., a voltage spike causing the clock buffers to momentarily speed up, shortening a clock cycle). These effects are collectively called *clock skew*.

Fig. 6.4 redraws the setup time constraint accounting for clock skew. Each clock edge has hashed lines indicating the range of times that it might

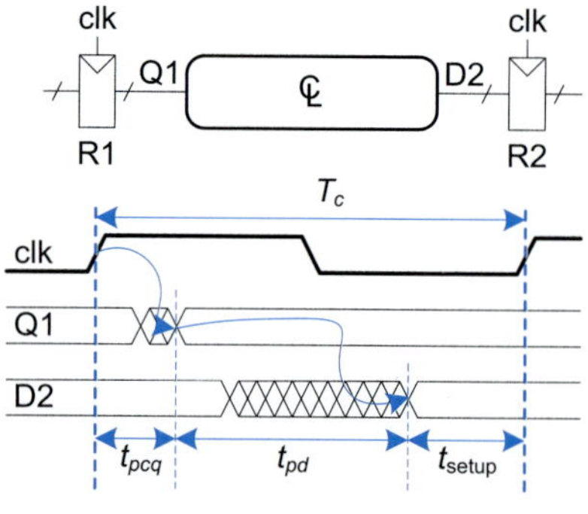

Fig. 6.2 Setup time constraint.

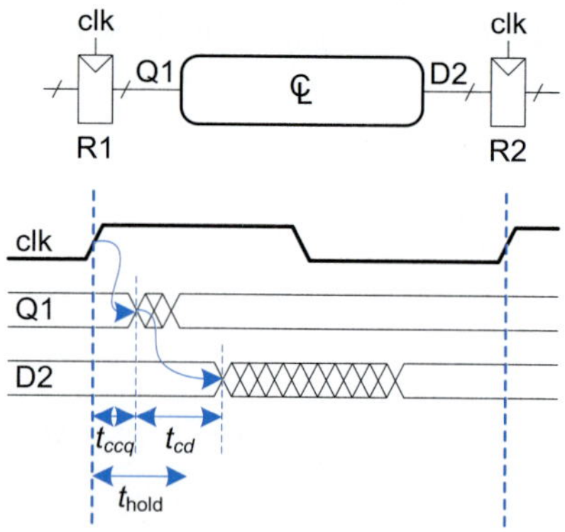

Fig. 6.3 Hold time constraint.

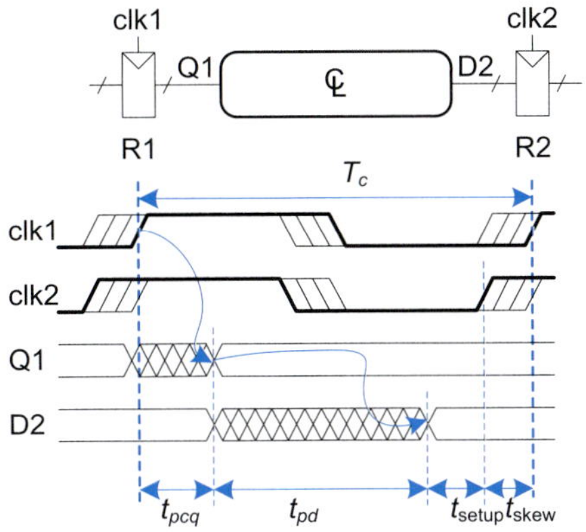

Fig. 6.4 Setup time constraint with clock skew.

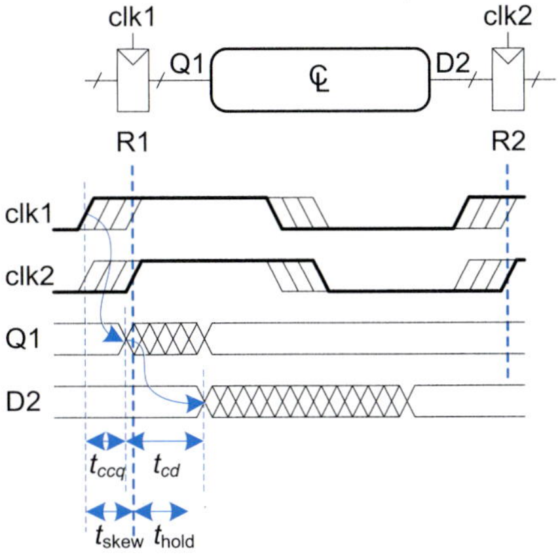

Fig. 6.5 Hold time constraint with clock skew.

switch. The clock skew t_{skew} is defined as the range between the earliest and latest that a clock edge might switch. In the worst case, *clk1* rises on the latest skewed clock edge, and *clk2* rises on the earliest skewed edge. After propagation delays through R1 and the combinational logic, *D2* must set up in time. The clock skew makes the setup constraint more difficult to satisfy.

$$T_c \geq t_{pcq} + t_{pd} + t_{\text{setup}} + t_{\text{skew}} \tag{6.3}$$

Fig. 6.5 redraws the hold time constraint accounting for clock skew. In the worst case, *clk1* rises early and *clk2* rises late. The skew effectively increases the hold time.

$$t_{ccq} + t_{cd} > t_{\text{hold}} + t_{\text{skew}} \tag{6.4}$$

Setup and hold times and clock skew on an SoC are all measured in picoseconds (10^{-12} s). Nevertheless, they are vitally important to achieve multigigahertz operation.

6.1.1.2 Static Timing Analysis

Synthesis performs *static timing analysis* (*STA*) to determine the worst-case timing by adding up the propagation delays along each path between clocked elements, which starts at the rising edge of the clock (*startpoint*), considered time = 0, passes through the *Q* output of the flip-flop, and propagates through a chain of combinational logic. The path must arrive at the *D* input of the next flip-flop (*endpoint*) some setup time before the next rising edge of the clock. The *arrival time* is the time at which the endpoint settles ($t_{cq} + t_{pd}$), and the *required time* is the clock period less the setup time ($T_c - t_{\text{setup}}$). The *slack* is the difference between the required time and the arrival time. If the slack is positive, the circuit operates properly at the desired frequency. If it is negative, the circuit cannot achieve the desired frequency.

In static timing analysis, t_{cq} is the maximum clock-to-q delay referred to earlier as t_{pcq}.

The *worst negative slack* (*WNS*) is the most negative slack of all paths, or zero if none are negative. The *total negative slack* (*TNS*) is the sum of all the negative slacks at each endpoint.

The *critical path* is the one with the WNS, limiting the clock period. The synthesis tool strives to reduce this critical path to improve performance. For example, if a gate must drive a high fanout (e.g., a select signal driving a 64-bit multiplexer), synthesis can choose a cell with a high *drive strength* that is larger but faster at driving the heavy load. If one signal arrives late on the critical path, the synthesis tool might precompute the output assuming that the signal is 0 and again assuming it is 1, then select the output from these two possibilities when the late signal arrives. Similarly, whereas a critical path might use a fast prefix adder, a noncritical path might use a smaller, cheaper ripple carry adder. Hence, critical paths tend to use faster structures that cost more area, and noncritical paths use the smallest implementations. Some processes offer low threshold voltage (low-V_t) transistors with better speed but exponentially more leakage power. Synthesis inserts low-V_t cells selectively on the critical paths. In a well-balanced system, many paths are nearly equally long, and there is no single outlier that could be optimized to speed up the whole system.

If synthesis cannot meet the target clock period, the *logic design engineer* may modify the RTL, simplifying the critical path or moving logic between pipeline stages. The *physical design engineer* may tweak the synthesis constraints, such as flattening the module hierarchy to enable more optimizations, preplacing modules so that critical cells are near each other, tuning the clock network, or enabling low-V_t cells. If none of these efforts succeed, the chip will still operate correctly but at a reduced clock frequency. Moreover, static timing analysis assumes worst-case environment and manufacturing. The chip will often run faster than analysis predicted.

PC enthusiasts know they can often overclock their CPU (central processing unit), especially if they have a good cooling system.

For hold times, a similar analysis is performed using contamination delays (the minimum time through the combinational logic), but this analysis is independent of clock frequency. Hold time analysis is essential to ensure signal transitions do not compromise the endpoint flip-flop's sampled value. Hold time failures are catastrophic because they cause the chip to run unreliably at any speed; thus, these failures cannot be fixed by simply changing the clock speed. The hardest case to satisfy is back-to-back flip-flops with no logic between them ($t_{cd} = 0$). Synthesis sometimes must insert delay buffers to ensure hold time is satisfied. Alternatively, FPGAs often prevent hold time problems by providing a flip-flop that is so slow (big t_{ccq}) that the system is guaranteed to work, albeit slowly.

External inputs and outputs are assigned required arrival times relative to the clock edge to satisfy setup and hold time constraints.

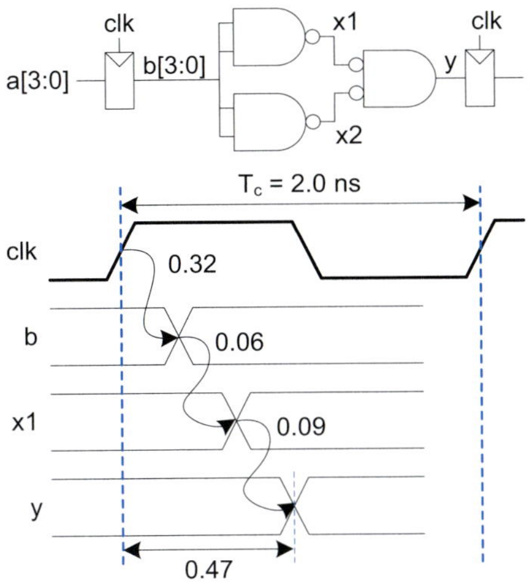

Fig. 6.6 **Schematic annotated with timing.**

Code Example 6.1 describes a synchronous sequential system with a 4-input AND gate between two flip-flops. Fig. 6.6 shows the schematic, where the AND gate is implemented with 2-input NAND and NOR gates. Suppose the flip-flops have a $t_{cq} = 0.32$ ns and a $t_{setup} = 0.13$ ns, and the NAND and NOR gates have propagation delays of 0.06 and 0.09 ns, respectively, as annotated in the schematic. If the target clock period is 2 ns, y has a required time of $2.0 - 0.13 = 1.87$ ns and an arrival time of $0.32 + 0.06 + 0.09 = 0.47$ ns. The path has slack of $1.87 - 0.47 = 1.4$ ns.

Code Example 6.1 EXAMPLE OF TIMING ANALYSIS

```
module example (
  input  logic       clk,
  input  logic [3:0] a,
  output logic       q);

  logic [3:0] b;
  logic       y;

  flop #(4) f1(clk, a, b);
  assign y = &b;           // 4-input AND
  flop #(1) f2(clk, y, q);
endmodule
```

6.1.2 Power

As digital circuits toggle between 0 and 1, they consume *dynamic* power to charge and discharge capacitance on the nodes. Even when they are idle, they draw some *leakage* power because transistors never turn completely OFF.

Power is energy per time. If a system consumes energy E per clock cycle and the clock period is T_c, the power is

$$P = E/T_c \tag{6.5}$$

The clock frequency f is the reciprocal of the period T_c, so power is proportional to frequency:

$$P = Ef \tag{6.6}$$

All transistors and wires on a chip have some capacitance, generally on the order of a femtofarad for micron-scale structures (1 fF = 10^{-15} F). As you may recall from a physics course, the energy required to charge a capacitor from 0 to V_{DD} is proportional to the capacitance and quadratic with the voltage. This is called *dynamic energy* because it is caused by signals changing. Half of this energy is stored in the capacitor, and the other half is dissipated as heat during the charging process. When the capacitor discharges from V_{DD} back to 0, the remaining energy stored in the capacitor is dissipated as heat.

$$E_{dyn} = CV_{\mathrm{DD}}^2 \tag{6.7}$$

A clock signal that rises and falls once every clock cycle with frequency f therefore consumes

$$P_{\mathrm{dyn}} = E_{\mathrm{dyn}}f = CV_{\mathrm{DD}}^2 f \tag{6.8}$$

In general, the dynamic power is proportional to the *activity factor* α, which measures the number of times the signal rises and falls each cycle. A clock has an activity factor of $\alpha = 1$ because it rises and falls once a cycle. A data signal that rises or falls only once in a clock cycle switches half as often and thus draws half the power, so $\alpha = ½$. A truly random data signal that rises or falls in half the cycles and keeps its old value in the other half has $\alpha = ¼$. Most signals switch even less often; for example, only one or a few bits of the program counter change in a typical cycle, and the upper bits of an integer are often all 0s. Empirically, $\alpha = 0.1$ is a reasonable estimate for the activity of data signals, corresponding to a signal that rises or falls in 20% of the cycles. Activity factors are also called *toggle rates*.

$$P_{\mathrm{dyn}} = \alpha C V_{\mathrm{DD}^2} f \tag{6.9}$$

Dynamic power is computed for each component based on that component's activity factor and capacitance. The power of an entire block or chip is the sum of the power consumed by each component.

Transistors that are supposed to be OFF actually draw some leakage current I_{leak}. This leakage is predominantly subthreshold leakage between the source and drain, but tiny transistors also have some tunneling current through the gate as electronics quantum mechanically teleport across the ultrathin gate insulator. The leakage power is proportional to the leakage current and the supply voltage

$$P_{\mathrm{leak}} = I_{\mathrm{leak}} V_{\mathrm{DD}} \tag{6.10}$$

In some processes and cells, the leakage current differs for HIGH vs. LOW outputs. However, over a large circuit, it is reasonable to estimate leakage as the average of these two cases.

When a chip is operating at a high frequency f, the dynamic power is usually at least an order of magnitude greater than the leakage power, so leakage can be ignored. However, when the chip is idle, it often stops the clock, so $f = 0$, and leakage accounts for all of the power. Mobile devices that interact with a user spend most of their time waiting for the user, and hence their standby battery life is determined by the leakage power.

Energy depends on the number and size of components, so it correlates with area. For example, the Wally floating-point divide/square root unit consumes an average of 0.28 fJ/μm^2/cycle in the TSMC 28HPC+ (TSMC28) process under typical conditions. Hence, minimizing area is a rough proxy for minimizing power.

Modern SoCs use other knobs to optimize power [Keating07, Weste12]. Cell libraries with multiple drive sizes and threshold voltages allow the designer to trade dynamic or leakage power for speed on the critical path. Clock gating and power gating turn off dynamic and leakage power, respectively, of idle subsystems. Dynamic voltage and frequency scaling reduce the supply voltage and clock frequency to save power when a component does not need to operate at peak performance. Adaptive voltage scaling measures the real-time chip performance and adjusts the supply voltage to reduce the power wasted by unnecessarily large guard bands for manufacturing and environmental variations.

6.1.3 Area

Synthesis makes many design decisions trading performance against area. For example, whereas a ripple carry adder is slow but small, a prefix adder is larger but faster. Synopsys Design Compiler provides *DesignWare* arithmetic components that automatically select a good implementation based on timing constraints. As another example, cell libraries contain flavors of cells with different *drive strengths*. A cell with a larger drive strength has wider transistors that deliver more current and can drive a given capacitive load faster. However, the cell also takes more area and power and presents a larger capacitive load on the previous stage.

6.2 FABRICATION TECHNOLOGIES

Integrated circuit fabrication technologies are characterized by their *process node* or *feature size*. The 10-μm node through the 28-nm node, which were developed from 1972 until 2009, are known as *planar processes*, in which transistors are built from layers on a flat silicon wafer, including the silicon source, drain, and channel, an insulating gate dielectric, and a conductive gate. The process node of a planar process refers to the drawn length of the transistor gate. Processes typically scaled length and width by a factor of $1/\sqrt{2} \approx 0.7$ from one node to the next, such that each node packed twice as many transistors in the same area [Dennard74]. Smaller transistors were not only cheaper but also faster and used less energy, leading to a cycle that drove the semiconductor revolution for four decades. According to Dennard scaling theory, delay scales proportionally to feature size and power (per gate) scales quadratically with feature size.

Vigorous worldwide competition led to new nodes every 2 to 3 years. Fig. 6.7 shows the progression of Intel process nodes from 10 μm to 10 nm and beyond, indicating a thousand-fold feature size reduction and a million-fold increase in transistor density over five decades or a doubling of transistor density every 2.5 years.

As transistors shrunk, the power supply voltage also scaled down to avoid damaging the tiny devices. To maintain performance, the transistor threshold voltage also decreased. Unfortunately, the leakage of OFF transistors increases exponentially as threshold voltage decreases. Below the 28-nm node, planar processes struggled to improve performance without excessive leakage power. Chip manufacturers turned to FinFETs (fin-shaped field-effect transistors), in which transistors are turned on their side and the gate is wrapped around three sides to reduce leakage. The process node is no longer linked to any physical dimension, but smaller nodes are now a marketing term for general advances in materials and fabrication that improve performance or power. For example, TSMC describes that, compared to its 5-nm process, its 3-nm process has 33% better transistor density and is 10% to 15% faster at the same power or 25% to 30% lower power at the same speed [Shilov20].

The technical details of more advanced nodes are closely guarded trade secrets. The 28-nm node is the latest node that universities can presently access at moderate cost with reasonable amounts of paperwork. For

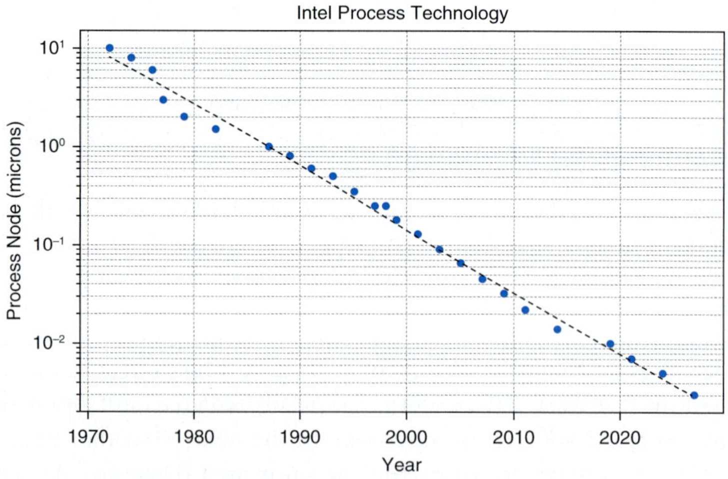

Fig. 6.7 Planar process feature size progression. (Data from [WikichipProcTech].)

example, IMEC supports universities and companies prototyping on the TSMC 28HPC+ process; the nondisclosure agreements require about three months to complete, and a batch of a hundred unpackaged 1-mm^2 chips costs about $12,000. The NDA (nondisclosure agreement) forbids sharing certain details, so this book only shows overall results, not detailed 28-nm timing reports.

This book also illustrates synthesis in an older node with open-source access so that users can more easily reproduce the book's results. Cypress Semiconductor spun off its foundry business, which Skywater Technology acquired in 2017. Skywater partnered with Efabless and Google to provide open-source access and free chip fabrication for universities. The Skywater 130-nm process design kit (SKY130 PDK) is available at github.com/google/skywater-pdk. The Oklahoma State University cell library for Skywater 130 is available at foss-eda-tools.googlesource.com/skywater-pdk/libs/sky130_osu_sc_t12.

SKY130 has a discontinuity in the transistor I-V curve model that sometimes produces simulation artifacts. Skywater has a more advanced 90-nm process (SKY90) with more robust transistor models that was intended for open-source access in 2022. The scripts in this chapter support both SKY90 and SKY130. However, all results shown in this chapter are from a prerelease version of SKY90. Unfortunately, it appears that Google has abandoned the open-source release of the SKY90 PDK.

6.2.1 Inverter Characteristics

Table 6.1 summarizes key characteristics of these processes. Power and speed are given under typical process conditions at nominal supply voltage and room temperature (25°C). The track pitch is the width plus spacing of

Table 6.1 Typical inverter characteristics.

Characteristic	Units	SKY130	SKY90	TSMC28
Nominal supply voltage	V	1.8	1.2	0.9
Track pitch	μm	0.37	0.35	0.090
Cell height	tracks	12	8	9
INVX1 FO4 delay	ps	99.5	43.3	11.6
INVX1 area	μm^2	4.4	1.96	0.252
INVX1 leakage	nW	0.13	2.0	1.1
INVX1 input capacitance	fF	2.8	3.49	0.54

a layer 1 or 2 metal wire (metal1 or metal2), which can usually be found in the technology LEF (.tlef) file in the PDK. The cells in a library are typically the same height so that they snap together easily; this height is usually 7-12 metal tracks. An INVX1 has unit drive strength, while an INVX2 has twice the drive strength. An FO4 inverter is an inverter driving four identical copies of itself. The FO4 delay is the delay of such an inverter and a handy metric for the overall performance of a process. The theory of logical effort [Sutherland99] indicates that gates in well-designed fast circuits have delays comparable to an FO4 inverter.

6.2.1.1 Process Variation

All results in this book are reported under typical process, voltage, and temperature (PVT) conditions defined earlier. Real circuits must operate over a range of temperatures; for example, commercial-grade parts are tested in an environment from 0° to 70°C, while military-grade parts are tested from −55° to 125°C. The transistor junction temperatures can be substantially higher when the chip is drawing a lot of power. The supply voltage often varies by 10% or more because of voltage regulator error and power supply distribution noise. Random fluctuations in the manufacturing process produce transistors that may be faster or slower than typical (TT). Fast (FF) transistors have lower threshold voltages and hence more leakage, while slow (SS) transistors have the opposite. Leakage increases exponentially with temperature.

Wally's `bin/libppa.pl` script parses Liberty files to extract delay, leakage, area, and capacitance.

Table 6.2 summarizes INVX1 delay and leakage across these process, voltage, and temperature (PVT) corners, showing that leakage increases by two orders of magnitude from typical to hot and fast. Delay varies by roughly a factor of two between fast and slow extremes. Products must leave *margin* so that they achieve specifications despite these variations.

6.2.2 Datapath Units

Designers typically estimate the delay, area, and energy of functional units and memories when planning a design. Moreover, because fabrication technologies are constantly advancing, it is useful to express delay and area in units that are independent of technology so that insights from one design remain relevant in the next design. A typical metric is to list all delays relative to an FO4 inverter. Functional unit delay tends to track FO4 inverter delay remarkably well, although paths dominated by long wires do not.

Table 6.2 INVX1 delay and leakage variations.

			SKY130		SKY90		TSMC28	
Process	Voltage (V)	Temperature (°C)	FO4 Delay (ps)	INVX1 Leakage (nW)	FO4 Delay (ps)	INVX1 Leakage (nW)	FO4 Delay (ps)	INVX1 Leakage (nW)
TT	Nominal	−40			41.1	0.1		
TT	Nominal	25	99.5	0.013	43.3	2.0	11.6	1.1
TT	Nominal	85			44.4	15.5	11.6	12.2
TT	Nominal	125			45.1	43.3		
TT	−10%	25	118.7	0.008	50.6	1.3	14.4	0.7
TT	+10%	25	92.6	0.017	38.2	3.0	9.9	1.6
SS	Nominal	−40					13.0	0.0
SS	Nominal	25			60.7	0.3		
SS	Nominal	125					12.8	23.0
SS	−10%	−40	206	0.000	68.4	0.0	16.7	0.0
SS	−10%	25			70.3	0.2		
SS	−10%	125	182	0.28	71.0	8.1	15.1	17.6
FF	Nominal	−40					10.8	0.1
FF	Nominal	25			31.0	13.4		
FF	Nominal	125					10.8	80.5
FF	+10%	−40	64.6	0.025	26.7	1.4	9.0	0.2
FF	+10%	25			27.9	20.9		
FF	+10%	125	68.5	211.2	29.2	268.2	9.4	111.8

To characterize the delay, area, power, and energy for each block, many synthesis runs are performed, sweeping the target frequency around the maximum achievable. Fig. 6.8 shows the results for a 32-bit adder in SKY90. As target frequency increases, the achieved delay initially decreases, and the area increases (gray dots). Beyond some target frequency, the delay ceases improving significantly, and the area and delay become noisy (blue dots) as Design Compiler struggles to achieve an impossible target. Selecting the rightmost gray dot gives a reasonably stable design

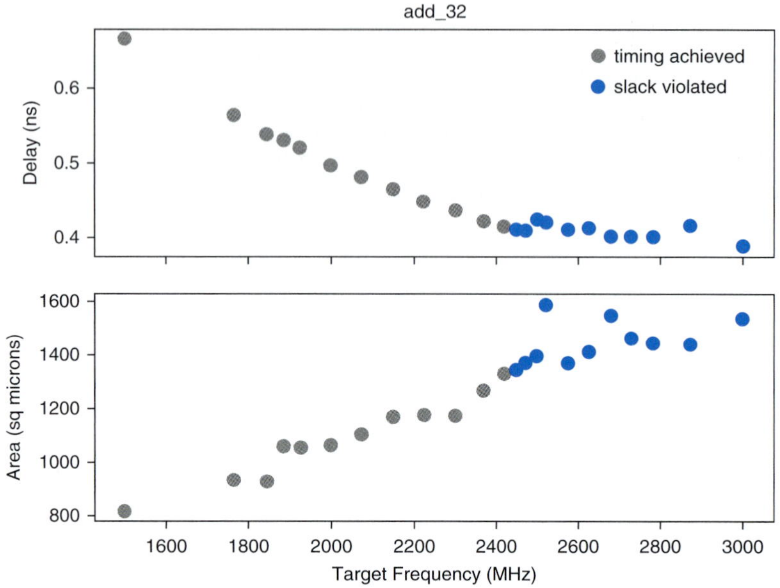

Fig. 6.8 **32-bit adder area and delay in SKY90.**

Table 6.3 **32-bit fast adder characteristics at TT 25°C.**

Characteristic	Units	SKY130	SKY90	TSMC28
Delay	ps	891	449	100
	FO4	9.0	10.4	8.6
Area	μm^2	2,581	1,441	209
Leakage Power	nW	18	714	1,060
Dynamic Energy	fJ/op	685	658	82

It is suspicious that SKY130 and SKY90 have similar dynamic energies (see Table 6.3) because energy is usually lower at a smaller feature size. It raises questions about whether the libraries are characterized accurately.

point to measure the best achievable delay and the resulting area and energy given that delay, listed in Table 6.3. We will refer to these as *fast add32 PPA*.

Dynamic energy is the total energy for the unit to perform one operation. It is measured using an activity factor of 1 for clocked nets and 0.1 for all other nets. Adders are combinational, so none of their nets are clocked. Dynamic energy is computed as the dynamic power times the clock period. Leakage power is measured with a static probability of 0.5, meaning each net has an equal chance of being 0 or 1. These activity factors are the Synopsys Design Compiler default values, listed as `power_default_toggle_rate` and `power_default_static_probability`.

Fig. 6.9 explores the PPA of other datapath building blocks, including comparators, priority encoders, simple left shifters, multipliers,

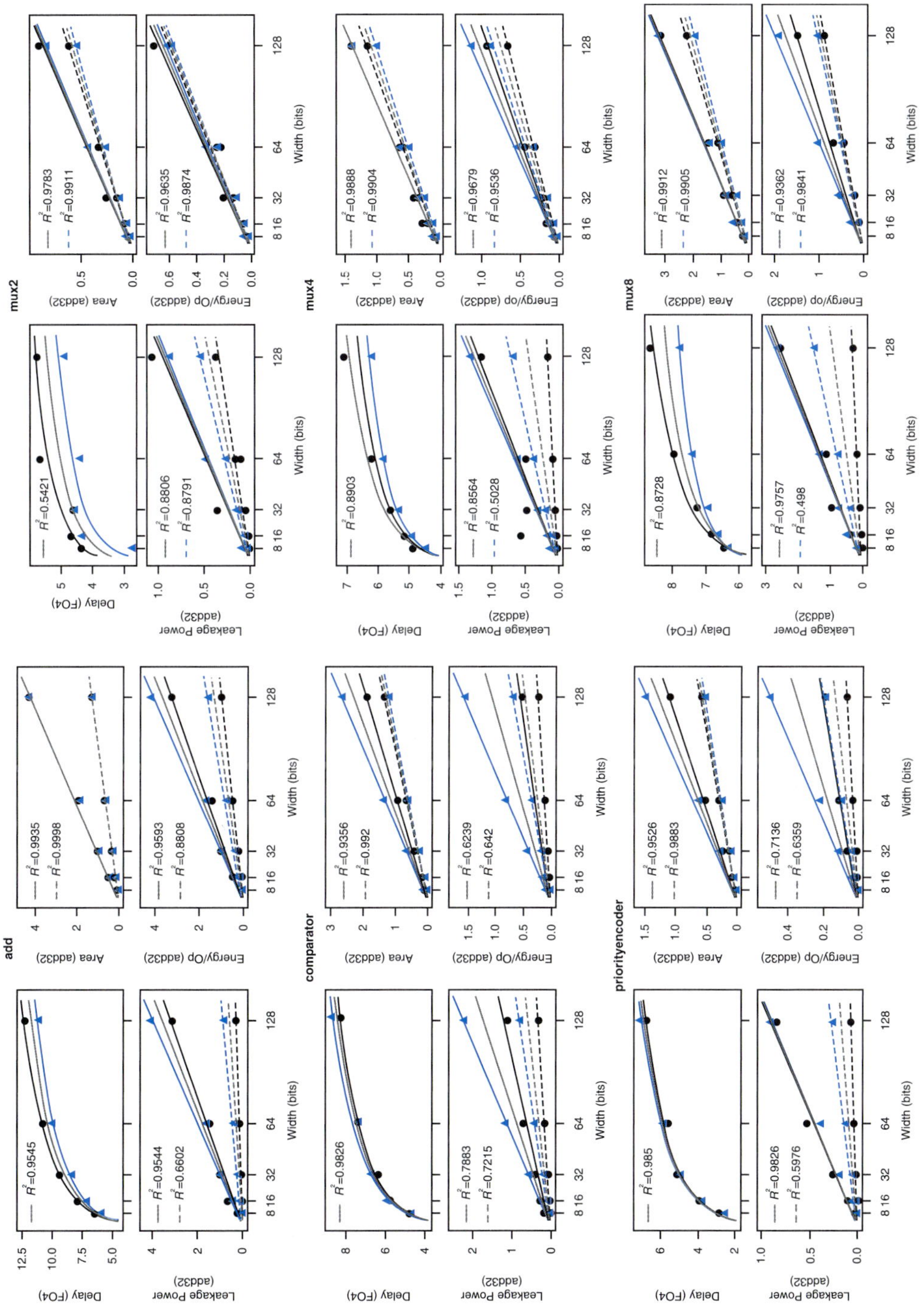

Fig. 6.9 Datapath component PPA.

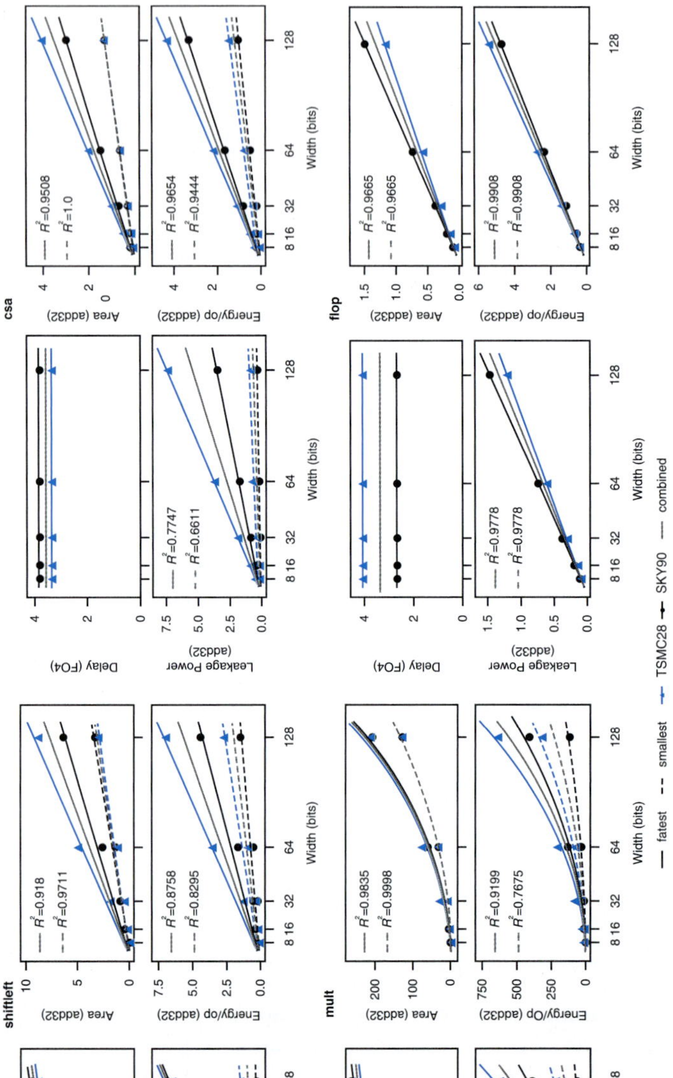

Fig. 6.9 Datapath component PPA. Cont'd

multiplexers, carry-save adders, and flip-flops. Each block has four graph panels, each containing synthesis results for 90- and 28-nm technologies, along with an average of the two. The x-axis lists size (8–128 bits). The delay panel shows the best achievable delay, as a function of size, expressed in FO4 inverter delays. The area panel shows the area as a function of size. The solid lines show the area for the design with the best achievable delay, and the dashed lines show the smallest achievable area when delay doesn't matter. For example, an adder uses a prefix structure to achieve best delay, but it uses a much more compact ripple structure if delay doesn't matter. Similarly, the leakage power and energy panels show results for best achievable delay and for smallest achievable area.

The R^2 fit values for the average delay, area, leakage, and dynamic energy are shown on the plots. For most results, they are close to 1, indicating that the average curve fit is a good match to both 90- and 28-nm results from 8 to 128 bits.

Table 6.4 summarizes the fits as a function of the number of inputs N. Most components have a best achievable delay involving a constant term and a term growing logarithmically with the component size. Carry-save adders (CSAs) and flip-flops naturally have a delay independent of the number of inputs. Area, leakage, and energy typically scale linearly with the normalized size $S = N/32$ except that they are quadratic for multipliers.

Table 6.4 Datapath element performance, power, and area (PPA).

		At Best Delay (Relative to Fast add32)			At Lowest Area (Relative to Fast add32)		
Element	Best Delay (FO4)	Area	Leakage	Energy	Area	Leakage	Energy
add	$1.9 + 1.4 \log_2 N$	$1.0S$	$0.89S$	$0.91S$	$0.32S$	$0.15S$	$0.33S$
comparator	$2.1 + 0.91 \log_2 N$	$0.58S$	$0.44S$	$0.27S$	$0.33S$	$0.15S$	$0.12S$
priorityencoder	$0.0 + 0.98 \log_2 N$	$0.32S$	$0.22S$	$0.087S$	$0.14S$	$0.044S$	$0.033S$
shiftleft	$0.46 + 1.60 \log_2 N$	$1.9S$	$2.1S$	$1.4S$	$0.77S$	$0.28S$	$0.48S$
mult	$0.0 + 6.0 \log_2 N$	$13S^2$	$13S^2$	$33S^2$	$7.8S^2$	$3.5S^2$	$14S^2$
mux2	$2.6 + 0.41 \log_2 N$	$0.21S$	$0.23S$	$0.16S$	$0.14S$	$0.11S$	$0.14S$
mux4	$3.1 + 0.50 \log_2 N$	$0.35S$	$0.31S$	$0.26S$	$0.27S$	$0.11S$	$0.19S$
mux8	$4.9 + 0.46 \log_2 N$	$0.80S$	$0.66S$	$0.43S$	$0.53S$	$0.23S$	$0.24S$
csa	3.6	$0.87S$	$1.3S$	$0.95S$	$0.33S$	$0.15S$	$0.31S$
flop	3.3	$0.33S$	$0.33S$	$1.3S$	$0.33S$	$0.33S$	$1.3S$

Note that multiplexer delay is measured from the select input and grows with the number of inputs and bus width (N) because the select signal must drive many bits. The delay from the data input is essentially independent of the number of bits.

Key takeaways are that most datapath components have a delay that grows with the log of the width, except that flip-flops and carry-save adders have a delay of 3-4 FO4 independent of the number of inputs. Microarchitects can make early critical path estimates from these component delays. Datapath component area grows linearly with width (surprisingly, even shifters), except that multipliers grow quadratically. Microarchitects can also make early size estimates to help evaluate tradeoffs; for example, a comparator is about half the size of a fast adder, and a shifter is about twice the size.

6.2.3 Memories

On-chip RAM can be built from flip-flops or static RAM (SRAM) cells. Flip-flops are easy to use and are readily available with standard logic synthesis, but their area and power consumption are far greater than those of SRAM. For example, in TSMC28, the flip-flop cell area is 1.89 μm^2, while the SRAM bit cell area is 0.13 μm^2. However, SRAM requires additional *peripheral circuitry*, including a precharge circuit for bitlines, sense amplifiers, and pitch-matched decoders. It is created with SRAM generators, provided by the fab or third-party IP vendors.

Similarly, on-chip ROM can be synthesized from combinational logic gates that provide the desired input/output relationship, or it can be created with ROM generators. We find that the TSMC28 ROM generators have so much overhead that logic synthesis normally produces smaller and faster results for a 128 × 32-bit boot ROM.

Wally by default describes memories with `always_ff` statements as in Section 4.2.3.3.2. When the `USE_SRAM` configuration is set, memories of specific sizes from Table 6.5 are replaced with a memory wrapper module instantiation. The wrapper in turn should instantiate a compiled SRAM macro for the target process. The names of these cells may be considered confidential by the vendor, so they are not specified in this textbook or the code.

Table 6.5 compares the flip-flop and memory modules used in Wally. The flop-based designs are much larger and more power hungry but are faster. SRAM has greatest density and power benefits for large arrays where the peripheral area is modest compared to the bit cell array. As shown in Fig. 4.23, flop-based memories can have either an asynchronous or synchronous read. The SRAM modules are inherently synchronous with a read time involving an address setup t_{setup} before the clock edge and a clock-to-Q delay t_{cq} to the read data output. Wally uses memories with

> Obtaining SRAM generators can be difficult, especially for academic designers. No Skywater generator is available. The TSMC generator entails various licensing hassles, and it produces a high aspect ratio with large column circuits and poor area utilization, so the area advantage is lower than expected. At the time of this writing, OpenRAM did not produce satisfactory RAM decoders in our processes of interest.

> If a multiport RAM is unavailable, one can synthesize a register file from latches rather than flip-flops to save area. For example, the RISC-V Ibex core optionally uses latches in the register file. They do not play nicely with FPGAs or Verilator, which are optimized for flop-based designs (see ibex-core.readthedocs.io/en/latest/03_reference/register_file.html).

Table 6.5 Memory block characteristics.

RAM	Application	Process	Style	Size (μm^2)	t_{setup} addr (ps)	t_{cq} read (ps)	Leakage (μW)	Dynamic Energy (pJ/access)
ram1p1rwbe_64x128 64 × 128 single port	Cache Data	SKY130	Flop	465k	152	1412	1.4	177
		SKY90	Flop	196k	46	536	38	74
		TSMC28	Flop	33k	13	212	162	26
			SRAM	7k	107	365	15	9
ram1p1rwbe_64x44 64 ×44 single port	Cache Tags	SKY130	Flop	164k	152	1354	0.56	61
		SKY90	Flop	69k	46	506	14	26
		TSMC28	Flop	12k	13	192	58	8
			SRAM	3k	107	305	6	4
ram2p1r1wbe_64x32 64 ×32 dual port	Branch history table (folded from 1024 × 2)	SKY130	Flop	117k	152	1287	0.41	45
		SKY90	Flop	49k	46	490	10	19
		TSMC28	Flop	9k	13	187	43	6
			SRAM	8k	50	459	20	4
ram2p1r1wbe_1024x68 1024 × 68 dual port	Branch target buffer	SKY130	Flop	2,245k	152	1,946	12	1270
		SKY90	Flop	1,653k	46	740	332	626
		TSMC28	Flop	261k	13	380	1,390	217
			SRAM	48k	72	550	84	10

synchronous read, as given in Code Example 4.22, allowing either type of memory to be used in the same pipeline.

The SRAM numbers in Table 6.5 come from the memory compiler datasheet directory. The size comes from the Area entry. t_{cq} is the access time (t_{cd}) entry. t_{setup} is the address setup (t_{as}) entry. Leakage is the total leakage current (in μA) multiplied by the 0.9 V supply voltage. Dynamic energy is the read dynamic current (in μA/MHz) multiplied by V_{DD} times the period of 1 MHz (1 μs). For example, the 64 × 128-bit RAM has leakage current of 16.77 μA and thus leakage power of 16.77 μA × 0.9 V = 15.1 μW. It has dynamic read current of 11.5 μA/MHz and thus dynamic energy of 11.5 μA/MHz × 0.9 V × 10^{-6} s = 10.35 pJ/read.

Dynamic energy per read is actually multiplied by the read frequency (in MHz), but that cancels with the read period (in μs). For example, if a read occurs at a frequency of 2 MHz, the read period is 1/(2 MHz) = 1/2 μs. So, the energy per read is: 11.5 μA/MHz × 2 MHz/read × (0.9 V) × (1/2 μs), which is still 10.35 pJ/read.

The flop-based numbers in Table 6.5 come from Design Compiler synthesis reports. Edit the RAM module default parameters to specify the size. The design is synthesized with an impossibly fast target (10 GHz) to estimate the best speed.

```
$ make synth DESIGN=ram1p1rwbe TECH=sky90 CONFIG=syn_rv64gc FREQ=10000 USESRAM=0
```

The area and leakage numbers come straight from synthesis reports. The dynamic energy is the reported dynamic power (in mW) multiplied by the 100-ps target period. The t_{cq} read delay comes from the timing report. t_{setup} is obtained as the setup time of a flip-flop from another synthesis run such as `wallypipelinedcore` (see Section 6.3).

6.3 RUNNING SYNTHESIS

Run synthesis with:

```
$ cd $WALLY/synthDC
$ make synth DESIGN=wallypipelinedcore TECH=sky130 CONFIG=syn_rv32e FREQ=330
```

The `CONFIG` options are described in Section 6.5.1.

The Makefile uses `.synopsys_dc.setup` and `scripts/synth.tcl`, both of which are necessary to read to follow this section. The Makefile sets the top-level design to `wallypipelinedcore` and configuration to rv32e. The target technology is SKY130. The target frequency is 330 MHz, a little faster than achievable. Synthesis takes only a few minutes to run because the rv32e configuration is simple. In contrast, synthesizing the rv64gc configuration with 16-KiB caches can take more than an hour. Watch for warnings and errors in the messages printed.

The run results can be large. You should delete the contents of the run directory (`rm -rf runs`) periodically to save disk space unless you need to refer to old runs.

The synthesis results are stored in `runs` within a directory named for the design, configuration, technology, frequency, date, and git hash for future reference. The scripts copy all the Verilog code into a subdirectory called `hdl`. The first time synthesis is run, it analyzes the Liberty (`.lib`) file, which describes the cell library, and caches the analyzed library in a directory called `alib-52`.

The scripts copy the specified configuration into the `config` subdirectory. As discussed in Section 4.2.5.6, the parameterized top-level module requires a wrapper to load the `config.vh` file and `parameter-defs.vh`. This file (e.g., `wallypipelinedcorewrapper`) is automatically generated by `synth.tcl` using `wrapperGen.py` in the run `hdl` subdirectory if the top-level design module contains the keyword `cvw_t`. `synth.tcl` synthesizes the wrapper to pass the parameters to the top-level module but generates reports for the top-level module (e.g., `wallypipelinedcore`).

The synth.tcl script mostly uses commands that are custom to Design Compiler, but it also uses native Tcl commands to read environment variables, copy files, and perform actions conditionally. The key Design Compiler commands are analyze, elaborate, and compile_ultra, which read the Verilog and run synthesis.

If you have access to TSMC28 libraries, run synthesis to get much higher frequency and lower area than in the SKY130 process. Invoke with:

```
$ make synth DESIGN=wallypipelinedcore TECH=tsmc28 CONFIG=syn_rv32e FREQ=3000
```

With suitable technology files (called Milkyway and TLU+), Design Compiler can run *physical synthesis*, creating an internal model of cell placement at the same time it maps Verilog onto logic gates. This allows Design Compiler to estimate wire lengths and optimize appropriately, producing area and delay that agree with final placement and routing results to within about 5%. This synthesis is also called *topographical mode*. TSMC28 results in this chapter are produced using topographical mode for best accuracy. The Makefile specifies TECH=tsmc28 for no physical synthesis or TECH=tsmc28psyn for physical synthesis.

The Design Compiler User Guide is proprietary. Licensed institutions can download it from solvnet.synopsys.com. If you are serious about optimizing your synthesis results, it is worth reading.

When the USESRAM configuration is set, the synthesis scripts will also load the memory libraries from Section 6.2.3 for most compact cache RAM, and branch predictor tables. Invoke with:

```
$ make synth DESIGN=wallypipelinedcore TECH=tsmc28psyn CONFIG=syn_sram_rv32gc
FREQ=3000 USESRAM=1
```

6.4 REVIEWING LOG FILES

cd into $WALLY/synthDC/runs/RUN, where RUN is the name of the synthesis you just completed. The directory contents are shown in Fig. 6.10. The rest of this section describes the logs from wally-pipelinedcore_rv32e_orig_sky130nm_330_MHz... generated in Section 6.3.

The config directory contains the configuration file (e.g., config/rv32e/config.vh) used in synthesis. The other files shown in Fig. 6.10 are described in the following sections.

To check out the Verilog code used in a previous run, use git checkout HASH, where HASH is a code such as b4ff90e1 at the end of the name of the run directory. The code is in src. You can reproduce the simulation by running make synth again with the parameters from the run directory name, for example, make synth b4ff90e1. Run git checkout main to return to the main branch.

6.4.1 synth.out

synth.out is the output log file that gives warnings and error messages. Fig. 6.11 shows some of the messages in the log file for each of the stages of compilation. Mapping and optimization complete many passes to improve WNS and area. See the added comments in blue to indicate what is occurring at each step.

```
├── config
│   ├── config.vh
│   ├── config-shared.vh
│   ├── BranchPredictorType.vh
│   └── parameter-defs.vh
├── hdl
│   ├── wallypipelinedcore.sv
│   ├── wallypipelinedcorewrapper.sv
│   ├── <other hdl files>
├── mapped
│   ├── wallypipelinedcore.ddc
│   ├── wallypipelinedcore.sdc
│   ├── wallypipelinedcore.sdf
│   └── wallypipelinedcore.sv
├── reports
│   ├── area.rep
│   ├── cell.rep
│   ├── check_design.rpt
│   ├── constraint_all_violators.rpt
│   ├── constraint.rep
│   ├── mdu_timing.rep
│   ├── mindelay.rep
│   ├── per_module_timing.rep
│   ├── power.rep
│   ├── qor.rep
│   └── timing.rep
├── synth.out
```

Fig. 6.10 **run directory contents.**

6.4.2 Mapped Output Files

The mapped directory has several outputs of synthesis.

- **DESIGN.sdc:** Synopsys Design Constraints file; uses Synopsys Tcl commands to provide the design constraints, including the clock period, driver and load cells, delays from primary inputs and outputs, and maximum fanout
- **DESIGN.sdf:** Standard Delay Format file; gives the design timing suitable for further timing analysis
- **DESIGN.ddc:** an encrypted binary database file; contains the synthesis results that Synopsys tools load for further analysis
- **DESIGN.sv:** contains the synthesized Verilog netlist. By default, our scripts run hierarchical synthesis, preserving the same module hierarchy and signal names as in the original Verilog. This makes interpreting the results feasible, but it also prevents cross-module optimizations, which is described in Section 6.5.2.

The reports in this section were generated around January 2024. In spring 2024, a Flush signal slightly increased the length of the critical path to the external bus unit (EBU). We anticipate that path may be fixed later. We also anticipate rv64gc area will increase slightly as extensions such as Zfa are implemented.

This section uses DESIGN to indicate the top-level module name, which, in this example, is wallypipelinedcore.

Design Compiler warns that clk is a high-fanout net because it drives so many flip-flops. Placement tools will handle it with a special clock-routing algorithm designed to balance delay.

```
# Synthesis Synopsys Flow
. . .
analyze -f sverilog -lib WORK $my_verilog_files    # Compile all source files
Compiling source file ./runs/wallypipelinedcore_rv32e_orig_sky130nm_330_MHz_2023-11-03-07-
38__5e84d5e6/hdl/wallypipelinedcorewrapper.sv
Opening include file ./runs/wallypipelinedcore_rv32e_orig_sky130nm_330_MHz_2023-11-03-07-
38__5e84d5e6/hdl/config.vh
Compiling source file ./runs/wallypipelinedcore_rv32e_orig_sky130nm_330_MHz_2023-11-03-07-
38__5e84d5e6/hdl/wallypipelinedcore.sv
. . .                                          # Target SKY130 (130 nm library), TT, 1.08 V, 25 C
Presto compilation completed successfully.
Loading db file
'/opt/riscv/cad/lib/sky130_osu_sc_t12/12T_ms/lib/sky130_osu_sc_12T_ms_TT_1P8_25C.ccs.db'
Loading db file '/cad/synopsys/SYN/libraries/syn/dw_foundation.sldb'
elaborate $my_toplevel -lib WORK
Loading db file '/cad/synopsys/SYN/libraries/syn/gtech.db'
Loading db file '/cad/synopsys/SYN/libraries/syn/standard.sldb'
  Loading link library 'sky130_osu_sc_12T_ms_TT_1P8_25C.ccs '
  Loading link library 'gtech'
Running PRESTO HDLC
Presto compilation completed successfully. (wallypipelinedcore)
Elaborated 1 design.
Current design is now 'wallypipelinedcorewrapper'.
Information: Building the design 'wallypipelinedcore' instantiated from design
'wallypipelinedcorewrapper' with the parameters
. . .
Inferred memory devices in process             # Found flip-flops in design
in routine flopenl_WIDTH32 line 33 in file
'./runs/wallypipelinedcore_rv32e_orig_sky130nm_330_MHz_2023-11-03-07-38__5e84d5e6/hdl/flopenl.sv'.
===============================================================================
|    Register Name    |   Type    | Width | Bus | MB | AR | AS | SR | SS | ST |
===============================================================================
|        q_reg        | Flip-flop |  32   |  Y  | N  | N  | N  | N  | N  | N  |
===============================================================================
. . .
  Linking design 'wallypipelinedcorewrapper'
  Using the following designs and libraries:
  --------------------------------------------------------------------------
  sky130_osu_sc_12T_ms_TT_1P8_25C.ccs (library)
/opt/riscv/cad/lib/sky130_osu_sc_t12/12T_ms/lib/sky130_osu_sc_12T_ms_TT_1P8_25C.ccs.db
  dw_foundation.sldb (library) /cad/synopsys/SYN/libraries/syn/dw_foundation.sldb
. . .
============================================================================
| DesignWare Building Block Library  |         Version         | Available |
============================================================================
| Basic DW Building Blocks           | R-2020.09-DWBB_202009.5 |     *     |
| Licensed DW Building Blocks        | R-2020.09-DWBB_202009.5 |     *     |
============================================================================
# Problems are related to unused signals in the configurable design
Information: There are 2472 potential problems in your design. Please run 'check_design' for more
information. (LINT-99)
. . .
Information: Uniquified 2 instances of design 'flopenl_WIDTH32'. (OPT-1056)
# Uniquify instances to optimize each one separately
  Beginning Pass 1 Mapping
  ------------------------
  Processing 'ifu'                             # Map modules onto cell library
. . .
```

Fig. 6.11 `synth.out` **log.**

```
  Updating timing information
Information: Updating design information... (UID-85)
Information: Removing unused design 'mux2_WIDTH3_1'. (OPT-1055)
Information: The register 'ifu/PCPDReg/q_reg[0]' is a constant and will be removed. (OPT-1206)
                                                                  # Remove unused registers
  Beginning Mapping Optimizations  (Ultra High effort)            # Start synthesis
  -------------------------------
  Mapping Optimization (Phase 1)
  Mapping Optimization (Phase 2)
  Mapping Optimization (Phase 3)                                  # First try is big and slow
                                      TOTAL
   ELAPSED              WORST NEG     SETUP    DESIGN                                  LEAKAGE
    TIME      AREA        SLACK       COST    RULE COST           ENDPOINT              POWER
  --------- --------- --------- --------- --------- ------------------------- ---------
    0:00:11  104050.7       0.37     196.9        0.0                                420.8164
    0:00:11  103514.9       0.80     460.9        0.0                                419.0923
  Beginning Constant Register Removal
  ----------------------------------
Information: The register 'ieu/c/controlregW/q_reg[1]' is a constant and will be removed. (OPT-
1206)                                                             # No reason for flops on constant values.
. . .
  Beginning Global Optimizations
  ------------------------------
  Numerical Synthesis (Phase 1)
  Numerical Synthesis (Phase 2)
  Global Optimization (Phase 1)
  Global Optimization (Phase 2)
. . .
Information: Added key list 'DesignWare' to design 'alu_32_165949. (DDB-72)
# Designware Libraries have optimized components
  Beginning Delay Optimization                                    # Get faster
  ----------------------------
    0:00:12   96295.4       0.76     406.6        0.0                                384.4135
    0:00:14   97328.8       0.26     124.8        0.0                                389.7814
    0:00:14   97328.8       0.26     124.8        0.0                                389.7814
    0:00:14   97328.8       0.26     124.8        0.0                                389.7814
Information: The register 'dut/ieu/c/controlregW/q_reg[2]' is a constant and will be removed.
(OPT-1206)
Information: The register 'dut/ieu/c/controlregW/q_reg[0]' is a constant and will be removed.
(OPT-1206)
    0:00:14   97102.2       0.26     124.8        0.0                                389.5597
    0:00:14   96738.9       0.43     172.3        0.0                                388.6815
  Beginning WLM Backend Optimization                              # WLM is wire load model
  --------------------------------------
  Beginning Leakage Power Optimization  (max_leakage_power 0)
  -----------------------------------
                                      TOTAL
   ELAPSED              WORST NEG     SETUP    DESIGN                                  LEAKAGE
    TIME      AREA        SLACK       COST    RULE COST           ENDPOINT              POWER
  --------- --------- --------- --------- --------- ------------------------- ---------
    0:00:28   96382.3       0.14      55.4        0.0                                472.3618
. . .
  Optimization Complete
  ---------------------
Memory usage for this session 165 Mbytes.
Memory usage for this session including child processes 165 Mbytes.
CPU usage for this session 40 seconds ( 0.01 hours ).
Elapsed time for this session 42 seconds ( 0.01 hours ).
# Elapsedtime < CPU usage if using multicore processing
```

Fig. 6.11 `synth.out` **log. Cont'd**

Code Example 6.2 shows the hazard unit's behavioral Verilog (on page 274), and Code Example 6.3 shows the synthesized netlist (on page 275). Many of the signals have been optimized out of the netlist because the `rv32e` configuration does not have a multiply and divide unit (MDU), floating-point unit (FPU), or privileged unit. Hence many inputs, including `MDUStallD`, `FPUStallD`, and all the trap and control and status register (CSR) signals, are tied to 0 and can be ignored.

The #1 delays in the Verilog in Code Example 6.2 are included for simulation to help see causality. They are ignored by synthesis.

Design Compiler has a Design Vision tool that displays netlists as schematics. Invoke it using:

```
$ design_vision-xg
```

Use File → Read and open the DDC file (e.g., in `runs/<current_run>/mapped/wallypipelinedcore.ddc`). Navigate to the hazard unit (`hazard`) in the Logical Hierarchy pane, then use Schematic → New Schematic View. Double click on the symbol in the schematic to see the gates inside. Fig. 6.12 shows the synthesized schematic. Observe that it exactly matches the netlist from Code Example 6.3.

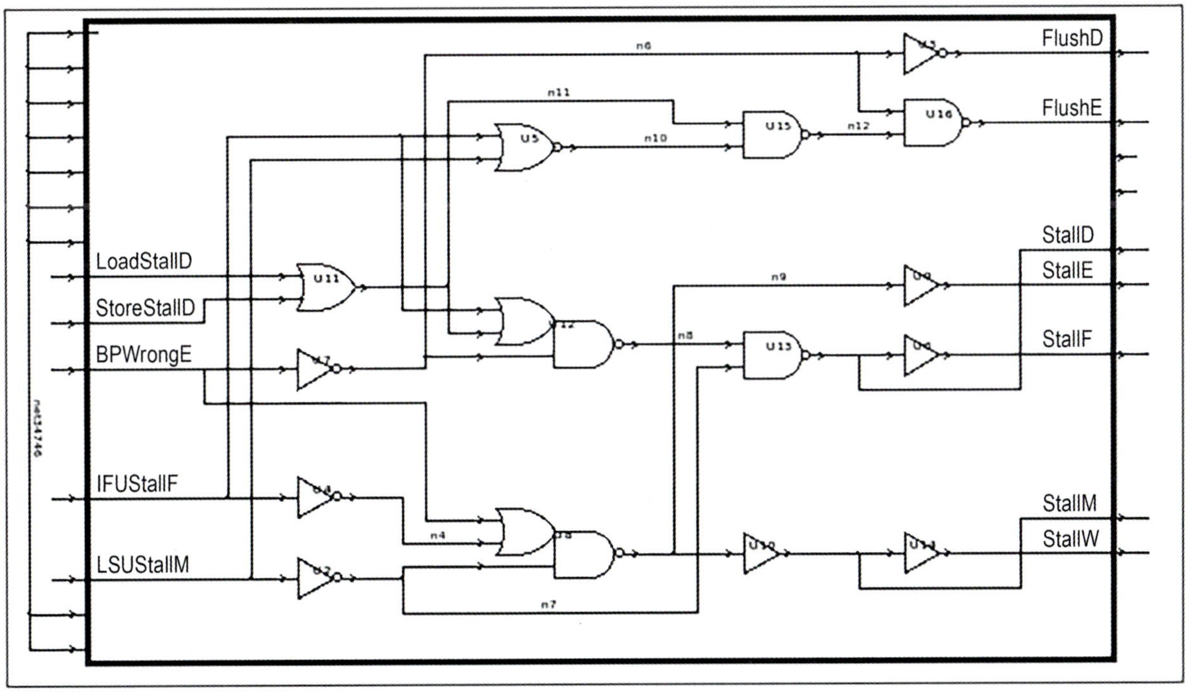

Fig. 6.12 **Hazard unit (`hazard`) synthesized schematic.**

Code Example 6.2 hazard.sv

```
module hazard (
  // Detect hazards
  input  logic  BPWrongE, CSRWriteFenceM, RetM, TrapM,
  input  logic  LoadStallD, StoreStallD, MDUStallD, CSRRdStallD,
  input  logic  LSUStallM, IFUStallF,
  input  logic  FCvtIntStallD, FPUStallD,
  input  logic  DivBusyE, FDivBusyE,
  input  logic  EcallFaultM, BreakpointFaultM,
  input  logic  wfiM, IntPendingM,
  // Stall & flush outputs
  output logic StallF, StallD, StallE, StallM, StallW,
  output logic FlushD, FlushE, FlushM, FlushW
);

  logic StallFCause, StallDCause, StallECause, StallMCause, StallWCause;
  logic LatestUnstalledD, LatestUnstalledE, LatestUnstalledM, LatestUnstalledW;
  logic FlushDCause, FlushECause, FlushMCause, FlushWCause;

  logic WFIStallM, WFIInterruptedM;

  // WFI logic
  assign WFIStallM = wfiM & ~IntPendingM;         // WFI waiting for an interrupt or timeout
  assign WFIInterruptedM = wfiM & IntPendingM;    // WFI detects a pending interrupt.

  // Stalls and flushes
  assign FlushDCause = TrapM | RetM | CSRWriteFenceM | BPWrongE;
  assign FlushECause = TrapM | RetM | CSRWriteFenceM |(BPWrongE & ~(DivBusyE | FDivBusyE));
  assign FlushMCause = TrapM | RetM | CSRWriteFenceM;
  assign FlushWCause = TrapM & ~WFIInterruptedM;

  // Stall causes
  assign StallFCause = '0;
  assign StallDCause = (LoadStallD | StoreStallD | MDUStallD | CSRRdStallD | FCvtIntStallD |
                        FPUStallD) & ~FlushDCause;
  assign StallECause = (DivBusyE | FDivBusyE) & ~FlushECause;
  assign StallMCause = WFIStallM & ~FlushMCause;
  assign StallWCause = (IFUStallF & ~FlushDCause) | (LSUStallM & ~FlushWCause);

  // Stall each stage for cause or if the next stage is stalled
  assign #1 StallF = StallFCause | StallD;
  assign #1 StallD = StallDCause | StallE;
  assign #1 StallE = StallECause | StallM;
  assign #1 StallM = StallMCause | StallW;
  assign #1 StallW = StallWCause;
  // Detect the first stage that is not stalled
  assign LatestUnstalledD = ~StallD & StallF;
  assign LatestUnstalledE = ~StallE & StallD;
  assign LatestUnstalledM = ~StallM & StallE;
  assign LatestUnstalledW = ~StallW & StallM;

  // Flush stage if the previous stage is the last one stalled or the system has reason to flush
  assign #1 FlushD = LatestUnstalledD | FlushDCause;
  assign #1 FlushE = LatestUnstalledE | FlushECause;
  assign #1 FlushM = LatestUnstalledM | FlushMCause;
  assign #1 FlushW = LatestUnstalledW | FlushWCause;
endmodule
```

Code Example 6.3 `wallypipelinedcore.sv` NETLIST HAZARD MODULE

```
module hazard ( BPWrongE, CSRWriteFenceM, RetM, TrapM, LoadStallD, StoreStallD,
        MDUStallD, CSRRdStallD, LSUStallM, IFUStallF, FCvtIntStallD, FPUStallD,
        DivBusyE, FDivBusyE, EcallFaultM, BreakpointFaultM, wfiM, IntPendingM,
        StallF, StallD, StallE, StallM, StallW, FlushD, FlushE, FlushM, FlushW
);
  input BPWrongE, CSRWriteFenceM, RetM, TrapM, LoadStallD, StoreStallD,
        MDUStallD, CSRRdStallD, LSUStallM, IFUStallF, FCvtIntStallD,
        FPUStallD, DivBusyE, FDivBusyE, EcallFaultM, BreakpointFaultM, wfiM,
        IntPendingM;
  output StallF, StallD, StallE, StallM, StallW, FlushD, FlushE, FlushM,
        FlushW;
  wire  n4, n6, n7, n8, n9, n10, n11, n12;

  sky130_osu_sc_12T_ms__inv_1 U2 ( .A(LSUStallM), .Y(n7) );
  sky130_osu_sc_12T_ms__inv_3 U3 ( .A(n6), .Y(FlushD) );
  sky130_osu_sc_12T_ms__inv_1 U4 ( .A(IFUStallF), .Y(n4) );
  sky130_osu_sc_12T_ms__nor2_1 U5 ( .A(IFUStallF), .B(LSUStallM), .Y(n10) );
  sky130_osu_sc_12T_ms__buf_1 U6 ( .A(StallD), .Y(StallF) );
  sky130_osu_sc_12T_ms__inv_2 U7 ( .A(BPWrongE), .Y(n6) );
  sky130_osu_sc_12T_ms__oai21_1 U8 ( .A0(BPWrongE), .A1(n4), .B0(n7), .Y(n9) );
  sky130_osu_sc_12T_ms__buf_4 U9 ( .A(n9), .Y(StallE) );
  sky130_osu_sc_12T_ms__buf_2 U10 ( .A(n9), .Y(StallM) );
  sky130_osu_sc_12T_ms__or2_1 U11 ( .A(LoadStallD), .B(StoreStallD), .Y(n11) );
  sky130_osu_sc_12T_ms__oai21_1 U12 ( .A0(IFUStallF), .A1(n11), .B0(n6), .Y(n8) );
  sky130_osu_sc_12T_ms__nand2_1 U13 ( .A(n8), .B(n7), .Y(StallD) );
  sky130_osu_sc_12T_ms__buf_2 U14 ( .A(StallM), .Y(StallW) );
  sky130_osu_sc_12T_ms__nand2_1 U15 ( .A(n11), .B(n10), .Y(n12) );
  sky130_osu_sc_12T_ms__nand2_1 U16 ( .A(n6), .B(n12), .Y(FlushE) );
endmodule
```

6.4.3 Quality of Results (QoR) Reports

Quality of results (QoR) reports summarize key timing and area results. Timing metrics include the critical path slack (WNS), TNS, and any hold time violations. Area metrics include the number and area of combinational, sequential, and buffer/inverter cells. For example, `qor.rep` in Fig. 6.13 shows timing and area results for Wally `rv32e` in the SKY130 process. As highlighted in blue text, the critical path goes through 37 levels of logic. The target clock period is 3.03 ns, but the design has 0.15 ns of negative slack, so it does not meet timing and has a best achievable clock period of 3.18 ns. The area is 95,313 μm^2, or about 0.3 mm on a side. The processor is built from 8,226 cells, of which 1,173 are sequential (flip-flops), 2,069 are inverters or buffers, and the rest are logic gates. Synthesis took about 2 minutes. Synthesis involves a certain amount of random choice, and thus results vary slightly from run to run.

```
****************************************
Report : qor
Design : wallypipelinedcore__165949
Version: R-2020.09-SP5-5
Date   : Fri Nov  3 07:39:11 2023
****************************************

  Timing Path Group 'clk'
  -----------------------------------
  Levels of Logic:              37.000000
  Critical Path Length:          3.030967
  Critical Path Slack:          -0.152755
  Critical Path Clk Period:      3.030303
  Total Negative Slack:        -57.774506
  No. of Violating Paths:      710.000000
  Worst Hold Violation:          0.000000
  Total Hold Violation:          0.000000
  No. of Hold Violations:        0.000000
  -----------------------------------

  Cell Count
  -----------------------------------
  Hierarchical Cell Count:           78
  Hierarchical Port Count:         6697
  Leaf Cell Count:                 8226
  Buf/Inv Cell Count:              2069
  Buf Cell Count:                   335
  Inv Cell Count:                  1734
  CT Buf/Inv Cell Count:              0
  Combinational Cell Count:        7053
  Sequential Cell Count:           1173
  Macro Count:                        0
  -----------------------------------

  Area
  -----------------------------------
  Combinational Area:     52579.678278
  Noncombinational Area:  42734.023876
  Buf/Inv Area:           10614.396889
  Total Buffer Area:       2400.485956
  Total Inverter Area:     8213.910933
  Macro/Black Box Area:       0.000000
  Net Area:                   0.000000
  -----------------------------------
  Cell Area:              95313.702154
  Design Area:            95313.702154

  Design Rules
  -----------------------------------
  Total Number of Nets:            8614
  Nets With Violations:               0
  Max Trans Violations:               0
  Max Cap Violations:                 0
  -----------------------------------

  --------------------------------------------------------------------
  Design  WNS: 0.152755  TNS: 57.774506  Number of Violating Paths: 710
  Design (Hold)  WNS: 0.000000  TNS: 0.000000  Number of Violating Paths: 0
  ---------------------------------------------------------------
```

Fig. 6.13 `qor.rep.`

6.4.4 Performance Reports

The synthesis script uses the `report_timing` command to generate `timing.rep` that lists the worst overall timing path and `per_module_timing.rep` that lists the longest path through each of the various modules. For rv32e, the longest path involves a branch comparison in the Execute stage triggering a branch that leads to a stall or flush. In SKY90 and TSMC28, the stall endpoint is longest. In SKY130, the flush path is 2 ps slower than the stall path, but the flush path is more complicated to describe because it involves canceling an EBU transaction. Therefore, this section describes the stall critical path.

Fig. 6.14 annotates the stall path listed in the `timing.rep` report in Fig. 6.15 using SKY130. Fig. 6.14 is a diagram of the rv32e core adapted from Fig. 7.13. All times are in nanoseconds and capacitances in picofarads, as

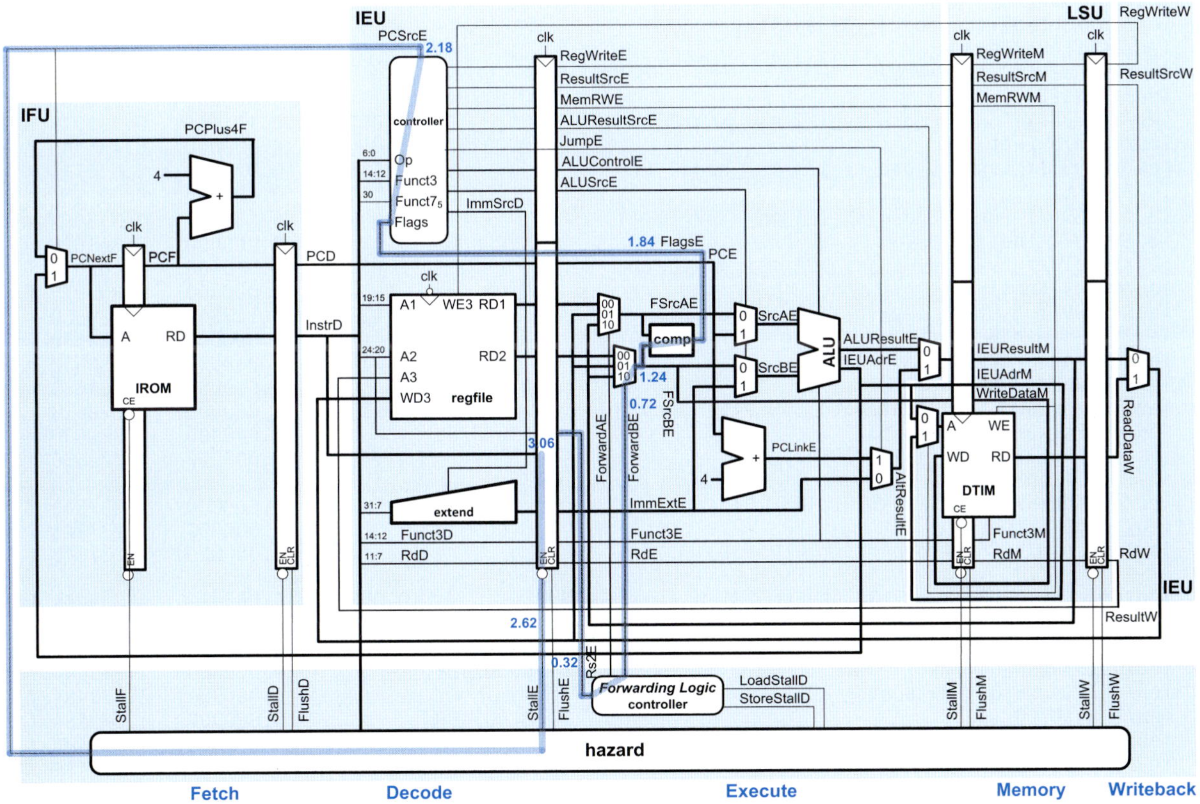

Fig. 6.14 rv32e core with critical path annotated in blue.

```
# A fanout number of 1000 was used for high fanout net computations.

Operating Conditions: PVT_1P8V_25C   Library: sky130_osu_sc_12T_ms_TT_1P8_25C.ccs
Wire Load Model Mode: top

  Startpoint: ieu/dp/Rs2EReg/q_reg[0]
              (rising edge-triggered flip-flop clocked by clk)
  Endpoint: ifu/InstrEReg/q_reg[31]
            (rising edge-triggered flip-flop clocked by clk)
  Path Group: clk
  Path Type: max

  Point                                                   Fanout    Cap        Trans      Incr       Path
  -----------------------------------------------------------------------------------------------------------
  clock clk (rise edge)                                                                   0.000000   0.000000
  clock network delay (ideal)                                                             0.000000   0.000000
  ieu/dp/Rs2EReg/q_reg[0]/CK (sky130_osu_sc_12T_ms__dff_1)                     0.000000   0.000000 # 0.000000 r
  ieu/dp/Rs2EReg/q_reg[0]/Q (sky130_osu_sc_12T_ms__dff_1)                      0.080838   0.320341   0.320341 f
  ieu/dp/Rs2EReg/q[0] (net)                                4        0.013949              0.000000   0.320341 f
  ieu/dp/Rs2EReg/q[0] (flopenrc_WIDTH5_3)                                                 0.000000   0.320341 f
  ieu/dp/Rs2E[0] (net)                                              0.013949              0.000000   0.320341 f
  ieu/dp/Rs2E[0] (datapath__395949)                                                       0.000000   0.320341 f
  ieu/Rs2E[0] (net)                                                 0.013949              0.000000   0.320341 f
  ieu/fw/Rs2E[0] (forward)                                                                0.000000   0.320341 f
  ieu/fw/Rs2E[0] (net)                                              0.013949              0.000000   0.320341 f
  ieu/fw/U36/Y (sky130_osu_sc_12T_ms__xnor2_1)                                 0.089581   0.111006   0.431348 r
  ieu/fw/n19 (net)                                         1        0.002578              0.000000   0.431348 r
  ieu/fw/U35/Y (sky130_osu_sc_12T_ms__nand2_1)                                 0.056313   0.070386   0.501734 f
  ieu/fw/n54 (net)                                         2        0.004396              0.000000   0.501734 f
  ieu/fw/U34/Y (sky130_osu_sc_12T_ms__inv_1)                                   0.038368   0.051942   0.553675 r
  ieu/fw/n21 (net)                                         1        0.002665              0.000000   0.553675 r
  ieu/fw/U10/Y (sky130_osu_sc_12T_ms__nand2_1)                                 0.035893   0.048400   0.602075 f
  ieu/fw/n4 (net)                                          1        0.002499              0.000000   0.602075 f
  ieu/fw/U9/Y (sky130_osu_sc_12T_ms__nor2_1)                                   0.119280   0.118555   0.720630 r
  ieu/fw/ForwardBE[1] (net)                                2        0.005788              0.000000   0.720630 r
  ieu/fw/ForwardBE[1] (forward)                                                           0.000000   0.720630 r
  ieu/ForwardBE[1] (net)                                            0.005788              0.000000   0.720630 r
  ieu/dp/ForwardBE[1] (datapath__395949)                                                  0.000000   0.720630 r
  ieu/dp/ForwardBE[1] (net)                                         0.005788              0.000000   0.720630 r
  ieu/dp/fbemux/s[1] (mux3_WIDTH32_0)                                                     0.000000   0.720630 r
  . . .
  ieu/dp/fbemux/y[26] (mux3_WIDTH32_0)                                                    0.000000   1.237343 f
  ieu/dp/n126 (net)                                                 0.009574              0.000000   1.237343 f
  ieu/dp/comp/b[26] (comparator_WIDTH32)                                                  0.000000   1.237343 f
  . . .
  ieu/dp/comp/U272/Y (sky130_osu_sc_12T_ms__oai21_1)                           0.052593   0.063912   1.839484 f
  ieu/dp/comp/flags[0] (net)                               1        0.002488              0.000000   1.839484 f
  ieu/dp/comp/flags[0] (comparator_WIDTH32)                                               0.000000   1.839484 f
  ieu/dp/FlagsE[0] (net)                                            0.002488              0.000000   1.839484 f
  ieu/dp/FlagsE[0] (datapath__395949)                                                     0.000000   1.839484 f
  ieu/FlagsE[0] (net)                                               0.002488              0.000000   1.839484 f
  ieu/c/FlagsE[0] (controller__395949)                                                    0.000000   1.839484 f
  ieu/c/FlagsE[0] (net)                                             0.002488              0.000000   1.839484 f
  ieu/c/branchflagmux/d1[0] (mux2_WIDTH1)                                                 0.000000   1.839484 f
  ieu/c/branchflagmux/d1[0] (net)                                   0.002488              0.000000   1.839484 f
  ieu/c/branchflagmux/U1/Y (sky130_osu_sc_12T_ms__nand2_1)                     0.055557   0.051331   1.890814 r
  ieu/c/branchflagmux/n3 (net)                             1        0.002578              0.000000   1.890814 r
```

Fig. 6.15 **timing.rep.**

```
ieu/c/branchflagmux/U4/Y (sky130_osu_sc_12T_ms__nand2_1)      0.052831  0.063409  1.954223 f
ieu/c/branchflagmux/y[0] (net)              1    0.004722               0.000000  1.954223 f
ieu/c/branchflagmux/y[0] (mux2_WIDTH1)                                  0.000000  1.954223 f
ieu/c/BranchFlagE (net)                          0.004722               0.000000  1.954223 f
ieu/c/U33/Y (sky130_osu_sc_12T_ms__xnor2_1)                   0.056597  0.100257  2.054480 f
ieu/c/n22 (net)                             1    0.002418               0.000000  2.054480 f
ieu/c/U34/Y (sky130_osu_sc_12T_ms__oai21_1)                   0.128619  0.122617  2.177096 r
ieu/c/PCSrcE (net)                          2    0.004966               0.000000  2.177096 r
ieu/c/PCSrcE (controller__395949)                                       0.000000  2.177096 r
ieu/PCSrcE (net)                                 0.004966               0.000000  2.177096 r
ieu/PCSrcE (ieu__395949)                                                0.000000  2.177096 r
PCSrcE (net)                                     0.004966               0.000000  2.177096 r
ifu/PCSrcE (ifu__395949)                                                0.000000  2.177096 r
ifu/PCSrcE (net)                                 0.004966               0.000000  2.177096 r
ifu/U3/Y (sky130_osu_sc_12T_ms__buf_1)                        0.054478  0.107858  2.284954 r
ifu/BPWrongE (net)                          1    0.004894               0.000000  2.284954 r
ifu/BPWrongE (ifu__395949)                                              0.000000  2.284954 r
BPWrongE (net)                                   0.004894               0.000000  2.284954 r
hzu/BPWrongE (hazard__395949)                                           0.000000  2.284954 r
hzu/BPWrongE (net)                               0.004894               0.000000  2.284954 r
hzu/U6/Y (sky130_osu_sc_12T_ms__inv_2)                        0.037596  0.048914  2.333868 f
hzu/n5 (net)                                4    0.011301               0.000000  2.333868 f
hzu/U7/Y (sky130_osu_sc_12T_ms__nand2_1)                      0.037941  0.046295  2.380163 r
hzu/n6 (net)                                1    0.002578               0.000000  2.380163 r
hzu/U8/Y (sky130_osu_sc_12T_ms__nand2_1)                      0.100068  0.096502  2.476665 f
hzu/StallM (net)                            4    0.011132               0.000000  2.476665 f
hzu/U13/Y (sky130_osu_sc_12T_ms__buf_2)                       0.054383  0.144729  2.621394 f
hzu/StallE (net)                            4    0.016288               0.000000  2.621394 f
hzu/StallE (hazard__395949) <-                                          0.000000  2.621394 f
StallE (net)                                     0.016288               0.000000  2.621394 f
ifu/StallE (ifu__395949)                                                0.000000  2.621394 f
ifu/StallE (net)                                 0.016288               0.000000  2.621394 f
ifu/U6/Y (sky130_osu_sc_12T_ms__inv_2)                        0.073454  0.077434  2.698829 r
ifu/n256 (net)                              6    0.015610               0.000000  2.698829 r
ifu/InstrEReg/en (flopenr_WIDTH32_1)                                    0.000000  2.698829 r
ifu/InstrEReg/en (net)                           0.015610               0.000000  2.698829 r
ifu/InstrEReg/U8/Y (sky130_osu_sc_12T_ms__nand2_1)            0.055165  0.067654  2.766483 f
ifu/InstrEReg/n3 (net)                      1    0.004619               0.000000  2.766483 f
ifu/InstrEReg/U9/Y (sky130_osu_sc_12T_ms__inv_2)              0.065370  0.071717  2.838200 r
ifu/InstrEReg/n21 (net)                     1    0.013498               0.000000  2.838200 r
ifu/InstrEReg/U4/Y (sky130_osu_sc_12T_ms__inv_6)              0.078840  0.089298  2.927498 f
ifu/InstrEReg/n106 (net)                   32    0.077421               0.000000  2.927498 f
ifu/InstrEReg/U108/Y (sky130_osu_sc_12T_ms__oai22_1)          0.120045  0.128082  3.055580 r
ifu/InstrEReg/n66 (net)                     1    0.002852               0.000000  3.055580 r
ifu/InstrEReg/q_reg[31]/D (sky130_osu_sc_12T_ms__dff_1)       0.120045  0.000000  3.055580 r
data arrival time                                                                 3.055580

clock clk (rise edge)                                                   3.030303  3.030303
clock network delay (ideal)                                             0.000000  3.030303
ifu/InstrEReg/q_reg[31]/CK (sky130_osu_sc_12T_ms__dff_1)                0.000000  3.030303 r
library setup time                                                     -0.152149  2.878155
data required time                                                                2.878155
-----------------------------------------------------------------------------------------------
data required time                                                                2.878155
data arrival time                                                                -3.055580
-----------------------------------------------------------------------------------------------
slack (VIOLATED)                                                                 -0.177425
```

Fig. 6.15 `timing.rep`. Cont'd

```
ieu/dp/ImmExtEReg/q_reg[18]/D (sky130_osu_sc_12T_ms__dff_1)   0.121576  0.000000   3.081019 r
data arrival time                                                                   3.081019

clock clk (rise edge)                                                   3.030303   3.030303
clock network delay (ideal)                                             0.000000   3.030303
clock uncertainty                                                      -0.100000   2.930303
ieu/dp/ImmExtEReg/q_reg[18]/CK (sky130_osu_sc_12T_ms__dff_1)            0.000000   2.930303 r
library setup time                                                     -0.152505   2.777798
data required time                                                                 2.777798
-----------------------------------------------------------------------------------------------
data required time                                                                 2.777798
data arrival time                                                                 -3.081019
-----------------------------------------------------------------------------------------------
slack (VIOLATED)                                                                  -0.303221
```

Fig. 6.17 `timing.rep` excerpt showing setup time with uncertainty.

```
Point                                                Fanout      Cap      Trans      Incr        Path
-----------------------------------------------------------------------------------------------------
clock clk (rise edge)                                                              0.000000    0.000000
clock network delay (ideal)                                                        0.000000    0.000000
lsu/ReadDataMWReg/q_reg[2]/CK (sky130_osu_sc_12T_ms__dff_1)            0.000000   0.000000 #  0.000000 r
lsu/ReadDataMWReg/q_reg[2]/Q (sky130_osu_sc_12T_ms__dff_1)             0.041536   0.191974    0.191974 r
lsu/ReadDataMWReg/q[2] (net)                           2     0.003163              0.000000    0.191974 r
lsu/ReadDataMWReg/U8/Y (sky130_osu_sc_12T_ms__mux2_1)                  0.034672   0.023973    0.215947 r
lsu/ReadDataMWReg/n3 (net)                             1     0.002566              0.000000    0.215947 r
lsu/ReadDataMWReg/q_reg[2]/D (sky130_osu_sc_12T_ms__dff_1)             0.034672   0.000000    0.215947 r
data arrival time                                                                              0.215947

clock clk (rise edge)                                                              0.000000    0.000000
clock network delay (ideal)                                                        0.000000    0.000000
clock uncertainty                                                                  0.100000    0.100000
lsu/ReadDataMWReg/q_reg[2]/CK (sky130_osu_sc_12T_ms__dff_1)                        0.000000    0.100000 r
library hold time                                                                 -0.050201    0.049799
data required time                                                                             0.049799
-----------------------------------------------------------------------------------------------------
data required time                                                                             0.049799
data arrival time                                                                             -0.215947
-----------------------------------------------------------------------------------------------------
slack (MET)                                                                                    0.166148
```

Fig. 6.18 `mindelay.rep` showing hold time with uncertainty.

the synthesis process. Fig. 6.18 shows that hold slack has also decreased by 0.105 ns compared to Fig. 6.16 (from 0.271 to 0.166 ns).

During placement and routing, CAD tools may insert delay buffers to fix hold time violations.

6.4.5 Power Reports

The `power.rep` file in Fig. 6.19 shows the power in each module. Dynamic power is categorized as *switching power* and *internal power*. Switching power accounts for the load capacitance (gates and wires driven by each cell),

```
****************************************
Report : power
        -hier
        -analysis_effort low
Design : wallypipelinedcore__395949
Version: R-2020.09-SP5-5
Date   : Thu Nov 16 07:18:39 2023
****************************************


Library(s) Used:

    sky130_osu_sc_12T_ms_TT_1P8_25C.ccs (File:
/opt/riscv/cad/lib/sky130_osu_sc_t12/12T_ms/lib/sky130_osu_sc_12T_ms_TT_1P8_25C.ccs.db)


Operating Conditions: PVT_1P8V_25C   Library: sky130_osu_sc_12T_ms_TT_1P8_25C.ccs
Wire Load Model Mode: top


Global Operating Voltage = 1.8
Power-specific unit information :
    Voltage Units = 1V
    Capacitance Units = 1.000000pf
    Time Units = 1ns
    Dynamic Power Units = 1mW    (derived from V,C,T units)
    Leakage Power Units = 1nW


--------------------------------------------------------------------------------
                                       Switch   Int      Leak     Total
Hierarchy                              Power    Power    Power    Power    %
--------------------------------------------------------------------------------
wallypipelinedcore__395949                0.820    5.302  715.939    6.122 100.0
  hzu (hazard__395949)                 7.04e-03 1.90e-03    1.674 8.95e-03   0.1
  ebu.ebu (ebu_XLEN32_PA_BITS34_AHBW32)
                                       3.79e-02    0.177   10.097    0.215   3.5
  lsu (lsu__395949)                    5.16e-02    0.609   32.021    0.661  10.8
  ieu (ieu__395949)                       0.458    3.299  602.129    3.758  61.4
  ifu (ifu__395949)                       0.264    1.213   69.435    1.477  24.1
```

Fig. 6.19 `power.rep`.

and internal power accounts for the capacitance inside each cell. Both are expressed in units of mW. The leakage power is given in units of nW. The power is calculated at the target clock frequency even if the circuit has negative slack.

Observe that the leakage power is 715 nW compared to dynamic power of (5.302 + 0.820) = 6.122 mW. In other words, leakage power is 0.012% of the dynamic power. When a system is operating at full speed, leakage is typically negligible. However, leakage influences battery life when the system is quiescent.

Power is estimated using Design Compiler's default activity factor of 1.0 (100%) for clocks and 0.1 (10%) for non-clock signals. It also assumes equal probability of 0 and 1 states on each node to compute leakage power. Activity factors can be determined more accurately by running a simulation of the synthesized netlist on a real workload and recording the number of times each node toggles in a VCD (Verilog Change Dump) file. See `wsim --vcd` and the `MAKE_VCD` mode in `testbench.sv`. However, the results are highly sensitive to the workload; for example, the CoreMark benchmark has no floating-point instructions and thus may predict no activity in the FPU.

Power is computed at the target clock frequency and hence changes with the frequency target. Energy per cycle or operation is a more insightful metric because it is independent of frequency. Energy per cycle is obtained by multiplying the power by the target clock period (the reciprocal of frequency). In this case, the power is 6.122 mW, and the target frequency is 333 MHz, so the energy is $6.122 \times 10^{-3}/333 \times 10^{6} = 18$ pJ/cycle.

6.4.6 Area Reports

Fig. 6.20 shows an area report that gives a breakdown for each module. As highlighted in blue, the total area is 95,064 μm^2, or about 0.1 mm^2; 71% of the area is in the integer execution unit (`ieu`), and more than half of that area is the register file (`ieu/dp/regf`). The total area is based on cell area, assuming wires can run over top. Place and route tools typically need an extra 10%–30% area to have the flexibility to place critical cells near each other and avoid routing congestion.

6.4.7 Other Reports

The reports directory has a number of other handy outputs.

- **`check_design.rpt`** lists potential problems with the design. Because Wally is configurable, it lists hundreds of unconnected ports. Many inputs are unused in rv32e because the features that would use them do not exist. For example, the `TimerIntM` timer interrupt pin is unconnected because there is no privileged unit to accept interrupts.
- **`constraint_all_violators.rep`** lists all endpoints that have negative slack.
- **`constraint.rep`** summarizes unsatisfied constraints, particularly maximum delay.

```
****************************************
Report : area
Design : wallypipelinedcore__395949
Version: R-2020.09-SP5-5
Date   : Thu Nov 16 07:18:39 2023
****************************************
Library(s) Used:
    sky130_osu_sc_12T_ms_TT_1P8_25C.ccs (File:
/opt/riscv/cad/lib/sky130_osu_sc_t12/12T_ms/lib/sky130_osu_sc_12T_ms_TT_1P8_25C.ccs.db)
Number of ports:                         6891
Number of nets:                         13479
Number of cells:                         8265
Number of combinational cells:           7014
Number of sequential cells:              1173
Number of macros/black boxes:               0
Number of buf/inv:                       2060
Number of references:                      11
Combinational area:              52330.105886
Buf/Inv area:                    10525.996487
Noncombinational area:           42734.023876
Macro/Black Box area:                0.000000
Net Interconnect area:      undefined  (No wire load specified)
Total cell area:                 95064.129763
Total area:                 undefined
The above information was reported from the logical library. The following are from the physical
library:
Hierarchical area distribution
------------------------------
                                   Global cell area          Local cell area
                                   -------------------  ------------------------------
Hierarchical cell                  Absolute    Percent  Combi-      Noncombi-   Black-
                                   Total       Total    national    national    boxes   Design
--------------------------------   ----------  -------  ----------  ----------  ------  ----------
wallypipelinedcore__395949         95064.1298   100.0     40.0488      0.0000   0.0000
wallypipelinedcore__395949
ebu.ebu                             4124.5380     4.3    949.9380      0.0000   0.0000
ebu_XLEN32_PA_BITS34_AHBW32
ebu.ebu/IFUInput                    2252.9892     2.4     30.2808      0.0000   0.0000
controllerinput_PA_BITS34
ebu.ebu/IFUInput/genblk1.RestorMux   412.6980     0.4    412.6980      0.0000   0.0000
mux2_WIDTH43
ebu.ebu/IFUInput/genblk1.SaveReg    1810.0104     1.9    746.2752   1063.7353   0.0000
flopenr_WIDTH43
ebu.ebu/LSUInput                     252.0144     0.3    252.0144      0.0000   0.0000
controllerinput_PA_BITS34_SAVE_ENABLED0
ebu.ebu/ebufsmarb                    669.5964     0.7    228.5712      0.0000   0.0000  ebufsmarb
ebu.ebu/ebufsmarb/BeatCounter        295.4820     0.3     60.5616      0.0000   0.0000
counter_WIDTH4
. . .
hzu                                   98.6568     0.1     98.6568      0.0000   0.0000
hazard__395949
ieu                                67129.1145    70.6     25.3968      0.0000   0.0000
ieu__395949
ieu/c                               2178.2640     2.3    555.3108      0.0000   0.0000
controller__395949
ieu/c/branchflagmux                   23.4432     0.0     23.4432      0.0000   0.0000
mux2_WIDTH1
ieu/c/controlregD                     59.5848     0.1     27.3504     32.2344   0.0000
```

Fig. 6.20 area.rep.

```
flopenrc_WIDTH1_0
ieu/c/controlregE              984.1260    1.0     403.9068     580.2192  0.0000
flopenrc_WIDTH35
ieu/c/controlregM              431.7456    0.5     173.8704     257.8752  0.0000
flopenrc_WIDTH25
ieu/c/controlregW              124.0536    0.1      59.5848      64.4688  0.0000
flopenrc_WIDTH5_5
ieu/dp                       64108.8488   67.4     336.5076       0.0000  0.0000
datapath__395949
ieu/dp/IEUResultMReg          1695.7248    1.8     664.2240    1031.5009  0.0000
flopenrc_WIDTH32_2
. . .
ieu/dp/WriteDataMReg          1714.2840    1.8     682.7832    1031.5009  0.0000
flopenrc_WIDTH32_1
ieu/dp/altresultmux            389.2548    0.4     389.2548       0.0000  0.0000
mux2_WIDTH32_1
ieu/dp/alu                    8880.0887    9.3    4587.5412       0.0000  0.0000
alu_32_395949
ieu/dp/alu/sh                 4292.5476    4.5    4292.5476       0.0000  0.0000
shifter__395949
ieu/dp/comp                   1995.1140    2.1    1995.1140       0.0000  0.0000
comparator_WIDTH32
ieu/dp/ext                     465.4452    0.5     465.4452       0.0000  0.0000
extend__395949
ieu/dp/faemux                  902.5632    0.9     902.5632       0.0000  0.0000
mux3_WIDTH32_1
ieu/dp/fbemux                  904.0284    1.0     904.0284       0.0000  0.0000
mux3_WIDTH32_0
ieu/dp/ieuresultmux            615.8724    0.6     615.8724       0.0000  0.0000
mux2_WIDTH32_0
ieu/dp/regf                  34373.1031   36.2   13977.5196   20395.5835  0.0000
regfile_32_1
ieu/dp/resultmuxW              601.2204    0.6     601.2204       0.0000  0.0000
mux5_WIDTH32
ieu/dp/srcamux                 593.4060    0.6     593.4060       0.0000  0.0000
mux2_WIDTH32_3
ieu/dp/srcbmux                 615.8724    0.6     615.8724       0.0000  0.0000
mux2_WIDTH32_2
ieu/fw                         816.6048    0.9     816.6048       0.0000  0.0000  forward
ifu                          16242.7191   17.1    2312.0856       0.0000  0.0000
ifu__395949
ifu/AlignedInstrRawDFlop      1701.0972    1.8     669.5964    1031.5009  0.0000
flopenl_WIDTH32_1
ifu/FlushInstrEMux             278.3880    0.3     278.3880       0.0000  0.0000
mux2_WIDTH32_5
ifu/InstrEReg                 1710.3768    1.8     678.8760    1031.5009  0.0000
. . .
ifu/pcreg                     1641.5124    1.7     642.2460     999.2665  0.0000
flopenl_WIDTH32_0
lsu                           7429.0526    7.8     243.7116       0.0000  0.0000
lsu__395949
lsu/AddressMReg               1720.1448    1.8     688.6440    1031.5009  0.0000
flopenrc_WIDTH32_7
lsu/ReadDataMWReg             1474.9681    1.6     443.4672    1031.5009  0.0000
. . .
```

Fig. 6.20 `area.rep`. **Cont'd**

```
lsu/subwordread                       713.0640       0.8     713.0640       0.0000  0.0000
subwordread_LLEN32
lsu/subwordwrite                      450.7932       0.5     450.7932       0.0000  0.0000
subwordwrite_LLEN32
lsu/swbytemask                         58.6080       0.1      58.6080       0.0000  0.0000
swbytemask_32_0
------------------------------ ---------- ------- ---------- ---------- ------ ----------
Total                                                     52330.1059 42734.0239  0.0000
```

Fig. 6.20 `area.rep`. **Cont'd**

6.5 WALLY SYNTHESIS RESULTS

This section describes area and delay results from synthesizing various configurations of `wallypipelinedcore` with different constraints and synthesis options. The results are presented for SKY130, SKY90, and TSMC 28SKY130. The data is gathered under typical conditions (TT process, nominal voltage, 25°C), as in Section 6.2.1.

6.5.1 Synthesis Configurations

The primary configurations have large memories, so they can run arbitrary test programs during logic verification. For example, rv32i has an 8-MiB DTIM. Design Compiler does not gracefully build such a large memory from flip-flops. The runtime and memory are huge, and the synthesis usually fails. Therefore, we use derived configurations, that is, configurations with some tweaks such as smaller memories, summarized in Table 6.6. For Skywater, the derived syn_* configurations reduce the size of the DTIM, IROM, and cache and branch predictor memories. For TSMC, the derived syn_sram_* configurations only reduce the DTIM/IROM size and build the cache and branch memories from SRAM. Other derived configurations are used to turn off features to explore critical paths, as described in Section 6.5.4.

Table 6.7 summarizes the core memory configurations for each case. rv32e has only a register file. rv32i, rv64i, and rv32imc have register files, DTIM, and IROM. rv32gc has caches, branch predictors, TLBs, and PMP. Note that uncore memories and peripherals are not part of these results.

6.5.2 Area and Delay vs. Target Clock Period

Fig. 6.21(a) shows how the area and delay vary with target clock period for the rv32e configuration in the SKY130 technology. Synthesis achieves its timing goals at long target cycle times, as indicated by the gray dots. As the cycle time decreases, the area increases slightly because synthesis trades area for faster and more parallel logic. However, Fig. 6.20 shows that 38% of the area is consumed by the register file, with flip-flop area being independent of cycle time, so overall area does not increase too much with

Table 6.6 Derived configurations for synthesis.

Derived Configuration	Base Configuration	Tweaks
syn_rv32e	rv32e	Reduced memory size per Table 6.7 to keep runtime and memory reasonable for processes without SRAM
syn_rv32i	rv32i	
syn_rv32imc	rv32imc	
syn_rv32gc	rv32gc	
syn_rv64i	rv64i	
syn_rv64gc	rv64gc	
syn_sram_rv32e	rv32e	Use SRAM library for TSMC28
syn_sram_rv32i	rv32i	
syn_sram_rv32imc	rv32imc	
syn_sram_rv32gc	rv32gc	
syn_sram_rv64i	rv64i	
syn_sram_rv64gc	rv64gc	
syn_rv64gc_pmp0	rv64gc	Gradually disable features to explore critical paths
syn_rv64gc_noPriv	syn_rv64gc_pmp0	
syn_rv64gc_noFPU	syn_rv64gc_noPriv	
syn_rv64gc_noMulDiv	syn_rv64gc_noFPU	
syn_rv64gc_noAtomic	syn_rv64gc_noMulDiv	
syn_sram_rv64gc_pmp0	syn_sram_rv64gc	Gradually disable features to explore critical paths while using SRAM library
syn_ sram_rv64gc_noPriv	syn_sram_rv64gc_pmp0	
syn_ sram_rv64gc_noFPU	syn_ sram_rv64gc_noPriv	
syn_ sram_rv64gc_noMulDiv	syn_ sram_rv64gc_noFPU	
syn_ sram_rv64gc_noAtomic	syn_ sram_rv64gc_noMulDiv	

frequency. At shorter cycle times, synthesis is unable to achieve the target cycle time, as indicated by the blue dots. The achieved cycle time and area plateau at a point nearly independent of the target cycle time, although there is some synthesis noise resulting in slight variations from one point

Table 6.7 **Core memory configurations for synthesis.**

Element	Configurations	USESRAM	Size	Implementation
D$, I$	rv{32/64}gc	0	1 way each 512 B/way 8 64B lines 8 44 or 22-bit tags	[rv32gc: 8 × (512 + 22) = 4,272 flops rv64gc: 8 × (512 + 44) = 4,448 flops] × 2 caches
		1	4 ways each 4 KiB/way 64 512-bit lines 64 22/44-bit tags	[4 ram1p1rwbe_64x128 1 ram1p1rwbe_64x44] × 4 ways × 2 caches
BHT	rv{32/64}gc	0	2^5 entries × 2 bits	64 flops
		1	2^{10} entries × 2 bits	1 ram2p1r1wbe_64x32
BTB	rv{32/64}gc	0	2^5 entries × 36 or 68 bits	1,152 − 2,176 flops
		1	2^{10} entries × 68 bits	1 ram2p1r1wbe_1024x68
DTIM	rv32i, rv32imc	N/A	128 words × 32 bits	4,096 flops
	rv64i	N/A	64 words × 64 bits	4,096 flops
DTLB, ITLB	rv{32/64}gc	N/A	32 entries each x 61 or 106 bits	1,952 − 3,392 flops
PMP	rv{32/64}gc	N/A	16 entries × 64 bits	1,024 flops
Register File	rv32e	N/A	16 entries × 32 bits	512 flops
	rv32{i/imc/gc}	N/A	32 entries × 32 bits	1,024 flops
	rv64{i/gc}	N/A	32 entries × 64 bits	2,048 flops

to another. Normalizing delay in terms of fanout-of-4 (FO4) inverter delays helps compare designs across processes. The best achievable cycle time is 28-32 FO4.

Fig. 6.21(b and c) show area and delay versus clock period for SKY90 and TSMC28. The TSMC28 feature size is about 5 times smaller than SKY130. The area is about 15 times smaller, and the clock period is about 8 times shorter. The best achievable cycle time is similar for all designs when normalized to FO4 delays.

Ideal Dennard scaling would predict that a 5x feature size reduction makes a design 25x smaller and 5x faster; market requirements favor improving performance, typically at a cost of more area.

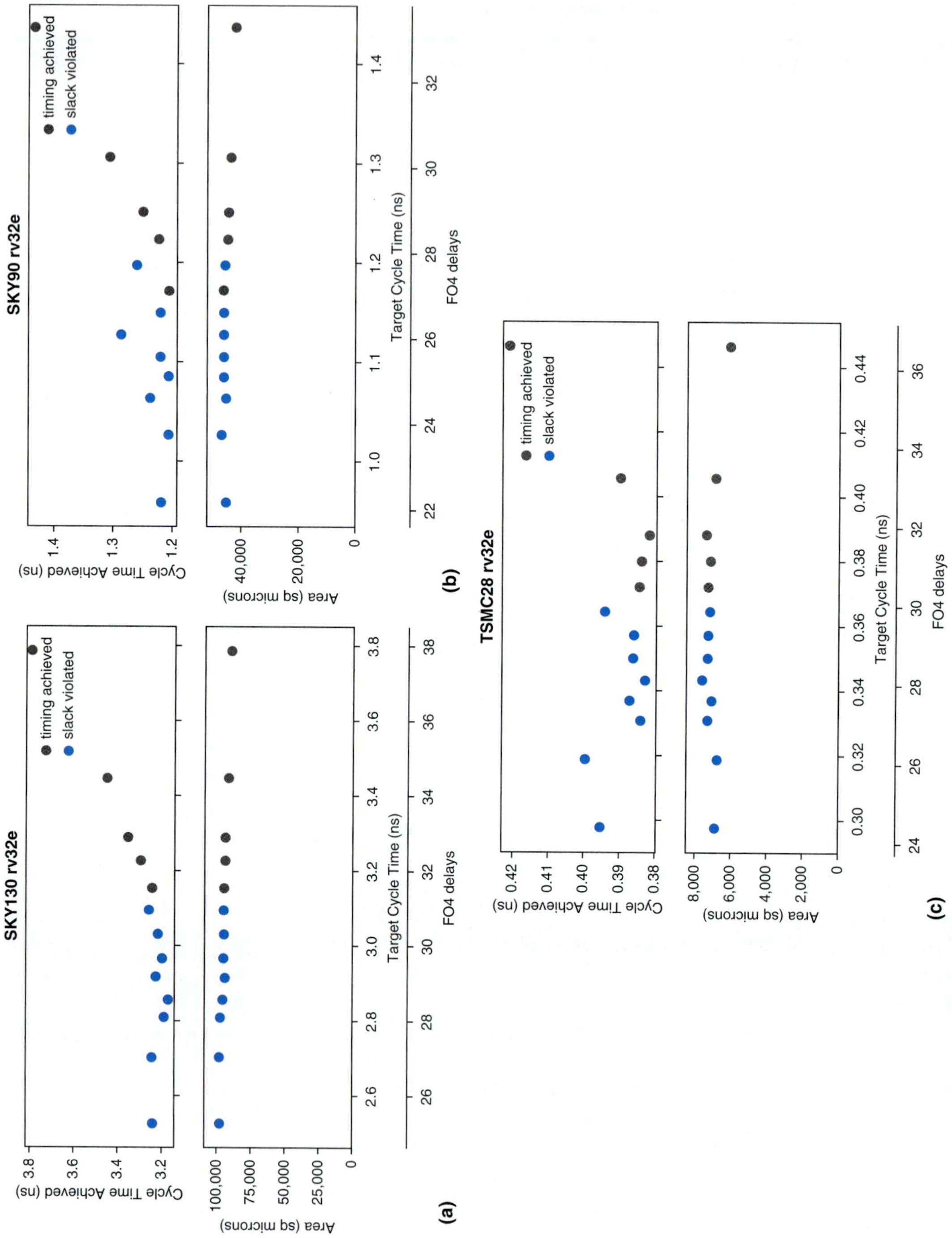

Fig. 6.21 `wallypipelinedcore` **rv32e area and delay vs. target clock period for (a) SKY130, (b) SKY90, and (c) TSMC28.**

The results and plots are generated with the following commands:

```
$ cd $WALLY/synthDC
$ ./wallySynth.py --freqsweep 330 --tech sky130
$ ./wallySynth.py --freqsweep 870 --tech sky90
$ ./wallySynth.py --freqsweep 2800 --tech tsmc28psyn
$ ./extractSummary.py
```

`wallySynthAll.py` automates running all of the simulations shown here.

The `--freqsweep` option requests a series of target frequencies centered around the specified value (e.g., 870 MHz = 1.15 ns). Use `wallySynth.py -h` to get more information on wallySynth options. The synthesis logs, reports, and outputs are placed in the `runs` subdirectory. `extractSummary.py` generates plots in `wallyplots/freqsweep*` from all of the runs, including these sweeps and other experiments described later in this section. Expect some messages because the other experiments have not yet been run. Linux users may display the plots using:

`extractSummary.py` requires several Python libraries to be installed: `sudo`, `pip3`, `install`, `matplotlib`, `scipy`, `sklearn`, and `adjustText`.

```
$ feh wallyplots/freqSweep_sky90_rv32e.png
```

Fig. 6.22 shows the same results when the `--maxopt` flag is added to the `wallySynth.py` invocation. Curiously, TSMC28 gets markedly faster, but SKY90 does not improve much. The `--maxopt` flag turns on the following Design Compiler optimizations to get better area and delay at the cost of more runtime:

- `flatten`: Allow optimizations across all modules but lose module names
- `optimize_registers`: Move logic across pipeline stages to improve critical paths

```
$ <remove or archive old files in the runs directory so they don't interfere>
$ ./wallySynth.py --freqsweep 350 --tech sky130 --maxopt
$ ./wallySynth.py --freqsweep 870 --tech sky90 --maxopt
$ ./wallySynth.py --freqsweep 3000 --tech tsmc28psyn --maxopt
$ ./extractSummary.py
```

6.5.3 Area and Delay vs. Configurations

Fig. 6.23 shows area and delay for the various configurations. Note that TSMC28 uses full-sized RAMs, while Skywater uses reduced-size flop-based memories. The results, plots, and area tables are generated with the following commands. The target frequency is set beyond what is achievable, to estimate the minimum cycle time.

```
$ ./wallySynth.py --configsweep --tech sky130 --targetfreq 1000
$ ./wallySynth.py --configsweep --tech sky90 --targetfreq 3000
$ ./wallySynth.py --configsweep --tech tsmc28psyn --targetfreq 10000 --usesram
$ ./extractSummary.py --sky130freq 1000 --sky90freq 3000 --tsmcfreq 10000
```

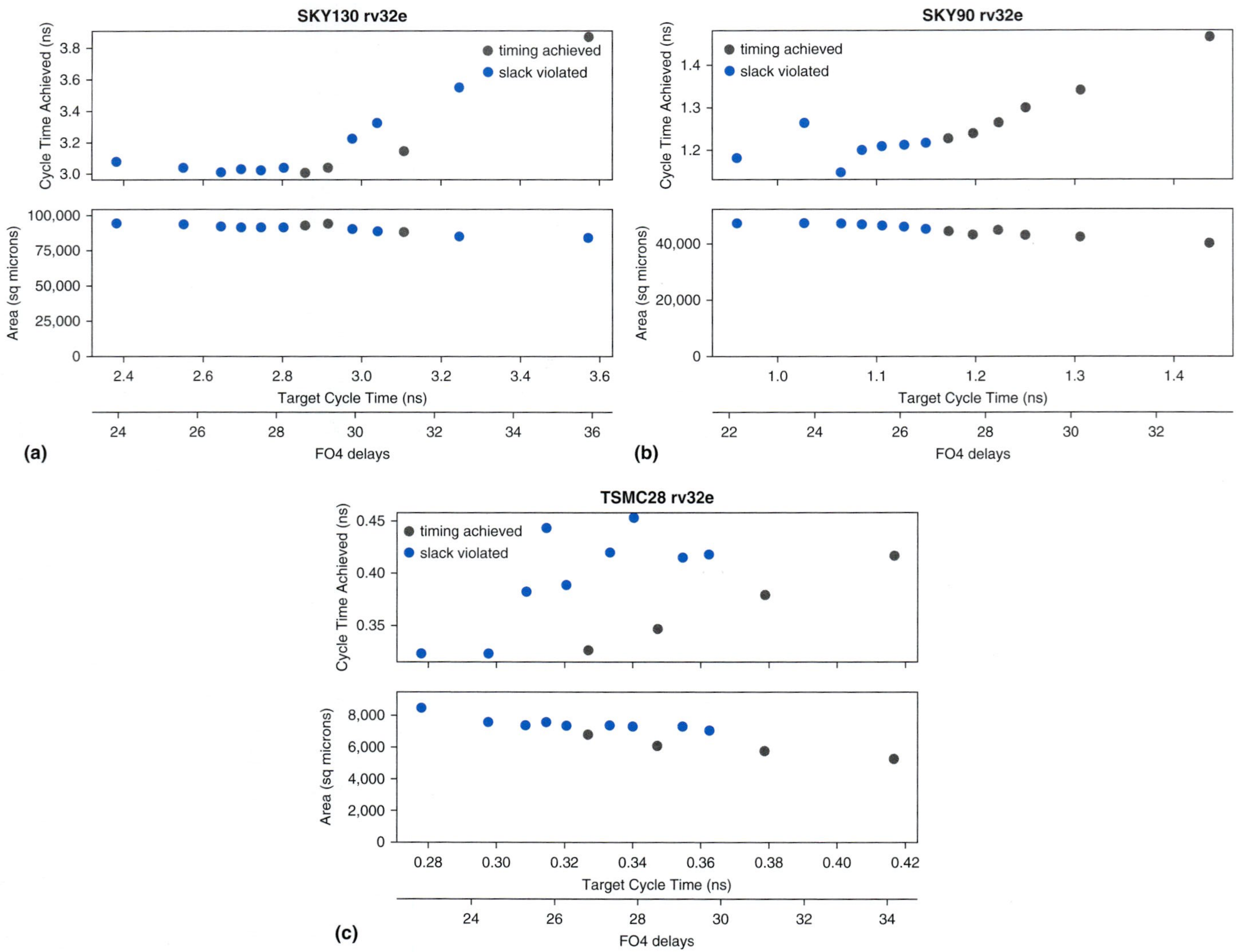

Fig. 6.22 `wallypipelinedcore` **rv32e area and delay vs. target clock period with maximum optimization for (a) SKY130, (b) SKY90, and (c) TSMC28.**

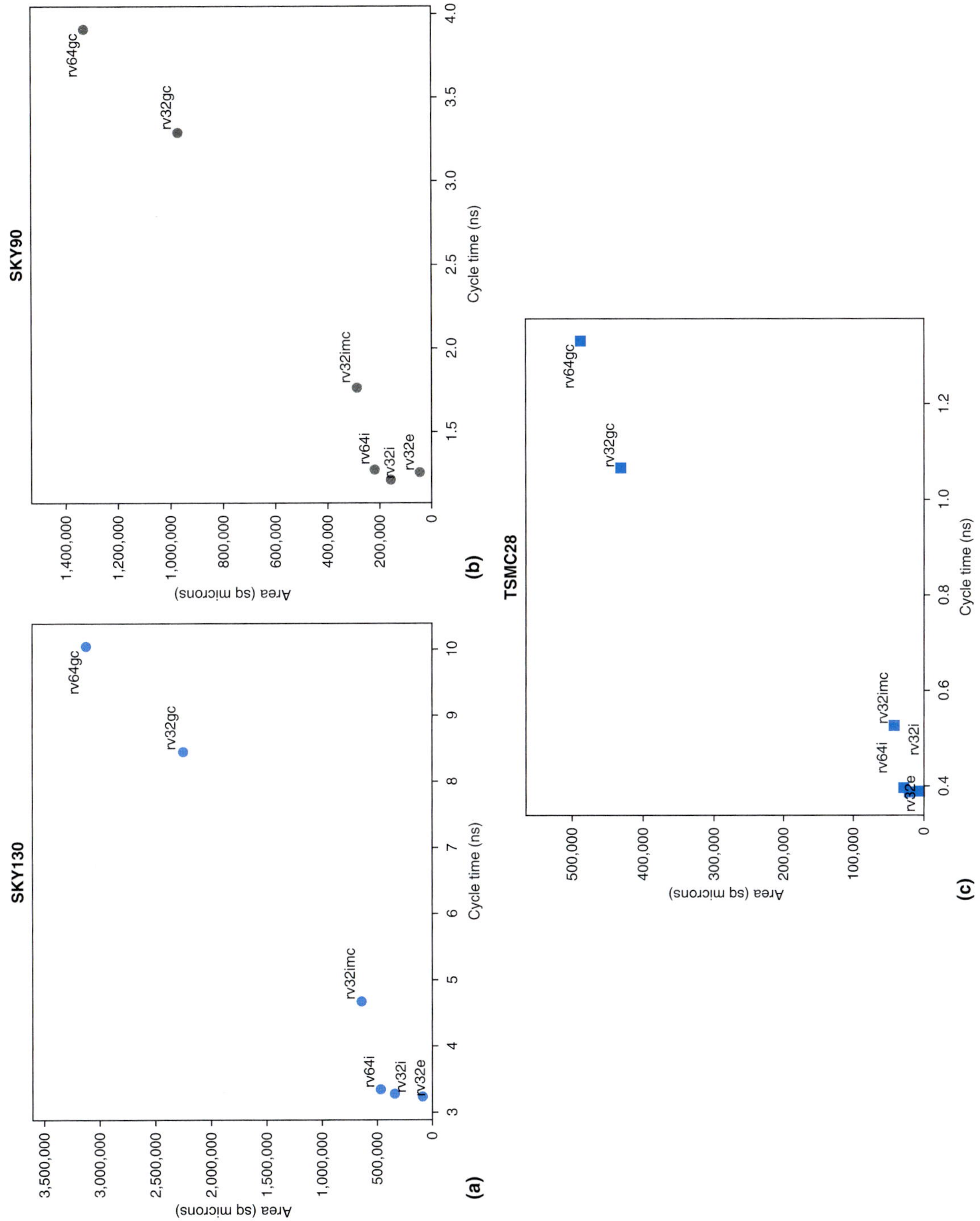

Fig. 6.23 **`wallypipelinedcore` area and delay vs. configuration for (a) SKY130, (b) SKY90, and (c) TSMC28.**

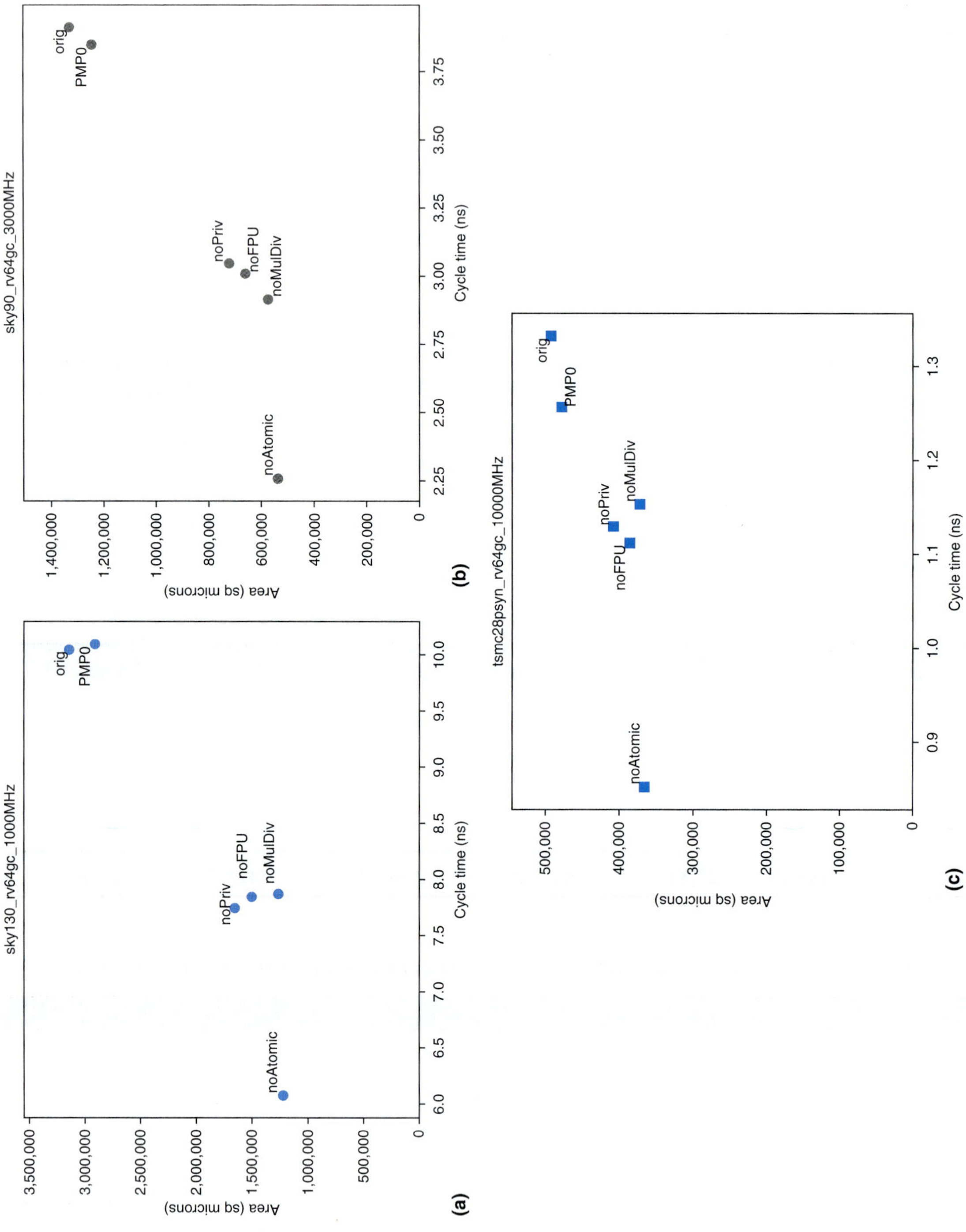

Fig. 6.24 **rv64gc `wallypipelinedcore` area and delay vs. features for (a) SKY130, (b) SKY90, and (c) TSMC28.**

Table 6.12 `wallypipelinedcore` **rv64gc feature deactivation experiment (TSMC28,** `USESRAM` **= 1, no** `MAXOPT`**).**

Feature	PMP Entries	Privileged Unit	FPU	MDU	Atomic Ops
orig	16	Y	Y	Y	Y
PMP0	0	Y	Y	Y	Y
noPriv	0	N	Y	Y	Y
noFPU	0	N	N	Y	Y
noMulDiv	0	N	N	N	Y
noAtomic	0	N	N	N	N

In SKY130, it is unexpected that PMP0 is slower than the baseline design (orig) and that noFPU and noMulDiv are slower than noPriv, calling into question the quality of the library models. Sometimes, you get what you pay for.

The SKY90 and TSMC28 results generally show the cycle time improving and area shrinking as features turn off. The area benefits are more pronounced on SKY130 and SKY90 because they use tiny flop-based memories; in TSMC28, the large RAMs account for most of the area.

6.6 SUMMARY

Logic synthesis maps synthesizable HDL code onto a cell library, minimizing area subject to a clock frequency constraint. This chapter examines the capabilities of three CMOS technologies. The older Skywater Technology 130- and 90-nm processes (SKY130 and SKY90) are low-cost technologies suited to microcontrollers and other chips of modest complexity. The newer TSMC 28 nm process (TSMC28) is the last of the mainstream planar processes before the advent of FinFETs, costly multipatterning, and draconian nondisclosure agreements (NDAs) that appeared starting with the 16-nm process. TSMC 28 is available for academic research and modern enough to be relevant for some commercial products. It also has an industry-strength set of libraries, including a memory compiler that can produce the SRAM memories needed for Wally's caches and branch predictor. The 28-nm process is 3-4x faster and offers about 8x better density and dynamic energy than the 90-nm process.

This chapter has illustrated how to use a Make-based script to synthesize Wally using the industry-standard Synopsys Design Compiler and how to interpret performance, power, and area (PPA) reports. In the 28-nm process, rv32e consumes about 6,500 logic cells (0.007 mm^2) and operates up to 2.8 GHz, limited by the Execute stage branch logic. rv64gc consumes about 200,000 logic cells and 42 RAM macros (0.5 mm^2) and operates at 800 MHz, limited by the time for the TLB and PMP to detect an illegal physical address, generate a fault, and update the program counter to fetch the next instruction from the trap handler.

Pipelined Core

7

7.1 Single-Cycle Core
7.2 Pipelining the RV32I Core
7.3 SystemVerilog
7.4 Test Plan
7.5 Summary

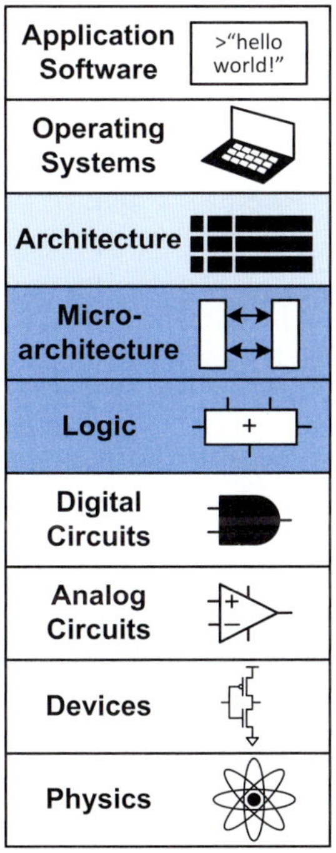

This chapter builds a simple configurable pipelined core capable of running RV32I or RV64I, integer instructions. Compared to the toy microarchitectures in Section 2.7, it implements the full integer instruction set. Fig. 7.1 shows the high-level block diagram with the unimplemented portions grayed out. The core contains ideal instruction and data memories (instruction read-only memory [IROM], data tightly integrated memory [DTIM]) and has no bus interface for external memory or peripherals. The trivial branch predictor always predicts branch not taken. There is no privileged unit, so the processor always runs in machine mode and does not take traps, support control and status registers (CSRs), or provide virtual memory. None of the other extensions, such as multiply/divide or floating-point, are supported in the core described in this chapter. All these features will be added to the core in subsequent chapters.

Tightly integrated memories (*TIMs*) are located near the processor and can typically be accessed within one processor clock cycle. They are also referred to as *tightly coupled* (*TCMs*), *closely-coupled* (*CCMs*), or *scratchpad* memories.

Recall that standard RISC-V instructions are 32 bits long. Operands are XLEN bits, where XLEN = 32 in RV32 and XLEN = 64 in RV64. Table 7.1 lists the various integer instructions common to the RV32I and RV64I instruction set as well as the extra instructions in RV64I. Tables I.1 and I.2 (on the last pages of the book) provide more details for these instructions. The processor in this chapter supports all these instructions. All instructions in a category are handled in approximately the same way.

Section 7.1 begins by building a single-cycle RV{32/64}I core that supports all these base instructions and executes each instruction in one cycle. Section 7.2 then converts the single-cycle design into a five-stage pipeline by adding pipeline registers and hazard logic. Section 7.3 explores how this core is implemented in SystemVerilog, and Section 7.4 defines the test plan for these configurations. Subsequent chapters fill in the remaining features:

- **8:** Privileged operations
- **9–13:** Memory, bus, and cache systems
- **14:** Compressed instructions
- **15:** Multiply and divide instructions

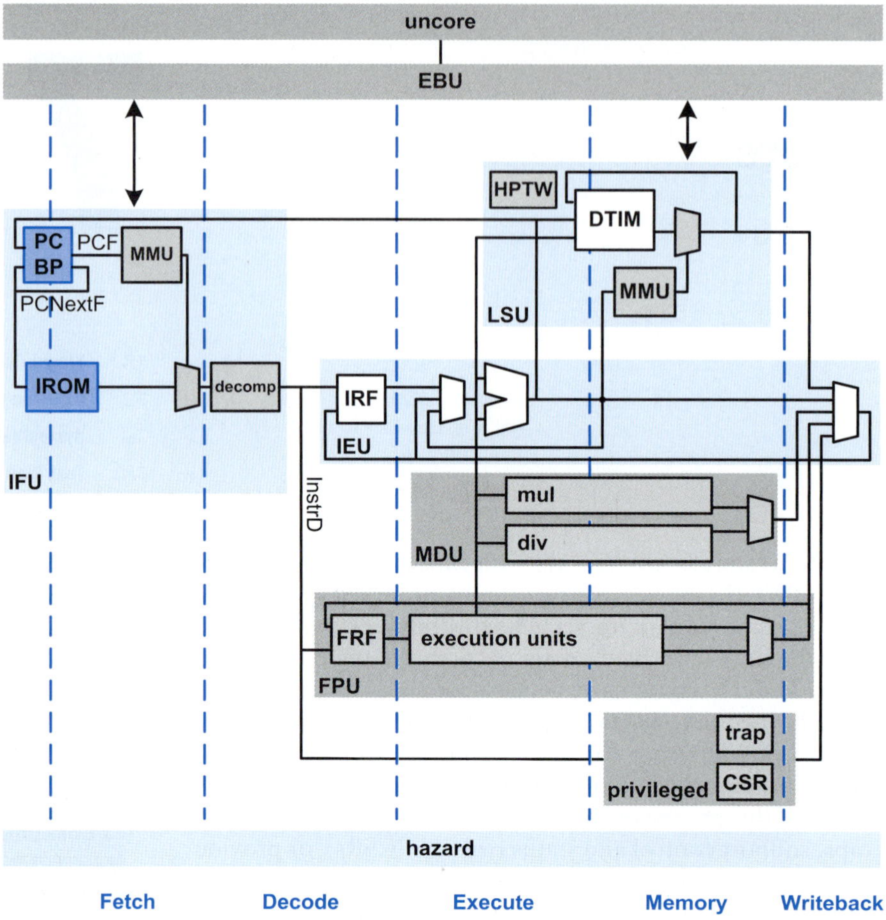

Fig. 7.1 RV{32/64}I high-level block diagram.

Table 7.1 Integer instructions.

Category	Op	Format	RV{32/64}I Integer Instructions	RV64I Extra Instructions
R-Type ALU	0x33/0x3B	R	add, sub, and, or, xor, sll, srl, sra, slt, sltu	addw, subw, sllw, srlw, sraw
I-Type ALU	0x13/0x1B	I	addi, andi, ori, xori, slli, srli, srai, slti, sltiu	addiw, slliw, srliw, sraiw
Loads	0x03	I	lw, lh, lhu, lb, lbu	ld, lwu
Stores	0x23	S	sw, sh, sb	sd
Branches	0x63	B	beq, bne, blt, bge, bltu, bgeu	
Jump and Link	0x6F	J	jal	
Jump and Link Register	0x67	I	jalr	
Load Upper Immediate	0x37	U	lui	
Add Upper Immediate PC	0x17	U	auipc	

- **16:** Floating-point instructions
- **17:** Bit-manipulation and cryptography instructions
- **18:** Atomic instructions
- **19:** Other extensions
- **20:** Peripherals

Chapters generally start with the principles applicable to many computer architectures and then examine the practices specific to RISC-V. They then define the test plan and finally provide a microarchitecture specification with the implementation details.

7.1 SINGLE-CYCLE CORE

Recall the simplified RV32I single-cycle core from Fig. 2.9, which supported only the `add`, `sub`, `and`, `or`, `slt`, `addi`, `andi`, `ori`, `slti`, `lw`, `sw`, `beq`, and `jal` instructions. This section details a full-featured single-cycle RV{32/64}I core shown in Fig. 7.2 that adds support for the rest of the instructions in Table 7.1, including the following:

- More R- and I-type ALU instructions, including RV64's W instructions, which operate on 32-bit values
- Byte, half-word, and double-word loads and stores
- More flavors of branches
- Upper immediate instructions: `lui` and `auipc`
- Jump and link register: `jalr`

In Chapter 9, we add a bus to access peripherals and off-chip memory. Then in Chapters 10–13, we replace the instruction and data memories with caches holding the most recently used portions of the off-chip memory, as well as add the `fence.i` instruction to synchronize the instruction cache after stores to program memory.

On the surface, Fig. 7.2 closely resembles the simplified core from Fig. 2.9. The prominent structural change is a new `AltResult` multiplexer, which chooses an alternate value to write to the register file: either the immediate for `lui` or `PCPlus4` for jump return addresses. Under the hood are several other changes:

- The datapath is configurable to be either 32 or 64 bits for RV32 or RV64.
- The ALU is enhanced to support the additional operations (xor, set less than unsigned, shifts). It takes an extra bit of control to specify RV64's RW/IW instructions, such as `addw` and `addiw`.
- The LSU adds circuitry to perform subword loads and stores.
- The comparator generates more flags (equal, less than, less than unsigned) to handle all the flavors of branches, and the controller picks the appropriate flag.
- The extender is enhanced to support U-format instructions.
- The controller adds entries for the new instruction categories: `jalr`, `lui`, and `auipc`.

Multiplexers and registers are named by their output signal. For example, the `AltResult` multiplexer outputs the `AltResult` signal.

This design is slightly different than the processor in *Digital Design and Computer Architecture* [Harris21], which uses the ALU for branch comparisons and a separate adder for branch address calculations. The revised organization, which uses the ALU for address calculation and a dedicated branch comparator, saves a bit of hardware when supporting all instructions and decreases critical path delay in small configurations.

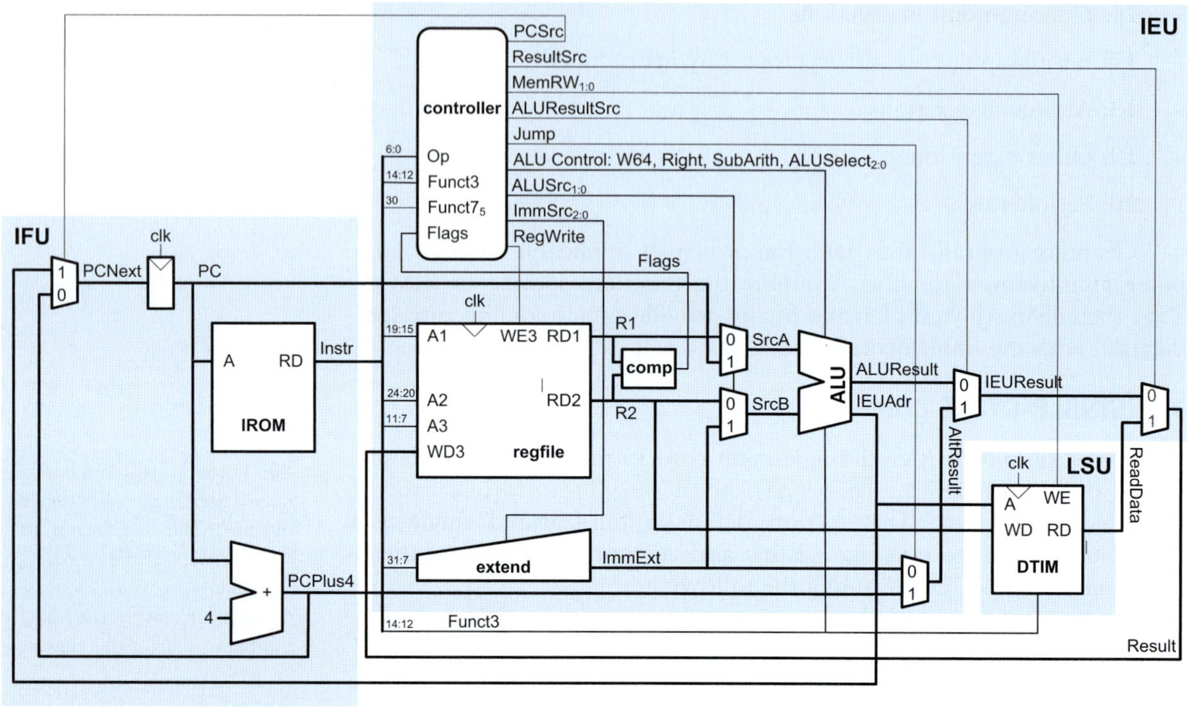

Fig. 7.2 **Single-cycle core.**

This section first discusses the single-cycle core by defining the architectural state that must exist in any RISC-V implementation: the program counter (PC) and registers. We then explore the `extend` block that extracts and extends the immediate from various formats of instructions. Next, we describe the arithmetic circuitry including the ALU and comparator. We then discuss the controller, which computes the multiplexer select and write enable signals to steer the datapath. Finally, we examine how to support subword loads and stores in the DTIM.

7.1.1 Architectural State

Any processor must have a certain amount of *architectural state* to remember the information visible to the programmer. In RISC-V, the architectural state comprises the PC, which contains the address of the current instruction, and the integer register file (IRF), which contains registers `x1`-`x31`. The single-cycle microarchitecture has this architectural state, instruction and data memories, and no other memory elements.

Because this chapter only deals with integer operations, we also refer to the IRF as simply the register file.

The IRF has three ports, so it can read two source registers and write a destination register in one cycle. Ports 1 and 2 are read ports, each with a 5-bit address (`A1`/`A2`) that specifies which of the 32 registers to read, and an XLEN-bit read data port (`RD1`/`RD2`), which delivers the contents of the specified register. Port 3 is a write port where `A3` specifies the register and `WD3` specifies the data to write; `WE3` specifies whether to write and a clock specifies when to write (on the rising edge). Register 0 is hardwired to 0, and writes to it are ignored.

In this implementation, the instruction memory is read-only and is preloaded with the program before the processor starts executing. The data memory is readable and writable.

Throughout this book, core diagrams such as Fig. 7.2 use the following conventions:

- **Module names:** bold font. These match the HDL module names (e.g., `hazard` for hazard unit, `lsu` for the load/store unit [LSU]).
- **Signal names:** UpperCamelCase. They match the HDL signal names.
- ***Descriptions:*** bold italic font.
- **Highlighted hardware/signals:** shown in bright blue.
- **Unused or optional hardware/signals:** shown in gray.
- **Bus widths:** single wires use thin lines, multibit buses (<XLEN) use medium lines, XLEN buses use thick lines, and >XLEN buses use very thick lines.

7.1.2 Immediate Extension

Another source of operands for many instructions comes from a 12-, 13-, 21-, or 32-bit immediate that is encoded in a 12- or 20-bit immediate field that is scattered around the instruction, as described in Section 2.2. The extender block extracts this immediate and appropriately sign-extends its most significant bit, which is always in instruction bit 31, to XLEN bits to create an extended immediate, `ImmExt`, as shown in Table 7.2 and Figs. 2.3 and 2.4. In RV64, the immediate is also sign-extended into bits 63:32 of `ImmExt`. The extender is just a big multiplexer controlled by an `ImmSrc` signal that depends on the instruction format, as will be described in Section 7.1.4.1.

RISC-V immediate encodings are strange to look at but are optimized to minimize the cost and delay of the sign extension multiplexers and to avoid interfering with the other instruction fields.

7.1.3 Arithmetic Circuitry

The processor has three arithmetic blocks:

- **ALU:** Performs an arithmetic, shift, or logical operation and address calculation
- **Comparator:** Compares two registers to produce flags for branches
- **Incrementer:** Increments the PC by 4 to calculate the next PC

Table 7.2 Extend operation.

ImmSrc	ImmExt	Type	Description
000	`{{20{Instr[31]}}, Instr[31:20]}`	I	12-bit signed
001	`{{20{Instr[31]}}, Instr[31:25], Instr[11:7]}`	S	12-bit signed
010	`{{20{Instr[31]}}, Instr[7], Instr[30:25], Instr[11:8], 1'b0}`	B	13-bit signed
011	`{{12{Instr[31]}}, Instr[19:12], Instr[20], Instr[30:21], 1'b0}`	J	21-bit signed
100	`{Instr[31:12], 12'b0}`	U	20-bit signed

The ALU has two outputs: ALUResult (for integer arithmetic/logical operations) and IEUAdr (for address calculations). The ALU can perform any of the RISC-V integer arithmetic/logical operations, including add, sub, and, or, xor, set-less-than (slt or sltu), and shifts (sra, srl, or sll). It also adds a base register to an immediate offset for memory, branch, and jump address calculations.

While the ALU is computing a branch target address, the branch comparator produces equal (eq) and less than (lt) flags for branches. The lt flag may come from signed or unsigned comparison, depending on the flavor of branch.

Table 7.3 lists the operations performed by the arithmetic circuitry for each type of instruction. The ALU computes ALUResult by performing addition or another operation (op) that depends on the ALU instruction being executed. The first source is either R1 from the register file or PC, and the second source is either R2 from the register file or the sign-extended immediate ImmExt. The integer execution unit (IEU) Result multiplexer (shown as two multiplexers in Fig. 7.2) chooses the ALUResult, the immediate, or PC + 4 (PCPlus4) to write to the register file. To minimize delay, the ALU also provides its adder output early (before the ALU's internal multiplexer, see Fig. 7.3) as IEUAdr, which is sent to the LSU and IFU (instruction fetch unit) as an address for memory operations, branches, and jumps.

7.1.3.1 ALU Design

Fig. 7.3 shows a diagram of the ALU. The ALU computes all operations in parallel and selects ALUResult from among them with a large multiplexer. The adder optionally inverts the B input to perform sub, slt, or sltu. The

Table 7.3 Arithmetic circuits summary.

Category	ALUResult	IEUResult	IEUAdr (and Destination)	Comparator
R-Type ALU	R1 op R2	ALUResult	x	
I-Type ALU	R1 op ImmExt	ALUResult	x	
Loads/Stores	X	x	R1 + ImmExt → LSU	
Branches	X	x	PC + ImmExt → IFU	R1 cmp R2
jal	X	PCPlus4	PC + ImmExt → IFU	
jalr	X	PCPlus4	R1 + ImmExt → IFU	
lui	X	ImmExt	x	
auipc	PC + ImmExt	ALUResult	x	

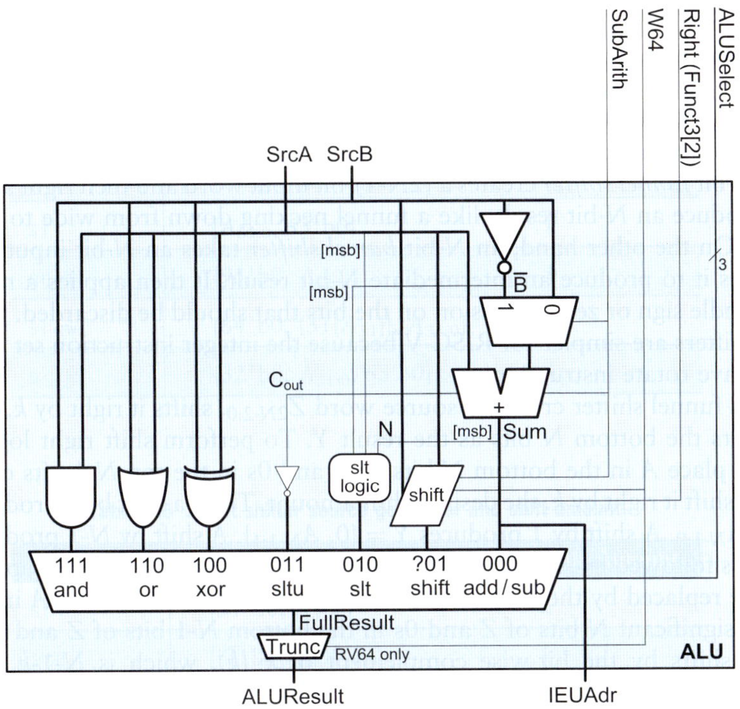

Fig. 7.3 **ALU schematic.**

The IEUAdr output is faster than ALUResult because it skips the multiplexer delay. This reduces the critical path when accessing memory.

shifter performs logical left shifts and logical and arithmetic right shifts (sll, srl, and sra). In RV64, it also handles W-type shifts on 32-bit words. The slt and sltu functions compute the lt and ltu flags and place them in the least significant bit of the result, with 0s in the upper bits. Bitwise gates perform and, or, and xor. ALU components are discussed further in the next sections.

7.1.3.2 Add/Subtract

The adder in the ALU performs addition or subtraction based on the control signal SubArith. When SubArith = 0, it adds, computing Sum = SrcA + SrcB. When SubArith = 1, it subtracts, computing Sum = SrcA + $\overline{\text{SrcB}}$ + 1 = SrcA − SrcB, using the two's complement identity that $-X = \overline{X} + 1$. Adder architectures and their area/delay tradeoffs have been heavily researched; see [Weste12] for an overview. It is usually expedient to describe the adder with + in the HDL and allow the logic synthesis tool to select the best adder given the constraints. If the adder is timing critical, synthesis generally produces a parallel prefix tree with $O(log_2 N)$ delay.

7.1.3.4 Set Less Than

slt and sltu instructions compute $\{C, Sum\} = A - B = A + (\overline{B} + 1)$ and check if the result is negative, which indicates that $A < B$. They can make the sign determination based on the two most significant bits of the difference $\{C, Sum_{N-1}\}$, which are called the C (*carry out* or, simply *carry*) and N (*negative*) flags: $N = Sum_{N-1}$.

Beware that N for negative flag and N for the number of bits in a word are unrelated uses of the same variable name.

For unsigned comparison, A and B are both positive. If the carry out is 0, $A - B$ is negative (so A is less than B). Thus, sltu $= \overline{C}$.

For signed comparison, we must consider overflow. When overflow occurs, the result's sign bit (and thus, N) is wrong. When subtracting, overflow occur when two conditions are true: (1) A and B have opposite signs, which, when subtracting, is effectively adding values of the same sign: $(A_{N-1} \oplus B_{N-1})$, and (2) the result's sign ($N = Sum_{N-1}$) differs from A's sign: $(A_{N-1} \oplus N)$. Thus, the *overflow* flag $V = (A_{N-1} \oplus B_{N-1})(A_{N-1} \oplus N)$. The true result is negative when $N \oplus V$. In other words, when overflow occurs, the N flag gave the wrong value, so it must invert. This simplifies to slt $= N \oplus V = A_{N-1}\overline{B_{N-1}} + A_{N-1}N + \overline{B_{N-1}}N$.

7.1.3.5 W-Type Instructions

For W-type RV64 instructions, the full 64-bit result should be truncated to 32 bits and then sign extended into the upper half. The 32-bit Trunc multiplexer shown in Fig. 7.5 performs this operation by choosing the upper half of ALUResult from either the full result or by sign-extending bit 31. In RV32, no W-type instructions exist, and this Trunc multiplexer is omitted.

7.1.3.6 Comparator

The dedicated branch comparator produces the eq and lt/ltu flags for branches based on two source registers. Wally uses a dedicated comparator rather than reusing the adder and slt logic because comparison is on the critical path of a simple pipelined processor and because the adder is also needed to compute the branch destination. The comparator is faster than the regular ALU for several reasons:

- The branch comparator inputs come from the register file, rather than the Src{A/B} and SubArith muxes.

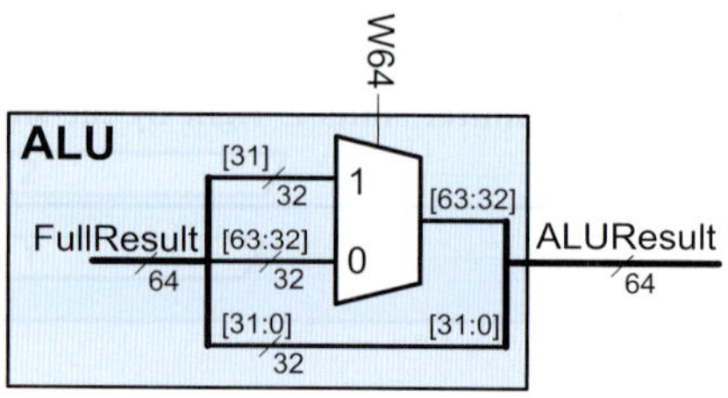

Fig. 7.5 Trunc multiplexer.

- Computing flags is less expensive than producing a full XLEN-bit result.
- The flags are available without waiting for ALU multiplexing after the adder.

A brute force method is to subtract $A - B$ and compute `lt` and `ltu` as in Section 7.1.3.4. The `eq` flag should also be set when the difference is 0. This method is overkill because the actual answer of the subtraction is unnecessary.

The design can be optimized using a parallel prefix circuit, as described by Equations 7.1–7.6. The circuit first computes whether a column's inputs are equal (E_i) or unsigned less than (L_i). It then merges these bitwise results to obtain group results for columns i through j ($E_{i:j}$, $L_{i:j}$). Columns $i{:}j$ are equal ($E_{i:j} = 1$), when both the upper group of bits ($E_{i:k}$) AND the lower group of bits ($E_{k\text{-}1:j}$) are equal (Equation 7.3). Columns $i{:}j$ of A are less than columns $i{:}j$ of B when the upper bits are less ($L_{i:j}$) OR the upper bits are equal AND the lower bits are less ($E_{i:k}L_{k\text{-}1:j}$); see Equation 7.4. After computing these signals for all columns, the prefix circuit outputs the equal (`eq`) and less than unsigned (`ltu`) flags.

The prefix signals, $Ei{:}j$ and $Li{:}j$, are each 1-bit signals indicating whether bits $i{:}j$ of A and B are equal or less than unsigned, respectively: $E_{i:j} = (A_{i:j} == B_{i:j})$ and $L_{i:j} = (A_{i:j} < B_{i:j})$.

Precomputation

(7.1) $E_i = \overline{A_i \oplus B_i}$ — Bits i of A and B are equal

(7.2) $L_i = \overline{A_i} B_i$ — Bit i of A < bit i of B (viewed as unsigned numbers)

Prefix Recursion

(7.3) $E_{i:j} = E_{i:k} E_{k\text{-}1:j}$ — Bits $i{:}j$ are equal

(7.4) $L_{i:j} = L_{i:k} + E_{i:k} L_{k\text{-}1:j}$ — Bits $i{:}j$ are less than unsigned

Output Logic

(7.5) `eq` $= E_{N\text{-}1:0}$ — $A == B$

(7.6) `ltu` $= L_{N\text{-}1:0}$ — $A < B$ unsigned

Signed comparison can be performed with the same hardware as unsigned comparison by flipping the most significant bit of A and B. The principle is that unsigned numbers are in the range $[0, 2^N - 1]$, and signed numbers are in the range $[-2^{N\text{-}1}, 2^{N\text{-}1} - 1]$. Flipping the most significant bit is equivalent to adding $2^{N\text{-}1}$ to both positive and negative two's complement numbers and converts them to unsigned inputs (i.e., shifts them to the range $[0, 2^N - 1]$) without changing which one is less.

Example 7.2 COMPARING 32-BIT TWO'S COMPLEMENT NUMBERS

Compare the following 32-bit two's complement numbers by inverting the most significant bit (msb) of each one and then performing unsigned comparison: -7 and 3, -5 and -10.

Solution: -7 is 0xFFFFFFF9, and 3 is 0x00000003. Flipping the most significant bit of each number and performing unsigned comparison results in: 0x7FFFFFF9 < 0x80000003, which is correct ($-7 < 3$). -5 is 0xFFFFFFFB, and -10 is 0xFFFFFFF6. Flipping the most significant bit of each number and performing unsigned comparison results in: 0x7FFFFFFB $\geq$ 0x7FFFFFF6, which is correct ($-5 \geq -10$).

The comparator can use bubble-pushing and alternating stages to perform the prefix recursion with only 1 level of AOI/OAI (AND-OR-INVERT/ OR-AND-INVERT) gates.

A prefix adder that performs $A + B$ calculates the precomputation signals (column propagate and generate) as $P_i = A_i \oplus B_i$ and $G_i = A_i B_i$. Because this prefix subtractor is performing $B + \overline{A}$, $P_i = \overline{A_i} \oplus B_i = \overline{A_i \oplus B_i}$ and $G_i = \overline{A_i} B_i$.

Thus, the comparator handles both signed and unsigned comparison by adding two XOR gates to conditionally flip the most significant bits of each input for signed comparison and then using the unsigned comparison described earlier. The `lt`/`ltu` $= L_{N\text{-}1:0}$ output is now correct for either signed or unsigned comparison. Thus, an $N = 2^{\ell}$-bit comparator's critical path involves an XOR to handle signed/unsigned comparison, two 2-input gates for precomputation, and ℓ levels of AND and AND-OR logic to compute $L_{N\text{-}1:0}$ and $E_{N\text{-}1:0}$ (`lt`/`ltu` and `eq`).

The comparator recursion closely resembles subtraction. Consider computing $\{C, Sum\} = B - A - 1 = B + \overline{A}$. For unsigned numbers, $C = 1$ if and only if $A < B$. The prefix recursion for $B + \overline{A}$ is shown in Equations 7.7–7.14. It is identical to the comparator recursion from Equations 7.1–7.6 except P corresponds to E and G corresponds to L.

Precomputation

(7.7) $P_i = \overline{A_i \oplus B_i}$ Column i propagates a carry

(7.8) $G_i = \overline{A_i} B_i$ Column i generates a carry

Prefix Recursion

(7.9) $P_{i:j} = P_{i:k} P_{k\text{-}1:j}$ Bits $i{:}j$ propagate a carry

(7.10) $G_{i:j} = G_{i:k} + P_{i:k} P_{k\text{-}1:0}$ Bits $i{:}j$ generate a carry

Output Logic

(7.11) $Sum_i = P_i \oplus G_{i-1:0}$

(7.12) $C = G_{N\text{-}1:0}$

(7.13) `eq` $= (Sum == 0) = P_{N\text{-}1:0}$ $A == B$

(7.14) `ltu` $= G_{N\text{-}1:0}$ $A < B$ unsigned

To highlight the similarities and differences between the subtractor and comparator, Fig. 7.6 shows diagrams of the subtractor-based (Equations 7.7–7.14) and optimized (Equations 7.1–7.6) prefix network unsigned comparator designs, also referred to as simply *prefix comparators*. For

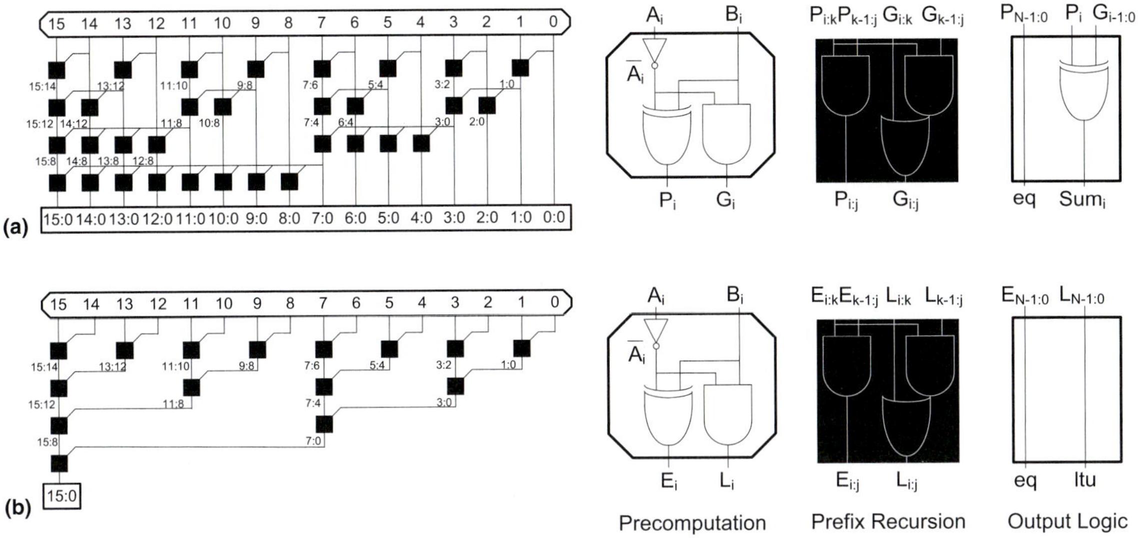

Fig. 7.6 **Prefix comparators: (a) subtract, (b) optimized.**

both networks, the precomputation is the same, with *N* AND, XOR, and NOT gates. The optimized comparator produces only the final `eq` and `ltu` results, so it does not need to calculate the intermediate column results or the *Sum* output. Thus, whereas the subtractor has $N\log_2 N / 2$ recurrence cells (black boxes), the comparator only has *N* recurrence cells, and the *N* final XORs to produce the *Sum* in the subtractor are not needed in the comparator.

Despite these design optimizations, Design Compiler is smart enough to synthesize high-level operations onto well-optimized DesignWare blocks, so we do not need to write prefix equations to get satisfactory results. Thus, Wally's `comparator` module in Code Example 7.1 expresses the comparison operations in Verilog using == and < statements rather than prefix equations. The Wally `comparator` logic shown below calls the output `lt`, whether it is signed or unsigned.

Code Example 7.1 COMPARATOR LOGIC USING == AND <

```
// For signed numbers, flip msb before comparing
assign af = {a[WIDTH-1] ^ sgnd, a[WIDTH-2:0]};
assign bf = {b[WIDTH-1] ^ sgnd, b[WIDTH-2:0]};
assign eq = (a == b);    // Equality
assign lt = (af < bf);   // Signed or unsigned less than
```

7.1.4 Controller Design

Fig. 7.7 shows a block diagram of the controller. The controller has a main decoder that computes most of the control signals based on an instruction's

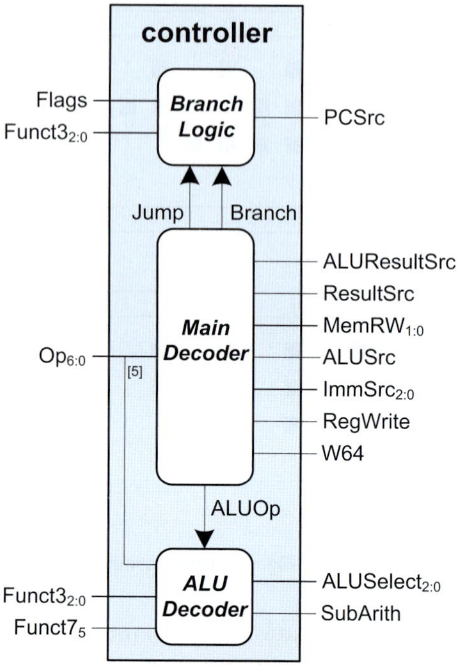

Fig. 7.7 **Controller.**

`op` field. The controller also has branch logic to compute `PCSrc` based on the branch/jump type and flags, and an ALU decoder.

7.1.4.1 Main Decoder

The don't care (x) entries in Table 7.6 do not matter but are treated as 0 in Wally to avoid illegal values in simulation.

The main decoder truth table is given in Table 7.6. Each row corresponds to a category of instructions specified by the `op` field (`Instr[6:0]`). The rows in blue text are only required for RV64I. The columns indicate the output signals. `ALUSrc` is a 2-bit value `{ALUSrcA, ALUSrcB}` driving the two source multiplexers. `MemRW` is also a two bit value {`MemRead`, `MemWrite`} that determines whether to read, write, or leave the memory idle.

7.1.4.2 Branch Logic

The comparator implementation from Section 7.1.3.6 handles signed and unsigned comparison with the same logic, so `lt` is also used for `ltu` in the branch logic.

The controller's branch logic, shown in Fig. 7.8, outputs the PC source signal `PCSrc`, which chooses the next PC as PC + 4 when `PCSrc` = 0 and the branch or jump target address (`IEUAdr`) otherwise. The main controller sends the branch logic the `Branch` and `Jump` signals, which indicate whether the instruction is a branch or jump, the `funct3` field from the instruction, which indicates the type of branch, and the {`eq`, `lt`, `ltu`} flags from the comparator. Given the `funct3` and flags, the branch logic determines

Table 7.6 Main decoder.

Category	op	Branch	Jump	ALUSrc	ImmSrc	ALUOp	W64	ALUResultSrc	ResultSrc	RegWrite	MemRW
R-type ALU	0x33	0	0	00	xxx	1	0	0	0	1	00
I-type ALU	0x13	0	0	01	000	1	0	0	0	1	00
RW-type ALU	0x3B	0	0	00	xxx	1	1	0	0	1	00
IW-type ALU	0x1B	0	0	01	000	1	1	0	0	1	00
Loads	0x03	0	0	01	000	0	0	x	1	1	10
Stores	0x23	0	0	01	001	0	0	x	x	0	01
Branches	0x63	1	0	11	010	0	0	x	x	0	00
`jal`	0x6F	0	1	11	011	0	0	1	0	1	00
`jalr`	0x67	0	1	01	000	0	0	1	0	1	00
`lui`	0x37	0	0	xx	100	x	0	1	0	1	00
`auipc`	0x17	0	0	11	100	0	0	0	0	1	00

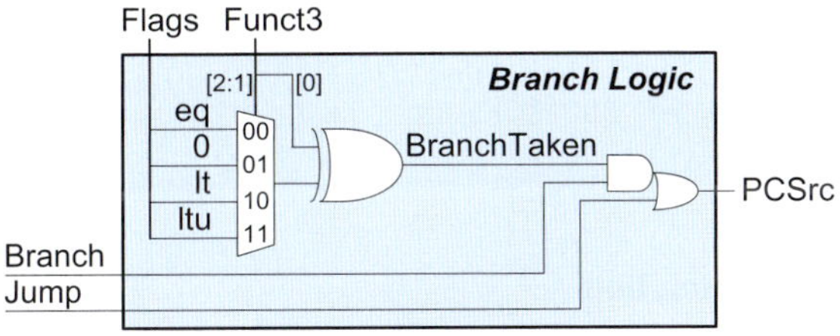

Fig. 7.8 Branch logic.

whether the branch should be taken according to Table 7.7. For example, a `bne` will be taken if `R1` ≠ `R2`, indicated by an `eq` flag of 0. This table simplifies to a 4:1 multiplexer to select the flag of interest followed by an XOR to invert the flag if necessary. The branch logic then sets `PCSrc` if the instruction is either a jump or a conditional branch that should be taken.

7.1.4.3 ALU Decoder

The controller's ALU decoder determines the ALU's operation by producing `ALUSelect` and `SubArith`, which is 1 when subtracting or arithmetic right shifting. Recall from Fig. 7.3 that the ALU also requires `W64` and `Right` control signals. `W64` comes from the main decoder, and `Right = Funct3[2]`, which is 1 when right shifting.

Table 7.7 BranchTaken determination.

Branch Type	Funct3	BranchTaken
beq	000	eq
bne	001	$\overline{\text{eq}}$
<undefined>	01x	0
blt	100	lt
bge	101	$\overline{\text{lt}}$
bltu	110	ltu
bgeu	111	$\overline{\text{ltu}}$

This decoder is *lazy* in that it only looks at bit 5 of Funct7 rather than checking Funct7 == 0100000. When traps are supported, RISC-V strongly recommends throwing an illegal-instruction exception if Funct7 ≠ 0100000. Thus, when ZICSR_SUPPORTED = 1, the decoder is stricter, at the cost of a few extra gates.

Note that sub also needs to look at bit 5 of the opcode to distinguish sub from addi, which has the same Funct3 field and no Funct7.

ALUSelect is Funct3 for ALU instructions or 000 for auipc to add the PC and the upper-immediate.

SubArith indicates the ALU is doing a subtraction (including set less than) or arithmetic right shift. This is determined from the funct fields (Funct3 and Funct7) when ALUOp is 1:

```
slt      = (Funct3 == 010)
sltu     = (Funct3 == 011)
sra      = (Funct3 == 101 & Funct7[5])
sub      = (Funct3 == 000 & Funct7[5] & Op[5])
SubArith = ALUOp & (slt | sltu | sra | sub)
```

7.1.5 Instruction Fetch Unit

The *instruction fetch unit* (IFU) in Fig. 7.2 comprises the PC register, IROM, PC incrementer, and PCNext multiplexer. The IROM is preloaded with the program, and stores to program addresses are disregarded. The incrementer adds 4 to find the address of the next instruction. The multiplexer picks between PC + 4 and the branch or jump destination in IEUAdr.

Chapter 9 describes how to fetch instructions from an external memory over a bus interface, and Chapter 13 describes how to add an instruction cache and branch predictor to improve performance.

7.1.6 Load/Store Unit

The LSU contains a DTIM with single-cycle access. In this section, we assume that memory reads occur asynchronously and that writes occur on the rising edge of the clock. Section 7.2.1 introduces more practical memories with synchronous reads.

The DTIM internally accesses entire XLEN-bit words or double words. To handle an instruction such as `lb`, `lbu`, `lh`, or `lhu`, the DTIM reads the full word, extracts the appropriate byte or halfword, places it in the least significant bits of the result, and then sign extends to full length. To handle `sb` or `sh`, the DTIM replicates the subword into all positions of the data write signal's word. It asserts byte write signals (`ByteMask`) to indicate which bytes within the word should be written based on the size of the store (indicated by `Funct3`) and the offset within the word (in the least significant bits of `Adr`). These operations are referred to as *subword loads and stores*. For RV64, `lw`, `lwu`, and `sw` are also subword operations because the native word size is 64 bits.

Another way to perform a subword write to memory is to read the word from memory, modify it, and write it back. This is slower and consumes more power. The hardware to implement bit or byte enables is modest, so most static random access memory (SRAM) generators and field programmable gate array (FPGA) memories support such enables.

Fig. 7.9 shows the LSU, including subword load and store blocks for RV64. The RV32 design is similar but drops support for `sd`, `ld`, and `lwu`, and only produces a 4-bit `ByteMask` and 32-bit `WriteData` and `ReadData`.

The LSU supports subword operations, but the random access memory (RAM) always reads and writes full-length *native* words. When the core doesn't have floating-point hardware, the native length is LLEN = XLEN. When floating-point is added in Chapter 16, the native length

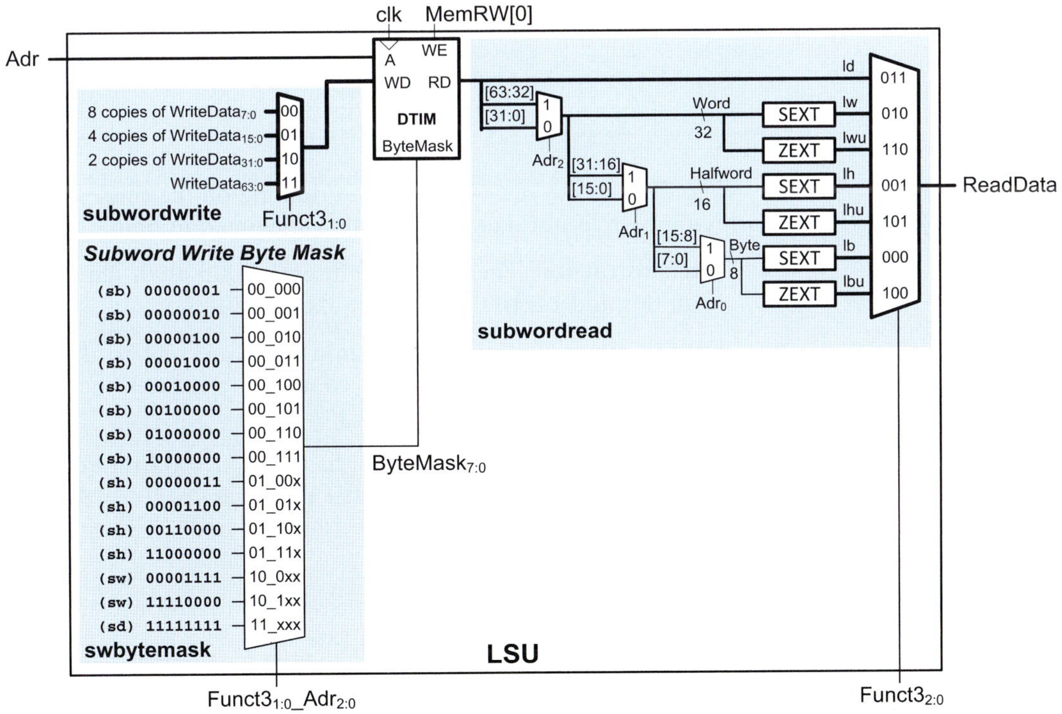

Fig. 7.9 RV64 LSU.

becomes LLEN = max(XLEN, FLEN). Read LLEN as the "longer" of XLEN and FLEN.

For example, a RV32GC configuration has XLEN = 32 but FLEN = 64, so the native accesses are 64-bit words. The `ReadData` and `WriteData` bus widths are this native length.

Section 8.4.7 adds big-endian support. Section 9.5 describes how to access data from an external memory over a bus interface. Chapter 12 covers a fully-featured LSU with a data cache and memory management unit (MMU). Chapter 17 describes the atomic memory extensions.

7.1.7 Sample Program

To visualize the processor in action, consider running the program in Code Example 7.2, starting at address 0x80000000, and assume address 4 contains 42.

Code Example 7.2 `ch7.s` SAMPLE PROGRAM

```
start:
    0x80000000: lw x1, 4(x0)            00402083
    0x80000004: sw x1, 8(x0)            00102423
    0x80000008: add x2, x1, x1          00108133
    0x8000000c: beq x1, x2, done        00208463
loop:
    0x80000010: jal x0, loop            0000006f
done:
```

Fig. 7.10 shows the `lw` instruction running on the single-cycle processor, with the relevant datapath signals highlighted in blue.

- **Fetch instruction:** The PC is initially at 0x80000000. The IROM reads from this address to find `Instr` = 0x00402083.
- **Decode instruction:** The instruction fields are sent to the controller, register file, and extender. The controller produces the appropriate control signals.
- **Get source operands:** The register file reads `R1` = 0 from register `x0` and reads garbage for `R2`. The extender produces `ImmExt` = 4. The source multiplexers choose `SrcA` = 0 from `x0` and `SrcB` = 4 from the immediate.
- **Execute:** The ALU adds these two sources to produce `ALUResult` and `IEUAdr` = 4.
- **Read memory:** The DTIM reads from this address to find `ReadData` = 42.

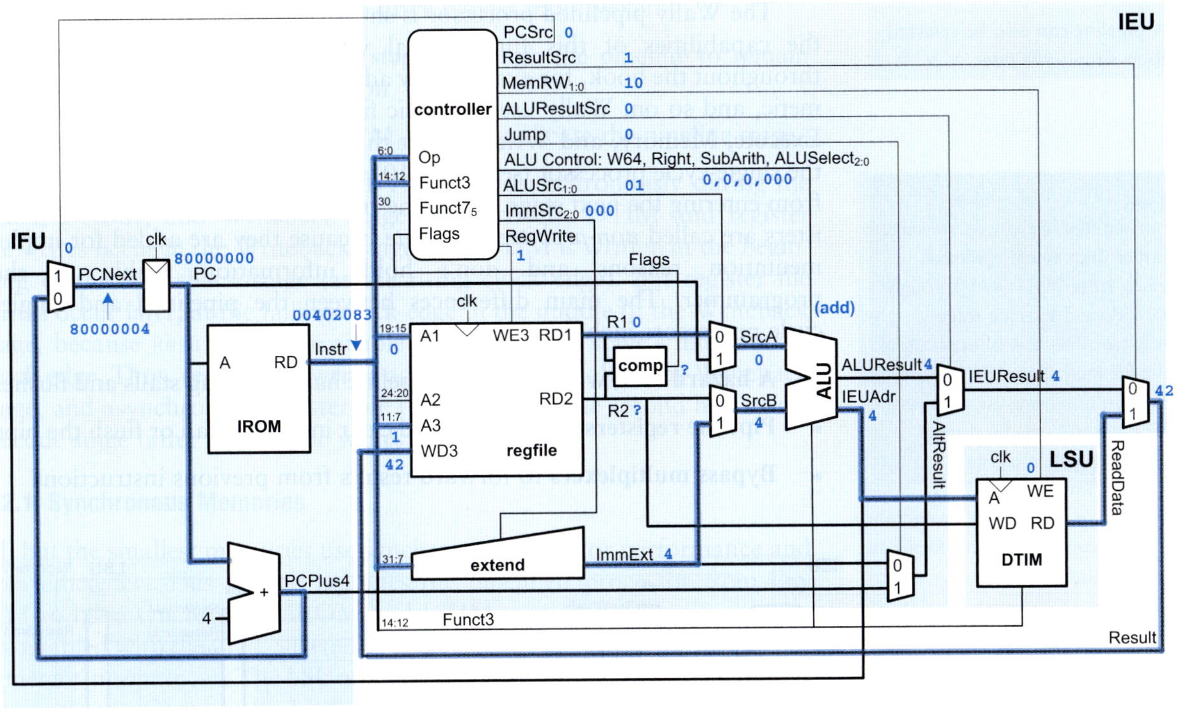

Fig. 7.10 **Single-cycle processor executing `lw` instruction.**

- **Write result:** The result multiplexer chooses `Result` = 42. The register file writes 42 into `x1`.
- **Increment PC:** Meanwhile, the PC incrementer computes `PCPlus4` = 80000004, which is selected as `PCNext`.

On the next cycle, `PC` is 80000004, so the IROM reads `Instr` = 00102423. Predict the rest of the results until you are comfortable with the processor operation.

7.2 PIPELINING THE RV32I CORE

The cycle time of the single-cycle processor is limited by the longest path to execute an instruction, including fetching the instruction, reading from the register file, performing an ALU operation, accessing data memory, and writing the result back to the register file. Pipelined processors shorten the cycle time and improve throughput by separating the processor into multiple pipeline stages; hence, the next instruction can start before the current one completes. This introduces a set of *hazards* when one instruction depends on the results of a previous instruction that has not yet completed.

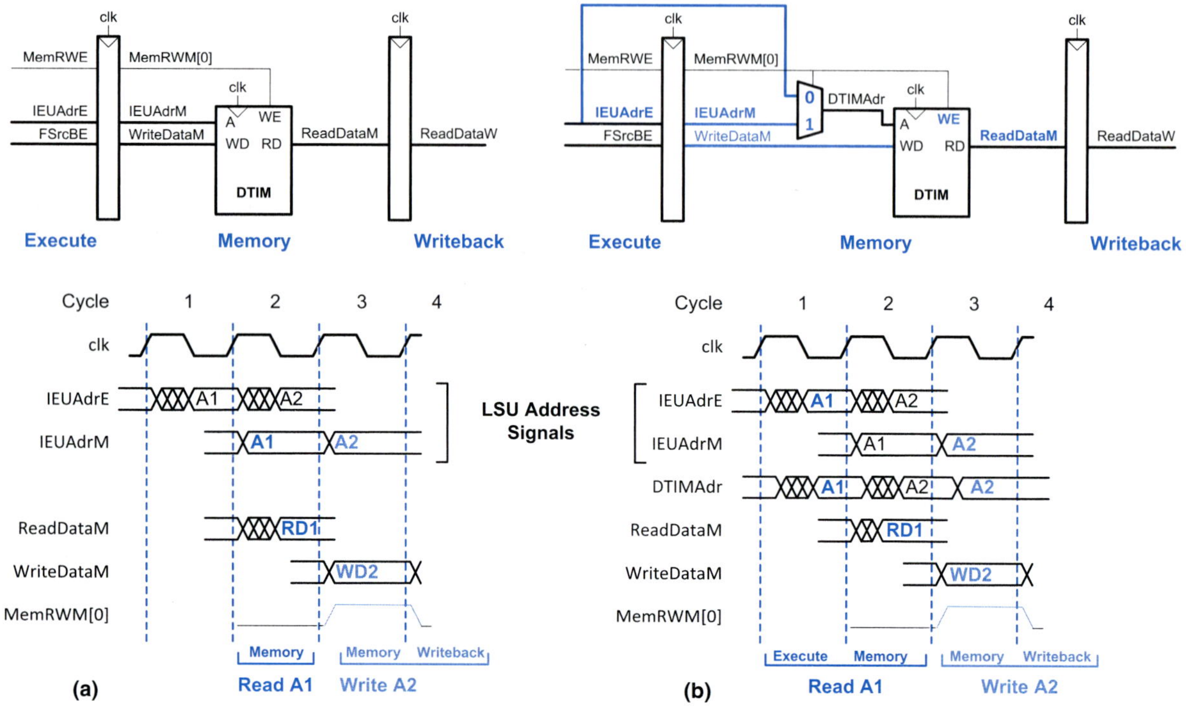

Fig. 7.12 DTIM timing: (a) asynchronous read and (b) synchronous read.

same cycle – that is, while the store is in the Memory stage and the load is in the Execute stage. This is discussed further in Section 7.2.2.

Fig. 7.13 shows the improved pipeline diagram using synchronous memories. Observe that the IROM and DTIM are shifted left to indicate that the read address must set up before the clock edge. The DTIM adds a `DTIMAdr` mux. Both memories add a chip enable to hold the old read value during a stall.

In summary, an access to the DTIM involves work in the Execute, Memory, and Writeback stages, as shown in Table 7.8.

Table 7.8 LSU DTIM access steps.

Stage	DTIM Read	DTIM Write
Execute	Compute and set up memory address: `IEUAdrE` = base address + offset	Compute and set up memory address: `IEUAdrE` = base address + offset
Memory	Read `ReadDataM` synchronously from `IEUAdrE`	Set up `IEUAdrM`, `WriteDataM`, `MemRWM`
Writeback	Write `ReadDataW` into register file	Write `WriteDataM` synchronously to DTIM

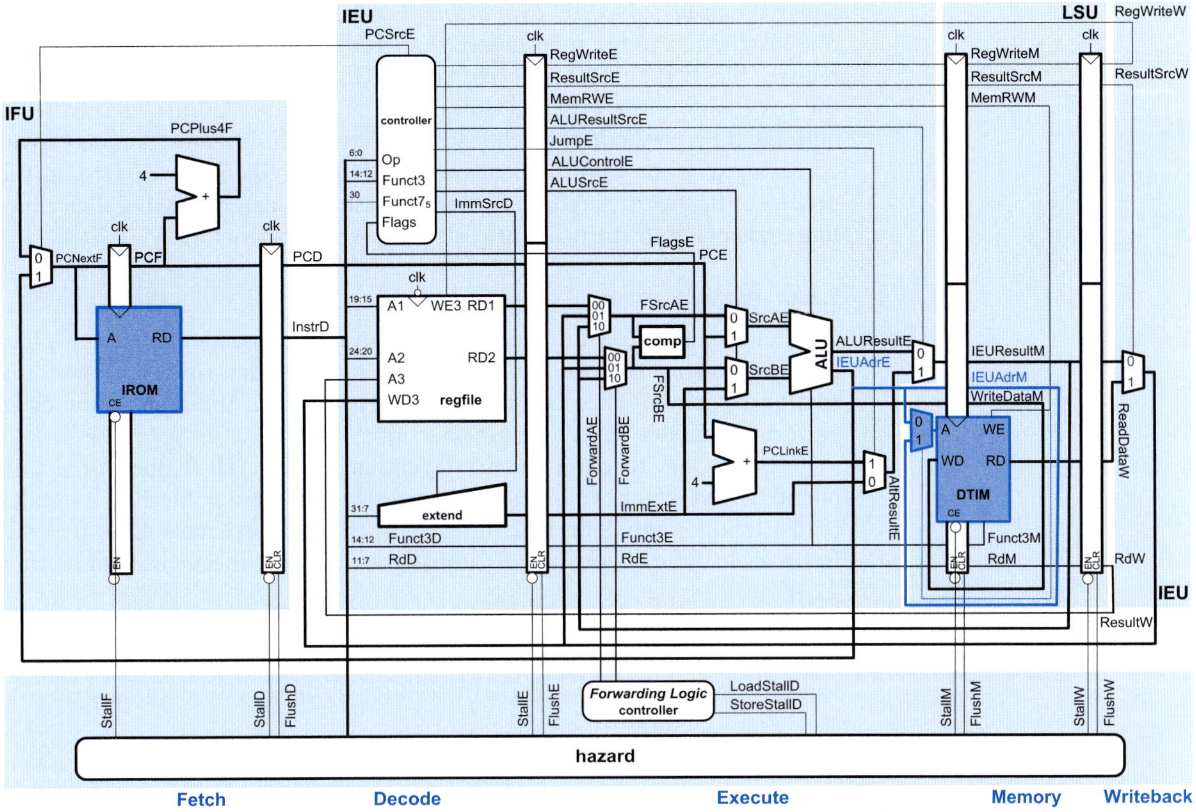

Fig. 7.13 Improved Wally pipelined processor with synchronous memories.

The IROM has a similar issue as the DTIM: the address for synchronous reads must setup before the Fetch stage. Hence, the IROM must be read with `PCNextF` rather than `PCF`. Therefore, reset is applied to the `PCNextF` mux rather than the `PCF` register, as shown in Fig. 7.14. To reduce clutter, the reset to `PCNextF` is not shown in other figures, such as Fig. 7.13.

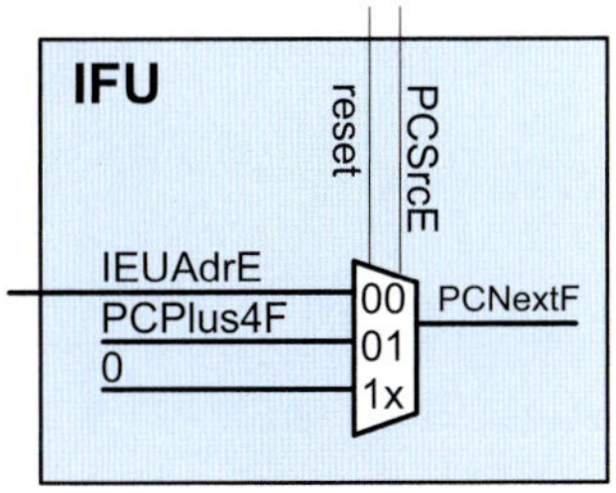

Fig. 7.14 `PCNextF` with reset.

7.2.2 Hazards

Pipeline hazards occur when an instruction depends on the results of another instruction that has not yet completed. They can be classified as data hazards and control hazards. In a *data hazard*, an instruction needs a value from a previous instruction that has not yet completed. In a *control hazard*, caused by branches and jumps, the decision of which instruction to fetch next has not been made by the time the fetch should occur. Some data hazards can be solved by *forwarding* (also called *bypassing*) a result from one instruction to the next before that result has been written back to the register file. If the result isn't available to forward by the time it's needed, the processor *stalls* instead, waiting for the result. Control hazards are addressed by predicting which instruction to fetch. If the prediction proves correct, execution continues normally. If the prediction was wrong, the processor flushes the incorrectly fetched instructions, by clearing their pipeline registers, and starts again with the correct instruction.

7.2.2.1 Hazard Example

Fig. 7.15 shows an example of a program that causes hazards on a pipelined processor. The program is on the left, with each row corresponding to an instruction. The pipeline stages are shown on the right in the cycle each one operates for a given instruction.

The `addi` instruction has no dependencies. In cycle 1, the instruction memory (IM) fetches the `addi`. In cycle 2, the register file/sign extender reads the operands 0 and 42. In cycle 3, the ALU executes $42 + 0 = 42$. In cycle 4, the data memory is unused. In cycle 5, the register file writes 42 to `t0`.

The `ori` instruction depends on `t0`. In cycle 3, it reads `t0` from the register file but gets an outdated value because the `addi` instruction won't write `t0` until cycle 5. This is called a *read after write* (*RAW*) data hazard. The

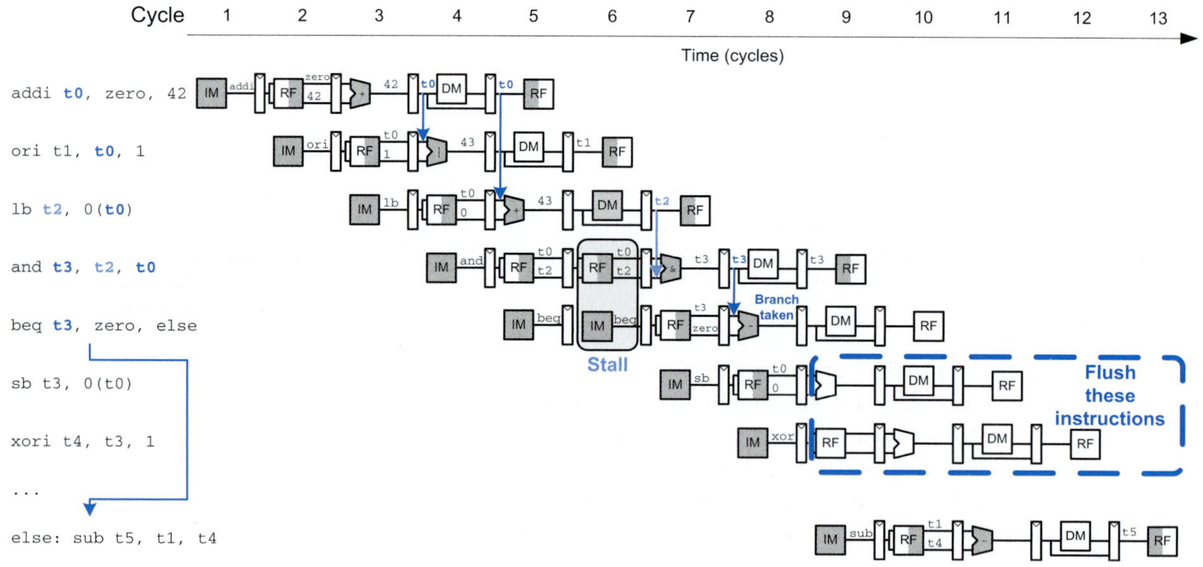

Fig. 7.15 Program with hazards.

pipeline solves the hazard in cycle 4 by forwarding t0 from the Memory stage of the addi to the Execute stage of ori so the correct value is used.

The lb instruction also depends on t0, so there is another read after write hazard. This time, the pipeline solves the hazard in cycle 5 by forwarding t0 from the Writeback stage of the addi to the Execute stage of lb.

The and instruction depends on both t0 from addi and t2 from lb. addi writes t0 to the register file in the first half of cycle 5 (i.e., at the falling edge of clock), so and reads t0 from the register file correctly in the second half of cycle 5, and no forwarding is necessary. Unfortunately, t2 comes from the data memory and is not yet available when and wants it by the start of cycle 6. Thus, the hazard unit must stall the pipeline during cycle 6: and remains in the Decode stage, and beq remains in the Fetch stage. On cycle 7, t2 is now available and is forwarded from the Writeback stage of lb to the Execute stage of and.

The beq instruction must predict whether to take the branch. Branch predictors are discussed in depth in Section 13.1, but our simple pipelined processor always predicts that the branch is not taken and adds 4 to PC as usual. Thus, it fetches sb in cycle 7 and xori in cycle 8. By cycle 8, the beq executes and determines if t3 = 0. If so, as we assume in Fig. 7.15, the branch should have been taken. The hazard unit flushes the sb and xori instructions and fetches sub in cycle 9. Flush is performed by clearing the flushed instruction's pipeline registers or at least all those that contain write enables such as RegWrite or MemWrite, so that the flushed instruction changes no state and behaves as a nop.

7.2.2.2 Pipeline Registers

As shown in Fig. 7.13, each pipeline register takes a stall and flush signal. For example, the register between the Decode and Execute stages takes StallE and FlushE to stall or flush the instruction entering the Execute stage. Implicitly, all registers are also synchronously resettable to begin at a known initial state. Thus, pipeline registers typically use flip-flops with enable, reset, and clear (flopenrc) as shown in Fig. 7.16. $\overline{\text{Stall}}$ drives the enable (en) pin. Flush drives the clear pin. Synchronous reset has highest priority, initializing Q to 0 even if en = 0. Enable has next priority, recirculating Q when disabled.

Clear has the lowest priority in this implementation, although clear and reset could be ORed together in an alternate register design.

Enable uses a multiplexer in this implementation, although it could be ANDed with clk in an alternate register design. Such clock gating requires a latch on the enable to prevent clock glitching, as well as CAD tools to model the gated clock delay for setup and hold time analysis.

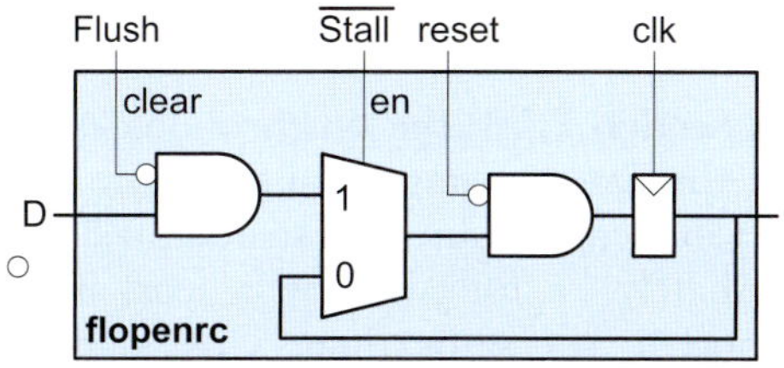

Fig. 7.16 Pipelined register: flopenrc register with enable, reset, and clear.

7.2.2.3 Forwarding Logic

The pipelined processor adds forwarding logic to the controller, as shown in Fig. 7.17, to forward results from later stages (Memory and Writeback) to the Execute stage. The forwarding logic outputs are combinational. The pipeline forwards a result to the Execute stage if a source register in the Execute stage (`Rs1E` or `Rs2E`) matches the valid destination register in a later stage (`RdM` or `RdW`). A destination register is valid if it is nonzero (writes to register 0 are ignored) and will be written. If the source register matches destinations in multiple stages, the pipeline picks the earliest one, which corresponds to the most recent instruction. The forwarding logic also detects hazards that it cannot solve by forwarding. These hazards, such as load and store dependencies, are sent to the hazard unit to stall or flush the pipeline.

Forwarding is done with multiplexers for `SrcA` and `SrcB` as shown in Fig. 7.13. The forwarding logic computes the forwarding multiplexer select signals `Forward{A/B}E`. Specifically, the `ForwardAE` logic for `SrcA` is:

```
if      ((Rs1E == RdM) & RegWriteM) & (Rs1E != 0) # Mem stage
        ForwardAE = 10
else if ((Rs1E == RdW) & RegWriteW) & (Rs1E != 0) # WB stage
        ForwardAE = 01
else    ForwardAE = 00                            # No forwarding
```

The `ForwardBE` logic for `SrcB` is analogous but involves `Rs2E` instead of `Rs1E`.

The result of a load instruction is not available until the end of the Memory stage. A dependent instruction requires the result by the start of the Execute stage. Therefore, a load immediately followed by use requires a one-cycle stall. This is called a *structural hazard* because the structure of the pipeline cannot forward the load result to the next instruction. This can be detected in the Decode stage (`LoadStallD`) by checking if the instruction in the Execute stage is a load (indicated by bit 1 of the memory read/write enable signal, `MemRWE[1]` = 1) and if the load destination register matches either source register in the Decode stage (`MatchDE` = 1).

```
MatchDE = (Rs1D == RdE | Rs2D = RdE)
LoadStallD = MemRWE[1] & MatchDE
```

As discussed in Section 7.2.1, the synchronous DTIM needs the read address to setup in the Execute stage and the write address to setup in the Memory stage. This creates another structural hazard that a DTIM store cannot immediately be followed by a load because both would need the address port at the same time. This `StoreStallD` is detected

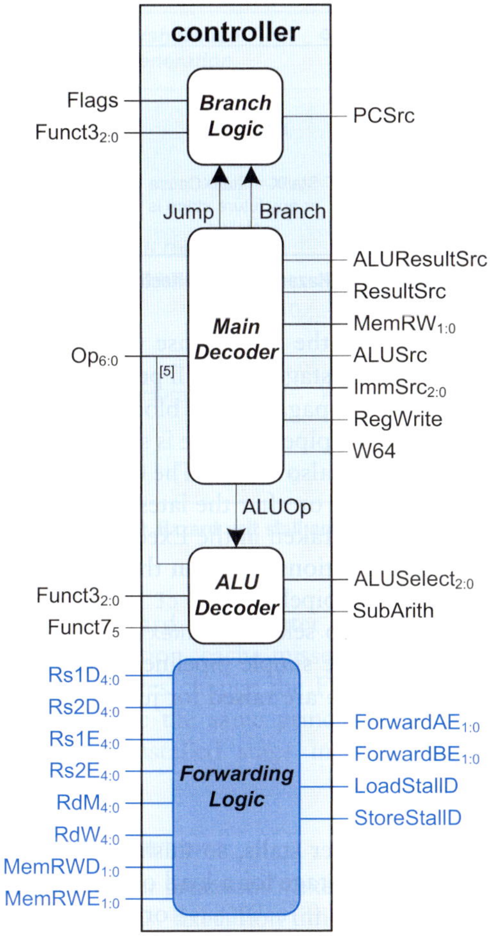

Fig. 7.17 Forwarding logic in controller.

when the instruction in the Execute stage is a store (`MemRWE[0]`) and the instruction in the Decode stage is a load (`MemRWD[1]`).

```
StoreStallD = MemRWE[0] & MemRWD[1]
```

7.2.2.4 Hazard Unit

The hazard unit monitors various hazards that cannot be resolved by forwarding, such as load-use dependencies and taken branches. Fig. 7.18 shows the general architecture of the hazard unit. The unit receives various hazard signals from the IEU such as `PCSrc`, which is 1 when a branch is taken, and `LoadStall` or `StoreStall`, which are from the forwarding logic within the controller. Using these hazard inputs, the Flush Cause Generation block computes `FlushXCause` signals for each pipeline stage X that

Table 7.9 CoreMark instruction mix.

Type	Number of Instructions	Percent of Total Instructions	CPI
R- or I-type ALU	1,320,259	49.7	1
Jumps / returns	46,750	1.8	3
Branches	588,905	22.2	1.84
Loads / stores	698,738	26.3	1.41
Total	**2,654,652**	**100**	**1.32**

when running on Wally. The benchmark is compiled for rv32im with no branch prediction support (`BPRED_SUPPORTED` = 0) to match the static branch predictor from this chapter.

```
$ cd $WALLY/benchmarks/coremark
$ make ARCH=rv32im_zicsr CONFIG=nobpred_rv32gc run
```

Adding a branch predictor in Section 13.3.3 improves the branch CPI. Adding bit manipulation instructions reduces the instruction count. Together, these improve the overall CPI to 1.14 as described in Section 21.1.

The instruction mix comes from the performance counter results. Branches are taken 41.9% of the time. Our static branch predictor predicts not taken; therefore, misprediction rate is 41.9%. Each misprediction requires an additional two cycles to flush the two instructions in the Fetch and Decode stages after the branch target address is calculated in the Execute stage. Jumps always require three cycles: one to execute the jump and two to flush the next two instructions. The performance counters indicate that 40.1% of loads cause a use after load stall. Thus, the CPI for R-/I-type ALU operations is 1, jumps is 3, branches is $1 + 0.419 \times 2 = 1.84$, and loads/stores is $1 + 0.401 \times 1 = 1.40$. The overall CPI is then a weighted sum of these CPIs: CPI $= (0.497 \times 1) + (0.018 \times 3) + (0.222 \times 1.84) + (0.263 \times 1.41) = 1.32$. This approximates CoreMark's reported CPI: # clock cycles / # instructions = 3,36,604 / 2,654,652 = 1.27.

7.3 SYSTEMVERILOG

The Wally configurable pipelined processor is implemented in SystemVerilog. It uses `$WALLY/config/rv64i/config.vh` to specify the set of supported features described in this chapter for a 64-bit processor. The rv32i configuration is similar. Wally's rv32e configuration is also similar, but minimizes size by shrinking the register file from 32 to 16 entries and replacing the IROM and DTIM with an AHB interface to off-chip memory described in Section 9.5. The configurable source code is in `$WALLY/src`.

The full source code contains conditional generate statements to produce only the configured hardware. However, it has modules and ports sufficient to support all configurations. It is worth reading and relating to the diagrams in this chapter and the coding guidelines from Section 4.3 to get an understanding of the processor design.

7.4 TEST PLAN

The instructions are tested using the riscv-arch-test suite [riscv-arch-test]. This is an evolving set of tests provided by RISC-V International to offer a basic level of confidence that an implementation conforms to the RISC-V specification.

The configuration in this chapter supports the rv32i_m/I, rv32e_m/E, or rv64i_m/I suite of tests. These include a collection of tests for each of the RV32I or RV64I unprivileged instructions. Each instruction has a series of tests produced by the RISC-V constrained random test generator, riscv-ctg. The tests perform an operation and then write a result called the *signature* to memory. If the full set of signatures produced by the device under test matches a reference signature, the test suite passes.

The constrained random test generator is driven by a configuration file that lists the constraints for each instruction [crtg]. Code Example 7.3 lists the constraints for the `addi` instruction. It tests all the different registers being used for `rs1` and `rd`, including cases where `rs1` and `rd` are the same register. To test a range of operand values, it selects values including 0, 1, the most negative, the most positive, walking ones, walking zeros, alternating bits, and random values, including operand pairs that are equal, unequal, and that have each of the four patterns of signs. Walking ones start with a value that is all 0s except the least significant bit. The 1 then shifts (walks) to the left (i.e., 00...001, 00..010, 00...100, etc.). Walking zeros are the similar but start with all 1s and the 0 walks to the left.

Code Example 7.3 `addi` TEST VECTOR CONSTRAINTS

```
addi:
            config:
              - check ISA:=regex(.*I.*)
            opcode:
              addi: 0
            rs1:
              <<: *all_regs
            rd:
              <<: *all_regs
            op_comb:
              <<: *ifmt_op_comb
            val_comb:
              <<: [ *ifmt_val_comb_sgn, *base_rs1val_sgn, *ifmt_base_immval_sgn]
              abstract_comb:
                'sp_dataset(xlen,["rs1_val",("imm_val",12)])': 0
                <<: [*rs1val_walking, *ifmt_immval_walking]

ifmt_val_comb_sgn: &ifmt_val_comb_sgn
  'rs1_val == imm_val': 0
  'rs1_val != imm_val': 0
  'rs1_val > 0 and imm_val > 0': 0
  'rs1_val > 0 and imm_val < 0': 0
  'rs1_val < 0 and imm_val > 0': 0
  'rs1_val < 0 and imm_val < 0': 0

base_rs1val_sgn: &base_rs1val_sgn
  'rs1_val == (-2**(xlen-1))': 0
  'rs1_val == 0': 0
  'rs1_val == (2**(xlen-1)-1)': 0
  'rs1_val == 1': 0

ifmt_base_immval_sgn: &ifmt_base_immval_sgn
  'imm_val == (-2**(12-1))': 0
  'imm_val == 0': 0
  'imm_val == (2**(12-1)-1)': 0
  'imm_val == 1': 0

rs1val_walking: &rs1val_walking
  'walking_ones("rs1_val", xlen)': 0
  'walking_zeros("rs1_val", xlen)': 0
  'alternate("rs1_val",xlen)': 0
```

Table 7.10 lists the number of test cases per instruction for RV32I, RV32E, and RV64I. These statistics can be found using `$WALLY/bin/testcount.pl`. The tests are continually evolving, and the test count has increased slightly since this data was last collected in 2022. All these load and store tests are aligned so they do not cause misalignment faults (see Section 8.2.5). RV64 has more test cases because there are more operand values to cover.

Table 7.10 riscv-arch-test I cases.

Instruction	RV32I Tests	RV32E Tests	RV64I Tests
add-01	588	583	679
addi-01	561	561	619
addiw-01			625
addw-01			678
and-01	584	588	677
andi-01	562	554	619
auipc-01	64	63	63
beq-01	583	590	679
bge-01	590	586	674
bgeu-01	726	728	821
blt-01	580	586	674
bltu-01	726	727	814
bne-01	584	585	678
fence-01	1		1
jal-01	32	16	32
jalr-01	33	27	33
lb-align-01	33	18	32
lbu-align-01	32	17	33
ld-align-01			32
lh-align-01	32	16	32
lhu-align-01	32	16	32
lui-01	63	63	63
lw-align-01	32	16	32
lwu-align-01			32
misalign1-jalr	1		1
or-01	587	589	678

(Continued)

Table 7.10 riscv-arch-test I cases. Cont'd

Instruction	RV32I Tests	RV32E Tests	RV64I Tests
ori-01	560	557	622
sb-align-01	70	78	135
sd-align-01			132
sh-align-01	71	71	133
sll-01	89	88	153
slli-01	88	89	151
slliw-01			150
sllw-01			153
slt-01	584	581	676
slti-01	560	561	617
sltiu-01	699	697	759
sltu-01	722	722	819
sra-01	90	90	154
srai-01	87	87	152
sraiw-01			152
sraw-01			154
srl-01	92	89	153
srli-01	90	88	153
srliw-01			152
srlw-01			154
sub-01	592	584	670
subw-01			678
sw-align-01	68	68	132
xor-01	588	583	675
xori-01	566	559	624
Total	**12,642**	**12,521**	**17,836**

The riscv-arch-test suite has several limitations. It does not test interactions between instructions, such as those that cause pipeline hazards. It also has limited coverage of the instruction operands. Wally uses other functional coverage tests to fill these gaps.

See Section 5.11.6 for instructions of how to run a regression including these tests.

7.5 SUMMARY

This chapter has developed a processor that runs the RV{32/64}I base instruction set. The processor is partitioned into an instruction fetch unit (IFU), integer execution unit (IEU), and load/store unit (LSU). The processor has no external memory interface, so it uses an instruction ROM (IROM) in the IFU to hold instructions and data tightly-integrated memory (DTIM) in the LSU to hold data. The architectural state comprises the program counter (PC) and the 31 registers in the integer register file (IRF).

The IFU contains the IROM, organized as a 32-bit wide read-only memory. It also contains an incrementer to add 4 to the PC and a mux to select PC + 4 or the branch/jump target address from the IEU.

The IEU contains a datapath and a controller. The datapath has the IRF, extension logic to assemble an immediate from various bits of the instruction, and an ALU capable of addition/subtraction, shifts, and logical operations. Branch comparison is on the critical path for simple processors, so the datapath has a fast branch comparator separate from the ALU. The datapath also has a variety of multiplexers to choose sources and results appropriately. The controller contains combinational logic that decodes the `op` and `funct` fields of the instruction to determine write enables and multiplexer selects.

The LSU contains the DTIM, organized as an XLEN-bit wide RAM with a single read/write port and byte write enables. It has subword read multiplexers controlled by the least significant address bits and the access type to choose a subset of the word and sign or zero-extend the subword to XLEN-sized ReadData. It also has subword write logic to replicate the smaller WriteData value and turn on the appropriate byte enables.

We begin with a single-cycle core, then convert it into a pipelined design by adding pipeline registers and hazard/forwarding logic and adapting timing for synchronous memories (DTIM and IROM). The five stages of the pipeline are:

Fetch: Read instruction from IROM and update PC

Decode: Decode instruction to compute control signals, read registers from IRF

Execute: Execute ALU, branch, and jump instructions and calculate an address for loads/stores

Memory: Read or write DTIM if applicable

Writeback: Update IRF with new result

Pipelining greatly increases the clock frequency at a modest increase in complexity. Hazards are the main challenge for a pipelined processor. When one instruction depends on the result of a previous instruction that is not yet written back to the register file, the forwarding logic, in many cases, resolves the problem by forwarding the result of the previous instruction to multiplexers at the start of the Execute stage for the current instruction. Unfortunately, load instructions have a two-cycle latency, and their result is not available until the end of the Memory stage. Hence, the hazard unit stalls the pipeline for a cycle when a load is immediately followed by a dependent instruction. A store followed by a load also causes a stall because of a structural hazard with the memory's address port. Finally, the direction of a branch is not known until the comparison is in the Execute stage. The processor continues fetching consecutive instructions as if the branch is not taken, so it must flush the next two instructions when it discovers that a branch should be taken.

The Wally pipelined processor is described using SystemVerilog in the Wally repository. To instantiate only the hardware needed for the simple processor described in this chapter, use the rv32i, rv32e, or rv64i configurations. These configurations have been verified using the integer tests in the riscv-arch-test suite. The next chapters add features to this pipelined processor.

8 Privileged Operations

8.1 Principles

8.2 RISC-V Practices

8.3 Test Plan

8.4 Wally Implementation

8.5 Summary

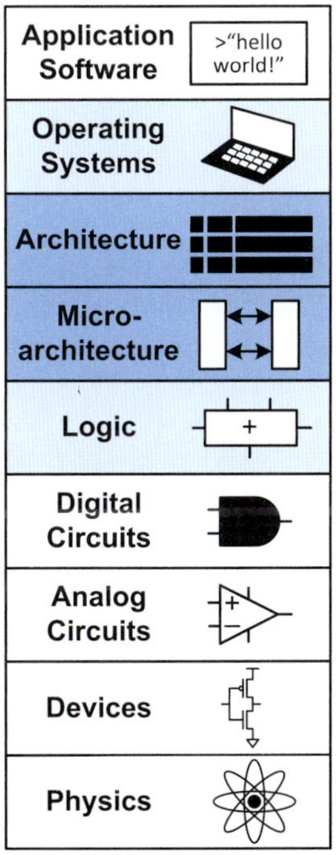

In the basic integer core described so far, a program can access any feature and any location in memory. This raises two problems: a faulty program can crash the machine, and a malicious program can compromise the machine. Processors add privileged modes to provide protection and security to prevent these problems.

Processors also must be able to respond to unusual program events called *exceptions* or *faults* (e.g., trying to access nonexistent memory) and external events called *interrupts* (e.g., a keystroke or periodic timer). The processor responds with a *trap* that saves the current program counter (PC), raises the privilege mode if necessary, and jumps to a *trap handler* that is located at a specified memory address. The trap handler may save other state, respond to the trap, restore the saved state, and restore the PC to continue the program.

This chapter introduces privileged operations and traps as well as the *control and status registers* (*CSRs*) and related instructions needed to support them. Section 8.1 gives an overview of the principles and motivations for privileged operations. Section 8.2 then discusses RISC-V privileged features, including privileged modes, privileged instructions, CSRs, traps, and a sample trap handler. Section 8.3 describes the test plan. Section 8.4 explores an implementation of the privileged features. Memory management, including virtual memory and physical memory protection (PMP), is such a large and important part of privileged operations that it is mostly covered in Chapter 11.

Other privileged features include debug mode and hypervisors. *Debug mode* allows a debugger to step through a program one instruction at a time and inspect all of the architectural state. This is inherently a privileged operation because the debugger could access sensitive information. A *hypervisor* switches among multiple guest operating systems running on *virtual machines*, giving each the illusion of owning the entire machine. For example, I am writing this book on a Mac but drawing the illustrations in Visio, which only runs on Windows. I use Parallels to create a virtual Windows machine on the Mac, with a hypervisor switching between the

A trap handler can also be referred to as an *exception handler*, *interrupt handler*, or simply *handler*.

Control and status registers (CSRs) are sometimes referred to simply as registers for brevity.

The RISC-V privileged specification is nearly as long as the unprivileged instruction set architecture (ISA) specification, and much more intricate. It is the authoritative source and covers privileged operations in more detail than this chapter could; it is essential reading for developers concerned about these features (see riscv.org/specifications/ratified/).

two. The privileged architecture supports trapping on certain instructions and CSR accesses to support virtual machines. The Hypervisor (H) extension improves virtual machine performance by avoiding the need for some of these traps. Neither debug mode nor hypervisor features are covered in this book.

8.1 PRINCIPLES

The functions of a privileged unit are to provide security and protection and to handle traps. A set of privileged instructions and CSRs supports these functions.

8.1.1 Security and Protection

Security and protection are closely related but distinct terms. In this chapter, we define *security* as the general problem of preventing unauthorized agents from accessing a computer system and detecting and mitigating damage if access does occur. *Protection* refers to specific hardware and software mechanisms to prevent both malicious and accidental access to resources.

Early computers only ran one program at a time, but developers soon discovered a need to share the computer among multiple programs and users. The first general-purpose time sharing was the Compatible Time Sharing System (CTSS) running on a modified IBM 7094 [Corbato62]. Designers immediately discovered the need to protect one program from another and added architectural features to address this need. Intel introduced memory management and protection in its 80286 processor in 1982, but developers of operating systems were slow to extend their code to take advantage of these features because of the challenge of maintaining backward compatibility. For example, Microsoft operating systems such as Windows 95 were notoriously unstable because a bug in one program could bring down the whole system. Almost twenty years after the hardware support for protection on PCs became available, Microsoft Windows 2000 took advantage of it to give mainstream users a moderately stable computing experience.

"There are only two types of companies: those that have been hacked and those that don't know they have been hacked."—**John Chambers, CEO, Cisco**

Malware has become an enormous economic drain and national security threat. Cybercriminals and governments use malware to steal personal information, hold corporate data ransom, and attack critical infrastructure such as power plants and military labs. Around 2010, a computer worm called Stuxnet, attributed to Israel and the United States, reportedly wiped out 20% of Iran's nuclear centrifuges by causing them to spin out of control, slowing its weapons program. In 2015 and 2016, Russian hackers repeatedly attacked Ukraine's power grid and shut off power in the capital city Kyiv. The FBI recorded $16B of cybercrime losses reported by US businesses and individuals in 2024 [IC3-21]; this is likely just the tip of the iceberg. In 2024, the US National Security Agency (NSA) announced that Chinese hackers had gained access to control portions of US critical infrastructure, including energy, transportation, communication, and water treatment for at least the past five years, and had stepped up cyberattacks on US critical infrastructure ahead of potential conflict in the Pacific or over Taiwan.

Malware threats include confidentiality, integrity, and availability attacks. A *confidentiality* attack steals sensitive information such as credit card numbers. An *integrity* attack tampers with data, such as changing grades on a transcript or encrypting a hospital's patient records until a ransom is paid. An *availability* attack denies service, such as bringing down a website.

Any device connected to the internet is now under nearly constant attack by automated scripts probing for access. Most attacks target known vulnerabilities, such as weak passwords, bugs in outdated programs, or gullible employees. Rigorous information technology (IT) policies of using unique secure passwords, deploying firewalls and antivirus software, quickly patching security holes, and training employees about suspicious messages can mitigate some of these attacks. The most sophisticated attackers, often state sponsored, also seek out new bugs and vulnerabilities that are unknown to vendors and users. Such *zero-day* exploits are used to take control of phones, computers, and critical infrastructure for spying and economic warfare. Zombie armies of compromised computers are also used to probe other computers or launch distributed denial of service attacks.

Securing a large amount of code, especially code from untrusted sources, is very difficult. Larger code bases have more bugs that can be exploited. A first step is to divide a system into at least two *privilege modes*. Most processes operate in user mode (U-mode). U-mode programs can only access resources such as memory and files granted to them by the OS kernel. They are protected from accessing other resources, so even if a U-mode program is malicious or buggy, it cannot read or corrupt resources owned by another program or the OS. When a process needs to request more resources, it must call the OS, which runs in supervisor mode (S-mode). The OS verifies that the request is valid and assigns the resources, then returns to the U-mode portion of the OS kernel. Now, the attack surface is limited to the smaller OS kernel rather than the entire set of programs on the machine.

8.1.2 Traps

Recall that a processor takes a trap to stop normal program execution and jump to a trap handler when an interrupt or exceptional program event occurs. Traps are an important protection mechanism, providing a secure method to invoke system calls at a higher privilege level and a way to stop programs that attempt to access unauthorized data or code.

RISC-V only supports precise traps.

Traps are categorized as precise or imprecise. A precise trap is one associated with a specific instruction such that all previous instructions have retired (completed), and no later instructions have modified architectural state or memory. In contrast, an imprecise trap is one in which the pipeline cannot be restarted at a specific instruction, for example, if some instructions have retired out of order when the trap occurs.

To illustrate imprecise programs, consider the following program:

```
0x1000: div t1, t2, t3
0x1004: add s0, s0, s1
```

Suppose the `div` instruction takes 20 cycles, and the independent `add` instruction takes 1 cycle. Suppose the core allows the `add` to write its result to the register file and retire before the `div` retires. Then suppose a timer interrupt occurs after the `add` but before the `div` retires, causing an imprecise trap. After the trap handler completes, to what PC should it return? If it returns to 0x1000, the `div` will reissue properly, but `add` will execute a second time and produce an incorrect result in `s0`. If it returns to x01008, the `div` instruction will not have executed at all, leaving an incorrect result in `t1`.

Precise traps allow a program to resume from a well-defined place. To precisely handle traps, a pipeline should have a retire or commit stage at the end to delay modifying architectural state until all previous instructions have retired. For in-order processors, the writeback stage at the end of the pipeline serves this role. Out-of-order processors are more complicated and may use a reorder buffer to delay committing results to architectural state until all previous instructions have completed.

For example, if a load causes a virtual memory page fault, the trap handler should fetch the missing page and then resume the load without having other partially completed instructions leaving the system in an inconsistent state. Similarly, if a debugger takes a trap when reaching a breakpoint, all the instructions before the breakpoint should be complete, and none of the instructions after the breakpoint should appear to have been started. Handling *imprecise* traps in a way other than terminating the program is complicated and sometimes impossible, so most processors generate precise traps.

8.1.3 Control and Status Registers (CSRs)

Most processors have CSRs that report or change the behavior of the processor. Examples of CSRs are:

- Registers to control traps, such as enabling/disabling interrupts or saving the program counter
- Registers to enable/disable optional features
- Performance counters to assist profiling programs

8.2 RISC-V PRACTICES

RISC-V provides protection and security by supporting up to three privilege modes: *machine*, *supervisor*, and *user*, also referred to as *M-mode*, *S-mode*, and *U-mode*, respectively. As described in Section 8.2.1, these modes enable protection by allowing only higher privilege modes access to some operations, CSRs, and portions of memory so that malicious programs at lower privilege levels cannot access or corrupt critical data or code. The privileged specification also adds a set of instructions (see Section 8.2.2) to support traps and memory synchronization and a set of CSRs (see Section 8.2.3), some of which are only accessible in a given privilege mode. Section 8.2.4 defines the state of the system on reset.

The base RISC-V ISA has only M-mode and lacks CSRs. The Zicsr extension provides CSRs, and the U and S extensions provide the additional privilege modes. For quick reference on the last pages of this book, Table I.8 summarizes privileged instructions, and Tables I.9–I.13 summarize the CSRs and traps.

The privileged specification also defines the causes and handling of traps, discussed further in Sections 8.2.5 and 8.2.6. As an overview, traps to machine mode operate as follows:

1. Elevate the privilege to M-mode.
2. Save the current program counter in the `mepc` (machine exception PC) CSR.
3. Place a code indicating the reason for the trap in `mcause`.
4. Optionally record additional information about the trap in `mtval`.
5. Record the previous privilege mode and interrupt enable state in `mstatus`.
6. Disable interrupts.
7. Jump to a *trap vector* address (address of the trap handler code) stored in `mtvec`.
8. The trap handler responds to the trap (Section 8.2.6).
9. The trap handler then uses the `mret` (M-mode exception return) instruction to return to the address in `mepc` and to restore the privilege mode and interrupt enable.

Fig. 8.1 shows the CSRs and trap return instruction (`mret`) used in an M-mode trap.

Most traps in Linux are handled by the S-mode OS. An M-mode trap handler could reroute traps to S-mode (in software), but it is faster for the hardware to directly *delegate* the trap. The `medeleg` and `mideleg` CSRs (machine exception and interrupt delegation registers) indicate which

The RISC-V privileged specification has evolved over time and is still evolving, so it is important to know which version of the extensions are supported. The Sm1p12 and Ss1p12 extensions indicate that Version 1.12 of the machine- and supervisor-mode specifications are supported.

Note that each flavor of the `epc` is set to the PC of the instruction that traps. For some traps such as interrupts or page faults, the trap handler returns to that PC to rerun that instruction. For other traps, such as an environment call (`ecall`), the trap handler adds 4 to the `epc` so that it returns to the next instruction.

CSRs

mepc	PC where trap occurred
mcause	Trap cause
mtval	More info about trap (optional)
mstatus	Previous privilege mode and interrupt enable state
mtvec	Instruction address of trap handler

Trap return instruction

mret	Return to mepc and Restore previous privilege mode and interrupt enable

Fig. 8.1 **CSRs and trap return instruction (`mret`) used in M-mode trap.**

exceptions and interrupts should be delegated to S-mode. This way, when such a trap occurs, it can immediately go to the S-mode (OS) trap handler instead of the M-mode handler. S-mode traps are similar to M-mode traps but use a corresponding set of registers, such as `sepc` and `stvec`, and a supervisor trap handler that returns using `sret`.

The current privilege mode is stored in hardware, as described in Section 8.4.1, but it is not directly visible to a program. However, the privilege mode may be changed by the `mret` and `sret` instructions, as discussed in this chapter. The mode is generally obvious: whereas user programs run in U-mode, the OS runs in S-mode and makes SBI calls to M-mode.

8.2.1 Privilege Modes

Table 8.1 lists the RISC-V privilege modes and their encodings. The processor boots in M-mode. Simple embedded microcontrollers ("bare metal") only use M-mode. Secure microcontrollers add U-mode to separate untrusted application code from a smaller trusted secure kernel. Processors running sophisticated operating systems such as Linux also add a supervisor mode (S-mode), which supports virtual memory and provides security while allowing an M-mode supervisor binary interface (SBI) to abstract platform-specific details from the OS, such as sending interprocessor interrupts or rebooting the computer. Section 2.3.6 described the SBI.

Table 8.1 **RISC-V privilege modes.**

Code	Mode
00	User
01	Supervisor
11	Machine

8.2.2 Privileged Instructions

Both the RISC-V privileged and unprivileged specification define instructions to interact with privileged features. The unprivileged specification defines the following instructions:

- `csr*` CSR access instructions (including `csrrw`, `csrrs`, and `csrrc`)
- `ecall` Environment call
- `ebreak` Debugger breakpoint

The RISC-V privileged specification introduces the following instructions:

- `mret` Return from M-mode trap
- `sret` Return from S-mode trap
- `wfi` Wait for interrupt
- `sfence.vma` Synchronize virtual memory by flushing TLB entries

`sfence.vma` is described in Section 11.2.4, and the other instructions are described here.

8.2.2.1 CSR Access Instructions

RISC-V defines a family of instructions to read and modify the CSRs: `csrrw`, `csrrs`, and `csrrc`. Each reads the value of the CSR into `rd`. Then the instructions, respectively, write, set, or clear bits of the CSR specified by `rs1`.

The instructions also come in immediate forms `csrrwi`, `csrrsi`, and `csrrci`, that use a 5-bit unsigned immediate instead of `rs1` to indicate which bit to write, set, or clear.

Programs commonly want to just read or just write the CSRs. `csrr rd, csr` is a pseudoinstruction, *CSR read*, for reading `csr` into `rd`; it translates into the actual RISC-V instruction `csrrs rd, csr, x0`. To write a CSR, pseudoinstruction *CSR write* may be used. `csrw csr, rs1` writes `rs1` into `csr`; it translates into the actual RISC-V instruction `csrrw x0, csr, rs1`.

These and other common pseudoinstructions are listed in Table I.7 (on the last pages of the book).

Pseudoinstructions `csrs`/`csrc` are also commonly used. They set/clear bits in the CSR as specified by bits that are high in `rs1`. `csrs csr, rs1` makes `csr = csr | rs1`, and `csrc csr, rs1` makes `csr = csr & ~rs1`.

8.2.2.2 Exception Generating Instructions

Two instructions purposefully trigger an exception: `ecall` or `ebreak`. `ecall` triggers an exception to call the OS. For example, in a system with all three privilege modes, a U-mode program uses `ecall` to invoke the S-mode OS through the application binary interface (ABI), and the OS uses `ecall` to invoke the firmware through the supervisor binary interface (SBI).

`ebreak` triggers a breakpoint exception to transfer control to a debugger. A debugger sets a breakpoint by replacing the corresponding instruction with an `ebreak` and recording the replaced instruction. When the program reaches the breakpoint, the `ebreak` trap occurs, entering the debugger code. The debugger may display the current PC and register values. To single-step through the instruction at the breakpoint, the debugger restores the instruction at the breakpoint, inserts an ebreak at the next instruction, and returns to the program.

8.2.2.3 Trap Return Instructions

The mret and sret instructions return from M- and S-mode trap handlers, respectively. mret sets the program counter to the return value stored in mepc and restores the previous privilege mode and interrupt enable state from the mstatus register. Specifically, it changes the privilege mode to mstatus.MPP (machine previous privilege mode) and mstatus.MIE (machine interrupt enable) to mstatus.MPIE (machine previous interrupt enables). These status bits are described in Section 8.2.3.1.

Bitfields of a register are indicated by .<bitfield name> after the register. For example, mstatus.MPP indicates the MPP bitfield within the mstatus CSR.

mret then sets mstatus.MPP to the least-privileged supported mode (U-mode, if supported) and sets mstatus.MPIE. If mstatus.MPP was not M-mode, mret also clears mstatus.MPRV (the Modify PRiVilege mode bit), as discussed in Section 11.2.1.5. mret causes an illegal-instruction exception if attempted from S- or U-mode.

sret is analogous to mret but returns to sepc and manipulates the SPP, SIE, and SPIE bits of the sstatus register. sret causes an illegal-instruction exception if attempted from U-mode or while mstatus.TSR = 1 (TSR is an acronym for trap on sret). sret then sets mstatus.SPP to U-mode and mstatus.SPIE to 1 and clears mstatus.MPRV.

8.2.2.4 Wait for Interrupt

Most processors spend most of their time in an *idle loop*, waiting for an interrupt that indicates the user has pressed a key, moved the mouse, or tapped the screen, or a timer is scheduling the program's next action. The wait for interrupt instruction (wfi) is used to save power while the processor is in such an idle loop. After a wfi instruction executes, the processor stops executing instructions until an interrupt is pending. wfi preferably puts the processor in a low power state, stopping clocks to most of the core and possibly lowering the power supply voltage. To facilitate a hypervisor switching between guest operating systems, if wfi is executed in U- or S-mode and mstatus.TW = 1 (TW = timeout wait) or wfi is executed in U-mode and S-mode is supported, wfi will time out after an implementation-specific delay (possibly 0) by triggering an illegal-instruction exception.

wfi will resume program execution when an interrupt is pending even if interrupts are globally disabled. In this case, the pending interrupt is not taken, but the code after wfi can poll interrupt sources and respond right away, without needing to save and restore registers as is done in a trap handler. Thus, the system can improve response time by globally disabling interrupts before a wfi.

8.2.3 Control and Status Registers (CSRs)

RISC-V supports up to 4,096 CSRs, which are XLEN bits wide. Some CSRs require 64 bits of information and have an associated h upper half for RV32. CSRs are specified by 12-bit addresses:

- **The two most significant bits** of the address are 00 or 01 for standard read/write registers and 11 for read-only registers.

- **The next two bits** are 00 for U-mode, 01 for S-mode, and 11 for M-mode registers. CSRs can only be accessed from a privilege level greater than or equal to their mode.

Table I.9 (see the last pages of the book) summarizes the standard RISC-V CSRs, excluding those associated with hypervisors and debug/trace. The CSR type is M/S/U (machine/supervisor/user) for the mode and RO or RW for read-only/read-write.

Trap registers are described in Section 8.2.5. The supervisor address translation and protection (`satp`) register for virtual memory is described in Section 11.2.1.1. Physical memory protection (PMP) registers are described in Section 11.2.2.2. Floating-point CSRs are described in Section 16.4.1. The remaining CSRs are described next.

This book does not discuss or support CSRs associated with hypervisors or debug/trace, including `dcsr`, `dpc`, `dscratch0`, `dscratch1`, `scontext`, `mtinst`, and `mtval2`.

The privileged specification provides an elaborate description of what occurs when a program writes to reserved CSRs or writes illegal values. This chapter glosses over these special cases. However, occasionally acronyms will appear:

- **WPRI:** Writes Preserve values, Reads Ignore values (for reserved fields)
- **WLRL:** Write Legal/Read Legal values only (programs must write/read legal values)
- **WARL:** Write Any/Read Legal values (writing illegal values is allowed, but they won't stick)

8.2.3.1 Status Register

The 64-bit status register (`mstatus` or `sstatus`) holds information such as the previous privilege level, interrupt enables, floating-point/vector enables, timeout wait, and XLEN for U- and S-mode. The status register is the most complex CSR. All the bits of the status register are accessible in M-mode using the `mstatus` register, and a subset is accessible in S-mode using the `sstatus` register. Fig. 8.2 shows the bitfields of the 64-bit `mstatus` register for RV64.

Fig. 8.3 shows the same bitfields packed into two 32-bit registers, `mstatus` and `mstatush`, for RV32; the relative positions are the same except that SD (state dirty) is shifted into bit 31 to make checking dirty state easy and `SXL`/`UXL` are not supported.

Fig. 8.4 shows `sstatus` for RV64; the `MIE`, `MPIE`, `MPP`, `MPRV`, `TVM`, `TW`, `TSR`, `SXL`, `SBE`, and `MBE` fields are inaccessible from S-mode, but the other fields are the same as `mstatus`. `sstatus/h` for RV32 is the same as `sstatus` except `SD` is again shifted into bit 31. The bitfields are described next.

`MPP`/`SPP`: Previous Privilege

The 2-bit `MPP` field holds the previous privilege mode from before a trap into M-mode (`MPP` = 00, 01, or 11 for U-, S-, or M-mode, respectively). The `mret` instruction restores the privilege mode to the value saved in `MPP`. Likewise, the 1-bit `SPP` field holds the previous privilege mode before a trap into S-mode (`SPP` = 0 or =1 for U- or S-mode, respectively), and `sret` restores the privilege mode to the value saved in `SPP`.

`MIE`/`SIE` and `MPIE`/`SPIE`: Global Interrupt Enables and Previous Interrupt Enables

A trap handler must disable interrupts, at least initially, so that it is not interrupted while processing an interrupt. The `mstatus.MIE` field enables (1) mstatus

SD	WPRI	MBE	SBE	SXL[1:0]	UXL[1:0]
63	62:38	37	36	35:34	33:32

WPRI	TSR	TW	TVM	MXR	SUM	MPRV	XS[1:0]	FS[1:0]	MPP[1:0]	VS[1:0]	SPP	MPIE	UBE	SPIE	WPRI	MIE	WPRI	SIE	WPRI
31:23	22	21	20	19	18	17	16:15	14:13	12:11	10:9	8	7	6	5	4	3	2	1	0

Fig. 8.2 RV64 `mstatus` register.

mstatush

WPRI	MBE	SBE	WPRI
31:6	5	4	3:0

mstatus

SD	WPRI	TSR	TW	TVM	MXR	SUM	MPRV	XS[1:0]	FS[1:0]	MPP[1:0]	VS[1:0]	SPP	MPIE	UBE	SPIE	WPRI	MIE	WPRI	SIE	WPRI
31	30:23	22	21	20	19	18	17	16:15	14:13	12:11	10:9	8	7	6	5	4	3	2	1	0

Fig. 8.3 RV32 `mstatush` and `mstatus` registers.

sstatus

SD	WPRI	UXL[1:0]
63	62:34	33:32

WPRI	MXR	SUM	WPRI	XS[1:0]	FS[1:0]	WPRI	VS[1:0]	SPP	WPRI	UBE	SPIE	WPRI	SIE	WPRI
31:20	19	18	17	16:15	14:13	12:11	10:9	8	7	6	5	4:2	1	0

Fig. 8.4 RV64 `sstatus` register.

Do not confuse the `mie` machine interrupt enable register (containing per-interrupt enables, see Fig. 8.7) with the similarly named `MIE` global interrupt enable field of the `mstatus` (machine status) register (see Figs. 8.1–8.3 and Table I.10).

or disables (0) all M-mode interrupts for a hart operating in M-mode. In M-mode, interrupts delegated to S-mode are always disabled regardless of `SIE`.

When a trap to M-mode occurs, the old value of `MIE` is copied to `MPIE`, and `MIE` is cleared, as described earlier, so that a recursive trap cannot immediately occur. The `mret` instruction restores `MIE` to the value in `MPIE` to resume normal operation. Therefore, the trap handler generally does not need to manipulate the `mstatus.MIE` bit in software.

Likewise, the `SIE` field enables (1) or disables (0) all interrupts delegated to S-mode for a hart operating in S-mode. In S-mode, M-mode interrupts are always globally enabled regardless of `MIE`, although the Machine Interrupt Enable register (`mie`) may be used to disable individual interrupt sources. When a trap to S-mode occurs, `SIE` is copied to `SPIE` and is then cleared. `sret` restores `SIE` to the value in `SPIE`.

MBE/SBE/UBE: Endianness

These bits control whether loads and stores are treated as little-endian (0) or big-endian (1) in M-/S-/U-mode. Instructions are always little-endian so that a disassembler can unambiguously interpret the instruction. An implementation may hardwire these bits to specific values or allow them to change under software control.

Endianness is defined in Section 2.1.10.

FS/VS/XS/SD: Floating-Point/Vector/Extension/State Status

Floating-point and vector extensions introduce new registers. If these registers are in use, the trap handler must save and restore them during context switches, which is a costly operation. The FS and VS fields indicate whether floating-point or vector registers are *dirty* (different from the last saved version) and need saving. The XS field can do the same for a user-defined extension.

Table 8.2 describes the status codes. For example, when FS = 00 (off), the floating-point unit is disabled, and operations accessing floating-point state cause an illegal instruction trap. Enable floating-point by setting FS = 01 (initial) in software.

```
lui t0, 2          # 1 in bit 13
csrs mstatus, t0   # Set FS = 01 = initial
```

On reset, typically mstatus.FS = 00, and floating-point is disabled. The OS may enable floating-point during the boot process. When writing a bare-metal M-mode program with floating-point instructions, don't forget to turn on floating-point, lest your program take an illegal-instruction trap.

The hardware sets FS = 11 (dirty) when the processor writes a floating-point register or fcsr. When FS indicates a dirty state, the OS must save the floating-point registers and fcsr during a context switch. Once the state is saved, FS can be marked as clean (FS = 10), and future context switches do not need to save the floating-point state unless it is written again.

SD (state dirty) is set if FS, VS, or XS is 11. When SD is 0, no state is dirty and context switches do not need to save or recall any floating-point, vector, or other extension state.

Table 8.2 **State status codes.**

Status	FS/VS Meaning
00	Off
01	Initial
10	Clean
11	Dirty

TW: Timeout Wait

When TW = 1 and the system is in U- or S-mode, a wfi instruction will produce an illegal-instruction exception if it does not complete within an implementation-defined bounded time. This allows the hypervisor to switch to a different guest OS rather than wasting time sleeping in the current OS.

MPRV, MXR, SUM: Modify Effective PRiVilege, Make eXecutable Readable, Supervisor User Memory

These bits influence address translation and protection as described in Section 11.2.1.5.

TVM, TSR: Trap Virtual Memory, Trap on sret

Both of these bits facilitate efficient hypervisors, which are not discussed in detail in this book. But generally, when TVM = 1, accessing satp or executing sfence.vma in S-mode raises an illegal-instruction exception. When TSR = 1, executing sret in S-mode raises an illegal-instruction exception. A hypervisor trap handler could receive these instructions and manage virtual memory or operating systems returns appropriately.

Table 8.3 XLEN codes.

Code	XLEN
01	32
10	64
11	128

UXL/SXL: User/Supervisor XLEN

For RV64, these fields control the effective XLEN in U- or S-mode according to Table 8.3, allowing the processor to emulate RV32. These fields are often hardwired to 10 for processors that do not need to change their width on the fly. The M-mode XLEN is stored in the two most significant bits of the misa register, which is discussed next.

8.2.3.2 misa Register

The machine ISA (misa) register bitfields indicate which standard extensions are supported by the processor, as given in Table I.12 (on the last pages of the book). The corresponding bit is set for each supported extension.

The association of letters to features is a bit too cute, and RISC-V has long ago run out of letters. Additional extensions have longer names starting with Z (for any mode) or S (for supervisor mode), described in Table A.6. At the time of this writing, a standard way to discover which extensions are available in a hart is still being defined. Linux tests for the presence of some extensions by trying them during boot, taking an illegal instruction trap if the extension is unavailable, and generating a table of which extensions are available.

8.2.3.3 Machine Information Registers

The read-only machine information registers include mvendorid, marchid, mimpid, mhartid, and mconfigptr. All these registers can be 0 in a simple single-core processor or may have more information to identify the design, hart number, and configuration.

JEDEC assigns the Vendor ID for a fee. RISC-V International assigns a unique Arch ID to a processor architecture upon request. A list of all the RISC-V architecture IDs is at github.com/riscv/riscv-isa-manual/.

8.2.3.4 Counters

RISC-V has up to 32 64-bit performance counters to keep track of various events. The first three are standard, measuring the number of clock cycles, number of timer ticks, and number of instructions retired. The others are platform dependent and optionally configurable; they could track events such as cache accesses and misses, branch predictions and mispredictions, pipeline stalls, and so on. These counters are helpful to profile program behavior and identify program or microarchitectural areas for optimization.

Performance monitoring CSRs (see Table I.9) include the following:

- Up to 32 performance counters: mcycle/h, minstret/h, and mhpmcounter3/h — mhpmcounter31/h
- 32-bit mcounteren and scounteren that determine which privilege modes can read each of the counters
- mhpmevent3/h—mhpmevent31/h (hardware performance monitor event) registers to optionally select which events are observed by which counters
- 32-bit mcountinhibit (machine counter inhibit) register to disable counters and, thus, reduce energy consumption of unused counters

In RV64, each performance counter holds a single 64-bit value (i.e., mhpmcounter3). In RV32, each performance counter is divided into low and high 32-bit halves (i.e., mhpmcounter3 and mhpmcounter3h).

RISC-V also provides read-only views of the performance counters for less-privileged access: cycle/h, time/h, instret/h, and hpmcounter3/h—hpmcounter31/h. The first three counters are special. Counter 0 (cycle) tracks the number of clock cycles elapsed on the current hart. Counter 1 (time) tracks a system timer that may operate at a constant speed independent of each hart's clock. Counter 2 (instret) tracks the number of instructions retired (completed) by a hart. The other 29 counters may be hardwired to specific events, configurable among many events based on mhpmevent3—31, or simply not implemented.

time/h is a memory-mapped register, not a CSR.

Each counter can be read and written from M-mode (i.e., with mcycle, mhpmcounter) or read-only from any mode (i.e., with cycle, hpmcounter). Performance counters raise security risks such as timing attacks, in which information is deduced by measuring the speed of variable latency operations such as cache accesses or division. Thus, the mcounteren and scounteren bits can selectively turn off access to counters from S- and U-mode, respectively. These enables could be read-only as 0 if the counters are only accessible in M-mode.

The time register comes from a system timer rather than a per-hart timer, so there is no mtime register to which a user could write the time. The system timer is often in the Core Local Interruptor (CLINT) peripheral, which will be discussed further in Section 20.2.

The rdcycle rd pseudoinstruction is equivalent to csrr rd, cycle to read the current cycle count. Similarly, rdtime and rdinstret pseudoinstructions read the other special timers. In RV32, they also have an h form (rdcycleh, rdtimeh, and rdinstreth) to read the upper 32 bits.

> Overflow of 64-bit counters is not a practical concern in present systems. If a counter increments every cycle at 1 GHz, it would have to run continuously for 585 years before overflowing.

The base RISC-V instruction set has no counters, even with the Zicsr extensions. The Zicntr extension provides the standard `cycle`, `time`, and `instret` family of counters. The Zihpm extension provides one or more platform-specific hardware performance monitor counters. The Sscounterenw extension indicates that `scounteren` is writable (as opposed to read-only 0), and thus counters can be accessible from U-mode.

8.2.3.5 CSR Interaction with Peripherals

The timer value and most interrupts come from peripherals outside the RISC-V core. In many systems, these peripherals are called the *Core Local Interruptor* (CLINT) and the *Platform-Level Interrupt Controller (PLIC)*. The CLINT contains the system timer, which may run at a constant rate independent of any hart's sleeping or variable frequency. It also contains timer compare registers and software interrupt registers on a per-hart basis to generate machine timer and software interrupts for each hart. The PLIC prioritizes and routes interrupts from other peripherals to the external machine and supervisor interrupt inputs of each core. These peripherals are accessed using memory-mapped input/output (I/O) rather than through CSRs. However, reading the `time` register returns a value from the CLINT timer. Each hart's `mip` and `sip` register fields are affected by the interrupt inputs from peripherals. The CLINT and PLIC are described in more depth in Chapter 20.

8.2.3.6 Environment Configuration Registers

Some RISC-V extensions raise security risks, especially when the OS or firmware is not designed to support them. For example, the Sstc extension defines a `stimercmp` CSR. If the M-mode firmware has not been enhanced to save and restore `stimercmp` when switching between S-mode processes, such processes could exchange covert information by writing and reading `stimercmp`. Similarly, with Zicbom, a user-level process could use `cbo.inval` on a dirty cache line to leave memory in an inconsistent state. The `{m/s}envcfg` registers allow the system to disable such extensions at lower privilege levels when they are not safely supported.

The machine and supervisor environment configuration registers `menvcfg` and `senvcfg` enable the availability of certain features in lower privilege modes. Fig. 8.5 shows the bitfields in these registers. `menvcfg` is always 64 bits, so `menvcfgh` is used to access the upper half in RV32. The `STCE`, `PBMTE`, and `ADUE` bits control S-mode features, so they only exist in `menvcfg`. The other bits apply to both S- and U-modes and exist in both `menvcfg` and `senvcfg`. In a system with S- and U-modes, the enables must be set in both `menvcfg` and `senvcfg` to enable the feature for U-mode. The bitfields of unsupported extensions are hardwired to 0.

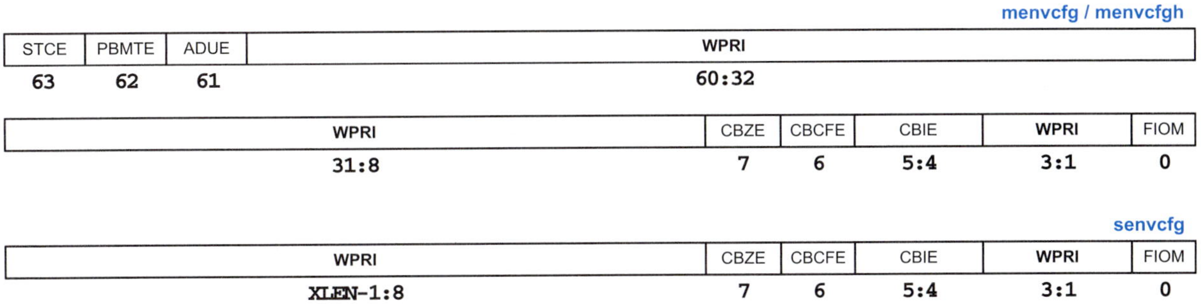

Fig. 8.5 **`menvcfg/h` and `senvcfg` registers.**

When the Sstc Supervisor Timer Compare extension is supported (see Section 8.2.5.4), `menvcfg.STCE` enables access to the `stimecmp` register in S-mode. If `menvcfg.STCE` = 0, accessing `stimecmp` in S-mode causes an illegal-instruction exception, and the `mip.STIP` and `sip.STIP` bits revert to their behavior as if Sstc were not implemented.

When the Svpbmt Page Based Memory Types extension is supported (see Section 11.2.2.1.1), `menvcfg.PBMT` enables this feature. If `menvcfg.PBMT` = 0, the core behaves as if Svbpmt is not supported.

When the Svadu Hardware A/D Update extension is supported (see Section 11.2.1.4), `menvcfg.ADUE` (A/D Update Enable) enables this feature. If `menvcfg.ADUE` = 0, the core takes a page fault when it needs to update the `A` or `D` bits of a page table entry.

When Zicbo* Cache Management Operation extensions are supported, the `CBZE`, `CBCFE`, and `CBIE` bits enable support for the `cbo.zero`, `cbo.flush`, and `cbo.inval` instructions as discussed in Section 19.1.

When the `FIOM` bit is set, a `fence` instruction in a privilege level below M-mode is modified so that I/O fences also imply order for main memory accesses, as described in Section 17.2.2.

The RISC-V privileged specification also defines the `mseccfg` CSR for security configuration. It controls the Zkr random number generation and Smepmp Supervisor Memory Execution Prevention extensions. Wally does not support either of these extensions, so the register is hardwired to 0.

8.2.4 Reset Behavior

On reset, a core begins in M-mode. The PC is set to an implementation-defined reset vector, typically 0x1000 or 0x80000000. `mcause` is set to a value indicating the cause of the reset or 0 if the system does not distinguish reset conditions. `mstatus.MIE` and `mstatus.MPRV` are cleared, and the `pmp` registers are all disabled by setting the `A` and `L` fields to 0. Other hart state is undefined, so the system must initialize all registers and CSRs before using them.

8.2.5 Traps

RISC-V traps are classified as either *exceptions*, which are caused by instructions, or *interrupts*, which are caused by external events such as a key press. Some exceptions are also referred to as *faults*. Recall that traps by default are handled in M-mode but can be delegated in hardware to an S-mode handler, instead of the M-mode handler needing to route the trap to the OS.

Table 8.4 summarizes the standard RISC-V traps, listed in priority from highest to lowest. If multiple traps occur on the same instruction, the highest priority is triggered. For a trap into mode *x* (**m**achine or **s**upervisor), the `xcause`, `xepc`, and `xtval` registers are written. The most significant bit of the `xcause` register is 1 for an interrupt and 0 for an exception. The trap code is written into the low bits of `xcause`. The exception program counter, `xepc`, is set to the address (PC) of the current instruction, the one that generated the exception or that was interrupted. The trap value register `xtval` can either be cleared to 0 or written with trap-specific information to help the software trap handler (see Table 8.4). The processor then changes PC to a trap handler address specified in `xtvec`. All addresses are virtual when virtual memory is enabled (see Chapter 11).

Traps that are handled in the same privilege mode where they were caused (machine to machine or supervisor to supervisor) are called *horizontal* traps. Traps to a higher privilege level are *vertical* traps. Traps never go to a lower privilege level.

When a trap to M-mode occurs, the privilege mode and machine interrupt enable (`mstatus.MIE`) are stored to `mstatus.MPP` and `mstatus.MPIE`, respectively, so that `mret` can later return to the correct state. The privilege mode is set to M-mode, and `mstatus.MIE` is cleared so that the trap won't be immediately interrupted again. The CSRs associated with M-mode traps are as follows:

> `mstatus.MIE` is also known as a *global interrupt enable* because clearing it disables all M-mode interrupts. For example, the hardware clears this bit when taking a trap to disable all interrupts during the trap handler. The `mie` register provides per-interrupt enable bits.

- `mstatus`: Interrupt enable, previous interrupt enable, previous privilege mode bits
- `mcause`: Cause of interrupt or exception (see Table 8.4)
- `mtval`: Value that gives extra trap information (see Table 8.4)
- `mtvec`: Address (also called vector) of trap handler
- `mepc`: Exception PC, holding return address from trap
- `mscratch`: Typically holds the exception stack pointer
- `mip`: Interrupt pending that lists which interrupts are currently pending
- `mie`: Interrupt enable that lists which interrupts are taken (enabled) or ignored (disabled)

Table 8.4 Trap cause (`xcause`) and value (`xtval`) in priority order.[a]

		Cause (`xcause`)		Value (`xtval`)
	Trap	Interrupt Bit (msb)	Trap Code	
Interrupts	M-mode external interrupt	1	11	0
	M-mode software interrupt	1	3	0
	M-mode timer interrupt	1	7	0
	S-mode external interrupt	1	9	0
	S-mode software interrupt	1	1	0
	S-mode timer interrupt	1	5	0
Exceptions	Instruction page fault	0	12	PC
	Instruction access fault	0	1	PC
	Illegal instruction	0	2	Instruction
	Instruction address misaligned	0	0	Target Address
	Breakpoint	0	3	PC
	`ecall` from U-mode	0	8	0 (`a0` = arg)
	`ecall` from S-mode	0	9	0 (`a0` = arg)
	`ecall` from M-mode	0	11	0 (`a0` = arg)
	Store/AMO address misaligned	0	6	Store Address
	Load address misaligned	0	4	Load Address
	Store/AMO page fault	0	15	Store Address
	Load page fault	0	13	Load Address
	Store/AMO access fault	0	7	Store Address
	Load access fault	0	5	Load Address

[a]Store and load misaligned faults become lowest priority when misaligned cacheable accesses are handled in hardware using the Zicclsm extension (see Section 19.2).

Similarly, when a trap to S-mode occurs, the privilege mode and supervisor interrupt enable `mstatus.SIE` are stored to `mstatus.SPP` and `SPIE`, respectively; the privilege mode is set to S-mode; and `mstatus.SIE` is cleared. S-mode traps use a similar set of CSRs but lack views of the M-mode state.

To precisely handle traps, a pipeline should have a *retire* or *commit* stage at the end to delay modifying architectural state until all previous instructions have retired. For in-order processors, the writeback stage at the end of the pipeline serves this role. Out-of-order processors are more complicated and may use a *reorder buffer* to delay committing results to architectural state until all previous instructions have completed.

8.2.5.1 Precise Traps

RISC-V traps are precise. Recall that a *precise* trap is one associated with a specific instruction such that all previous instructions are retired, and no later instructions are retired.

RISC-V exceptions are precise: all instructions prior to the exception complete, and neither the instruction causing the exception nor any subsequent instructions change the architectural state of the system before the exception is handled. This is accomplished by flushing the instruction that caused the exception and any subsequent instructions that are in-flight in the pipeline.

RISC-V interrupts are also precise. Although interrupts are not caused by a particular instruction, they are handled between instructions, with all instructions up to some point retired and no instructions after that point modifying architectural state. Interrupts are handled within a bounded amount of time from the triggering condition.

8.2.5.2 Exceptions

Exceptions are caused by instructions that encounter a special situation. The causes include illegal instructions, accesses to unavailable or misaligned addresses, `ecall`/`ebreak` instructions, and virtual memory page faults (see Table 8.4), as described in this section.

The `medeleg` (exception delegation) register controls which exceptions are delegated to S-mode. It contains a bit for each of the standard exception causes from Table 8.4, as shown in Fig. 8.6, where each grey bit corresponds to an exception cause (unused bits, such as 10 and 14, are set to 0). A one in a given bit delegates that exception to the S-mode handler. For example, if bit 2 of `medeleg` is 1, illegal instruction exceptions call the S-mode handler instead of the M-mode handler. However, traps never reduce privilege levels, so an exception occurring in M-mode will always trap to the M-mode handler.

For example, in a system with a Linux-like S-mode OS, `medeleg` normally delegates at least page faults, U-mode OS calls, and breakpoints to S-mode because these exceptions should be handled by the OS. If they weren't delegated, the M-mode handler would have to pass them along to the OS.

Illegal Instructions

A hart throws an *illegal-instruction exception* when it tries to execute an instruction that either doesn't exist or is unavailable in the current privilege

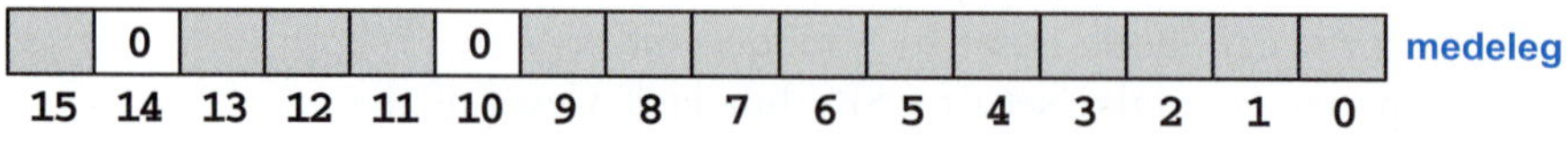

Fig. 8.6 `medeleg` **exception delegation register.**

mode. The 0x00000000 instruction is always illegal, and many other instructions are not implemented in the base ISA or standard extensions. Some instructions such as `mret` are unavailable in lower privilege levels. Accessing an unimplemented CSR or a privileged CSR from a lower privilege level also throws an illegal-instruction exception.

Access Faults

A hart throws an *access fault* when it tries to fetch, read, or write from an inaccessible memory location. Examples include reading from a nonexistent memory location, writing to a read-only memory (ROM), making word accesses to a peripheral that only handles byte accesses, or accessing a protected region of memory from an insufficient privilege level. Accessible regions are defined by *physical memory attributes* (*PMA*, essentially generalized memory address decoders) and *physical memory protection* (*PMA*) registers described in Section 11.2.2. A PMA checker checks for allowed accesses.

Address Misaligned

Most memories can only directly handle accesses from an address that is *naturally aligned*, that is, a multiple of the access size. For example, loading a 4-byte word from 0x1000 is fine, but loading that word from 0x1001 is not because it involves bytes from the words at both 0x1000 and 0x1004. Compilers appreciate this limitation and normally align data correctly, but bad or unusual code might attempt a misaligned access.

A hart may throw a *load* or *store address-misaligned* exception when it attempts a memory operation to an address that is not a multiple of the access size. For example, a load word instruction to address 0x1002 is misaligned because 0x1002 is not a multiple of 4 bytes. The trap handler could kill the program or it could handle the exception by issuing a pair of aligned accesses and then shift, mask, and merge to deliver the misaligned result and return to the program.

A hart will throw an *instruction address-misaligned* exception when it attempts to branch or jump to an illegal address. Note that branch and jump offsets are implicitly a multiple of 2 and that `jalr` sets the least significant bit of the destination address to 0, so instruction addresses are inherently a multiple of 2. When the Compressed extension is not supported (see Chapter 14), instructions are 4 bytes long and must be word aligned, so a branch or jump to an odd multiple of 2 causes an exception. When the Compressed extension is supported, instructions only require halfword alignment so instruction address-misaligned exceptions cannot occur.

With the Zicclsm extension (`ZICCLSM_SUPPORTED`), cacheable main memory regions of memory support misaligned accesses in hardware, as described in Section 19.2. Such accesses may take longer but do not throw a load or store address-misaligned exception.

`ecall` and `ebreak`

`ecall` and `ebreak` instructions are used for OS calls and to invoke debugger breakpoints. They throw the corresponding exception to elevate the privilege and transfer control.

Page Faults

When virtual memory is enabled, attempting to access a virtual page that is not in physical memory throws a *page fault*. The OS can then fetch the virtual page from storage and place it in physical memory before reattempting the access. Virtual memory also allows pages to be designated as user-only or supervisor-only, and code-only or data-only; violating these protections causes a page fault that might terminate the program. Virtual memory is discussed further in Section 11.2.1.

8.2.5.3 Interrupts

RISC-V defines three types of interrupts (timer, software, and external—TI, SI, and EI, respectively) at two privilege levels, machine and supervisor. Interrupts are managed with the `mip` and `mie` *machine interrupt pending* and *enable* registers shown in Fig. 8.7. For example, `mip.MEIP` is set when a machine external interrupt is pending. `mie.MEIE` is set to enable machine external interrupts.

Only 6 bits of the interrupt registers are not hardwired to 0 (Figs. 8.7 and 8.8). All bits tied to 0 are optimized away.

Like exceptions, interrupts can also be delegated to the OS (S-mode handler) by setting bits in the `mideleg` (interrupt delegation) register. `mideleg` contains a bit for each of the standard interrupt causes from Table 8.4, as shown in Fig. 8.8. For example, bits 1, 5, and 9 correspond to S-mode software, timer, and external interrupts (`SSI`, `STI`, `SEI`).

15:12	11	10	9	8	7	6	5	4	3	2	1	0	
0	MEIP	0	SEIP	0	MTIP	0	STIP	0	MSIP	0	SSIP	0	mip
0	MEIE	0	SEIE	0	MTIE	0	STIE	0	MSIE	0	SSIE	0	mie

Fig. 8.7 `mip` **and** `mie` **registers.**

15:12	11	10	9	8	7	6	5	4	3	2	1	0	
0	MEI	0	SEI	0	MTI	0	STI	0	MSI	0	SSI	0	mideleg

Fig. 8.8 `mideleg` **interrupt delegation register.**

The machine and supervisor interrupt nomenclature is confusing. Machine and supervisor interrupts have no fundamental hardware differences, but supervisor interrupts are conventionally delegated to the S-mode handler, while machine interrupts go to the M-mode handler. Thus, in a system with an S-mode OS, `mideleg` is set to 0x222 to delegate supervisor external, timer, and software interrupts to the S-mode handler. Section 22.2.11 describes Linux interrupt handling further.

S-mode code uses the `sip` and `sie` registers which are subsets of `mip` and `mie` that contain only the S-mode bits. Only the bits of `sip` that are delegated are visible: `sip = mip & mideleg`. `sip.SSIP` (S-mode software interrupt pending) is writable in order to clear the interrupt after it is handled; the other `sip` bits are read-only because they are caused by external agents.

When multiple interrupts are pending, RISC-V processors prioritize M-mode interrupts before S-mode and external, then software, and then timer interrupts. This generally causes the most important interrupts to be serviced first.

In a system with no S-mode, S-mode interrupts are typically not produced but would go to an M-mode trap handler if they occurred.

RISC-V implementations may choose which interrupts can be delegated. Like most implementations, Wally only permits delegating supervisor interrupts to S-mode.

8.2.5.4 Interrupt Triggering and Generation

Interrupts can be triggered by peripherals or by writing bits in the `mip` CSR, as shown in Fig. 8.9 and as described further below. `mip.SEIP`, `SSIP`, and `STIP` are writable, allowing M-mode code to send interrupts to S-mode. All six bits are readable, returning the values on the right side of the figure.

External interrupts are produced by peripherals to indicate that an event has occurred, such as data being ready from a serial port. A RISC-V core normally has two external interrupt pins (`MExtInt` and `SExtInt`) for the two privilege levels. External circuitry arbitrates among interrupts from all the various peripherals to drive these two pins. In a typical system, the PLIC performs this arbitration and drives the interrupt pins for

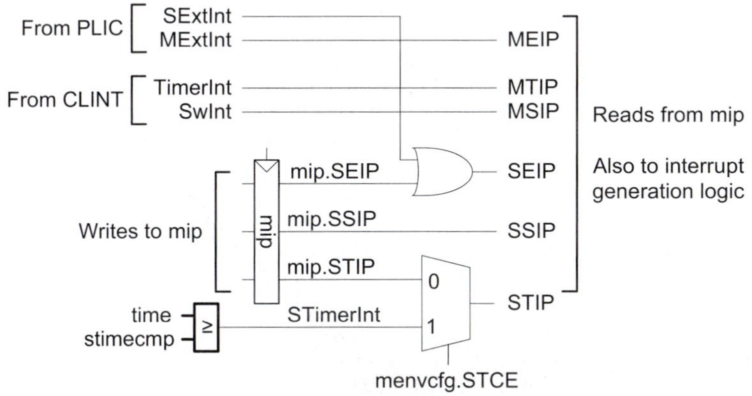

Fig. 8.9 Interrupt sources.

each hart. The trap handler queries the PLIC to determine which peripheral caused the interrupt, then handles the interrupt from that peripheral and writes a peripheral register to clear the interrupt. S-mode external interrupts can also be triggered by M-mode software making `mip.SEIP` = 1, as shown in Fig. 8.9.

Software interrupts are produced by programs to communicate between harts. In a typical system, the CLINT generates an M-mode software interrupt `SwInt` for a hart when another hart writes to a memory-mapped bit in the CLINT. The trap handler then writes to that memory-mapped bit in the CLINT to clear the machine software interrupt. If the interrupt should be handled by the supervisor, the M-mode handler sets `mip.SSIP` to trigger the S-mode interrupt, which is delegated to the supervisor handler, which then clears `sip.SSIP`.

It is cumbersome for the timer interrupt to trap to a machine mode handler that sets STIP that causes a trap to the supervisor mode handler. The RISC-V Sstc extension adds a `stimecmp` register to produce supervisor timer interrupts directly from hardware when `time` $\geq$ `stimecmp`. `menvcfg.STCE` must be set to 1 to enable this capability.

Timer interrupts provide a heartbeat for the OS scheduler to switch between tasks. The CLINT produces a machine timer interrupt `TimerInt` when the timer exceeds some `mtimecmp` compare value. Some timer interrupts are handled directly by the M-mode firmware, and others are delegated to the supervisor by writing `mip.STIP`. The trap handler then writes the CLINT `mtimecmp` register to a value exceeding the current time to clear the interrupt and prepare for the next event needed by the M-mode firmware or S-mode OS.

Fig. 8.10, shows the logic to generate an interrupt, `InterruptM`. A hart takes an interrupt if interrupts are globally enabled and if any interrupt is both pending in `mip` and enabled in `mie`. The global enable depends on whether the interrupt is delegated to S-mode or not. Interrupts are always globally enabled in U-mode. Delegated interrupts are globally enabled in

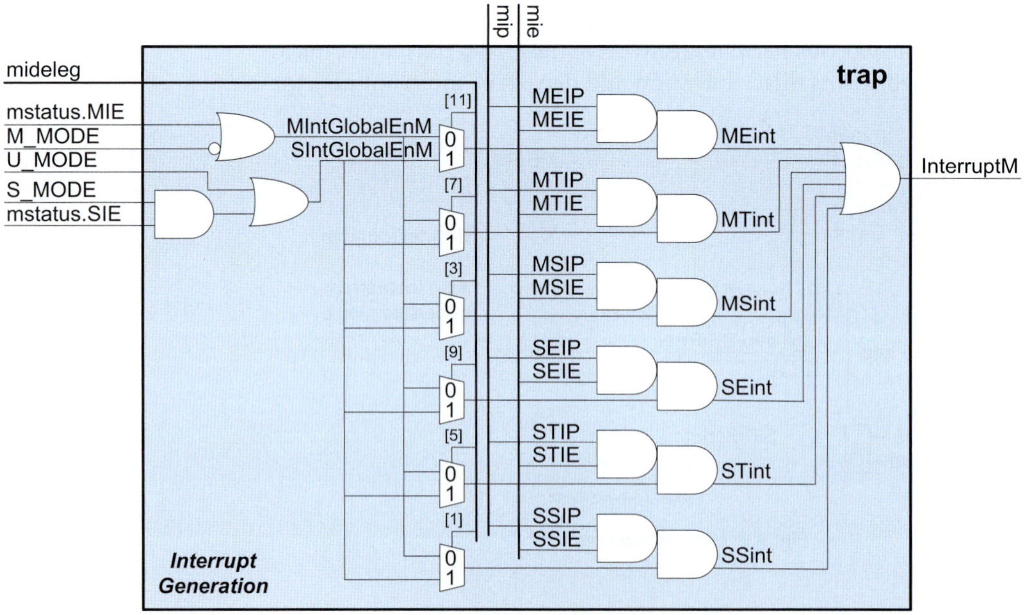

Fig. 8.10 Interrupt generation.

S-mode when `mstatus.SIE` is set and are globally disabled in M-mode. Undelegated interrupts are globally enabled in S-mode and in M-mode when `mstatus.MIE` is set. In other words, undelegated interrupts are enabled when `mstatus.MIE` = 1 (regardless of mode) or in any mode other than M-mode.

8.2.5.5 Interrupt Vectoring

The addresses in `mtvec` or `stvec` are the trap vector (i.e., trap handler address) for M- or S-mode. Recall that the hart jumps to the trap vector when a trap occurs. This trap vector must be a multiple of 4, so the bottom two bits are ignored and treated as 0.

To handle interrupts faster, RISC-V optionally allows the core to jump to a different address depending on the interrupt source. This is called *vectored interrupts*. RISC-V enables interrupt vectoring by making the bottom two bits of the trap vector be 01. Exceptions still jump directly to the trap vector (discarding the least significant bit), but interrupts jump to the trap vector plus four times the trap cause (i.e., the trap cause shifted left by 2 bits). For example, a machine software interrupt has a cause of 3, so it jumps to the trap vector +12. Processors with vectored interrupts typically require that the trap vector is aligned to a multiple of 64 (i.e., its lower 6 bits are 0) so that the vector addition reduces to concatenating the shifted cause with the upper bits of the trap vector address.

8.2.6 Trap Handler

Code Example 8.1 illustrates a sample RV64 M-mode trap handler, which handles two types of traps: timer interrupts and OS calls (`ecall`). It ignores any other traps. If the trap is a timer interrupt, the trap handler adds more time to the timer compare register. If the trap is an `ecall`, it looks for an argument in `a0`. If the argument is 0–3, it changes to that privilege mode (0 = user, 1 = supervisor, 3 = machine) by changing `mstatus.MPP` before issuing `mret`. If the argument is 4, it terminates the program by calling `write_tohost`. For testing purposes, the handler also logs each trap by writing `mcause` and `mtval` to signature memory each time a trap occurs.

Run the trap using:

```
$ cd $WALLY/examples/asm/trap
$ make
$ make sim
```

The RISC-V Advanced Interrupt Architecture (AIA) extension (Smaia/Ssaia) was ratified in June 2023, with additional interrupt features focused on the needs of large, high-performance RISC-V systems. It replaces the PLIC with a more general interrupt controller that supports message-signaled interrupts from PCI Express and similar devices and reduces the interrupt overhead for systems with hypervisors.

There is only room for one instruction at the trap vector, so this instruction is normally a jump to the actual trap handler.

Exceptions are not vectored because they are less sensitive to latency. For example, a timer interrupt must be handled quickly, but an illegal-instruction exception does not.

As introduced in Chapter 5, *signature memory* is memory set aside for the test program to write. After tests are completed, this memory should have a specific *signature*, that is, memory values that show that the tests were completed successfully. The signature depends on the test being run.

```
.section .tohost
tohost:
# Write to HTIF (host target interface) to terminate program
    .dword 0
fromhost:
    .dword 0
.EQU XLEN, 64
begin_signature:
    .fill 6*(XLEN/32),4,0xdeadbeef
end_signature:
# Initialize stack with room for 512 bytes each for program and trap handler
.bss
    .space 512
topofstack:
# And another stack for the trap handler
.bss
    .space 512
topoftrapstack:
```

8.3 Test Plan

The system call (`ecall`) to change privilege mode in Code Example 8.1 is very useful in test code, but it would be dangerous in a real system!

The privileged test plan involves testing exceptions, interrupts, CSR accesses, and privileged instructions.

At the time of this writing in 2025, riscv-arch-test has a small set of privileged tests including misaligned loads, stores, and branches/jumps as well as `ecall` and `ebreak`. Wally has additional privileged tests in `tests/wally-riscv-arch-test`. The memory management tests related to virtual memory, PMP checking, and PMA checking are described in Chapter 11. Traps related to atomic instructions are described in Chapter 17. The other tests are described in this section. The test plan is described in detail in the cvw repository under `docs/testplans`.

Table 8.5 summarizes the privileged instructions and the tests that exercise them. Table 8.6 summarizes the CSRs and the tests that exercise

Table 8.5 Summary of privileged instruction tests.

Instruction	Testing
`ecall`	WALLY-trap-01
`ebreak`	WALLY-trap-01
`mret`	WALLY-trap-01
`sret`	WALLY-trap-s-01
`wfi`	WALLY-wfi-01 and WALLY-status-tw-01
`sfence.vma`	Not explicitly tested yet

Table 8.6 Summary of CSR tests.

CSR	Test(s)
Counters	Not tested
`fflags`	FPU tests in Chapter 16
`frm`	FPU tests in Chapter 16
`fcsr`	FPU tests in Chapter 16
`sstatus`	See Table 8.7
`sie`	WALLY-sie-01
`stvec`	WALLY-stvec-01 and WALLY-trap-s-01
`sscratch`	WALLY-trap-s-01
`sepc`	WALLY-trap-s-01, WALLY-trap-u-01
`scause`	WALLY-trap-s-01, WALLY-trap-u-01
`stval`	WALLY-trap-s-01, WALLY-trap-u-01
`sip`	Not explicitly tested
`satp`	WALLY-mmu-sv32-01, WALLY-mmu-sv39-01, WALLY-mmu-sv48-01
`senvcfg`	Bits tested in the context of features they enable
`mvendorid`	WALLY-minfo-01
`marchid`	WALLY-minfo-01
`mimpid`	WALLY-minfo-01
`mhartid`	WALLY-minfo-01
`mconfigptr`	WALLY-minfo-01
`mstatus`	See Table 8.7
`misa`	WALLY-misa-01
`medeleg`	WALLY-trap-s-01, WALLY-trap-u-01
`mideleg`	WALLY-trap-s-01, WALLY-trap-u-01
`mie`	WALLY-mie-01
`mtvec`	WALLY-mtvec-01, WALLY-trap-01
`mscratch`	WALLY-trap-01
`mepc`	WALLY-trap-01
`mcause`	WALLY-trap-01

(Continued)

Table 8.6 Summary of CSR tests. Cont'd

CSR	Test(s)
mtval	WALLY-trap-01
mip	Not explicitly tested
menvcfg	Bits tested in the context of features they enable
pmpcfg[]	WALLY-pmp-01
pmpaddr[]	WALLY-pmp-01

> The privileged tests are among the more difficult tests to write. Enumerating the cases to test is lengthy, and checking that the tests cover all of these cases requires care. Our reference model of Spike (see Section 3.3.8) lacks the peripherals to generate some of the interrupts, so the tests could not be checked against a reference model, increasing the chance of making the same error misinterpreting the specification in the test case as in the RTL. The privileged specification leaves many choices to the implementation, and these tests are only written to check Wally's implementation, not all valid implementations. Even for Wally, these tests only cover basic privileged functions. We believed these tests were better than anything else we could find at the time they were written in 2022 but expect more comprehensive tests will eventually be available from the RISC-V Architecture Test Special Interest Group.

them. The `mstatus/sstatus` CSRs contain many bitfields; the tests for each bitfield are summarized in Table 8.7.

All privileged tests are in `tests/wally-riscv-arch-test/riscv-test-suite/rv{32,64}i_m/privilege`.

riscv-arch-test intends to support U- and S-mode tests in directories named `rv{32,64}i_{u,s}`. This naming structure is not yet implemented, so the non-M-mode tests are placed in `rv{32,64}i_m/privilege/testname-{u,s}-01`. Configurations that do not support U- or S-mode should omit the associated tests.

The privileged unit had a high number of difficult bugs for a variety of reasons:

- Complex specification with many easily-overlooked corner cases
- Exceptions and CSR accesses disrupt pipeline behavior
- This unit interacts with peripherals, which were also lightly tested

The Wally privileged tests include the `WALLY-TEST-LIB-32/64.h` file. This file includes a generic trap handler that logs a signature for each trap, including `mcause` and optionally `mtval` and certain `mstatus` bits. It also contains a collection of functions and macros to cause traps and change privilege modes. Most tests call these functions directly. The MMU/PMA/PMP tests contain a table of tests, and calls a function called `run_test_loop` that applies each test. The following sections describe how test programs trigger the various traps.

8.3.1 Generating Exceptions

RISC-V defines 14 standard exceptions as listed in Table 8.4. Table 8.8 gives an example of how each exception is triggered. Note that exceptions caused by atomic instructions are covered in Section 17.2.3.1.

Table 8.7 Summary of `mstatus` bitfield tests.

Field	Test
SD	WALLY-status-fp-enabled-01
MBE	WALLY-endianness-01
SBE	WALLY-endianness-01
SXL	WALLY-status-xlen-01
UXL	WALLY-status-xlen-01
TSR	WALLY-trap-sret-01
TW	WALLY-status-tw-01
TVM	WALLY-mmu-svxx-01
MXR	WALLY-mmu-svxx-01
SUM	WALLY-mmu-svxx-01
MPRV	WALLY-mmu-svxx-01
XS	<not supported>
FS	WALLY-status-fp-enabled-01
MPP	WALLY-trap-01
VS	<not supported>
SPP	WALLY-trap-s-01
MPIE	WALLY-trap-01
UBE	WALLY-endianness-01
SPIE	WALLY-trap-s-01
MIE	WALLY-mie-01
SIE	WALLY-sie-01

With the Zicclsm extension (see Section 19.2), misaligned loads and stores to cachable memory are handled by hardware rather than trapping, but misaligned accesses to peripherals still generate an exception. Therefore, these exceptions are tested by accessing the CLINT.

8.3.2 Generating Interrupts

Six interrupts exist, as listed in Table 8.4: timer, software, and external interrupts, each handled in M- or S-mode. Table 8.9 gives examples of how each interrupt can be triggered by a test program. Recall from Section 8.2.5.4 that some interrupts are caused by peripherals and others are caused by writing to the `mip` or `sip` registers.

Table 8.8 Methods of causing exceptions.

Exception	Examples of Exception Generation
Instruction address misaligned	This exception is not tested because privileged tests run with the Compressed extension, which can never generate misaligned instruction addresses.
Instruction access fault	`jr zero // Jump to unimplemented address`
Illegal instruction	`.word 0 // 0x00000000 is an illegal instruction`
Breakpoint	`ebreak`
Load address misaligned	`li t3, 0x2000001 // Peripheral access must be aligned` `lw t4, 0(t3) // Load from a misaligned address`
Load access fault	`lw t1, 0(zero) // Load from unimplemented address`
Store address misaligned	`li t3, 0x2000001 // Peripheral access must be aligned` `sw t4, 0(t3) // Store to a misaligned address`
Store access fault	`sw t4, 0(zero) // Store to unimplemented address`
`ecall` from U-mode	`GOTO_U_MODE // A macro that switches to U-mode` `ecall`
`ecall` from S-mode	`GOTO_S_MODE // A macro that switches to S-mode` `ecall`
`ecall` from M-mode	`GOTO_M_MODE // A macro that switches to M-mode` `ecall`
Instruction page fault	See MMU tests in Section 11.3
Load page fault	See MMU tests in Section 11.3
Store page fault	See MMU tests in Section 11.3

8.4 WALLY IMPLEMENTATION

Fig. 8.12 shows the Wally core with privileged support. Compared to Fig. 7.1, it adds the privileged unit, which includes CSRs and trap logic.

The privileged unit adds architectural state: the CSRs and the 2-bit privilege mode. Privileged instructions operate in the Memory stage and update the architectural state on the rising edge of the clock at the start of the Writeback stage.

8.4.1 Wally Privileged Unit

Fig. 8.13 shows a block diagram of the privileged unit, including major busses connecting components. The major components are the `csr` unit, which holds the CSRs; the `trap` unit that issues traps; the `privmode` unit that tracks the privilege mode; and `privdec`, which decodes privileged instructions. These blocks all operate in the Memory stage, and the `csr` and `privmode` blocks update state at the start of the Writeback stage. The privileged pipeline registers (`privpiperegs`) delay faults that occur in the Decode and Execute stages so that they are processed in the Memory stage instead.

The `privdec` module decodes the instruction `InstrM` and sends appropriate control signals to the other modules.

Table 8.9 Examples of causing interrupts.

Interrupt	Program to Generate an Interrupt

S-mode software	Write a 1 to `sip.SSIP`.
	`li t3, 0x2              # SSIP is bit 1`
	`csrs sip, t3             # Set supervisor software interrupt pending`

M-mode software	Write the `MSIP` bit of the CLINT.
	`la t3, 0x02000000        # MSIP register in CLINT`
	`li t4, 1                 # 1 in the lsb`
	`sw t4, 0(t3)             # Write MSIP bit of the mip register`

S-mode timer	If Sstc is not supported, write a 1 to the `mip.STIP`.
	`li t3, 0x20             # STIP is bit 5`
	`csrs mip, t3             # Set supervisor time interrupt pending`
	If Sstc is supported, set `menvcfg.STCE` and write `stimecmp` (like M-mode timer below)

M-mode timer	Set `mtimecmp` to `mtime` + <some offset> by writing to the CLINT and wait.
	`# The following code works for both RV32 and RV64.`
	`# RV64 alone would be easier using double-word adds and stores`
	`    li t3, 0x30           # Desired offset from the present time`
	`    mv a3, t3             # Stop waiting for interrupt after this many loops`
	`    la t4, 0x02004000     # mtimecmp register in CLINT`
	`    la t5, 0x0200BFF8     # mtime register in CLINT`
	`    lw t2, 0(t5)          # Low word of mtime`
	`    lw t6, 4(t5)          # High word of mtime`
	`    add t3, t2, t3        # Add desired offset to the current time`
	`    bgtu t3, t2, nowrap   # Check new time exceeds current time (no wraparound)`
	`    addi t6, t6, 1        # If wrap, increment most significant word`
	`    sw t6,4(t4)           # Store into most significant word of mtimecmp`
	`nowrap:`
	`    sw t3, 0(t4)          # Store into least significant word of mtimecmp`
	`time_loop:`
	`    addi a3, a3, -1`
	`    bnez a3, time_loop    # Wait for interrupt to occur or loop to time out`

(Continued)

Table 8.9 Examples of causing interrupts. Cont'd

<table>
<tr><th>Interrupt</th><th>Program to Generate an Interrupt</th></tr>
<tr><td>S-mode external</td><td>Use the same code as Machine external interrupts (see below) except set the S priority threshold to 0, set the M priority threshold to 7, and enable source 3 in S-mode at address 0x0C002080.</td></tr>
<tr><td>M-mode external</td><td>External interrupts depend on a peripheral to cause them. Wally presently can receive external interrupts from the GPIO and UART peripherals, routed through the PLIC to the <code>ExtInt</code> pin. This set of tests relies on the GPIO to cause an interrupt.
The first section of this code configures both the PLIC and GPIO to allow for an external interrupt. This involves raising GPIO priority above the M-mode priority threshold, enabling the GPIO, and enabling input pins as interrupters within the GPIO.<pre>
========== Configure PLIC ==========
Assume a3 is set to number of iterations to wait for interrupt
M priority threshold = 0
  la t3, 0xC200000   # PLIC IntThreshold[M-mode]
  li t4, 0
  sw t4, 0(t3)   # PLIC IntThreshold[M-mode] = 0
# S priority threshold = 7
  la t3, 0xC201000   # PLIC IntThreshold[S-mode]
  li t4, 7
  sw t4, 0(t3)   # PLIC IntThreshold[S-mode] = 7
# source 3 (GPIO) priority = 1
  la t3, 0xC000000   # PLIC IntPriority base address
  li t4, 1
  sw t4, 0x0C(t3)   # PLIC IntPriority[3] 1
# enable source 3 interrupts in M -mode
  la t3, 0x0C002000   # PLIC IntEnable[M-mode]
  li t4, 0b1000
  sw t4, 0(t3)   # PLIC IntEnable[M-mode][3] = 1
  la t3, 0x10060000   # GPIO base address
  li t4, 0x1
  sw t4, 0x08(t3)   # GPIO.output_en = 1 enable pin 1 as an output
  sw t4, 0x04(t3)   # GPIO.input_en = 1 also enable pin 1 as an input
 # for loopback test
  sw zero, 0x1C(t3)   # Clear GPIO.rise_ip
  sw zero, 0x24(t3)   # Clear GPIO.fall_ip
  sw zero, 0x2C(t3)   # Clear GPIO.high_ip
  sw zero, 0x34(t3)   # Clear GPIO.low_ip
  sw t4, 0x28(t3)   # GPIO.high_en = 1 enable interrupt on high input
  sw t4, 0x0C(t3)   # GPIO.output_val = 1 raise pin 1 (cause interrupt)
m_ext_loop:
  addi a3, a3, -1
  bnez a3, m_ext_loop   # Wait a3 iterations to see if interrupt fires
</pre></td></tr>
</table>

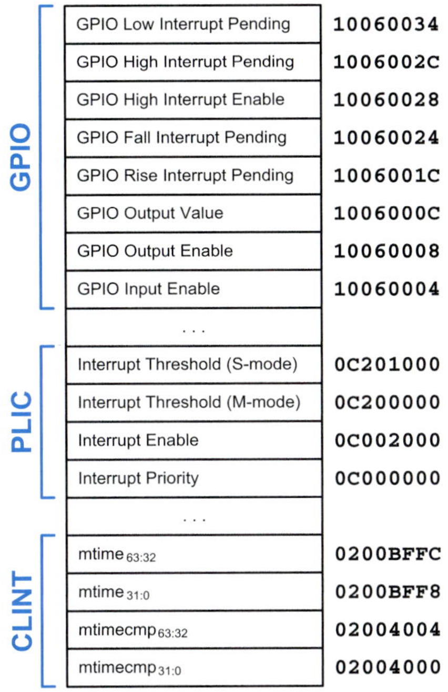

Fig. 8.11 Memory-mapped registers used for interrupt generation.

The CSR unit (`csr`) contains all the CSRs and has the following functions:

- Reads and writes the CSRs over the `CSRReadValW` and `SrcAM` signals, respectively, connecting to the IEU.
- During traps, it does the following:
  - o Sends delegation and interrupt register values (`mideleg`/`medeleg`, and `mip`/`mie`) to the trap unit (`trap`) and reads the trap cause from the trap unit
  - o Sends *x*`status` bits to `trap` and `privmode`, the privileged mode unit (`privmode`), which holds the current privilege mode
  - o May record `InstrM` (in *x*`tval`) and `PCM` (in *x*`epc`) from the instruction fetch unit (IFU) and `IEUAdrM` (in *x*`tval`) from the IEU
  - o Uses the CSRs (*x*`tvect`) to determine the trap destination, `UnalignedPCNextF`
- Sends or receives the remaining signals (discussed in future chapters).
  - o Sends `pmp` and `satp` CSRs to the LSU to control memory management (see Chapter 12)
  - o Sends the `frm` (floating-point rounding mode) CSR to control the FPU, which returns the `SetFFlagsM` signals (see Chapter 16)

The code must change if the peripheral base addresses move or the generally-purpose I/O (GPIO) or platform-level interrupt controller (PLIC) are not available to produce and route external interrupts.

Note that interrupt tests are not simulated in Sail because Sail does not have the same peripherals to generate interrupts. Chapter 20 describes Wally's peripherals.

Recall from the Wally memory map in Fig. 2.8 that the PLIC and CLINT registers are mapped to addresses starting at 0x0C000000 and 0x02000000, respectively. Fig. 8.11 shows the memory-mapped registers used in the Table 8.9 examples. A full list of registers is found in Chapter 20.

Additions to diagrams are highlighted in bright blue with a blue outline. For example, Fig. 8.12 shows the addition of the privileged unit. Modules in gray have not yet been added.

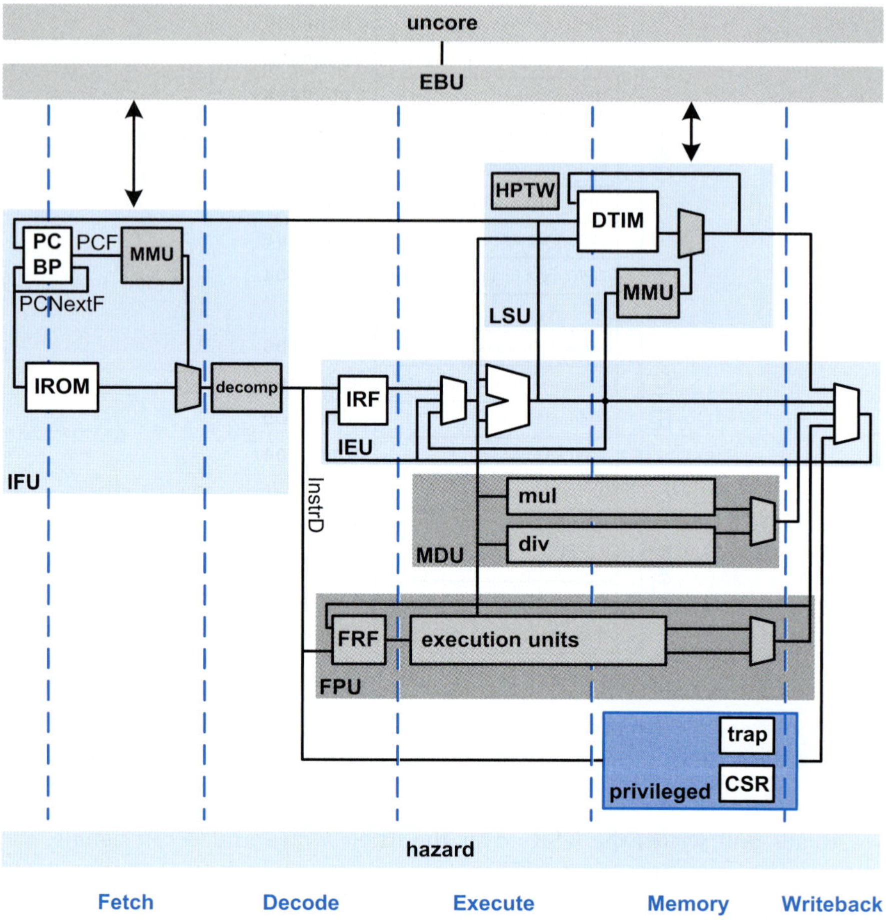

Fig. 8.12 **Wally high-level block diagram, highlighting privileged support.**

- o The CLINT provides the `MTIME_CLINT` signal for the timer (see Section 20.2)

The `trap` module receives exceptions and interrupts; determines what to take, considering the current privilege mode (`PrivilegeModeW`); and responds with `TrapM`, `DelegateM`, and `CauseM`.

The `privmode` module changes the privilege in response to traps and `mret`/`sret` instructions.

Recall that Wally is configurable, and all features beyond the basic instruction set can be turned on or off at compile time using configuration options. Privileged support is enabled by setting the `ZICSR_SUPPORTED` flag to 1 in `config.vh`. When `ZICSR_SUPPORTED` = 0, all the privileged signals and architectural state are tied to 0 and optimized away by logic synthesis.

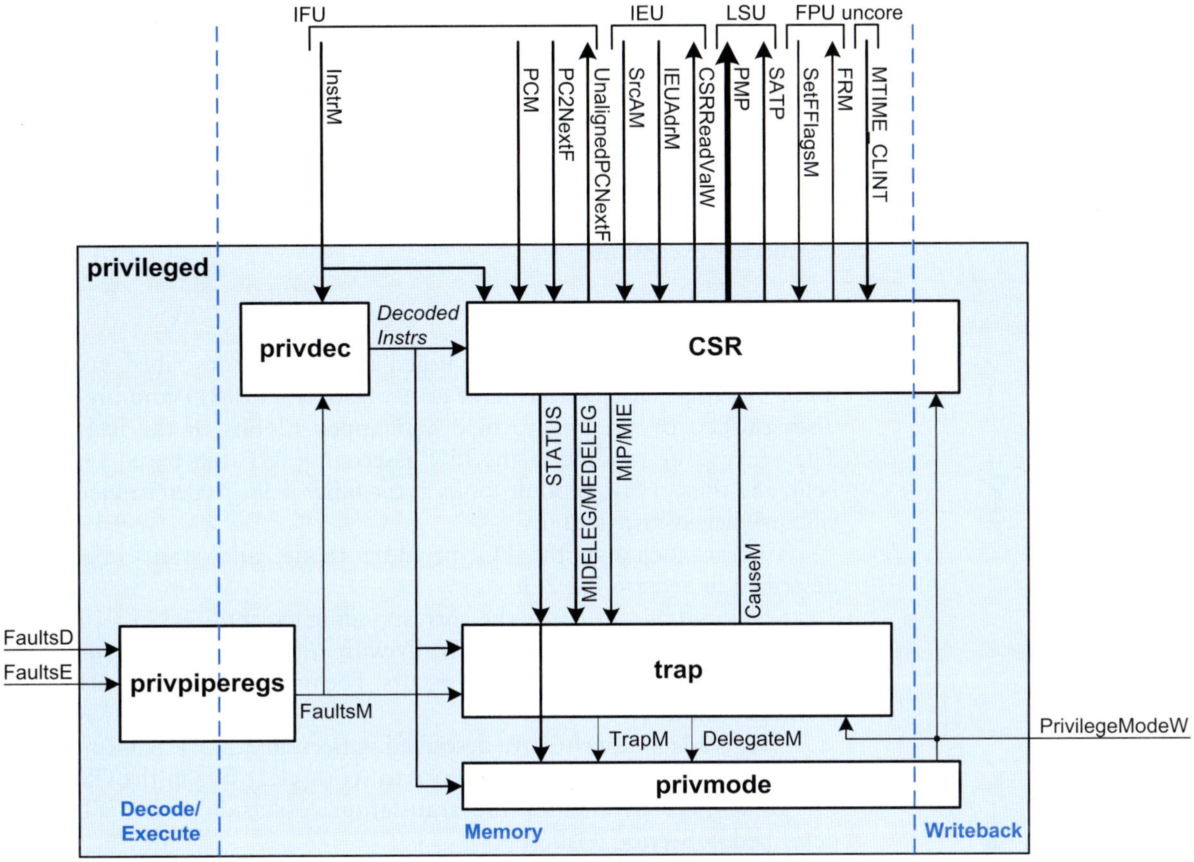

Fig. 8.13 **Privileged unit block diagram.**

Various features within the privileged unit are further enabled with flags, including:

- `S_SUPPORTED` Supervisor mode
- `ZICOUNTERS_SUPPORTED` Performance counters
- `VECTORED_INTERRUPTS_SUPPORTED` Vectored interrupts
- `BIGENDIAN_SUPPOPRTED` Big-endian loads and stores
- `WFI_TIMEOUT_BIT` WFI times out after $2^{\text{WFI\_TIMEOUT\_BIT}}$ cycles

The Sstvecd extension indicates that an implementation also supports direct (unvectored) interrupts with the bottom two bits of the trap vector being 00. RVA20S64 and later profiles expect Sstvecd, and Wally always supports it, independent of `VECTORED_INTERRUPTS_SUPPORTED`.

Several more privileged configurations affect the memory management unit described in Section 11.4.

- `VIRTMEM_SUPPORTED` Virtual memory
- `PMP_ENTRIES` 0, 16, or 64

Privileged instructions and CSR accesses that are either unsupported or not permitted in the current privilege mode cause an illegal-instruction exception.

8.4.2 Privileged Instructions

The privileged unit decodes privileged instructions described in Section 8.2.2. All these instructions have an `op` field of 1110011 and are further distinguished by the `funct3` field and upper 12 bits of the instruction. The `controller` module in the IEU asserts `PrivilegedM` based on the `op` field, and the `privdec` module looks at the other fields to determine the specific privileged instruction.

`mret/sret` change the PC, privilege mode, and status register as described in Section 8.2.2.3.

`ecall` and `ebreak` cause the corresponding exceptions.

`wfi` (wait for interrupt) stalls Wally with `wfi` in the Memory stage until an interrupt is pending or timeout occurs. The timeout counter is also in the privileged unit.

CSR reads and writes are described in Section 8.2.2.1. Sets and clears involve a read-modify-write sequence to set or clear bits in the CSR.

`sfence.vma` invalidates the translation lookaside buffer (TLB) as described in Section 11.2.4.

`privdec` also generates an illegal-instruction exception when the instruction is not defined or supported, or `wfi` times out.

8.4.3 Control and Status Registers

The `csr` module contains the CSRs along with logic to read, write, set, and clear bits in the CSRs. It also contains logic to compute the next `mtval` (see Table 8.4) and PC (Fig. 8.17) during traps; the logic is here rather than in the `trap` module to preserve locality because it depends on *x*`tvec` (`mtvec`, `stvec`) and *x*`epc` (`mepc`, `sepc`). The CSRs are partitioned into submodules:

- `csrsr mstatus / sstatus` Status registers
- `csri mip / sip / mie / sie` Interrupt pending and enable registers
- `csrc` Performance counters
- `csrm` Other M-mode CSRs
- `csrs` Other S-mode CSRs
- `csru` U-mode CSRs (floating-point)

Each CSR submodule contains a register for each CSR, address decoders, write enables, and read multiplexers. If none of the submodules detect a legal access, the `csr` module causes an illegal-instruction exception.

8.4.3.1 Performance Counters

When `ZICSR_SUPPORTED` and `ZICNTR_SUPPORTED` are both 1, Wally provides 64-bit performance counters to count events. The number of counters is specified by the `COUNTERS` configuration variable, to a maximum of 32. The active counters are defined in Table 8.10; any others are hardwired to 0.

`testbench.sv` prints the performance counters at the end of a test if the `PrintHPMCounters` parameter is 1. Benchmarks in Chapter 21 parse the log file and extract performance numbers.

8.4.3.2 Machine Information Registers

Wally is assigned the OpenHW Foundation `mvendorid` of 0x602, the RISC-V International `marchid` of 36, and an `mimpid` of 0x100 (for Wally version 1.0.0). The single hart has `mhartid` = 0 and `mconfigptr` = 0.

8.4.3.3 `misa` and XLEN

Wally treats `misa` as a read-only register, so features cannot be disabled on the fly. In particular, `misa.XLEN` is hardwired, so an RV64 machine cannot run RV32I code. Similarly, `mstatus.SXL` and `mstatus.MXL` are hardwired to `misa.XLEN`.

The Ssu64xl extension indicates that 64-bit U-mode is available. Wally inherently supports this extension on any RV64 configuration.

8.4.4 Traps

The `trap` module manages pending exceptions and interrupts, global and local interrupt enables, and delegation registers. Key outputs include:

- `TrapM` Trap should occur
- `CauseM` Cause of highest-priority trap
- `DelegateM` Trap should be delegated to S-mode

Fig. 8.14 shows a schematic of the trap unit. `mie`, `mip`, and `mideleg` are 12-bit inputs from the CSRs. Based on the privilege mode and global interrupt enable signals, the unit determines whether M-mode and S-mode interrupts are enabled. Considering whether a particular interrupt is delegated to S-mode, the trap unit determines which interrupt types are enabled. Precise interrupts are always associated with a specific PC so that the handler can return properly. Therefore, when the instruction in the Memory stage is not valid, interrupts are disabled until the next valid instruction arrives (`InstrValidM` = 1). Meanwhile, the trap unit determines which interrupts are pending (`PendingIntsM`) based on the `mie` and `mip` registers. The interrupts that are both pending and enabled are valid

Table 8.10 Performance counters.

CSR	Count	Enable	Section with Description
B00	Cycles elapsed	<always>	—
B01	Unused		—
B02	Instructions retired	`InstrValidM`	8.2.3.4
B03	Total branches	`BranchM`	13.3.3.3
B04	Total jumps	`JumpM & ~ReturnM`	13.3.3.3
B05	Total returns	`ReturnM`	13.3.3.3
B06	Branch predictor fetched wrong instruction	`BPWrongM`	13.1.4, 13.3.3
B07	Branch prediction wrong direction	`BPDirPredWrongM`	13.1.2, 13.3.3.1
B08	Branch/jump prediction wrong target address	`BTAWrongM`	13.3.3.2
B09	Return address stack wrong address	`RASPredPCWrongM`	13.3.3.4
B0A	Instruction class predictor wrong	`IClassWrongM`	13.3.3.3
B0B	Load stalls	`LoadStallM`	7.2.2.3
B0C	<Deprecated>		
B0D	Data cache accesses	`DCacheAccess`	10.4.3
B0E	Data cache misses	`DCacheMiss`	10.4.3
B0F	Data cache miss cycles	`DCacheStallM`	10.4.1
B10	Instruction cache accesses	`ICacheAccess`	10.4.3
B11	Instruction cache misses	`ICacheMiss`	10.4.3
B12	Instruction cache miss cycles	`ICacheStallF`	13.3.5
B13	CSR writes executed	`CSRWriteM`	8.4.5.4
B14	`fence.I` executed	`InvalidateICacheM`	13.3.1.1
B15	`sfence.VMA` executed	`sfencevmaM`	11.4.4
B16	Number of interrupts taken	`InterruptM`	8.4.4
B17	Number of exceptions taken	`ExceptionM`	8.4.4
B18	Division stall cycles (Integer & FP)	`DivBusyE | FDivBusyE`	15.4.2 & 16.6.4

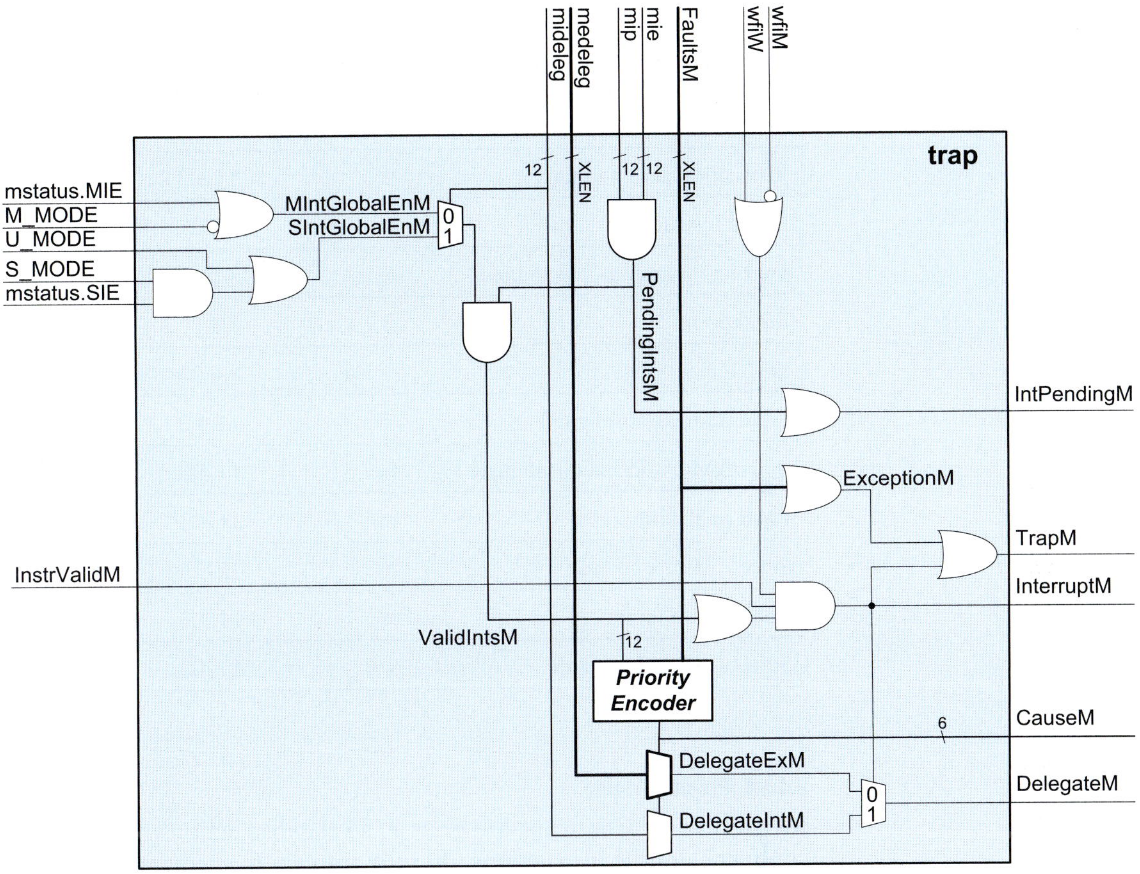

Fig. 8.14 **Trap unit.**

(`ValidIntsM`). A priority encoder considers all the valid traps to determine a `CauseM` code for the highest priority trap. If any trap occurs, `TrapM` is asserted. Depending on the cause and type of the trap, the unit examines the corresponding bit of the `mideleg` or `medeleg` register to assert `DelegateM` if the trap should be delegated to S-mode. `IntPendingM` is also raised when any interrupt is pending (to end a `wfi`), even if it is not enabled.

Exceptions are detected according to Table 8.11.

Fence instructions, described in Sections 8.4.5.5 and 17.2.2, ensure an ordering of memory accesses.

8.4.5 Pipeline Hazards

The privileged unit flushes the pipeline on traps, CSR writes, *xret* (`mret`, `sret`), and fences to ensure subsequent instructions see the correct state. It also stalls the pipeline during CSR reads and `wfi`. Fig. 8.15 shows the hazard unit. Compared to the baseline unit in Fig. 7.19, new signals are highlighted in blue. Each new flush and stall cause is discussed in the following subsections.

The hazard unit will continue to be expanded in future chapters to support new features.

Table 8.11 Exception sources.

Exception	Source	Section with Description
Instruction page fault	`mmu`	11.4.3
Instruction access fault	`mmu`	11.4.3
Illegal instruction	`privdec`	8.4.2
Instruction address misaligned	`ifu`	8.2.5.1
Breakpoint	`privdec`	8.4.2
`ecall`	`privdec`	8.4.2
Load address misaligned	`mmu`	11.4.3
Store/AMO address misaligned	`mmu`	11.4.3
Load page fault	`mmu`	11.4.3
Store/AMO page fault	`mmu`	11.4.3
Load access fault	`mmu`	11.4.3
Store/AMO access fault	`mmu`	11.4.3

8.4.5.1 Precise Traps

Recall that RISC-V traps are precise, so the instructions before the trap complete and the instructions after the trap are flushed and do not affect architectural state. The simplest way to ensure consistency in the architectural state is to update all state at the same time. The register file is written in the Writeback stage, so Wally updates all other state at this time. In particular, instructions changing the privilege mode or CSRs write this state on the rising edge of the clock at the start of the Writeback stage.

Exceptions may occur as late as the Memory stage; load/store faults are not known until this time. Again, it is simplest to have all traps take place at the same place in the pipeline and as late as possible before architectural state is updated, meaning traps occur in the Memory stage and change state at the rising edge of the clock at the start of the Writeback stage. When a trap occurs, the instruction in the Memory stage and all subsequent instructions in earlier stages are flushed. In Fig. 8.15, `TrapM` causes the full pipeline flush. On the rising edge of the clock at the start of the Writeback stage, CSRs are updated (including `epc`, `tval`, `cause`, and `status`), and the privilege mode and PC are changed to take the trap.

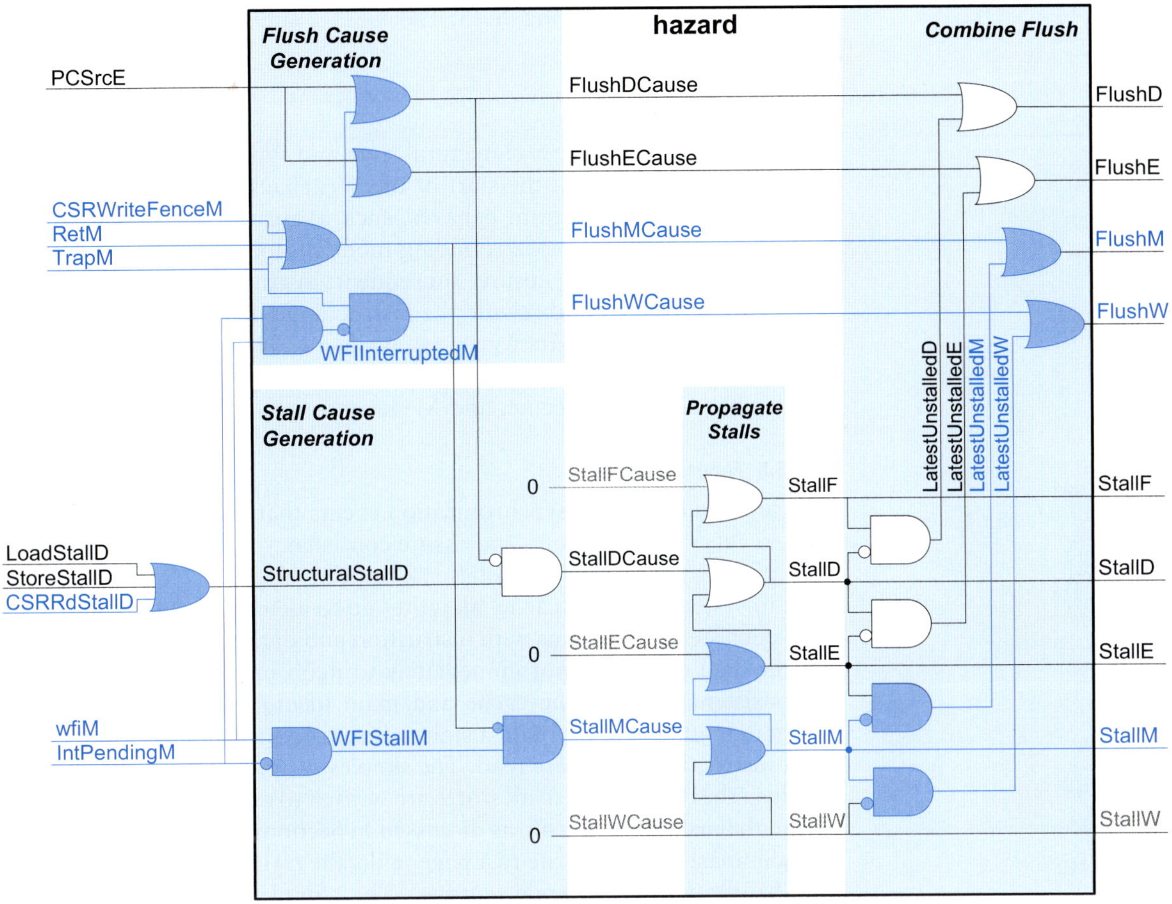

Fig. 8.15 Hazard unit, with privileged changes highlighted in blue.

8.4.5.2 `mret` and `sret`

Like traps, `mret` and `sret` change the PC and privilege mode on the rising edge of the clock at the start of the Writeback stage. Hence, they must flush all subsequent instructions in the pipeline. In Fig. 8.15, `RetM` causes a flush of all but the Writeback stage.

8.4.5.3 CSR Reads

CSR reads occur in the Memory stage so they obtain the latest CSR values, including any changes that may have been triggered by the previous instruction, such as CSR writes or floating-point operations that change flags. Therefore, like load and store structural hazards, CSR reads have a 2-cycle latency from issue to use, so the pipeline must stall in the Decode stage when the next instruction source register depends on the CSR read destination register. In Fig. 8.15, `CSRRdStallD` stalls the Decode stage for a CSR read hazard.

8.4.5.4 CSR Writes

CSR writes also change architectural state and thus must occur on the rising edge of the clock at the start of the Writeback stage. Writes to CSRs change the way a program behaves, such as turning on virtual memory translation or changing the rounding mode of floating-point instructions. The simplest way to ensure subsequent instructions reflect the new behavior is to flush and reissue the instructions after a CSR write.

The controller asserts `CSRWriteM` while a CSR write is in the Memory stage and in turn raises `CSRWriteFenceM`. In response, the hazard unit flushes the Decode, Execute, and Memory stages, as shown in Fig. 8.15.

8.4.5.5 Fences

Fence instructions are synchronization events that ensure memory consistency. `fence.i` and `sfence.vma` ensure consistency of the instruction memory and virtual memory translations, respectively.

In systems with no cache, all writes go to memory, which is inherently self-consistent. In systems with instruction and data caches, a store may go to the data cache but not immediately to main memory, so data may be inconsistent between the cache and main memory. An instruction read from that address will obtain a stale value. `fence.i` makes sure that consistent instruction values are read. The simplest way to implement `fence.i` is to flush the data cache (so all stores are written to main memory) and invalidate the instruction cache, as discussed in Sections 10.4.2.6 and 10.4.5.4.

Similarly, after a write to a page table, the translation lookaside buffer (TLB) may contain stale entries. The simplest way to implement `sfence.vma` is to invalidate the entire TLB as discussed in Section 11.2.4.

Like CSR writes, fences assert `CSRWriteFenceM` to flush and refetch all instructions following the fence.

> `fence.i` is defined in the Zifencei extension (configured when `ZIFENCEI_SUPPORTED` = 1 in `config.vh`). `fence.i` (or `cbo.flush` in Zicbom, see Section 19.1.1.1) is required for coherence in a system with a cache. However, it can be treated as a `nop` in cacheless systems.
>
> `sfence.vma` is part of the privileged specification and is supported when virtual memory is enabled (configured when `VIRTMEM_SUPPORTED` = 1 in `config.vh`).

8.4.5.6 `wfi`

The wait for interrupt module (`wfi`) waits until an interrupt is pending or until it times out and causes an illegal-instruction exception. `WFIStallM` in the hazard unit (Fig. 8.15) is 1 while the `wfi` is active and no interrupts are pending. In Fig. 8.15, `WFIStallM` drives `StallM` to keep `wfi` in the Memory stage until an interrupt is pending or a timeout raises `TrapM`. When an interrupt becomes pending, `WFIStallM` deasserts and `WFIInterruptedM` asserts, to keep the `wfi` instruction from flushing as it advances to the Writeback stage and commits. `InterruptM` and `TrapM` are asserted, flushing the instructions after `wfi` and jumping to the trap handler. `WFIStallM` would preferably gate clocks to the unused parts of the core, but Wally does not yet implement clock gating.

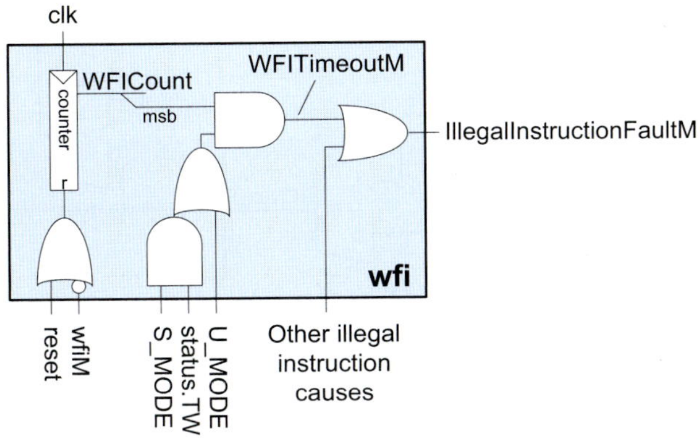

Fig. 8.16 wfi **timeout logic.**

Fig. 8.16 shows the wfi timeout logic in the privdec module for a system supporting S-mode. When wfi reaches the Memory stage, the hazard unit asserts StallM to stall the pipeline until an interrupt is pending or a timeout occurs. The WFICounter begins counting from zero. When the most significant bit is set and the system is in either U- or S-mode with status.TW set, it asserts WFITimeoutM, which causes an illegal-instruction exception. As a result, the processor flushes wfi from the pipeline and enters the trap handler.

In a system that does not support S-mode, the timeout only occurs in U-mode when status.TW is set.

The WFI_TIMEOUT_BIT configuration parameter determines the number of cycles to wait before timing out. It has a default value of 16, meaning wfi times out after 2^{16} cycles.

8.4.6 Instruction Fetch Unit (IFU) and Privileged Changes

Fig. 8.17 enhances the PC logic to support privileged operations: traps, mret/sret, and refetching instructions after a pipeline flush. The privileged changes in the privileged module are highlighted in blue. The multiplexer in gray is instantiated only when vectored interrupts are supported.

In the original Wally pipelined processor (Fig. 7.11), the PCSrcE mux selected PCNextF from either PCPlus4F or IEUAdrE (the branch or jump target address). In Fig. 8.17, the output of the PCSrcE mux is now called PC1NextF. If fences or CSRs are supported, the CSRWriteFenceM signal flushes the pipeline as described in Sections 8.4.5.4 and 8.4.5.6 and reloads the program counter with the instruction after the CSR write or fence. The flush is triggered in the Memory stage, so the next instruction is in the Execute stage and its address is in PCE. Thus, the PCWriteFenceM multiplexer chooses PC2NextF from either PC1NextF or PCE.

The next multiplexer handles traps and trap returns. During a trap, the PC is loaded from the TrapVectorM, which is either mtvec

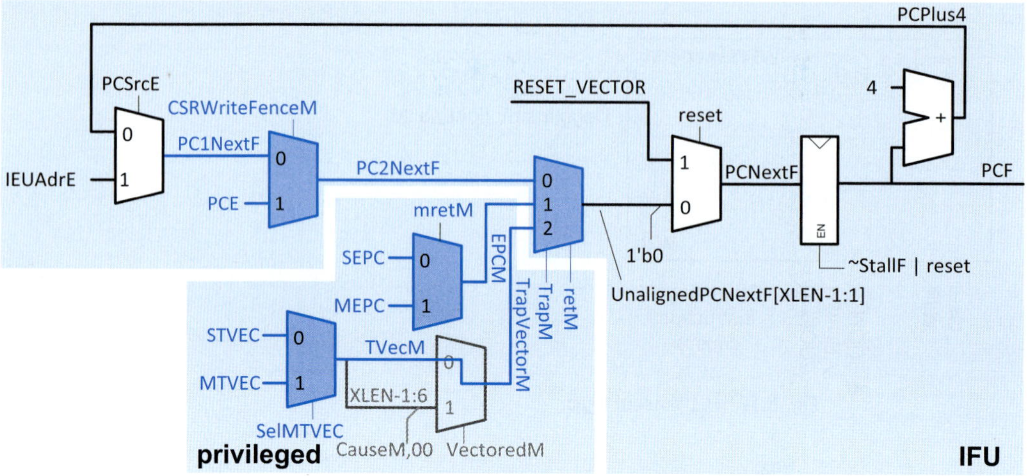

Fig. 8.17 **PC logic, with privileged changes highlighted in blue.**

or `stvec` depending on the privilege mode after the trap. If `VECTORED_INTERRUPTS_SUPPORTED` = 1 (in `config.vh`), interrupts add an offset based on the interrupt cause, as shown by the gray mux. When vectored interrupts are supported, `mtvec` and `stvec` must be aligned to a multiple of 64 so that adding the cause offset is a trivial concatenation. During an `mret` or `sret`, the PC is loaded from `mepc` or `sepc`, respectively. The least significant bit (lsb) of the PC is always zero, so any nonzero lsb of the trap vector or EPC is discarded to form `PCNextF`.

8.4.7 Endian Swapper

Wally processes all data internally as little-endian. When the RISC-V data endianness modes are supported in configuration (`BIGENDIAN_SUPPORTED` = 1), Wally adds endian swappers to reverse big-endian data being loaded and stored if the `mstatus` register enables big-endian.

Example 8.1 BIG- AND LITTLE-ENDIAN MEMORY

Consider the following RV64 program. What are the memory and register contents upon completion in little-endian mode? In big-endian?

```
li  x1, 0xFEDCBA98_76543210
sd  x1, 1000(x0)
ld  x2, 1000(x0)
lwu x3, 1000(x0)
lwu x4, 1004(x0)
lbu x5, 1001(x0)
```

Solution: The memory contents for each mode are shown in Fig. 8.18. In little-endian, the least significant byte of `x1` is stored in the lowest address. In big-endian, the least significant byte is stored in the highest address. The contents of the registers are shown in Table 8.12.

Little-Endian Memory Contents	Byte Address	Big-Endian Memory Contents
FE	1007	10
DC	1006	32
BA	1005	54
98	1004	76
76	1003	98
54	1002	BA
32	1001	DC
10	1000	FE

Fig. 8.18 Memory contents for endian example.

Table 8.12 Register contents for endian example.

Load	Register	Little Endian	Big Endian
`ld x2, 1000(x0)`	`x2`	`FEDCBA98_76543210`	`FEDCBA98_76543210`
`lwu x3, 1000(x0)`	`x3`	`00000000_76543210`	`00000000_FEDCBA98`
`lwu x4, 1004(x0)`	`x4`	`00000000_FEDCBA98`	`00000000_76543210`
`lbu x5, 1001(x0)`	`x5`	`00000000_00000032`	`00000000_000000DC`

Observe that a full-length write in either endianness followed by a read in the same endianness returns the original value. However, subword loads return different portions of the full word depending on the endianness. A program can use this behavior to deduce the endianness of the machine.

Fig. 8.19 shows the RV64 LSU with the optional `endianswap` modules in gray. It is similar to the LSU from Fig. 7.9 except that it swaps endianness before writing to memory and after reading from memory. Big-endian subword reads must also swap the order of the subword by inverting the address. The `BigEndianM` signal is asserted when the `mstatus.MBE`, `SBE`, or `UBE` bit is set corresponding to the current privilege mode. The `endianswap` module exchanges byte 0 with byte 7, 1 with 6, and so forth. RV32 is the same except that `endianswap` only exchanges byte 0 with byte 3 and byte 1 with byte 2 in a word. The endian logic adds a multiplexer delay to the Memory stage read and write paths, which might impact cycle time.

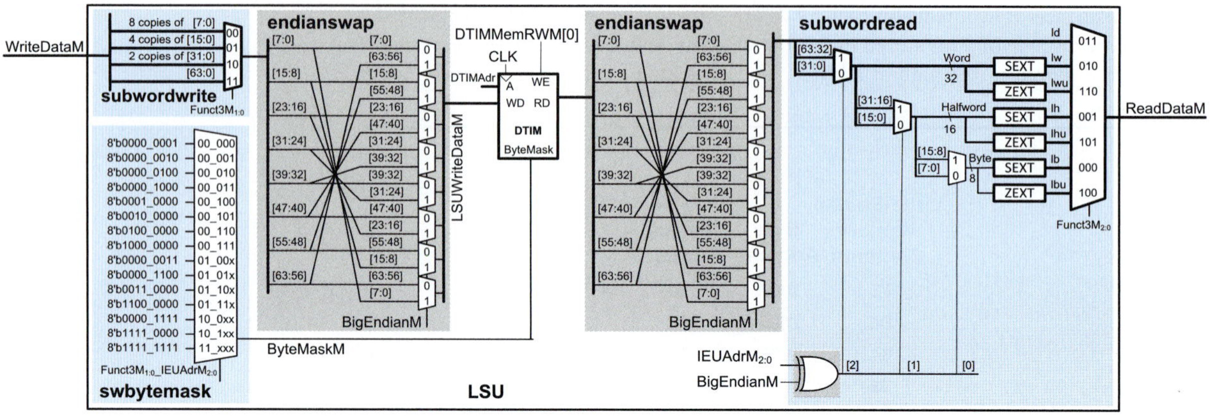

Fig. 8.19 RV64 LSU with **endianswap** modules.

8.5 SUMMARY

A robust microprocessor needs to protect itself from crashes or compromise caused by buggy or malicious programs. Bad instructions cause exceptions, such as illegal-instruction, access-fault, address-misaligned, or page-fault exceptions. System calls (`ecall`) and debugger breakpoints (`ebreak`) are also exceptions. The microprocessor also receives interrupts from asychronous events, such as timers, interprocessor communication, or external peripherals. Exceptions and interrupts are collectively called *traps*. Traps leave the current program and jump to a trap handler. Processors add multiple privilege modes for robustness and security. Programs run in user mode, in which mode they cannot interfere with each other. The trap handler is part of the firmware that runs in machine mode, in which it has full power to allocate machine resources and kill faulty programs. Application processors add an intermediate supervisor mode, to run an OS such as Linux without giving the entire OS kernel full M-mode powers. Each privilege mode is associated with a set of CSRs to configure traps, enable features, or monitor performance counters.

The `mret` and `sret` instructions return from M- and S-mode trap handlers. `csrrw`, `csrrs`, and `csrrc` instructions read and write, set, or clear CSRs, respectively. The `wfi` instruction can be used to save power by placing the processor in a low-power state until an interrupt occurs.

In a system without S-mode, traps always elevate the privilege to M-mode. In a system with S-mode, traps default to M-mode, but traps can be delegated to the S-mode OS via the `medeleg` and `mideleg` registers. A set of CSRs communicates information about the trap. Let x = `m` (machine) or `s` (supervisor), the privilege mode of the trap handler. When a trap occurs, the core enters x mode. It saves the old program counter in `xepc`, disables interrupts, places a code in `xcause` that indicates the reason for the trap, places additional information such as the bad address or instruction in `xtval`, and jumps to a trap handler at `xtvec`. The trap handler typically swaps the stack pointer (`sp`) with an exception stack pointer in the `xscratch` register, then saves the integer registers onto the exception stack so that it can use the registers without corrupting the program that was running. It then reads `xcause` to determine the cause of the trap and respond accordingly. For example, it may perform a system call, respond to the timer or external interrupt, or kill a program that made a bad memory access. The trap handler then restores the registers and stack pointer. Finally, it performs the `xret` instruction, which reenables interrupts, restores the old privilege level, and returns to the old program counter stored in `xepc`.

CSRs with names beginning with `m` can only be accessed from M-mode, so lower privilege programs cannot seize control of the processor. The `mstatus` register is particularly important; it contains bitfields to globally enable interrupts, specify the privilege mode to which the trap handler should return, configure endianness, and so forth. Other machine CSRs contain codes defining the vendor, architecture, implementation, hart IDs, and information about the supported extensions. `mcycle` and `minstret` are standard performance monitor counters that track the number of cycles and instructions that have elapsed. The platform-dependent `mhpmcounter` hardware performance monitor counters keep track of the number of times other events occur, such as cache misses or branch mispredicts.

Similarly, CSRs starting with `s` can only be accessed from S-mode or above. The `satp` supervisor address translation and protection CSR controls virtual memory, which is described in depth in Chapter 11.

Wally adds a privileged unit to track the privilege level, store the CSRs, and handle traps and privileged instructions. Traps, `xret` instructions, CSR writes, and fences flush the pipeline to discard partially completed instructions and ensure subsequent instructions see the updated privilege and CSRs.

If big-endian mode is supported, the LSU adds endianswap muxes to reverse the bytes written and read from the data memory. Instructions are always stored in little-endian order.

Wally defines an extensive set of wally-riscv-arch-test programs to test privilege modes, traps, and CSRs.

9

Bus Interface

9.1 Principles
9.2 AMBA Buses
9.3 Other SoC Buses
9.4 Test Plan
9.5 Wally Implementation
9.6 Summary

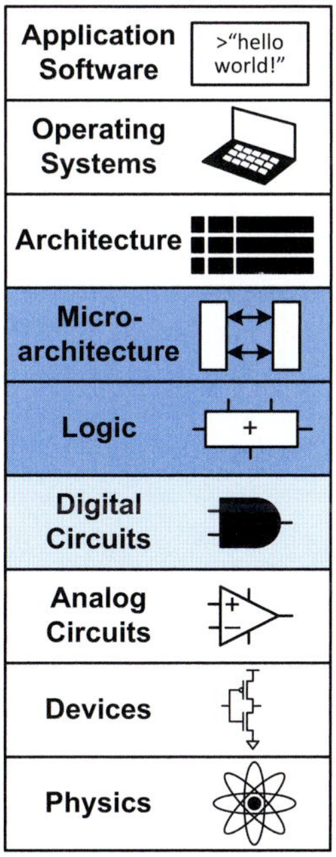

A processor's capabilities are extended by interfacing it with external memory and peripherals such as a keyboard, mouse, or display. Digital systems use *bus interfaces* to connect components with standardized communication protocols. For example, processor cores use one or more standard SoC buses to communicate with memory and peripherals. Using a standard SoC bus makes it practical to attach off-the-shelf peripherals designed for that bus.

This chapter begins by exploring the principles driving SoC bus design. It then surveys a variety of standard SoC buses, including AHB (Advanced High-performance Bus) and APB (Advanced Peripheral Bus), which are used in Wally, as well as AXI (Advanced eXtensible Interface), TileLink, and others.

The chapter then extends the Wally processor by adding an external bus unit (EBU) that controls an AHB bus. Whereas the Wally processor from Chapter 7 used idealized instruction and data memories (IROM [instruction read-only memory] and DTIM [data tightly integrated memory]) located inside the core, the EBU added in this chapter provides access to RAM (random access memory) and ROM (read-only memory) located outside the processor in the uncore portion of the SoC, as highlighted in bright blue at the top of Fig. 9.1.

9.1 PRINCIPLES

A bus connects one or more *controllers* to one or more *peripherals* or *responders*. The peripheral might be a memory, such as SRAM (static RAM), DRAM (dynamic RAM), or ROM, or an I/O (input/output) device such as a timer, interrupt controller, general-purpose input/output (GPIO) device, serial port, or data converter. In most bus protocols, the controller sends an *address* to specify which peripheral and which memory location or register

Chapter 20 describes off-chip peripheral interfaces using SPI and UARTs. High bandwidth off-chip interfaces such as DDR, PCIe, USB, and Ethernet are beyond the scope of this book.

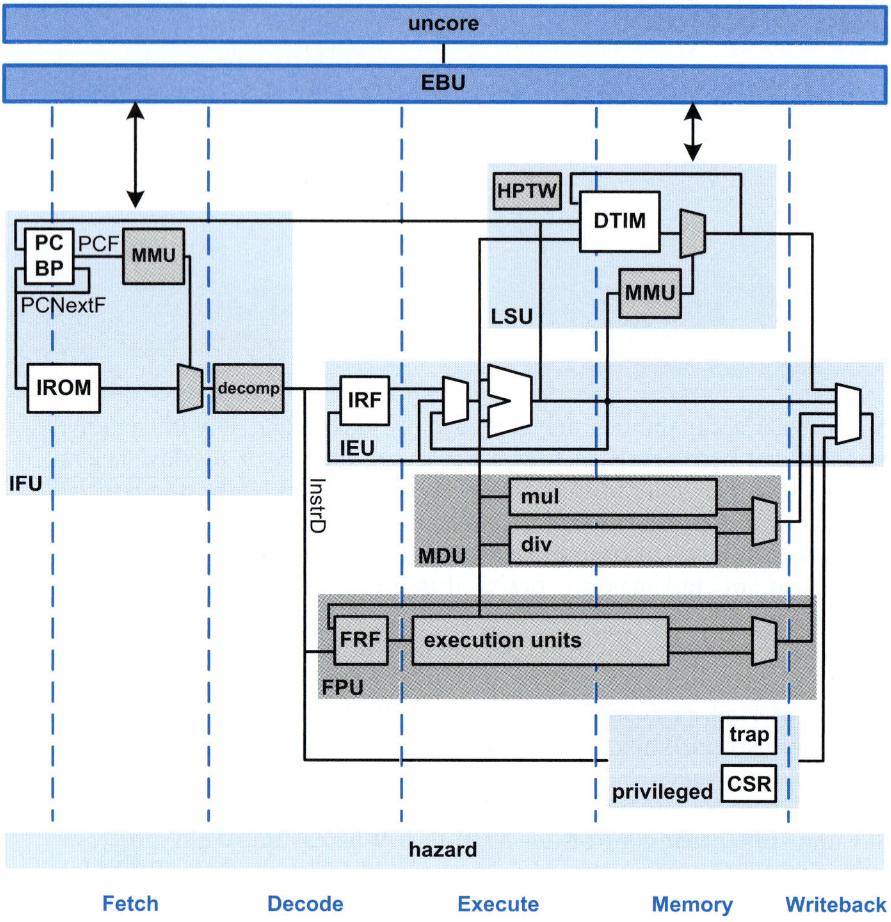

Fig. 9.1 **RV{32/64}I high-level block diagram highlighting the EBU and uncore.**

The *uncore* part of an SoC is, as expected, everything except the processor core. It includes the bus interfaces, memory, and peripherals.

within that peripheral to access. The bus also transmits *data* that the controller reads from or writes to the peripheral.

Fig. 9.2 builds up a sequence of progressively more practical buses, showing the interface signals and timing. The simplest example of a bus is an *asynchronous read-only bus*, like the instruction memory interface from Fig. 7.2. As shown in Fig. 9.2(a), the controller sends an *address* (A), and a peripheral responds with *read data* (RD) within a bounded amount of time. However, most memories and peripherals are *synchronous* for ease of implementation: they need the address to set up before a clock edge, and they return the read data after that clock edge, as shown in Fig. 9.2(b). Peripherals other than ROMs and sensors need a read/write interface. They need to know what to write (*write data*, WD), where to read/write (*address*, A), whether to write (*write enable*, WE), and when to read/

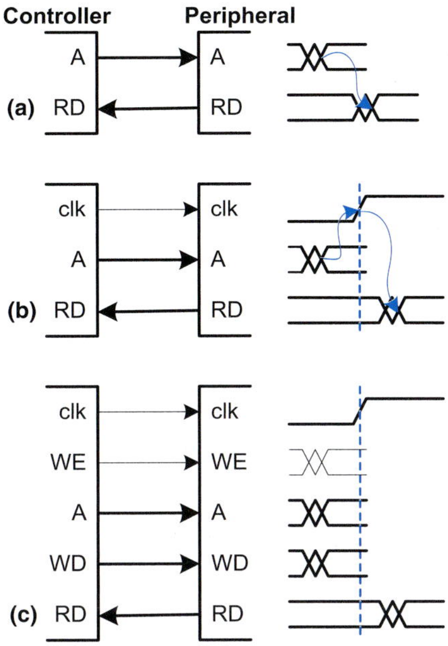

Fig. 9.2 Basic bus interfaces: (a) asynchronous read-only, (b) synchronous read-only, and (c) synchronous read/write

write (`clk`). Hence, in the *synchronous read/write* bus shown in Fig. 9.2(c), the controller sends the address (`A`), write data (`WD`), and write enable (`WE`) before the clock edge. The peripheral then returns read data (`RD`) after the clock edge. When `WE` is asserted, the controller also writes to the peripheral just after the clock edge.

9.1.1 Bidirectional vs. Unidirectional Buses

The bus in Fig. 9.2(c) has separate unidirectional `RD` and `WD` (read and write data) ports. Pins on a chip and wires on a circuit board are relatively expensive, so designers were historically motivated to combine these two ports into a single bidirectional port, as shown in Fig. 9.3. A tristate buffer, with alternating enable input polarity, is added on both the controller (left) and peripheral (right) sides. When `WE` is 1, the controller writes the data from signal `D` to the peripheral; otherwise, the controller reads data from the peripheral on signal `D`.

In an SoC, far more wires are available. Moreover, long on-chip wires need repeaters (buffers) to achieve reasonable speed, making bidirectional operation impractical. Hence, SoC buses universally favor unidirectional

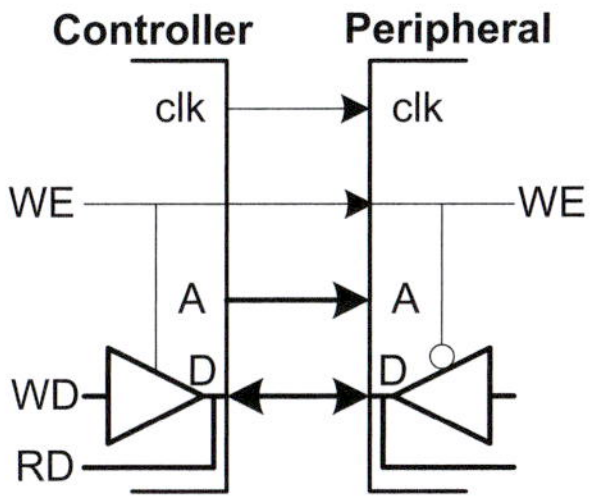

Fig. 9.3 Synchronous read/write bus with a bidirectional data port.

Buses are also referred to as links.

Buses formerly used the nomenclature *master* and *slave*, and you will still see these terms in standards documents. These terms carry dehumanizing and derogatory connotations, so the industry is moving toward deprecating this language [Ellis21, OSHA22]. There is not yet agreement on new terminology. In 2021, ARM began using the terms *manager* and *subordinate* or *requester* and *completer*. Other applications, such as SCSI buses, use *initiator* and *target* for these two roles. In the absence of a new standard, this book uses the terms *controller* and *peripheral* (or *device*).

RISC-V generalizes the address decoder as a *physical memory attributes* (*PMA*) unit that also determines whether the address is cacheable, what transfer widths are allowed, and so on, as described in Section 11.2.2.1.

buses. Tristate buffers driving on-chip bidirectional buses are typically forbidden because of their poor performance and tool support.

Unidirectional point-to-point off-chip buses also can run at a far higher speed than bidirectional multidrop buses, which have multiple peripherals that connect to the bus at different points along the bus wires. It is now cheaper to achieve a required bandwidth by running a moderate number of wires that transmit data serially at high speed in each of several unidirectional links than to run a larger number of wires at lower speed in a single bidirectional link. Section 9.1.6 discusses off-chip buses further.

9.1.2 Handshaking

A controller only needs to access the bus some of the time. To save power, it can provide a *request* (REQ) line that is only asserted when it puts a valid address on the bus.

Some peripherals take more than one bus cycle to respond. For example, DRAM is fast for sequential reads but slow for reads to random addresses. A bus often provides an *acknowledge* (ACK) or *ready* (RDY) signal back from the peripheral indicating that the transfer is complete. This request/acknowledge sequence is known as *handshaking*. Fig. 9.4 shows the timing for a peripheral with a 2-cycle read access time. Observe how RDY remains low for a cycle before rising on the second cycle when the read data (RD) is available.

9.1.3 Multiple Peripherals

In a practical system, the controller communicates with multiple peripherals. SoCs typically use *memory-mapped I/O* to select a particular peripheral based on the address range. The SoC contains an *address decoder* to produce select signals to choose which peripheral or memory to read based on the address. The SoC *memory map* defines the addresses of each peripheral.

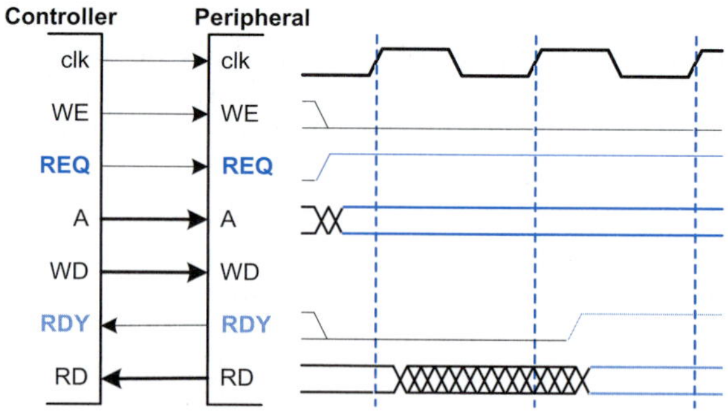

Fig. 9.4 Handshake timing for multicycle access.

Fig. 9.5 shows buses that link multiple peripherals with *shared* and *point-to-point* data ports. Both versions require an address decoder. The shared bus allows multiple peripherals to drive the RD and RDY lines using tristates, as was shown in Fig. 9.3. In contrast, the point-to-point design requires a read multiplexer to choose the read data (RD) and ready (RDY) signals from the selected peripheral. Shared buses minimize the number of wires, but point-to-point buses are much faster and eliminate potential contention among peripherals. Shared buses have many problems in SoCs, including accurately modeling delay, verifying contention, and lacking repeaters to drive long wires. Therefore, SoC on-chip buses use point-to-point wiring.

9.1.4 Multiple Controllers

Larger systems also have multiple controllers that may share the same bus. One or more controllers may place a request on any given cycle by

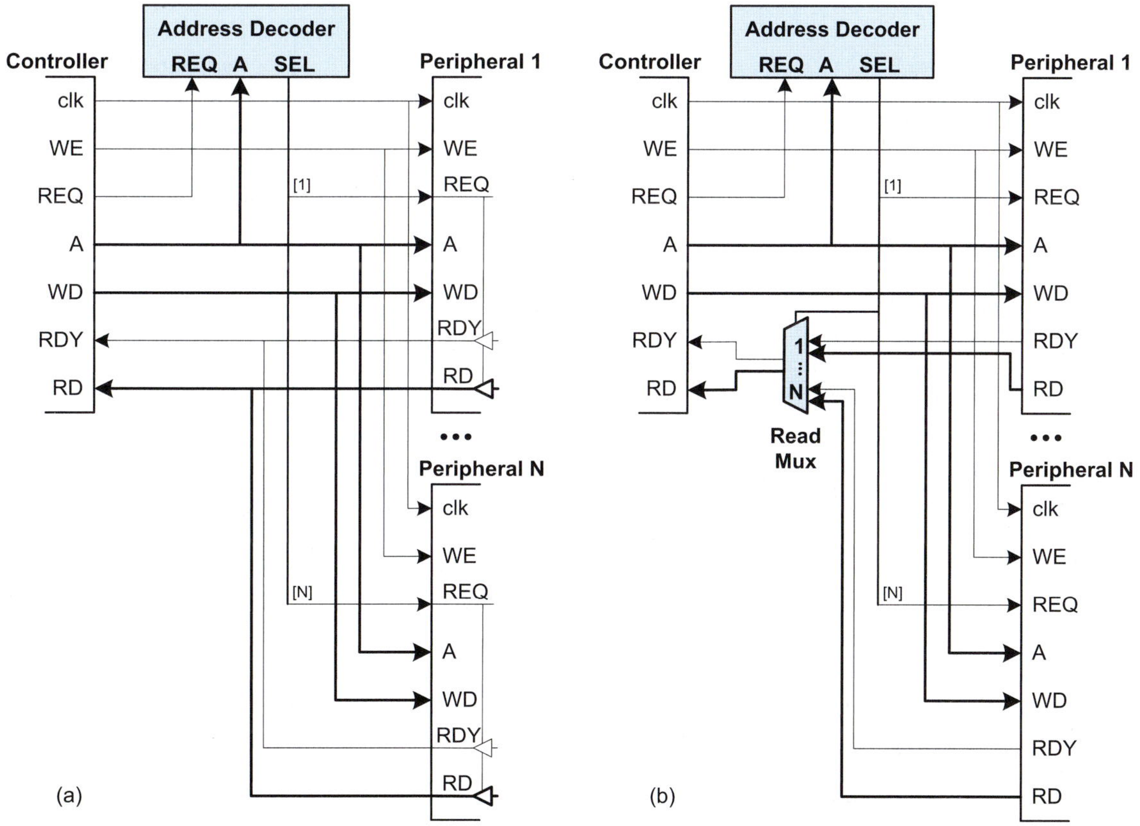

Fig. 9.5 Bus with multiple peripherals: (a) shared data ports and (b) point-to-point data ports with read multiplexer.

asserting their REQ. An *arbiter* picks among the requests by granting access to at most one controller and chooses the address (A), write data (WD), and write enable (WE) from that controller, as shown by the Grant Mux in Fig. 9.6. This synchronous bus uses a common clock for all components, which is not shown.

Arbiters can use various methods to pick which request to grant, including static, round-robin, time-division multiple access, and dynamic priority algorithms. In *static priority*, each controller has a fixed priority, and the requesting controller with the highest priority is granted access. This could starve a low-priority controller if a high-priority controller is making continuous requests. In *round robin*, access is granted in a circular manner, so every controller will eventually get access. This is fair but could cause long delays on a critical request. In *time division multiple access (TDMA)*, controllers are assigned time slots of varying length depending on their bandwidth requirements. In *dynamic priority*, the priority is adjusted in real time based on the system needs.

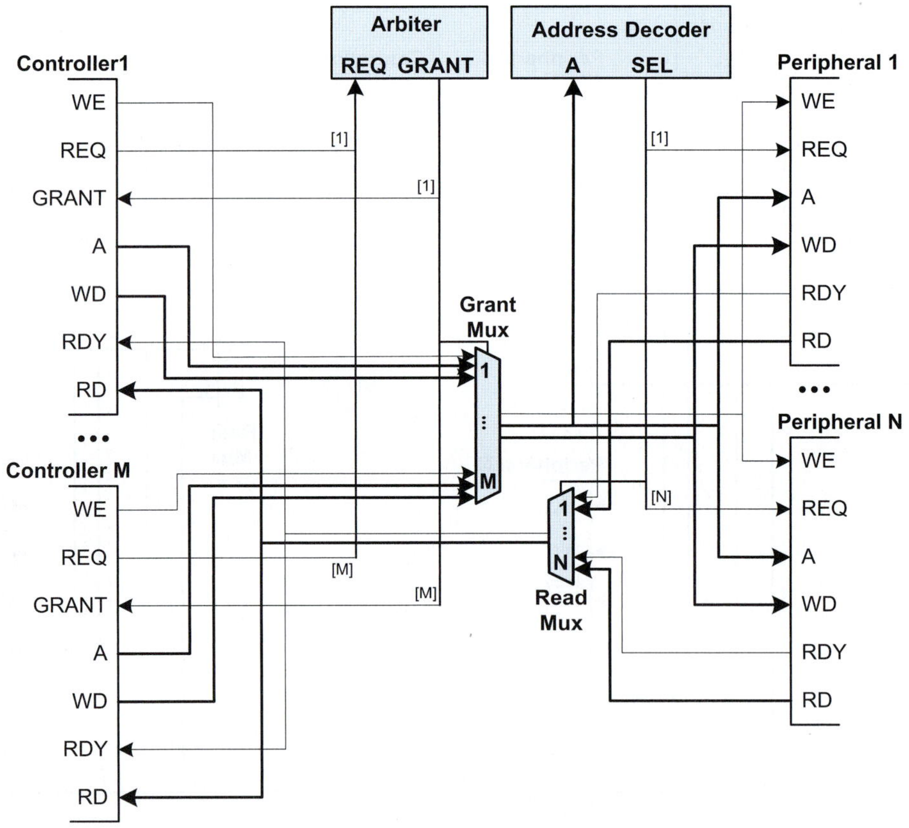

Fig. 9.6 Bus with multiple controllers and arbitration.

Systems with multiple controllers may need to *synchronize* between controllers so that only one can access a resource at a time. Synchronization is performed with *atomic* operations described in Chapter 17. These operations often read, modify, and then write a location in memory in such a way that no other controller can interrupt the operation. Thus, buses often have a *locking* signal so that a controller that is granted access can complete the atomic operation before another controller gains access.

In a multicontroller system, a device may act as both a peripheral and controller. When the device is a controller, it may perform *direct memory access* (*DMA*), reading or writing memory without going through the processor. DMA introduces challenges maintaining *coherence* between memory and caches such that any store to an address will eventually be seen by subsequent loads from the same address. For example, if a device writes directly to an address in DRAM but a cache already holds an older value of the contents of that address, reads to the cache will not see the new value written by the device, and the memory system is incoherent. Section 19.1 describes the RISC-V cache management operation (CMO) extension, which includes additional RISC-V instructions to invalidate and clean the data cache to maintain coherence with a DMA controller. A more detailed discussion of DMA coherence is beyond the scope of this book.

9.1.5 Other Bus Features

Additional bus features include support for multiple transfer modes (e.g., subword writes, burst transfers, and split transfers), cache coherency, and bridges between buses. High-performance systems may abandon buses altogether in favor of higher-speed networks. This section gives an overview of these features.

Buses often support various transfer modes. Most buses can write data smaller than the full word size. They provide either a `Size` or `ByteEn` signal that indicates which bytes to write. To reduce the overhead of repeated arbitration, buses also often define *burst* transfers consisting of a sequence of reads or writes, often to sequential addresses. Finally, when a data transfer is slow, it may tie up a bus for many processor cycles. For example, when a DRAM read is made to a new row, the bus may take many tens of cycles to respond. *Split transfer* buses tag the transfer with a unique ID and then open the bus for other operations. Sometime later, the bus returns the data and tag to the controller that initiated the request.

Complex systems often use more than one type of bus interface. For example, a system may have a high-speed memory bus and a low-speed bus for other (including legacy) peripherals. A *bridge* interfaces two buses to translate the interaction from one bus to another and so that their protocols are each satisfied.

Some buses, such as ACE and TileLink Cached, incorporate *cache coherence* capabilities directly into the bus protocols to support *shared-memory multiprocessors* in maintaining *coherence* across their cache memories.

High-performance systems often need more bandwidth than a single bus can provide. For example, they may need to allow multiple processors to concurrently access multiple banks of level 3 cache memory. A *Network on Chip* (NoC) provides a flexible and high-performance interconnect. Types of NoC architectures include *crossbars*, where every component can communicate with any other component; *indirect networks*, in which components communicate through a series of switches; and *meshes* or *rings*, in which every component can communicate with its neighbors.

9.1.6 High-Speed Off-Chip Links

The bandwidth of off-chip links has grown along with processor performance. AT&T developed the first commercial modem in 1958 for the SAGE system that connected radar systems across the United States to detect intrusions by Soviet nuclear bombers. The modem ran over the phone lines at 110 baud (symbols per second), transmitting 80 bits per second (bps), which is 10 characters per second, between mainframe computers and teletype terminals.

Bandwidth has increased rapidly since those first links. In 2025, personal computers are equipped with USB-C ports running at 40 Gbps to external devices and HDMI 2.2 ports supporting 96 Gbps using inexpensive cables. On the motherboard, a dual-channel DDR5-6400 memory system transmits approximately 400 Gbps between DRAM and a CPU (central processing unit), and PCIe 5.0 x16 transmits up to 1 Tbps between a graphics card and CPU. In data centers, Ethernet runs at 800 Gbps between cabinets. Wireless networks have also become very fast, with Wi-Fi 7 and 5G cellular theoretically capable of more than 10 Gbps.

Although long-distance networks use fiber-optic cables driven by lasers, shorter links are primarily electrical over copper wires. As the data moves along the copper at nearly the speed of light, transmission line effects are critical, and the wires must be properly *terminated* with a *matched impedance*, making point-to-point links essential. The links use *differential signaling*, in which the data is represented as a difference in voltage between a pair of wires, rather than as a single-ended signal relative to a ground line. Differential signaling reduces *crosstalk*, *common-mode noise*, and *electromagnetic interference* because current is balanced, flowing out and back along the pair rather than returning over a shared ground line.

Some high-speed links such as DDR (double-data-rate) memory use *source-synchronous clocking*, in which a clock is sent along with a parallel bundle of data bits. The clock rises and falls each cycle, generally at a rate as fast as the transmission line can reliably support. If the memory bits only switched once per clock cycle, they would only use half of the available bandwidth of the line. Hence, *DDR links* send data on both rising and falling edges of the clock (thus, its name: double-data-rate) to use the same toggle rate as the clock and fully saturate the transmission line's capabilities.

Source-synchronous speeds are limited by mismatches in flight time between bits in the bundle. These mismatches are influenced by the line lengths, impedances, and noise. *High-speed serial links* avoid these mismatches by sending a serial bitstream over a differential pair encoded in such a way that the bitstream conveys both clock and data. The receiver performs *clock/data recovery* (*CDR*) to determine when to sample the serial stream, then decodes the data bits. For example, *8b/10b* encoding sends 8 bits of data in a 10-bit symbol such that the transmitted bitstream toggles at least once every 5 bits even when the unencoded bitstream is a continuous pattern of 0s or 1s. The redundant bits can also provide error detection and correction. A phase-locked loop (PLL) in the CDR detects these transitions to align the sampling clock. Similarly, 10-Gb Ethernet uses *64b/66b* encoding, PCIe 5 uses *128/130b* encoding, and USB 3.1 uses *128b/132b* encoding to guarantee transitions while reducing the overhead. PCIe 5 x16 uses 16 serial *lanes* each operating at 32 GHz (64 Gbps).

PCIe (PCI Express), DDR memory, USB, and Ethernet each have specifications that are many hundreds of pages long, and they require a complex controller and *physical interface* (*PHY*) beyond the scope of this book. SoC designers purchase these controllers and PHYs as *IP* (*intellectual property*) blocks. The controllers may be integrated onto an ASIC (application-specific integrated circuit) or FPGA (field programmable gate array) and connected to the rest of the SoC via AXI or another standard on-chip bus. The SoC physical interface involves custom mixed-signal circuitry such as impedance-controlled differential output drivers and phase-locked loops. These are also sold by IP vendors.

[Dally98] provides a good introduction to high-speed communication. All these sophisticated techniques for fast off-chip links cost chip area and power. Within the SoC, it remains more cost-effective to use wide buses, instead of serial buses, to deliver high local bandwidth.

9.2 AMBA BUSES

ARM's Advanced Microcontroller Bus Architecture (AMBA) has been a leading family of open-standard on-chip buses used in both ASICs and FPGAs since 1996. The three major AMBA flavors are:

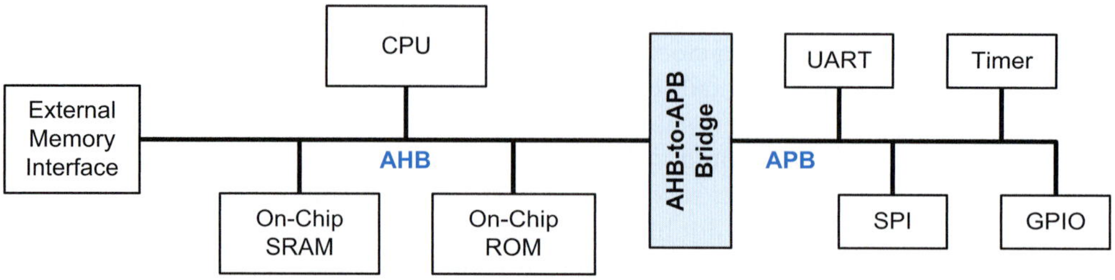

Fig. 9.7 **AHB/APB system.**

David Flynn was the original architect of ARM's synthesizable processors and the AMBA on-chip interconnect. He received the IEEE/RSE James Clerk Maxwell Medal in 2019 for these contributions as well as for his contributions to microprocessor technology and deployment. He also coauthored the *Low Power Methodology Manual* [Keating07], which defined industry-standard methods of clock and power gating. Flynn was an ARM fellow and worked at the company from 1991 to 2018. He earned his BSc in computer science from Hatfield Polytechnic and his doctorate in electrical engineering from Loughborough University.

- **APB** (Advanced Peripheral Bus): Simple unpipelined
- **AHB** (Advanced High-performance Bus): Pipelined
- **AXI** (Advanced eXtensible Bus): Multichannel, high-speed

Fig. 9.7 shows an example system that uses AHB and APB. Whereas the high-speed CPU and on-chip memories use AHB, the slower peripherals use the simpler APB bus. The AHB-to-APB bridge translates signals from one protocol to the other. AHB can also be used as an interface to off-chip memory.

The following subsections describe APB [APB21], AHB [AHB21], and AXI [AXI21]. Both APB and AHB use two phases for address and data transfer, but AHB is faster because the phases of consecutive operations can overlap. AXI is now the dominant standard for systems with multiple controllers or high-speed DRAM interfaces, but the 500-page specification [AXI21] is beyond the scope of this chapter.

9.2.1 APB

Fig. 9.8 shows an APB bus connecting a controller to one or more peripherals. The controller is an AHB-to-APB bridge. The standard APB signal names are similar to the generic ones from Fig. 9.5 but are prefixed by `P`, as listed in Table 9.1 for the main APB signals except `PCLK` and `PRESETn`. The `PRESETn` signal resets all peripherals when asserted low. The address decoder, which is part of the controller, generates the `PSEL[N:1]` signal for the various peripherals and the mux. The APB mux returns the data and ready signals from the selected peripheral. `PADDR` may be up to 32 bits, and `PRDATA`/`PWDATA` may be 8, 16, or 32 bits.

Each transfer has two phases: *Setup* and *Access*. The Setup phase takes one clock cycle, during which the controller drives the `PADDR`, `PWRITE`, and `PSEL` signals onto the bus along with `PWDATA` for a write. The Access phase takes one or more clock cycles, during which the controller holds these Setup phase signals steady and asserts `PENABLE`; the peripheral performs the read or write. Upon a read, the peripheral responds with `PRDATA`. The peripheral raises `PREADY` when it is done reading or writing. Typically,

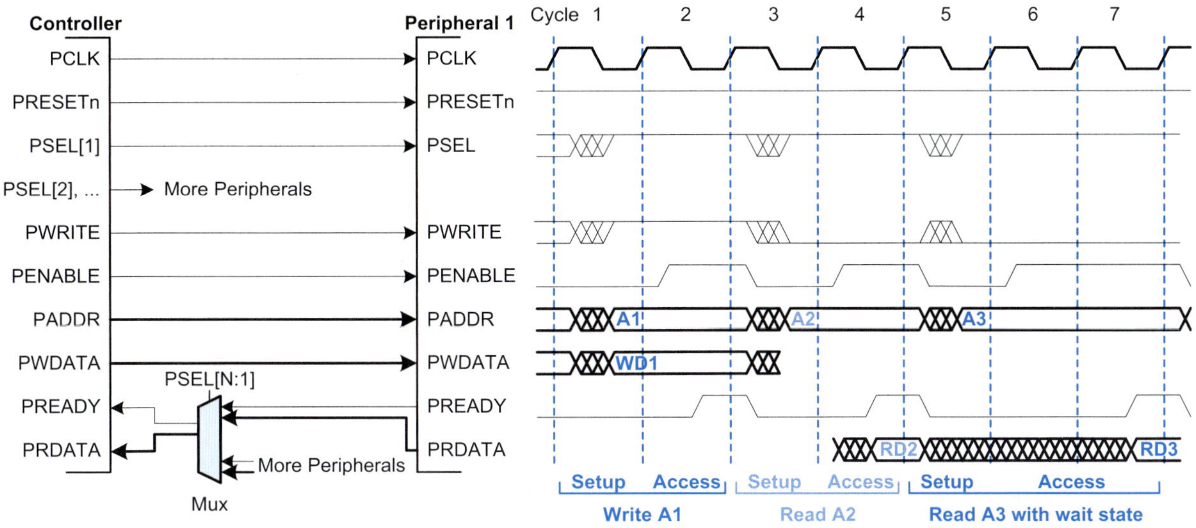

Fig. 9.8 APB interface.

the Access phase is one cycle long, but a slow peripheral could hold PREADY low for one or more *wait states* before responding. A peripheral that always responds in a single Access phase cycle can tie its PREADY high. If no peripheral is selected, the APB mux asserts PREADY so the bus remains in a ready state.

The waveforms in Fig. 9.8 illustrate read and write transfers. In cycles 1 and 2, the APB bus performs a write transfer with no wait states. In cycles

Table 9.1 Main APB signals.

Signal	Description	Width (bits)	Phase	Source
PADDR	Address to read or write	32	Setup	Controller
PRDATA	Read data from peripheral	32	Access	Peripheral
PWDATA	Write data to peripheral	32	Setup	Controller
PWRITE	Write enable	1	Setup	Controller
PENABLE	Enable read or write	1	Access	Controller
PREADY	Peripheral asserts when Access phase completes	1	Access	Peripheral
PSTRB	Write byte enable	4	Setup	Controller

3 and 4, the bus performs a read with no wait states. In cycles 5–7, the bus performs a read with one wait state.

Not shown in Fig. 9.8 are the optional PSTRB (APB strobe) and PSLVERR (APB peripheral error) signals. PSTRB[3:0] is a byte write enable, with one bit per byte. A peripheral may assert PSLVERR when an error occurs.

ASIC and FPGA implementations use specially designed tristate buffers (also called *drivers*) capable of driving large capacitive PCB (printed circuit board) wire loads. They are instantiated in the top-level module, and therefore the gpio_apb module has three interface signals: GPIOEN, GPIOOUT, and GPIOIN.

Fig. 20.3 shows a full-featured GPIO peripheral with an APB interface.

Example 9.1 GPIO USING APB

Design a 32-bit GPIO peripheral that uses an APB interface. The peripheral has three 32-bit memory-mapped I/O registers: GPIOOUT, GPIOIN, and GPIOEN. GPIOOUT is the output register (for writing to the GPIO pins); GPIOIN is the read-only input register (for reading the values on the GPIO pins); and GPIOEN configures the GPIO pins as outputs or inputs (1 = output, 0 = input), with each bit controlling a tristate enable connected to the GPIO pins. These registers are mapped to a 2-bit address: GPIOOUT is at address 00, GPIOEN at 01, and GPIOIN at 10.

Solution: Fig. 9.9 shows the 32-bit GPIO peripheral. The core of the peripheral is a tristate buffer that enables writing GPIOOUT onto the GPIO pins for each bit of GPIOEN[31:0] that is asserted. The pins can always be read onto GPIOIN. A 2:3 address decoder determines which register is being accessed. When PWRITE, PSEL, and PENABLE are asserted, the GPIOOUT or GPIOEN register is written when PADDR [1:0] is 00 or 01, respectively. The GPIO multiplexer is controlled by the address (PADDR[1:0]) to read the appropriate register. The peripheral has no wait states; thus, it always asserts PREADY.

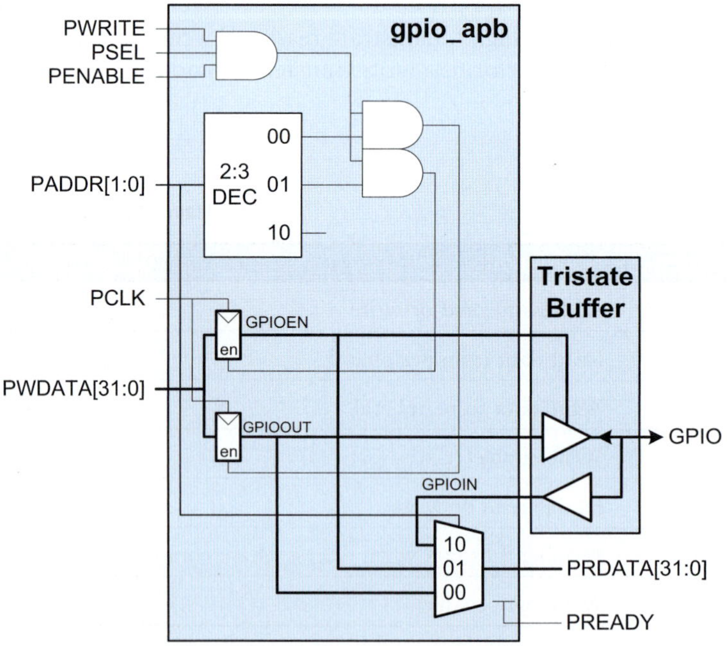

Fig. 9.9 GPIO peripheral with APB interface.

9.2.2 AHB

Like APB, AHB uses two phases of operation and takes at least two cycles for a transfer, with the option of additional wait states. However, AHB doubles the throughput by pipelining, that is, overlapping the two phases. AHB refers to the two phases as Address (Adr) and Data phases, which are similar to APB's Setup and Access phases, as listed in Table 9.2.

Fig. 9.10 shows an AHB bus connecting a controller to one or more peripherals. The address decoder generates `HSEL[N:1]` for *N* peripherals and the AHB mux. `HADDR` may be up to 64 bits, and `HRDATA`/`HWDATA` are typically 32, 64, 128, or 256 bits, permitting wider transfers than APB. AHB adds an `HTRANS` signal to indicate the type of transfer, as summarized in Table 9.3. Typically, `HTRANS` is IDLE when the bus is not in use, or NONSEQ for ordinary transfers. Section 9.2.2.5 introduces SEQ burst transfers.

Fig. 9.10 illustrates a write to address A1; then a read from A2; then another read from A3, which adds a wait state. In the Address phase, the controller places the address `HADDR` and control signals `HWRITE` and `HTRANS` on the bus, and the address decoder drives `HSEL` to the appropriate peripheral. In the Data phase, the controller sends `HWDATA`, or the peripheral

ARM uses inconsistent terminology between APB and AHB.

AHB was originally developed to support multiple controllers and split transactions [AMBA99], but a simpler flavor called AHB-Lite omits this support. Systems requiring these additional features soon migrated to the more capable AXI bus, leaving AHB-Lite as the favored flavor of AHB. ARM recognized this in the recent AHB5 specification, which is based on AHB-Lite with optional extra features related to security and exclusive transfers. This book uses the term *AHB* to refer to AHB5 and focuses on the required features that are common to AHB-Lite and AHB5.

Table 9.2 Comparison of APB and AHB terminology.

Term	APB	AHB
Controller	Requester	Manager
Peripheral	Completer	Subordinate
Phase 1	Setup	Address (or Adr)
Phase 2	Access	Data
Signals	Use prefix `P`	Use prefix `H`

Table 9.3 HTRANS transfer types.

HTRANS	Type	Description
00	IDLE	No data transfer
01	BUSY	Like IDLE, but in the middle of a burst
10	NONSEQ	Single transfer or first transfer of a burst
11	SEQ	Subsequent transfers of a burst

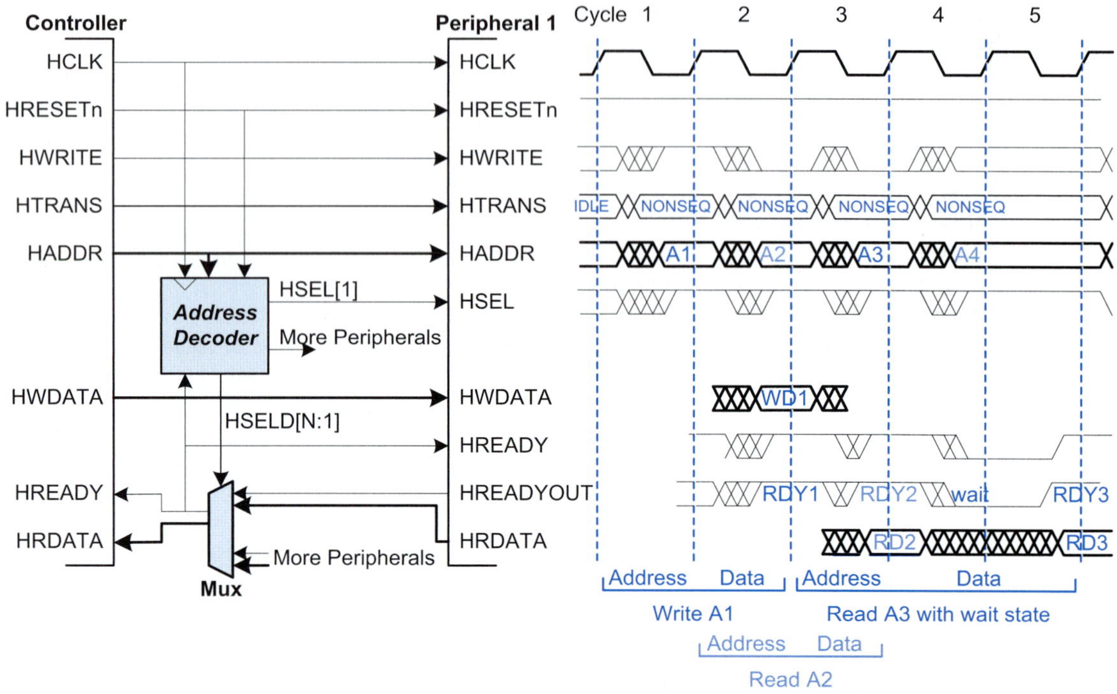

Fig. 9.10 **AHB interface in the uncore.**

responds with HRDATA. The peripheral also raises HREADYOUT to signal that the Data phase is complete, or the peripheral may keep it low to insert one or more wait states before completing. The Address phase of the next transfer begins when the previous transfer is in its Data phase. This requires an extra register to pipeline HADDR and the control signals. Because of this pipelining, AHB has nearly twice the throughput of APB.

The AHB multiplexer chooses the HREADYOUT and, on a read, HRDATA from the selected peripheral. It chooses based on HSELD, a version of HSEL delayed to align with the Data phase. As in APB, when no HSEL signals are asserted, the AHB mux asserts HREADY to prevent the bus from hanging. This is called the *default subordinate* behavior. AHB also sends HREADY to the peripherals to mark the completion of one transaction and start of the next. When HTRANS is NONSEQ or SEQ and a peripheral receives HREADY and HSEL, the peripheral must capture the Address phase inputs so that it can generate the addresses for the next transactions.

Fig. 9.11 shows a schematic of the address decoder and multiplexer, including the comparators to check HADDR against a peripheral's address range to find HSEL, the AND-OR muxing for HRDATA and HREADY, the HSELD delay register, and the default HREADY logic. The Wally PMA (physical memory attribute) checker, pmachecker, located in the memory

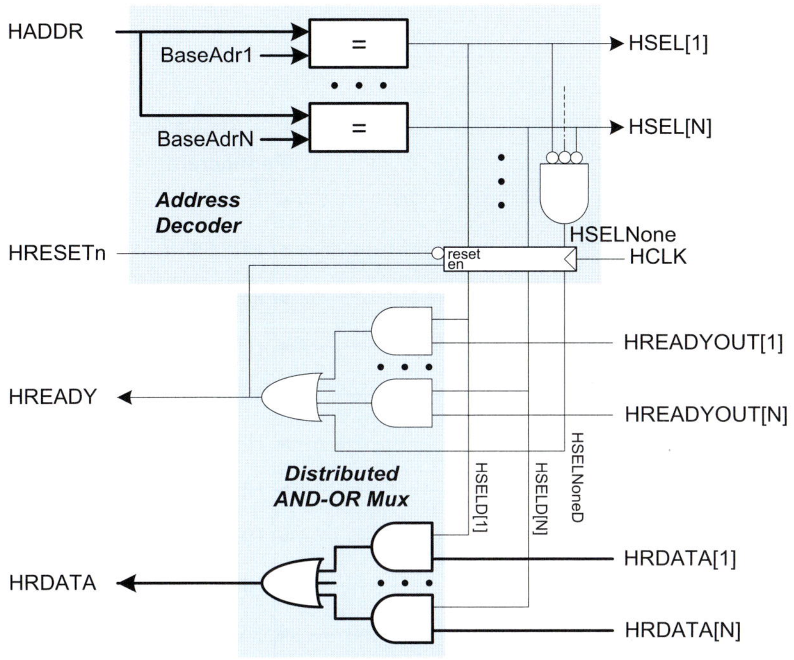

Fig. 9.11 **AHB address decoder and multiplexer.**

management unit (MMU) (Fig. 11.16), expands the address decoders to also check access types and sizes.

9.2.2.1 Additional AHB Signals

A full AHB link has several other signals shown in Fig. 9.12 that were left out of Fig. 9.10 for simplicity. Table 9.4 lists the main AHB signals other than `HCLK` and `HRESETn`. The controller sends `HSIZE`, `HWSTRB`, `HBURST`, `HPROT`, and `HMASTLOCK` and optionally `HEXCL`, `HMASTER`, and `HNONSEC`. All of these signals are transmitted in the Address phase except `HWSTRB`, which is sent in the Data phase with `HWDATA`. The peripheral replies with `HRESP` and, optionally, `HEXOKAY` to indicate success of an exclusive transfer. `HRESP` is a 2-bit signal that shows the status of the transfer (00: OKAY, 01: ERROR, 10: RETRY, 11: SPLIT).

`HSIZE` indicates the size of the transfer, as defined in Table 9.5, not to exceed the width of the data buses. `HWSTRB` (write strobe) conveys byte write enables to write only certain bytes within a larger transfer. For example, Table 9.6 shows two ways to write a single byte using an AHB bus. It assumes a little-endian memory system with a 32-bit bus and writes 0x42 to address 0x1002. The first option sends a whole word and only enables 1 byte in the word to be written, and the second option makes a single-byte transfer.

For Wally, `HADDR` width (see Table 9.4) is `ADDR_WIDTH` = `PA_BITS` (physical address bits), which is 34 bits for RV32 and 56 bits for RV64, as will be discussed in Chapter 11.

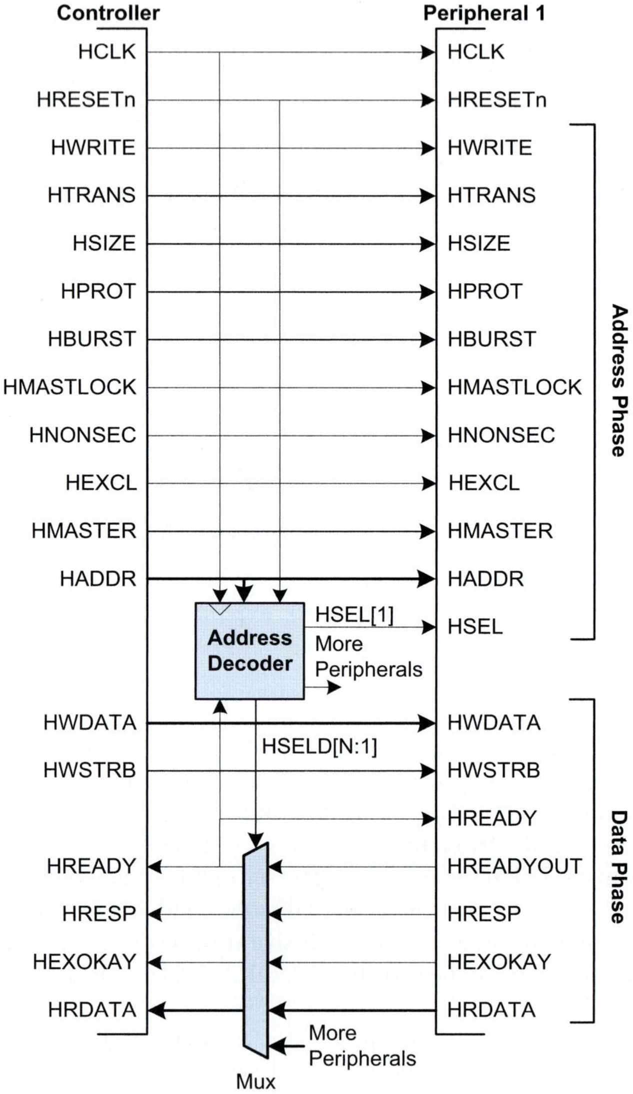

Fig. 9.12 **Full AHB interface.**

Table 9.4 Main AHB signals.

Signal	Description	Width (bits)	Phase	Source
HADDR	Address to read or write	ADDR_WIDTH	Address	Controller
HRDATA	Read data from peripheral	AHBW	Data	Peripheral
HWDATA	Write data to peripheral	AHBW	Data	Controller
HWRITE	Write enable	1	Address	Controller
HREADY	Peripheral asserts when Data phase completes	1	Data	Peripheral
HTRANS	Type of transaction (see Table 9.3)	2	Address	Controller
HSIZE	Size of transaction (see Table 9.5)	3	Address	Controller
HWSTRB	Write strobe: indicates which bytes to write	AHBW/8	Data	Controller
HBURST	Information about bursts (see Section 9.2.2.5)	3	Address	Controller
HPROT	Protection information	4	Address	Controller
HMASTLOCK	Master lock for atomic instructions	1	Address	Controller
HEXCL	Exclusive transfer for atomic instructions (optional)	1	Address	Controller
HMASTER	Master for atomic instructions (optional)	0-8	Address	Controller
HNONSEC	Identifies nonsecure transactions (optional)	1	Address	Controller
HRESP	Peripheral transmits status of transfer	1	Data	Peripheral
HEXOKAY	Exclusive transfer okay: is 1 when successful (optional)	1	Data	Peripheral

Table 9.5 **HSIZE transfer sizes.**

HSIZE[2:0]	Number of Bits	Size
0	8	Byte
1	16	Half word
2	32	Word
3	64	Double word
4	128	
5	256	
6	512	
7	1024	

Table 9.6 Two options for a 1-byte transfer on an AHB bus.

Signal	Option 1	Option 2
HADDR	0x1000	0x1002
HSIZE	2 (4 bytes)	0 (1 byte)
HWSTRB	0b0100	0b0001
HWDATA	0x00420000	0x00000042

HPROT conveys optional information about protection, including cacheability, bufferability, and privilege/user and data/instruction modes. HBURST gives hints about burst transfers as will be discussed in Section 9.2.2.5. HMASTLOCK (master lock) and the optional HEXCL/HMASTER (exclusive/master) signals can be used for atomic operations. The optional HNONSEC signal identifies nonsecure transactions for systems with security extensions.

Many systems don't need all the signals listed in Table 9.4. The controller can tie HSIZE to the bus size (e.g., 2 for a 32-bit bus), HPROT = 3 (ignored by most peripherals), HBURST = 0 (no burst), and HMASTLOCK = 0, and omit optional signals. In a system where peripherals don't generate errors, the peripheral can tie HRESP to 0 and the controller can ignore HRESP.

9.2.2.2 AHB Peripherals

An AHB peripheral must perform several functions to interface with the bus:

- Detect new transactions.
- Retime writes, so that HWDATA arrives in the Data phase rather than in the Address phase.
- Read or write registers in the peripheral.
- Produce HRDATA in the Data phase.
- Generate the HREADYOUT, HRESP, and HEXOKAY acknowledgement.

A peripheral starts a new transaction when the bus contains a valid transaction, the peripheral is selected, and the previous transaction is completing. The peripheral detects these conditions by asserting an internal signal, initTrans, when HTRANS is NONSEQ or SEQ, HSEL is asserted, and HREADY is asserted from the previous transaction. When initTrans asserts, the peripheral either reads or writes depending on the value of HWRITE.

HTRANS[1] indicates that the transaction is NONSEQ or SEQ.

When interacting with a peripheral that has no wait states, a controller reads from the peripheral on the rising edge after the Address phase and HRDATA is available during the Data phase. However, a write to the peripheral occurs on the rising edge *after* the Data phase because HWDATA is not available until the Data phase. Hence, when HREADY is asserted, HADDR and HWRITE must be captured at the end of the Address phase to retain their values for writes in the Data phase.

A peripheral may tie HREADYOUT to 1 if it never needs wait states. Otherwise, the peripheral only asserts HREADYOUT when the access is complete.

9.2.2.3 Example: AHB SRAM Interface

Fig. 9.13 shows a single-port synchronous SRAM with an AHB interface. It asserts initTrans when the RAM is selected (HSEL = 1), a transaction is pending (HTRANS[1] = 1), and the previous transaction is completing (HREADY = 1). memRead or memWrite is asserted when initTrans is 1. A register delays HADDR and memWrite by a cycle to align with HWDATA arriving in the Data phase for writes and to retain the values for transactions with wait states. A multiplexer chooses the RAM address from HADDR on reads with

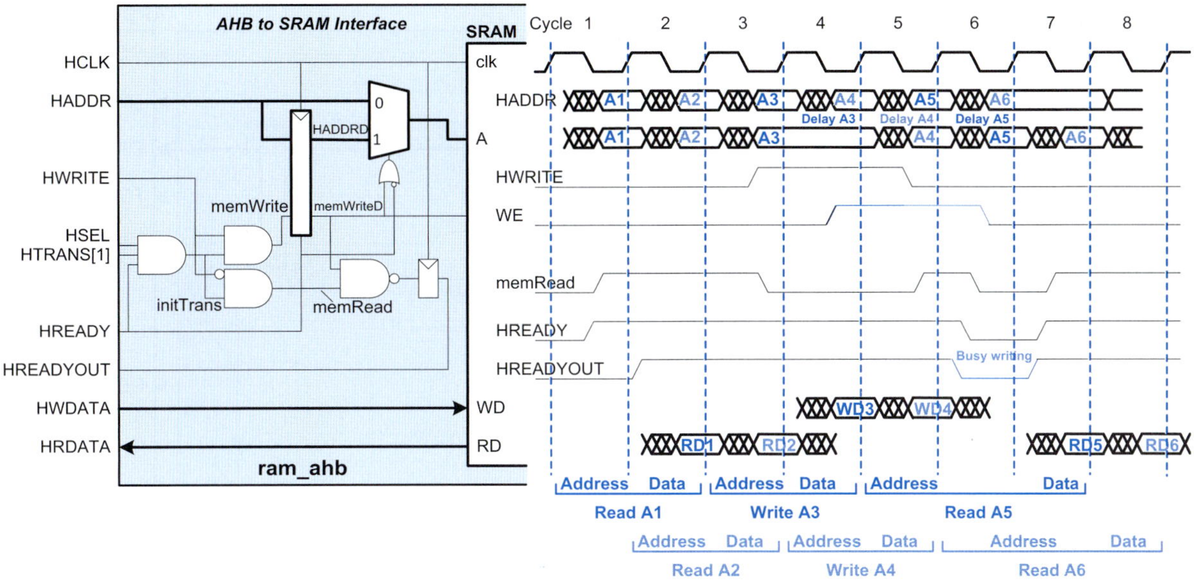

Fig. 9.13 AHB SRAM interface.

no wait state or from the delayed HADDRD for writes and for reads with one or more wait states.

Reads occur on the rising edge at the end of the Address phase. Writes occur on the rising edge at the end of the Data phase, so their address and write enable must be delayed a cycle. Thus, the SRAM can accommodate consecutive reads and consecutive writes and a read followed by write – but not a write immediately followed by a read because the SRAM has a single read/write port and cannot accept the read and write addresses on the same cycle. Therefore, the SRAM interface inserts a single wait state by lowering HREADYOUT for one cycle for a write followed by read.

The waveforms in Fig. 9.13 show two consecutive reads to A1 and A2, two consecutive writes to A3 and A4, and then two more reads to A5 and A6. The read after write has a wait state created by lowering HREADYOUT. As a result, HREADY and memRead must wait, delaying the read from A5.

A simple SRAM expects the write data at the same time as the address and control signals. The AHB protocol complicates the interface because write data arrives a cycle late. This timing is an artifact of AHB's roots as the native memory interface of early ARM processors [Flynn97]. For example, the ARM6 had a 3-stage pipeline in which the address arrived a cycle before the write data.

9.2.2.4 Example: AHB-to-APB Bridge

Fig. 9.14 shows an AHB to APB bridge. The left side behaves as a peripheral on an AHB bus. The right side behaves as a controller on an APB bus. The bridge translates AHB transactions into APB transactions, adjusting the phases of the signals before returning HREADYOUT. The bridge supports a configurable number of peripherals. It expects select signals for each peripheral from the AHB address decoder. The bridge contains the APB read mux to choose the results from the selected peripheral. The bridge accepts

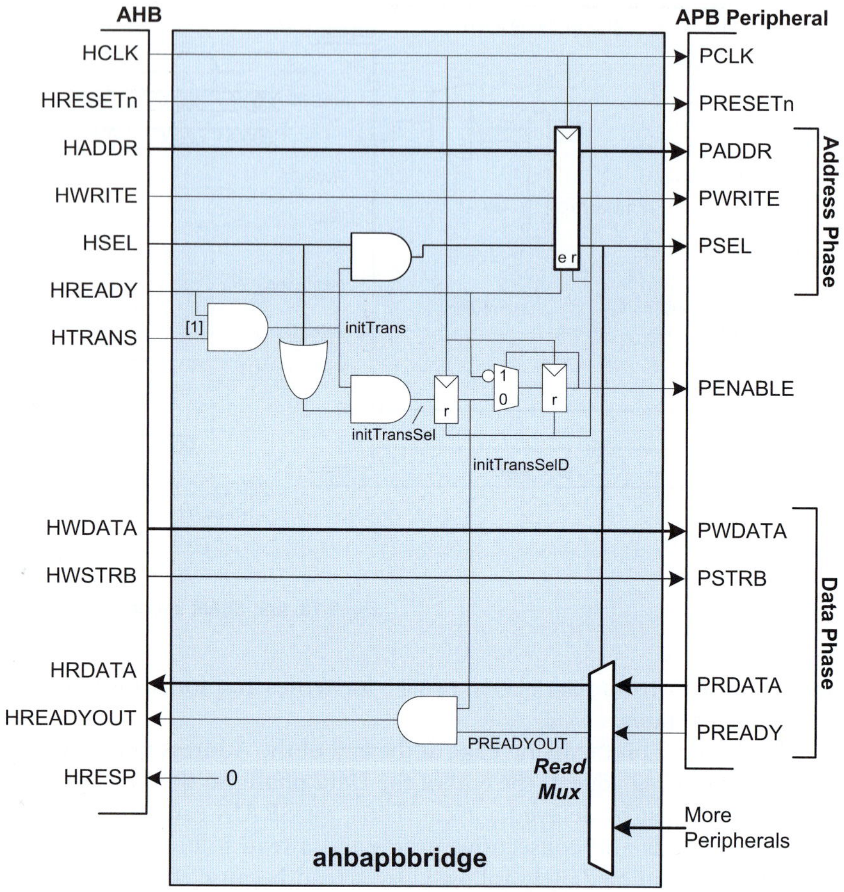

Fig. 9.14 AHB-to-APB bridge.

the full width of the data buses and uses `HWSTRB` byte enables for subword writes.

Fig. 9.15 shows how an AHB read and write without wait states would be translated to APB signals. `PRDATA` (APB read data) is available in the cycle after `PADDR` changes, so no adjustments need to be made on reads from the APB peripheral. On a write, because APB peripherals are not pipelined, `HADDR`, `HWRITE`, and `HSEL` are delayed by a cycle to align with `HWDATA`. The resulting delay in `PREADY` would also cause `HREADYOUT` to be delayed by a cycle (not shown).

9.2.2.5 Burst Transfers

DRAM is organized into rows called *pages*, which are typically 4–16 Kib in size. DRAM reads a full page of memory into its sense amplifiers and

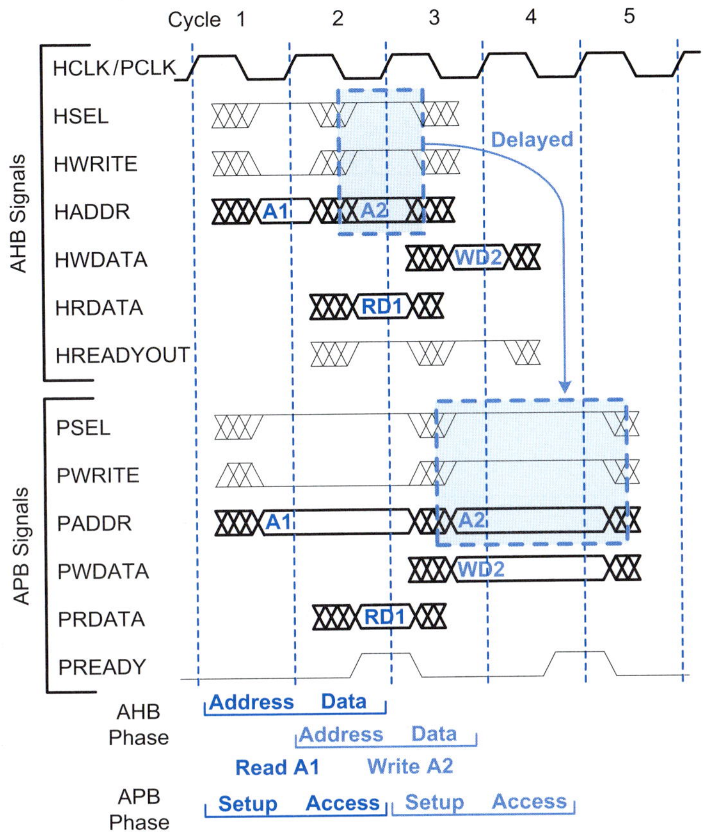

Fig. 9.15 Translation of AHB to APB signals.

returns words much faster from that open page than from a new page. Thus, a DRAM controller is much faster at bursts of accesses to consecutive memory locations than accesses to random memory locations.

AHB uses SEQ transactions and the HBURST signal to tell a peripheral that such a burst is occurring so that some peripherals can respond faster. The AHB standard defines various HBURST encodings for fixed and variable length, with addresses that increment or wrap around an aligned line. Simple peripherals ignore the SEQ and HBURST hints and just respond to HADDR.

For example, Fig. 9.16 shows a 4-beat write burst followed by a single read. The HBURST signal is set to INCR during the write burst to indicate addresses are incrementing. On the first beat, HTRANS is NONSEQ, and on subsequent beats, it is SEQ. HSIZE is 2 (010), so each beat transfers one 32-bit word, and the 4 beats transfer 128 bits all together.

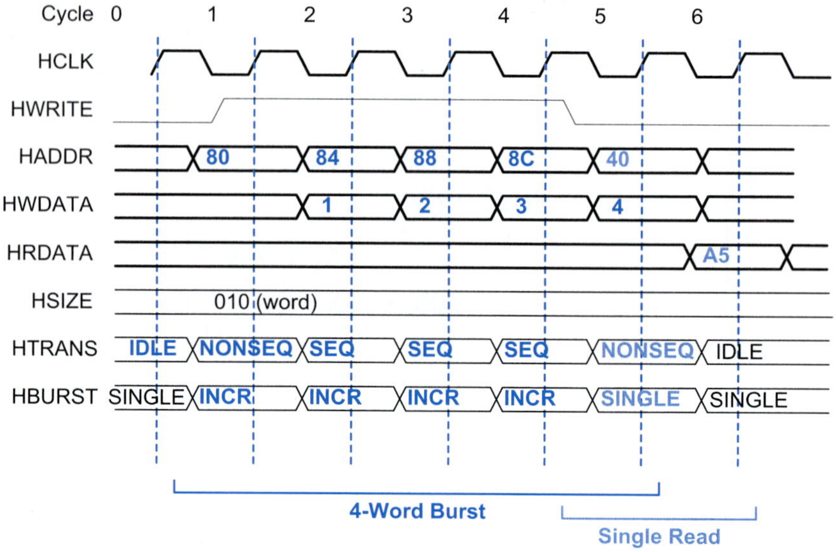

Fig. 9.16 AHB burst transfer.

9.2.2.6 Multilayer AHB

Multicontroller AHB systems use independent AHB buses called *layers* for each controller. The layers are connected to peripherals through an interconnect matrix with decoders for each controller and arbiters managing access to each peripheral, as shown in Fig. 9.17.

9.2.3 AXI

AHB is designed for the address phase to take a single clock cycle. The response may take multiple cycles, but one transaction must complete before the next begins. However, modern SoCs must handle multicycle

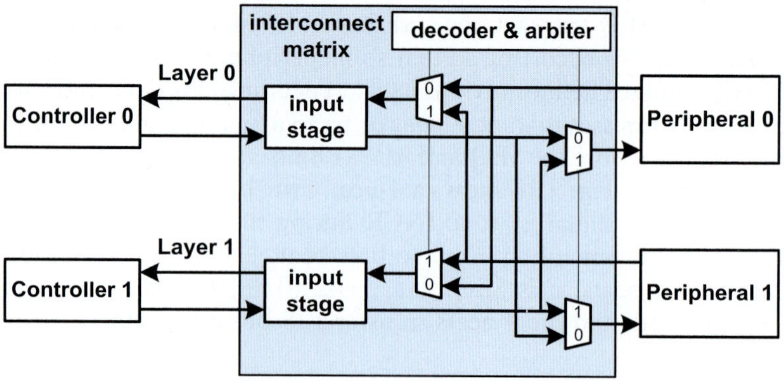

Fig. 9.17 Multilayer AHB. (Adapted from [MultilayerAHB04].)

on-chip delays, multiple cores accessing the same peripherals and memory, cache coherence, and devices acting as both a controller and peripheral. As clock frequencies have increased, the flight time to send information across an SoC has become multiple clock cycles. An SoC also often has many heterogeneous cores communicating with one or more DDR DRAM controllers as well as accelerators and other on-chip peripherals. The DRAM can service multiple concurrent requests, and access to an open page of DRAM is much faster than access to a new page. To maintain cache coherence, one core's cache may need to snoop transactions from other cores. Direct memory access (DMA) devices may act as a peripheral when being configured by a core, then as a controller when issuing memory requests.

The Advanced eXtensible Interface (AXI) addresses all these issues by providing independent pipelined address and data channels operating at high clock rates that handle multiple pending requests and out-of-order responses [AXI21]. AXI uses the terms *manager* and *subordinate* in place of controller and peripheral. It was designed for multiple managers and subordinates connected by flexible interconnect matrices that support cache coherency and DMA devices. Each channel can have arbitrary latency, so pipeline registers can be added to meet clock speed targets. AXI handles large bursts, such as transferring entire cache lines between memory and a core; the manager only sends the burst's start address, and the subordinate autoincrements the address for subsequent beats. Byte strobes enable unaligned data transfers. Simple bridges connect AXI to AHB or APB. AXI is now the leading standard for SoCs with high-performance multicore processors. It is also widely used on FPGAs, especially for high-bandwidth IP blocks such as memory controllers. This section briefly outlines the AXI4 protocol.

AXI4 is based on five independent transaction channels. Table 9.7 summarizes the main AXI signals on the five channels [AXI21]. For a read transaction, shown in Fig. 9.18(a), a manager sends address and control information on the *read address channel* to a subordinate. The subordinate responds with a burst of data on the *read data channel*, possibly taking multiple beats (data transfers) to convey the information. For a write transaction, shown in Fig. 9.18(b), a manager sends address and control information on the *write address channel* to a subordinate. It also sends a burst of data on the *write data channel*. The write data burst can come after the address. The subordinate acknowledges on the *write response channel*. Each of the device's five channels may be routed through an interconnection matrix arbitrating among multiple managers and subordinates.

AXI uses a global clock called `ACLK` and an active low reset `ARESETn`. Each channel has a *source* that sends most of the information and asserts `VALID` and a *destination*, which replies with a `READY` signal. Every

Table 9.7 AXI channel signals.

Signal	Description	Width (bits)	Source
	Read Address Channel		
ARID	Identification tag	ILEN	Manager
ARADDR	Starting address	ALEN	Manager
ARLEN	Burst length in beats: ARLEN + 1	8	Manager
ARSIZE	Bytes in a beat: $2^{ARSIZE} = 1...128$	3	Manager
ARBURST	Burst mode (01 = increment)	2	Manager
ARVALID	Channel signals are valid	1	Manager
ARREADY	Transfer on channel can be accepted	1	Subordinate
	Read Data Channel		
RID	Identification tag	ILEN	Subordinate
RDATA	Read data	$2^{3\text{-}10}$ (8-1,024)	Subordinate
RRESP	Status (00 = normal successful transfer)	2	Subordinate
RLAST	Last beat in the transaction	1	Subordinate
RVALID	Channel signals are valid	1	Subordinate
RREADY	Transfer on channel can be accepted	1	Manager
	Write Address Channel		
AWID	Identification tag	ILEN	Manager
AWADDR	Starting address	ALEN	Manager
AWLEN	Burst length in beats: AWLEN + 1	8	Manager
AWSIZE	Bytes in a beat: $2^{AWSIZE} = 1...128$	3	Manager
AWBURST	Burst mode (01 = increment)	2	Manager
AWVALID	Channel signals are valid	1	Manager
AWREADY	Transfer can be accepted	1	Subordinate
	Write Data Channel		
WID	Identification tag	ILEN	Manager
WDATA	Write data	$2^{3\text{-}10}$ (8-1024)	Manager

(Continued)

Table 9.7 AXI channel signals. Cont'd

Signal	Description	Width (bits)	Source
WSTRB	Write strobe: which bytes hold valid data	WDATA/8	Manager
WLAST	Last beat in the transaction	1	Manager
WVALID	Channel signals are valid	1	Manager
WREADY	Transfer on channel can be accepted	1	Subordinate
	Write Response Channel		
BID	Identification tag	ILEN	Subordinate
BRESP	Status (00 = normal successful transfer)	2	Subordinate
BVALID	Channel signals are valid	1	Subordinate
BREADY	Transfer on channel can be accepted	1	Manager

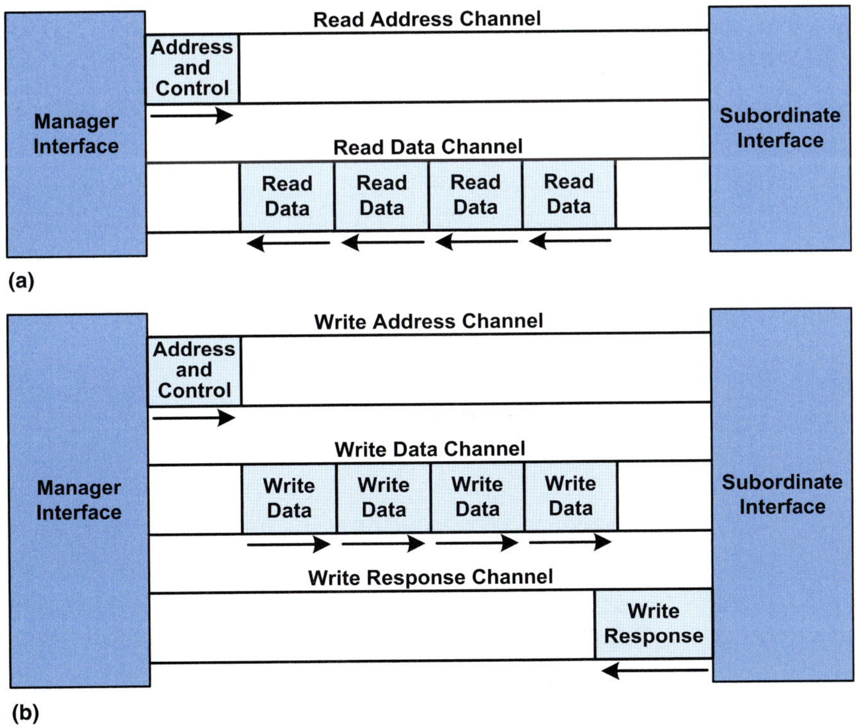

Fig. 9.18 AXI transactions: (a) read and (b) write. (Adapted from [AXI21])

transaction includes an identification tag to distinguish out-of-order responses. AXI also supports other optional control signals that convey atomic transactions, cache coherence, security, quality of service, regions, and user-defined control. The widths of the ID, address, and data buses are system-dependent (ILEN, ALEN, and 8−1024 bits).

Every channel has VALID and READY handshaking signals, thus providing two-way flow control so both the manager and subordinate can control the rate of information movement. The signals are active high, so they are asserted when 1. The channel source places information on the channel and raises the VALID signal to begin a transaction. The destination asserts its READY signal to indicate that it is able to accept a transaction; it may do so even before seeing VALID. The transaction occurs on the next rising ACLK edge when both VALID and READY are asserted. The source can then lower VALID.

Fig. 9.19 illustrates the handshaking; signals sent by the manager are blue, and those sent by the subordinate are black. Single-headed arrows point to signals that may be asserted before or after the signal at the start of the arrow. Double-headed arrows point to signals that must be asserted only after the signal at the start of the arrow asserts.

Fig. 9.19 (a) shows the read transaction handshake involving the Read Address and Read Data channels:

- The manager places information on the Read Address channel and asserts ARVALID without waiting for ARREADY.
- The subordinate raises ARREADY when it is ready to accept the address. This could be before or after ARVALID.
- When the subordinate sees both ARVALID and ARREADY, it accepts the Read Address transaction. When information is available on the

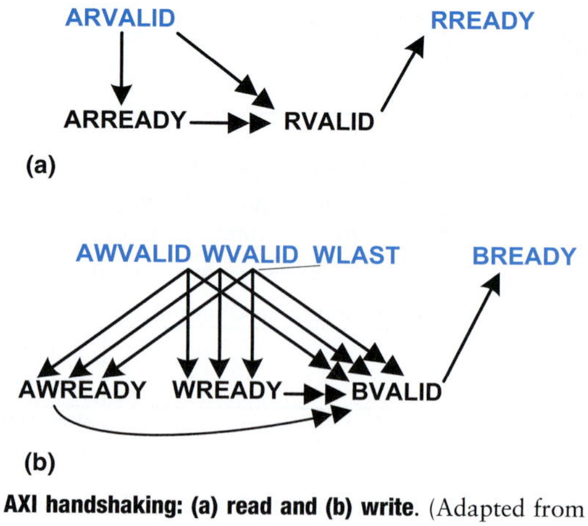

Fig. 9.19 AXI handshaking: (a) read and (b) write. (Adapted from [AXI21].)

Read Data channel, which may take some time, the subordinate asserts RVALID without waiting for RREADY. The subordinate sets RRESP = 0 for a successful normal transaction, or nonzero otherwise.

- The manager raises RREADY when it is ready to accept the Read Data. This could be before or after RVALID.

Fig. 9.19(b) shows the write transaction handshake involving the Write Address, Write Data, and Write Response channels. Note that the subordinate may receive information on the Write Address and Write Data channels in either order because of different numbers of register stages in the channel or uncertain delays through the interconnect matrix.

- The manager places information on the Write Address channel and asserts AWVALID. It also places the first beat of a burst on the Write Data channel and asserts WVALID. The manager continues sending beats on consecutive cycles and asserts WLAST on the final cycle. The manager does not wait for AWREADY or WREADY to drive the valid (AWVALID/WVALID) signals.
- The subordinate raises AWREADY and WREADY when it is ready to accept the address and data, respectively. This could be before or after AWVALID and WVALID.
- When the subordinate sees AWVALID and AWREADY, it accepts the Write Address transaction.
- When the subordinate sees WVALID and WREADY, and WLAST arrives, it accepts the Write Data transaction.
- When the subordinate has accepted both Write Address and Write Data transactions for the same ID tag, it places BRESP for that ID on the Write Response channel and raises BVALID.
- The manager raises BREADY when it is ready to accept the response. This could be before or after BVALID.

If sending information on a channel takes multiple cycles, a lag occurs between the time a subordinate deasserts READY and the time the manager sees that the subordinate cannot receive more transactions. The subordinate either needs an input buffer to accept transactions during this lag or must handshake by waiting to assert READY until a VALID transaction arrives on the channel. Waiting on handshakes over a multicycle channel keeps the bus idle most of the time, so buffering is necessary for good performance.

AXI5 adds various optional capabilities (e.g., loopback, poison bits, and ARM-specific features) to the AXI4 protocol described earlier. AXI4-Lite is a subset of AXI limited to in-order transactions comprising single-beat bursts of 32 or 64 bits. A simple bridge can connect AXI-Lite subordinates to an AXI bus. The AXI Coherence Extensions (ACE) protocol provides more hardware support for cache coherency. The Coherent Hub Interface (CHI) extends ACE for large, scalable communication networks with directory-based cache coherence.

AXI is ideal for high-performance on-chip communication because it is easy to support wide buses on chip, and they can be pipelined for speed. AXI's bursts and out-of-order capabilities are well-suited for linking cores to on-chip memory controllers. However, a 32-bit AXI5 bus involves 224 interface pins [AXI21], which is far too many for a chip-to-chip interface on a simple microcontroller. AXI also entails greater implementation complexity, such as buffering out-of-order responses. AHB is a better fit for small cores with simple peripherals.

9.3 OTHER SoC Buses

Several other buses are found on SoCs.

Wishbone is an open-standard SoC bus developed by Silicore Corporation and maintained by OpenCores for portable IP cores [Wishbone10]. It has similarities to AHB but doesn't delay write data by a cycle. It lacks the cache coherency features of AXI and other buses. Wishbone is compatible with many free OpenCores, but it is not used as much in commercial systems.

The OpenHW group uses the Silicon Labs Open Bus Interface [OBI22] in several of its RISC-V processors. It resembles a simplified AXI interface, with an address and response channel.

CoreConnect, developed by IBM, is another commercial SoC bus. It includes a high-speed processor local bus (PLB) with a bridge to a simpler on-chip peripheral bus (OPB). It is available for free but requires a license from IBM. Xilinx FPGAs used CoreConnect IPs before switching to AXI; little CoreConnect IP is now available.

SiFive developed an open-standard SoC bus called TileLink for ease of implementation and to provide high speed and cache coherence [TileLink20]. A key difference between AHB and TileLink is that AHB allows one controller to directly communicate with multiple peripherals, but TileLink allows only point-to-point links between a single controller and a single peripheral. TileLink adds a switch component when it needs to connect multiple agents or peripherals. TileLink's point-to-point links facilitate high-speed SoC networks in which the link is pipelined to permit multiple messages in-flight concurrently. However, SiFive has found that customers preferred the broadly accepted AHB or AXI bus rather than a novel bus.

9.4 TEST PLAN

The AHB bus has primarily been tested through normal use of memory and peripherals.

The `ahb64_ram_{0/1/2}_{0/1}_rv64gc` tests stress the variable latency AHB interface and burst mode operation. The first number gives the

RAM latency (0, 1, or 2 extra cycles of memory latency), and the second number indicates whether burst mode cache line fills are enabled (1) or not (0). These tests run fadd_b11-01, where the active data exceeds the cache sizes and, thus, generates memory traffic. A more comprehensive test plan would be beneficial.

Chapter 10 describes caches.

9.5 WALLY IMPLEMENTATION

The basic pipelined processor core from Fig. 7.13 has no bus interface because its memories (DTIM and IROM) are inside the core and it does not access peripherals. This section adds an *uncore* module with memory and peripherals that communicate with the core over an AHB bus. The memory and peripherals are collectively called AHB *devices*. When the BUS_SUPPORTED configuration variable in config.vh is 1, Wally adds an AHB interface to the load/store unit (LSU) and instruction fetch unit (IFU) to access peripherals or memory other than the DTIM or IROM. Wally also adds an *external bus unit* (EBU) that arbitrates between requests from the LSU and IFU and passes the selected request onto the uncore's AHB bus.

Wally's AHB interface is shown in Fig. 9.20. Wally includes AHB, EBU, and uncore modules when BUS_SUPPORTED = 1. The wallypipelinedsoc top-level module comprises the wallypipelinedcore and uncore modules. wallypipelinedcore is the processor core, which includes the LSU, IFU, and EBU (modules lsu, ifu, and ebu). The uncore includes on-chip RAM (ram_ahb) and ROM (rom_ahb) and an optional AHB

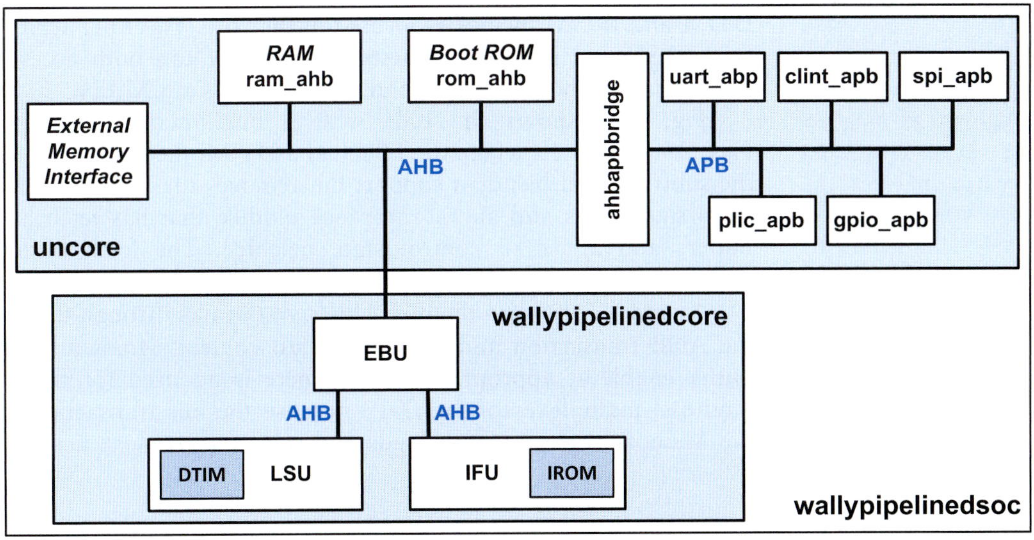

Fig. 9.20 Wally AHB interface and top-level system diagram.

Table 9.8 Wally bus memory configuration options.

Configuration Parameter	Meaning	Address
UNCORE_RAM_SUPPORTED	RAM in the uncore on AHB	0x80000000
BOOTROM_SUPPORTED	ROM in the uncore on AHB for bootloader	0x1000
EXT_MEM_SUPPORTED	AHB interface to RAM outside the uncore (e.g., via a DDR memory controller on an FPGA)	0x80000000

interface to off-chip memory. Table 9.8 summarizes the configuration parameters for these bus-based memories. The uncore also has an AHB-to-APB bridge (ahbapbbridge) to various peripherals described in Chapter 20. In the rv{32/64}i configuration (or to increase performance in configurations with a bus), the LSU and IFU contain a DTIM and IROM, as described in Chapter 7. Chapters 12 and 13 add optional caches.

To enable bus access, the LSU and IFU add AHB interfaces that include an FSM to sequence the AHB phases and a register to capture HRDATA. The core adds an EBU to arbitrate between LSU and IFU requests. Wally adds a new top-level wallypipelinedsoc module that contains the wallypiplinedcore and the uncore memories and peripherals.

The rv32imc configuration has a DTIM and IROM as well as a bus to uncore peripherals. It does not include external (uncore) memories.

Fig. 9.21 omits several AHB control signals for brevity. HBURST is hardwired to 0 for now; Section 12.2 introduces burst-mode cache line fetches and writebacks. HPROT is tied to 4'b0011 and HMASTLOCK is tied to 0 because protection and bus locking are not supported. HRESP is ignored because the memory system does not produce any error conditions. The uncore module generates HSEL based on HADDR and configured memory ranges for memories and peripherals, as discussed in Section 9.5.3.1.

9.5.1 LSU and IFU AHB Interface

This section describes a configuration where the LSU and IFU add AHB bus interfaces to access uncore memory instead of or in addition to the DTIM and IROM memories located in the core. The EBU arbitrates between LSU and IFU bus requests so that they can both access uncore AHB devices. The bus HRDATA and HWDATA widths are XLEN.

Fig. 9.21 shows the LSU with a bus interface. Compared to Figs. 7.13 and 8.19 where the LSU had a DTIM, this updated LSU keeps the subword and biendian support but removes the DTIM and adds the AHB signal pins and an ahbinterface module that has fetch and write buffers and a busfsm control state machine. The fetch buffer holds the HRDATA read from the bus, and the write buffer holds the values driving HWDATA and HWSTRB onto the bus. The busfsm walks through the phases of the AHB transaction and generates AHB control signals and the fetch buffer enable at appropriate times. There is no need for the DTIM A (address) multiplexer of Fig. 7.13 because the bus transaction starts in the Memory stage for both loads and stores; so IEUAdrM ties directly to LSUHADDR.

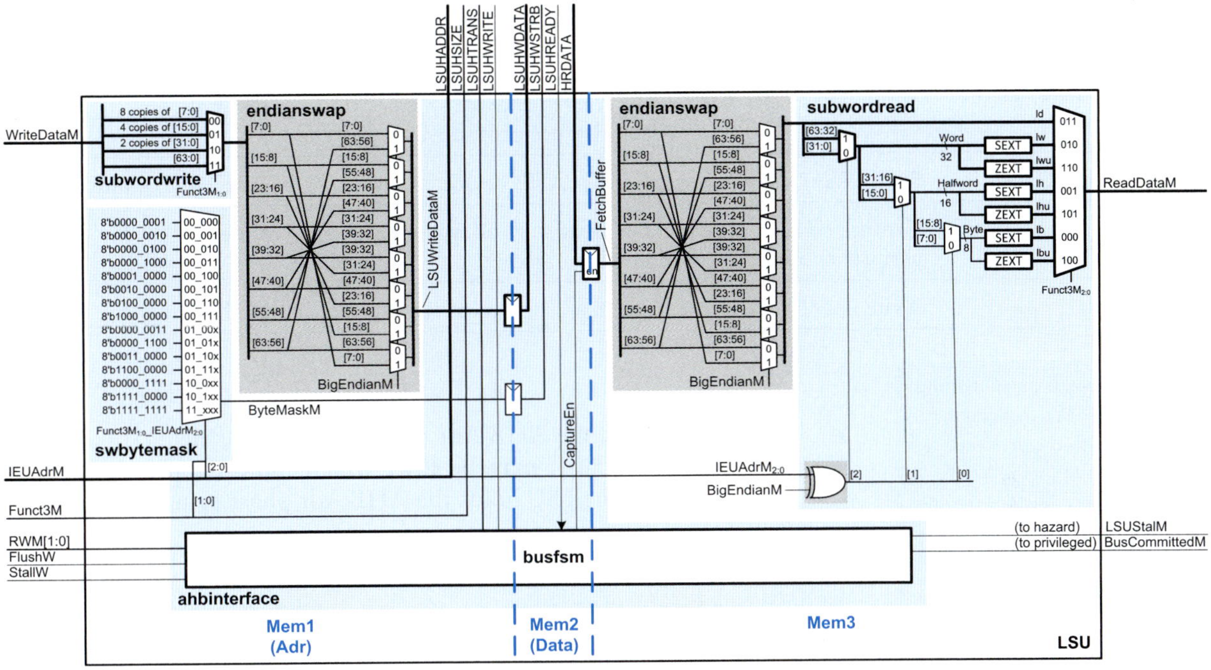

Fig. 9.21 **LSU with AHB interface (`lsu` module configured with `` `BUS_SUPPORTED `` = 1, `` `DTIM_SUPORTED `` = 0, `` `DCACHE_SUPPORTED `` = 0).**

Recall from Table 7.8 that an LSU DTIM access involves work in the Execute, Memory, and Writeback stages but only stalls for hazards. In contrast, an access to bus-based memory always takes at least two extra cycles, as shown in Fig. 9.22 for `ld` followed by additional instructions (`i1`, `i2`, ...). The LSU stalls the CPU during these extra cycles by setting `BusStall` = 1, so the LSU with an AHB interface behaves as a multicycle (nonpipelined) system. Table 9.9 summarizes the steps the LSU takes to access the AHB bus. The Memory stage is stalled while it takes at least three cycles:

The Mem2 and Mem3 phases could be merged, but this might increase the cycle time of the processor. In a processor with a DTIM/IROM or cache, bus accesses are infrequent, so it is better to have a fast cycle time than a lower bus latency.

Cycle	F	D	E	M	W	State/AHB	BusStall
0	ld						0
1	i1	ld					0
2	i2	i1	ld				0
3	i3	i2	i1	ld		Adr	1
4	i3	i2	i1	ld		Data	1
5	i3	i2	i1	ld		Mem3	0
6	i4	i3	i2	i1	ld		0

Fig. 9.22 **`ld` instruction using the AHB bus.**

Table 9.9 **LSU AHB access steps.**

Stage	Read	Write
Execute	Compute memory address IEUAdrE = base address + offset	
Mem1/AHB Adr	Drive HADDR = IEUAdrM; generate other AHB control signals Writes also perform subword write muxing	
Mem2/AHB Data	Read HRDATA synchronously from AHB device	Drive HWDATA onto AHB
Mem3	Capture HRDATA into ReadDataM at start of cycle, then perform subword read and endian muxing	Write HWDATA into device
Writeback	Write ReadDataW into register file	(Nothing)

The AHB bus is pipelined: the current request's Adr phase overlaps with the previous request's Data phase. Wally could leverage this pipelining to eliminate the two bus stall cycles and, thus, reduce AHB memory access latency. But this would require modifying the standard 5-stage pipeline so that memory accesses occur in 2 cycles instead of 1. Right now, the Fetch or Memory stage receives its address at the beginning of the stage and produces its read or write data at the end. To take advantage of AHB's pipelining, the address would need to be calculated a cycle earlier (which would increase the LSU's IEUAdrE critical path) or the data would need to be read or written a cycle later (which would add an extra cycle to the LSU's Load Delay Hazard). These modifications could be made, but bus transactions are infrequent because most accesses are filtered by the cache, and Amdahl's Law says to make the common case fast. So, it is better to maintain a lower cycle time than pay the price for a lower bus latency.

- In the first phase, the AHB Adr phase (also called Mem1), HREADY (in this case, LSUHREADY), may be low, which indicates the bus cannot take a request. The LSU holds the AHB Adr phase signals steady until LSUHREADY is high.
- In the second phase, the AHB Data phase (also called Mem2), the device reads or writes. A slow device may stretch this phase beyond one cycle by keeping LSUHREADY low.
- The third phase (Mem3) captures HRDATA in the FetchBuffer register and muxes the FetchBuffer bits for either subword accesses or, if needed, to reorder bytes for endianness to produce ReadDataM.

The LSU's AHB bus interface has a three-state FSM controller (busfsm) shown in Fig. 9.23 that sequences the steps from Table 9.9 for both stores and loads. State names match the AHB specification's Adr and Data phases with the added third state, Mem3. All memory requests start in the Adr phase. If read or write is requested, busfsm asserts BusStall to stall the pipeline while the bus services the request. When the bus is ready, busfsm asserts HTRANS = 2 (NONSEQ) to initiate a single AHB read or write and advances to the Data state. While in the Data state, the FSM asserts BusCommitted to prevent the privileged unit from taking interrupts until the bus transaction completes. When the bus raises HREADY to complete the Data phase, busfsm advances to Mem3 and raises CaptureEn on reads to enable the fetch buffer to sample HRDATA. While in Mem3, BusStall is lowered, so the pipeline can advance. The Stall, BusStall, and BusCommitted signals are discussed further in Section 9.5.5.

The IFU bus interface is similar to the LSU interface but is 32-bit little-endian read-only. Thus, it drops HWDATA, HWRITE, HWSTRB, and the subword and endian logic, as shown in Fig. 9.24, and HSIZE is 010 to always fetch 4-byte instructions. The IFU busfsm receives StallD instead of StallW. In the

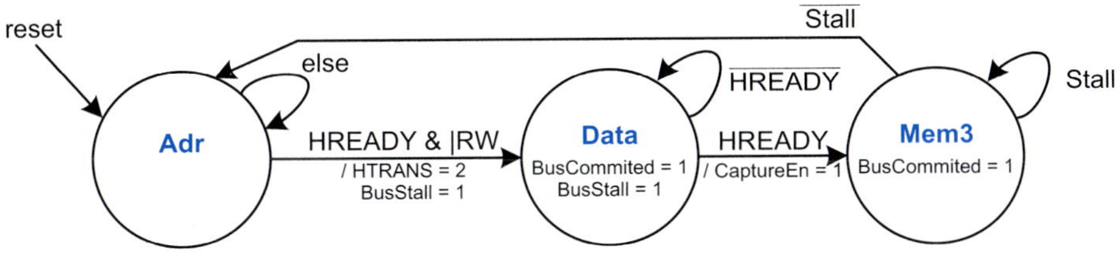

Fig. 9.23 `busfsm` **state machine;** `Stall` = `StallW` **for the LSU and** `StallD` **for the IFU.**

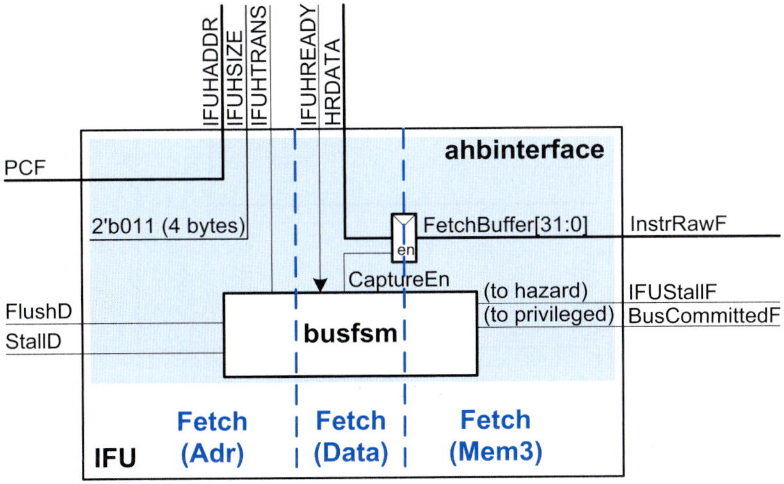

Fig. 9.24 IFU AHB interface (`ifu` **module configured with** `` `BUS_SUPPORTED `` = **1,** `` `IROM_SUPORTED `` = **0,** `` `ICACHE_SUPPORTED `` = **0).**

Without some sort of local memory, the processor runs slowly because each instruction or data access takes at least 3 cycles over the bus. Chapter 12 describes the full LSU with a cache and bus. The cache services most accesses, while the bus fetches cache misses from RAM and accesses peripherals in the uncore.

Mem3 stage, the IFU outputs `InstrRawF` from the fetch buffer. The AHB signals from each unit have corresponding `IFU` or `LSU` prefixes, and the EBU must arbitrate between requests from the two units, as discussed next.

9.5.2 External Bus Unit

The external bus unit (EBU) implements a subset of the multilayer AHB protocol to arbitrate between the IFU and LSU AHB interfaces (also called controllers). LSU requests get priority. The selected controller's request is placed on the bus to the uncore. As shown in Fig. 9.25, the EBU consists of an FSM arbiter (`ebufsmarb`), muxes to choose the selected controller, and registers to retain the other controller's request. Section 12.2.3 extends the EBU to handle burst transactions for fast cache line fills.

When `LSUSelect` is 1, the muxes send the LSU signals to the bus. Otherwise, if `IFUSelect` is 1, the mux sends the IFU signals. If neither is selected, the bus receives 0, making it idle. The read data bus (`HRDATA`) goes to both the IFU and LSU. Only the LSU writes to external memory,

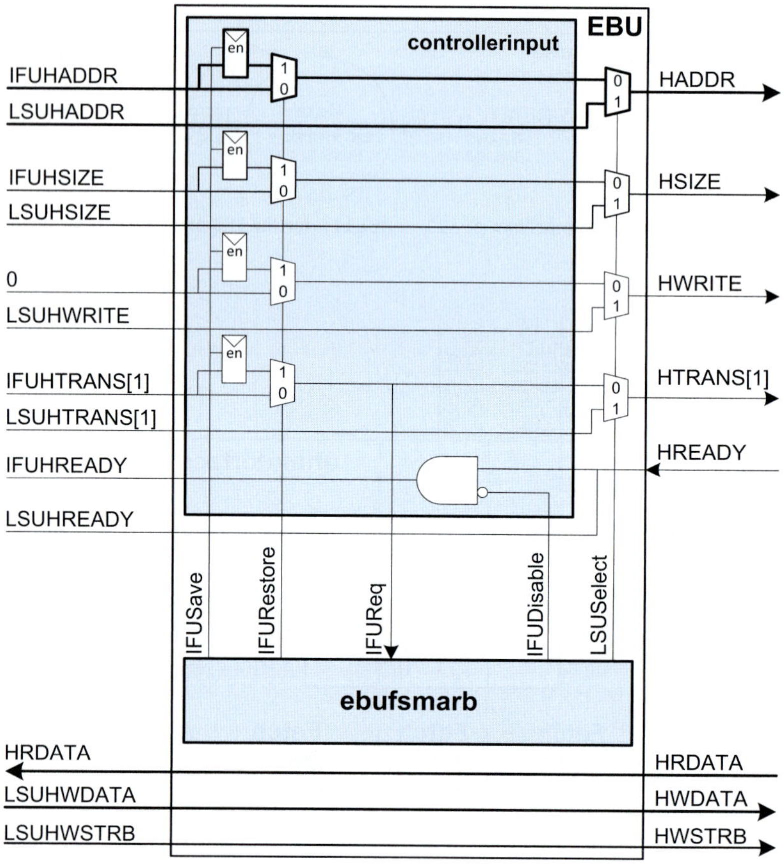

Fig. 9.25 External bus unit (EBU).

so HWDATA only connects to the LSU, and the IFU never asserts IFUHWRITE, so that signal (just above LSUHWRITE) is tied to 0.

The ebufsmarb, shown in Fig. 9.26, arbitrates when both the LSU and IFU make requests at the same time. It always gives priority to the LSU to prevent deadlocking the pipeline. Only two states are required, an Idle state to indicate either 1 or 0 requests are active and an Arbitrate state, which selects the LSU. It is a Mealy machine with outputs that are a function of the current state and inputs. When two requests occur in the Idle state, the FSM transitions to Arbitrate and selects the LSU. Because the IFU detects HREADY = 1, it assumes the Adr phase of the request was taken and moves on to the Data phase. So, the ControllerInput module must save the IFU's Adr phase HADDR and control signals for later. Data phase signals are inherently delayed, so they need not be stored and replayed. When the device completes the memory request, HREADY asserts, and the FSM returns

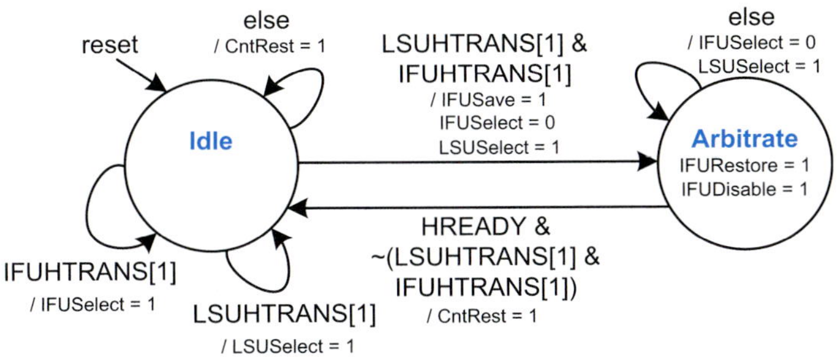

Fig. 9.26 EBU FSM arbiter (ebufsmarb), with the LSU having priority over the IFU.

to Idle. Upon returning to Idle, the IFU Adr phase is replayed by driving the stored IFU request onto the bus.

Fig. 9.27 shows the timing of a concurrent IFU and LSU read followed by two more IFU reads:

- **In cycle 0**, both the IFU and LSU make requests by driving `IFUHTRANS` and `LSUHTRANS` to NONSEQ (2). The FSM chooses the LSU, raising `LSUSelect` to grant the LSU bus access and drive the LSU Adr phase signals onto the bus: `HADDR` = `LA1`. The IFU request is saved inside the `controllerinput` register with `IFUSave` = 1.
- **In cycle 1**, the FSM transitions to the Arbitrate state. `HREADY` is 1, indicating that the device has returned `HRDATA`. `LSUHREADY` asserts to capture `HRDATA` into the LSU. Meanwhile, the EBU selects the saved IFU request's Adr phase signals (IA1) to drive onto the AHB bus. `IFUHREADY` is 0 to continue stalling the IFU.
- **In cycle 2**, the FSM transitions to the IDLE state. The device responds to the IFU request with `HRDATA` for IA1 and `HREADY` = 1.
- **In cycles 3 and 4**, two more requests originate from the IFU but, because the LSU is no longer making requests, no arbitration is needed.

When the data cache is added to the LSU, cache misses will generate AHB burst requests. Burst operations require the EBU to count each word of the transaction and ensure the lower priority IFU is not interrupted during a burst.

9.5.3 System-on-Chip

As shown in Fig. 9.20, the Wally SoC (`wallypiplinedsoc`) includes both the processor core and the uncore module that contains memory, peripherals, and an external memory interface connected to the core via the bus. Chapter 20 describes the peripherals further. The external memory interface could be an off-chip AHB device such as a DDR4 memory controller. The uncore module also contains address decoding and device multiplexing. Fig. 2.5 shows Wally's default memory map.

HRDATA and HWDATA bus widths are AHBW, which is set to the natural word size XLEN in Wally configurations. Section 11.2.1 discusses virtual and physical addressing and explains that RISC-V uses 34-bit physical addresses for RV32 and 56-bit physical addresses for RV64. HADDR is a physical address of width PA_BITS (34 or 56). Recall that programs use 32- or 64-bit addresses. In systems with no MMU, RV32 prepends two leading 0s to create 34-bit addresses, and RV64 drops the 8 most significant bits to create 56-bit addresses. Systems with an MMU that supports virtual memory translate XLEN-bit virtual addresses to PA_BITS physical addresses (see Chapter 11).

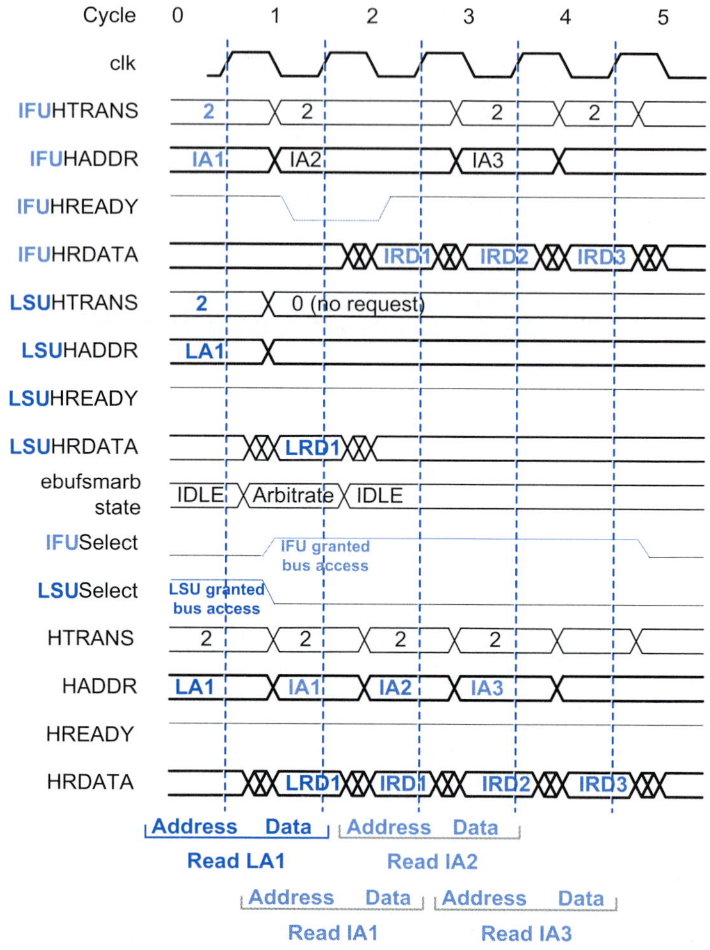

Fig. 9.27 Timing diagram for IFU and LSU arbitration of concurrent requests.

9.5.3.1 Address Decoding and Device Mux

As shown in Fig. 9.12, AHB requires an address decoder and multiplexer to select and choose from among multiple devices. Fig. 9.28 shows the uncore, which includes the address decoder, mux, and several peripherals, emphasizing the HADDR, HREADY, HREADYOUT, and HSEL connections.

Recall that AHB is pipelined into Adr and Data phases and that a transaction ends when the selected device asserts its HREADY signal in its Data phase. Because of pipelining, the HREADY signal in the Adr phase of one transaction comes from the Data phase of the previous transaction using the delayed HSEL signal, HSELD.

The decoder uses the configuration parameters <DEVICE>_SUPPORTED, <DEVICE>_BASE, and <DEVICE>_RANGE to determine which devices (e.g.,

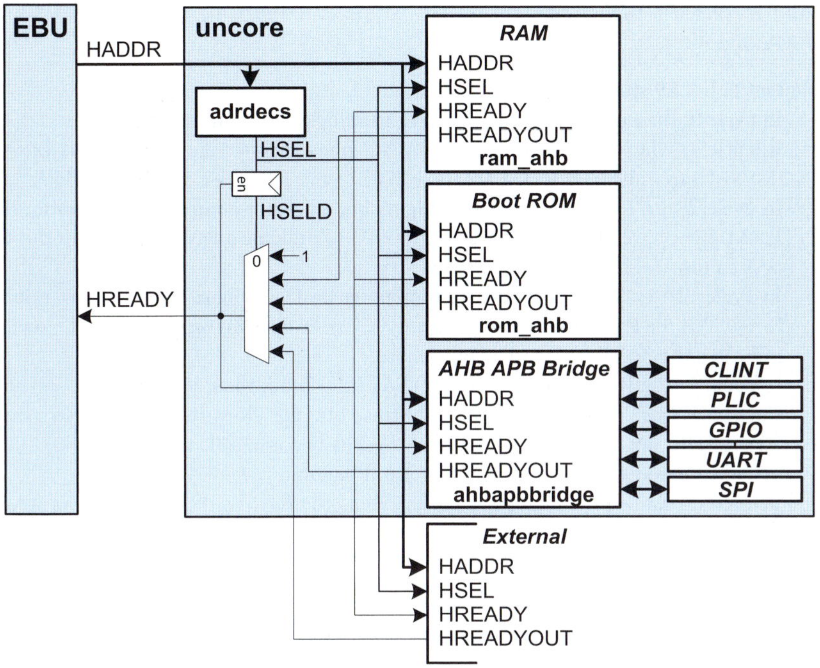

Fig. 9.28 **Uncore interface from EBU to devices.**

Boot ROM, uncore RAM, GPIO) are supported and where they appear in the memory map.

The address decoders module (adrdecs) receives HADDR in the Adr phase and produces a one-hot HSEL signal, with 1 bit per device. The module contains an address comparator for each device; an implementation is shown in Section 11.4.1.1. HSEL[0] is asserted when HADDR doesn't match any device address. When the previous transaction's Data phase completes, as indicated by that device asserting HREADY, the current transaction enters the Data phase and latches HSEL to HSELD. The currently selected device reads or writes during the Data phase and raises its HREADYOUT when complete. The HREADY multiplexer picks HREADYOUT from the selected device (based on HSELD) and drives HREADY back to the EBU and to all the devices. If no device is selected, the bus is idle and immediately asserts HREADY to avoid locking up. The result multiplexer, not shown in this diagram, also selects HRDATA and HRESP from the selected device.

Example 9.2 ROM AND RAM READS WITH WAIT STATES

Suppose an AHB bus is initially idle, then reads from the ROM immediately followed by a read from the RAM, and then returns to idle. Suppose the ROM and RAM are slow, with 2 and 1 extra wait states, respectively. Let the ROM and RAM be devices

1 and 2. Sketch a timing diagram of HSEL, HSELD, HREADYOUT (from the ROM and RAM), and HREADY.

Solution: Fig. 9.29 shows a timing diagram.

- **In cycle 0,** the bus is idle: HSEL[0] = 1 to indicate no requests.
- **In cycle 1,** when the ROM transaction Adr phase begins, the controller asserts HSEL[1] to indicate a ROM access. HREADY was 1 because the bus was idle.
- **In cycle 2,** the ROM's Data phase begins but does not complete. In addition, the RAM Adr phase begins and asserts HSEL[2] to indicate a RAM access, but the access cannot begin until HREADY rises.
- **In cycles 2–4,** the slow ROM is completing its Data phase and the mux selects HREADY as the ROM's HREADYOUT using HSELD (a delayed version of HSEL), which indicates the peripheral in the Data phase.
- **In cycle 4,** the ROM finally asserts HREADYOUT (and, thus, HREADY).
- **In cycle 5,** the RAM's Data phase begins, but the slow RAM does not assert HREADYOUT until cycle 6. Meanwhile, the bus becomes idle and asserts HSEL[0] to indicate no new requests in cycles 5 and beyond.
- **In cycles 5 and 6,** the mux selects HREADY as the RAM's HREADYOUT. The RAM finally asserts HREADYOUT in cycle 6.

Upon reset, both HSEL[0] and HSELD[0] must be set to 1 to assert HREADY and, thus, prevent the bus from locking up.

9.5.4 Combined Bus and Tightly Integrated Memories

An embedded processor typically doesn't have caches but instead has a small IROM and DTIM in the core as well as a bus to peripherals or external memories. For example, the Wally rv32imc configuration is organized this way. The address decoder must distinguish between accesses to these TIMs versus bus accesses and use a mux to access them accordingly.

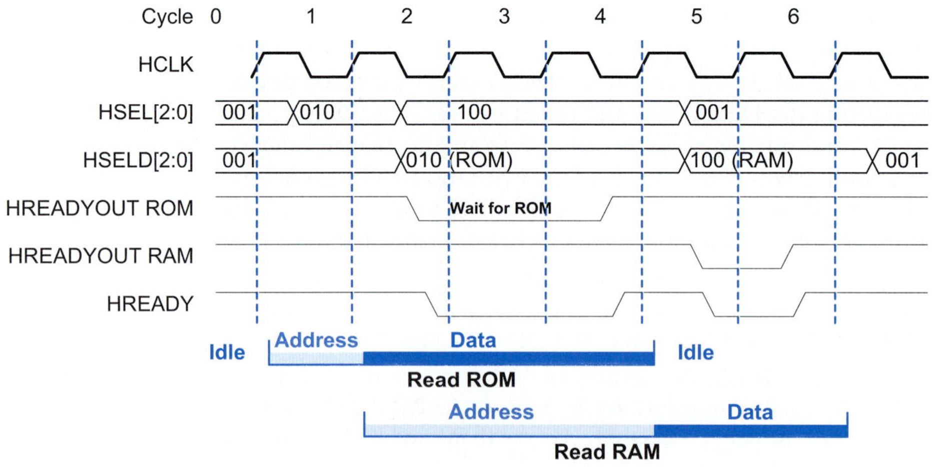

Fig. 9.29 AHB timing diagram of consecutive reads to different devices with wait states.

Wally handles this decoding using a PMA checker in the MMU, as described in Sections 11.4.1.1 and 12.3.1. Fig. 9.30 shows a diagram of the LSU with both a DTIM and AHB bus. The PMA checker's address decoder outputs `SelDTIM`, which controls a multiplexer to select between the DTIM's read data (`DTIMReadDataM`) or the AHB's fetch buffer (`FetchBuffer`). `SelDTIM` also demuxes the read/write signal `RWM` to either the DTIM or `busfsm`.

Fig. 9.31 abstracts the LSU to emphasize the data and address paths. The `subwordwrite`, `subwordread`, and `endianswap` modules are abbreviated as SWW, SWR, and ES. The `busfsm` is hidden in the `ahbinterface`. The processor stalls in the Memory stage during the multicycle bus access. The three (or more) AHB cycles (Adr, Data, Mem3) are not shown, and the core resumes in the Memory stage when the stall is released during Mem3.

The IFU uses the same approach to select between the IROM and the Bus, as shown in Fig. 9.32. A separate PMA checker in the IFU's MMU outputs `SelIROM` to select the IROM or the Bus. Unlike the LSU's DTIM, the IROM only reads instructions. `PCNextF` directly drives the IROM read address port.

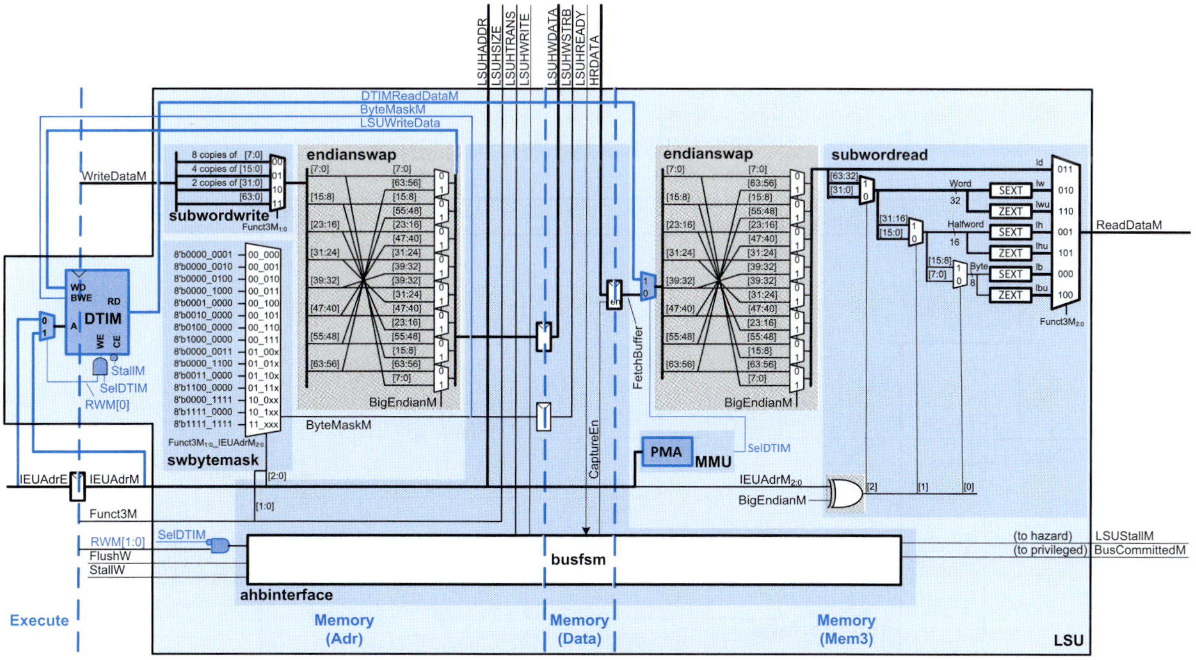

Fig. 9.30 **LSU with DTIM and bus.**

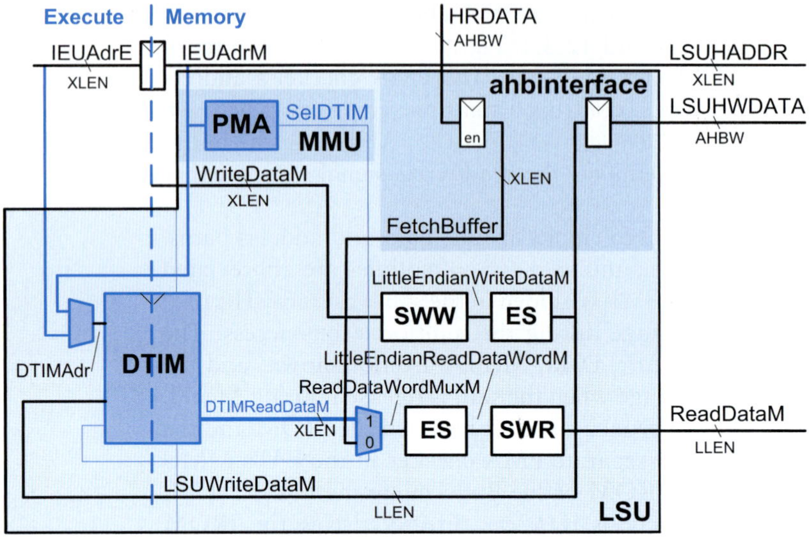

Fig. 9.31 Abstracted LSU with DTIM and bus.

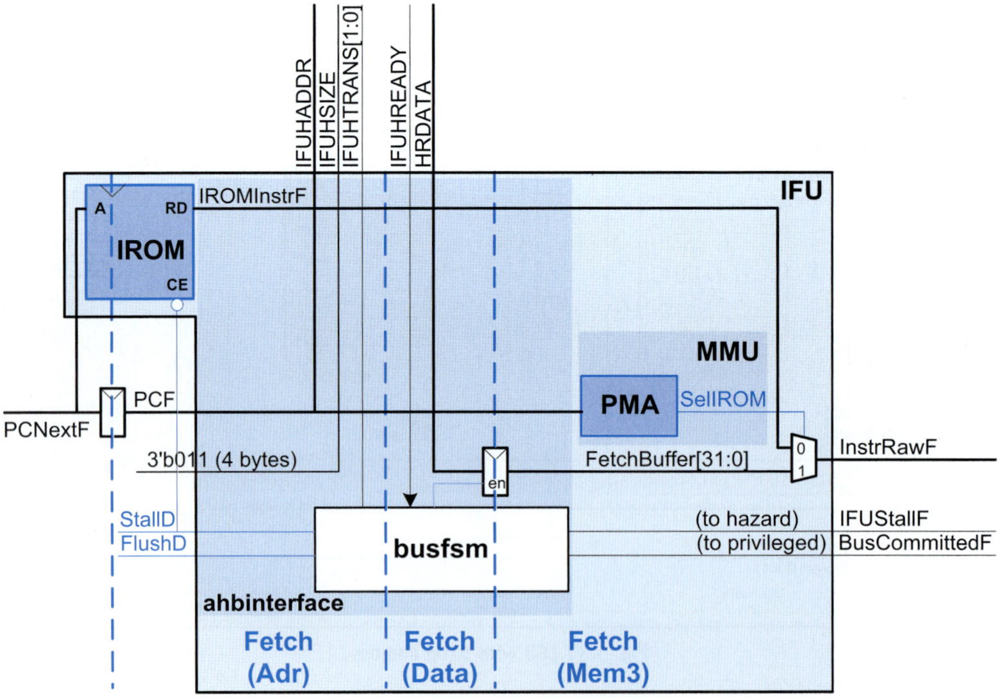

Fig. 9.32 IFU with IROM and bus.

9.5.5 Bus Hazards

Each AHB interface in the IFU and LSU takes multiple cycles to complete a bus transaction. The `busfsm` from Fig. 9.23 must *interlock* with the hazard unit to correctly stall the pipeline and delay interrupts occurring in the middle of a bus transaction.

Because each bus transaction takes multiple cycles, the LSU or IFU `busfsm` generates `BusStall` while in the midst of a transaction. Specifically, `BusStall` asserts when the bus transitions from the Adr phase into the Data phase and remains high until the Data phase completes. When `BusStall` is lowered, the hazard unit unstalls the pipeline, and the IFU or LSU enters its Mem3 stage, producing a valid instruction `InstrRawF` or capturing the load result in `ReadDataM`.

In the LSU, `LSUStallM` = `BusStall` from its `busfsm`. Similarly, the IFU's `IFUStallF` = `BusStall` from its `busfsm`. The hazard unit raises `StallW` to stall the entire pipeline when either unit stalls, as shown in Fig. 9.33. Traps take priority and should flush the pipeline rather than processing the LSU/IFU stall. However, traps are only allowed at the start (i.e., in the Adr phase) of a bus transaction. The `CommittedM/F` signals, discussed later in this section, postpone interrupts arriving in the midst (in the Data phase) of a bus transaction.

Recall that when both the IFU and LSU need to access the bus, the EBU prioritizes the LSU. In that case, when the LSU completes its request and lowers `LSUStallM`, the pipeline remains stalled until the IFU's request is finished. Because `StallW` is still asserted, the LSU's `busfsm` remains in Mem3 to prevent reissuing the bus request. After the IFU's `busfsm` transitions to Mem3, both units lower their `BusStall` signals and the pipeline resumes. Other hazards such as CSR writes may also stall the pipeline, keeping the busfsm in Mem3 to prevent reissuing the bus request.

The `busfsm`'s `BusCommitted` signal delays pending interrupts until a bus transaction completes. `BusCommitted` must remain high while waiting in Mem3 so that the interrupt does not occur until the next instruction. `BusCommitted` is zero only during the Adr phase, which allows interrupts to occur before either the LSU or IFU starts a bus access. If a trap occurs while the IFU or LSU is in the Adr phase, the bus transaction is canceled by driving `HTRANS` to 0.

An *interlock* disallows certain states to occur that would conflict. For example, an interrupt that occurs during a bus transaction must be delayed until after the bus transaction completes.

`LSUStallM` also asserts on cache and hardware page table walker stalls (see Section 12.4.1). Similarly, `IFUStallF` also asserts on cache stalls (see Section 13.3.5) and spills (see Section 14.3.3).

Recall that a stall at a given stage stalls all previous stages. Thus, a stall in the Writeback stage stalls the entire pipeline.

The bus stall could be expressed as follows:

```
StallWCause = (IFUStallF &
~FlushFCause | LSUStallM)
& ~FlushWCause
```

A Fetch flush (`FlushFCause`) has priority over an IFU stall (`IFUStallF`), and a Writeback flush (`FlushWCause`) has priority over an LSU stall. `FlushFCause` is not explicitly defined but is equivalent to `FlushDCause` because a flush in one stage also flushes all previous pipeline stages. Similarly, `FlushDCause` always asserts whenever `FlushWCause` asserts. Thus, the bus stall simplifies to the following, as also shown in Fig. 9.33:

```
StallWCause = (IFUStallF &
~FlushDCause | LSUStallM)
& ~FlushWCause
```

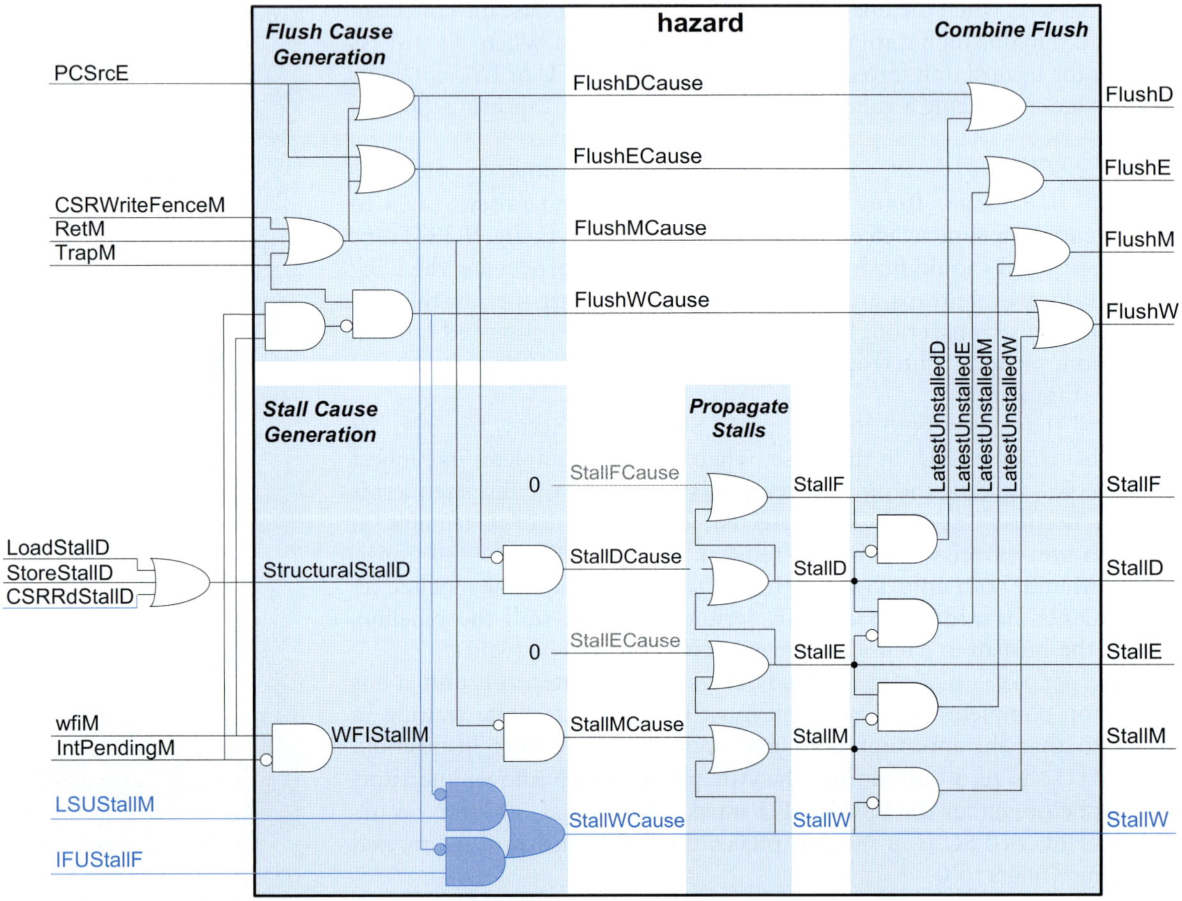

Fig. 9.33 Hazard unit, with changes due to the AHB highlighted in blue.

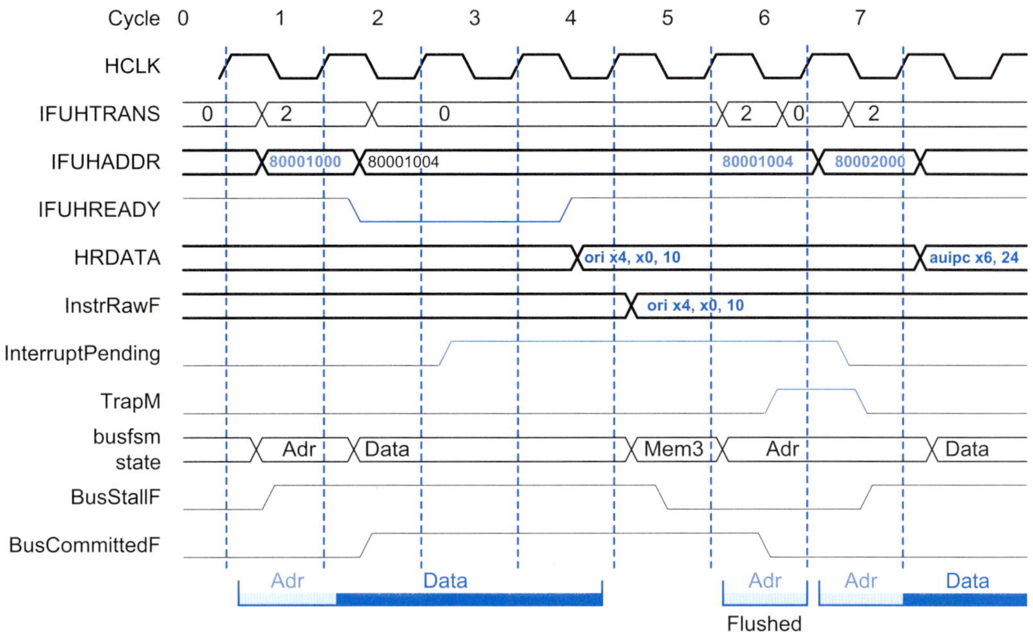

Fig. 9.34 Privileged unit interlock with the IFU's `busfsm`.

Fig. 9.34 shows how the IFU `busfsm` delays interrupts until the current bus transaction completes during the execution of Code Example 9.1:

- **In cycle 1**, the CPU begins fetching the `ori` instruction from 0x80001000 by entering the Adr phase of the bus.
- **In cycle 2**, the `busfsm` asserts `BusCommittedF` when entering the Data phase of the bus.
- **In cycle 3**, an interrupt becomes pending (`InterruptPending` = 1), but because `BusCommittedF` = 1, the interrupt is delayed.
- **In cycle 5**, the bus completes fetching `ori` (enters Mem3 phase).
- **In cycle 6**, the CPU begins fetching the `addi` instruction from 0x80001004 by entering the Adr phase of the bus. `BusCommittedF` = 0 in the Adr phase, which allows the pending interrupt to be handled. `TrapM` flushes the pipeline, including the `addi` fetch that was in the Adr phase.
- **In cycle 7**, the trap handler begins fetching the `auipc` from 0x80002000, which enters the Adr phase of the bus.

The IFU could allow `TrapM` in cycle 5 to flush the completed fetch because the IFU only reads instructions. In contrast, when an instruction begins to write to memory, it must commit. Section 11.2.2.1 defines *nonidempotent* reads that have side effects and also must commit. For example, reading the Receive Buffer Register (RBR) in a 16550 UART also empties the buffer. Therefore, the RBR is nonidempotent because consecutive reads to the same address return different values.

The hazard unit could choose to stall just the Decode stage when `IFUStallF` asserts. However, this would mean that an exception could occur in the Memory stage while an instruction is being fetched. To ensure the exception is taken, the hazard unit would have to stall the Memory stage until the IFU lowers `BusStall`. Caches complicate this issue further. Thus, Wally instead stalls the entire pipeline during both IFU and LSU bus accesses.

Code Example 9.1 DELAY INTERRUPTS DURING BUS TRANSACTION

```
80001000:      ori  x4, x0, 10
80001004:      addi x5, x4, 10
. . .
TrapHandler:
80002000:      auipc x6, 24
```

In this chapter, `CommittedF` = `BusCommittedF`. Later chapters will add other cases when `CommitedF` asserts. Similar to the IFU, the LSU asserts `CommitedM` when its `BusCommitedM` asserts. The `trap` module in the privileged unit asserts `Committed` when either the LSU's `CommitedM` or the IFU's `CommitedF` assert. `Committed` delays interrupts until the instruction completes. Fig. 9.35 shows the `trap` unit from Fig. 8.14 with the new `Committed` logic highlighted in blue.

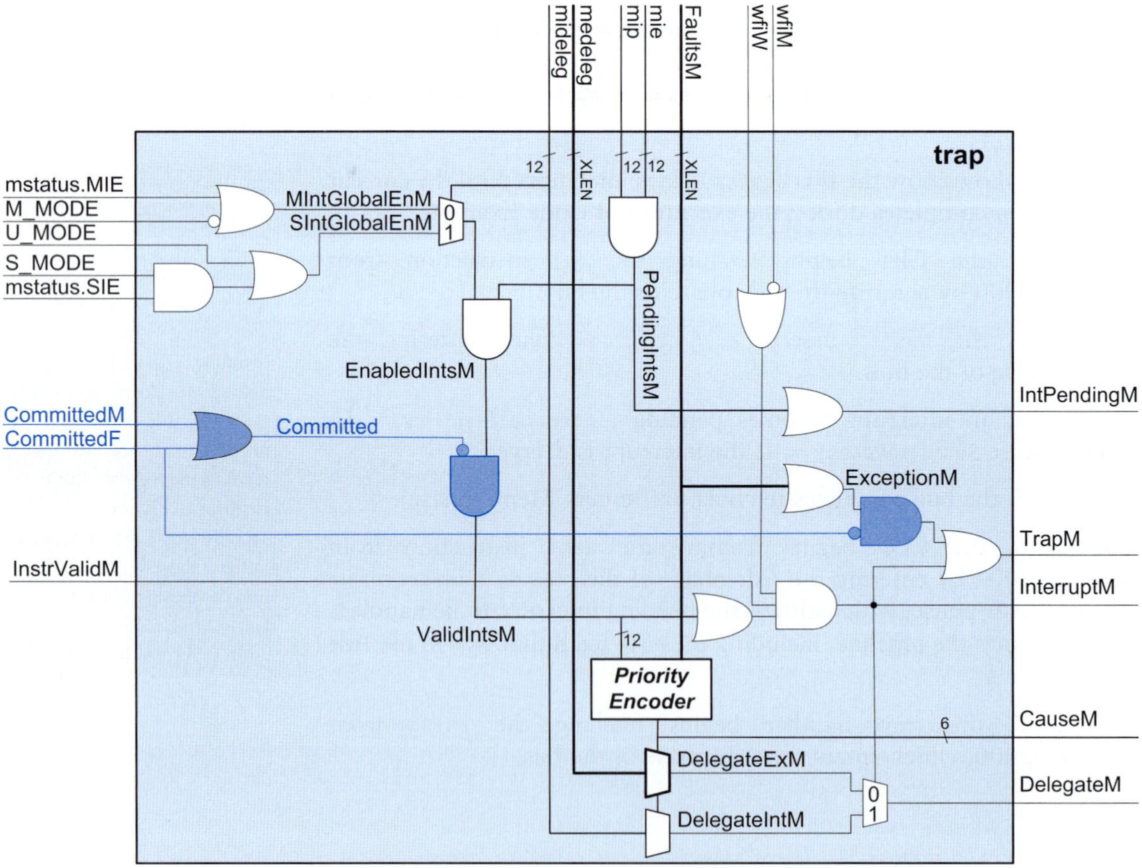

Fig. 9.35 `trap` **unit delays pending interrupts while committed instructions complete.**

9.6 SUMMARY

A bus interface provides a standardized way for a system to communicate with external memory or peripherals. Modern on-chip buses normally use synchronous point-to-point links between a controller and one or more peripherals. The controller generates a clock signal and sends an address and some control information to each peripheral. During a write operation, the controller also sends the write data. During a read operation, the peripheral responds with the read data. The peripheral may insert wait states if it needs longer to respond. Because of the point-to-point links, the controller uses a read multiplexer to choose the read data from the selected peripheral. High-performance buses are normally pipelined to overlap consecutive accesses. They support features such as burst and subword transfers and bus locking for atomic operations. Many bus protocols are used in industry, so a bridge is often needed to adapt between protocols.

Wally uses an EBU to connect the IFU and LSU to a single AHB bus so they can access peripherals and external memory, beyond the capacity of an IROM or DTIM. When both the IFU and LSU want to use the bus at the same time, the EBU grants priority to the LSU so the pipeline can make forward progress. Bus transfers take multiple cycles and must finish once they have started, so the EBU interlocks with the hazard and privileged units to stall the pipeline and prevent interrupts during in-flight transfers until the transfer completes.

Caches

10

10.1 **Principles**
10.2 **RISC-V Practices**
10.3 **Test Plan**
10.4 **Wally Implementation**
10.5 **Summary**

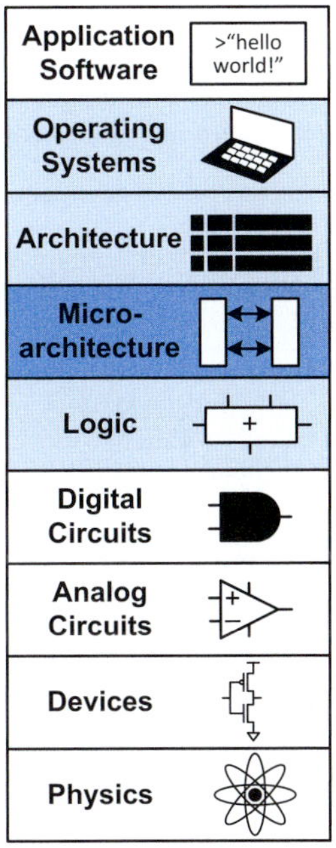

An ideal processor would have a large, fast, cheap memory. Unfortunately, the memory available on-chip is usually static random access memory (SRAM), which is fast but expensive per bit and has relatively low density. Off-chip dynamic random access memory (DRAM) provides gigabytes of memory at low cost and high density but is significantly slower than a CPU (central processing unit). Processors provide the illusion of a large, fast, cheap memory by storing most data in the large, cheap DRAM but *caching* frequently accessed information in the smaller but fast on-chip SRAM. Pipelined CPU designs typically use separate first-level (L1) caches for instructions and data so that both can be accessed concurrently without stalling the pipeline. The *instruction cache* (*I$*) is generally only readable, but the *data cache* (*D$*) must support both reads and writes. In systems with multiple cache levels, these are referred to as *level 1* caches (L1I$ and L1D$). Larger systems usually also have a *second-level cache* (L2$), with larger capacity but longer access times, that holds both instructions and data. Multicore systems often have an even larger *third-level cache* (L3$) shared across cores. Some server SoCs, such as IBM Power 8 and 9, add an off chip L4$ using embedded DRAM (eDRAM) that is still faster than going to the main system DRAM.

The levels of cache and memory in the system are called the *memory hierarchy*. Fig. 10.1 shows a memory hierarchy with three levels of cache, a main memory, and *virtual memory*. Virtual memory uses a slow but inexpensive hard drive to hold more information than fits in DRAM.

This chapter begins with a short review of cache principles. Then it examines the cache architecture used in the Wally instruction and data caches. Section 9.5 described how the processor accesses main memory using the system bus. Chapter 11 discusses virtual memory, and Chapters 12 and 13 discuss the rest of the load/store unit (LSU) and instruction fetch unit (IFU), which contain these caches.

The *lowest* level of the memory hierarchy is an L1$; the *higher* levels are the L2$ and L3$ followed by main memory and then virtual memory.

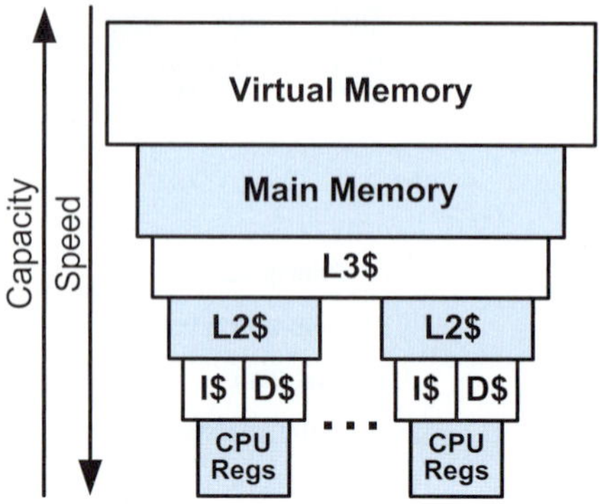

Technology	Capacity	Access Time	Bandwidth (GB/s)	Price / GiB
SSD	128 GiB - 8 TiB	100 - 1000 us	0.05 - 15	$0.05
HDD	1 - 20 TiB	5 - 20 ms	0.001 - 0.1	$0.02
DRAM	16 - 128 GiB	50 - 150 ns	30 - 50	$3
SRAM	3 - 100 MiB	10 - 25 ns	> 100	$100s
SRAM	128 KiB - 2 MiB	2 - 10 ns	> 100	$100s
SRAM	16 - 64 KiB	0.5 - 4 ns	> 100	$100s
Flip-Flops	100 - 1000 Bytes	100 - 500 ps	250 - 1000	$1000s

Fig. 10.1 Memory hierarchy.

See [Hennessy19] for a more detailed introduction to caches.

In March 2025 AMD released the Ryzen 9 9950X3D with a massive 128 MiB L3$ using a 3D stacked memory. Two dice, the 9950X with a 64 MiB L3$ and an extra 64 MiB L3$ cache, are stacked and connected with through-silicon vias (TSV). Previous generations of X3D chips, the 5800X3D and 7800X3D stacked the cache die on top of the commute die to simplify power delivery. However, the 9950X3D moved the cache die below the commute die to improve device cooling achieving higher clock frequencies with both designs boosting to 5.7 GHz. The 9950X3D has 20 − 30 % higher frames rates than the 9950X in many video games, which highlights the benefits of a large cache.

10.1 PRINCIPLES

Caches improve the effective latency and bandwidth of the overall memory system by taking advantage of repetitive patterns in program behavior. For some data-intensive applications, large caches improve performance more than CPU clock speed does.

A memory access *hits* in the cache if it finds the information there. If it *misses* the cache, it must go to the next-level cache or to main memory to find the information, which increases the access latency.

This section first examines how caches affect performance, then explores how caches exploit *locality* to increase hit rates, and finally describes principles to build a cache.

10.1.1 Cache Performance

An ideal memory system would introduce no delays, but a practical system slows down when the cache misses and needs to wait for data from the next level of the memory hierarchy. The *miss rate* is the fraction of cache accesses that miss. The *miss penalty* is the average number of cycles required to handle the miss. In a simple pipelined processor, misses to the L1$ stall the pipeline until it can fetch the information from a higher-level cache or main memory. The number of stall cycles is:

$$\text{Memory stall cycles} = \text{Number of instructions} \times (\text{Memory accesses / instruction}) \times \text{Miss rate} \times \text{Miss penalty} \quad (10.1)$$

Example 10.1 EFFECT OF CACHES ON CPI AND PERFORMANCE

Suppose a pipelined processor has a CPI (cycles per instruction) of 1.2 when all memory accesses hit in the cache. 30% of instructions are loads or stores. The miss rate is 2%, and the miss penalty is 40 cycles. How much slower is the processor because of cache misses? How much slower would it be if there were no cache?

Solution: The program has 1.3 memory accesses/instruction (1 instruction fetch and 0.3 loads/stores). So, the number of memory stall CPI, which is Equation 10.1 divided by the number of instructions, is $1.3 \times 0.02 \times 40 = 1.04$. This adds to CPI, so overall CPI is $1.2 + 1.04 = 2.24$. This increases program execution time by $(2.24 - 1.2)/1.2 = 87\%$. If there were no cache, every memory access would incur a miss penalty, and CPI would be $1.2 + (1.3 \times 1 \times 40) = 53.2$, or about 24x (53.2/2.24) slower.

A measure of overall memory performance is the *average memory access time* (*AMAT*), which is defined as the average time memory operations take to complete. AMAT is:

$$\text{AMAT} = \text{A}_1 + \text{MR}_1(\text{A}_2 + \text{MR}_2(\text{A}_3 + \text{MR}_3 \ldots)) \qquad (10.2)$$

A_i is access time, and MR_i is miss rate for level i of the memory hierarchy. The miss penalty for level i is the AMAT starting at level $i+1$.

Example 10.2 AMAT

If 95% of instructions hit in the L1$, which has a 1-cycle access time and 15-cycle miss penalty, what is the AMAT?

Solution: AMAT $= 1 + 0.05 \times 15 = 1.75$ cycles.

Cache misses are classified as *compulsory*, *capacity*, or *conflict*. The first time data is accessed, a compulsory miss occurs because the data must be loaded into the cache. Capacity misses occur when the cache is too small to hold the concurrently used data, which is called the *working set*. Conflict misses occur when too many addresses map to the same set (location in the cache), causing active cache lines to be *evicted*. Section 10.1.2 addresses design methods to reduce each class of miss.

The miss penalty can be reduced by using a faster, more expensive memory or by adding intermediate levels of cache. The miss penalty increases as the processor cycle time gets faster relative to the main memory.

Example 10.3 AMAT AND CPI

A 5-stage pipelined CPU has 2 levels of caches and DRAM main memory. The L1I$ has a hit rate of 98% and a 1-cycle access time. The L1D$ behaves the same except the hit rate is 96% and The L2$ has a hit rate of 80% (for accesses that missed the L1$) and a 3-cycle access time. DRAM has a 30-cycle access time. What is the AMAT for loads/stores and for instructions? Suppose the running application has 40% loads/stores, and all other instructions take a single cycle to execute if the L1I$ hits. Use AMAT to estimate overall CPI.

Solution: AMAT for fetching instructions is $0.98 \times 1 + 0.02(3 + 0.2(30)) = 1.16$. AMAT for accessing data is $0.96 \times 1 + 0.04(3 + 0.2(30)) = 1.32$. The CPI of a non-load/store is the instruction fetch AMAT, 1.16. To compute the load/store CPI, add the two AMATs and subtract 1 to account for the overlap of one instruction fetched while 1 load or store is read or written. $\text{CPI}_{\text{load/store}} = 1.16 + 1.32 - 1 = 1.48$. So, $\text{CPI} = 0.6 \times 1.16 + 0.4 \times 1.48 = 1.29$ cycles.

A mainstream PC in 2022 typically has 16-64 GiB of DDR4 DRAM costing $3-$6/GiB with a bandwidth of about 19.2-35.2 GB/s and a latency of nearly 100 ns to access a random word.

A Core i7 12900K Alder Lake processor costs about $600 and has 30 MiB of SRAM in the L3$. The L3$ has a 17.6 ns latency at the 3.2-GHz base core frequency, and the 48 KiB L1$ has a latency of under 1 ns [RyzenVi9]. The L1 caches deliver about 1 TB/s (1000 GB/s) of bandwidth. Fig. 1.24 shows a Raptor Lake die with the L2$ and L3$ highlighted.

Out-of-order processors can hide part of the miss penalty by executing other instructions while waiting, though data dependencies in the program limit this parallelism. Such caches are called *nonblocking* because the CPU is not stalled on a miss. Multithreaded processors hide the miss penalty by switching to another thread when a cache misses.

Memory access time is an increasing concern as processor clock rate increases. If main memory takes 50 ns to return data, a processor running at 20 MHz (= 1/50 ns) takes 1 processor clock cycle to access main memory, whereas a processor running at 2 GHz takes 100 clock cycles.

10.1.2 Locality

The information held in a cache is generically referred to as data, even though for an I$ the information held is instructions.

Caches reduce the AMAT for programs that reuse the same or adjacent data. Code Example 10.1 shows a `for` loop that adds the loop index to each element in an array. This code exhibits both *temporal* and *spatial* locality, which are defined as follows:

- **Temporal locality:** Recently accessed memory is likely to be used again.
- **Spatial locality:** Nearby memory addresses are likely to be used.

In Code Example 10.1, the I$ sees temporal locality because the same instructions (in the loop) are executed repeatedly and spatial locality because consecutive instructions are executed. The D$ sees temporal locality because each array element is read and then written. It also sees spatial locality because consecutive elements are accessed.

Code Example 10.1 LOOP DEMONSTRATING LOCALITY

```
for (int index = 0; index < 100; index++)
  array[index] = array[index] + index;
```

Caches exploit both types of localities by holding a copy of recently accessed instructions and data. When the CPU reads or writes an address, it first checks the cache; if the instruction or data is present, the CPU accesses it immediately. However, if the item is missing, the cache fetches this item and nearby consecutive items into a *cache line*. Spatial locality tells us the nearby addresses have a high probability of being used soon, so it is beneficial to fetch data from these nearby addresses into the cache. Each access to the next-level cache incurs some latency independent of data size plus additional time to transfer each word in the line, depending on line size. So, making the line too long may hurt performance.

Some texts refer to a cache line as a cache *block*.

10.1.3 Direct-Mapped Cache

The simplest kind of cache is a *direct-mapped* cache, in which the least significant bits of the address specify the location of an item in the cache. Fig. 10.2 shows a 4-KiB direct-mapped cache organized as 128 *sets*, each containing a single 256-bit (32-byte) cache line and an associated *tag* holding the upper bits of the line's address. Suppose addresses are 64 bits. Bits 11:5 of the address specify the set and bits 4:0 specify the bytes within the set. If the cache reads 8-byte *double words* aligned to 8-byte boundaries, bits 2:0 are always 000, and bits 4:3 specify one of the four 8-byte words in the line, called the *word offset*, indicated by "ww" in the address of Fig. 10.2.

The word offset is also referred to as the *block offset*.

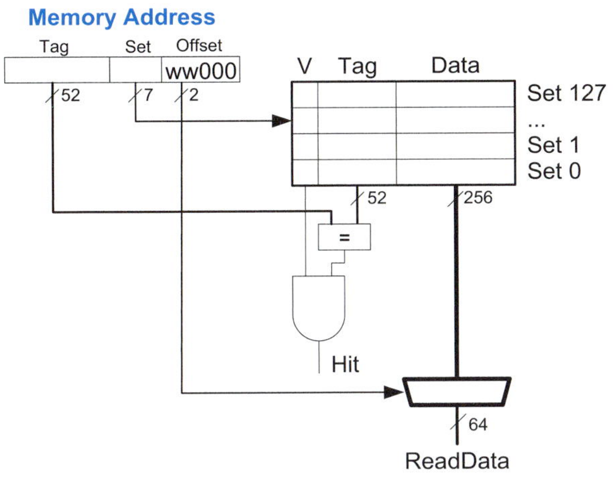

Fig. 10.2 **4-KiB direct-mapped cache with 128 256-bit cache lines.**

Any memory address with the same 11:5 bits will map to the same set. Therefore, the cache must store a *tag* that indicates which address is in the set, as well as a valid (V) bit indicating that the set contains an item at all. During a memory read, the cache compares the tag (upper bits of the requested address) with the tag in the set. Address bits 11:5 determine the set. If the line is valid (V = 1) and the tags match, the cache *hits* and returns the selected data word from the cache line. The word offset, address bits 4:3, determines the word within the line. Subword reads (`lb`, `lbu`, `lh`, `lhu`, `lw`, `lwu`) would then use bits 2:0, called the *byte offset*, to select the byte(s) within the word.

Because multiple addresses map to the same set, the cache must hold both the data and its address. However, the set and word offset bits already encode the lower bits of the address, so only the remaining upper bits, the tag, must be held in the cache, not the entire address.

Example 10.4 DETERMINE SET AND TAG

In which set would the cache from Fig. 10.2 hold an 8-byte double-word located at memory address 0x2138? What would the tag be?

Solution: Fig. 10.3 shows the cache organization. With 128 lines, this cache needs $\log_2 128 = 7$ bits of the address to indicate the set. The least significant 3 bits are 000 for each 8-byte word, and the next 2 bits indicate which of the four words (00−11) is requested within the 32-byte line. The next 7 bits indicate the set. The most significant bits are the tag. In binary, the memory address 0x2138 is 0010 0001 001 **11** 000, with the set bits highlighted in blue. Hence, the 8-byte word is stored in set 9 with a tag of 2, indicated by the bits to the left of the set bits. The stored word is the most-significant 8-byte word of the four in the set, as indicated by the bolded "**11**" in the address.

The size or capacity of a cache refers to the number of data bits and does not count tag or valid bits.

Multiple addresses map to the same cache set. Using the cache from Example 10.4, an 8-byte word at address 0x4120 (0100 0001 001 **00** 000) would also be held in set 9 but with a tag of 4. A *conflict* occurs when the cache accesses two different addresses with the same set but different tags. The direct-mapped cache must evict one line to make room for the other. This increases the cache miss rate.

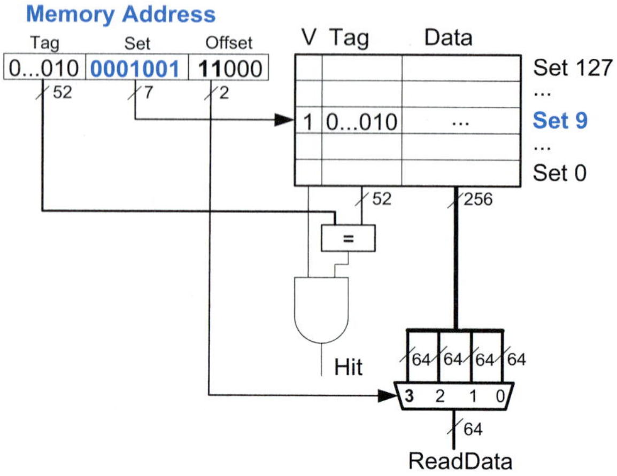

Fig. 10.3 Mapping of address 0x2138 to a 4-KiB direct-mapped cache with 32-byte lines.

If the tag does not match or the set is invalid, the cache *misses*. The cache controller must send a request to the next level of the memory hierarchy to fetch the line. The cache then writes the new data along with the tag of the fetched line and sets the valid bit to 1. If the set previously contained a valid line, the cache must first *evict* the old line to make room for the new one. Read-only caches simply replace the old line with the new one. Writable caches may first have to write a line that has been changed (called a *dirty line*) to the higher level of the memory hierarchy as discussed in Section 10.1.4. Each cache line adds a *dirty bit* (*D*) to indicate a dirty line. The valid and dirty bits are collectively called *status bits*.

10.1.4 Cache Writes

Instruction caches only need to read data, but data caches also need to write data from the CPU. This involves adding a write port, supporting partial line writes to a multiword cache line, and updating the written data in higher levels of the memory hierarchy.

Fig. 10.4 shows the interface of a generic read/write cache. The write port includes `WriteData` and read/write enable (`RW`) inputs as well as a `ByteMask` input that indicates which bytes in the word should be written.

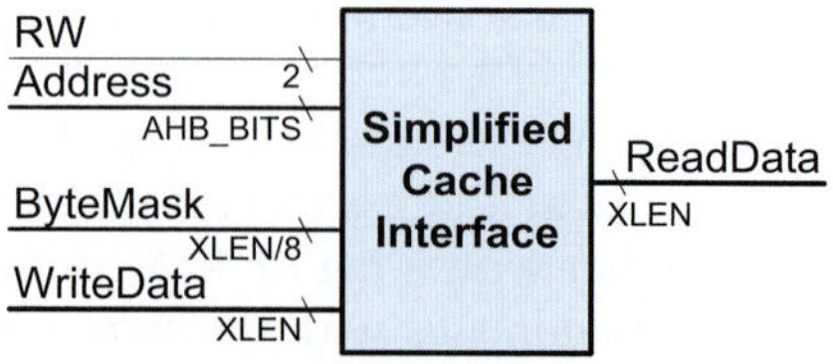

Fig. 10.4 Read/write cache interface.

For example, ByteMask is 0b11111111 for an sd instruction that writes all 8 bytes of a double-word, 0b00000001 for a sb instruction writing the least significant byte, or 0b11000000 for a sh instruction writing the most significant half-word. Read and write enables are grouped together in the RW[1:0] signal; read enable is the msb, RW[1], and write enable is the lsb, RW[0].

A store begins the same way as a cache read: the cache reads the tag and valid fields of the selected set to detect a hit. On a hit, WriteData is written to the correct word offset within the line, and the dirty bit (D) for that line is set so the cache knows that the line will need to be written back to the next level of the memory hierarchy when it is evicted. The valid and dirty bits are collectively called *status bits*.

Fig. 10.5 shows an example of writing 0xAABBCCDD to address 0x14. It assumes a 16-bit address, XLEN = 32 bits, 16-byte (4-word) cache lines, and 64 sets. First, WriteData is replicated four times in the line-length (4-word) signal LineWriteData. A demultiplexer shifts ByteMask (0xF) left by the word offset (0b01) into the correct position within LineByteMask (0x00F0) and zeros the other bits. LineByteMask controls which bytes are written within the line, so only bytes 7:4 are updated in the cache line. The dirty bit is also set.

If a write misses, the cache must first fetch the line, as is done for a read miss. After the data arrives from the next level of the memory hierarchy, the cache then merges the new write data into the line at the correct word offset before writing the line into the cache memory. On read or write misses, LineByteMask is 0xFFFF because all bytes need to be written.

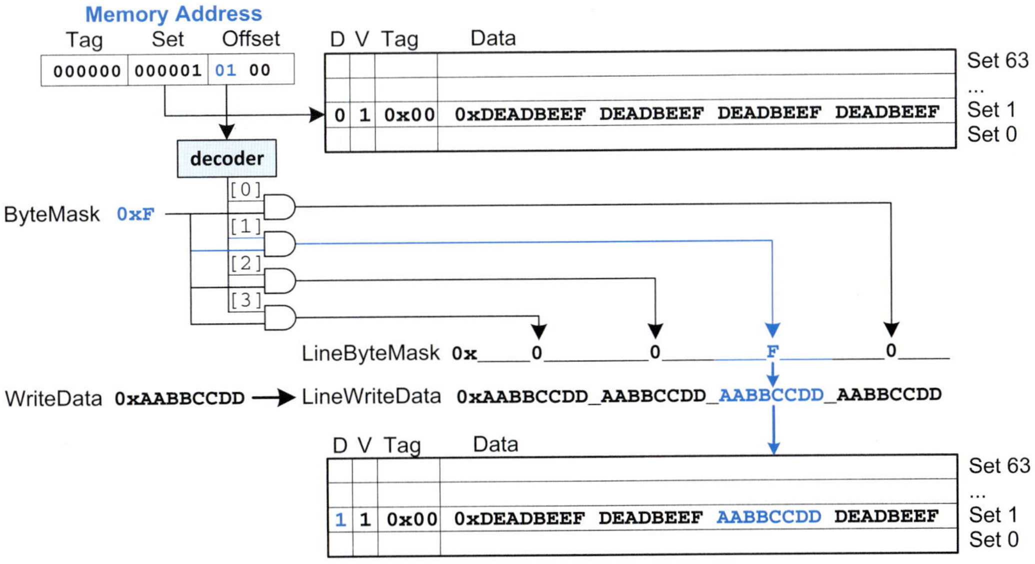

Fig. 10.5 Direct-mapped cache write.

The memory hierarchy becomes inconsistent when data is written to a cache but not to the higher levels of the memory hierarchy. The two strategies for handling this inconsistency are called *write-through* and *write-back*.

In a write-through cache, each write is also sent to the next level(s) of the memory hierarchy, so all memories remain consistent. This seriously limits write bandwidth and costs substantial power, so it is only viable for simple low-performance systems.

Practical processors now use write-back caches, which use the dirty bit to indicate that the cache line has changed and, thus, is inconsistent with the next level of the memory hierarchy. When a line is initially fetched on a read miss, the dirty bit is cleared. Cache writes set the dirty bit. When a dirty line is *evicted*, it must be written back to the next level of the hierarchy.

Multicore systems add another degree of complexity because different cores may have different dirty lines in their local caches. The programmer expects memory *coherency*, meaning that all cores will eventually see the value most recently stored to an address. Thus, the memory system uses a *coherency protocol* such as snooping or directories to maintain coherency between the caches [Hennessy19].

10.1.5 Associative Caches

With direct-mapped caches, conflicting addresses compete for the same set. This can lead to conflict misses, as shown in Code Example 10.2, which alternately accesses two 8-KiB arrays. If run on a direct-mapped 8-KiB cache, every access will miss in the cache because each `a[i]` and `b[i]` map to the same set. Thus, the cache has a 100% miss rate and does not benefit from spatial locality. In fact, the system will run more slowly than it would without a cache because it wastes time fetching each cache line and then immediately evicting it. This is called *thrashing*.

Code Example 10.2 LOOP CAUSING THRASHING

```
int64_t a[1024], b[1024];
int i;
for (i = 0; i < 1024; i++)
  b[i] = a[i];
```

An *N*-way *set associative cache* (or *associative cache*, for brevity) reduces conflicts by providing *N* separate *ways* in each set. A line can be placed in any of the ways. Each way has its own tag, valid, and dirty bits. Fig. 10.6 shows an 8-KiB 2-way associate cache with 128 sets, each containing 256 bits (32 bytes = 4 8-byte words) of data in each way.

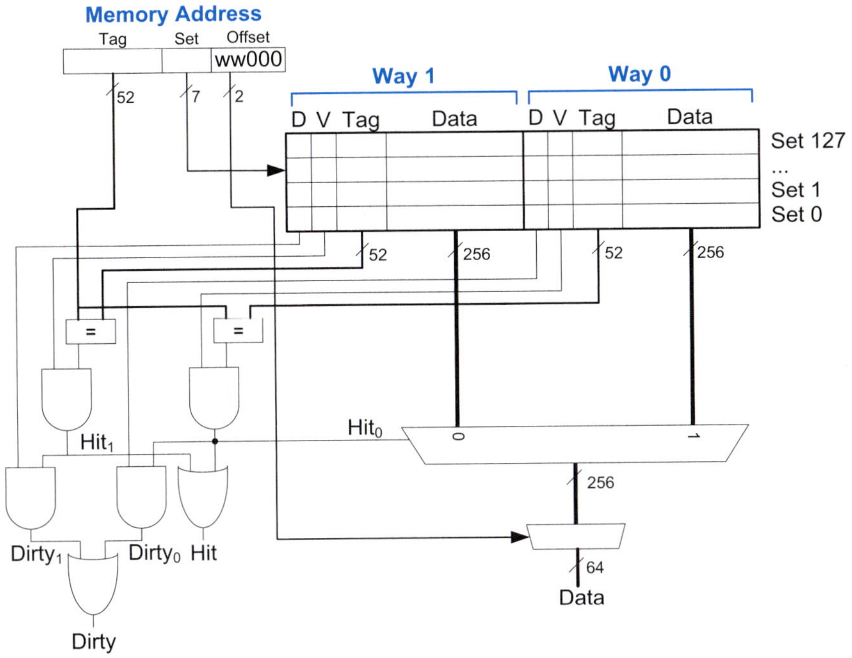

Fig. 10.6 **Associative cache read access.**

Like the direct-mapped cache discussed earlier, this cache reads 8-byte words, so the three least significant bits of the address are always 0. The next two bits of the address (ww) specify one of the four words in the line. The next seven bits indicate the set, and the upper bits are compared against the tag in both ways. If either way hits, the matching way reads the data.

When an associative cache misses and all ways are valid in the set, the cache must select a *victim* way to evict to make room for the new line. The cache *replacement policy*, discussed in Section 10.1.7, describes how to pick the victim way.

Considering Code Example 10.2 again with the 2-way associative cache, no conflict misses occur anymore because `a[i]` and `b[i]` each reside in a different way. The cache misses on the first word in each line but then hits on the next three words because of spatial locality. Even though the program has no temporal locality (repeated access to the same data), it still has a 75% hit rate because of spatial locality.

A direct-mapped cache can be viewed as a 1-way associative cache. The opposite extreme is a *fully associative* cache, which has a single set containing all the lines in different ways. Fully associative caches never have conflict misses, although they still have capacity misses. The fully associative hit comparators and hit multiplexer are expensive and grow with the size of the cache, so full associativity is only practical for modest-sized caches. Translation lookaside buffers (TLBs), described in Section 11.4.1.3, are often fully associative to achieve the best hit rate for a limited capacity.

10.1.6 Cache Design Tradeoffs

Cache designers must pick the line size, associativity, and total capacity to achieve the best performance given a limited area and power budget.

Section 10.1.2 describes how longer cache lines improve hit rate due to spatial locality but take longer to fill or evict. For a given cache size, very long lines also increase the miss rate because fewer lines are available altogether, increasing conflict misses. Fig. 10.7 shows these tradeoffs for various cache sizes (4–256 KiB) when running the SPEC92 benchmark [SPEC92].

Fig. 10.8 plots cache miss rate versus size and associativity when running the SPEC2000 benchmark [SPEC2000]. Capacity misses dominate, especially for typical 16- to 32-KiB L1 caches. Capacity misses can be alleviated by providing a larger L2$. Conflict misses can be significant in direct-mapped caches but become rare in caches with at least 4-way associativity. As we will see in Section 11.1.1.7, in systems with virtual memory, such associativity is also used to ensure that each way is no larger than the page size, which moves address translation off the critical path. In nontrivial programs, compulsory misses are negligible because they only occur at the start of a program.

Higher-level caches may be *inclusive* (containing all contents of the lower-level cache), *exclusive* (containing none), or *non-inclusive non-exclusive* (*NINE*, possibly containing some). Exclusive caches make better use of limited cache space, but inclusive top-level caches are needed to maintain cache coherence in multiprocessor systems. To provide significant miss rate reduction, the next-level cache should be significantly (at least 4-8

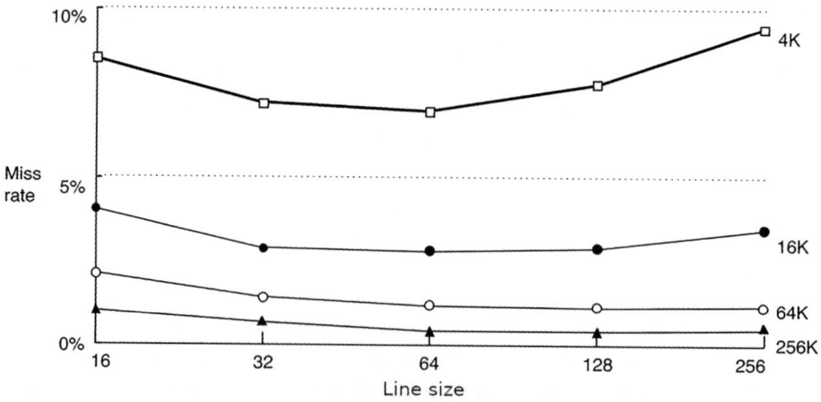

Fig. 10.7 Cache miss rate vs. line size for various cache sizes. (Figure courtesy of Elsevier [Hennessy19, Figure B.10]. Adapted and reprinted with permission.)

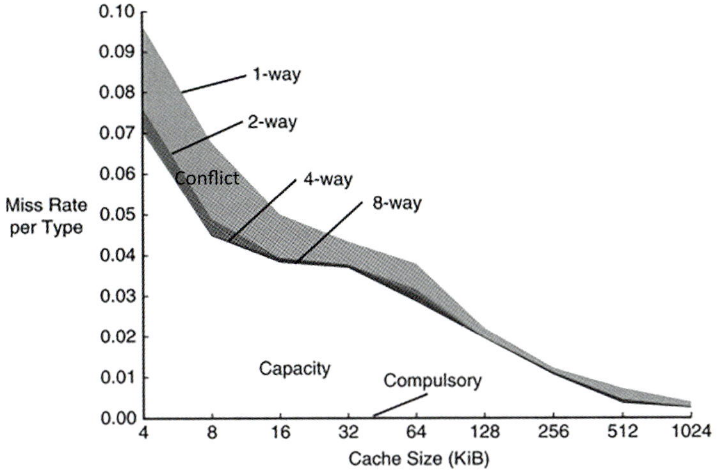

Fig. 10.8 Cache miss vs. size and associativity. (Figure courtesy of Elsevier [Hennessy19, Figure B.9]. Adapted and reprinted with permission.)

times) bigger than the one below it. So, in a system with 16-KiB L1I\$ and L1D\$, the combined L2\$ should be at least $8(16 + 16) = 256$ KiB. L2 caches with up to several MiB are sometimes used, but they can have long access times when they get too big. Hence, the system may add an L3\$, commonly with 4- to 64-MiB of capacity. The highest-level cache is shared across all cores of a multicore system to support shared-memory communication between cores.

L1 caches are typically 8-32 KiB to achieve 1-2 cycle latency, depending on clock speeds and quality of circuit design. Each way is usually 4 KiB to work nicely with 4-KiB virtual memory pages, so they have 2-8 ways. L1 caches often have 64-byte lines, which offer a good tradeoff between hit rate and miss penalty.

10.1.7 Replacement Policy

As mentioned in Section 10.1.5, an associative cache needs a replacement policy to select a victim way to evict during a cache miss when all ways are in use (all valid bits in the set are 1). An ideal replacement policy selects the way with the lowest probability of being used in the near future. However, program behavior is not known ahead of time, so we use past behavior to predict the future. The *least recently used* (*LRU*) replacement policy records the order in which the ways are accessed (read or written) and selects the oldest accessed way as the victim.

In a 2-way associative cache, LRU uses 1 extra bit in each set to indicate which of the two ways was used least recently. So, each set's LRU bit indicates the *victim way*, which way to replace on a miss, and is updated any time that set is accessed. For example, suppose a load misses and that set's LRU bit is 0, indicating way 0 is the victim. The cache writes back the set's way 0 if dirty and then fetches the missing line into way 0 and updates the LRU bit to 1.

LRU does not scale efficiently beyond two ways because recording the order of accesses and selecting the oldest becomes expensive. For example,

Despite pLRU being an approximation of LRU replacement, the way to replace is still called the LRU way, and the bits to keep track of that way are still called the LRU bits.

an N-way associate cache can keep a list of N entries for each set indicating the relative order of accesses. When the cache accesses a way, it also moves the entry for that way to the head of the list. This requires $N\log_2 N$ LRU bits per set and hardware to search and update the list.

Tree pseudo-LRU, also referred to simply as *pseudo-LRU (pLRU)*, approximates LRU with a less-costly implementation that uses only N-1 LRU bits and still achieves a hit rate that is nearly as good as LRU. The bits are organized as a binary tree, with $\log_2 N$ levels, to approximate which side of the tree was least recently accessed. pLRU works by grouping ways into powers of two: 2 groups, then 4 groups, and so on until only one way is in each group.

Fig. 10.9(a) shows an example of a tree for an 8-way associative cache in which way 4 was least recently used (LRU). The LRU bits N0–N6 in the tree are shown in circles and the ways are squares. A zero (0) in the tree indicates the right branch was more recently accessed and the left branch is older, a one (1) indicates the opposite. Traversing the tree along the arrows to the leaf box finds the pLRU (labeled LRU) victim way. In Fig. 10.9(a), the root LRU bit N6 is 1, so the right half (ways 4–7) were not used in the last access. Following the arrow to the right half leads to a 0 bit at N4, so the left half (ways 4–5) were not used the last time an access occurred to ways 4–7. Following the next arrow left to N1 reveals another 0, so way 4 was not used the last time an access occurred to ways 4–5. When the set is full and a new entry is needed, pLRU traverses the binary tree by following the arrows from N6 to N4 to N1 to select way 4 as the victim way to replace.

After a cache access uses a particular way, pLRU updates the tree to reflect this way's recent use. Every node between the root and the accessed way is updated to point away from the accessed way. Fig. 10.9(b) illustrates this pLRU update after accessing way 6, with the affected nodes labeled in blue. The path to traverse the tree from the root to the accessed

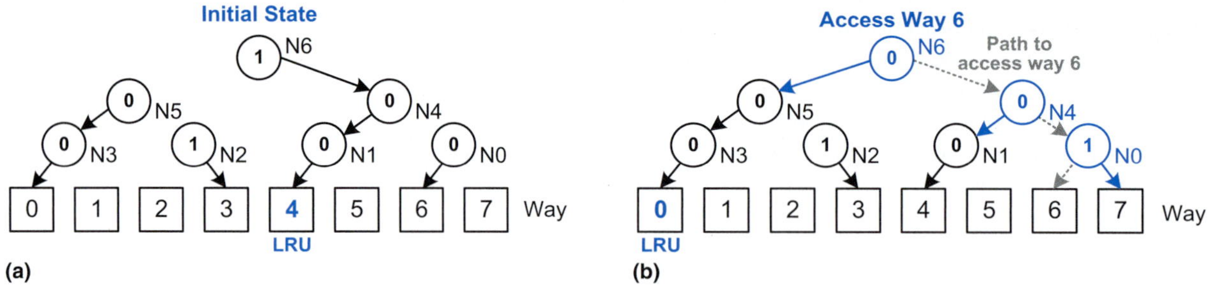

Fig. 10.9 **Pseudo-LRU examples: (a) initial and (b) after accessing way 6.**

way (6) is N6 to N4 to N0, shown by the gray dashed arrows. Each traversed node's LRU bit is updated to point away from the accessed way, as shown by the blue arrows: The root node N6 changes to 0 to point away from ways 4–7 (toward ways 0–3). The next node N4 is already 0, pointing away from ways 6–7 (toward ways 4–5). The final node N0 becomes 1 to point away from way 6 (toward way 7). Now the arrows indicate way 0 is least recently used.

Example 10.5 LRU TREE UPDATE

How would the tree of Fig. 10.9(b) be updated after accessing way 3? What way is now least recently used?

Solution: The path from the root to way 3 traverses nodes N6, N5, and N2, as shown by the gray dashed arrows in Fig. 10.10. Each node along this path is updated to point *away* from way 3, as shown by the blue arrows. Now the arrows indicate way 4 is least recently used.

Both LRU and pLRU can behave poorly under pathological access patterns. Thrashing occurs if the working set is larger than the number of ways. For example, if a program's working set maps five addresses to the same set and then accesses them in the same order repeatedly, a 4-way associative cache will conflict miss on every access.

Tree pLRU was first used in 1989 in the Intel 486 CPU [Intel89].

10.1.8 Cache Flush and Invalidation

Sometimes the cache may need to be *flushed* (have dirty lines written back to the next level of the memory hierarchy) or *invalidated* (valid bit V = 0 for all cache lines). For example, on reset, the cache must be invalidated. Also, because an instruction cache is read-only, anytime a program needs to write instructions, the writes go to the data cache instead. Hence, for the new instructions to be visible in the instruction cache, the processor must flush the dirty lines from the data cache back to the next level of the memory hierarchy and invalidate the instruction cache so that it reads the new instructions from the next level of the memory hierarchy.

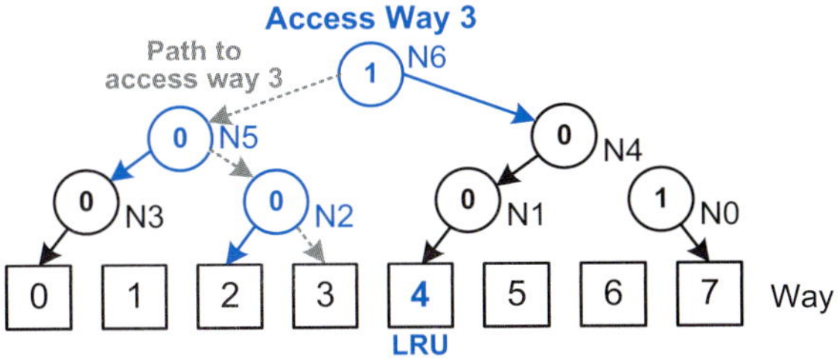

Fig. 10.10 Pseudo-LRU example after accessing way 3.

A flush operation walks through all data cache lines and writes the dirty lines to memory, then clears the dirty bits. An invalidate operation walks through cache lines and clears the valid bits.

Many processors support a sleep mode in which the cache array is powered off to save energy. Before entering sleep, write-back caches should be flushed. All caches should be invalidated when waking up.

Beware that the term flush has multiple meanings. A *cache flush* writes back dirty lines. A *pipeline flush* clears an instruction from a pipeline stage. In this chapter, *flush* refers to a cache flush except when discussing pipeline hazards.

10.2 RISC-V PRACTICES

In addition to handling fetches, loads, and stores, RISC-V caches have several design issues specific to the architecture:

- Atomic memory operations (AMOs) read a memory location, modify it, and write back to the same location in one indivisible, uninterruptable operation such that no other operation can access the memory between the read and write. This is discussed in Chapter 17.
- Coherency between the I$ and D$ is maintained by flushing the D$ and invalidating the I$ using a `fence.i` instruction. Cache invalidation and flushes are discussed in Sections 10.4.2.6 and 10.4.5.4.
- A physical memory attributes (PMA) checker indicates whether an address is cacheable. For example, memory-mapped peripherals cannot be cached because their value may change because of external events, such as a voltage changing on a general-purpose input/output (GPIO) pin.
- Section 11.4.1.1 describes PMAs, and Section 12.3.1 describes the LSU logic that handles uncacheable addresses.
- When compressed instructions are supported, instructions may wrap across cache lines. Such spills will be handled in Section 14.3.3.
- The operating system (OS) likes to know the size of a cache line so it can allocate memory sensibly aligned to the cache. The Zi64b RISC-V extension indicates that cache lines are 64 bytes. This size has become an industry standard because it is a good tradeoff between spatial locality and time to fetch a missing cache line. Wally defaults to 64-byte (512-bit) cache lines and thus implements Zi64b by default.

10.3 TEST PLAN

The Wally caches are currently exercised through the ordinary instruction tests and Linux boot. Section 21.3.1 describes cache performance verification.

10.4 WALLY IMPLEMENTATION

Wally has optional configurable level 1 instruction and data caches (L1I$/L1D$). Currently, it does not have any higher-level caches. Fig. 10.11 shows a simplified top-level block diagram of the `cache` module. A more detailed version is shown in Fig. 10.24. The cache interfaces to the core, bus, privileged unit, and hazard unit. It contains three main submodules: `cacheway`, `cachefsm`, and `cacheLRU`. An *N*-way associative cache stores lines in *N* `cacheway` modules, each of which contains data, tag, and status bits for each line. The `{I/D}CACHE_NUMWAYS` configuration parameter specifies *N*. Each `cacheway` handles load, store, invalidate, and flush operations. The `cachefsm` sequences multicycle operations such as line fetches caused by cache misses. An associative cache also contains a `cacheLRU` unit to select a victim way for replacement. The shaded boxes show other top-level cache

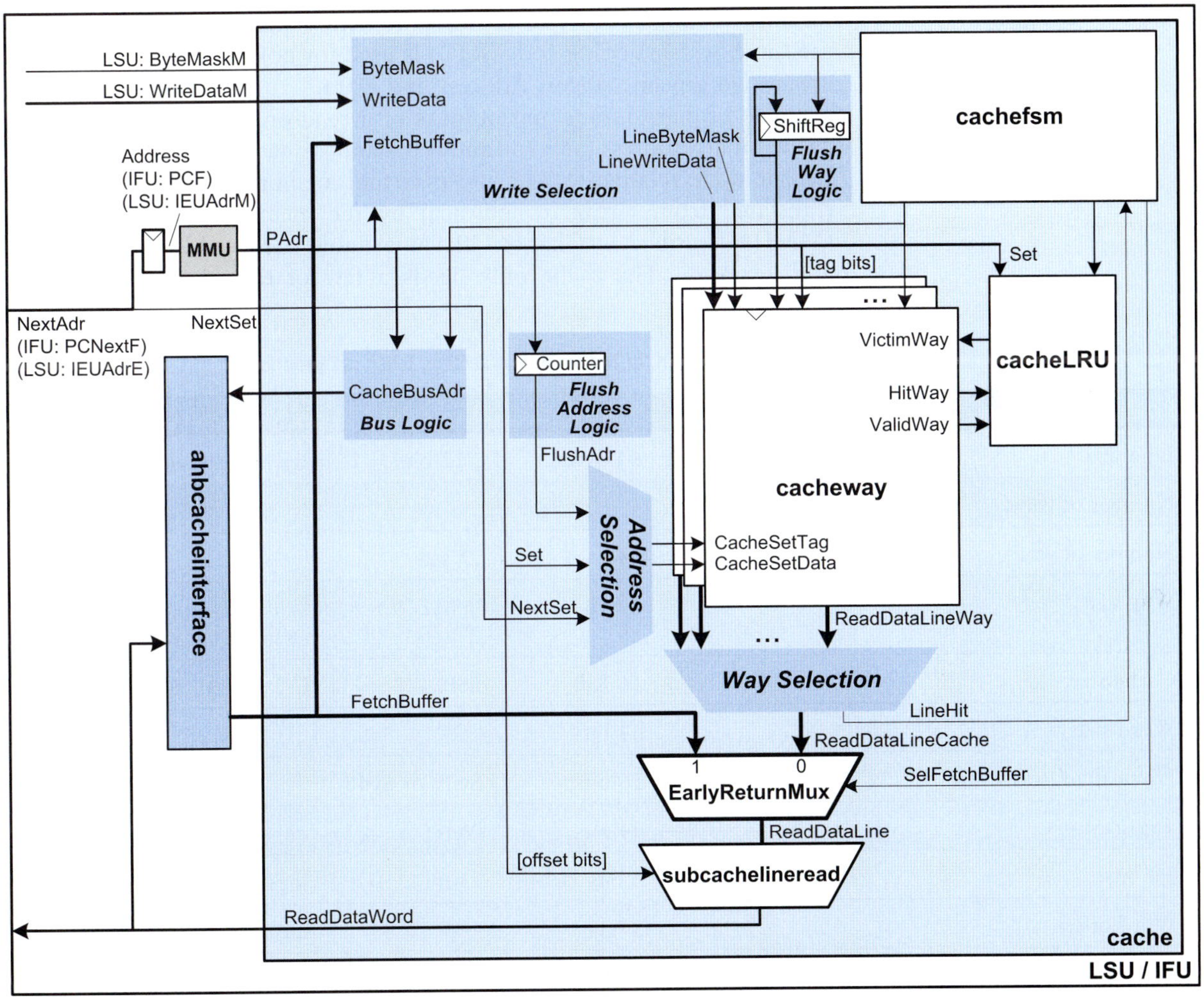

Fig. 10.11 **Cache top-level block diagram.**

logic. The cache and hazard unit interact to stall the pipeline during cache misses, and the cache interlocks with the privileged unit when traps occur. Table 10.1 lists the cache parameters and common values. This section explains the cache implementation common to instruction and data caches, and Chapters 12 and 13 explore the design issues specific to integrating the caches into the LSU and IFU.

The I$ and D$ both use the same configurable SystemVerilog `cache` module. The module has a `READ_ONLY_CACHE` parameter, which is 0 for the D$ and 1 for the I$. Both types of caches support reads, but only data caches support writes, writebacks, and flushing dirty lines (`READ_ONLY_CACHE` = 0). Both instruction and data caches support invalidation.

The caches use synchronous memories that expect the address, control, and write data to set up before the rising edge of the clock. They read or write on the rising edge. Unlike the rest of the CPU core, the caches do not use suffixes F, D, E, M and W to denote the pipeline stage because the same SystemVerilog code (RTL) is used for both caches. Caches operate in two pipeline stages, Address and Data.

The primary inputs and outputs of a read-only cache used during a cache hit are address (`NextAdr` and `PAdr`) and `ReadDataWord`. `PAdr` is delayed by one cycle relative to `NextAdr`; if virtual memory is enabled, it is also translated from virtual to physical by the memory management unit (MMU) described in Chapter 11. A writable cache adds the `WriteData` and `ByteMask`. The cache only needs a few of the address bits, called

Table 10.1 Cache configuration for figures in this section, when not generic.

Cache Parameter	Parameter Name	Value	Description	
Line size	LINELEN	64 bytes	512 bits = 2^6 bytes = 8 doublewords	
Number of ways	NUMWAYS	4		
Number of sets	NUMSETS	64	2^6 sets	
Way size		4 KiB	NUMSETS × LINELEN = 2^6 sets × 2^6 bytes/set = 2^{12} bytes	
Cache size		16 KiB	NUMWAYS × NUMSETS × LINELEN	
Word size	WORDLEN	64 bits		
Physical address size	PA_BITS	56 bits	Sv39 or Sv48	
Tag size		44 bits	PAdr[55:12]	PA_BITS − $\log_2$(Way size)
Set bits		6 bits	PAdr[11:6]	$\log_2$(NUMSETS)
Word offset		3 bits	PAdr[5:3]	$\log_2$(LINELEN/WORDLEN)
Byte offset		3 bits	PAdr[2:0]	$\log_2$(WORDLEN)

NextSet or Set, to specify which set to access. These bits are the same in the virtual and physical addresss, so accessing the cache does not have to wait for translation.

During the Address stage, the least significant bits of NextAdr (NextSet) set up before the clock edge, indicating which set to read. During the Data stage, the cache reads the ways. It checks the valid bits and compares the tags to the upper bits of PAdr to determine if each way hits. For a read hit, the cache returns a word of data from that way on ReadDataWord. For a write hit, the cache uses PAdr (Set), WriteData, and ByteMask to write part of a line to the way on the next clock edge.

10.4.1 Interfaces

Fig. 10.12 shows the Wally cache interfaces to the core (lsu and ifu), privileged, and hazard units.

The previous section described the primary interface signals used during a cache hit: address (NextAdr and PAdr), ReadDataWord, WriteData, and ByteMask. The cache also takes the CacheRW[1:0] control signal, with bit 1 indicating a read and bit 0 indicating a write.

On a cache miss or eviction, the cache fetches or writes back a line from memory over the EBU via the AHB cache interface (ahbcacheinterface).

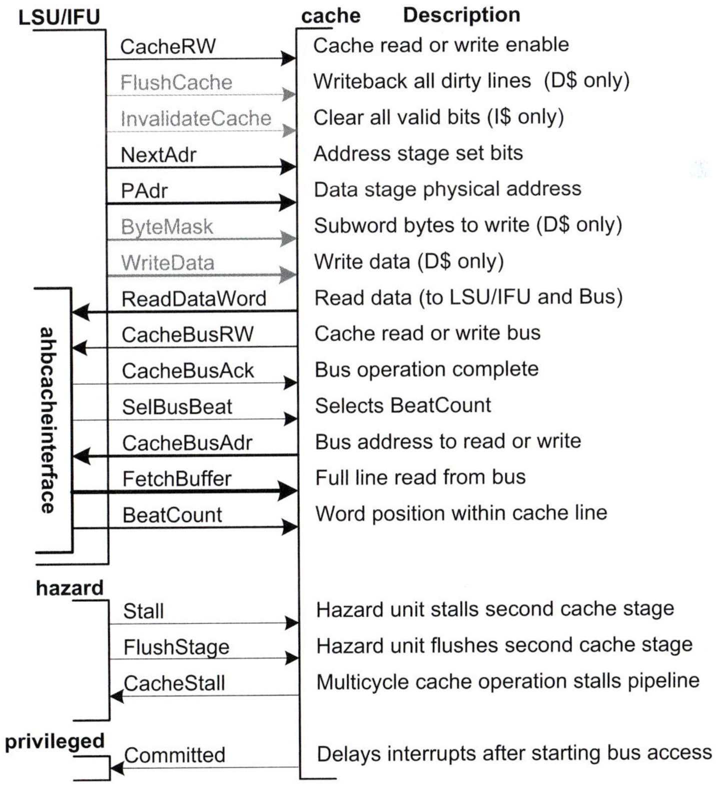

Fig. 10.12 Cache interfaces to CPU and memory bus.

Although Fig. 10.12 shows the cache outside the LSU/IFU for ease of drawing, the cache is actually located inside the LSU/IFU and communicates with other components within these units. The black signals are major busses or control signals, and the gray signals only exist for specific types of caches (I$ or D$). Chapters 12 and 13 delve into the LSU and IFU interfaces.

The CacheBusAdr and CacheBusRW[1:0] signals indicate the address to access and whether to read or write. The EBU uses burst transfers for performance, with BeatCount and SelBusBeat tracking and selecting the word in the burst and CacheBusAck signaling completion. Words read from the bus accumulate in the fetch buffer until the full cache line is ready to write to the cache. Words are written to the bus one at a time over ReadDataWord.

The caches may stall the pipeline or postpone an interrupt (using their CacheStall output), and the pipeline may stall the I$ or D$ or flush a D$ operation using a cache's Stall input. During a cache miss or other multicycle operation, the cache asserts CacheStall, which causes the hazard unit to assert StallW and stall the entire pipeline. Conversely, when the hazard unit asserts StallD or StallW, which are connected to the Stall input of an I$ or D$, respectively, the given cache stalls. When the cache's Data stage needs to be flushed, the hazard unit asserts FlushStage. For example, TrapM flushes the pipeline's Memory stage so a D$ operation should be flushed. The cache sends Committed to the privileged unit to postpone interrupts while a bus operation is in progress.

FlushStage is a pipeline flush, not a cache flush. It resets the cache FSM and cancels pending D$ writes.

10.4.2 Cache Way Module

The cacheway module implements a single way of an *N*-way associative cache. Fig. 10.13 shows a block diagram of cacheway configured according to Table 10.1. cacheway includes storage arrays to hold data, tags, and status bits for each line in a way. It uses the following:

- Separate SRAM arrays to hold the data and tags
- Arrays of flip-flops to hold the status (valid and dirty) bits for each cache line so these bits can be easily written or reset

The overall way size (4 KiB) matches the virtual memory page size, which is described in Section 11.1.1.7. The line length of 64 bytes is a typical length in commercial processors and offers a tradeoff between spatial locality and miss penalty. The number of lines in each cache way is NUMSETS = CACHE_WAYSIZEINBYTES/(CACHE_LINELENINBITS/8) = (4,096 bytes/way)/([512/8]bytes/line) = 64 lines/way.

The physical address is 56 bits in rv64gc. As shown in Table 10.1, the 12 least significant bits (the set, word offset, and byte offset bits) are used to select a byte within a 4-KiB way,... so the tag is the remaining 44 bits. Thus, the tag array size is 64 lines × 44 bits/line.

The cache way reads the tag specified by CacheSetTag. If ReadTag matches the upper bits of the physical address PAdr and the line is valid, the cache hits and returns the data line as ReadDataLineWay. The control logic in Fig. 10.13 handles pipeline flushes, cache invalidation, and updating status bits. This section details how cacheway reads, writes, fetches, writes back, and invalidates cache lines. It also describes a structural hazard caused by synchronous SRAM timing.

cacheway is one of the largest modules in Wally, so it is designed for area efficiency. The large data and tag arrays are built from SRAM for density and energy efficiency. SRAM generators have a maximum width, typically around 128 bits. Therefore, the 64-line × 512-bit wide data array is partitioned into four 64 × 128-bit banks of SRAM. The arrays have byte enables to support writing words or bytes within the line. They also have chip enable pins to save power and retain their old data output when disabled. When the hazard unit stalls the pipeline, the chip enable is turned off, so the cache retains its output. The status bits may be modified

Table 6.5 shows how SRAM is roughly four times denser and has two times lower energy than flop-based memories.

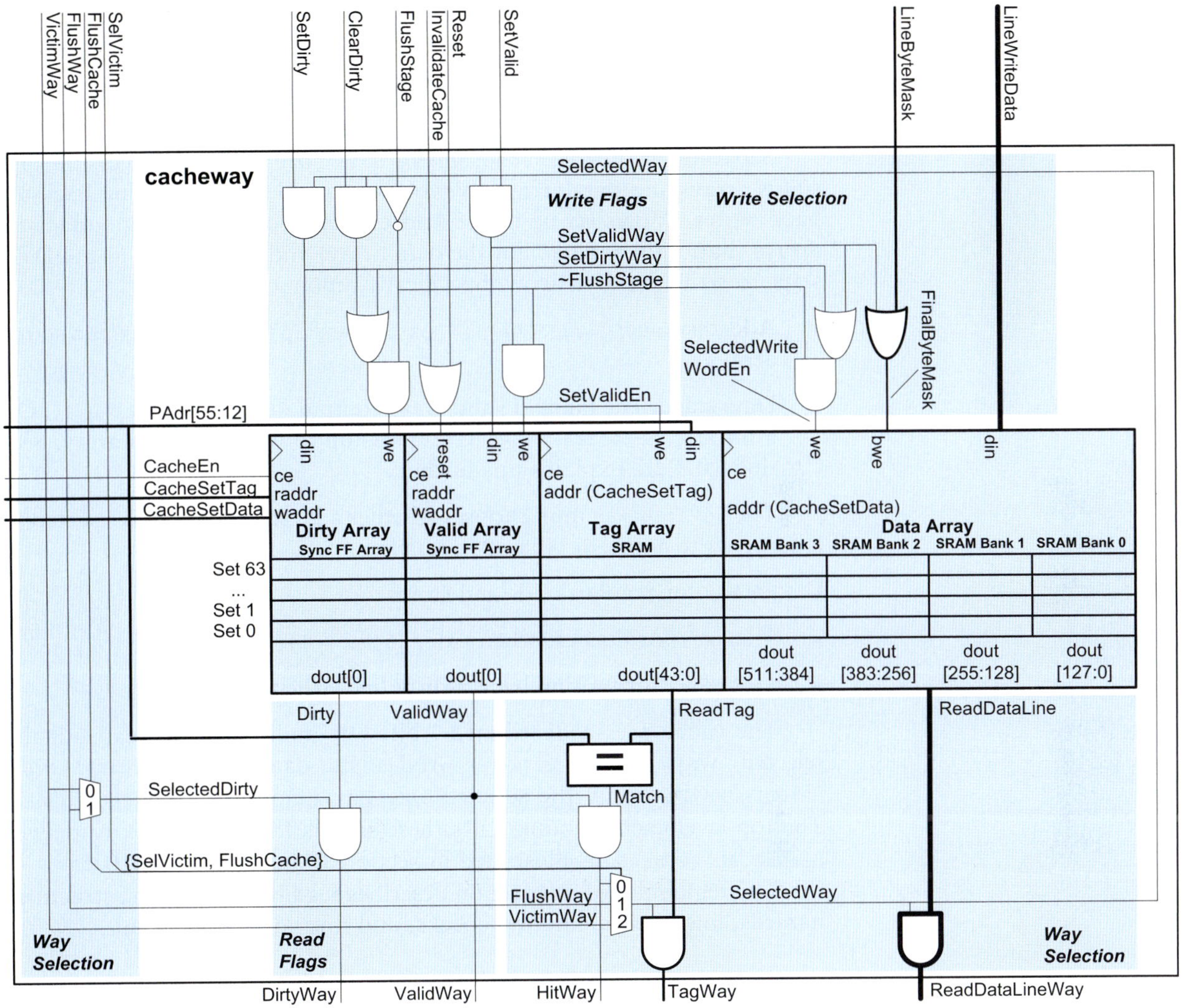

Fig. 10.13 `cacheway` **block diagram.**

independent of writing the data and tag arrays, so they are most conveniently built from flip-flops. The valid bits are reset on startup or invalidation. The data, tag, and dirty bits are ignored if the valid bit is 0, so they do not require the overhead of reset circuitry.

The synchronous memories require an address to setup before the rising clock edge and read or write on the clock edge. So, the cache operates in two stages: Address and Data stages. For cache reads that hit, this operates as follows:

- **Address stage:** `CacheSetTag` sets up to read the tag and status arrays. `CacheSetData` sets up to read the data array. `CacheSetTag = CacheSetData = NextSet`.

- **Data stage:** The SRAM arrays output the tag and data bits, ReadTag and ReadDataLine, and the valid array outputs the valid bit, ValidWay. A hit occurs when the requesting address tag and the line's tag match and that line is valid.

Separate set signals, CacheSet{Tag/Data}, are needed to access the tag/status arrays and data arrays to avoid structural hazards during back-to-back writes. Writes set up in the Address stage to read the tag and status arrays (but, unlike reads, not the data array) and then set up in the Data stage to write the data array. Specifically, upon a write hit:

- **Address stage:** CacheSetTag = NextSet sets up to read the tag and status arrays.
- **Data stage:** The tag and valid arrays output ReadTag and ValidWay. On a hit, CacheSetData = PAdr[11:6] sets up to write the data array. Section 10.4.2.2 explains this further.

The SelectedWay mux chooses whether a given cache way is selected to read or write a line. SelectedWay equals:

- HitWay for regular loads and stores
- FlushWay for flushes (when FlushCache = 1)
- VictimWay for writing back a dirty line (when SelVictim = 1)

The cache uses a distributed AND-OR multiplexer to read out the selected way. Each cacheway ANDs its data (ReadDataLine) with SelectedWay, so only the selected way has a nonzero ReadDataLineWay. The top-level cache module ORs these signals from each way to deliver ReadDataLineCache, as illustrated in Section 10.4.5.1 and Fig. 10.25.

The remainder of this section describes read hits, write hits, structural hazards, line fetches, line writebacks, and invalidation in detail.

10.4.2.1 Cache Way Read Hit

A cache way read (load or instruction fetch) begins in the Address stage with the set bits (CacheSetTag = CacheSetData = NextSet) specifying which set to read, as shown in Fig. 10.14(a). On the rising edge of the clock (start of the Data stage), cacheway reads ReadDataLine, ReadTag, and ValidWay. If ReadTag matches the address's tag (ReadTag == PAdr[55:12]) and the line is valid (ValidWay = 1), then Match = 1 and, thus, HitWay = 1, which indicates a hit in that way. SelectedWay = HitWay = 1, so the ReadDataLineWay AND gate outputs ReadDataLine.

Example 10.6 CACHE WAY HITS

A processor loads 4 unsigned bytes from address 0x00_0000_8000_0384. Sketch waveforms for a cache way hit using the rv64gc configuration of Fig. 10.13 and Table 10.1 and no virtual memory. The cache line contains 0x0...0_AAAA_5555_1111_2222.

Solution: Fig. 10.14(b) shows a timing diagram.

- **In cycle 0** (Address stage), set up to read data and tags from line 14: CacheSet{Tag/Data} = NextAdr[11:6] = 001110_2.
- **In cycle 1** (Data stage), the cache way reads ReadDataLine, ReadTag, and Valid from line 14. The line is valid and the tag matches PAdr, so HitWay = 1, SelectedWay = 1, and ReadDataLineWay outputs the whole line, 0x0...0_AAAA_5555_1111_2222. The subcachelineread module (see Fig. 10.11) then selects the word from that line using the word offset bits PAdr[5:2] = 0001 to select ReadDataWord = 0xAAAA_5555.

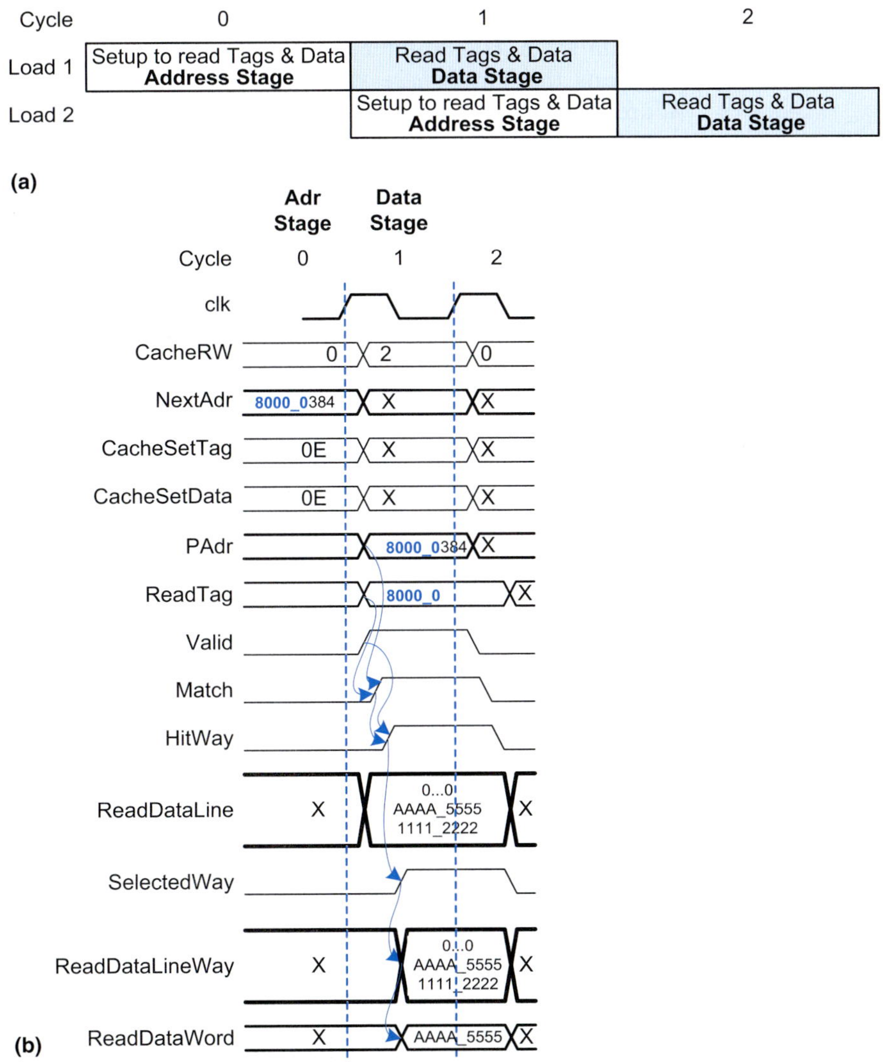

Fig. 10.14 **Cache read: (a) back-to-back reads and (b) detailed `cacheway` timing diagram for a single read.**

10.4.2.2 Cache Way Write Hit

A cache way write (store) begins in the Address stage by reading the tag and status arrays to check the tag and valid bits using CacheSetTag. However, the data array is not read. During the Data stage, SetDirty = 1, so upon a hit (SelectedWay = HitWay = 1), cacheway asserts SetDirtyWay. The cache drives CacheSetData with the set to write, LineWriteData with WriteData, and LineByteMask to specify the bytes within the line to write. On the rising clock edge at the end of the Data stage, the line is written and the dirty bit is set.

By not reading the data array in the Address stage (which is unnecessary for stores), back-to-back stores do not have to stall because only the tag array is used in the Address stage, and only the data array is used in the Data stage. For example, Fig. 10.15 shows three consecutive store instructions that hit, where the white and blue boxes show the cache's Address and Data stages, respectively. In cycle 0, store 1 sets up its address to read the tag and status arrays. In cycle 1, store 1 reads the tag and status bits, detects a hit, and sets up to write the status array and its word into the data array (at the end of the cycle). Also in cycle 1, store 2 sets up to read the tag and status arrays. In cycle 2, store 2 reads the tag and status bits, detects a hit, and sets up to write the status bits and its word into the cache's data array, and so on. Accessing the tag and data arrays in this way requires separate addresses, CacheSetTag and CacheSetData. For loads, these addresses are the same and are both driven (set up) in the Address stage. However, for stores, only CacheSetTag is used in the Address stage, and only CacheSetData is used in the Data stage. So, for writes, CacheSetData lags CacheSetTag by one cycle.

The status arrays may be used in both stages: they are read in the Address stage and, for writes, written in the Data stage. However, the status arrays are dual-ported, so they may be read and written in the same cycle. In contrast. the tag and data arrays are single-ported, so they can only be read *or* written in a given cycle.

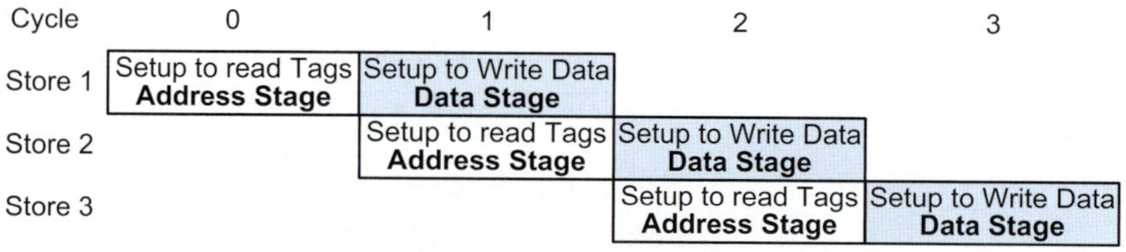

Fig. 10.15 Timing of back-to-back stores that hit.

Example 10.7 CACHE WRITE HITS

Sketch waveforms for a processor writing a 4-byte value 0x12345678 to address 0x00_0000_8000_0388 followed by a second 4-byte write of 0xAABBCCDD to address 0x00_000_8000_03C8. Assume the line is already in the cache.

Solution: Fig. 10.16 shows the waveforms:

- **In cycle 0** (Address stage of store 1), CacheSetTag = NextAdr[11:6] = 14, as in Example 10.6.
- **In cycle 1** (Data stage of store 1), the cache way reads the tag and valid bit from set 14. The cache way hits because the tag matches (ReadTag = PAdr[55:12]), and ValidWay is 1, so Match = 1, HitWay = 1, and SelectedWay = 1, and SetDirtyWay = 1. The cache drives CacheSetData = PAdr[11:6] = 14 to write

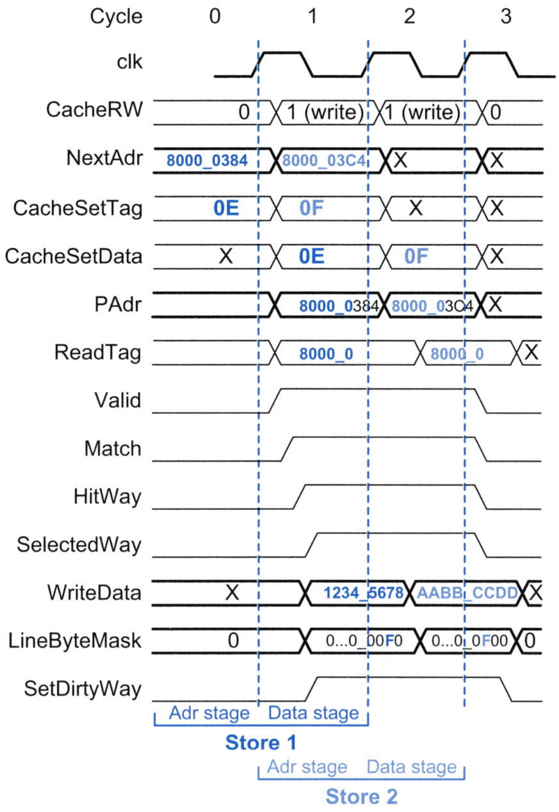

Fig. 10.16 `cacheway` timing diagram for back-to-back stores.

to the status and data arrays. The word offset is `PAdr[5:2]` $= 0001_2 = 1$, so the write selection logic writes word 1 (bytes 7:4) in the 16-word line by computing `LineByteMask` = 0x0000_0000_0000_00F0. The cache replicates the write data into `LineWriteData` = 0x12345678_..._12345678_12345678. The write completes on the next rising clock edge at the end of the Data stage.

- **Also in cycle 1** (Address stage of store 2), `CacheSetTag` = `NextAdr[11:6]` $= 001111_2 = 15$.
- **In cycle 2** (Data stage of store 2), the word offset is $0010_2 = 2$, which selects `LineByteMask` = 0x0000_0000_0000_0F00 to write word 2 (bytes 11:8) in the line. `LineWriteData`= 0xAABBCCDD...AABBCCDD, which is the desired `WriteData` replicated in each word (group of 4 bytes).
- Notice that, for store hits, `CacheSetData` lags `CacheSetTag` by one cycle.

10.4.2.3 Structural Hazards

Whereas back-to-back stores do not cause hazards, a store followed by a load does. Fig. 10.17 shows this scenario, with the cache's Address stage

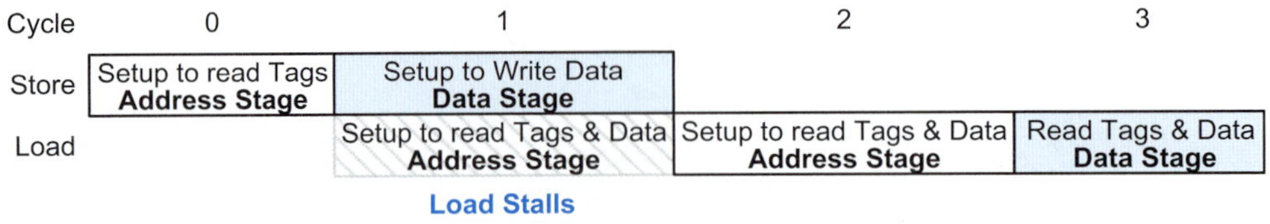

Fig. 10.17 Timing of a store followed by a load.

> This hazard is identical to the store delay hazard in the DTIM discussed in Section 7.2.2.3. The hazard logic in the integer execution unit (IEU) controller asserts LoadStallD when a load is followed by its use and StoreStallD when a store is followed by a load.

shown in white and its Data stage in blue, as before. The hazard occurs in cycle 1, where both the store and load want to access the data array: the store sets up to write the data array and the following load wants to read the data array, so the load must stall by one cycle.

10.4.2.4 Line Fetch

If no cache way hits during a memory access (HitWay for all cacheways is 0), the cache misses and must fetch the data from the next level of the memory hierarchy. This requires the following:

- Selecting a cache way for the new line, potentially needing to evict a line to make room (see Section 10.4.4).
- Writing back the evicted line if it is dirty (see Section 10.4.2.5).
- Fetching a new line using the AHB cache interface (ahbcacheinterface) word by word into a fetch buffer (see Sections 10.4.5.3 and 12.2.1).
- For a store miss, merging WriteData into the appropriate position in the fetch buffer (see Section 10.4.5.2).
- Writing the fetch buffer into the line of the selected way (see Section 10.4.5.2) while updating the tag from PAdr and setting the Valid bit. Dirty is set to 1 for a write or cleared to 0 for a read.

10.4.2.5 Line Writeback

If a cacheLRU selects a dirty line for eviction, the line must first be written back to the next level of the memory hierarchy. Fig. 10.18 shows how the victim way outputs its tag and data. When a cache way evicts a line, FlushCache = 0 and VictimWay = 1. The cache's controller drives SelVictim = 1, so SelectedDirty = 1. SelectedDirty is ANDed with the status array's dirty bit to check if the victim set is dirty. The SelectedWay mux chooses VictimWay, which is 1, so the cache way outputs TagWay and ReadDataLineWay. The cache writes the line in ReadDataLineWay back to the address indicated by TagWay and CacheSetTag. Section 12.2.2 describes writing the line over the bus interface one word at a time.

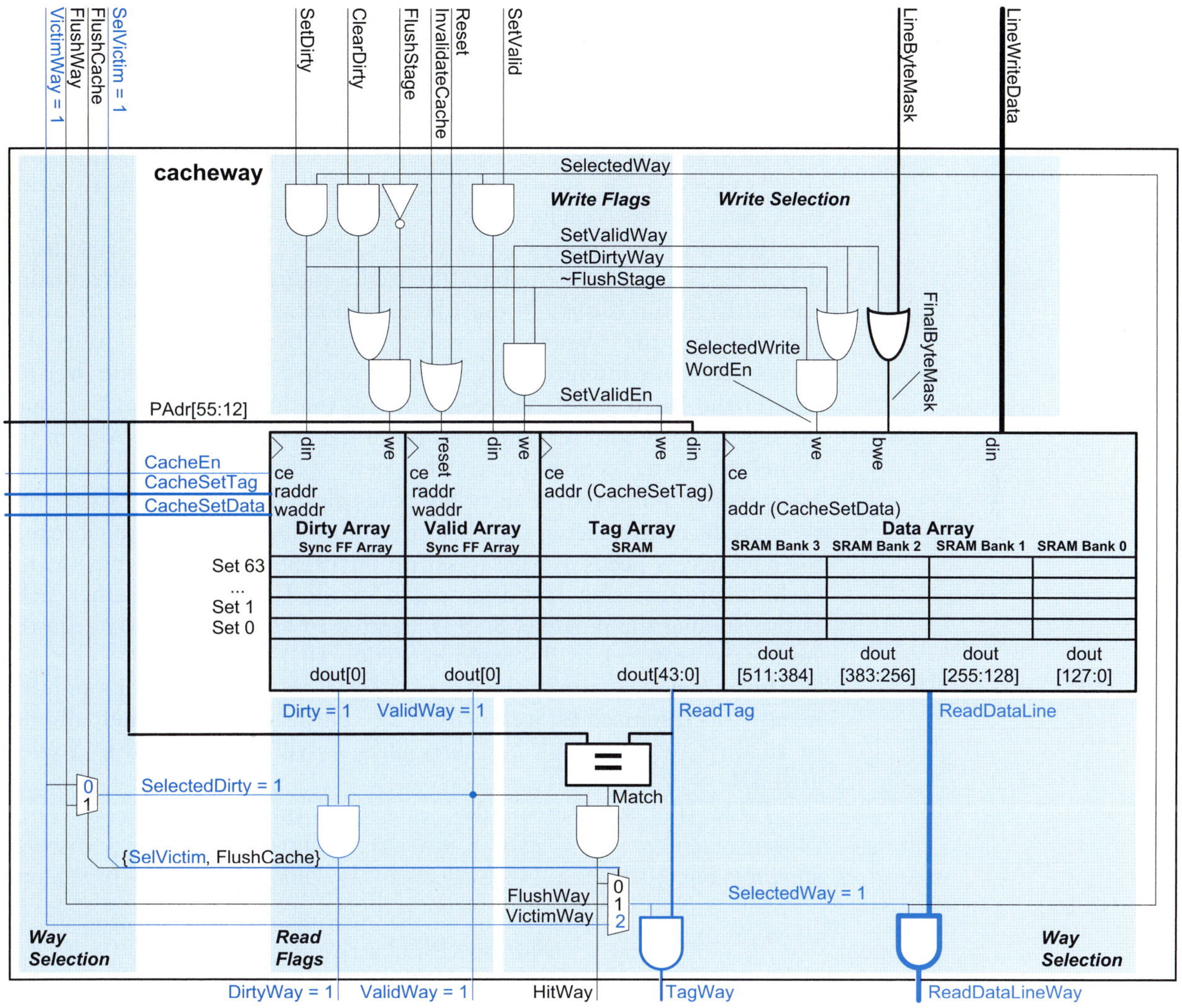

Fig. 10.18 `cacheway` **during line writeback.**

10.4.2.6 Way Invalidation

Recall from Section 10.1.8 that a cache is invalidated by clearing its valid bits. Both the D$ and I$ must be invalidated on reset. The Valid array contains resettable flip-flops. It receives a reset signal to clear all valid bits in parallel on the rising edge of the clock.

The Zifencei RISC-V extension defines `fence.i` to ensure coherence between the I$ and D$. For example, the OS may load a program or parts of a program into memory by copying them from other regions of memory. Because this is performed using loads and stores in the OS code, these writes go through the D$ rather than the I$, so the I$ may contain incorrect stale instructions. The `fence.i` instruction flushes the D$ to main memory and invalidates the I$ so that the I$ is forced to fetch the new instructions. `fence.i` is an expensive operation, so fences should be used sparingly.

10.4.3 Cache FSM

Fig. 10.19 shows the cache FSM (`cachefsm`), which handles cache misses and flushes. The FSM resets to the Access state, where it accepts read, write, invalidate, and flush requests from the CPU. If the cache hits on a read or write, it services the request immediately and remains in the Access state. Otherwise, `cachefsm` sequences the multicycle operations to write-back, fetch, or flush cache lines.

Read and write misses assert `AnyMiss`, select a victim line to evict using the `cachelru` module, write back the victim if dirty, and fetch the missing line. If the victim is dirty, it must be evicted by writing the line back over the bus (in the Writeback state). Each bus operation waits for `CacheBusAck` to continue. Then the cache fetches the missing line over the bus (in the Fetch state). The cache writes the missing line and tag into the selected way and sets the valid bit in the WriteLine state. Also in the WriteLine state, writes merge the new word from `WriteData` into `LineWriteData` and set the dirty bit; reads clear the dirty bit. Misses pass through the AddressSetup state before returning to the Access state because the next access needs a cycle to set up the cache address (set bits) to the cache before accessing it, as discussed in Section 10.4.2.1. If the other cache is also stalling the pipeline, Stall is asserted to hold the read data (in the AddressSetup state), as discussed in Section 10.4.6.

Invalidation occurs in the Access state. It resets all valid bits simultaneously, as discussed in Section 10.4.2.6. Flushes iterate through all lines of all ways, writing back any dirty lines as will be discussed in Section 10.4.5.4.

Table 10.2 summarizes the `cachefsm` outputs. `SetValid`, `SetDirty`, `ClearDirty`, `FlushCache`, and `SelWriteback` control writes to `cacheway`, as shown in Fig. 10.13. `SelVictim` selects a way when writing back or fetching

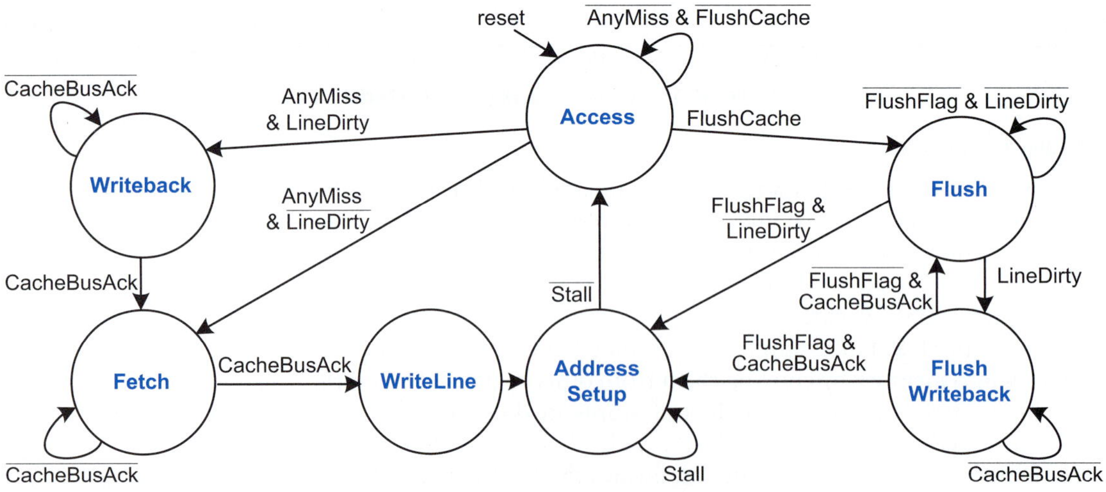

Fig. 10.19 **Cache FSM (`cachefsm`).**

Table 10.2 **cachefsm outputs.**

FSM Output	Destination	Meaning	Criteria to Assert
SetValid	cacheway, cachelru, cache	For a line in a way: set valid bit; write data and tag	WriteLine
SetDirty	cacheway, cache	For a line in a way: set dirty bit	Write & (Access & Hit | WriteLine)
ClearDirty	cacheway	For a line in a way: clear dirty bit	~Write & WriteLine | (Flush & ~LineDirty)
SelWriteback	cacheway	Writeback a cache line	NextState = Writeback
SelVictim	cacheway	A cache way outputs a victim line (used for writeback and fetch)	Access & Miss & LineDirty | WriteLine | WriteBack & ~CacheBusAck
LRUWriteEnable	cachelru	Update LRU after an access	Access & Hit | WriteLine
SelAdrData	cache	Select CacheSetData from PAdr (instead of NextAdr) during multicycle operation or store	Access & (Miss | Write) | Writeback | Fetch | WriteLine
SelAdrTag	cache	Select CacheSetTag from PAdr (instead of NextAdr) during multicycle operation	Access & Miss | Writeback | Fetch | WriteLine
SelFetchBuffer	cache	Pick word from FetchBuffer during line fetch	WriteLine | AddressSetup
CacheEn	cacheway, cachelru	Enable reads or writes	~Stall | FlushCache | AnyMiss | ~Access | InvalidateCache
CacheBusRW[1:0]	ahbcacheinterface	Request to read or write bus	[1]: Access & Miss & ~LineDirty | Writeback & CacheBusAck | Fetch & ~CacheBusAck [0]: Access & Miss & LineDirty | Writeback & ~CacheBusAck | FlushCheck & LineDirty | FlushWriteback & ~CachBusAck
FlushAdrCntEn	cache	Increment the flush line address counter	See Section 10.4.5.4
FlushWayCntEn	cache	Increment the flush way address counter	See Section 10.4.5.4
FlushCntRst	cache	Reset both set and way counters (that are used for flushes)	Reset | Flush & FlushFlag & ~LineDirty | FlushWritebakc & FlushFlag & CacheBusAck
CacheStall	hazard	Cache stalls the pipeline	NextState != (Access | AddressSetup)
CacheCommitted	trap	Cache delays interrupts	~Access
CacheAccess	Performance counter	Cache is accessed	Access & (AMO | CacheRW)
CacheMiss	Performance counter	Cache misses	Access & ~Hit

a line. LRUWriteEnable enables updated LRU bits to be written. SelAdrTag and SelAdrData control the address muxes for CacheSetTag and CacheSetData, respectively, and select among the current address (PAdr), the next address (NextAdr), or the flush address (FlushAdr), as discussed in Section 10.4.5.

After fetching a cache line, the cachefsm asserts SelFetchBuffer in the Writeline state to bypass the cacheway SRAMs and select the desired word from the fetch buffer (supplied by the AHB cache interface, see Fig. 10.12) instead. This saves a cycle as compared to waiting for the fetch buffer to be written to the cache and then reading the desired word from the cache. CacheEn asserts when reading or writing the cache, and CacheRW[1:0] is 1 when reading a cache line (line fetch) or 0 when writing a dirty line back, both using the AHB bus.

The flush counter controls are described in Section 10.4.5.4. CacheStall is described in Section 10.4.6, and CacheCommitted is described in Section 10.4.7.

The cachefsm pulses CacheAccess on each access, and CacheMiss on each miss. The performance counters use these signals to track cache accesses and misses.

10.4.4 pLRU Implementation

Section 10.1.7 introduced the theory of tree pseudo least recently used (pLRU) for selecting a victim way on a cache miss. This section maps the theory into an implementation suitable for a cache of any number of ways. An *N*-way associative cache puts the pLRU tree into an (*N*-1)-bit LRU signal associated with each set. On a miss, the cache selects the victim way from these LRU bits. The cache also updates the LRU bits after each access. This section uses an 8-way associative pLRU tree as an example and generalizes the result at the end.

Fig. 10.20(a) shows a pLRU tree for an N = 8-way associative cache. The node directions (left or right) are held in an (N-1)-bit LRU signal,

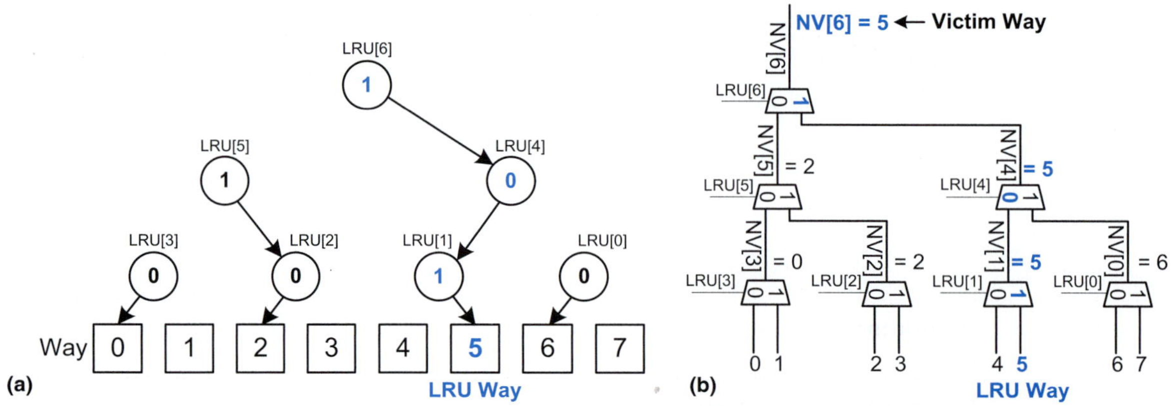

Fig. 10.20 8-way associative cache pLRU victim way implementation.

with 1 bit per node, that points to the least recently used (LRU) way. `LRU[N-2]` is the direction of the top node, and the remaining nodes map to the lower bits, top to bottom and left to right in the tree. In Fig. 10.20(a), `LRU` = 1100010.

Fig. 10.20(b) shows the tree of multiplexers used to find the *new victim* way. Each node calculates its new victim (LRU) way in its $\log_2$N-bit `NV` (new victim) signal and then passes that signal to the node above it. The overall victim way is the top node's new victim, `NV[N-2]`, in this case `NV[6]`. In this example, `LRU[6]` = 1 chooses the victim from ways 4–7. `LRU[4]` = 0 chooses the victim from ways 4 and 5, and `LRU[1]` = 1 selects the victim as way 5 (`NV[1]` = 5). The signal (`NV[1]` = 5) is passed up the tree using the multiplexers to make the new victim way 5 (`NV[6]` = 5).

Fig. 10.20(b) uses (`N-1`) $\log_2$N-bit 2:1 muxes to select the victim way. However, the way numbers (0–7) are constant, so the hardware simplifies dramatically, as shown in Fig. 10.21. The top multiplexer selects ways 0–3 (000-011) when `LRU[6]` = 0 and ways 4–7 (100-111) when `LRU[6]` = 1, so this reduces to the msb of `NV[6][2]` = `LRU[6]`. The lower multiplexers follow a similar pattern. The next most siginificant bit of the victim (`NV[6][1]`) will be either `LRU[5]` or `LRU[4]`, and the lsb will be `LRU[3]`, `LRU[2]`, `LRU[1]`, or `LRU[0]`. In the example figure, `NV[6][2:0]` = `{LRU[6], LRU[4], LRU[1]}` = 101.

Fig. 10.22 shows the `cacheLRU` module, which selects a victim way and updates the `LRU` bits held in the LRU Array. After a cache access, the hardware updates `LRU` to point nodes away from the accessed (i.e., most recently used) way. Thus, the LRU array is read at the end of the Address stage and written at the end of the Data stage.

The Victim Way logic in Fig. 10.22 selects either an invalid way, when available, or the victim calculated by the pLRU logic. `AllValid` is the bitwise AND of `ValidWay[N-1:0]`, a signal showing the valid way, using one bit per way.

> The `LRU` array is held in flip-flops rather than SRAM because it is small, requires reset, and is both read and written each cycle.

- When all ways are valid (`AllValid` = 1), the new victim is chosen as the pLRU output, (`VictimWayEnc` = `NV[6]`).

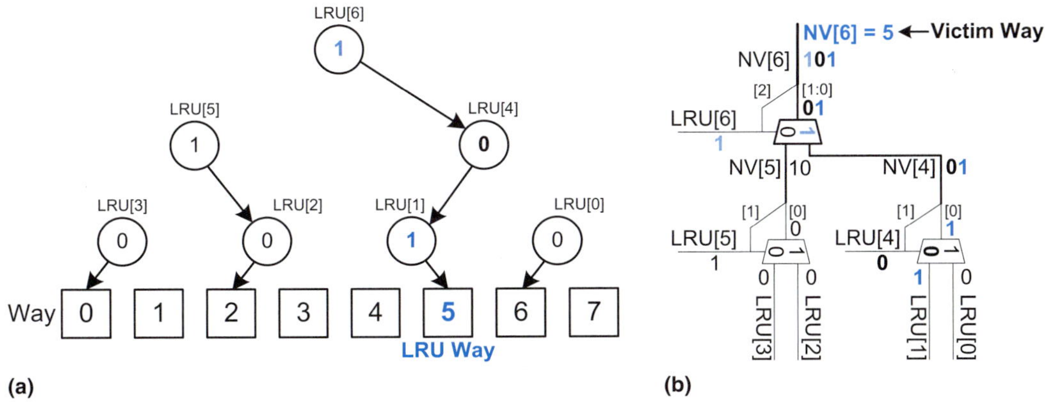

Fig. 10.21 **Simplified way selection hardware.**

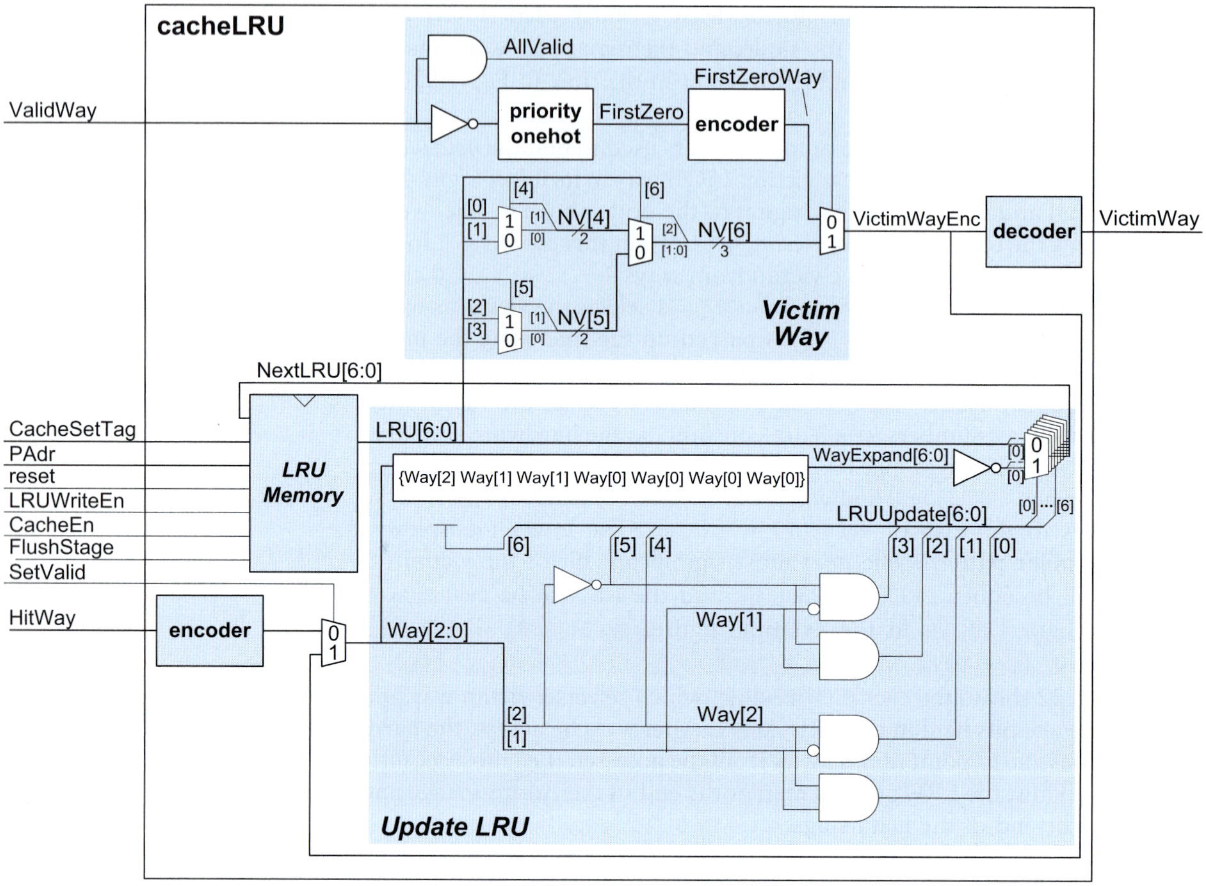

Fig. 10.22 **cacheLRU module.**

- Otherwise, the victim is an unused way; `priorityonehot` selects the lowest-numbered unused way (`FirstZero`), and the `encoder` converts the 1-hot output to binary (`FirstZeroWay`), which is then sent to the binary encoded victim way, `VictimWayEnc`.
- The `decoder` then converts the binary `VictimWayEnc` signal to the 1-hot `VictimWay` output. For example, if `VictimWayEnc` = 101 = 5, `VictimWay` = 0100000, with the fifth bit being 1.

The Update LRU logic at the bottom of Fig. 10.22 updates the LRU array to point *away* from the accessed way by calculating:

- **LRUUpdate:** An array with 1s for traversed nodes in the path to the accessed way; 1s indicate which nodes should be updated to point away from the accessed way.
- **Way:** The accessed way number.

- **WayExpand:** The direction each node points to reach the accessed way.
- **NextLRU:** Updated LRU bits, which are the original LRU bits for non-traversed nodes, and the inverse of WayExpand for traversed nodes, so that these nodes point away from the accessed way.

WayExpand replicates each bit of Way for each successive level of the pLRU tree. For example, if the accessed way (Way) is 5 (101), then WayExpand is 1001111. That is, to access way 5, the top-level node (LRU[6]) points right (1), the next-level nodes (LRU[5:4]) point left (00), and the lowest-level nodes (LRU[3:0]) point right (1111). Exactly one node at each pLRU tree level will be traversed to access a given way. Thus, because LRU[6] points right, only LRU[4]'s direction is relevant. And because LRU[4] points left, only LRU[1]'s direction is relevant.

Example 10.8 CALCULATE THE NEXT LRU BITS

Using Fig. 10.22, show the process for calculating NextLRU during an access to way 6 given LRU = 1100010.

Solution: Fig. 10.23 shows the pLRU tree values before and after an access to way 6. First, calculate LRUUpdate = **1010001**, which has 1s for traversed nodes (nodes 6, 4, and 0) along the path to way 6, shown with a gray dashed arrow. The other nodes are not involved in the access to way 6, so they will not be updated. Way = 110 (6), so WayExpand = 1110000. The NextLRU mux in Fig. 10.22 chooses LRU or WayExpand for each bit depending on each LRUUpdate bit (1010001). So, NextLRU equals the original LRU bits (LRU) for bits 5, 3, 2, and 1 (-1-001-) and WayExpand for bits 6, 4, and 0 (0-0---1), so that those nodes point away from way 6. Altogether, NextLRU = 0100011. The updated LRU bits (6, 4, and 1) are highlighted in blue in Fig. 10.23(b). Node 4 was already pointing away from node 6, so its direction happens to stay the same.

10.4.5 Top-Level Cache Logic

Fig. 10.24 shows the complete top-level cache module, providing more detail than the block diagram of Fig. 10.11, including the cacheway,

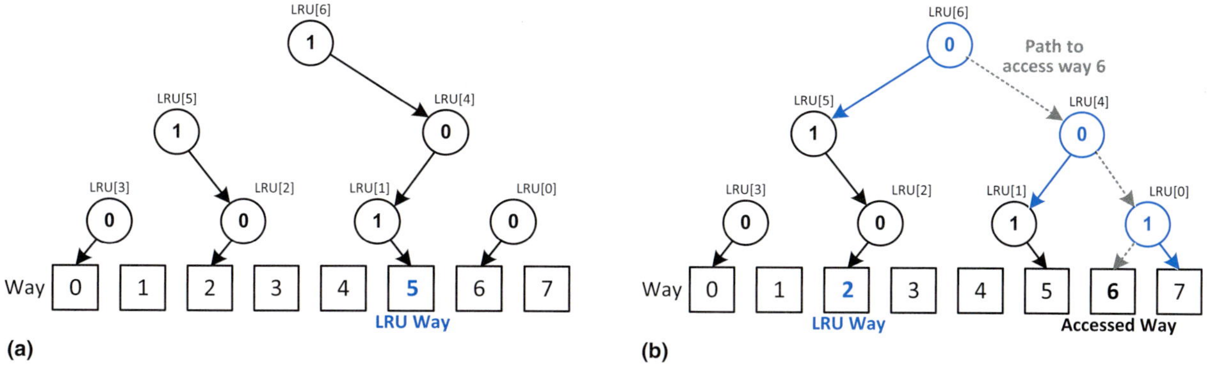

Fig. 10.23 pLRU tree (a) before and (b) after accessing way 6.

cachefsm, and cacheLRU modules described in previous sections. This section describes the shaded logic in Fig. 10.24.

- **Address selection:** Selects CacheSet{Tag/Data} from NextAdr, PAdr, or FlushAdr.
- **Way selection:** Delivers the tag, read cache line, dirty, hit, and valid bits from the selected way using a distributed AND-OR read mux.
- **Read selection:** Selects a word from the cache line for reads and writebacks.
- **Write selection:** Merges WriteData into FetchBuffer (using the bottom LineWriteData mux) during write misses and forms the byte mask for the line (DemuxedByteMask).
- **Bus logic:** Selects the address to send to the bus for fetches, writebacks, and flushes.
- **Flush logic:** Sequences through the lines and ways during a cache flush.

FetchBuffer, BeatCount, and SelBeatCount come from the AHB cache interface, which is described in Section 12.2. Section 12.5 gives examples of how the cache and bus transfer a burst of consecutive words during cache line fetches and writebacks.

10.4.5.1 Address, Way, and Cache Line Read

CacheSetData addresses the data array and CacheSetTag addresses the tag and status arrays. Independent addressing of the tag/status and data arrays allows for back-to-back stores without stalling as described in Section 10.4.2.2. During a flush, both are driven from FlushAdr to sequence through cache lines.

The read multiplexer that produces the cache line, ReadDataLineCache, is a distributed AND-OR multiplexer, as shown Fig. 10.25. Recall from Section 10.4.2 that, at most, one cache way outputs a nonzero cache line (ReadDataLineWay). These way outputs are ORed together to produce the cache's read line (ReadDataLineCache). The AND stage of the AND-OR mux is internal to each cacheway module (shown in Fig. 10.13), and the OR stage is part of the Way Selection logic.

Benefits of the distributed AND-OR mux over a traditional mux include its ability to directly use the one-hot HitWay signals (at most one HitWay signal is hot (1) across all ways) instead of having to encode that signal and its reduced power because at most one ReadDataLineWay bus is nonzero. The tag (Tag) and dirty bit (LineDirty) are similarly read using a distributed AND-OR multiplexer.

Distributed AND-OR muxes will appear again in the TLB (translation lookaside buffer) in Sections 11.4.1.3.1-11.4.1.3.3.

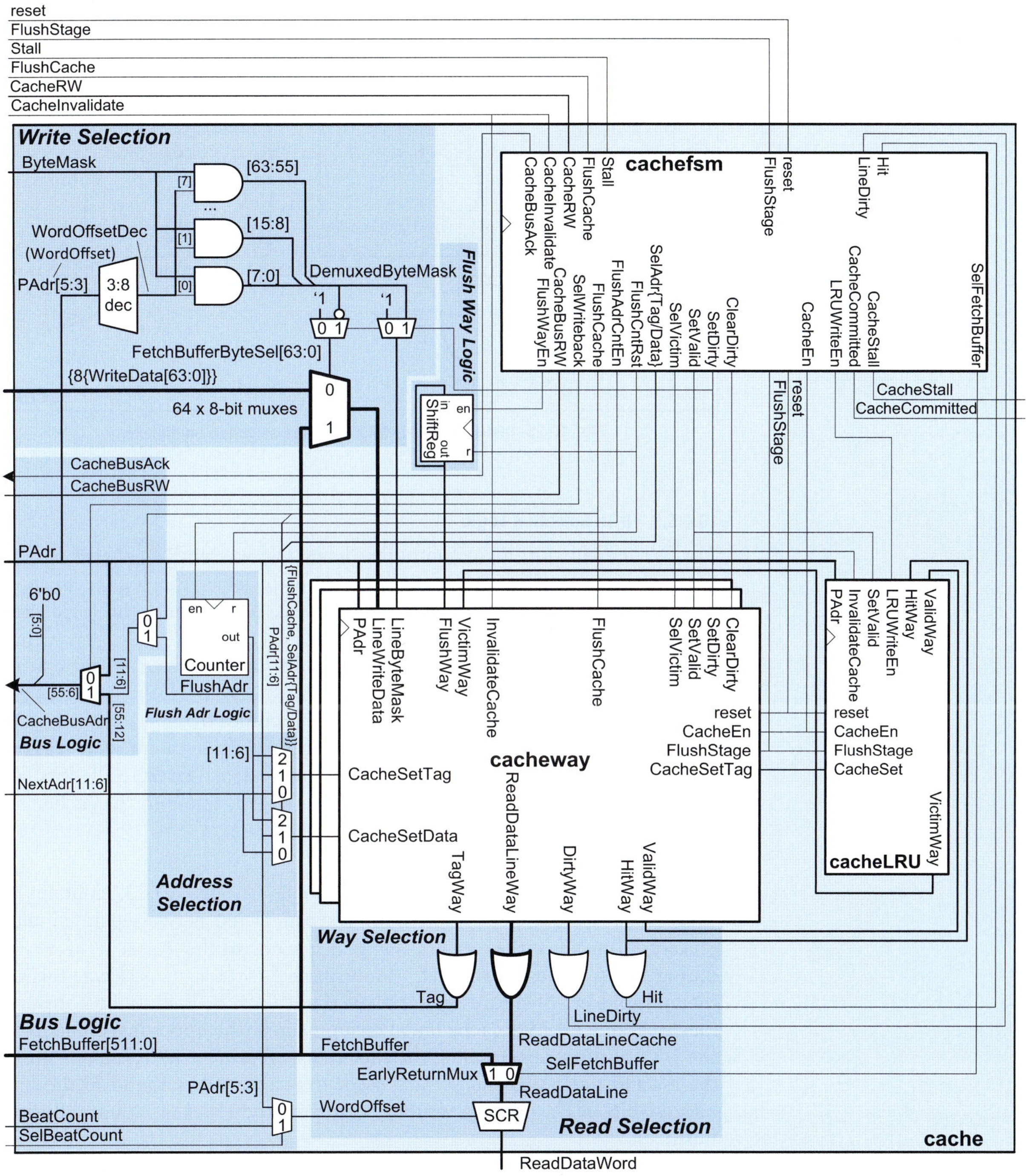

Fig. 10.24 Detailed cache module.

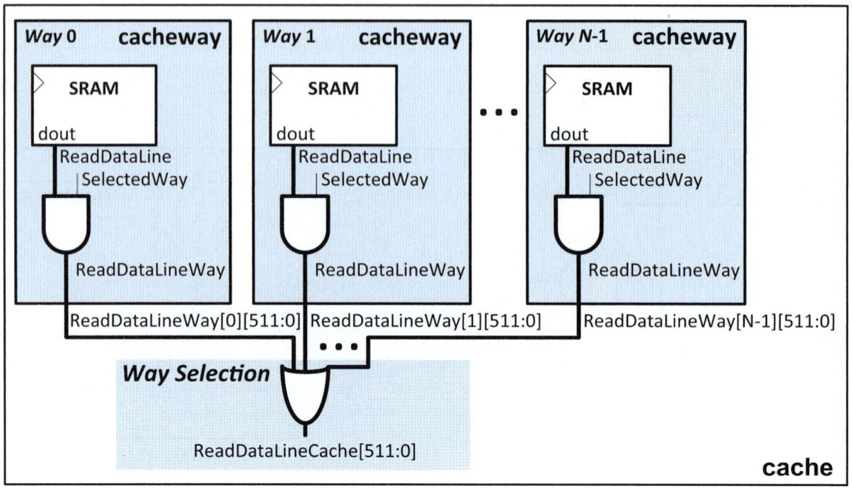

Fig. 10.25 Distributed AND-OR cache line selection multiplexer.

10.4.5.2 Write Selection Logic

The Write Selection logic computes `LineWriteData` and `LineByteMask` for the cache ways. During a store hit, the Write Selection logic replicates `WriteData` into each word of `LineWriteData`. Meanwhile, the logic uses the word offset bits of `PAdr` and the `ByteMask` to form `LineByteMask`, that indicates which bytes of the line should be written, as shown in Figs. 10.5 and 10.13.

During a cache line fetch, `LineWriteData` comes from `FetchBuffer`. On a write miss, `FetchBufferByteSelect` merges `WriteData` into the appropriate bytes of `LineWriteData`. The logic then asserts all bits of `LineByteMask` to write the entire (possibly merged) line into the cache.

10.4.5.3 Read Selection and Bus Interface Logic

The bus interface logic, labeled Bus Logic in Fig. 10.24, contains a `CacheBusAdr` address generation multiplexer (depicted as two separate multiplexers) to produce the `CacheBusAdr` that it sends to the `ahbcacheinterface` during a bus request. Table 10.3 shows how `CacheBusAdr` is selected using the multiplexer. All addresses are aligned to the cache line, so the bottom 6 bits are 0 in a 64-byte cache line. During a fetch, the upper bits come from the `PAdr` of the access causing the cache miss. During a writeback or flush, the evicted line's tag and set are used; the cache set is found in the line address bits of `PAdr` during writeback or from the `FlushAdr` counter during flush.

Table 10.3 **`CacheBusAdr` mux operation.**

{FlushCache, SelWriteback}	CacheBusAdr	Description
00	{PAdr[55:6], 000000}	Fetch
01	{Tag, PAdr[11:6], 000000}	Writeback
10	{Tag, FlushAdr, 000000}	Flush

The Read Selection logic selects the word within the cache line (see Fig. 10.24). This logic uses a sub-cache line read (SCR) mux and an early return mux to produce `ReadDataWord` for three cases:

- **Read hit:** The SCR mux selects a word within the cache line using the lower bits of the address: `WordOffset` = `PAdr[5:3]`.
- **Read miss:** The early return mux selects a word from the fetch buffer (`SelFetchBuffer` = 1) instead of the cache ways during a read miss. This saves a cycle at the end of the miss instead of waiting for the fetched cache line to be written to the cache before reading it.
- **Writeback:** The cache must writeback the cache line to external memory word by word. Thus, `ahbcacheinterface` asserts `SelBeatCount` so the SCR mux uses WordOffset = `BeatCount` to output each word of the cache line in sequence.

This is discussed further in Sections 12.2.1 and 12.2.2.

10.4.5.4 Cache Flush Logic

As mentioned in Section 10.4.2.6, `fence.i` flushes the D$. A cache flush writes all the dirty lines back to the next level of the memory hierarchy and clears the dirty bits but leaves the valid bits unchanged. This involves sequencing through each line of each way and reading the data, tag, valid, and dirty bits. If the line is valid and dirty, the data is written back to an address specified by the tag, and the dirty bit is cleared.

Fig. 10.26 highlights the flush logic, which contains two counters to sequence through the lines and ways. The `FlushAdr` counter counts through the lines in binary. The `FlushWay` counter uses a one-hot shift register to count through the ways. System reset initializes both counters, `FlushAdr` to 0 and `FlushWay` to 0...01. `FlushWayFlag` and `FlushAdrFlag` are inputs to the cache FSM. `FlushWayFlag` asserts when all ways in a given set have been flushed, and `FlushAdrFlag` asserts while flushing the last set in a cache. So, the entire cache has been flushed when `FlushFlag = FlushWayFlag & FlushAdrFlag` asserts.

During a flush, the cache FSM cycles through all ways of each successive set and writes back any dirty lines. A flush begins when the IEU

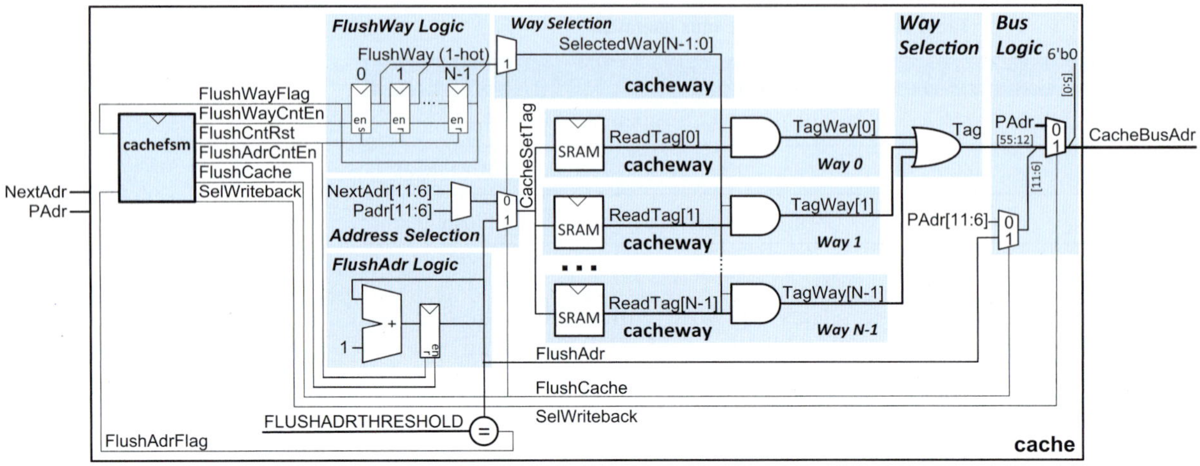

Fig. 10.26 Flush address generation.

asserts `FlushCache`, which causes the cache FSM (see Fig. 10.19) to transition from the Access to Flush state. During the transition, `FlushWay` = 1 and `FlushAdr` = 0, so the next cycle reads out way 0 set 0. The cache FSM asserts `FlushCache`, which chooses `FlushWay` (see Fig. 10.13) to select the 0th way. If that cache line is valid and dirty, `LineDirty` is asserted and `cachefsm` transitions to the FlushWriteback state. Concurrently, the cache computes the cache line's address by concatenating the line's `ReadTag` with the set bits from `FlushAdr` and appending zeros for the word and byte offset while asserting `CacheBusRW[0]` to initiate a bus write. The bus interface will increment the word offset to write back the dirty cache line word by word. When the bus completes the transaction by asserting `CacheBusAck`, `cachefsm` returns to the Flush state, where it increments `FlushWay` (by asserting `FlushWayCntEn`) to move onto flushing set 0 in way 1.

After the FlushWriteback state completes, that cache line will no longer be dirty, so `cachefsm` asserts `ClearDirty` on the transition from Flush to FlushWriteback. The dirty bit is cleared before data is written to the bus to minimize the number cycles spent on writeback. `ClearDirty` must be asserted for the current way and set (as indicated by `FlushWay` and `FlushAdr`).

For a single core, clearing the dirty bit before the writeback begins is safe. However, a multicore or multithreaded implementation may use a cache coherency mechanism, which requires the line to be cleared after writeback is completed.

If a line is clean (`LineDirty` = 0), `FlushWay` is incremented by asserting `FlushWayCntEn`. When `FlushWay` reaches the last way (10…0), it wraps back around to the first way (0…01), and `FlushAdr` is incremented to move to the next set by asserting `FlushAdrCntEn`.

The cache flush operation ends when the cache has iterated over all sets and ways, as indicated by `FlushAdr` and `FlushWay` hitting the maximum value. The check occurs in both the Flush and FlushWriteback states. `cachefsm` resets both counters by asserting `FlushCntRst` and then returns to the Access state.

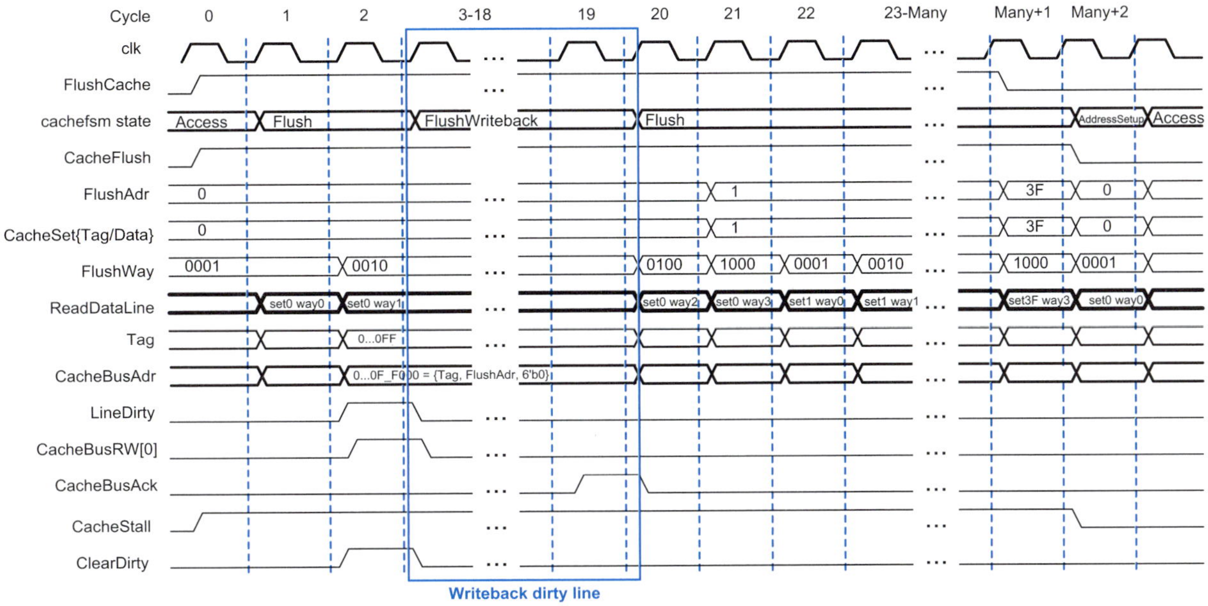

Fig. 10.27 Timing diagram for D$ flush, showing one dirty line written back.

Flushing the cache is time-consuming because the cache must check each cache line's dirty bit and write back any dirty lines. In the best case, all lines are clean so run time takes (the number of sets $\times$ the number of ways) cycles. The rv64gc configuration has 64 sets and 4 ways for a total of 256 cycles. If all lines are dirty and each writeback takes 15 cycles (in addition to the 1 cycle to check the dirty bit), the D$ flush will take $64 \times 4 \times (1+15)$ cycles $= 4{,}096$ cycles! Fortunately, D$ flushes are rare events.

Fig. 10.27 shows the timing of a D$ flush in a 4-way associative cache, showing the details of one dirty cache line written back over the bus.

- **In cycle 0,** the cache FSM receives `FlushCache`. It sets up `CacheSet{Tag/Data}` = `FlushAdr` in the Access state to read set 0.
- **In cycle 1**, set 0 way 0 is not dirty, so a writeback is not needed and `FlushWay`, which is a 1-hot signal, is incremented by shifting it to the left.
- **In cycle 2**, set 0 way 1 is dirty. The FSM asserts `ClearDirty` to reset set 0 way 1's dirty bit. The cache line's address, `CacheBusAdr` = 0x00_0000_000F_F000 is computed by concatenating the line's `Tag` with `FlushAdr` and six 0s. `CacheBusRW[0]` = 1 to begin a bus write.
- **In cycle 3**, `cachefsm` transitions to the FlushWriteback state where it holds `FlushAdr` and `FlushWay` constant until the bus transaction completes.

FlushAdr sets up which cache set to read (in the Address stage: CacheSet{Tag/Data} = FlushAdr), and FlushWay selects which way to read (in the Data stage). Recall that cache reads are pipelined, with one in the Data stage while the next is in the Address stage. As the last way is read (FlushWay = 1000), the set (FlushAdr) also increments, so that the cache sets up to read the new set (e.g., see cycle 21 of Fig. 10.27).

- **In cycle 19,** CacheBusAck rises to indicate the bus is finished and the cachefsm returns to the Flush state and enables FlushWay to increment, which occurs in the next cycle.
- **In the remaining cycles,** this process repeats until all cache lines are checked and written back if dirty. When cachefsm reaches the final set and way (set = 0x3F = 63, way = 3), the flush is complete, so the address (set) and way counters are reset (to set 0 way 0) and cachefsm transitions back to the Access state via the AddressSetup state.

10.4.6 Pipeline Stalls

Any time either the I$ or D$ misses or performs a cache flush, it asserts CacheStall to stall the entire pipeline while fetching or flushing. CacheStall remains asserted until the cachefsm returns to the Access or AddressSetup state. The LSU and IFU, which will integrate the caches in Chapters 12 and 13, assert LSUStallM or IFUStallF when their CacheStall occurs. These in turn stall the pipeline, as shown in Fig. 9.33.

Conversely, the hazard unit stalls the LSU or IFU by asserting StallW or StallD, respectively. These stalls are renamed to Stall as an input to the cache and ahbcacheinterface because these modules are instantiated by both the LSU and IFU.

Both caches can simultaneously miss, stalling the entire pipeline. Both cache FSMs concurrently request the missing line or writeback a dirty cache line. However, as discussed in Section 9.5.5, the external bus unit (EBU) gives priority to the LSU (D$ requests), so the D$ fetches its missing line first. The D$ then waits in the AddressSetup state, retaining ReadDataWord, until the I$ also fetches its missing line and the pipeline resumes. In AddressSetup, the D$ lowers CacheStall to signal the cache is no longer busy and reads from the FetchBuffer rather than cacheway. While the I$ is busy, the hazard unit asserts StallW, which is connected to the cache's Stall input. In the AddressSetup state the cachefsm only reads the next memory request when Stall asserts. This allows AddressSetup to behave like the Access State to set up CacheSetTag and CacheSetData in the Address stage.

10.4.7 Exceptions and Interrupts

At the beginning of a memory operation, the MMU, which is introduced in Chapter 11, checks the validity of the operation and generates an exception if the access is bad. Such exceptions, as well as other exceptions and interrupts, assert TrapM, which flushes the entire pipeline including pending cache operations.

A pipeline flush raises FlushStage in the I$ or D$, which is connected to receive FlushD and FlushW, respectively. Note that these signals indicate pipeline flushes, *not* cache flushes. `FlushStage` resets the given cache's `cachefsm` and aborts any ongoing writes to its `cacheway` and `cachelru`.

Interrupts are asynchronous, arriving at any time, even when the pipeline is stalled for a bus access or cache miss. When a bus transaction is in flight, the bus cannot be interrupted without potentially corrupting memory. Just as the `busfsm` asserts `BusCommitted` when a bus transaction is in progress (see Section 9.5.5), the `cache` asserts `CacheCommitted` when `cachefsm` is busy (i.e., in the middle of a fetch, flush, or write back). The LSU or IFU raises their `CommittedM` or `CommitedF` signal when either the bus or cache is busy. Pending interrupts are delayed until `CacheCommitted` falls, using the hardware from Fig. 9.35.

10.5 SUMMARY

Microprocessor designers approximate a large, fast, cheap memory with a large, slow, cheap main memory and a small, fast, cheap cache memory. If most memory accesses hit the cache, the AMAT is almost as fast as the cache, but the cost per total bit remains low. In a large system, this generalizes to a multilevel memory hierarchy with caches of varying size and speed.

Caches exploit temporal and spatial locality to obtain a high hit rate: memory locations that have been accessed recently are likely to be used again, and nearby memory locations are also likely to be used. To exploit spatial locality, caches are organized as a collection of lines, commonly 512 bits (64 bytes, or 8 double words) long.

The cache only holds a subset of memory. The middle bits of the requested address (the set bits) select the cache line, and the lower bits select the word within the line. The line also contains a tag as well as valid and dirty bits. If the tag bits match the upper bits of the address and the line is valid, the desired line is present, and the cache hits. Cache misses can be classified as compulsory (first time a line is accessed), capacity (cache is too small to hold the working set), or conflict (multiple addresses map to the same cache line). Long cache lines reduce compulsory misses at the expense of an increased miss penalty (time to transfer a cache line). Larger caches reduce capacity misses at the expense of cost and speed. Associative caches reduce conflict misses by providing multiple ways that can hold a given cache line, at the expense of complexity, cost, and access speed.

When a cache miss occurs, the entire line is transferred into the cache, so a subsequent access to another word in the line will hit. If the set is full, it must evict another line to make room. To exploit temporal locality by holding recently accessed lines, associative caches typically have an LRU or pLRU replacement policy to select the way to evict. If the evicted line is dirty, it must also be written back to memory.

A typical core makes most of its accesses to level 1 instruction and data caches (L1I$ and L1D$). The cache size is limited by the desired access time and is typically 8–32 KiB. Because of interactions with the virtual memory system described in the next chapter, each cache way is no bigger than the virtual memory page size, usually 4 KiB. Therefore, the cache generally has 2, 4, or 8 ways, which is sufficient to eliminate most conflict misses. Experience has shown that a 512-bit line size (64 bytes, or 8 double words) is a good tradeoff between spatial locality and cache line transfer time. Hence, a common L1$ organization is 16 KiB, built from four 4-KiB ways, each containing 64 64-byte lines, which achieves a miss rate of about 4% on Embench benchmarks. Wally rv32gc and rv64gc configurations have this cache topology. High-performance cores add an L2$ with about 256-1024 KiB capacity. Multicore systems add a L3$ with many megabytes of capacity that is also responsible for maintaining coherence when multiple cores are accessing the same data.

Wally uses a common parameterized module for both instruction and data caches. However, instruction caches are read-only, so they eliminate the write, dirty, and writeback logic. The heart of the cache is an array of cache ways. Each way is indexed by the set (middle bits of the address). The tags from each way are compared to the upper bits of the address. The hit logic checks for a valid way with a matching tag. If a hit occurs, a way selection mux picks the cache line from the matching way. For a read hit, the `subcachelineread` mux picks the word from the line based on the lower bits of the address. For a write hit, the cache writes the line on the next cycle and sets the line's dirty bit. The cache uses a byte mask to specify which bytes in the line to write, depending on the size of the access and the word offset in the low bits of the address.

When a cache miss occurs, the `cachefsm` module sequences through a series of steps. The `cacheLRU` module selects the way to replace using a pseudo-LRU algorithm. If the line is valid and dirty, it must be written back to main memory via the AHB bus. The line transfer involves a sequence of consecutive word transfers organized as a burst. The cache then fetches the missing line into a fetch buffer with another AHB bus burst and finally writes the line into the cache. An early return mux saves a cycle from the miss latency of a read data by returning the read data directly from the fetch buffer rather than waiting for it to be written into the cache.

The cache also has hardware support for flushes and invalidates. Flushes iterate over every line in every way, writing dirty lines back to memory. Cache invalidates clear the cache's valid bits. The `fence.i` instruction enforces coherence between instruction and data caches by flushing the D$ and invalidating the I$.

Memory Management Unit

11

11.1 **Principles**
11.2 **RISC-V Practices**
11.3 **Test Plan**
11.4 **Wally Implementation**
11.5 **Summary**

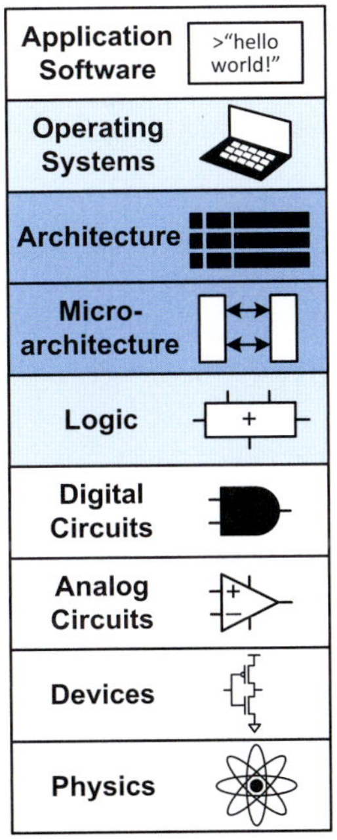

The *memory management unit* (*MMU*) is responsible for handling virtual memory and memory protection. This allows the programmer to write a program as if it has a large amount of memory that cannot conflict with memory from other programs and makes it more difficult for malicious programs to steal information from other programs. The RISC-V privileged specification defines the programmer's view of virtual memory, along with two memory protection checkers: the physical memory attributes (PMA) checker and the physical memory protection (PMP) checker. Wally implements these features using two MMUs, one in the instruction fetch unit (IFU) and one in the load/store unit (LSU). Wally also includes a single hardware page table walker (HPTW) to efficiently fetch page table entries (PTEs).

11.1 PRINCIPLES

This section summarizes the broad ideas of virtual memory and memory checking as they apply across computer architectures. Section 11.2 elaborates on RISC-V–specific issues.

11.1.1 Virtual Memory

Virtual memory enables each process to use its entire address space without being concerned about other processes trying to use the same addresses.

In a simple computer system without virtual memory, a process (i.e., a running program) directly specifies the address of main memory that it wishes to access; this is called *physical addressing*. Physical addressing presents a variety of problems for multitasking and large processes, as occurred in early operating systems. Each process must be loaded into a different place in memory. To allow an arbitrary set of processes to run concurrently, each process must be placed in an empty place in memory when it is loaded, and all the address references must be changed to reflect that location; this is known as *relocation*. Different processes need different amounts of memory, so when one process finishes, it may leave a chunk of free memory that is too small for another process; this is known as *fragmentation*. A process also might be too large to fit in memory at once. Early systems used burdensome *overlays*, in which the programmer might swap different parts of the program or data set between memory and disk as they were needed. Yet another challenge is that a process might accidentally write to another process's memory and cause both processes to crash, or a malicious process might read the password from another process's memory space. A better OS should provide *protection* to prevent processes from interfering with one another.

Virtual memory is an elegant solution to all these problems. With virtual memory, each process is given its own *virtual address (VA) space*, so the process has an illusion of a large, private, contiguous memory. Therefore, the process can be loaded at an address determined at compile/link time and does not require relocation. Because virtual addresses are private to the process, memory is protected from interference by other processes.

The virtual memory used by each process can be kept on the hard drive. Hard drives provide vast amounts of inexpensive storage, but are painfully slow, so computers use main memory (usually DRAM) to cache the commonly accessed virtual memory. This cache is called *physical memory*. Virtual memory is organized into fixed-sized blocks (conventionally 4 KiB) called *pages* that are analogous to cache lines. The MMU is a hardware block in the processor that performs *address translation*, converting virtual addresses used by the programmer into physical addresses (PAs) sent to the RAM. The OS manages moving pages between virtual (hard drive) and main (DRAM) memory as needed, so the programmer doesn't need to be concerned with overlays. Because pages are of fixed size, fragmentation does not occur. A process may have multiple threads that all share the same virtual address space, so communication between threads within a process is easier than communication between processes. Virtual memory is such a useful programming abstraction that all modern general-purpose operating systems use it.

Tom Kilburn at the Manchester Atlas console, 1971. (© 2014. This image is openly licensed under CC BY-SA 3.0.)

Virtual memory was invented by Tom Kilburn and his team for the Atlas computer at the University of Manchester [Kilburn62]. Atlas had 16,000 50-bit words of fast magnetic core memory with a 0.4-μs access time, 96,000 words of lower-cost drum memory on four rotating magnetic drums with a 4-μs access time, and a large amount of inexpensive storage on magnetic tapes with an average 88-μs transfer time per word. Now a typical computer has gigabytes of fast DRAM and terabytes of inexpensive hard drive storage.

In 1979, Ozalp Babaoglu, along with Sun Microsystems cofounder Bill Joy, added virtual memory support to Berkeley Unix on the VAX-11/780 computer that was distributed as 3BSD. The Intel 80286 introduced hardware support for virtual memory on x86 processors in 1982, but PC operating systems lagged nearly two decades. Operating systems such as Windows 3.11 (1992) and Mac MultiFinder (1987) used cooperative multitasking without protection where all programs shared the same address space. A bug in one program could crash the entire computer. This was not a rare occurrence. Windows NT supported virtual memory since 1993 but was incompatible with many popular programs. Windows 2000 and Mac OS X brought memory protection and preemptive multitasking to the mainstream personal computer world, making computing much more stable.

11.1.1.1 Pages

Pages of memory are kept in both physical (DRAM) and virtual (hard drive) memory. They are moved from virtual to physical memory as they are needed by a process. Let the *page size* be $P = 2^p$ bytes. Let a virtual address consist of n bits and a physical address consist of m bits, corresponding to $N = 2^n$ and $M = 2^m$ bytes of virtual and physical address space, respectively. n is typically 32 or 64 for 32- or 64-bit systems, but the upper bits may not all be used because, currently, few applications require $N = 2^{64} = 16$ exabytes of memory.

Hence, each process has up to $2^{n\text{-}p}$ virtual pages, and the system has $2^{m\text{-}p}$ physical pages. Like a larger cache line size, a larger page size takes advantage of spatial locality and has less overhead but greater penalties on a miss. Because disks have long latency but decent bandwidth, the page size should be large for highest performance. Most processors and operating systems have standardized on P = 4-KiB page sizes (p = 12 bits). Thus, a computer with M = 16 GiB of main memory (DRAM) would have m = 34 and would be able to hold $2^{34\text{-}12} = 2^{22}$ physical pages.

Each virtual page for a process is in one of three states:

- **Unallocated:** No data is associated with this page, so it does not use any disk space.
- **Cached:** An allocated page is currently cached in physical memory.
- **Uncached:** An allocated page is currently stored on disk.

Some systems use the terms *resident* and *nonresident* instead of cached and uncached pages.

Fig. 11.1 illustrates a mapping between virtual and physical pages. In this example, virtual page 0 is uncached and is only found somewhere on the hard drive. Virtual page 2 is cached in physical page 0 and so forth. Cached pages are found in physical memory and on disk, as indicated by the dashed blue lines, with the disk version potentially being out of date. Physical pages may also be empty, not holding any virtual page yet, as shown.

The area on the hard drive reserved for uncached (nonresident) virtual pages is called *swap space*.

The MMU uses a *page table* to perform address translation between virtual and physical pages, which are uniquely identified by virtual and

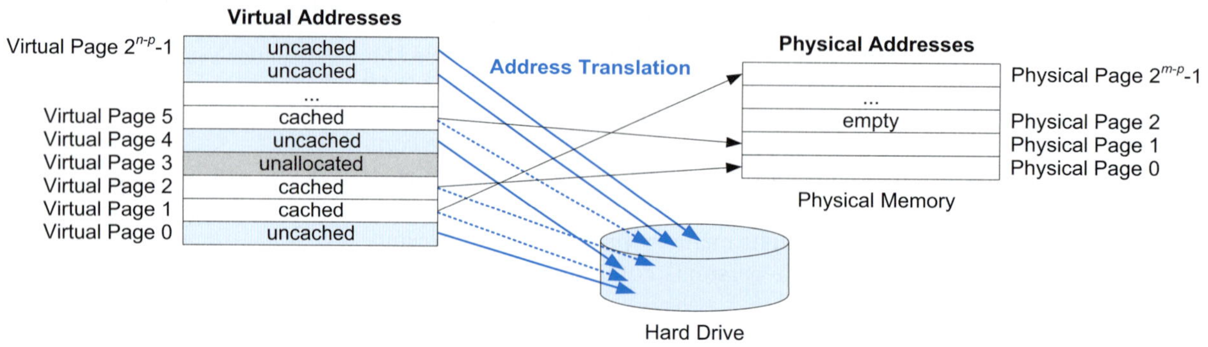

Fig. 11.1 Mapping between virtual and physical pages.

physical page numbers (VPNs and PPNs), respectively. The page table has one *page table entry* (*PTE*) for each virtual page. Each PTE contains status bits to indicate the state of the virtual page (unallocated, cached, or uncached) and address bits that specify the physical page number, if cached, or information for the OS to find the location on disk, if uncached. The upper *n-p* bits of the virtual address specify the *virtual page number* (*VPN*) and hence are used to index the page table. If the PTE indicates the cached state, then a *page hit* occurs and the MMU reads the (*m-p*)-bit *physical page number* (*PPN*). The bottom *p* bits, called the *page offset*, are the same for virtual and physical addresses and hence need no translation. The physical address is formed by concatenating the PPN and the page offset. Fig. 11.2 shows a page table for the mapping from Fig. 11.1, with 32-bit virtual addresses, 34-bit physical addresses, and 4-KiB pages.

Example 11.1 VIRTUAL TO PHYSICAL ADDRESS TRANSLATION

Suppose a program accesses virtual address 0x00005028. At what physical address will the data be found? Use the page table from Fig. 11.2, with $n = 32$, $m = 34$ and $p = 12$.

Solution: As shown in Fig. 11.3, the VPN is the upper $32 - 12 = 20$ bits of the virtual address: 0x00005. The page offset is the bottom 12 bits, 0x*028*. According to the page table, virtual page 5 maps to physical page **1**. Hence, the 34-bit physical address is 0x00000**1***028*.

11.1.1.2 Page Faults

If a process accesses a virtual page that is in the uncached or unallocated state, then a *page fault* occurs, and the system must fetch the page from disk. The processor takes a *page fault exception*. The OS handles the exception by reading the desired virtual page from disk, placing it into an empty physical page in main memory, and updating the PTE to the cached state with the proper PPN. The exception then returns to the

VPN	PPN	Status
$2^{n-p}-1$		Uncached
$2^{n-p}-2$		Uncached
...	...	...
5	000001	Cached
4		Uncached
3		Unallocated
2	000000	Cached
1	3FFFFF	Cached
0		Uncached

Fig. 11.2 Page table for mapping from Fig. 11.1.

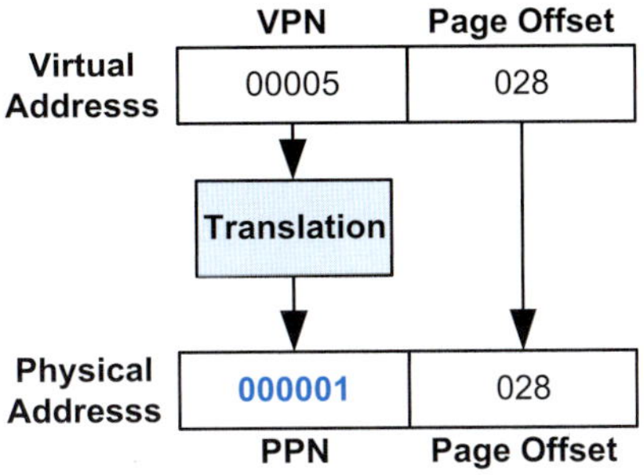

Fig. 11.3 Virtual to physical address translation.

interrupted instruction, which executes again and gets a page hit on the second try.

If the physical memory is completely full, the OS must choose a physical page to replace. If this *victim* page is *dirty*, the OS must write it back to disk. As in a regular cache, one might choose the least recently used (LRU) page. But determining true LRU is expensive, so the system may use a heuristic that approximates LRU. The PTEs typically contain additional state to keep track of whether the page is dirty or has been accessed recently.

11.1.1.3 Protection

Computers are constantly beset by malicious or buggy software. Virtual memory provides a convenient form of memory protection. Page table entries usually have additional bits to indicate whether a page is writable or executable and what privilege level(s) can access it. For example, code pages should not be writable, and data pages should not be executable so that a virus can't use a buffer overflow attack to inject bad instructions. The page tables must not be writable in user mode, lest the program rewrite its own page table to change this protection or to access another program's physical memory.

11.1.1.4 Multilevel Page Tables

The full page table described above can be unreasonably large. Because most processes use only a tiny fraction of the virtual address space, most PTEs contain unallocated pages and are unnecessary.

Example 11.2 1-LEVEL PAGE TABLE SIZE

Determine the size of the page table for $n = 32$-bit and $n = 64$-bit virtual addresses with 4-KiB pages.

Solution: The page table has 2^{n-p} entries, with $p = 12$. So, the number of entries is $2^{(32-12)} = 2^{20}$ for $n = 32$ and $2^{(64-12)} = 2^{52}$ for $n = 64$. Each PTE must be large enough to hold the PPN and some status bits and, as a practical matter, should be word aligned. Thus, each PTE is typically one word (4 bytes in a 32-bit system and 8 bytes in a 64-bit system). Hence, the page table size is 4 bytes $\times\ 2^{20} = 4$ MiB in the 32-bit system and 8 bytes $\times\ 2^{52} = 32$ PiB in the 64-bit system.

Note that the *root* (top-level) page table has the largest level number, whereas the lowest-level page table, which contains PPNs, is level 0.

Large page tables are inefficient or impossible to implement, so practical systems use multilevel page tables. In a 2-level page table, the upper bits of the virtual address are used to choose an entry from a level 1 page table, which points to a level 0 page table. The lower bits of the virtual address are used to index into the level 0 page table and choose a PTE, which contains the PPN and status bits. If only a small part of virtual memory is used, most of the level 0 page tables contain only unallocated pages. In such a case, the level 1 PTE is marked NULL (not valid), and the level 0 page table does not even exist.

For large virtual memory systems, this can be generalized to more levels. Each page table itself is typically chosen to fill one page. Only the top-level page table need be in main memory at all times. Some pages of the lower levels could themselves be uncached and loaded from secondary storage only upon demand.

Example 11.3 2-LEVEL PAGE TABLE SIZE

Consider the 2-level page table in Fig. 11.4 for $n = 32$ with 4-KiB pages. The virtual address is shown at the top: the most significant 10 bits index the level 1 page table, the next 10 bits index the level 0 page table, and the least significant 12 bits determine the page offset. Suppose the virtual memory has 2,048 (2^{11}) pages allocated at the bottom of the address space for code and an upward-growing heap, then a large gap, then one page for a downward-growing stack, and nothing above. Determine the size of the page table.

Solution: The level 1 page table occupies 1 page (4 KiB) because it contains 2^{10} 4-byte entries. Similarly, each level 0 page table occupies 1 page. With only 3 active level 0 page tables (2 for the 2^{11} virtual pages of code and heap and 1 for the single page allocated to the stack), the overall page table size is 4 pages $= 4 \times 4$ KiB $= 16$ KiB.

11.1.1.5 Translation Lookaside Buffer

A virtual memory system with a k-level page table adds k page table accesses to every memory access, which reduces performance dramatically. The MMU mostly avoids this penalty by using a small cache of recent virtual-to-physical page number translations, called a *translation lookaside buffer* (*TLB*). A small (e.g., 128-entry) fully associative TLB can achieve a

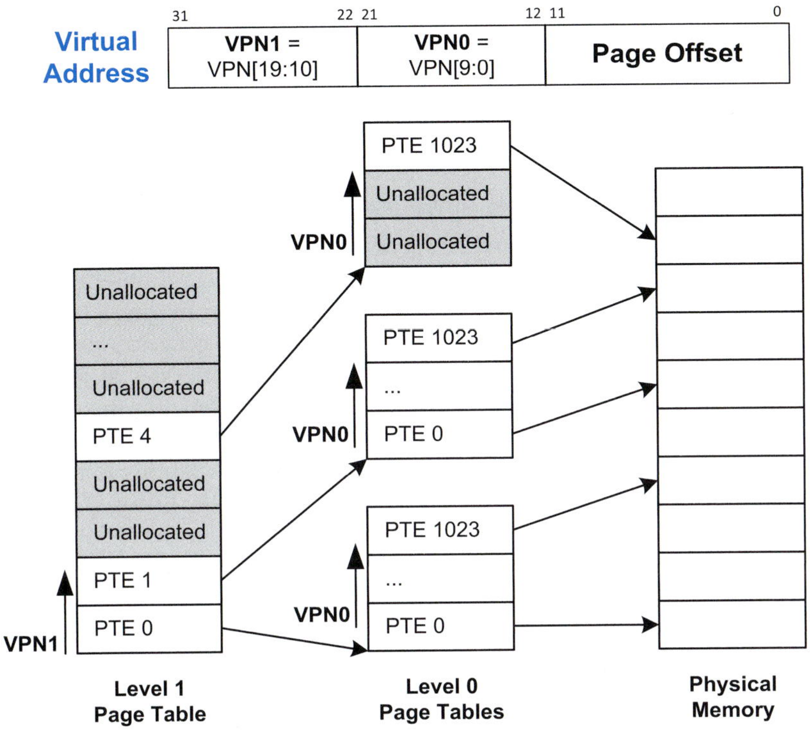

Fig. 11.4 **2-level page table.**

hit rate of better than 99% on many applications, and a larger TLB may achieve 99.99%. Some systems use a smaller first-level TLB with single-cycle access and a larger second-level TLB with high hit rate. For example, the AMD Opteron processor uses a 2-level TLB with 40 entries in the fast first level and 512 entries in the second level with 7-cycle latency [Hennessy19].

Historically, processors would take an exception on a TLB miss. The exception handler would then issue a sequence of page table reads to traverse the multilevel page table to reach the PTE and then put this entry in the TLB and return to the program. This is slow to do in software, so most modern processors have *hardware page table walkers* (*HPTWs*), which are state machines that issue the page table reads.

11.1.1.6 Variable Page Size

A program that accesses widely separated portions of a large data structure may touch many different pages, causing the TLB to run out of capacity and frequently miss. Thus, most architectures support more than one page size. For example, x86-64 supports 4-KiB, 2-MiB, and 1-GiB pages. Pages bigger than 4 KiB are called *large pages*, *huge pages*, or *superpages*.

These larger pages save space in the TLB and in the page tables. They make efficient use of disks or networks that achieve peak bandwidth when transferring large contiguous pages. However, the latency of paging these superpages in and out is high. Also, the OS must make a more complex decision of whether to allocate one superpage or many smaller pages.

11.1.1.7 Interaction of Processes, Virtual Memory, and Caches

Each process has its own virtual address space. Therefore, a particular virtual address in two different processes usually maps to a different physical address. The OS manages this by maintaining separate page tables for each process. The MMU keeps a pointer to the active process's page table.

The TLB must adjust translations for different processes as well. One solution is to flush the TLB on each context switch, but this incurs a performance penalty reloading the TLB. A higher-performance design adds a *process ID (PID)* field to each TLB entry so that translations for multiple processes can coexist. Some parts of memory, such as the OS or I/O peripherals, may belong to a *global address space* and thus be accessible by any process at the same virtual address. The TLB may include a global bit to indicate that these PTEs apply to all PIDs.

More than one process can access shared variables by configuring their page tables to point to a shared physical page. Thus, different virtual addresses can translate to the same physical address.

Programs use virtual addresses, so, ideally, cache accesses also use virtual addresses before they are translated to physical addresses. However, this is complicated because the same virtual address in different processes could correspond to different physical addresses, and different virtual addresses in different processes could *alias* to the same physical address. This is known as the *synonym problem*. The system could flush the cache on every context switch to prevent such aliasing, but this degrades performance. Although some solutions exist with various complications [Cekleov97], most systems avoid the problem entirely by using *virtually indexed, physically tagged* caches with a way size less than or equal to the page size (usually both 4 KiB). A processor accesses the L1 cache using the set bits from the page offset, which are not translated. The TLB is accessed concurrently with the L1 cache. After they are both read, the physical address from the TLB is checked against the physical address tag from the cache line. Page offsets are normally 12 bits, which limits each L1 cache way to 4 KiB. Larger L1 caches use associativity; for example, the Intel Haswell 32 KiB cache with 4-KiB ways is 8-way set associative. This limitation doesn't apply to higher-level caches because the physical address is available by the time an L1 cache miss is discovered.

Increasing associativity costs power (more RAMs to access concurrently) and time (a larger way mux). Some designs find 4-way associativity is a practical limit, restricting physically tagged virtually indexed L1

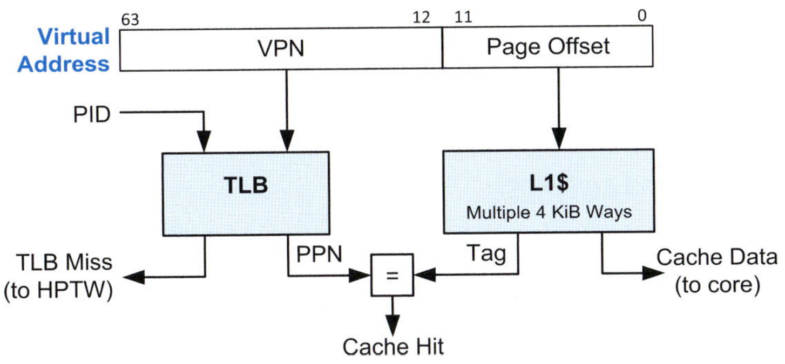

Fig. 11.5 Address translation and caching.

caches to 16 KiB. The AMD Opteron 64 KiB cache is 2-way associative with 32-KiB pages. It addresses the synonym problem with hardware *anti-aliasing* by checking for aliases in all eight possible locations during a cache miss, and invalidating an alias if it is in the cache. Alternatively, some operating systems use *page coloring* to allocate aliased pages to physical addresses with the same least significant bits (e.g., same 16 lsbs in a 2^{16}-byte cache).

Fig. 11.5 shows these interactions. The memory system receives a virtual address, which consists of a page offset and virtual page number. The fully associative TLB is accessed using the VPN and PID and produces the PPN and a TLB miss signal. If a TLB miss occurs, the hardware or software page table walker fetches the translation from the page tables in main memory and updates the TLB before continuing. Meanwhile, the level 1 cache is accessed using the page offset to produce the data and physical tag. The tag(s) are then compared against the PPN to determine whether the cache hit.

11.1.2 Physical Memory Checking

A processor with memory-mapped I/O needs to route physical memory accesses to the appropriate device. For example, a range of physical memory addresses correspond to main memory (DRAM), but other ranges could correspond to a UART (universal asynchronous receiver/transmitter), GPIO (general-purpose I/O), CLINT (Core-Local Interruptor), and PLIC (Platform-Level Interrupt Controller). Table A.2 in Appendix A lists the default memory map for Wally. Reserved addresses outside of mapped ranges should cause an illegal address exception. Moreover, the processor may need to know attributes about each of the devices, such as whether the values are cacheable (yes for memory, no for most peripherals) or whether only certain-sized memory accesses are supported. Some of these addresses or attributes may change when new devices are plugged into the

The DEC PDP-11 was a highly successful minicomputer in the 1970s with 16-bit addressing. Memories soon grew too large for 16 bits, and the PDP-11 architecture did not gracefully evolve to allow wider addresses, so the architecture rapidly became obsolete and was replaced by the even more successful VAX-11. The architects reflected on this error:

> *"There is only one mistake that can be made in computer design that is difficult to recover from – not having enough address bits for memory addressing and memory management. The PDP-11 followed the unbroken tradition of nearly every known computer"* [Bell76].

Extensions with the *Sv* previx refer to supervisor-mode features related to *v*irtual memory.

system. Moreover, certain physical addresses may only be accessed in a privileged mode for added security. Hence, most systems require some configurable checking of physical memory addresses in hardware. The details vary significantly between architectures.

11.2 RISC-V PRACTICES

RISC-V supports several flavors of virtual memory. It also supports memory checking through the PMA and PMP checkers. Tables I.14 to I.20 at the back of the book summarize CSRs and PTEs associated with these features. Bad memory accesses produce various types of faults. When a page table is modified, the OS uses `sfence.vma` to synchronize the TLB with the new page table. When a PTE is not in the TLB, a HPTW traverses the multilevel page table to find the missing entry.

11.2.1 RISC-V Virtual Memory

In RISC-V, machine-level processes exclusively use physical addressing. Virtual memory is an optional RISC-V feature. If supported, supervisor- and user-level processes can enable virtual addressing through the *supervisor address translation and protection* (`satp`) register. RV32 supports 32-bit virtual addresses and 34-bit physical addresses with a 2-level page table. Hence, each process may have up to a 4-GiB address space, but the system may have up to 16 GiB of physical memory. RV64 supports 39-, 48-, and 57-bit virtual addresses with 3-, 4-, and 5-level page tables, respectively, with 56-bit physical addresses. Future expansion is reserved for 64-bit virtual addresses with 6 levels of page tables, should 128 PiB (petabytes) of virtual memory prove insufficient. RV32 and RV64 systems can run in modes that support virtual memory called Sv32, Sv39, Sv48, and Sv57, with numbers corresponding to the virtual address size. Svbare mode, also called simply *bare mode*, indicates that virtual memory is off and address translation is disabled.

11.2.1.1 `satp` Register

The `satp` register enables the various virtual addressing modes and points to the root (level 1) page table for a process. It also contains the *address space identifier* (ASID), which acts as a short process ID (PID) assigned by the OS to differentiate multiple processes' virtual address spaces that might coexist in a TLB. Fig. 11.6 shows the RV32 and RV64 `satp` registers. PPN defines the PPN of the root page table. Shift it left by 12 to get the physical address (base address) of the root page table, which is always aligned to a page boundary. ASID is 9 or 16 bits. In RV32, `MODE` is 0 to disable virtual memory (Svbare mode) or 1 to enable it (Sv32 mode with

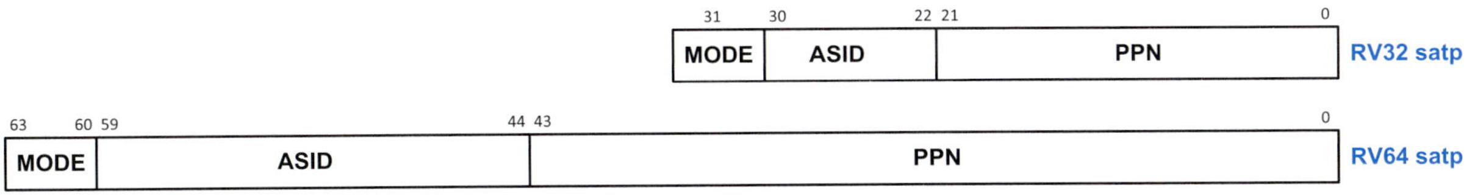

Fig. 11.6 Supervisor address translation and protection register (satp).

32-bit virtual addresses). In RV64, `MODE` is 0000 for Svbare, 1000 for Sv39, 1001 for Sv48, or 1010 for Sv57.

Each user or supervisor-mode process has its own virtual memory space and ASID. When the OS performs a context switch, it reloads `satp` with the page table address and ASID of the new process.

Older x86 processors without an ASID flushed the TLB on a context switch.

This chapter (and the Wally implementation) focus on virtual memory systems up to Sv48. The December 2021 release of the RISC-V specification added Sv57 and the implementation is a straightforward extension but only relevant to systems with exceptional memory requirements.

Example 11.4 2-LEVEL PAGE TABLE PHYSICAL ADDRESS ACCESSES

Consider again an Sv32 system with a 2-level page table organization shown in Fig. 11.4. Suppose `satp` = 0x80012345, and the virtual address being accessed is 0x013FF124. The PTE is not in the TLB. Determine all physical addresses that are accessed.

Solution: Fig. 11.7 depicts this access.

- The `satp` register's msb (`MODE` bit) is 1, so virtual memory is enabled. The next 9 bits, which are 0, give the ASID, which is unique to the process.
- The `satp`'s remaining 22 bits are the PPN of the root (level 1) page table: 0x012345. Append a page offset number of zeros (12 zeros) to obtain the 34-bit physical address (base address) of the level 1 page table: 0x012345000.
- From the virtual address, VPN1 = 4, which is the index into the level 1 page table. So, 0x012345010 is the address of the needed level 1 PTE. Specifically, 0x01234500 + 4 words (16 bytes) = 0x012345000 + 0x10 = 0x012345010.
- The level 1 PTE gives the PPN of the level 0 page table. In this case, it is 0x1AABBC. Append a page offset number of zeros (12 zeros) to get the physical address of the level 0 page table: 0x1AABBC000. VPN0 (0x3FF = 1023) is the index into the level 0 page table, so the level 0 PTE is at physical address 0x1AABBC000 + (0x3FF << 2) = 0x1AABBCFFC.
- That entry gives the PPN of the physical memory address: 0x377889.
- Append the page offset of 0x124 from the virtual address to get the final 34-bit physical address of 0x377889124.

11.2.1.2 Virtual Address and Page Table Structure

Fig. 11.8 shows the virtual address formats for Sv32, Sv39, and Sv48. In all formats, the bottom 12 bits, which are not translated, give the page offset for a 4-KiB page. Whereas in RV32 (Sv32), the PTEs are 32 bits = 4 bytes each, in RV64 (Sv39/48), the entries are 64 bits = 8 bytes each. Each page table fits on one page, so RV32 has $2^{12}/4 = 2^{10}$ entries in a page table, and RV64 has $2^{12}/8 = 2^9$ entries. Hence, in Sv32, the upper 10 bits of the

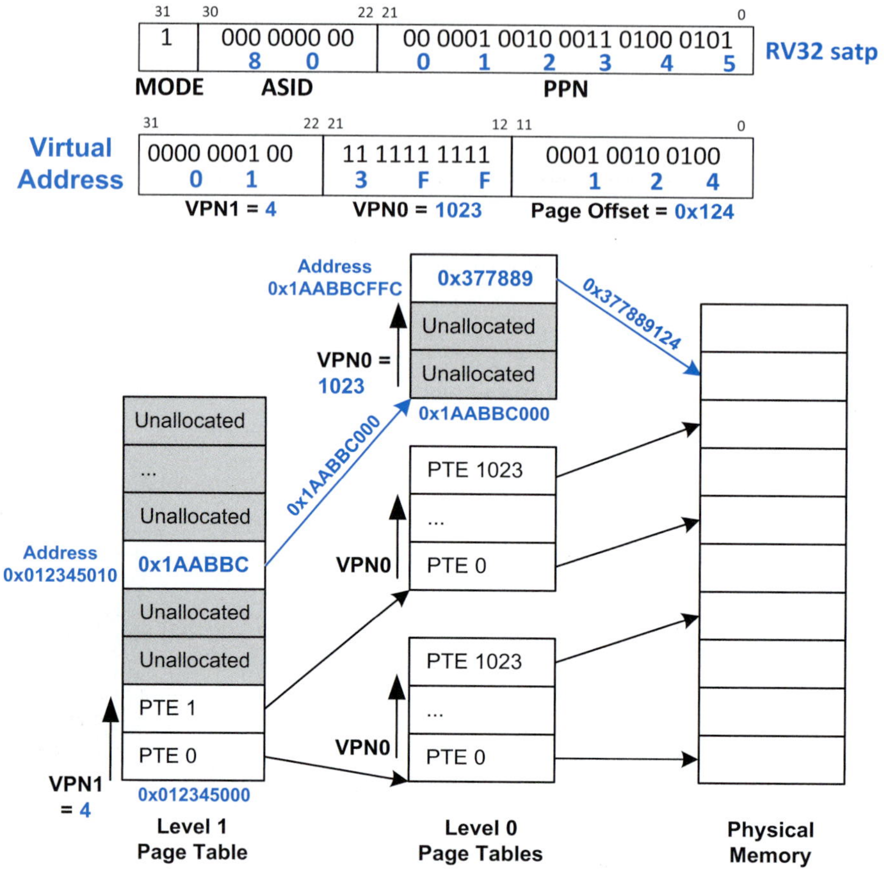

Fig. 11.7 Physical addresses of 2-level page table.

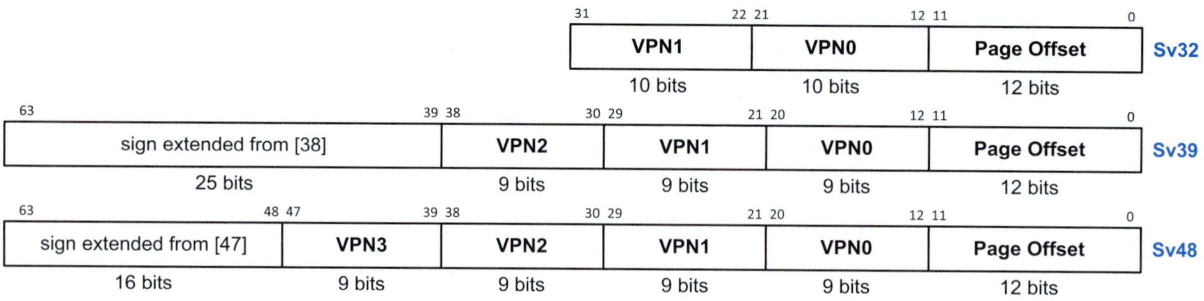

Fig. 11.8 RISC-V virtual address formats.

virtual address index into a level 1 page table, which then points to the level 0 page table. The remaining (middle 10) bits of the virtual address index into the level 0 page table, which holds the PPN. In Sv39, the upper 9 bits of the virtual address index into a level 2 page table, which points to the base address of the level 1 page table. The next 9 bits give the index into the level 1 page table, which then points to the level 0 page table. The next 9 bits index into the level 0 page table, which holds the PPN. Sv48 extends to a level 3 top-level page table. The virtual address space is $2^{32} = 4$ GiB for Sv32, $2^{39} = 512$ GiB for Sv39, and $2^{48} = 256$ TiB for Sv48. Sv39 is sufficient for many contemporary systems, uses less page table storage, and has faster page table walking, so it is preferred unless the larger address space is required. For RV64 (Sv39 and Sv48), the upper bits of the virtual address must all be the same (either all 1s or all 0s) so that the address space includes the bottom and top of the range but leaves nonexistent addresses in the middle; if the bits are not the same, a page fault occurs. This matches conventional practices, where user-mode programs occupy low addresses, while the supervisor-mode OS occupies high addresses, and the OS can easily distinguish between the two by looking at the most significant bit of the virtual address.

Fig. 11.9 shows the PTE formats for RV32 and RV64. Both formats contain the same 10 flag bits. In RV32, the PPN is in the upper 22 bits. Hence up to $2^{22+12} = 16$ GiB of physical memory is supported. In RV64, the PPN is 44 bits, so up to $2^{44+12} = 64$ PiB of physical memory is supported.

The Svnapot (naturally aligned power of two) and Svpbmt (page-based memory types) RISC-V extensions for RV64 use bits 63 and 62:61, respectively, to allow for larger page sizes (see Section 11.2.1.3.1) and to apply attributes to memory pages (see Section 11.2.2.1.1).

The flags are described in Table 11.1. The valid (`V`) bit is set if the page is in main memory. Otherwise, the MMU triggers a page fault, which causes the OS's *page fault handler* to load the page from disk. Collectively, `U`, `X`, `W`, and `R` provide protection. `X`, `W`, and `R` indicate if the page is executable, writable, and readable. A page fault occurs if the privilege or permissions are incorrect for the access attempted. In user mode, a fault occurs for any access if `U` $= 0$. In supervisor mode, if `U` $= 1$, a fault always occurs when trying to execute code; reads and writes are allowed only when `sstatus.SUM` (supervisor user memory) $= 1$. This protects the system from certain security vulnerabilities. The `G` bit indicates that a page is global and applies to all processes; the ASID is ignored for global pages. OS pages are typically global because the OS is called from many different user-mode processes.

The OS sets `sstatus.SUM` to read or write user-mode pages. But it normally keeps `SUM` low to avoid accidentally trampling user programs.

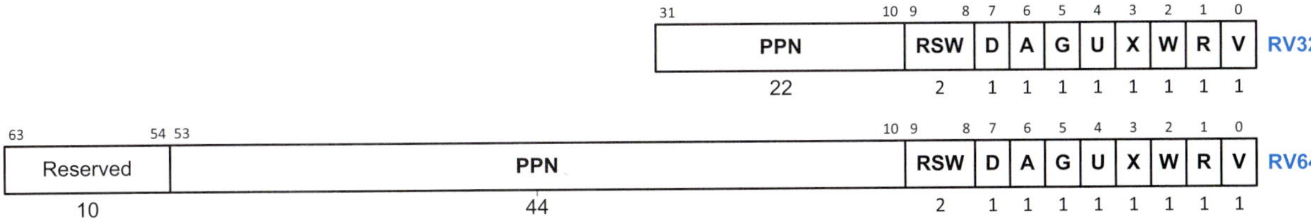

Fig. 11.9 RISC-V PTE formats.

Table 11.1 **PTE flags.**

Flag	Meaning
V	Valid: 1 if the page is in main memory, 0 if not
X, W, R	Execute, Write, Read Permissions: 000: Pointer to next level of page table 001: Read only 010: <illegal> 011: Read-write 100: Execute only 101: Read-execute 111: Read-write-execute
U	User mode
G	Global: page exists for all address spaces; ignore ASID in TLB matching
A	Accessed: page has been accessed at least once since all pages' A bits were last cleared
D	Dirty: at least one byte from the page has been written
RSW	Reserved for OS

The OS reserves part of the disk for *swap space* to hold pages that don't fit in physical memory. The OS uses the PPN field of the PTE to locate the page on disk. Paging from disk is slow, so if a computer doesn't have enough RAM to hold the working set, it will begin *thrashing*, frequently page faulting and reducing the system to a crawl. Mobile devices generally terminate an entire app that hasn't been used recently to reclaim its memory rather than enduring the delay and wear and tear of paging to flash memory.

When two processes wish to communicate through shared memory, they allocate a shared page through a system call such as `shmget()`. Each process has its own virtual addresses mapping to the shared physical addresses. Thus, each process also has its own PTE pointing to the shared PPN. The PTE flags could differ between processes; a producer-consumer buffer might be read-only for the consumer but read/write for the producer.

Table 11.2 RISC-V page sizes.

Format	Leaf Page Table Level	Page Offset Bits	Page Size	Page Type
Sv32	0	12	4 KiB	Kilopage (normal)
Sv32	1	$12 + 10 = 22$	4 MiB	Megapage
Sv39	0	12	4 KiB	Kilopage (normal)
Sv39	1	$12 + 9 = 21$	2 MiB	Megapage
Sv39	2	$12 + 9 \times 2 = 30$	1 GiB	Gigapage
Sv48	0	12	4 KiB	Kilopage (normal)
Sv48	1	$12 + 9 = 21$	2 MiB	Megapage
Sv48	2	$12 + 9 \times 2 = 30$	1 GiB	Gigapage
Sv48	3	$12 + 9 \times 3 = 39$	512 GiB	Terapage

11.2.1.3 Variable Page Sizes

RISC-V supports multiple page sizes. Usually, the level 0 page table is the leaf page table and points to normal 4-KiB pages (also called *kilopages*). The nonleaf PTEs use {`X`, `W`, `R`} = 000 to indicate that they point to the next level page table. (For these pointer PTEs, the `D`, `A`, and `U` bits are reserved and must be 0.) However, if `XWR` are nonzero for a higher-level PTE, then that entry points to a superpage, as shown in Table 11.2. The superpage must be aligned to its natural (page) boundary (with all the low-order bits being 0). The low-order untranslated bits of the virtual address form an offset into the superpage, which are also referred to by their more specific names of *megapage*, *gigapage*, or *terapage*. Megapages are 4 MiB in RV32 or 2 MiB in RV64. Gigapages are 1 GiB, and terapages are 512 GiB, and these are only available in RV64.

Recall that each page table fits into one page (4 KiB = 2^{12} bytes). In Sv32, each entry is 32 bits (4 bytes = 2^2 bytes), so each page table index is 10 bits ($2^{12}/2^2 = 2^{10}$). In Sv39 and Sv48, each entry is 64 bits (8 bytes = 2^3 bytes), so each page table index is 9 bits ($2^{12}/2^3 = 2^9$).

For this reason, megapages are larger in Sv32 because PTEs are 32 rather than 64 bits, giving room for twice as many PTEs in a fixed-size (4-KiB) page table.

In the page table examples in this chapter, white-shaded entries are good, gray entries are invalid (V = 0), and blue entries are malformed such that they should cause a page fault when used. For example, in Fig. 11.10, level 0 PTE 0 is blue because it is a pointer to the next level page table (XWR = 000), which is illegal for a level 0 PTE.

Example 11.5 ADDRESS TRANSLATION WITH VARIOUS PAGE SIZES

Consider an Sv48 virtual memory system with the page table in Fig. 11.10(a). (Fig. 11.10(b) shows the page table for an Sv32 virtual memory system.) Find the physical addresses corresponding to virtual addresses (a) 0000_0000_8020_1340, (b) FFFF_FFFF_FFFF_F238, (c) 0000_0000_8000_FFF8, and (d) 0000_0100_C0A0_ 7BC0.

Solution: Fig. 11.11 shows how the Sv48 virtual addresses are split into fields according to Fig. 11.8.

(a) VPN3 = 0, VPN2 = 2, VPN1 = 1, VPN0 = 1. Hence, entry 0 of the level 3 page table points to the bottom level 2 page table, whose entry 2 points to the bottom level 1 page table, whose entry 1 points to the bottom level 0 page table, whose entry 1 contains the 44-bit PPN 00_0000_8020_0. This is a normal kilopage because it is in the level 0 table. Thus, the page offset is the bottom 12 bits of the address (*340*), and the 56-bit PA is 00_0000_8020_*0340*.

(b) VPN3 = VPN2 = VPN1 = VPN0 = 511. Hence, entry 511 of the level 3 page table points to the top level 2 page table, whose entry 511 points to the top level 1 page table, whose entry 511 points to the top level 0 page table, whose entry 511 contains the 44-bit PPN 00_0000_8020_1. This is, again, a normal kilopage, so the page offset is *238*, and the 56-bit PA is 00_0000_8020_1*238*.

(c) VPN3 = 0, VPN2 = 2, VPN1 = 0. Hence, entry 0 of the level 3 page table points to the bottom level 2 page table, whose entry 2 points to the bottom level 1 page table, whose entry 0 contains the 44-bit PPN 00_0000_9000_0. This is a megapage because XWR is nonzero; thus, the leaf entry is in level 1, so the page offset comes from the bottom 21 bits of the VA (*00FFF8*). The bottom 9 bits of the megapage PPN are discarded before appending the 21-bit page offset, resulting in a 56-bit PA of 00_0000_*9000_FFF8*.

(d) VPN3 = 2. Hence, entry 2 of the level 3 page table contains the 44-bit PPN 00_0000_0000_0. This is a terapage because XWR is nonzero in this level 3 page table; thus, the leaf entry is on level 3 and the page offset comes from the bottom 39 bits of the VA (*00_C0A0_7BC0*). The bottom 27 bits of the PPN are discarded before appending the 39-bit page offset, resulting in a 56-bit PA of 00_*0000_C0A0_7BC0*.

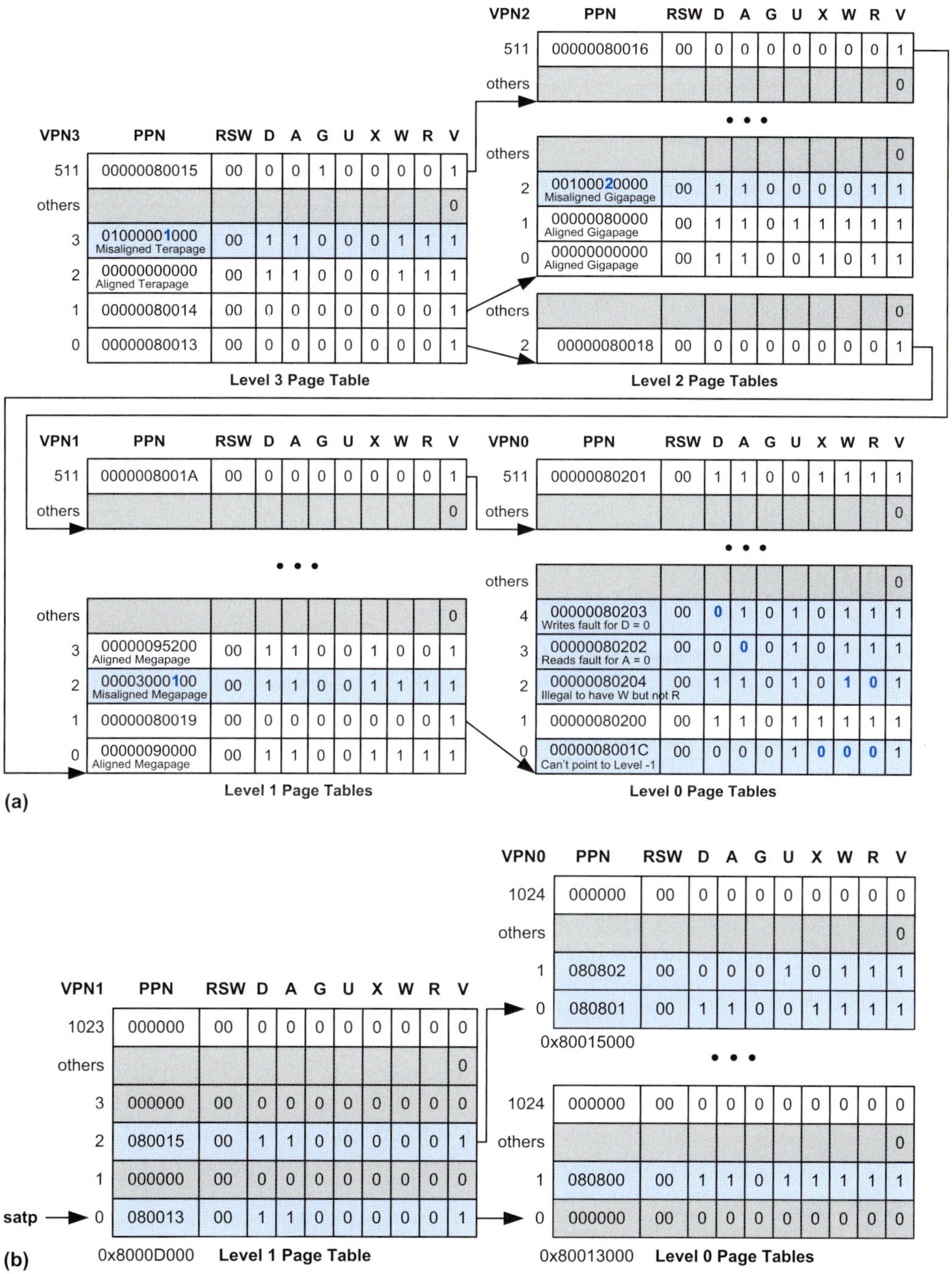

Fig. 11.10 Sample page tables: for an (a) Sv48 and (b) Sv32 system.

	Sign Extension (63–48)	VPN3 (47–39)	VPN2 (38–30)	VPN1 (29–21)	VPN0 (20–12)	Page Offset (11–0)
00000000_80201340	0000 0000 0000 0000	0000 0000 0 = 0	000 0000 10 = 2	00 0000 001 = 1	0 0000 0001 = 1	0x340
FFFFFFFF_FFFFF238	1111 1111 1111 1111	1111 1111 1 = 511	111 1111 11 = 511	11 1111 111 = 511	1 1111 1111 = 511	0x238
00000000_8000FFF8	0000 0000 0000 0000	0000 0000 0 = 0	000 0000 10 = 2	00 0000 000 = 0	0 0000 1111 = 15	0xFF8
00000100_C0A07BC0	0000 0000 0000 0000	0000 0001 0 = 2	000 0000 11 = 3	00 0000 101 = 5	0 0000 0111 = 7	0xBC0

Fig. 11.11 Virtual address fields.

Example 11.6 VIRTUAL AND PHYSICAL ADDRESS MAPPING

Consider again the Sv48 virtual memory system with the page table in Fig. 11.10(a). Sketch the virtual and physical address space and the mapping between them for the valid PTEs.

Solution: Fig. 11.12 shows the virtual and physical address spaces. Recall that virtual addresses are 48 bits, and physical addresses are 56 bits. The VPN3, 2, 1, and 0 fields are populated by tracing the page table entries, and their mapping to the PPN is extracted from the page tables. Note that larger pages can alias with smaller pages; for example, virtual addresses 0x0100_9531_A00C (on a terapage), 0x0080_5531_A00C (on a gigapage), and 0x0000_8071_A00C (on a megapage) all alias to physical address 0x00_0000_9531_A00C, as shown in Table 11.3; the page offsets are in blue, the VPNs are in italics, and the PPNs are in bold black.

11.2.1.3.1 Huge Pages

Many important programs use tens to hundreds of kilobytes of code and data. They are too small to make efficient use of a megapage yet clutter the TLB with large numbers of 4-KiB pages. Competing architectures such as ARMv8 have observed that 64-KiB *huge pages* are valuable, and Linux takes advantage of such pages when they are supported.

The optional RV64 Svnapot extension combines sixteen contiguous 4-KiB pages to create 64-KiB huge pages naturally aligned to 64-KiB address boundaries. In general, the extension anticipates any *naturally aligned power of two* (*NAPOT*) page size (i.e., 8, 16, 32, 64 KiB), but only 64-KiB pages are standardized in Version 1.0 of the extension.

When the Svnapot extension is supported, bit 63 of the PTE is the N bit. If a PTE has N = 1 and PPN[3:0] = 1000, the entry is considered NAPOT. The four least significant bits of the VPN are ignored when accessing the page table, so the 16 contiguous 4-KiB virtual pages all can share the same PTE. After translation, the four bottom bits of the PPN are replaced by these bits of the VPN. If PPN[3:0] has any

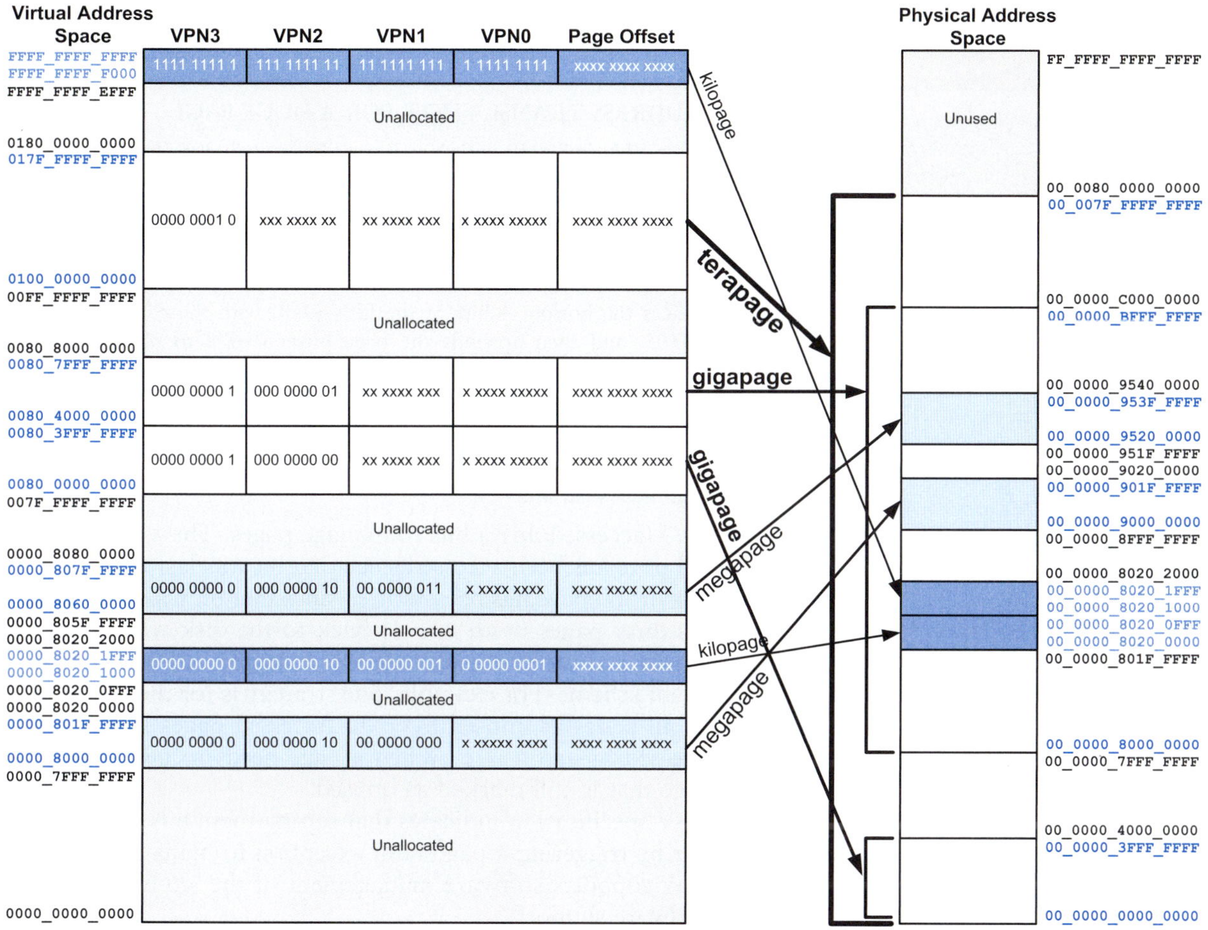

Fig. 11.12 Virtual and physical address spaces.

Table 11.3 Aliasing of virtual addresses to the same physical address.

Virtual and Physical Address Bits	
*0000 0001 0*000 0000 1001 0101 0011 0001 1010 0000 0000 1100	**VA:** 0x0100_9531_A00C
0000 0000 0000 0000 1001 0101 0011 0001 1010 0000 0000 1100	**PA:** 0x0000_9531_A00C
*0000 0000 1000 0000 01*01 0101 0011 0001 1010 0000 0000 1100	**VA:** 0x0080_5531_A00C
0000 0000 0000 0000 1001 0101 0011 0001 1010 0000 0000 1100	**PA:** 0x0000_9531_A00C
0000 0000 0000 0000 1000 0000 0111 0001 1010 0000 0000 1100	**VA:** 0x0000_8071_A00C
0000 0000 0000 0000 1001 0101 0011 0001 1010 0000 0000 1100	**PA:** 0x0000_9531_A00C

value other than 1000 in the PTE when PTE bit N = 1, a page fault occurs.

Example 11.7 ADDRESS TRANSLATION FOR A HUGE PAGE

Consider a PTE for VPN 0x80020 with N = 1 and PPN = 0x80308. Translate virtual address 0x80025678 to a physical address using this PTE.

Solution: This PPN's four least significant bits are 1000, so this is a NAPOT entry for a huge page. The address 0x80025678 has a VPN of 0x80025. The TLB matches against the VPN entry of 0x80020 (by ignoring the bottom four bits) and returns the PPN of 0x80308. It replaces the bottom 4 bits of the PPN (0x8) with those from the VPN (0x5), giving 80305, and then appends the page offset 0x678 to obtain a PA of 0x80305678. Observe that the bottom 16 bits of the virtual and PA are identical, as expected for a 2^{16} = 64-KiB huge page.

11.2.1.4 Dirty and Accessed Bits

The OS uses A/D (accessed/dirty) bits to manage pages. The A bit indicates that the virtual page has been accessed since that bit was last cleared. D indicates that the virtual page has been written since that bit was last cleared. The OS writes dirty pages (with D = 1) back to the disk when they are evicted and could use the A bit in a pseudo least recently used (pLRU) page replacement scheme. For example, one strategy is for the OS to periodically mark all pages as unused (A = 0). Accessed pages subsequently raise their accessed bits to A = 1. When a page needs to be evicted, the OS picks a page that is still marked as unused.

The RISC-V specification indicates that setting these bits can be done in hardware or by triggering a page fault exception to manage the bits in software. Linux supports software management of the A/D bits, so many cores lack hardware support.

The Svade (*supervisor virtual memory A/D exception*) extension indicates that the hart can take a page fault when accessing a page with A = 0 or when writing to a page with D = 0. The Svadu (*supervisor virtual memory A/D update*) extension indicates that the hart can update the A/D bits of the PTE in hardware without faulting. The Svadu extension also adds menvcfg.ADUE (bit 61). When ADUE = 0, the Svadu extension is disabled, and the hart takes a page fault to update the PTE in software rather than updating the PTE in hardware.

11.2.1.5 status Register

The mstatus/sstatus register (see Section 8.2.3.1) has three bits that affect address translation: MPRV, MXR, and SUM. These bits manage accesses across privilege levels.

The MPRV and MXR bits help write efficient trap handlers for misaligned loads and stores. Some RISC-V implementations perform misaligned loads and stores with a trap to a machine-mode handler rather than in hardware because the hardware is costly and rarely used. The handler needs to do address translation even though it is running in machine mode, which usually uses physical addressing only. When set, the mstatus.MPRV (**M**odify **PR**i**V**ilege) bit causes the system to do address translation and protection according to the privilege level specified in MPP (normally the user or supervisor mode that caused the trap). When set, mstatus.MXR (**M**ake e**X**ecutable **R**eadable) allows loads from executable nonreadable pages (with X = 1 but R = 0) so the trap handler can read instructions.

Supervisor code normally should not access information from user-mode pages, to reduce the risk of malicious user-mode programs affecting privileged operations. However, the supervisor occasionally needs to control user-mode data. The mstatus/sstatus.SUM (permit **S**upervisor **U**ser **M**emory access) bit allows these accesses when set; otherwise, supervisor-mode access to user memory causes a load or store page fault. The OS typically sets SUM only when necessary and then clears it for safety.

11.2.2 RISC-V Memory Protection

A key requirement for security is to control which processes have access to which parts of memory. RISC-V provides three, often-overlapping, memory protection mechanisms:

- The PMA checker defines the addresses and access limitations for memory ranges and memory-mapped I/O devices.
- The PMP checker restricts types and privilege of access to certain memory ranges.
- Virtual memory provides the X, W, R, and U access-permission flags and limits user-mode processes to only access pages explicitly allocated to them by the OS.

PMA and PMP checkers are also simply called the PMA and PMP.

11.2.2.1 PMA Checker

Most processors require an address decoder to access the appropriate bank of memory or I/O peripheral specified by a physical address. RISC-V uses a PMA checker to perform this address decoding. The PMA specifies and enforces the memory map and its attributes and is often hardwired at design time. However, it may be partially configurable (e.g., to allow for variable amounts of external memory to be connected to a processor).

RISC-V specifies that a core must contain a platform-dependent PMA that generates select signals for each memory bank or I/O peripheral and produces access faults if the memory access is unsuccessful. An unsuccessful access may be caused by a physical address located in a *vacant* portion of the memory map or an unsupported operation (e.g., a byte access to a peripheral that only supports word accesses).

Main memory is required to support all forms of accesses, but peripheral memory operations that might be constrained by the PMA checker include:

- Reads, writes, executes
- Access width, from byte to multiword burst
- Hardware page table accesses
- Cacheability (whether the value can be stored in the cache):
  - Normally yes for main memory and no for memory-mapped I/O
- Idempotency (whether the device produces the same outcome when accessed more than once)
  - Normally yes for main memory: reading or writing the same value twice is harmless
  - Normally no for peripherals that could have side effects
  - Instruction fetches and speculative or redundant accesses must be avoided for non-idempotent devices
- Atomic memory operations (AMOs); this depends on whether `lr-sc` (load reserved – store conditional) and AMOs are supported in hardware (see Chapter 17)
  - Often yes for all cacheable main memory

Table 11.4 lists the PMA attributes for the various memory regions of the Wally memory map from Fig. 2.8 and Table A.2. The base address is the beginning of the memory region, and the (base address + range) is the top of the region. It is prudent to place no memory at address 0 so that dereferencing a NULL pointer will cause an access fault. Some peripherals have fixed-sized registers and only permit accesses of that size. The size is commonly 32 bits, although it is 8 bits for the UART to match its byte interface.

11.2.2.1.1 Page-Based Memory Types (Svpbmt)

Some peripherals perform *direct memory access* to read or write from main memory, as introduced in Section 9.1.4. If the peripherals aren't specially designed to interact with the cache, they may face a coherence problem where the data cache contains valid data and the main memory contains different data. It is useful to mark certain portions of main memory as uncacheable to avoid this coherence problem. The portion of memory may be application dependent, so competing architectures such as x86 and ARMv8 allow the OS to mark certain PTEs as uncacheable pages. Linux relies on this capability, so RISC-V also offers it.

PMA ranges are non-overlapping because the memory map is non-overlapping.

Table 11.4 PMA for rv64gc Wally memory map.

Region	Base Address	Range	Access Width	ExecuteRead/ Write	Cacheable	Idempotent	Atomic Operations
RAM	80000000	7FFFFFF	Any	XWR	Yes	Yes	Yes
GPIO	10060000	FF	32-bit	RW	No	No	No
SPI	10040000	FFF	8-bit	RW	No	No	No
UART	10000000	07	8-bit	RW	No	No	No
PLIC	0C000000	3FF0000	32-bit	RW	No	No	No
CLINT	02000000	FFFF	Any	RW	No	No	No
SDC	00013000	FFF	{32/64}-bit	RW	No	No	No
Boot ROM	00001000	FFF	Any	XR	Yes	Yes	No

The optional RV64 RISC-V Svpbmt (*supervisor virtual memory page-based memory types*) extension allows a PTE to override the PMA for certain pages. The `menvcfg.PBMTE` bit must be set to enable page-based memory types (see Section 11.4.1.1). When enabled, bits 62:61 of the PTE are the `PBMT` field. If these bits are 00, the system uses the standard PMA checker. If the bits are 01 or 10, the system overrides the PMA checker to indicate uncacheable main memory and I/O, respectively. Whereas main memory is idempotent, I/O space is non-idempotent. If the `PBMT` field of the PTE is nonzero and Svpbmt is not supported or enabled, a page fault occurs. Linux automatically detects and uses Svpbmt.

11.2.2.2 PMP Checker

For security and to detect pointer errors in programs, it is helpful to limit which physical addresses software may access. An optional RISC-V PMP checker enforces protected address ranges concurrently with the PMA checker. By default, the PMP grants machine-mode access to all parts of physical memory and denies supervisor- and user-mode access to all of physical memory; the PMP is then configured in software to grant access to certain portions in S- or U-mode or to revoke access to certain portions in M-mode. For example, the OpenSBI M-mode firmware configures the PMP so that S- and U-mode software can access all memory except a region used by the firmware. Even if malicious software compromises the OS, the PMP prevents it from corrupting the firmware.

> Ordinarily, M-mode code has access to the entire memory. However, in secure systems, the M-mode boot loader might need to read cryptographic keys only during startup. The boot loader could then revoke access to ROM addresses containing the keys so that they are inaccessible to an attack that compromises other M-mode firmware.

Physical memory protection is specified by 0, 16, or 64 PMP address registers named `pmpaddr0-pmpaddr63`. These indicate the addresses associated with a corresponding region. Every PMP address register is also

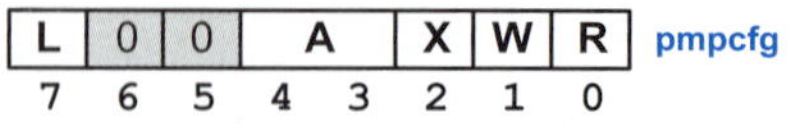

Fig. 11.13 **pmpcfg fields.**

associated with an 8-bit field in a pmpcfg (PMP configuration) register indicating the type of region and its access privileges (Fig. 11.13).

PMP addresses must be aligned to 4-byte boundaries, so the least significant bits of the addresses are 0 and are not stored in the pmpaddr registers. Thus, in RV32, the pmpaddr registers each specify bits [33:2] of a 34-bit physical address. In RV64, the registers each specify bits [55:2] of a 56-bit physical address.

pmpcfg registers hold multiple sets of the configuration fields shown in Fig. 11.13. The X, W, and R fields determine whether the region is executable, writable, and readable, respectively. The L field indicates if the region is locked. In a locked region (L = 1), the corresponding pmpaddr and pmpcfg registers cannot be written except when the system is reset. The XWR permissions always apply to S- and U-mode programs. When L is set, they also apply to M-mode programs. The two bits of A determine the type of PMP region, as described in Table 11.5.

The address of a PMP region is determined by its pmpaddr register and the type (A) field in its pmpcfg register. When the type is top of range (TOR), the region starts at the address in the previous pmpaddr and ends at the address in the current pmpaddr. When the type is NA4, the region is 4 bytes long starting at the address in pmpaddr. When the type is NAPOT, the region is a power of 2 long, whose base address and size are encoded in pmpaddr. Specifically, the string of 1s at the bottom of pmpaddr determines the size. If the least significant k bits of pmpaddr are 1s, the region size is 2^{k+3}. The base address is determined by treating these k bits as 0 and shifting the address left by 2.

The PMP checker examines each of the regions in sequential order to find the first region that matches the physical address. It applies the protection specified by the X, W, R, and L bits for that region.

Because the pmpcfg fields are only 8 bits, multiple sets of fields are packed into a single CSR. In RV32 or RV64, each pmpcfg register contains four (32/8 = 4) or eight (64/8 = 8) sets of 8-bit PMP configuration fields, respectively. Thus, each pmpcfg register supports 4 pmpaddr registers in RV32 and 8 in RV64. For example, in RV32, pmpcfg0 supports pmpaddr0-pmpaddr3, pmpcfg1 supports pmpaddr4-pmpaddr7, and so on. RV64 uses only even-numbered pmpcfg registers; so, pmpcfg0 supports pmpaddr0-pmpaddr7, pmpcfg2 supports pmpaddr8-pmpaddr15, and so on.

The PMP specification allows an implementation to require that PMP region sizes are a multiple of some *grain size* $g = 2^{(G+2)}$. If G > 0, then NA4 regions cannot exist.

Table 11.5 **PMP region types.**

A[1:0]	Name	Region Type
00	OFF	Disabled
01	TOR	Top of range
10	NA4	Naturally aligned 4-byte region
11	NAPOT	Naturally aligned power of two region ($\geq$ 8 bytes)

At reset, all PMP regions are OFF; so all accesses are granted to M-mode and denied to S- and U-mode (if PMP is supported).

Example 11.8 PMP ADDRESS RANGES

Considering Table 11.6, what is the range of addresses in region 6?

Solution: The `A` field of `pmpcfg` is 3 (11_2), so this is a NAPOT region. The region's `pmpaddr` = 0x2004013F has $k = 6$ trailing 1s, so its size is $2^9 = 512$ bytes. Discarding these ones gives 0x20040100 and shifting left by 2 gives 0x80100400. Hence, this region spans 512 bytes from 0x80100400 to 0x801005FF.

Table 11.6 **PMP regions example.**

Region	pmpaddr	pmpcfg	L	A	X	W	R	Comments
0	0FFFFFFF	1B	0	NAPOT	0	1	1	I/O 00000000-7FFFFFFF RW
1	20040000	00	0	OFF	0	0	0	NULL
2	2004003F	09	0	TOR	0	0	1	80100000-801000FC R
3	20040080	00	0	OFF	0	0	0	NULL
4	20040084	0C	0	TOR	1	0	0	80100200-80100210 X
5	200400C0	90	1	NA4	0	0	0	80100300-80100303 locked out
6	2004013F	18	0	NAPOT	0	0	0	80100400-801005FF no access
7	00000000	00	0	OFF	0	0	0	NULL
8	00000000	00	0	OFF	0	0	0	NULL
9	00000000	00	0	OFF	0	0	0	NULL
10	00000000	00	0	OFF	0	0	0	NULL
11	00000000	00	0	OFF	0	0	0	NULL
12	00000000	00	0	OFF	0	0	0	NULL
13	00000000	00	0	OFF	0	0	0	NULL
14	00000000	00	0	OFF	0	0	0	NULL
15	2FFFFFFF	1F	0	NAPOT	1	1	1	Main mem 80000000-FFFFFFFF RWX

Example 11.9 PMP MEMORY REGION CHARACTERISTICS

Describe the physical memory regions defined in Table 11.6.

Solution: The comments field describes the regions with their access permissions. For example, according to region 5 registers, the 4 bytes starting at 200400C0 << 2 = 80100300 is completely inaccessible from any privilege mode (L = 1). Region 2 registers define a TOR region that is read-only (R = 1) for U- and S-mode programs. The range starts at the address in the previous register (pmpcfg1 << 2 = 20040000 << 2 = 80010000) and ends at the address in its own register (pmpcfg2 << 2 = 2004003F << 2 = 801000FC). Register 15 makes all memory from 80000000-FFFFFFFF completely accessible. However, permission is evaluated in sequential order, so the lower-numbered regions 2, 4, 5, and 6 take precedence and are more restrictive.

If physical memory protection (PMP) is implemented, at least one region must be configured before entering supervisor or user mode; otherwise, all instruction fetches and data accesses will trap. A simple configuration is to make all memory accessible by setting pmpcfg0 = 0F and pmpaddr0 = all 1s.

11.2.3 Memory Exceptions

RISC-V has nine types of memory exceptions: three flavors (access, page, misaligned) for each of three types of accesses (instruction, load, store/AMO).

Access-fault exceptions occur when a process attempts an illegal access, such as to nonexistent memory, writing a read-only memory, or making an access in violation of PMP. The exception handler generally responds by printing an error and killing the process.

Page-fault exceptions occur when the virtual memory system detects an issue that cannot be handled in hardware. These exceptions are usually delegated to the S-mode exception handler because the OS typically deals with virtual memory. The handler must determine the cause. If the PTE is invalid, the OS must allocate a page of physical memory and fetch the missing page from disk. The OS may also set the access or dirty bits in the PTE. If the access permissions are inappropriate, the OS may kill the process.

Misaligned exceptions differ for instruction fetches and for data loads/stores. Instruction-misaligned exceptions only occur when the address is an odd multiple of 2 and compressed instructions are not supported. The handler should kill the process. Load-misaligned and Store/AMO-misaligned exceptions occur when the data address is not a multiple of the access size and the core does not have hardware support for misaligned accesses. Section 19.2 describes a hardware implementation of misaligned support for cacheable memory. When a misaligned access is not handled in hardware, the processor throws an exception, and the handler may issue a sequence of accesses to emulate the misaligned access. If the PMA Aligned attribute is 1, then a misaligned access can throw an access-fault instead. For example, peripherals may only support aligned accesses even though main memory supports misaligned accesses.

11.2.4 Fences

The `sfence.vma` instruction invalidates all TLB entries. The OS should use `sfence.vma` after modifying the page tables to ensure that programs see the new page tables rather than stale entries in the TLB.

If both PMP and virtual memory are implemented, the TLB may cache PMP permissions, even in M-mode, when virtual memory is not active. In this situation, the firmware must also issue a `sfence.vma` after updating PMP CSRs.

11.2.5 HPTW

An OS expects a hardware page table walker (HPTW) to handle TLB misses. The RISC-V Ssccptr (*supervisor cacheable coherent page table read*) extension indicates that the HPTW can access any main memory region. Wally's HPTW supports this feature. The HPTW is located in the LSU, but it handles misses from both the data and instruction TLBs (DTLB and ITLB). Section 12.3.4 shows how the HPTW is integrated into the LSU.

11.3 TEST PLAN

At the time of this writing, riscv-arch-test does not yet have MMU tests. Handwritten Wally MMU tests that exercise each feature at least once are in `tests/wally-riscv-arch-test/riscv_test_suite/rv{32/64}i_m/privilege`. More comprehensive functional verification tests are under development at the time of this writing.

The `WALLY-pma.S` and `WALLY-pmp.S` test files check basic operation of the PMA and PMP, including each type of normal and exceptional access. The virtual memory tests are in `WALLY-mmu-sv32` for RV32 and `WALLY-mmu-sv{39/48}` for RV64. If Svadu, Svnapot, and Svbpmt are supported, use `WALLY-mmu-sv*-svadu-svnapot-svpbmt` instead. These tests require a page table rich enough to exercise all the virtual memory features and exception conditions. SV48 uses the 4-level table from Fig. 11.10(a), and the other flavors use similar tables. No tests are yet written for `sfence.vma`.

The detailed MMU test plans are in the cvw repository under `docs/testplans`.

11.4 WALLY IMPLEMENTATION

Fig. 11.14 shows the Wally top-level block diagram, highlighting in blue the MMU and HPTW. Wally has two MMUs: an instruction MMU (IMMU) in the IFU and a data MMU (DMMU) in the LSU. Both have

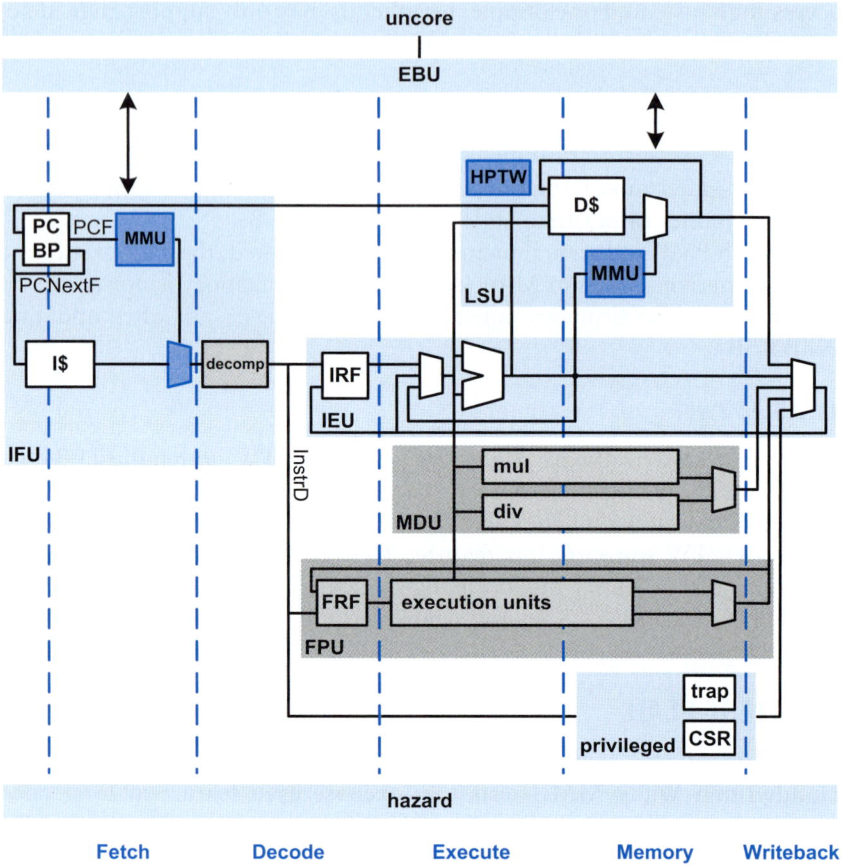

Fig. 11.14 Wally high-level block diagram, highlighting the instruction and data MMUs and HPTW.

identical SystemVerilog code, but the IMMU receives only execute requests (but no read or write requests), and the DMMU receives only read or write requests (but no execute requests), so logic synthesis optimizes out unnecessary access fault circuitry based on the parameters. The MMUs are configurable. At a minimum, the MMU must contain a small PMA checker to detect access and alignment faults for memory reads and writes. At a maximum, the MMU further contains a TLB for virtual memory, a PMP checker with 16 or 64 entries, and a PMA checker that supports many devices. The MMUs are present whenever `ZICSR_SUPPORTED` = 1 because they generate the access, misaligned, and page faults.

If virtual memory is supported, Wally adds an HPTW in the LSU to support TLB misses. The HPTW services both instruction and data TLB misses; only one HPTW is needed because only one miss can be serviced at a time through the bus interface anyway. All page table accesses are routed through the data cache (and on to external memory if the cache misses) and are checked by PMA and PMP checkers in the DMMU.

The MMUs are designed to be agnostic to the choice of bus interface; they send requests to the external bus unit (EBU), which directs the bus transactions. Figs. 12.2 and 13.12 show in more detail how the MMUs and HPTW are integrated into the LSU and IFU.

The configuration file (`configs/*/config.vh`) specifies the memory map for the PMA checker using the `BASE` and `RANGE` (size) parameters from Table A.4 corresponding to the typical memory map from Table A.2. Virtual addresses are always XLEN bits. Physical addresses are `PA_BITS` = 34 for RV32 and 56 for RV64. For example, the PLIC has `PLIC_BASE` = 0x0C000000 and `PLIC_RANGE` = 0x03FFFFFF, so the PLIC spans addresses 0x0C000000-0x0FFFFFFF. Other key MMU configuration parameters include:

- `VIRTMEM_SUPPORTED` — 1 to enable virtual memory
- `ITLB_ENTRIES/DTLB_ENTRIES` — TLB size, typically 16 each
- `PMP_ENTRIES` — 0, 16, or 64
- `SVNAPOT_SUPPORTED` — 64-KiB huge pages
- `SVPBMT_SUPPORTED` — Page-based memory types
- `SVADU_SUPPORTED` — HPTW updates A and D bits in PTE

11.4.1 MMU

Fig. 11.15 shows a block diagram of the MMU, focused on the major busses. The MMU always contains a PMA checker. If `VIRTMEM_SUPPORTED` = 1, it also contains a TLB, and if `PMP_ENTRIES` > 0, it also contains a PMP checker. The MMU receives the address `VAdr` used by the program. Note that `VAdr` is a virtual address if virtual memory is enabled or a physical address if not. If virtual memory is enabled, the MMU first translates the VPN to a PPN using a TLB. The MMU performs physical address decoding and checking to select the appropriate memory region and generate access faults. If the PTE is not in the TLB, the MMU asserts `TLBMiss`, which causes the HPTW to fetch the missing PTE.

All memory accesses have their addresses routed through the MMU so that it can perform PMA and PMP checking. The accesses may come from ordinary program execution (instruction fetch, or load/store operations) or from the HPTW. Addresses generated by the HPTW are physical, so they do not require translation but are still routed through the LSU MMU for PMA and PMP checking.

The MMU is on the critical path in Wally's 5-stage pipeline. The path occurs in the Memory stage, starting with the TLB translating a virtual

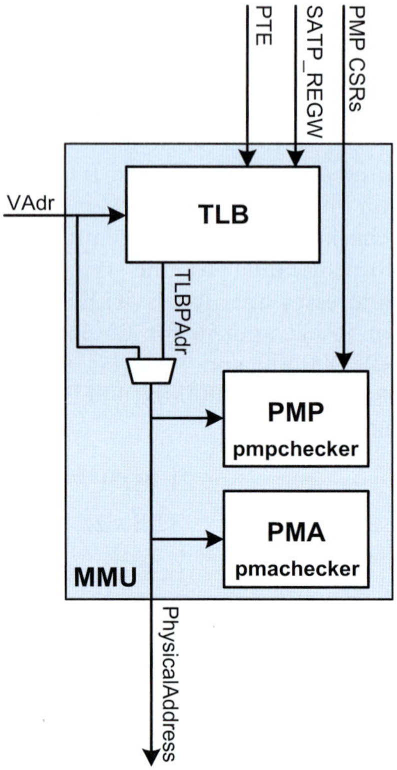

Fig. 11.15 Wally MMU (mmu).

address to a physical address. The PMP checker then compares this address to the PMP CSRs to determine if an access fault occurred. The privileged unit combines this information with other possible faults to cause a trap. The trap changes the program counter, which must set up in time to access the L1 I$ on the rising edge at the end of the cycle.

11.4.1.1 PMA Checker

The PMA checker uses the physical address to determine which memory region is accessed and the properties of that region. The PMA triggers an access fault if the region is vacant or does not support the type or size of access. Fig. 11.16 shows a block diagram of the RV64 PMA checker (`pmachecker`). It contains an array of address decoders (`adrdecs`) to check addresses in each region of the memory map, such as the tightly integrated on-chip memories, external memory, and the I/O peripherals. It checks which region the physical address matches and whether the access size and type are legal for that region, according to Table 11.4. `adrdecs`

> The RV32 PMA checker is nearly identical but uses shorter addresses and omits the leading 1 in the `SizeMask` to prohibit 64-bit accesses.

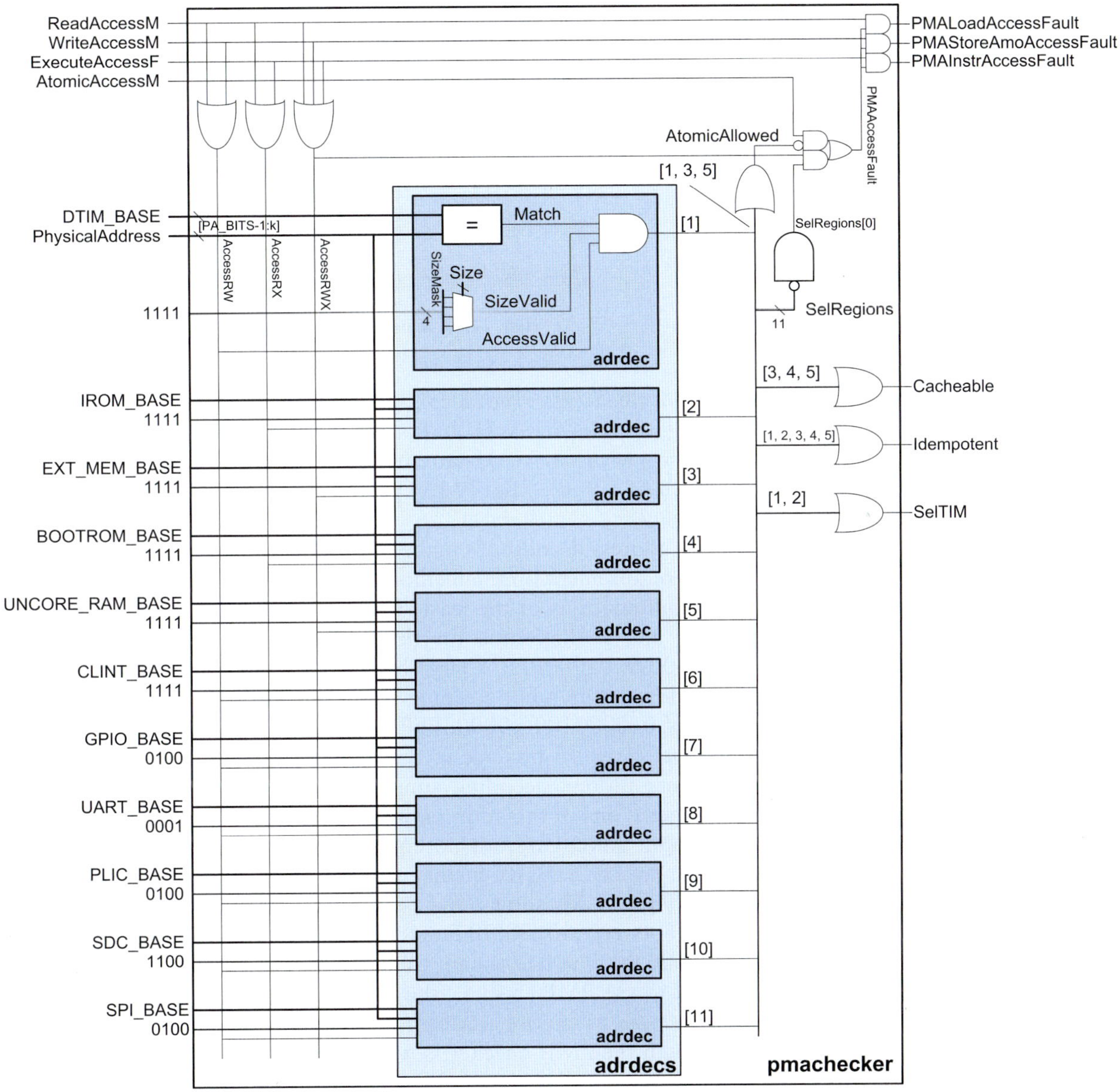

Fig. 11.16 **Wally PMA checker (`pmachecker`) for RV64.**

produces a one-hot select output (`SelRegions`) that indicates the accessed region or raises `SelRegions[0]` if the region is vacant. The PMA checker determines whether the access is to a cacheable, idempotent, and/or atomic region or to a tightly integrated memory in the core. For example, the uncore RAM, Boot ROM, and external memory (regions 3, 4, 5) are all

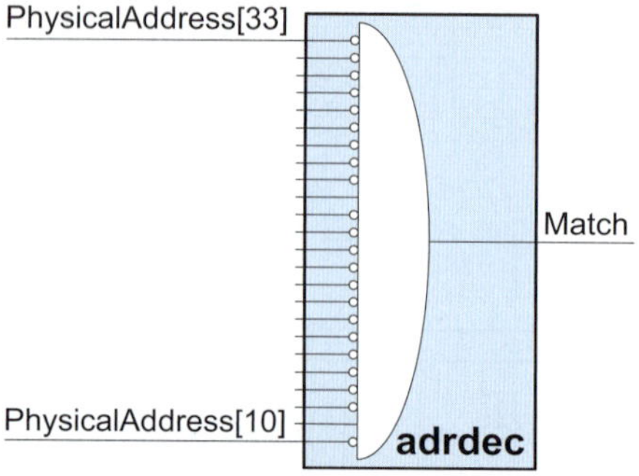

Fig. 11.17 Match logic example.

cacheable. If the access is not legal, the PMA checker produces an instruction-, load-, or store-access fault. For example, if an atomic memory operation (`AtomicAccessM` = 1) attempts to access a nonatomic region (`AtomicAllowed` = 0), the PMA produces a store/AMO-access fault (`PMAStoreAMOAccessFault` = 1).

The uncore contains the address decoders required to select one of the AHB devices, as discussed in Section 9.5.3.1. The `pmachecker` modules in both the instruction and data MMUs each contain another copy of the address decoders to check access faults. The EBU could choose `HSEL` signals coming from the instruction or data MMU instead of needing a third copy of the decoders. However, Wally supports configurations with a bus but no MMU, so it is simplest to have the extra decoders in the uncore.

Wally address decoding is made fast and cheap by imposing a natural alignment requirement on the memory map that each region covers all the possible combinations of the least significant bits while having the upper significant bits be the same for the region. For example, a region of 0x01000800-0x01000BFF is legal because it spans all values of the bottom 10 address bits while keeping the upper 24 address bits the same. A region of 0x00001000-00003FFF is not legal because not all combinations of the lower 13 bits are in the region. Because regions satisfy this requirement, address decoding simply involves an equality comparison on the upper *k* bits of the address, which involves *k* 2-input XNOR gates and a *k*-input AND gate but no costly subtraction. The base address is hardwired, so the XNORs further simplify to NOT gates or wires. For example, Fig. 11.17 shows the `Match` logic for the region spanning 0x001000800 to 0x001000BFF with 34-bit physical addresses. The upper 24 bits (0b00 0000 0001 0000 0000 0000 10) must match for that region. Compared to the PMP checker, the PMA checker does not need a costly priority encoder to choose among multiple matching regions.

Although the PMA checker is written to handle both instruction and data accesses, the `ReadAccess` and `WriteAccess` signals are tied low for the IMMU, and the `ExecuteAccess` signal is tied low for the DMMU, so logic synthesis can optimize away the unneeded access types in each PMA checker. If a given peripheral is not supported, its address decoder (`adrdec`) is also optimized out of the circuit.

When `SVPBMT_SUPPORTED` = 1 and the `PBMT` field of a PTE is nonzero, the PMA checker overrides the Cacheable and Idempotent attributes according to the page-based memory type.

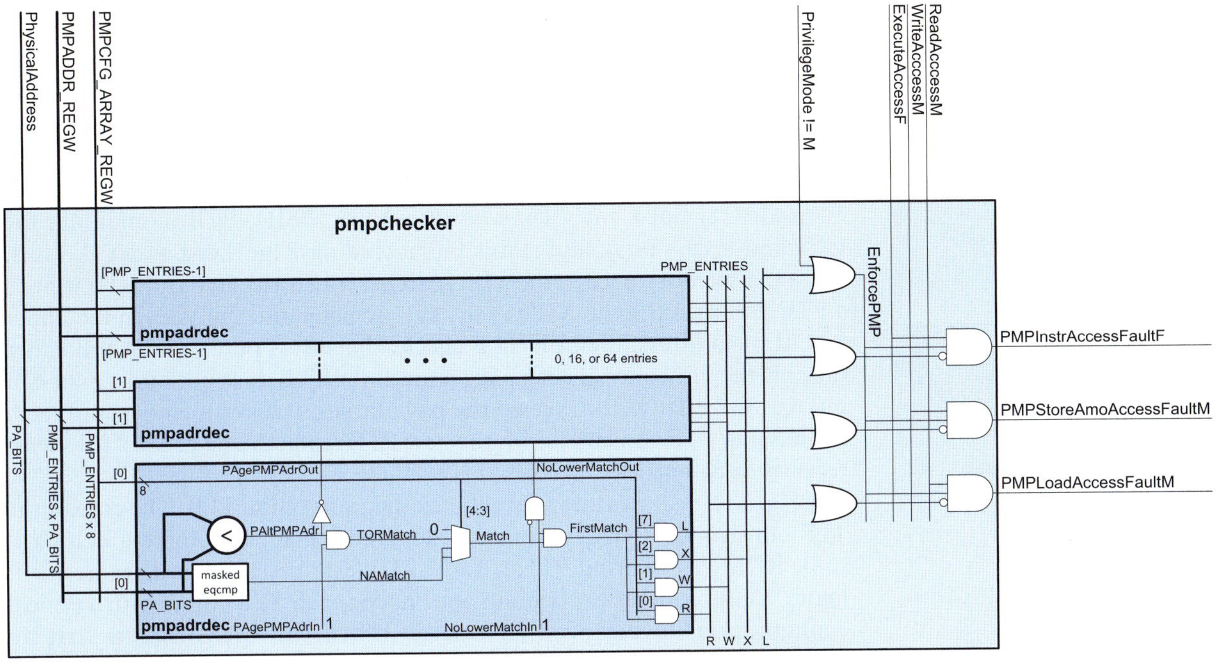

Fig. 11.18 Wally PMP checker (pmpchecker).

11.4.1.2 PMP Checker

The PMP checker causes access faults when a physical address violates PMP permission. Fig. 11.18 shows a block diagram of the PMP checker. It contains 16 or 64 PMP address decoder (pmpadrdec) blocks, each of which checks one PMP entry. Each pmpadrdec performs a less than comparison and a masked equality comparison to check for TOR and NAPOT/NA4 (naturally aligned) matches. The masked equality comparator tests only the more significant bits, disregarding trailing zeros in the PMP address. Based on the address mode (bits 4:3 of the pmpcfg field), the pmpadrdec selects the appropriate kind of match. More than one PMP entry could match a given address. A distributed priority encoder determines the first PMP entry that matches, FirstMatch. It uses daisy-chained AND gates to create a thermometer-encoded NoLowerMatch signal that is 1 for all of the entries up to the first match. The R, W, X, and L fields of pmpcfg for that entry are extracted using a distributed mux made of AND gates and a wide OR. pmpchecker enforces PMP rules when matching a locked region in M-mode, or anytime in S- or U-mode. Finally, the PMP checker generates an access fault based on the type of access occurring, privilege mode, and whether the PMP entry forbids that type of access.

When Wally supports compressed instructions or misaligned loads and stores, it also requires a grain size (PMP region size) at least as large as a cache line. Misaligned accesses that wrap a grain boundary are decomposed into multiple smaller accesses. PMP checks run independently on each access and trap if any access fails.

The critical path through the PMP checker includes the magnitude comparator, priority encoder, wide OR, and a handful of other gates.

The PMP check starts after the TLB delivers a physical address, and the check must complete in time to process a fault. This may limit the cycle time.

11.4.1.3 TLB

Recall that a fully associative cache has a single set and N ways, where N is the number of lines or, in the case of a TLB, entries.

A CAM compares its input with each of its entries. So, an N-entry CAM consists of N comparators. When a CAM's input matches an entry, it asserts a match for that entry.

A PTE is held in at most one TLB entry, so at most one entry matches.

The TLB quickly translates virtual to physical addresses using a cache of recent translations. It behaves as a fully associative cache with {I/D} TLB_ENTRIES. Each entry consists of a VPN, ASID, PTE (containing the PPN and access bits), and status bits including Page Type and Valid bits. During a memory access, the TLB searches all entries in parallel. If it finds a PTE that matches the ASID in the satp register and the VPN in the upper bits of the address, then the TLB checks the access permissions and either outputs the PPN or triggers a page fault. If the TLB does not find a match, it invokes the HPTW to traverse the page tables in main memory and fetch the missing PTE. An entry with the new PTE replaces one of the TLB's least recently used entries.

Fig. 11.19 shows a top-level block diagram of the TLB. The core of the TLB consists of a CAM (content-addressable memory: for tag comparison), RAM (for storing PTEs), and LRU (least-recently used replacement logic). These core blocks contain one line for each TLB entry, with a typical configuration having 32 TLB entries. When the CAM detects that an incoming tag matches the VPN and ASID, it raises the corresponding bit in Matches and reads that PTE from the RAM. When a new PTE needs to be written to the TLB, the LRU logic selects the line to be replaced and raises that line's bit in WriteEnables. The tlbmixer combines the PPN and page offset to produce the physical address. In the case of superpages, it also takes some address bits from the VPN. The controller tlbcontrol determines if the TLB hit, missed, or caused a page fault and whether translation should occur. The following subsections describe each block.

Unlike a conventional memory that accepts an address in binary and decodes it, the TLB has 1-hot Matches and WriteEnables signals to access an entry. The tlblru generates one WriteEnable for each entry, and the tlbcam generates one Match for each entry.

11.4.1.3.1 tlbcam

Fig. 11.20 shows a schematic of the TLB CAM, which holds the tags. The CAM is shown for systems that support Sv48 with a 36-bit VPN; those that only handle Sv39 or Sv32 use fewer bits and comparators. The CAM uses flops to hold the VPN, ASID, Page Type, and Valid bit. Each entry receives a write enable. The Valid bit is set to 1 on a write and cleared for all lines on a CAM flush. PageType is a two-bit code that indicates the type of page, as given in Table 11.7. The CAM matches if the VPN of the current access and the ASID field in satp match the values stored in a CAM line. Lower-order bits of the VPN are ignored for superpages, and the most

Table 11.7 Page types.

PageType	Type
00	kilopage
01	megapage
10	gigapage
11	terapage

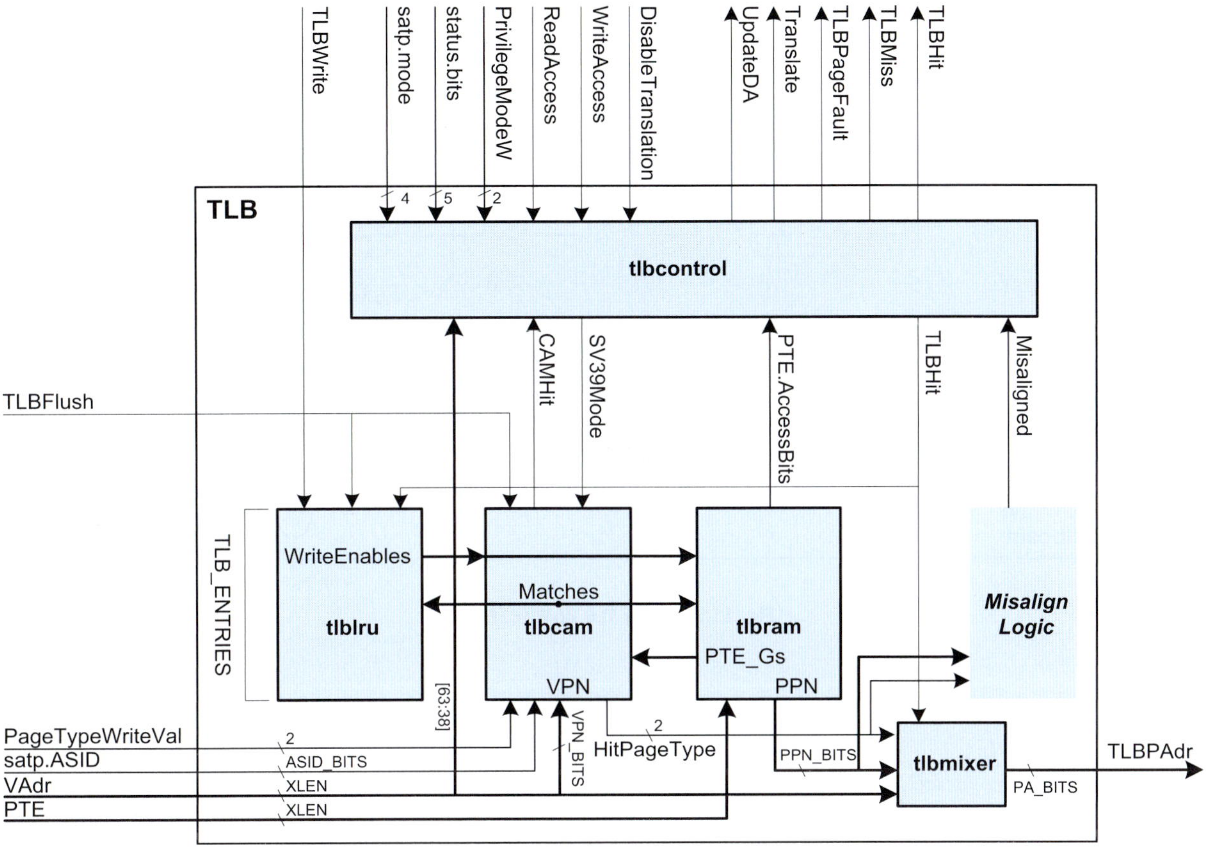

Fig. 11.19 **Wally TLB (`tlb`).**

significant bits are ignored when operating in Sv39 mode. The ASID is ignored if the PTE entry is marked global in the TLB RAM.

`CAMHit` asserts when any of the lines `Match` and `TLBFlush` is low. The CAM also outputs `HitPageType`, which indicates the type of the matched line. `CAMHit` and `HitPageType` are output using a distributed AND-OR multiplexer similar to Fig. 10.25 to combine the outputs of all the TLB lines (`tlbcamlines`) to determine if any of them hit and, if so, the page type. For example, each `tlbcamline` ANDs all the comparisons to create `Match` and then `tlbcam` ORs all the `Match` signals (and ANDs it with $\overline{\texttt{TLBFlush}}$) to produce `CAMHit`. Similarly, each `tlbcamline`'s `PageType` is ANDed with `Match` to produce `PageTypeRead` (so at most one `PageTypeRead` will be 1). Then `tlbcam` ORs `PageTypeRead` from all lines to produce `HitPageType`.

The Misalign Logic in Fig. 11.19 detects misalignment for superpages (mega-, giga-, or terapages) by detecting whether the correct PPN bits are 0s as determined by `PageType`.

A traditional custom CAM is built using a specialized cell that both holds a bit and also detects matches [Weste12]. Typical synthesized ASICs and FPGAs lack such CAM cells, so the TLB is implemented with flip-flops and gates to hold the tag and check the match.

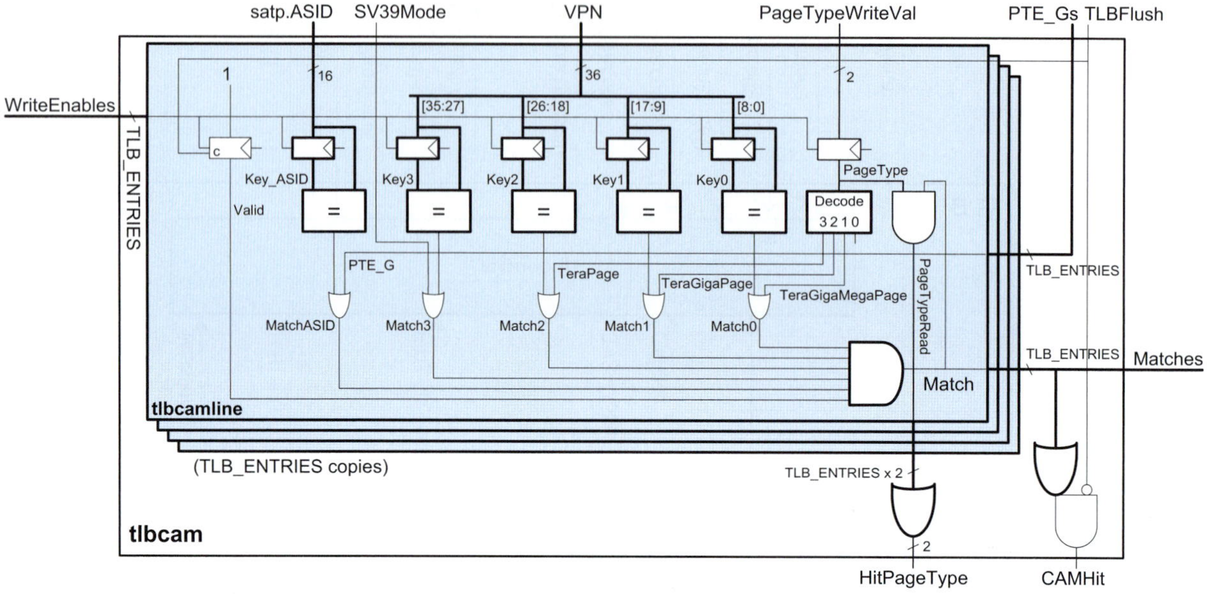

Fig. 11.20 Wally TLB CAM (`tlbcam`).

A PTE's G flag (`PTE_G`) is 1 when an entry is global (exists for all address spaces, see Table 11.1). The TLB RAM outputs `PTE_G` for all TLB entries (Fig. 11.21), and the TLB CAM then uses that signal to determine a match (Fig. 11.20).

11.4.1.3.2 `tlbram`

Fig. 11.21 shows a schematic of the TLB RAM (`tlbram`) that holds the PTEs. `tlbram`, like `tlbcam` is also built using flip-flops on an FPGA or ASIC. `tlbram` holds `TLB_ENTRIES` lines, each containing a PTE (`PageTableEntry`) which holds the PPN and 8 access bits (see Tables 11.1 and I.18). Like `tlbcam`, the `WriteEnables` bus determines which line to write, and a distributed AND-OR multiplexer reads out the line matched by the CAM. Essentially, the `Matches` signals act as read enables to select which entry, if any, to read out.

11.4.1.3.3 `tlbmixer`

Fig. 11.22 shows the RV64 `tlbmixer`. It forms the PPN from a combination of the VPN and PPN bits and the page offset. For kilopages, it concatenates the PPN with the page offset to form the physical address. For superpages, it selects some of the lower-order page number bits from the VPN rather than the PPN, as shown in Example 11.5. Because these low-order bits are always 0 in the PPN, the selection logic reduces from a multiplexer to an AND-OR gate. It then combines the `PPNMixed` signal with the page offset from the bottom 12 bits of the address. If the TLB misses, the physical address is forced to 0 (NULL); this isn't strictly necessary but will cause an access fault in the event that the TLB output is incorrectly used later. `TLBHit` is the `CAMHit` signal from `tlbcam` qualified by the access bits.

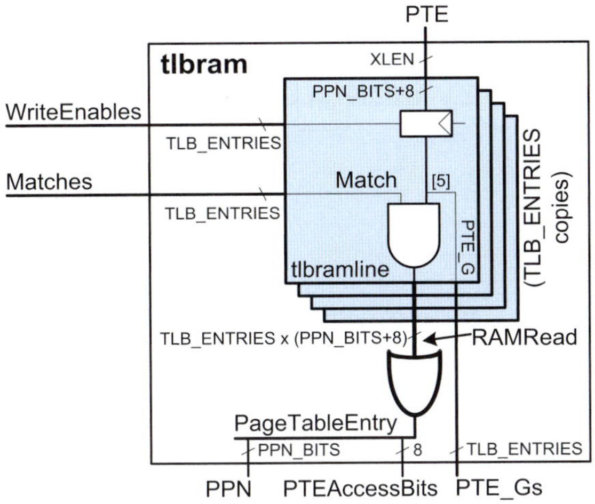

Fig. 11.21 Wally TLB RAM (`tlbram`).

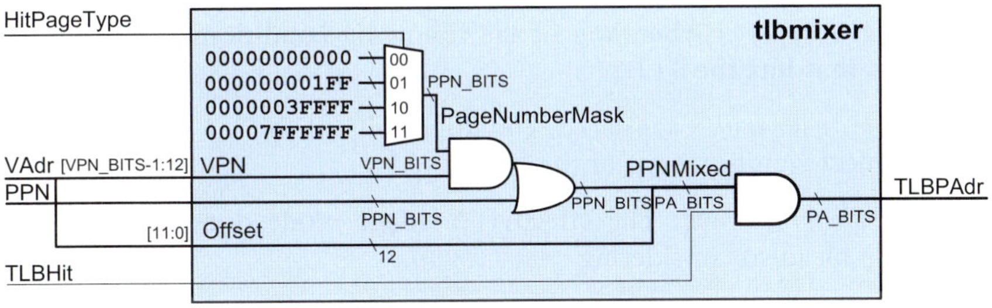

Fig. 11.22 Wally TLB mixer (`tlbmixer`).

11.4.1.3.4 `tlblru`

Fig. 11.23 shows a schematic of the TLB LRU unit that determines a least-recently used entry to replace. True LRU logic is too costly, so the block approximates it by keeping track of each entry that has been accessed and picking the first entry that hasn't been accessed. The recently used bits (`RUBits`) register keeps track of recently used entries. If all entries have been used recently, then all entries' `RUBits` are cleared. The priority unit `pri` picks the first entry that has not been recently used. On a write, the corresponding bit in the `WriteEnables` bus is raised to write the selected line of the CAM (tag) and RAM (PTE).

11.4.1.3.5 `tlbcontrol`

The `tlbcontrol` module determines the effective privilege mode; whether translation should occur; and whether the TLB hit, missed, or caused a page fault.

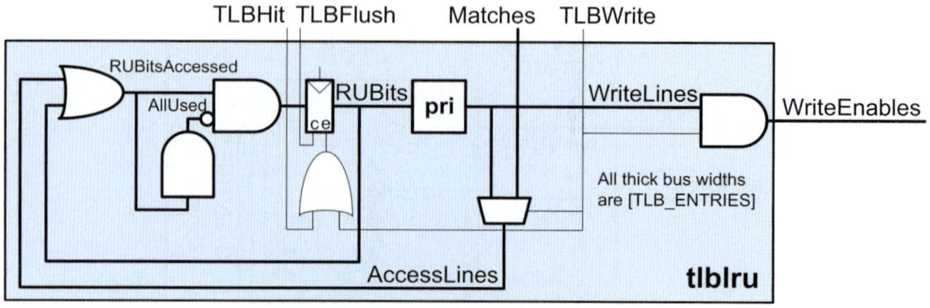

Fig. 11.23 Wally TLB LRU (`tlblru`).

Fig. 11.24 shows a schematic of `tlbcontrol` in the DTLB configuration. The TLB's `EffectivePrivilegeMode` is normally the core's `PrivilegeModeW`, but is `status.MPP` for the DTLB when the `status.MPRV` bit is set. Virtual address translation is enabled for S- and U-mode accesses when `satp.MODE`$\neq$ BARE but disabled during page table walks and cache flushes.

The TLB causes the appropriate kind of instruction/load/store page fault (`TLBPageFault`) when any of the following conditions occur during an access that hits the TLB.

- `UpperBitsUnequalPageFault`: The upper bits of the VA do not match the most significant bit of the VPN.
- `ImproperPrivilege`: The effective privilege mode doesn't match the `PTE.U` bit (also considering `status.SUM`).
- `InvalidRead/InvalidWrite`: The operation (execute/read/write) is prohibited by the `PTE.X/R/W` bits (also considering `status.MXR`).
- `UpdateDA`: The `PTE.A` bit is clear or the `PTE.D` bit is clear on a write. If `SVADU_SUPPORTED` = 0 or `menvcfg.ADUE` = 0, `TLBPageFault` is asserted to manage the `A` and `D` bits in software. Otherwise, `TLBPageFault` is not asserted, and the HPTW manages the bits.
- `PTE.V` = 0: The PTE is invalid.
- `Misaligned`: The PPN of a superpage is misaligned.

The `TLBHit` output is asserted when the CAM hits (`CAMhit`) during a read or write. The TLB misses when an access takes place, translation is enabled, no flush is occurring, and the CAM misses.

The ITLB is similar to the DTLB but is only used for instruction fetches. Hence, it checks `PTE.X` rather than `InvalidRead/InvalidWrite`.

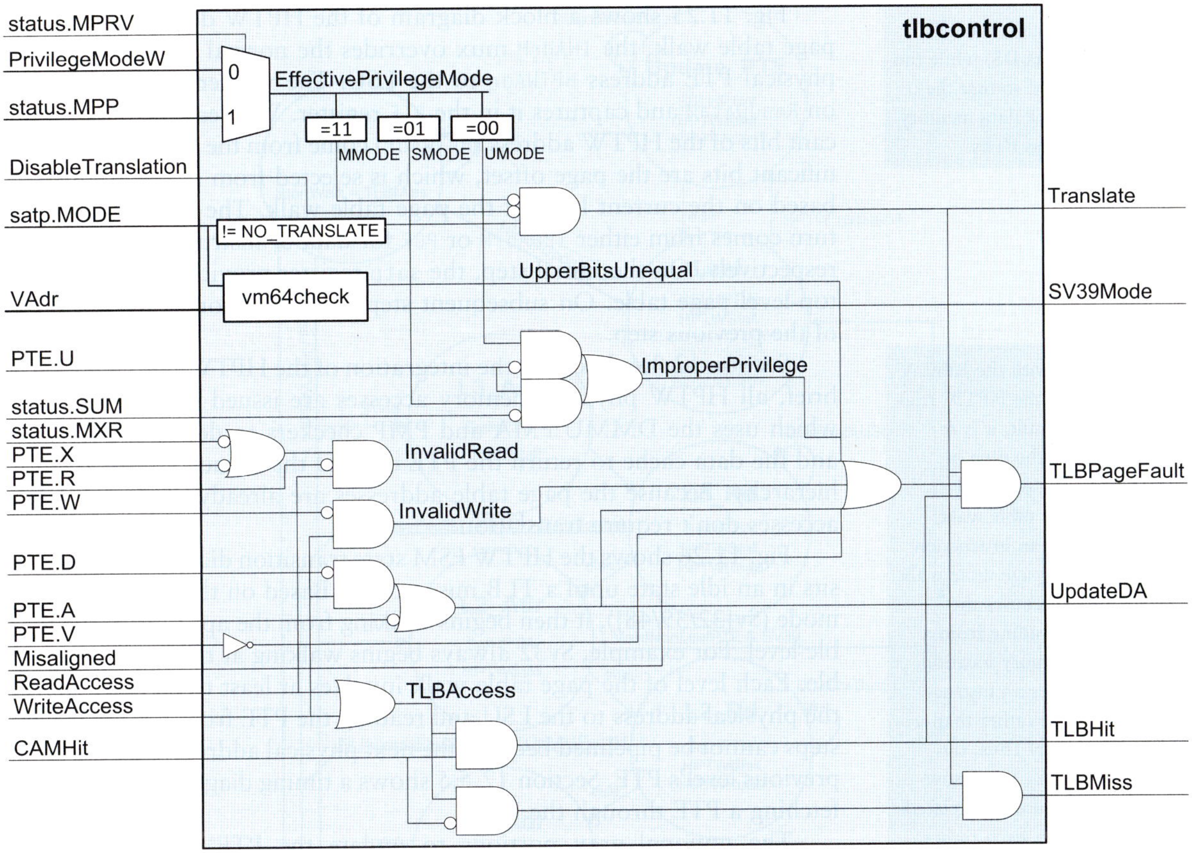

Fig. 11.24 Wally TLB control (`tlbcontrol`).

11.4.1.3.6 Svnapot Extension

When `SVNAPOT_SUPPORTED` = 1, the `tlbram` also asserts a `PTE_NAPOT` signal for each entry that is marked as a 64-KiB huge page by having the `PTE.N` bit set and `PPN[3:0]` = 1000. For huge pages, the `tlbcam` ignores the bottom four bits of the VPN, and the `tlbmixer` replaces the bottom four bits of the page number with the corresponding VPN bits. Hence, the cost of supporting huge pages is small. Only one TLB entry is used for the 64-KiB huge page compared to 16 entries to hold consecutive 4-KiB regular pages.

11.4.2 HPTW

The HPTW services a TLB miss by issuing a sequence of physical memory accesses to traverse the page table and fetch the missing PTE into the TLB. The HPTW consists of a datapath to insert page table accesses into the LSU, an FSM to sequence the walk through the multilevel page table, and control logic to detect page faults.

The HPTW is also simply called the *walker*.

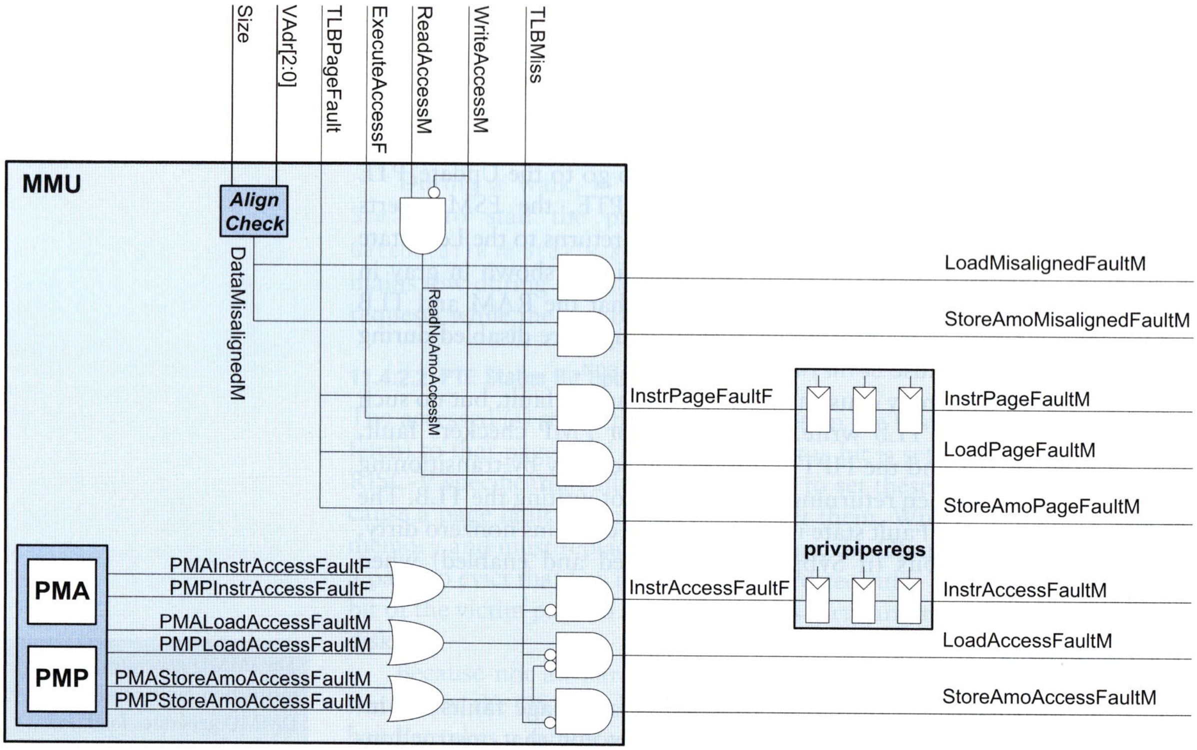

Fig. 11.28 MMU fault logic.

out-of-order and multicore machines. Wally simply treats them the same as `sfence.vma`.

11.5 SUMMARY

The MMU is responsible for virtual memory, PMA checking, and PMP checking.

Virtual memory elegantly addresses the problems of relocation, fragmentation, and protection for cores running multiple concurrent processes. It gives each process an independent virtual address space, so the process has the illusion of its own large and continuous memory. The memory is divided into pages, commonly 4 KiB, which can be placed anywhere in physical memory, or swapped out to disk when there is insufficient physical memory. Each process is protected from other processes because they can only access their own virtual addresses.

The MMU is responsible for translating the process's virtual address into a physical address to access main memory. Each process has a page table that identifies the (PPN) for each virtual page. In a practical system, the page table is too big, so it is organized hierarchically into 2-5 levels. The top-level page table is indexed with the most significant bits of the virtual address. It points to the next-level page table, which is indexed with the next most significant bits of the virtual address. This level in turn may point to another page table until the leaf page table entry (PTE) is reached. The leaf PTE contains the PPN and a set of flags such as valid, accessed, dirty, and access permission bits. If the PTE is invalid, has incorrect access permissions, or the page table is otherwise malformed, the MMU causes a page-fault exception, which fetches the missing page from disk into main memory, or kills a misbehaving process.

Traversing the page table on each memory access would greatly slow loads and stores, so the MMU contains a TLB, which is a cache of frequently used PTEs. Pages are large, so even a small TLB generally has a high hit rate because of spatial locality. If the TLB misses, the HPTW injects accesses into the LSU to walk the page table and find the missing PTE.

In RISC-V, machine-mode firmware uses physical addressing (except when `mstatus.MPRV` is enabled to facilitate address translation in an exception handler). The supervisor-mode OS enables virtual addressing with the `satp` CSR.

RISC-V also defines PMA and PMP checking. The PMA checker uses the physical address to determine which memory or peripheral is being accessed and the attributes of that memory. If the access goes to an unimplemented physical address or is of a type the peripheral cannot handle, the PMA checker causes an access-fault exception. The PMA is typically hardwired to reflect the system's memory map.

The PMP checker enhances security by limiting access to certain physical memory regions to machine or supervisor mode, triggering an access fault exception for disallowed accesses. For example, PMP is often used to restrict access to the firmware code to machine mode only, thus preventing a malicious user program from modifying the firmware. RISC-V has 0, 16, or 64 PMP CSRs to define PMP regions; 16 is typical for an application processor.

Wally's IFU and LSU each have their own MMU but share an HPTW in the LSU to handle TLB misses. Each MMU contains a TLB to translate virtual to physical addresses when virtual memory is enabled. The TLB is fully associative, using a CAM to identify which entry matches the VPN and then a RAM to read out the PPN and flags. The TLB operates concurrently with the cache access in the Fetch or Memory stage. Thus, the cache only knows the bottom 12 bits of the physical address when the access begins. By limiting cache ways to 2^{12} bytes = 4 KiB, the way is always indexed by a physical address. The rest of the physical address is available from the TLB by the time it is needed to compare against tags. Thus, the cache is physically addressed, avoiding tricky synonym problems. The `sfence.vma` instruction flushes the TLB by clearing its valid bits. The DMMU is on the critical path in Wally's 5-stage pipeline.

The wally-riscv-arch-test programs exercise the virtual memory system, PMA checkers, and PMP checkers.

Load/Store Unit

12

12.1 Wally Load/Store Unit Overview

12.2 Data Cache Integration

12.3 Memory Management and HPTW Integration

12.4 Stalls, Flushes, and Traps

12.5 Examples

12.6 Test Plan

12.7 Summary

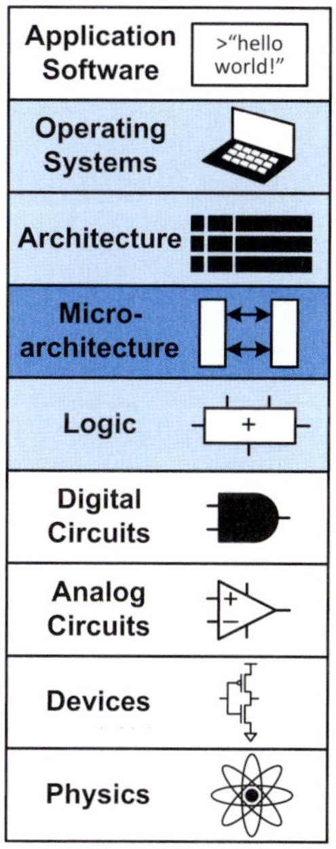

The load/store unit (LSU) handles data memory access for load and store instructions. It also fetches PTEs (page table entries) on behalf of the hardware page table walker (HPTW). Chapter 17 adds LSU support for AMOs (atomic memory operations).

Past chapters described various LSU elements. Chapter 7 introduced a simple LSU with a DTIM (data tightly integrated memory). Chapter 9 added support for off-chip memory and peripherals via the external bus unit (EBU). Chapter 10 described cache design, and Chapter 11 described the MMU (memory management unit), including address translation, the TLB (translation lookaside buffer), the HPTW (hardware page table walker), and PMP (physical memory protection) checker. This chapter combines all these capabilities into a fully-featured LSU.

Fig. 12.1 shows the Wally block diagram that highlights in blue the LSU features integrated in this chapter. The LSU receives memory addresses from the IEU (integer execution unit). When virtual memory is enabled, the TLB translates the addresses from virtual to physical. The PMP checker within the LSU faults on illegal accesses to protected memory. The LSU accesses the core memories: either the DTIM (not shown in Fig. 12.1) or the data cache (D$). When the LSU requests an address that is not in the core memories, it uses the external bus to access peripherals or memory in the uncore or off-chip. The HPTW is part of the LSU because it fetches PTEs through the D$ for TLB misses. The HPTW services both data and instruction TLBs and writes status bits back to the page tables in memory during page faults.

This chapter presents no new principles but focuses on the implementation of the LSU, especially the integration of the D$, MMU, and bus components previously described.

12.1 WALLY LOAD/STORE UNIT OVERVIEW

The Wally LSU can be configured to have a DTIM, a D$, or no core memory. Table A.1 in Appendix A lists various standard configurations and

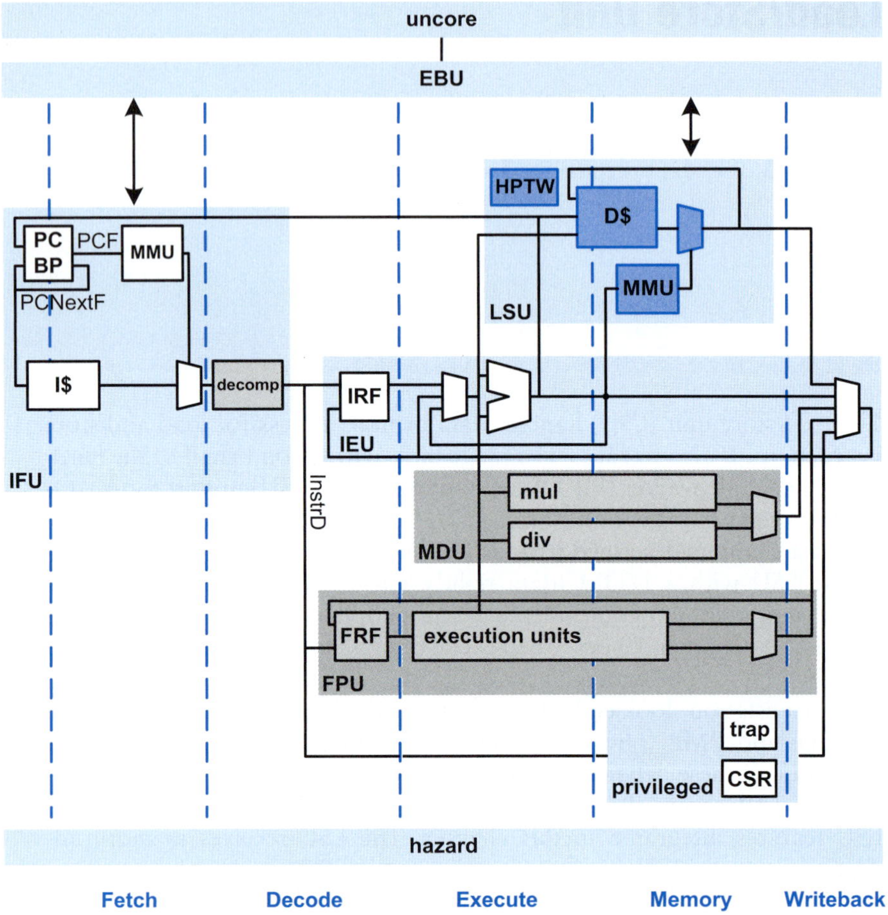

Fig. 12.1 Wally high-level block diagram, highlighting the LSU.

their LSU parameters. The embedded rv32e configuration has no core memory. The simple rv{32/64}i and microcontroller rv32imc configurations have tightly integrated memories. The apps processors rv{32/64}gc have caches. The microcontroller and apps processors have a privileged unit and MMU with a PMA (physical memory attributes) checker to catch illegal access and address misalignment faults; they also have a bus interface to uncore peripherals, external memory, or both. The apps processors also add virtual memory and a PMP checker to the MMU along with associated page faults.

Fig. 7.9 shows a basic LSU for the simple pipelined processor (rv{32/64}i). Fig. 8.19 adds optional endian swapping. The DTIM and IROM take up most of the area in the simple rv32i CPU, so for a small embedded core (rv32e), the DTIM is removed. Fig. 9.21 substitutes a bus interface to external memory and peripherals, which results in a tiny rv32e core area with no RAM. Figs. 9.30 and 9.31 combine the DTIM with the bus interface for an embedded microcontroller (rv32imc). This chapter adds the remaining LSU features, including integrating the D$, MMU, and HPTW for rv{32/64}gc.

Fig. 12.2 shows the fully featured (rv64gc) LSU for an application processor configuration capable of booting Linux. Control and byte write enables are omitted from top-level LSU diagrams for clarity. This configuration replaces the DTIM of Fig. 9.31 with a cache and AHB cache interface and adds an HPTW and a fully featured MMU. Details of the `endianswap` (ES), `subwordread` (SWR), and `subwordwrite` (SWW) modules are shown in Fig. 9.30 and are abstracted away here. Like the DTIM discussed in Section 7.1.6, the D$ supports XLEN-sized *native* words. The floating-point unit increases native word size to LLEN = max(XLEN, FLEN) and adds write multiplexing as described in Section 16.6.1. The atomic block is discussed in Section 17.4.1. These blocks are shown in gray because they are only needed when floating-point (F/D) or atomic (A) extension are enabled. The other components are explored in this chapter.

> Recall from Section 10.4.5.3 that the cache's sub-cache line read (SCR) module selects the word within the cache line (`ReadDataLine`) to produce `ReadDataWord`.

Fig. 12.2 is too complex to fully explain at once. Instead, this chapter builds up to the full LSU step by step. Section 12.2 integrates the D$ into

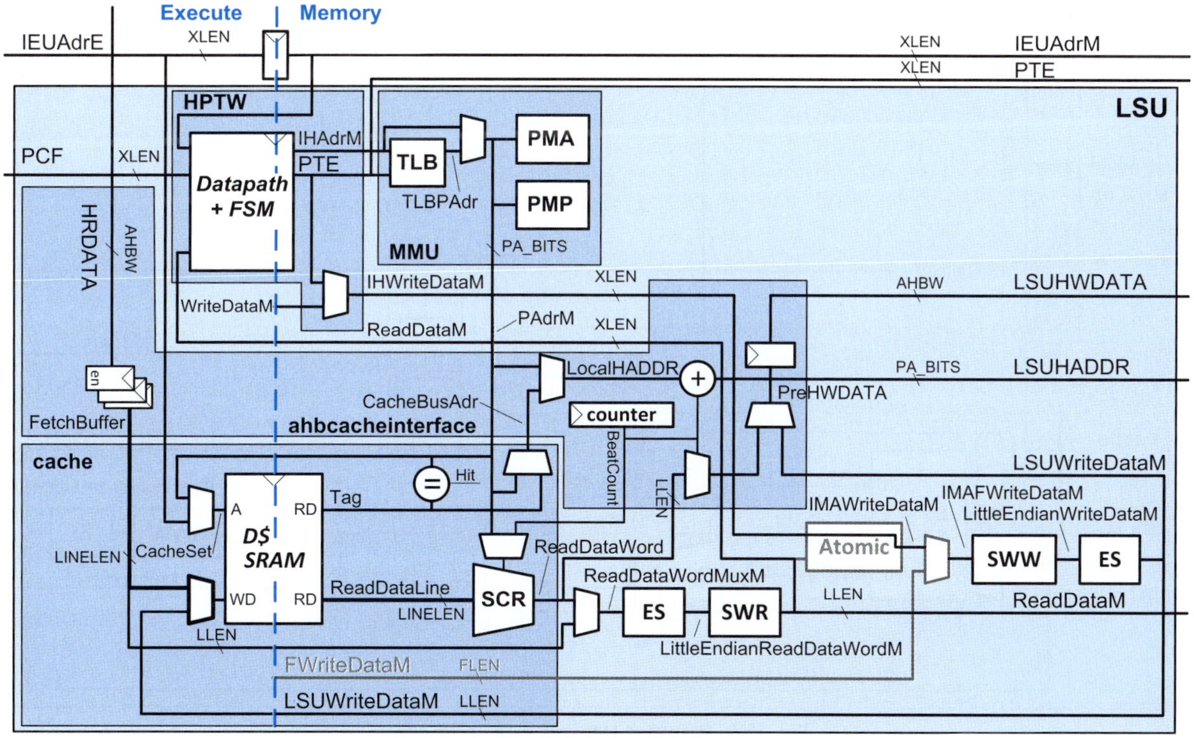

Fig. 12.2 **LSU block diagram.**

the LSU. Section 12.3 integrates the MMU. Section 12.4 describes LSU stalls, flushes, and traps. Section 12.5 presents timing diagrams explaining each of the use cases.

12.2 DATA CACHE INTEGRATION

This section integrates a D$ into the LSU, as shown in Fig. 12.3, if `DCACHE_SUPPORTED` = 1 in `config.vh`. Key features include servicing load and store hits, fetching lines from the external bus for cache misses, and writing back dirty evicted lines. Compared to Fig. 9.31, this LSU replaces the DTIM and `ahbinterface` modules with cache and `ahbcacheinterface` modules. It temporarily drops the PMA logic because this version of the LSU does not yet support uncached accesses; this capability will be restored in Section 12.3.1. The LSU retains the subword read/write (SWR/SWW) and endian swap (ES) modules.

The cache module was introduced in Fig. 10.24 in Section 10.4.5. All memory operations begin with the IEU calculating the address `IEUAdrE` in the Execute stage. On the rising edge starting the Memory stage, loads and stores both read the cache's ways using the lower (set) bits of `IEUAdrE`. Loads read both the tag and data using `CacheSetTag` and `CacheSetData`,

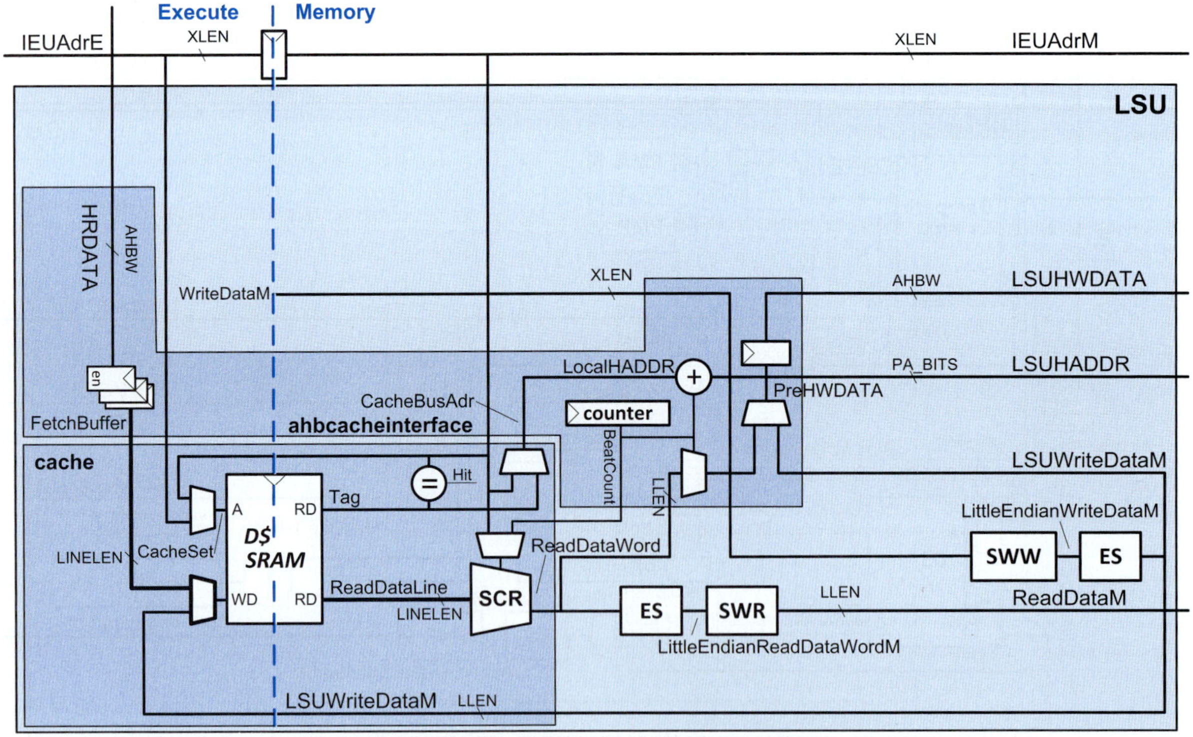

Fig. 12.3 Simplified LSU with data cache only.

but stores read only the tags, using CacheSetTag. If one of the tags is valid and matches the upper bits of IEUAdrM, the cache hits. Load hits output ReadDataM in the Memory stage, and store hits set up to write into the matching way (using LSUWriteDataW and CacheSetData = the lower (set) bits of IEUAdrM) on the next rising clock edge, at the start of the Writeback stage. Both loads and stores perform endian and subword muxing using the endianswap (ES) and subwordread/write (SWR/SWW) modules.

The CacheSet{Tag/Data} inputs to the cache, which access the cache's tag and data arrays separately, are simplified to CacheSet in LSU figures and timing diagrams, such as Figs. 12.2, 12.3, etc.

Cache misses stall the pipeline while fetching the missing line over the bus. Fig. 12.4 shows how the ahbcacheinterface manages communication between a cache and the EBU. At a high level, the cachefsm makes a bus request to writeback or fetch an entire cache line by placing the line address on CacheBusAdr and raising the write or read bit of CacheBusRW. For a writeback, the cache sends the evicted line ReadDataLine one word at a time over ReadDataWord, and the ahbcacheinterface generates an AHB write burst. For a line fetch, the ahbcacheinterface generates an AHB read burst and accumulates the words into FetchBuffer. ahbcacheinterface raises CacheBusAck when the burst completes.

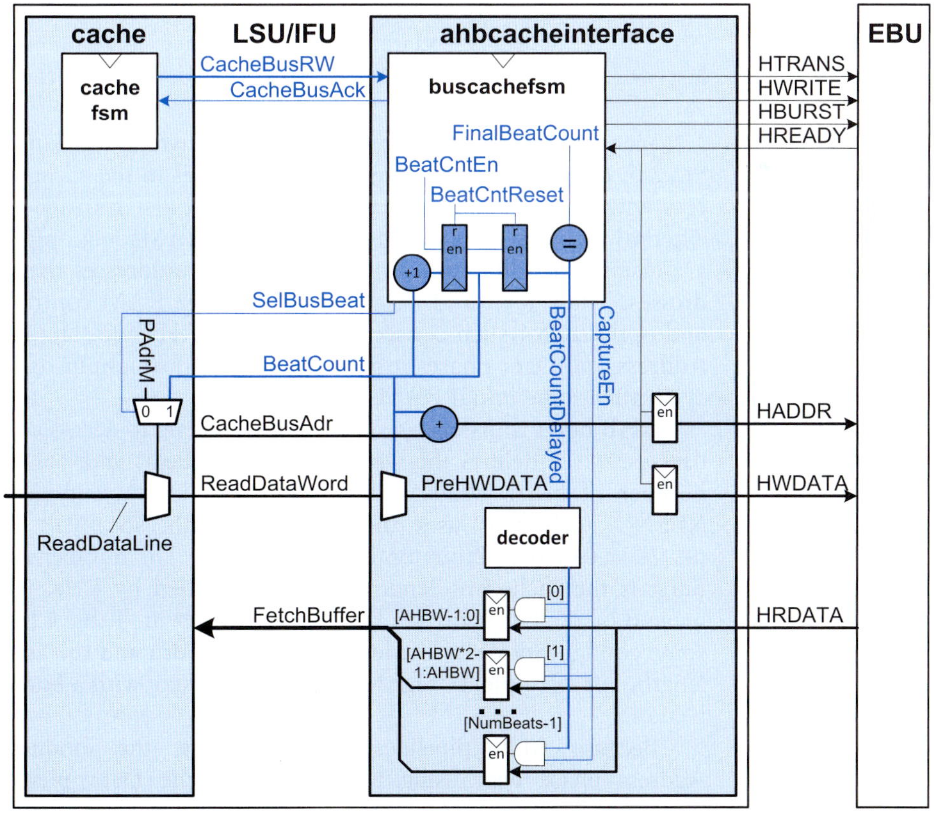

Fig. 12.4 **Cache, AHB cache interface, and EBU connections.**

The next subsections describe in more detail how the cache uses the ahbcacheinterface upon a miss to fetch and write back cache lines from noncore memory using burst mode.

12.2.1 Cache Line Fetches

During a cache miss, the cache selects a victim line to replace, as discussed in Sections 10.1.7 and 10.4.4. If the victim line is dirty, it must be written back to the next level of the memory hierarchy. Then the cache fetches the new line from the next level of the memory hierarchy (e.g., over the AHB bus), writes the line into its data array, and updates its tag array. This section discusses how the LSU handles cache line fetches, and Section 12.2.2 describes how it writes back dirty lines. Note that fetches of a new line occur on both load and store misses; store misses merge the stored word into the fetched line before writing the line into the cache.

To fetch a cache line, the cache places the starting address of the line on CacheBusAdr. This address is IEUAdrM with the offset bits zeroed. The cachefsm asserts CacheBusRW[1] to ask the ahbcacheinterface to perform the fetch. The ahbcacheinterface generates an AHB read burst to fetch all the words in the line into the FetchBuffer. BeatCount and BeatCountDelayed track the position within the burst to select which FetchBuffer word is updated by HRDATA. Once FetchBuffer receives the last word, ahbcacheinterface asserts CacheBusAck, and the cache writes FetchBuffer into the cache.

The AHB cache interface's bus cache FSM has a parameter BURST_EN that defaults to 1, which enables burst line transfers. It can be disabled by the testbench for testing nonburst transfers. Section 23.2.2.5 compares burst and nonburst performance of an FPGA implementation of Wally that uses DDR3 memory.

ahbcacheinterface implements burst transfers, introduced in Section 9.2.2.5, to fetch or write consecutive addresses in the cache line. Each access within the sequence is called a *beat*. A burst transfer can maximize external memory throughput because large RAMs (e.g., off-chip DRAM) are faster at reading and writing consecutive addresses than random addresses. The HBURST and HTRANS signals tell the RAM controller to expect and optimize for such consecutive accesses. AHB interfaces pipeline their Address and Data phases, obtaining a burst throughput of one word per cycle after some initial RAM latency.

AHB burst transfers use a sequence of beats to access consecutive addresses. HSIZE defines the word width for a beat, and HBURST defines the number of beats. Wally has a bus width of AHBW (typically equal to XLEN). Each beat uses the full bus width, so HSIZE encodes AHBW (HSIZE = 011 for 32 bits or 100 for 64 bits). The number of beats in the burst is the cache line length (LINELEN) divided by AHBW. HADDR is incremented by the bus width (in bytes) after each beat. Table 12.1 summarizes some configurations of line length and bus width and the associated burst length. For example, rv32gc fetches a 512-bit line with a burst of 16 32-bit beats.

Because AHB pipelines burst requests, the module tracks the Address phase beat using BeatCount and the Data phase beat using

Table 12.1 HBURST values.

Configuration	LINELEN (bits)	AHBW (bits)	HBURST	HSIZE
N/A	128	32	011 (4 beats)	011 (32 bits)
rv32gc	512	32	111 (16 beats)	011 (32 bits)
rv64gc	512	64	101 (8 beats)	100 (64 bits)

BeatCountDelayed. During the Address phase, ahbcacheinterface sends the memory address HADDR = CacheBusAdr + BeatCount << (AHBW/8). During the Data phase, the AHB cache interface stores HRDATA into FetchBuffer at BeatCountDelayed.

Fig. 9.23 shows the busfsm for a basic AHB interface, sequencing through Adr, Data, and Mem3 phases to read or write a single word over the bus. By adding a cache to the system, the bus must now support both single-word accesses (for peripheral accesses) and multiword bursts (for cache line fetches and writebacks). Fig. 12.5 shows the buscachefsm, an enhanced version of busfsm. It adds the CacheFetch and CacheWriteback states. This section describes cache fetches, and Section 12.2.2 describes cache writebacks. Section 12.3.1 then describes how the PMA checker detects uncacheable devices (e.g., peripherals) and issues single-word accesses; these are introduced in Section 9.1.3.

Cache bus requests start in the Adr state. If the cache requests a line fetch by asserting CacheBusRW[1] and the bus is free (HREADY = 1), the FSM enters the CacheFetch state and starts an AHB read burst by setting HTRANS = 2 (NON_SEQ) to indicate the first beat in the request. HBURST is set to the number of beats in the burst as shown in Table 12.1. HWRITE is set to 0. If the bus is not ready, the FSM remains in the Adr state.

The ahbcacheinterface uses a BeatCount counter to step through the beats in the burst each cycle the bus is ready (HREADY). For example, a DDR memory may have many wait states to service the first beat but then have single-cycle throughput for each subsequent beat. A comparator raises FinalBeatCount at the end of the burst.

The FSM remains in the CacheFetch state until the burst completes. After the first beat completes, HTRANS is set to 3 (SEQ) to signal sequential transfers in the burst. Each beat's HRDATA is written into the next word of FetchBuffer. When the burst completes its last beat, the FSM sends the cache CacheBusAck, and the FSM returns to the Adr state to wait for the next request.

If the memory operation is a load miss, the cache writes the cache line into its SRAM but also bypasses the SRAM to directly return

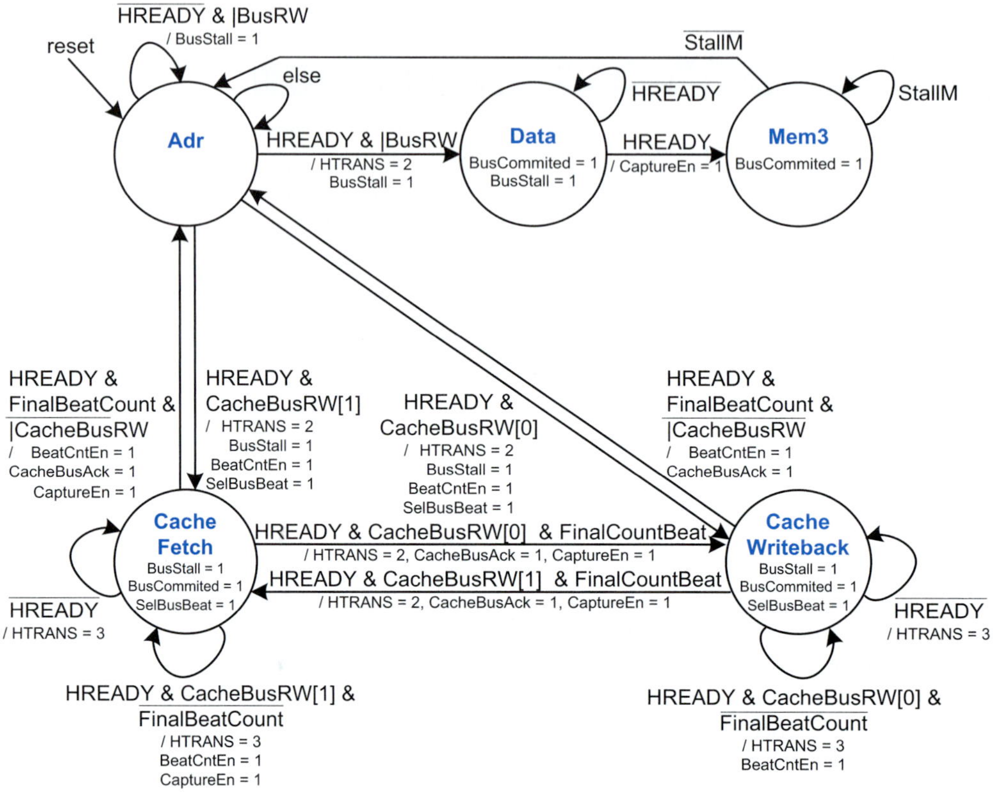

Fig. 12.5 The AHB cache interface's bus cache FSM (`buscachefsm`).

This path from the `FetchBuffer` to `ReadDataWord` is not shown in the LSU diagrams (Fig. 12.2, etc.).

Figs. 10.19 and 12.5 show the cache's FSM (`cachefsm`) and the bus cache FSM (`buscachefsm`), respectively.

`ReadDataWord` from the `FetchBuffer` to the core. Store misses merge the write data (`FinalWriteDataM`) with the `FetchBuffer` signal according to `ByteMask` before writing the line into the cache.

Code Example 12.1 illustrates cache misses and hits using a sequence of three memory instructions. Assume a word size (XLEN) of 64, a 32-byte (4-word) cache line, and a 64-bit (1-word) bus (`AHBW` = 64). Suppose the first load misses, and the cache fetches the line containing addresses 0x20-0x3F. Thus, the second load and store will hit. Fig. 12.6 shows the timing diagram for these instructions with reference to the CPU pipeline, an AHB bus read burst, and the `cachefsm` handshake with `buscachefsm`. The first load misses, which triggers a fetch burst of four 64-bit words. The first word has two wait states (`HREADY` = 0), and the next three words have single-cycle throughput. The next load and store hit the cache, so the bus is not involved.

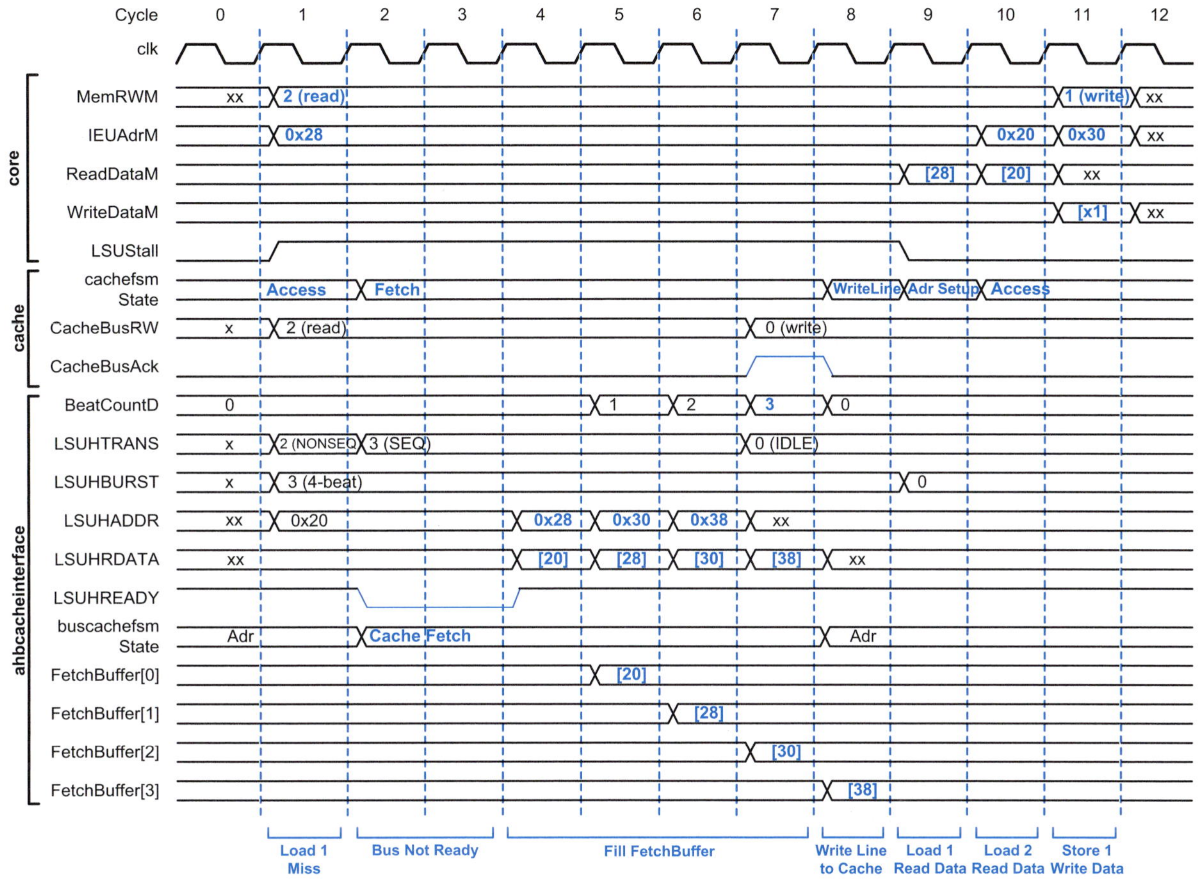

Fig. 12.6 Timing of a load miss followed by a load hit and store hit.

Code Example 12.1 LOAD MISS FOLLOWED BY LOAD AND STORE HITS

Sample code to demonstrate a load miss:

```
ld x1, 0x28(x0)
ld x2, 0x20(x0)
sd x1, 0x30(x0)
```

12.2.2 Cache Line Writeback

The `buscachefsm` handles cache line writebacks similar to how it handles fetches. The CacheWriteback state initiates a write burst similar to the fetch burst. Fig. 12.7 shows the timing for writing back an evicted line and then fetching the missing line. The data in each line is as follows:

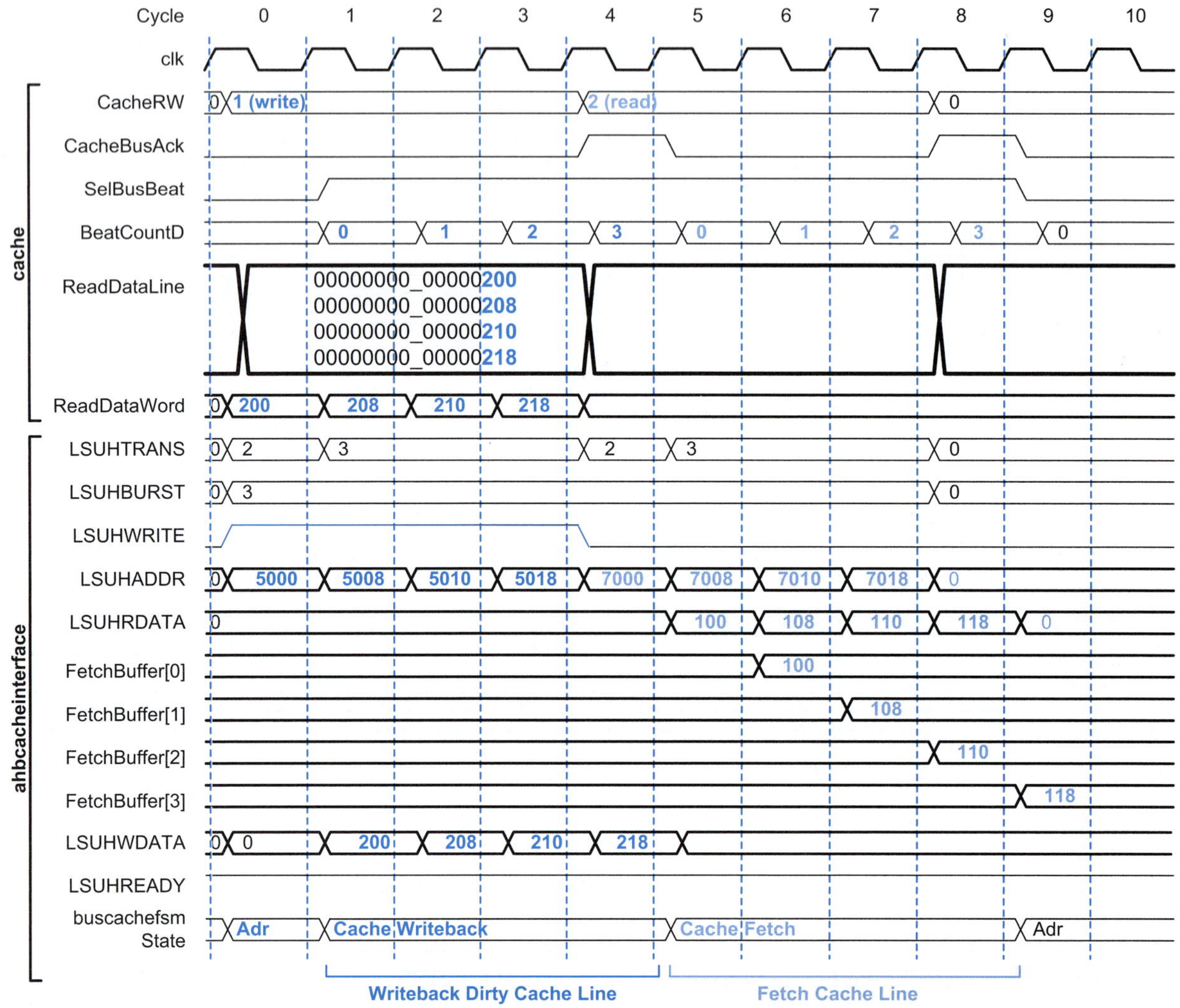

Fig. 12.7 Timing diagram of a fetch that requires a writeback.

- The evicted line's data (at address 0x5000):

0x00000000_00000218_00000000_00000210_00000000_00000208_00000000_00000200

- The missing line's data (at address 0x7000):

0x00000000_00000118_00000000_00000110_00000000_00000108_00000000_00000100

As with Fig. 12.6, the example assumes 8-byte words, a 32-byte (4-word) line, and a 64-bit (8-byte) bus, so each burst has four beats. Again, HTRANS is NON_SEQ on the first beat and SEQ on each of the subsequent beats. The BeatCount counter controls the SubCachelineRead (SCR) multiplexer (Fig. 12.4) by asserting SelBusBeat during writebacks to step through the words of the cache line to write a 64-bit word over the bus in each beat. This example assumes no wait states, so HREADY is always high. HWRITE is asserted for the line writeback (it was 0 for the line fetch). The CacheBusRW and CachBusAck signals handshake between the cachefsm (Fig. 10.19) and buscachefsm (see Fig. 12.5). CacheBusRW[1:0] = 1 (01_2) to writeback and 2 (10_2) to fetch.

Observe from Fig. 12.5 that the buscachefsm transitions directly from the CacheWriteback state to the CacheFetch state without an idle Adr state, pipelining the last beat Data phase of the writeback with the first beat Address phase of the fetch and saving a cycle of latency. Similarly, the buscachefsm can write back multiple lines consecutively without intervening idle states; this occurs during a cache flush, as described in Section 10.4.5.4.

12.2.3 Burst EBU

The EBU from Section 9.5.2 arbitrates between the IFU and LSU AHB requests, with the LSU having priority. Fig. 12.8 enhances the EBU to support burst transactions, with the changes shown in blue. The IFU's controllerinput module adds a register and multiplexers to save and restore the IFU burst request, IFUHBURST, during concurrent IFU and LSU requests. Both the IFU's and LSU's controllerinput modules add an AND gate to force their HREADY signals (IFUHREADY or LSUHREADY) to 0 when the other module has a burst in progress, which delays new requests until the in-flight request completes. The controllerinput module ORs the HTRANS bits to signal a request (IFUReq or LSUReq). HTRANS is 2 (NON_SEQ) for the first word of a burst request and 3 (SEQ) during the remainder of the request.

The EBU's FSM arbiter (ebufsmarb) (originally shown in Fig. 9.26) resolves requests between the buscachefsms of the IFU and LSU. It detects requests and produces control signals to grant AHB access to the LSU or IFU. ebufsmarb is enhanced in Fig. 12.9 to support burst mode. It adds a beat counter to track when the burst completes. The lookup table (LUT) decodes the number of beats from HBURST, according to Table 12.1 When an LSU request arrives while the IFU is in the midst of a burst, the LSUDisable signal asserts to delay the LSU request.

In its initial design, the EBU could disable IFU requests only, not LSU requests (see Fig. 9.25). In a system without virtual memory, this worked, because concurrent LSU and IFU requests only occurred in the same cycle, and the LSU always had precedence. However, in a system with virtual memory, LSU requests can occur after an IFU request is in flight because page table walks occur during a stalled pipeline.

For example, suppose a DTLB miss and I$ miss occur in the same cycle. The DTLB miss triggers a page table walk, and the I$ miss causes an IFU AHB request. While the IFU request is in flight, the HPTW traverses the page tables using the D$. The HPTW's first PTE access hits in the D$. However, its second access misses, which generates an LSU AHB request to fetch the missing cache line. Because LSU requests were not disabled and the LSU has precedence, the LSU request preempts the existing IFU cache line fetch, leaving stale instructions in the I$. This caused Wally to hang during Linux boot.

To fix this, the EBU added the `IFUReq` and `LSUDisable` signals. `LSUDisable` delays the LSU request in the example above until the IFU transaction completes. This was one of the most difficult bugs to figure out while designing Wally.

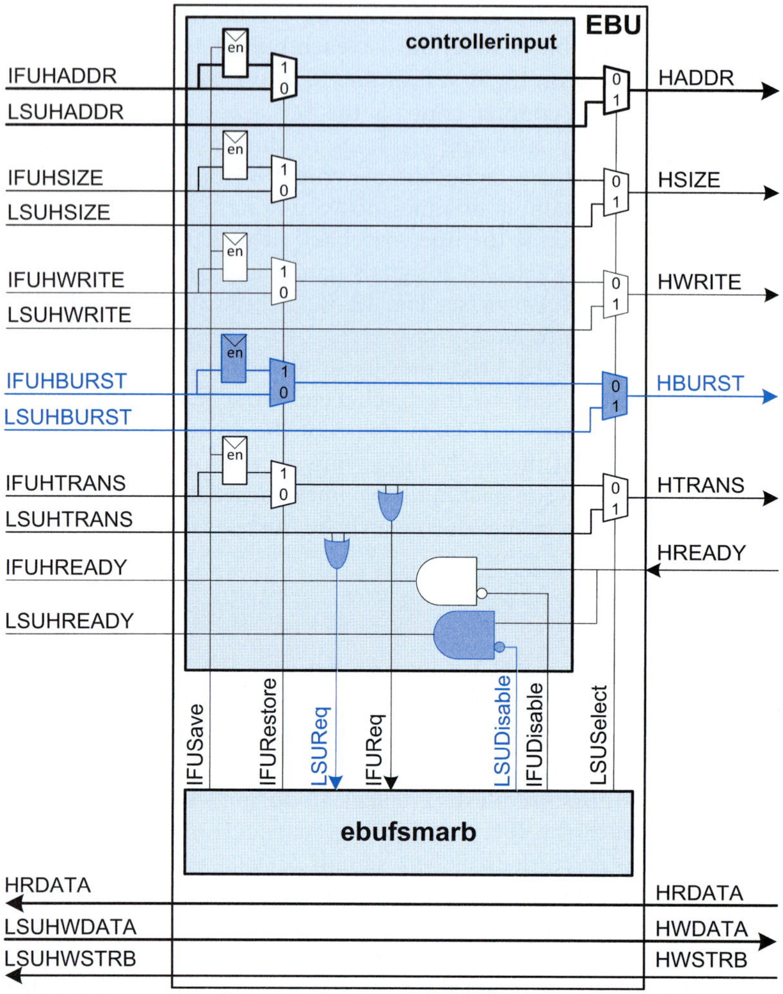

Fig. 12.8 **EBU with burst support.**

12.3 MEMORY MANAGEMENT AND HPTW INTEGRATION

This section integrates the main MMU components and the HPTW into the LSU, one by one. The three major MMU components are:

- PMA checker
- PMP checker
- TLB

These components rely on CSRs to handle memory exceptions and enable address translation, so the MMU is enabled when `ZICSR_SUPPORTED` = 1. The TLB also depends on `VIRTMEM_SUPPORTED` = 1,

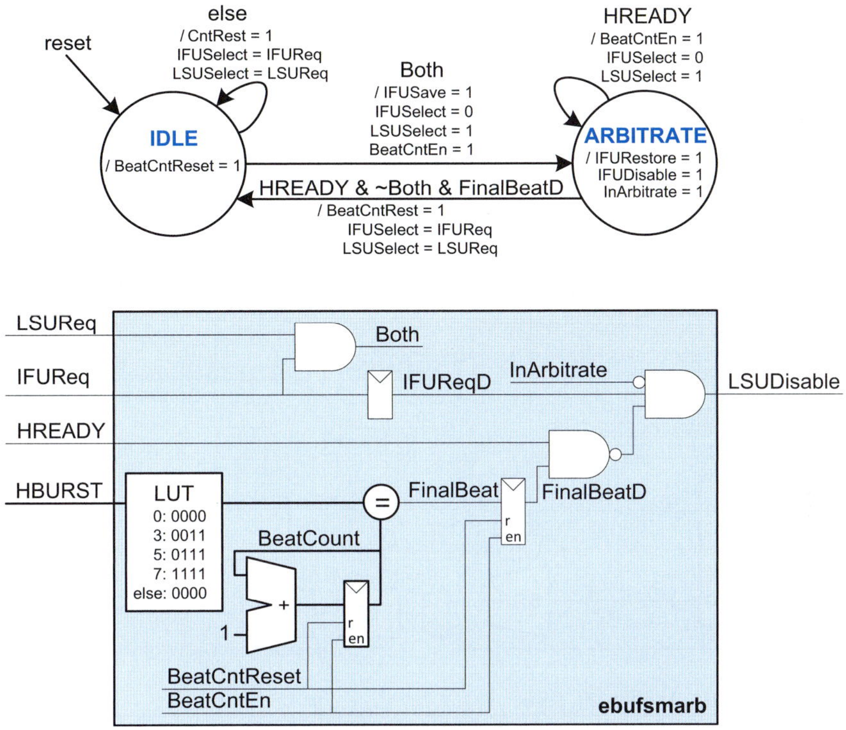

Fig. 12.9 **Fully featured `ebufsmarb` with burst support.**

and the PMP depends on `PMP_ENTRIES` > 0. With these MMU capabilities, Wally now has everything needed to boot Linux except atomic memory operations, which are described in Chapter 17.

12.3.1 PMA Checker

The PMA checker provides address decoding according to the memory map to select which memory device is accessed for an address range and what attributes that device has, as described in Section 11.2.2.1. Whereas main memory is cacheable, the peripherals are accessed directly over the bus. Fig. 12.10 enhances the LSU of Fig. 12.3 with PMA support to determine whether an access is cacheable. These new components are highlighted in blue. Cache flushes take priority and disregard the PMA by ORing `FlushDCacheM` with `CacheableM` to produce `CacheableOrFlushM`. Uncached accesses use AHB rather than the cache by driving `LocalHADDR` = `IEUAdrM` and `PreHWDATA` = `LSUWriteDataM` (for stores) or `ReadDataWordMuxM` = `FetchBuffer` (for loads).

The read and write control signals should go to either the cache or bus depending on whether the access is cacheable. Fig. 12.11 shows a

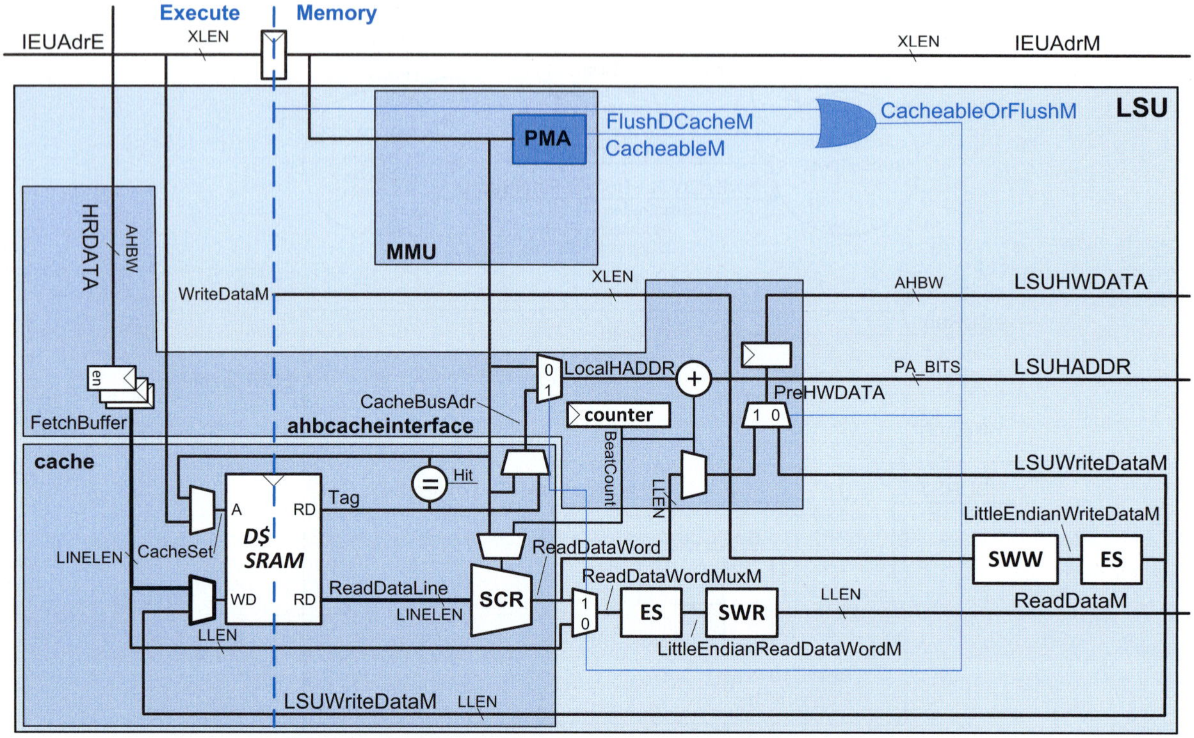

Fig. 12.10 Simplified LSU with data cache, bus interface, and PMA checker.

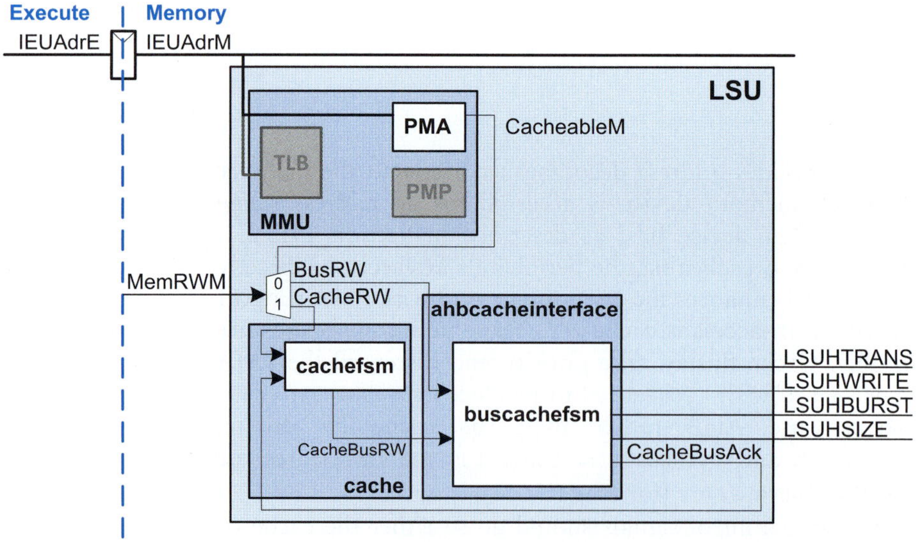

Fig. 12.11 LSU control path to select cached or uncached memory accesses.

demultiplexer added to the LSU to drive MemRWM from the IEU to either CacheRW or BusRW.

With the PMA integration, the LSU supports a fully featured set of seven types of memory operations, summarized in Table 12.2.

Table 12.2 Types of LSU memory operations.

Operation	Cacheable	Flush Cache	Access Type	LSUHADDR	ReadData WordMuxM	PreHWDATA	CacheWrite Data
Uncached load	0	0	BusRW = 10	cIEUAdrM	FetchBuffer [AHBW-1:0]	N/A	N/A
Uncached store	0	0	BusRW = 01	IEUAdrM	N/A	LSUWrite DataM	N/A
Cached load hit	1	0	CacheRW = 10	N/A	ReadData Word	N/A	N/A
Cached store hit	1	0	CacheRW = 01	N/A	N/A	N/A	LSUWrite DataM
Cache fetch	1	0	CacheBusRW = 10	CacheBusAdr + BeatCount	N/A	N/A	FetchBuffer
Cache writeback	1	0	CacheBusRW = 01	CacheBusAdr + BeatCount	N/A	ReadData Word	N/A
Cache flush	N/A	1	CacheBusRW = 01	CacheBusAdr + BeatCount	N/A	ReadData WordM	N/A

12.3.2 PMP Checker

The LSU's MMU contains an optional PMP checker described in Section 11.4.1.2. No LSU modifications are required. Together, the PMA and PMP checkers generate load- and store-access faults and misaligned faults, which are passed onto the privileged unit.

12.3.3 TLB and Cache Integration

Recall that when virtual memory is enabled, the program uses virtual addresses, but the memory system expects physical addresses. As discussed in Section 11.1.1, the processor translates virtual to physical addresses using a multilayer page table. A brute force implementation would slow down the processor with multiple page table memory accesses for each load or store, so a practical system adds a TLB to cache recently used translations.

The TLB is queried with the virtual address and returns a physical address and a hit signal. This section describes LSU accesses when the TLB hits, and Section 12.3.4 describes how the HPTW handles TLB misses.

As mentioned in Section 11.1.1.7, accessing a cache with a virtual address leads to synonym problems when multiple virtual addresses refer to the same physical address. These problems are avoided by indexing the cache with the least significant bits, which are the same for virtual and physical addresses. After translation, the tags (upper bits) are then checked against the physical tag. Therefore, the cache and TLB accesses occur in parallel, and the tag output of the cache is compared against the physical address output of the TLB. The size of a cache way is limited to the 4-KiB page size so that the cache set bits are in the physical page offset, which are available immediately without waiting for address translation.

Fig. 12.12 shows the LSU with a full MMU, including a TLB. When translation is off, the physical address PAdrM comes directly from IEUAdrM. When translation is on, the TLB looks up PAdrM concurrently with the cache read. In either case, the cache hit logic compares the tags to PAdrM.

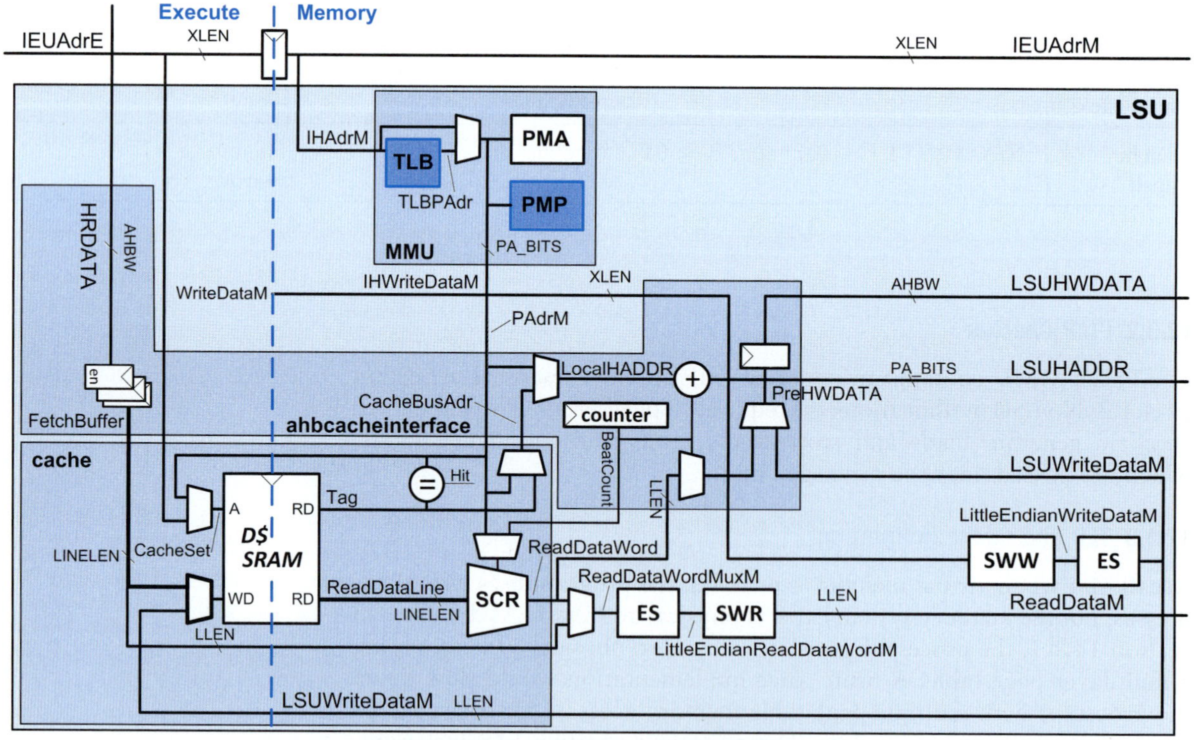

Fig. 12.12 LSU with full MMU and TLB address translation.

Like the cache, the ahbcacheinterface uses the physical address PAdrM in place of IEUAdrM for uncached memory operations. The cache and TLB are the boundary between virtual and physical addressing, which ensures that the bus always uses physical addresses.

With the inclusion of the full MMU, the LSU is nearly complete, but it lacks a mechanism to handle TLB misses. The next section integrates the HPTW from Section 11.4.2, which completes this last major LSU feature.

12.3.4 HPTW Integration

When virtual memory is enabled and the TLB misses, the processor must walk through the multilevel page table to *refill* the TLB with the missing page table entry (PTE). On a TLB miss, RISC-V may trigger a page-fault exception to ask the operating system to refill the TLB, but this can add hundreds of cycles to a TLB miss. Most application processors, including Wally, provide a HPTW to refill the TLB, as described in Section 11.4.2. TLB fill requests can come from either the ITLB (in the IFU) or DTLB (in the LSU), but the requests are all serviced by the single HPTW located in the LSU.

The HPTW issues a sequence of memory reads to each level of the multilevel page table, so it must have access to the memory system. The HPTW asserts SelHPTW to override other memory accesses and instead select the HPTW's address, control, and data during the page table walk.

The HPTW is located in the LSU and routes its memory reads through the LSU's D$ because this:

- Ensures coherency (recently changed PTEs may be in the D$ but not yet in main memory).
- Enables updating (writing) a PTE's accessed and dirty bits because the D$, unlike the I$, is writeable.
- Improves performance for page table walks because PTEs may be in the D$ instead of main memory.

Fig. 12.13 shows the LSU enhanced with the HPTW. Fig. 11.25 shows the simplified internal logic of the HPTW and Section 11.4.2 describes its operation. During a page table walk, the HPTW outputs physical addresses as IHAdrM so the PMA and PMP checker can verify the access is valid. The HPTW's IHAdrM mux (which is internal to the Datapath + FSM logic) chooses IEUAdrM (for regular loads and stores) or HPTWAdr (for a TLB miss and subsequent page table walk). Because the HPTW handles both ITLB and DTLB misses, the HPTW cobbles together the page table address (HPTWAdr) from either PCF or IEUAdrM (which are both virtual addresses), satp, and ReadDataM, which is the PTE read during the page table walk.

IHAdrM is a concatenation of names **I**EU**AdrM** and **H**PTWAdr.

The MMU drives IHAdrM onto PAdrM, and the LSU makes an untranslated PTE read from this physical address. It outputs the PTE via ReadDataM

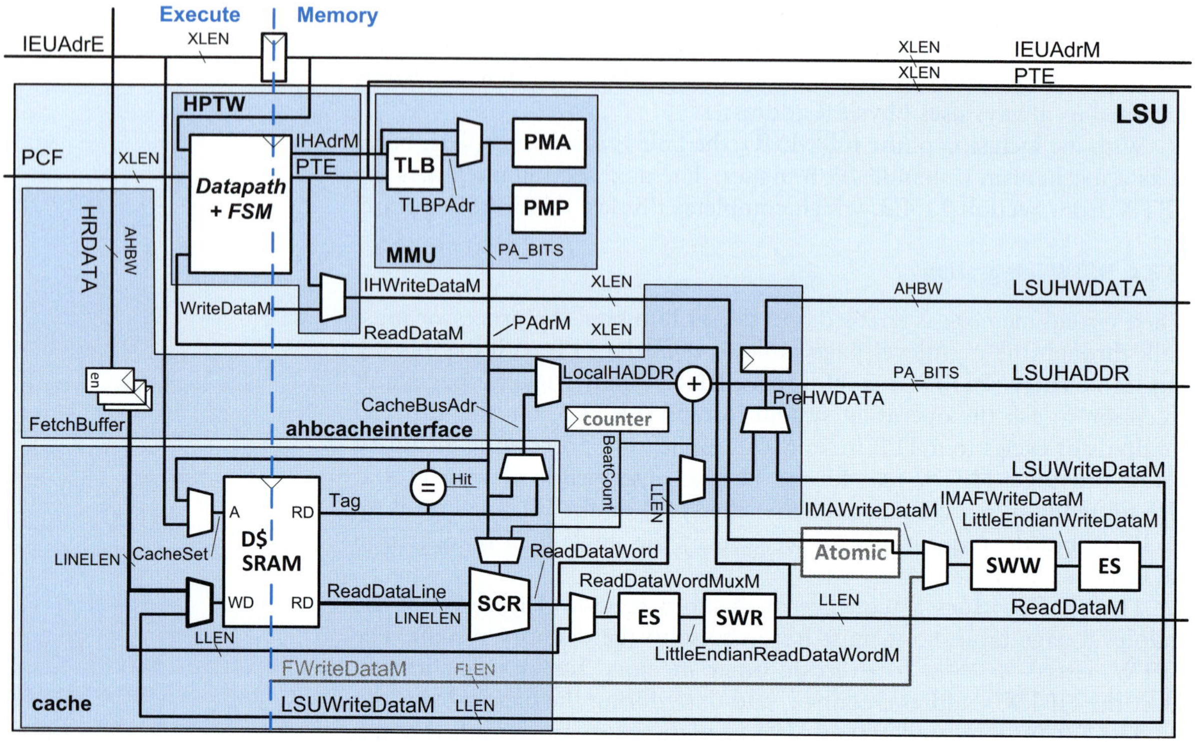

Fig. 12.13 LSU with HPTW.

> Fig. 12.13 abstracts away how the page table address (HPTWAdr) is generated but shows the main signals needed to produce it as inputs to the Datapath + FSM logic: IEUAdrM or PCF and ReadDataM. See Sections 11.2.1 and 11.4.2 for additional details.

to the HPTW. If the PTE is a leaf node, the HPTW FSM writes the entry to the TLB. Otherwise, it generates the next address of the page table walk and repeats the process.

12.4 STALLS, FLUSHES, AND TRAPS

The LSU stalls the pipeline during a bus access or hardware page table walk. It can generate traps due to access, misaligned, or page faults. Bus transactions take multiple cycles and should be atomic (i.e., complete once they start), so the LSU disables interrupts during bus transactions, including hardware page table walks. The LSU receives writeback-stage stall and flush signals, but the cache fetch/writeback and HPTW continue their operations even when the rest of the processor is stalled for the cache or TLB miss.

12.4.1 LSU Stall Generation

The LSU asserts LSUStallM during multicycle cache operations (including cache line fetch, writeback, and flush), during bus accesses, and when the HPTW is active. The hazard unit stalls all pipeline stages (StallF,

StallD, StallE, StallM, StallW) while LSUStallM is asserted. Specifically, LSUStallM asserts during a D$, HPTW, or bus stall, indicated by:

- DCacheStallM: asserted by the cachefsm during a cache miss or flush
- HPTWStall: asserted when the HPTW is processing a TLB miss
- BusStall: asserted by buscachefsm or busfsm when the bus is in use

On the last cycle of a memory operation, the respective stalls are lowered, which indicates that ReadDataM is valid or that the store has committed to memory.

The hazard unit also stalls the Decode stage for data dependencies and structural hazards. The forwarding logic in the controller raises LoadStallD when a load result is used by the next instruction and raises StoreStallD when a load follows a store (see Sections 7.2.2.3 and 10.4.2.3).

12.4.2 LSU Trap Generation

The TLB, PMA, and PMP generate page, access, and misaligned faults by asserting TrapM at the beginning of a memory access, before a cache miss could occur. Uncached memory operations are flushed if the access would generate a fault.

12.4.3 Disabling Interrupts

The LSU raises CommittedM to disable interrupts during a multicycle memory operation, as indicated by any of the following signals:

- BusCommittedM: a memory operation is in flight on the bus
- DcacheCommittedM: the cache has partially updated its state
- SelHPTW: HPTW is walking the page table

When CommittedM is asserted, the privileged unit suppresses pending interrupts, as shown in Fig. 8.14. Unlike LSUStall, CommittedM remains high during the last cycle of a memory operation.

12.4.4 LSU Response to Stalls and Flushes

The LSU's response to a stall depends on the current state of the D$, the bus, and the HPTW. Stalls are first gated by the HPTW with GatedStallW = StallW & ~SelHPTW. This allows the HPTW to issue loads and stores through the D$ while the rest of the pipeline is stalled. GatedStallW is renamed Stall in the cache. The cachefsm and buscachefsm only check Stall at the end of a multicycle memory operation, which allows the cache or bus to continue with the current operation while the pipeline is stalled. On the last cycle of a memory operation, if Stall is high, the cachefsm remains in the final, AddressSetup, state. This prevents the cache from executing the same memory operation over and over. Similarly, when Stall is asserted, the buscachefsm waits in its last state (Mem3, see Fig. 12.5) for uncached operations. However, buscachefsm does return to the Adr state after a cache line fetch or writeback so that the bus no longer asserts BusStall. This lowers LSUStallM, which allows the hazard unit to unstall the pipeline.

The Memory stage may be flushed in three cases: on the first cycle of a cache miss, on an uncached bus access due to a memory fault, or when another exception or interrupt flushes the pipeline. However, when a memory operation is in flight, Wally will not generate faults until that memory operation completes. In-flight cache fetch and writeback addresses were already checked by the PMA and PMP checkers, so they will not generate a fault when they are in flight. Specifically, cache fetches access PAdrM with the byte offset bits zeroed and cache writeback addresses are always the Victim Way's tag + the set bits from PAdrM, with zeros for the byte offset. These addresses were checked by the PMA and PMP when the cache line was first fetched.

12.5 EXAMPLES

This section illustrates the LSU in action, showing the active paths and timing diagrams for D$ hits and misses during loads and stores, uncached accesses, and TLB hits and misses.

12.5.1 Cache Load Hit

A load hit, highlighted in blue in Fig. 12.14, begins in the Execute stage using IEUAdrE, the address for accessing the D$'s SRAM. The cache uses the page offset bits, which are never translated, to choose the set in the cache. It reads the data, tags, and status bits. In the Memory stage, the TLB translates the virtual address IEUAdrM to the physical address PAdrM if address translation is enabled; otherwise, PAdrM = IEUAdrM. The cache compares its tags to PAdrM to indicate a hit and outputs all bytes in the cache line on ReadDataLine. The sub-cache line read (SCR) module uses the offset bits of PAdrM to select the word within ReadDataLine to produce ReadDataWord. Finally, because it is a hit, the mux passes ReadDataWord to ReadDataWordMuxM, optionally performs an endianswap (ES), and selects the final ReadDataM using the subwordread (SWR) module. The cachefsm (Fig. 10.19) remains in the Access state throughout the load hit.

12.5.2 Cache Store Hit

Like loads, a store begins by reading the cache SRAM and optionally translating the virtual address IEUAdrM to the physical address PAdrM. The

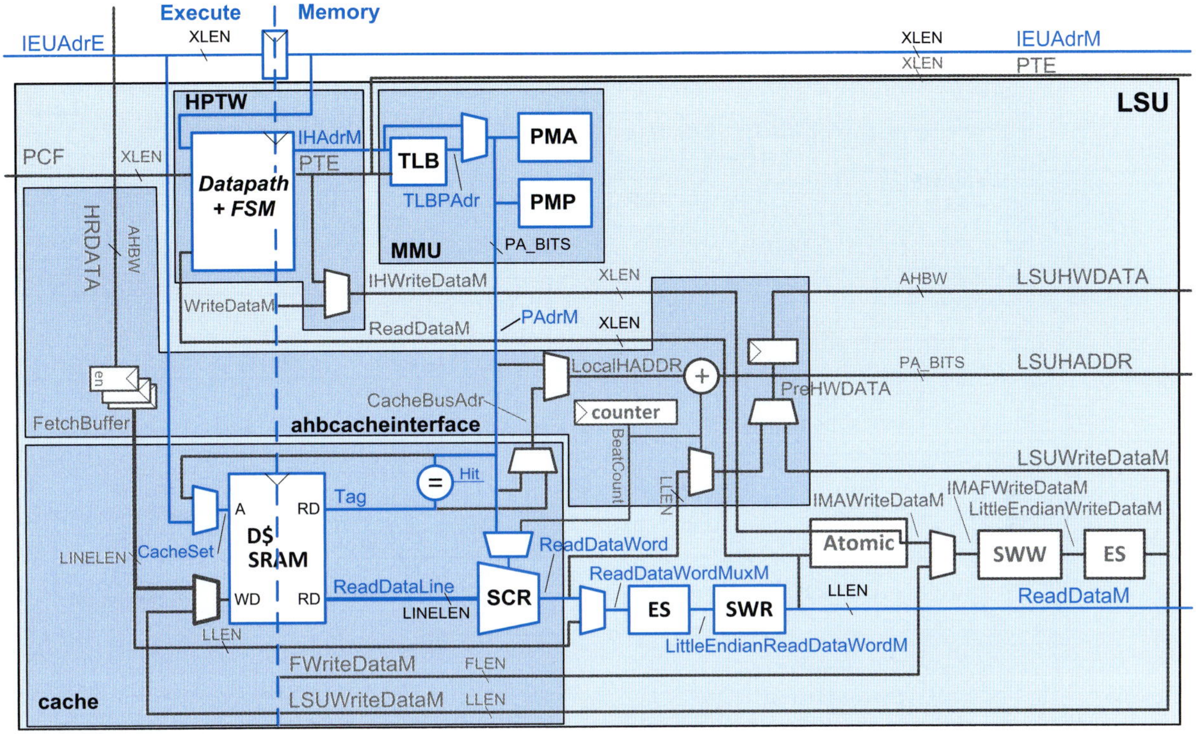

Fig. 12.14 **Example load hit.**

cache reads only the tags, not the data. It compares `PAdrM` to the `Tag` and checks the valid bit to detect a hit. On a hit, the cache sets up the `CacheSet` mux using `PAdrM`. In parallel, `WriteDataM` travels through the `subwordwrite` and `endianswap` muxes (labeled SWW and ES) and is written into the SRAM on the clock edge at the end of the Memory stage, as shown in Fig. 12.15. Not shown, the `subbytemask` (SBM) module sets the corresponding byte write enables for the subword and then the cache expands the byte write enables to the full width of the cache line in order to write `WriteDataM` into the correct location in the cache line.

Fig. 12.16 shows the timing diagram for consecutive load, store, load, and load hits to addresses 0x10, 0x18, 0x20, 0x08, which all hit in the same set. The first load and store have single-cycle throughput. The second load stalls for a cycle because of the store delay hazard caused by a store writing the cache data array during the Writeback stage and the following load attempting to read the cache data array in the same cycle in the Memory stage. In cycle 3, the controller detects the hazard in the Decode stage when a store is executing and a load is in the Decode stage. It asserts `StallD` to delay load 2 by a cycle and to insert a pipeline bubble between the two

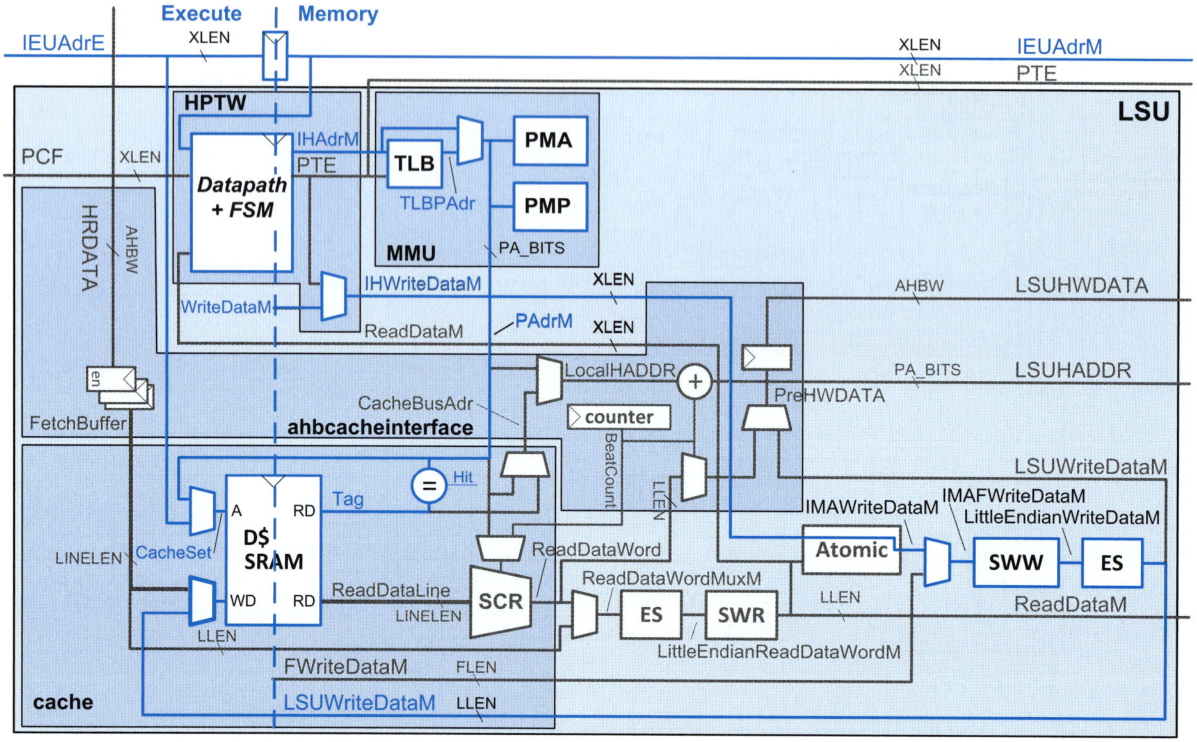

Fig. 12.15 **Example store hit.**

instructions. The last load has single-cycle throughput. All four accesses share the same cache line at set 0 (CacheSet = 0). The cache memory arrays are written on the rising clock edge at the beginning of cycle 5 (in the store's Writeback stage). On the rising clock edge at the beginning of cycle 6, the memory switches back to reading and ReadDataLine now contains the updated data from the store. An 8-byte word is written, so ByteMask = 11111111 (0xFF), and the fourth 8-byte word of the 8-word line is written, so LineByteMask expands this to 0x00000000FF000000 to indicate which 8 of the 64 bytes in the line should be written.

12.5.3 Cache Miss

On a load or store miss, the cache fetches the missing line from the next level of memory over the bus, as described in Section 12.2.1. Fig. 12.17 highlights the LSU path for fetching a cache line from the bus, and Fig. 12.6 shows the timing diagram. The cache outputs the cache line address, CacheBusAdr. ahbcacheinterface outputs BeatCount to sequence through the words of the cache line and to increment LSUHADDR, then raises CacheBusAck when FetchBuffer contains the cache line.

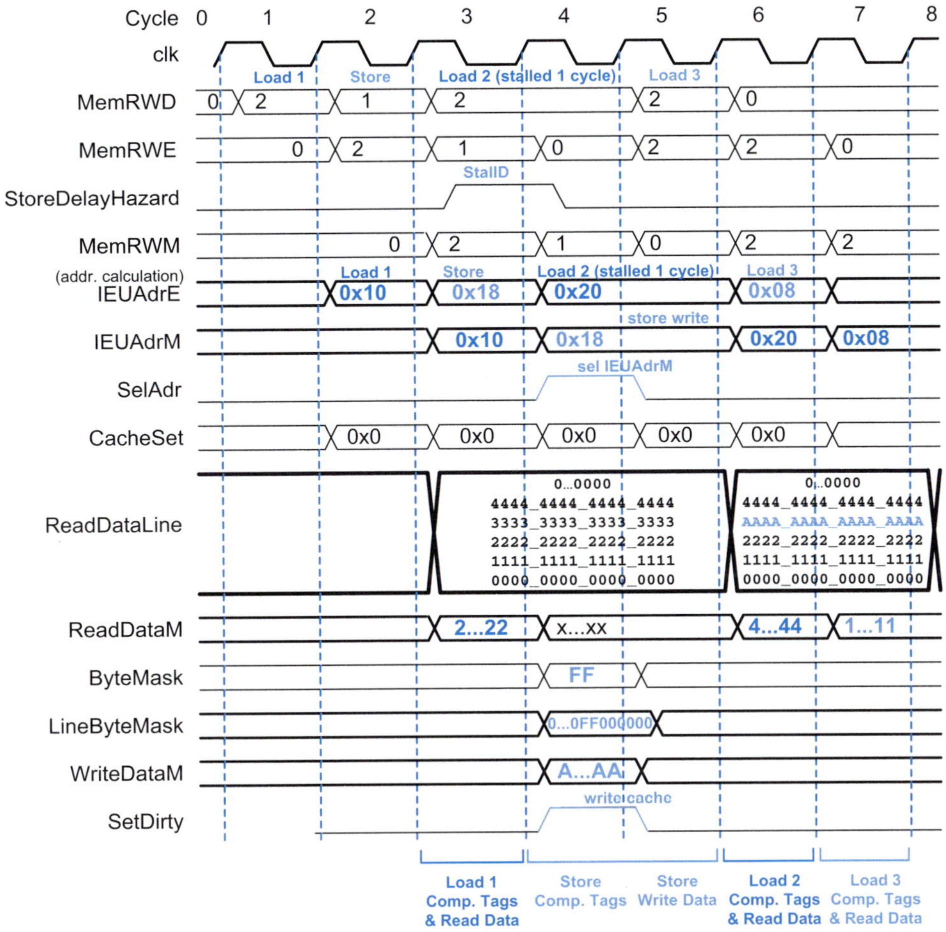

Fig. 12.16 Timing diagram of consecutive load, store, load, and load hits.

Section 12.2.2 describes a writeback of a dirty cache line. Fig. 12.18 highlights the cache line writeback paths in the LSU, and Fig. 12.7 shows the timing diagram.

12.5.4 Uncached Access

The LSU makes uncached accesses to uncacheable devices such as I/O peripherals and also in cacheless systems with external memory. Section 9.5.1 describes how the `ahbinterface` module makes single NON_SEQ bus requests. Section 12.3.1 reintroduces the PMA checker to select between cached and uncached bus access. Fig. 12.19 highlights the LSU uncached load path, and Fig. 12.20 highlights an uncached store.

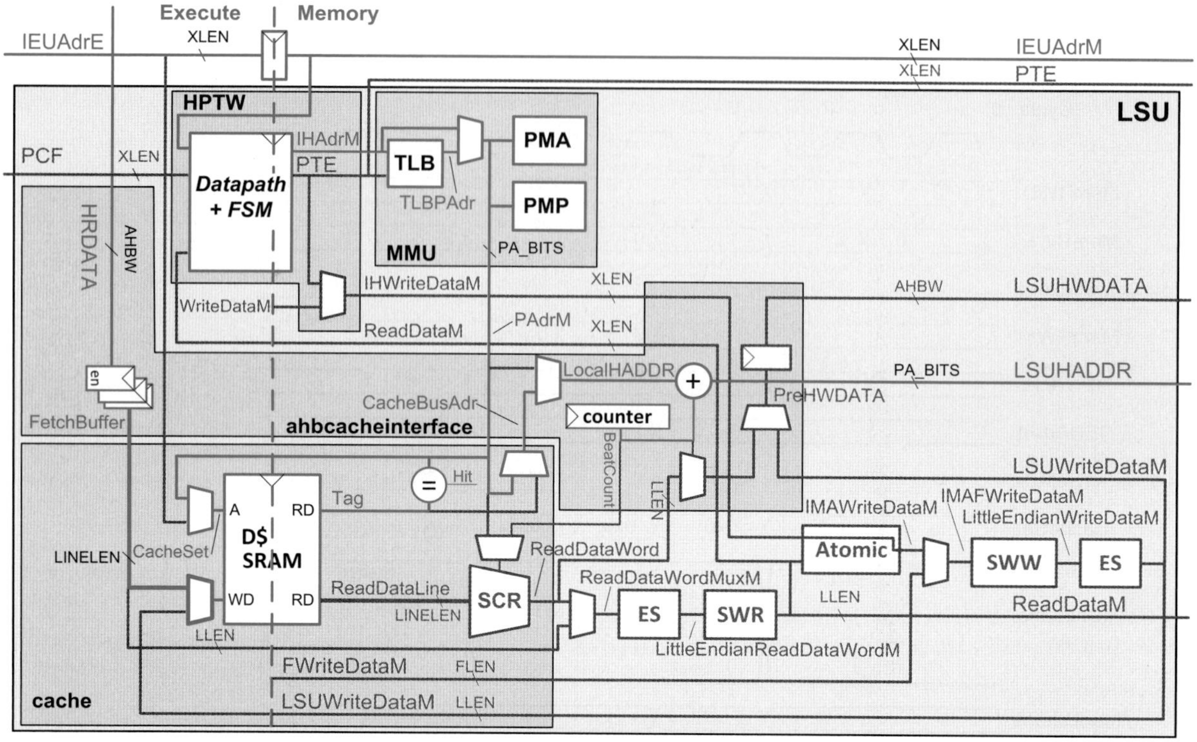

Fig. 12.17 Example cache line fetch.

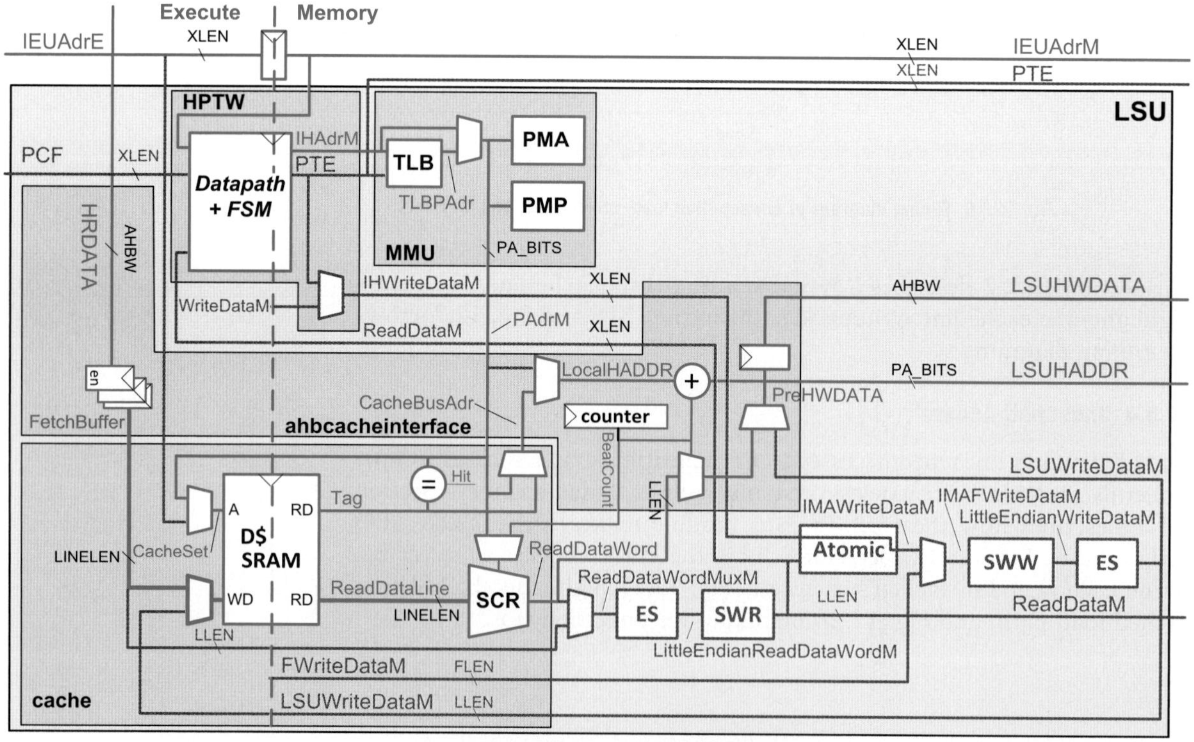

Fig. 12.18 Example cache line writeback.

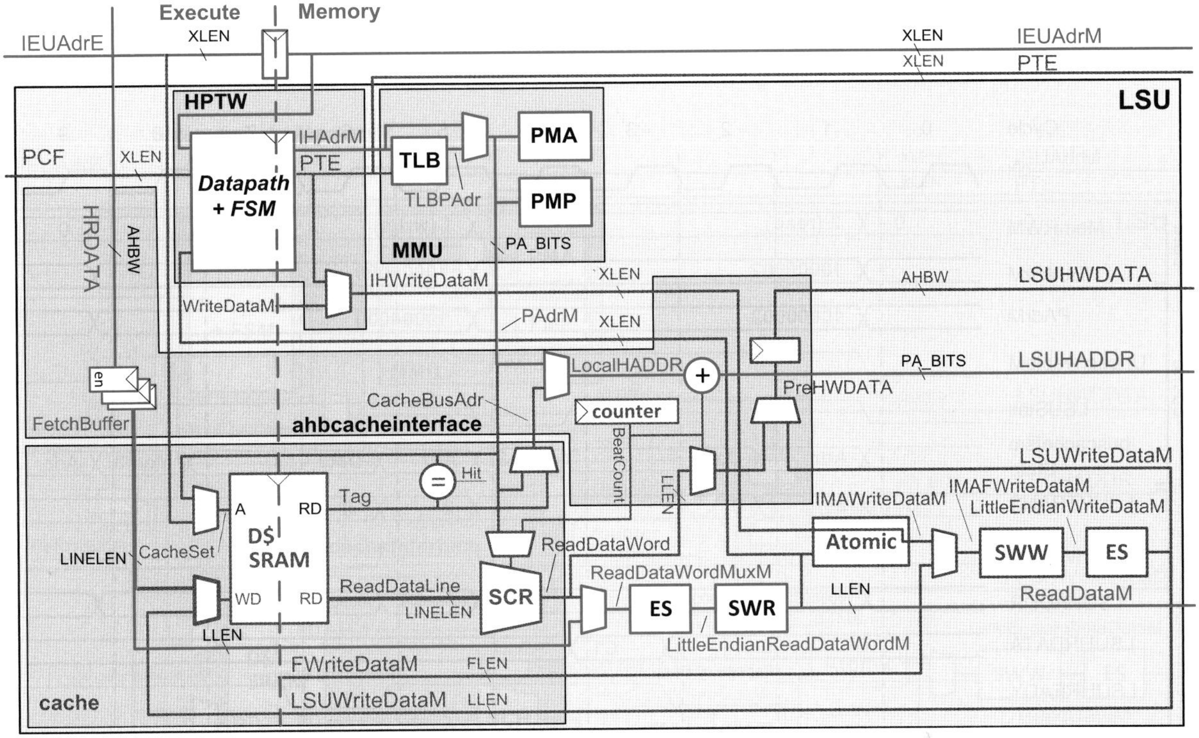

Fig. 12.19 **Example uncached load.**

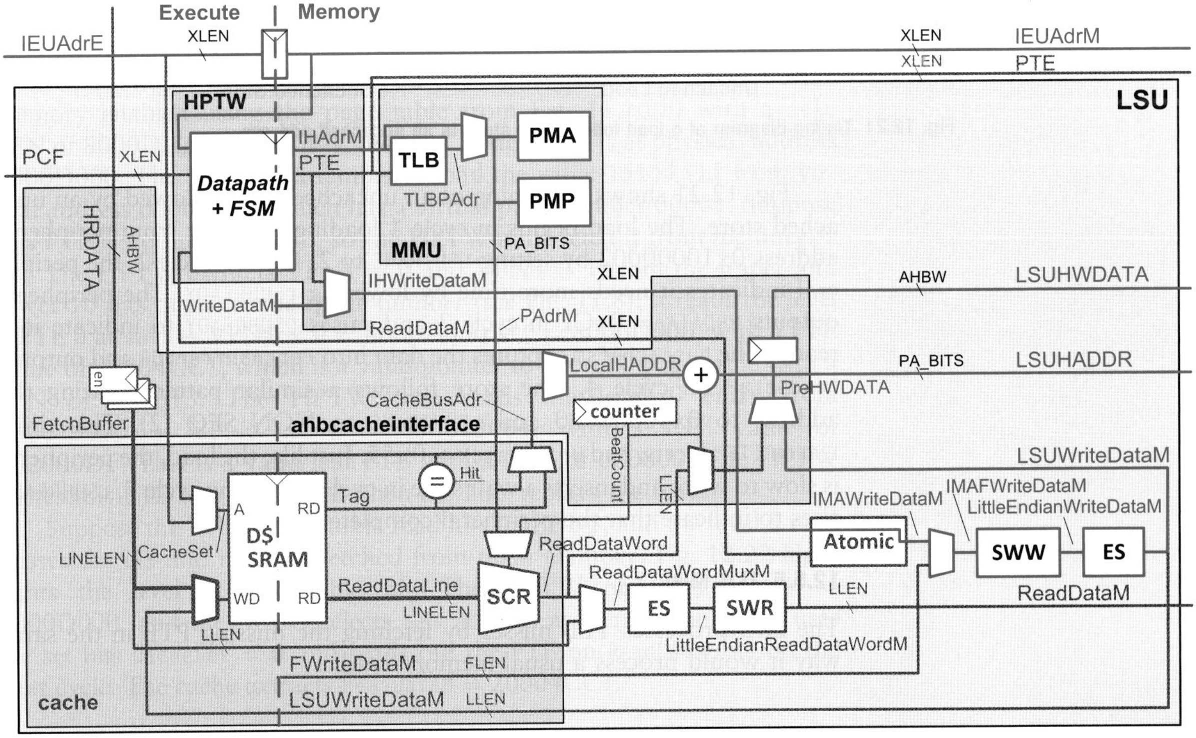

Fig. 12.20 **Example uncached store.**

On cycle 2, the cache detects a hit and sets up `LSUWriteDataM` = 202008D7. On cycle 3, the cache writes this value into its data array. Meanwhile, the HPTW updates the TLB with the new PTE.

12.6 TEST PLAN

The basic features of the LSU are tested with existing tests, including the load/store tests from Section 7.4, the privileged memory access fault tests from Section 8.3, the bus tests from Section 9.4, and the MMU tests from Section 11.3.

A more comprehensive test plan would stress the cache and its interaction with the HPTW.

12.7 SUMMARY

The LSU is a highly configurable module responsible for loads and stores. In the simplest configuration, it only has a DTIM. In the most sophisticated configuration, it interfaces to the EBU for main memory access, has a D$ to service most accesses, contains an MMU with a TLB for virtual to physical address translation, a PMP for physical memory protection, and a PMA for physical memory address checking; it also has an HPTW to service a TLB miss by the HPTW taking over the memory system to walk the page table. All these components have been described in earlier chapters. Some LSU features needed to coordinate among the components include the following:

- The `buscachefsm` to sequence cache line fetches and writebacks
- Multiplexers so that the HPTW can take over the address and write data signals during page table walks
- Stall logic for multicycle LSU operations and data dependencies
- Trap logic for page, access, and misaligned faults
- Disabling interrupts when a multicycle memory operation is in flight

Instruction Fetch Unit

13

13.1 **Branch Prediction Principles**

13.2 **Test Plan**

13.3 **Wally Implementation**

13.4 **Summary**

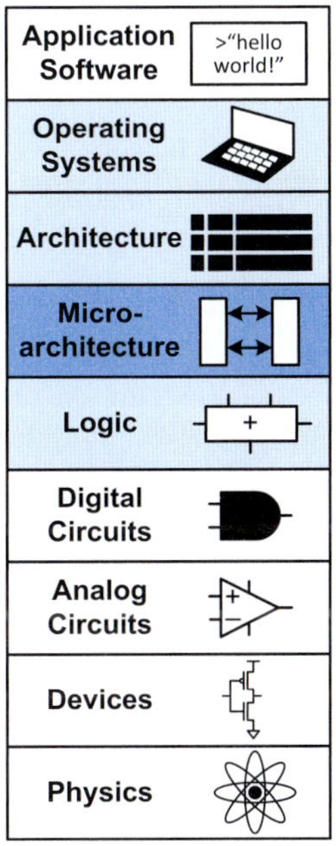

The *instruction fetch unit* (*IFU*) is the *front end* of the processor responsible for delivering instructions to the pipeline. It contains the program counter logic, instruction memory, and IMMU (instruction memory management unit). Chapter 14 adds IFU decompression logic for compressed instructions.

Like the LSU, the IFU also supports several memory options: a cache, tightly integrated memory (TIM), and bus interface to external memory. The memory system is somewhat simpler because the IFU holds instructions, which are read-only. The IMMU also resembles the LSU's MMU but relies on the LSU's hardware page table walker (HPTW). The major distinguishing feature of the IFU is the PC logic, which includes an optional branch predictor to improve performance.

Fig. 13.1 shows the Wally block diagram, with blue highlighting for the new IFU features integrated in this chapter. The PC logic generates the address of the next instruction. The IFU accesses the I$ (instruction cache) or the ITIM (instruction TIM) or uses the external bus to access external memory. Meanwhile, within the IMMU, the ITLB (instruction translation lookaside buffer) translates the PC from virtual to physical and the PMA and PMP checkers detect illegal accesses. ITLB misses are serviced by the LSU's HPTW.

This chapter introduces the principles of branch prediction. It then explores the Wally IFU implementation, including the I$, branch predictor, and HPTW interface.

An ITIM is also called an IROM (instruction read-only memory) because it is read-only. When included in a system, the IROM replaces (or works alongside) the I$ shown in Fig. 13.1.

13.1 BRANCH PREDICTION PRINCIPLES

Recall that the basic 5-stage pipeline from Section 7.2 predicts that branches are not taken and continues fetching subsequent instructions. In the Execute stage, it determines whether the branch should have been taken. If so, it flushes the next two instructions and updates the PC to the correct branch target address (BTA). So, the *misprediction penalty*, the number of cycles lost because of the misprediction, is 2 cycles. The flush increases the CPI (cycles per instruction) to 3 for taken branches. Branches

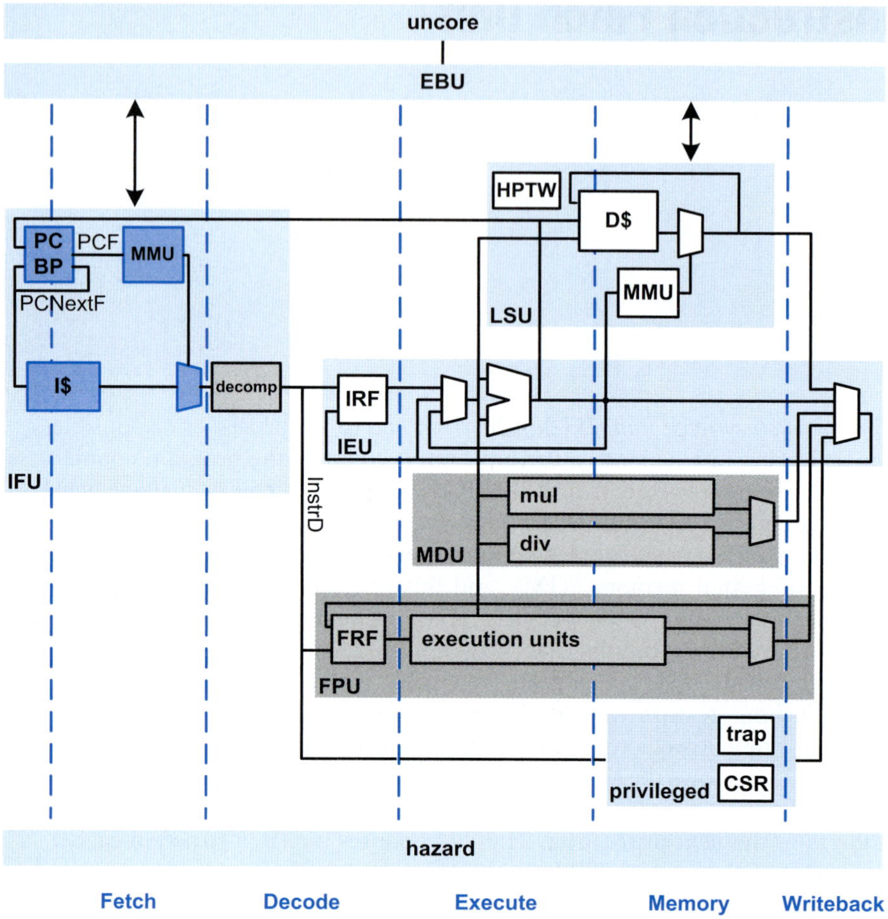

Fig. 13.1 **Wally high-level block diagram emphasizing the IFU.**

occur every 4-7 instructions in typical code, meaning that roughly 20% of executed instructions are branches.

Example 13.1 EFFECTS OF BRANCH MISPREDICTION ON CPI

In the CoreMark benchmark from Section 7.2.2.5, 22.2% of executed instructions are branches, and 41.9% of them are taken. Optimistically, assume that nonbranch instructions have a CPI of 1. What is the average CPI?

Solution: A taken branch requires 3 cycles: 1 to execute and 2 to flush the Fetch and Decode stages. A not taken branch requires 1 cycle.

$\text{CPI} = (1 - 0.168) \times 1 + 0.168 \times [3 \times 0.423 + 1 \times (1 - 0.423)] = 1.14$, which is 14% slower than the ideal CPI of 1.

Next, consider a 2-way superscalar processor (ideal CPI of 0.5), which predicts all branches as not taken and runs the same CoreMark code. A not-taken branch and other instructions require 0.5 cycles. A taken branch takes 3 cycles to execute, and it flushes 5 instructions: 2 from the Fetch stage, 2 from Decode, and 1 from Execute. CPI = $(1 - 0.168) \times 0.5 + 0.168 \times [3 \times 0.423 + 0.5 \times (1 - 0.423)] = 0.68$, which is 36% slower than the ideal CPI of 0.5.

Branch misprediction becomes much more significant in wide superscalar processors with deeper pipelines. For example, the Pentium 4 has a 20-cycle branch misprediction penalty and dispatches up to 4 instructions per cycle, meaning a misprediction could flush up to 80 instructions [Hinton01]. Even a misprediction rate of 1% would have a significant impact on CPI.

Branch prediction is an essential feature for high-performance processors. The simplest form is *static branch prediction*, such as predicting that all backward branches will be taken. *Dynamic branch prediction* improves accuracy by using recent branch history. Simple dynamic branch prediction uses 1 or 2 bits to remember the history of specific branches. *Gshare*, a widely used predictor, incorporates the history of multiple recent branches [McFarling93]. Branch predictors guess whether a branch is taken and work in conjunction with a *branch target buffer* (*BTB*) that holds the BTA. Other control instructions such as function returns and other indirect jumps use special predictors tuned to their behavior. Deeply pipelined superscalar processors devote even more resources to advanced branch predictors to minimize misprediction.

Branch prediction takes place in the Fetch stage to predict which instruction the IFU should fetch next. A dynamic branch predictor performs three main operations: it determines if the instruction is a branch or jump, predicts whether the branch is taken, and keeps track of target addresses for branches and jumps. To do this, the dynamic branch predictor uses:

- A *branch history table* (*BHT*) to track recent branch outcomes and use them to predict whether the branch will be taken.
- A BTB to record recent branches or jumps and their target addresses. The target buffer is essentially a direct-mapped cache indexed by the PC that produces the target address for taken branches and jumps.
- A *return address stack* (*RAS*) to track recent return addresses because function returns are poorly tracked with an ordinary BTB.

Each of these components is discussed further in the remainder of this section.

Although these principles apply to any microarchitecture, this section illustrates branch predictors using a 5-stage pipeline with branch conditions computed in the Execute stage, as is the case with Wally. This section shows abstract diagrams to illustrate the principles, and the detailed Wally diagrams in 13.3 expand on them. Signal and unit names in this section are the same as in Wally for consistency. When discussing branch predictors, all control flow instructions, including jumps, are referred to, generally, as branches for brevity, except when the difference matters.

> Direct jumps, such as branches and `jal`, use the immediate field to specify the target address relative to the PC. Indirect jumps, such as `jalr`, specify the *location* where the target address is held, for example, in a register. Direct jumps compute the same target address each time they are executed, but indirect jumps may not. For example, the `ret` instruction, which returns from a function call and is a pseudoinstruction for `jr ra`, is *indirect* because it specifies the target address in a register. So, the target address could be different for each execution of the same `jr`.

13.1.1 Static Branch Prediction

A static branch predictor relies on information known at compile time. The simplest static branch predictor guesses that a branch is not taken, as in Section 7.2. However, backward branches, such as those used in `for` and `while` loops, are taken about 90% of the time. Forward branches are taken about 50% of the time. A better static branch predictor guesses that backward branches are taken. The ARM 11 core used static branch prediction [ARM11].

13.1.2 Simple Dynamic Branch Prediction

When discussing branch prediction, direction refers to whether a branch is taken or not; it does not refer to the direction of the branch (forward or backward).

Dynamic branch prediction captures past program behavior to guess the *direction* of the current branch, that is, whether it is taken or not. A *1-bit dynamic predictor* tracks whether a branch was taken last time and predicts to do the same next time.

Consider the nested loops in Code Example 13.1. `branch_j` at the end of the `j` loop (address 0x134: `blt s1, t1, forj`) is executed 10 times for each iteration of the `i` loop. That `blt` is taken on the first nine times and not taken on the 10th. A 1-bit predictor mostly guesses taken. On the tenth time, it is wrong, and it now records not taken. However, on the next iteration of the `i` loop, `branch_j` should be taken, but it is predicted not taken. Thus, the branch is mispredicted on the first and last iteration of the `j` loop. In this example, the prediction accuracy is 80% for `branch_j`. Similarly, the `i` loop's first and last branches are also mispredicted, so that loop has a $(5 - 2)/5 = 60\%$ prediction accuracy.

Code Example 13.1 NESTED LOOPS

C Code

```
int i, j;



for (i=0; i<5; i++)
  for (j=0; j<10; j++)
    dostuff
```

Assembly Language

```
# i in s0, j in s1
0x100:        li t0, 5           # Bound on i loop
0x104:        li t1, 10          # Bound on j loop
0x108:        li s0, 0           # i = 0
0x10C: fori:  li s1, 0           # j = 0 (start of i loop body)
0x110: forj:                     # dostuff (start of j loop body)
...
0x130:        addi s1, s1, 1     # j++
0x134:        blt s1, t1, forj   # branch_j: Repeat if j < 10
0x138: donej: addi s0, s0, 1     # i++
0x13C:        blt s0, t0, fori   # branch_i: Repeat if i < 5
0x140: donei:
```

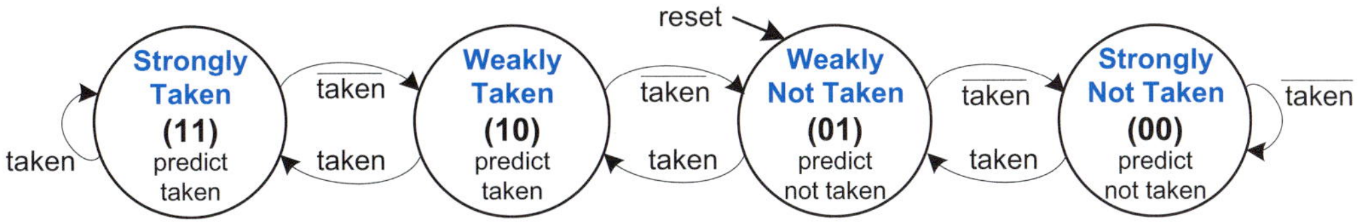

Fig. 13.2 **2-bit branch predictor state transition diagram.**

The *2-bit dynamic branch predictor* (or *2-bit predictor*) in Fig. 13.2 improves the accuracy by remembering more history and only changing the prediction after multiple incorrect guesses. The states are encoded such that the most significant bit indicates the prediction (1 for taken, 0 for not taken). During the `j` loop, the predictor for `branch_j` enters the Strongly Taken state, with an output of 11. When it mispredicts on the last time through the `j` loop, it moves to the Weakly Taken state (10). The next time it starts the `j` loop, it still predicts `branch_j` as taken. Thus, `branch_j` is only mispredicted the last time through the `j` loop. In this example, the prediction accuracy is 90% for `branch_j`.

The 2-bit branch predictor is also called a *bimodal* predictor.

The FSM's state bits are referred to as a *saturating counter*, with the counter saturating at 11 when a given branch is Strongly Taken and 00 when it is Strongly Not Taken.

The branch predictor stores the history information in a BHT (branch history table). The table has a finite number of entries and is typically indexed by the less significant *b* bits of the PC, typically excluding the two lsbs, which are 0 if all instructions are 32 bits. Fig. 13.3 shows a 2^b-entry BHT built from a dual-port synchronous RAM with one read and one write port for reading the predicted direction in the Fetch stage and writing the updated prediction in the Execute stage. Each 2-bit entry indicates the predictor state for any branch with the corresponding less significant address bits. Just before the Fetch stage, `PCNextF` indexes the BHT so that it outputs the predicted direction `BPDirF` just after the next clock edge, at the beginning of the Fetch stage. The predicted direction is pipelined to the Execute stage, where the true branch outcome is actually calculated. An XOR gate compares the predicted and actual direction, `PCSrcE`, which is 1 when the branch is taken. If the prediction is wrong, `BPDirWrongE` = 1, and the PC logic must flush the previous two instructions from the pipeline and fetch the correct instruction, as will be described in Section 13.1.4.

The BHT behaves similarly to a direct-mapped cache with no tags, which means conflicts could exist between branches at different program addresses. These conflicts reduce prediction accuracy but don't cause program failure because the consequence of a mispredicted branch is just to flush the incorrectly fetched instructions.

Throughout the branch predictor, predicted signals are prefixed with `BP`.

The branch predictor updates the BHT with the new predicted direction (`NewBPDirE`) in the Execute stage, according to Fig. 13.2, after the IEU determines the true branch direction. Even though the branch predictor of Fig. 13.2 is drawn as an FSM, only its state is stored in the BHT. The 2-bit saturating counter logic updates the BHT state based on the current state (`BPDirE`) and the direction of the branch (`PCSrcE`).

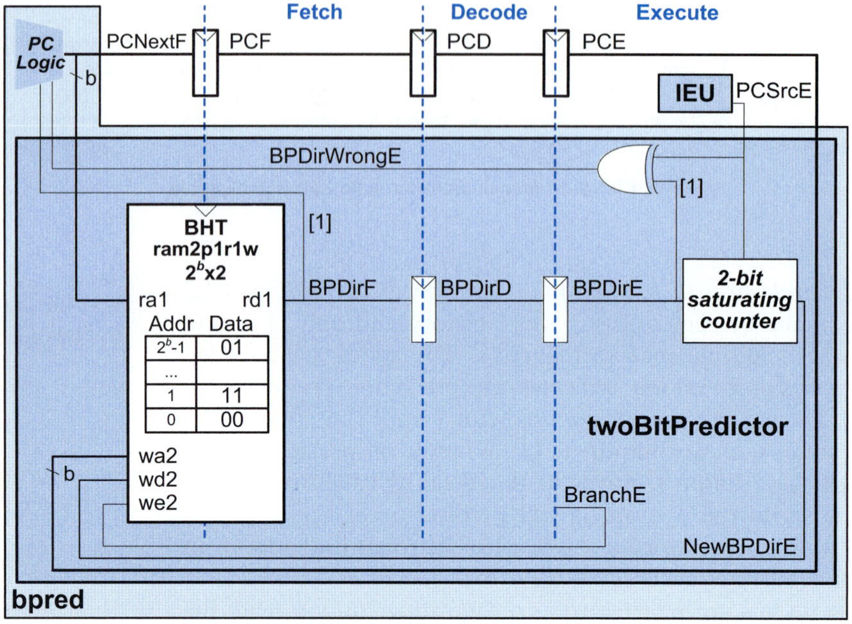

Fig. 13.3 BHT for 2-bit branch predictor.

When the prediction is wrong, `BPDirWrongE` = 1 triggers the hazard unit to flush the Fetch and Decode stages. The PC logic updates `PCNextF` to the correct PC, as described in Section 13.3.3.5.

Example 13.2 BHT ENTRIES

Suppose Code Example 13.1 is run on the 2-bit branch predictor from Fig. 13.3 with $b = 4$. The lower two bits of the PC are always 0, so PC[5:2] are used to index the BHT. At reset, the BHT entries are initialized to 01 (Weakly Not Taken). What are the values in the BHT after the program completes?

Solution: This first branch is at address 0x134 = 0b0001001**1101**00. Bits 5:2 of this address are **1101**, so the branch maps to BHT entry 13. The next branch is at address 0x13C (0b0001001**1111**00) and maps to BHT entry 15. The other BHT entries are unused. Each branch is repeatedly taken, moving to the Strongly Taken state. However, on the last iteration of each loop, the branch is not taken, moving back to the Weakly Taken (10) state. Thus, entries 13 and 15 both contain 10, and the other entries remain at their reset value 01.

Fig. 13.4 shows the 2-bit branch predictor misprediction rate on SPEC89 benchmarks for a 4,096-entry BHT. The accuracy (misprediction rate) is highly dependent on the application, ranging from 82% to 99%.

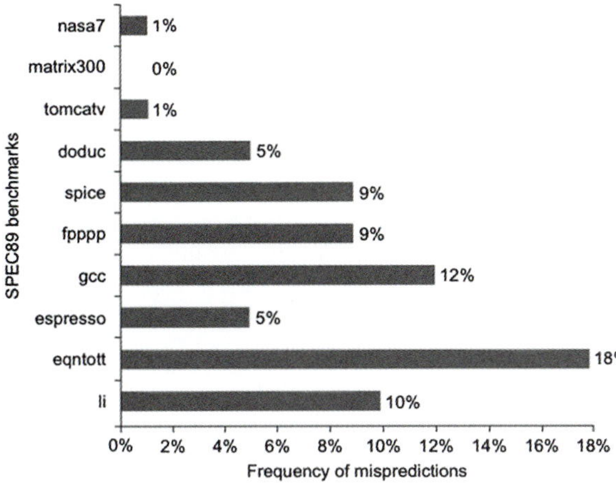

Fig. 13.4 2-bit branch predictor accuracy with 4,096 entries. (Figure courtesy of Elsevier [Hennessy19]. Reprinted with permission.)

Some complex programs, such as gcc, have relatively low accuracy, but scientific computing programs with highly repetitive loops (e.g., `nasa7`, `matrix300`, and `tomcatv`) are largely predictable and, thus, have higher accuracy. Increasing the BHT size to infinity improves results by less than 1%, so 4,096 entries is sufficient for many programs.

> A 4,096 × 2-bit SRAM has an extreme aspect ratio, making it impractical to build on a chip. It is normally folded into a more square-like shape. For example, a 32x fold gives a 128 × 64-bit SRAM, with the lower bits of the original address used as mux selects and bit write enables to read or write 2 of the 64 bits.

13.1.3 Gshare

The 2-bit predictor considers each branch independently. Careful program analysis shows that some branch outcomes are correlated with others. To improve accuracy, a *global history dynamic branch predictor* considers the history of the last several branches. The widely used *Gshare* predictor incorporates global history to achieve prediction accuracy greater than 96% with 4,096 entries [McFarling93].

Code Example 13.2 shows a snippet of the `eqntott` SPEC89 benchmark with a heavily used and difficult-to-predict branch. If `branch_aa` and `branch_bb` are both not taken (and hence make `aa` and `bb` = 0), then `branch_aabb` will be taken. Hence, `branch_aabb` partially depends on the outcome of previous branches. By separating the prediction of `branch_aabb` into four scenarios that depend on the outcomes of the previous two branches, the predictor can reduce the interference between the different scenarios and improve the prediction accuracy [Yeh91, Pan92].

Code Example 13.2 eqntott CORRELATED BRANCHES

C Code	Assembly Language

if (aa == 2)
  aa = 0;
if (bb == 2)
  bb = 0;
if (aa != bb) {
  dostuff
}
	# aa in s0, bb in s1
	0x1000: li t0, 2 # 2 for comparison
	0x1004: bne s0, t0, skipaa # branch_aa: Skip if aa != 2
	0x1008: li s0, 0 # aa = 0
	0x100C skipaa: bne s1, t0, skipbb # branch_bb: Skip if bb != 2
	0x1010: li s1, 0 # bb = 0
	0x1014 skipbb: beq s0, s1, skipaabb # branch_aabb: Skip if aa != bb
	0x1018: # dostuff
	0x101C skipaabb:

Fig. 13.5 shows a block diagram of a Gshare branch predictor. Like a 2-bit dynamic predictor, Gshare uses a history table, a dual-port synchronous SRAM with 2-bit entries that indicate the prediction direction (whether it is taken or not) and the prediction strength as given in Fig. 13.2. Gshare calls this a *pattern history table* (*PHT*) instead of a BHT because it reflects patterns of branch history.

Gshare and global history predictors are often referred to as *correlating predictors* because they consider how branch outcomes are interrelated with other branches. They are also frequently called *2-level adaptive predictors* because they record branch history using two levels: the GHR and PHT.

Gshare adds a *k*-bit *global history shift register* (*GHR*) that tracks the outcome of the last *k* branches. Instead of using the PC directly to index the PHT, Gshare indexes it using a *hash*, which is `PCNextF ⊕ GHR`. Therefore, sequences of branches with different histories each map to different locations and are separately predicted. Empirically, prediction accuracy is best when the number of table index bits *b* (extracted from the PC) equals the GHR length *k*.

Gshare updates the GHR and PHT after each branch outcome is known in the Execute stage. A given branch affects many entries of the PHT based on the GHR, so Gshare needs time to train all these entries before producing accurate predictions.

Like the 2-bit predictor, Gshare flushes the pipeline on misprediction and updates `PCNextF`. The PC logic is described in Sections 13.1.4, 13.3.2, and 13.3.3.5.

Fig. 13.6 shows the improved branch prediction accuracy of Gshare over the simple 2-bit saturating counter predictor for Embench. Gshare can reasonably achieve branch prediction accuracy approaching 97% after sufficient warmup. Gshare generally outperforms 2-bit branch prediction because it captures the correlation among branches. However, for tiny history tables, 2-bit outperforms Gshare because it has fewer conflicts. Fig. 13.6 also shows theoretical accuracy from a branch predictor simulator (CModel) described in Section 21.3.2.1. Wally matches the theoretical results well, with discrepancies caused by the pipeline delay between making predictions and updating the tables.

Several processors use Gshare such as the Shakti RISC-V core (configurable size) and the Sun UltraSPARC-III (with 16K entries [Horel99]).

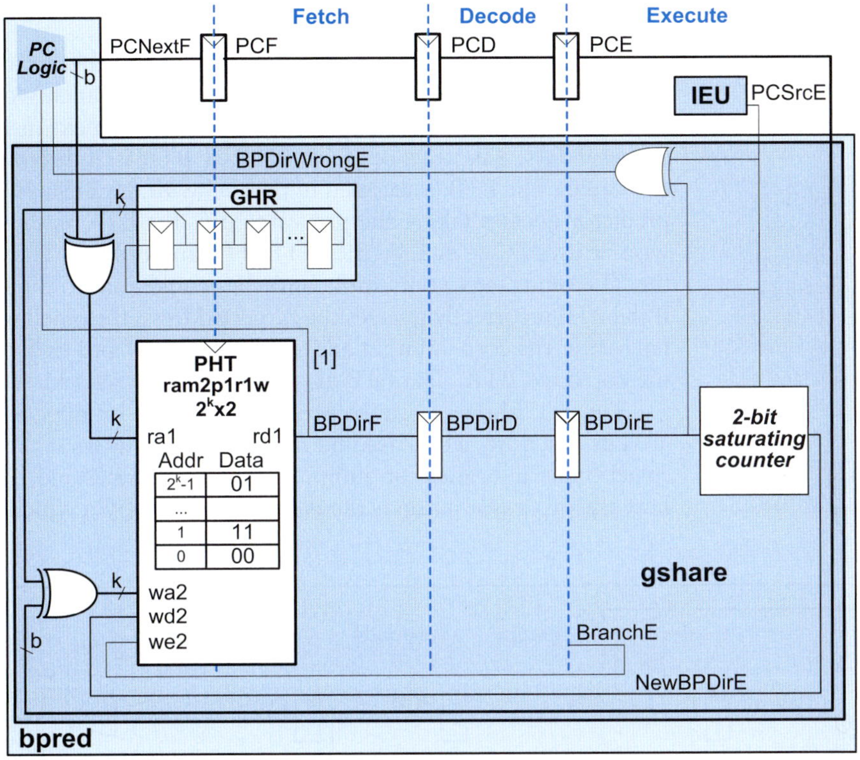

Fig. 13.5 Gshare dynamic branch predictor.

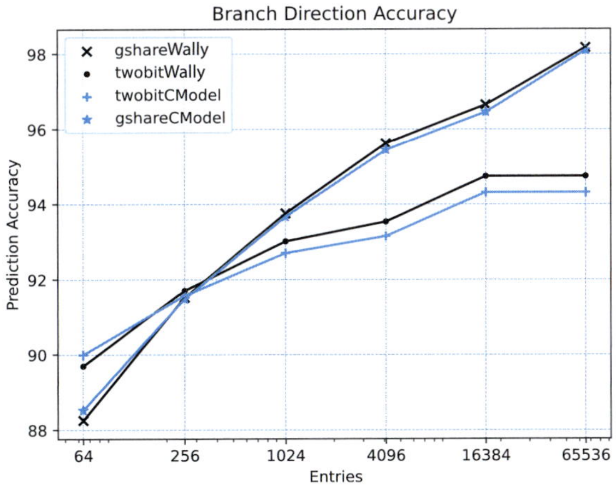

Fig. 13.6 Branch prediction accuracy for Embench, varying BHT and PHT size.

13.1.4 Branch Target Buffer

The branch predictor must guess not only whether a branch is taken but also the branch target address (BTA) where the branch would go if taken. Both guesses should be ready in time to fetch the next instruction. A BTB caches BTAs. Like the BHT or PHT, a BTB is indexed with the lower bits of PC. The BTB is accessed in parallel with the BHT/PHT. If the branch predictor guesses taken, the processor uses the BTA as the next PC; otherwise, it uses PC + 4. Like the BHT/PHT and unlike a conventional cache, the BTB omits tags, so multiple branches could alias to the same BTB entry. If the BTB incorrectly guesses the target address, the pipeline flushes the bad fetch after the correct target address is computed and updates the BTB with the corrected BTA. The BTB also caches the target address for jumps.

Because multiple instruction alias to the same BTB entry, the BTB may mispredict both the target address and instruction class. The data shows that this usually has a small effect on BTB accuracy, although certain benchmarks such as Embench's `nsichneu` experience notable performance loss because of this aliasing.

Fig. 13.7 shows a branch predictor with a 2-bit direction predictor and a BTB. For now, let us assume the hardware knows if the Fetch stage instruction is a branch or jump. Sections 13.3.3.2 and 13.3.3.3 explain how the processor predicts the *instruction class* (*IC*), which tells the branch

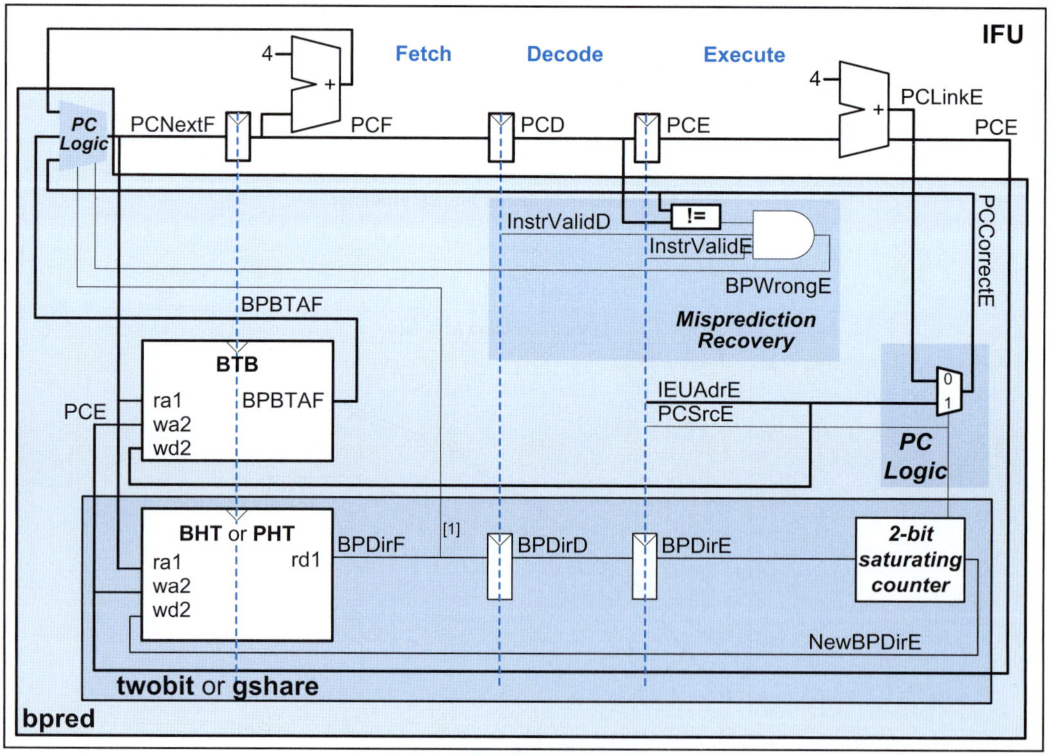

Fig. 13.7 Branch predictor with BTB.

predictor if the instruction is a branch, jump, function return, or jump and link. If the instruction is a branch and the direction predictor predicts taken or if the instruction is a jump, then the PC logic selects `PCNextF` = `BPBTAF` (branch predictor BTA from the BTB). However, if a branch is predicted not taken, then the PC logic selects the fall-through address, `PCNextF` = `PCF` + 4. Like the BTH/PHT, the BTB's memory array is synchronous SRAM so `PCNextF` sets up the read address before the rising clock edge at the start of the Fetch stage, and `BPBTAF` is valid at the beginning of the Fetch stage.

If the Execute stage determines the next PC was mispredicted, it asserts `BPWrongE` and PC logic selects `PCNextF` = `PCCorrectE`, the PC of the correct next instruction after the branch. If the Execute stage instruction is a taken branch or a jump, then `PCCorrectE` is the BTA, `IEUAdrE`. Otherwise, `PCCorrectE` is the fall through address `PCE` + 4. Unlike the branch direction predictors of Fig. 13.3 and Fig. 13.5, which checks if the branch predictor's direction matches, now the predictor needs to check if the PC of the next instruction (`PCD`) matches what the `IEU` computed (`PCCorrectE`). If they mismatch and both instructions are valid (`InstrValidD` = `InstrValidE` = 1), the branch predictor fetched the wrong instruction and must assert `BPWrongE` to flush the Decode and Execute stages.

Fig. 13.8 shows the BTB accuracy versus the number of BTB entries for Embench. Again, Wally's implementation matches the theoretical CModel closely. The working sets of Embench programs are typically small and comfortably fit inside both the I$ and branch predictor. Even with just 256 entries, the BTB hits 99% of all jump and branch instructions after

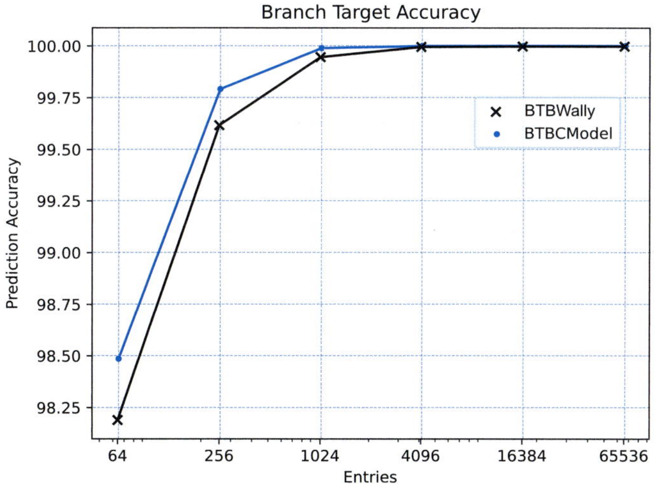

Fig. 13.8 BTB accuracy for Embench vs. number of entries.

a sufficient warmup time. All RISC-V branches and most jumps have *direct* branch/jump targets because they compute their target address relative to PC. A direct target never changes; therefore, the BTB readily achieves near 100% accuracy. The next section describes the most common type of *indirect* jumps, function returns.

Section 2.9 lists BHT and BTB sizes for several RISC-V processors. BTB sizes range from 4 to 256 entries in low-end processors. The smaller BTBs are typically associative and don't contain instruction class prediction.

13.1.5 Return Address Stack

The BTB works well for branches and `jal` because the destination is part of the instruction. However, function returns (`ret`) are difficult to predict because they have different return addresses depending on where the function was called. Because function calls occur in pairs with returns, a *return address stack* (*RAS*) reliably predicts the return address. A RAS is a small hardware stack that pushes the return address when calling a function and pops the address when returning.

Fig. 13.9 shows the block diagram of a RAS, which consists of a small memory and a pointer (`Ptr`) to the top of the stack. When the processor executes a function call, `jal ra, <dest>`, it computes the return address (`PCLinkE` = `PCE` + 4). It *pushes* this return address onto the stack by incrementing `Ptr` and then writing the return address to the RAS memory at `Ptr`. When the processor fetches a return instruction, it *pops* the return address off the stack by reading the RAS memory at `Ptr` and decrementing `Ptr`.

A RAS typically has 3 to 16 entries because the majority of programs have a limited call depth. However, when the RAS overflows, `Ptr` wraps

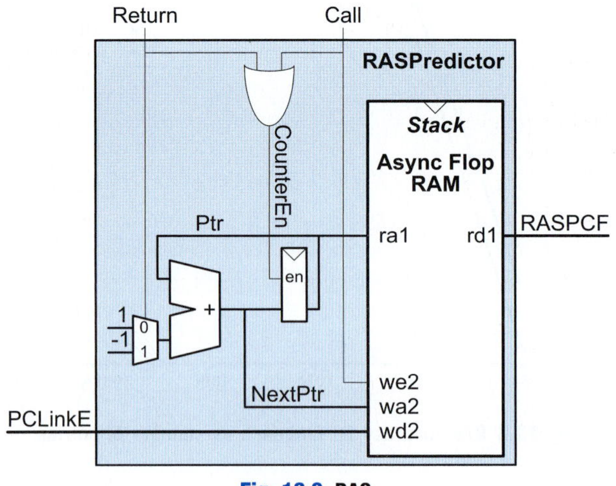

Fig. 13.9 RAS.

around, overwriting the oldest valid return address. In practice, this has a small effect on prediction accuracy because the recent pushes are still valid.

Fig. 13.10 shows a sequence of two function calls and returns using a RAS with 4 entries. In Fig. 13.10(a), `Ptr` is initialized to the bottom entry. In Fig. 13.10(b), a `jal` at 0x105c pushes its return address, 0x1060. In Fig. 13.10(c), the next `jal` also pushes its return address, 0x1144. In Fig. 13.10(d), a `ret` pops the stack and predicts `PCNextF` = 0x1144. In Fig. 13.10(e), the last `ret` pops `PCNextF` = 0x1060.

Fig. 13.11 integrates the RAS into the branch prediction hardware from Fig. 13.7. If the Fetch stage instruction is `ret`, the RAS pops the return address, and the branch predictor selects `BPPCF` = `RASPCF`. The IFU computes and pushes the return address `PCLinkE` = `PCE` + 4 onto the RAS during the Execute stage.

`jal` and `ret` instructions are usually paired together; however, buggy programs could execute `ret` without a preceding `jal`. Without a `jal` the RAS pops the stack as usual but will contain an invalid return address. The misprediction recovery logic of Fig. 13.11 detects the incorrect address and flushes the pipeline just like a branch misprediction. Additionally, program code could change the return address in `ra`, but the RAS won't be updated; this would cause `ret` to jump to (fetch) the wrong instruction. Again, the recovery logic detects the mistake and flushes the pipeline. In general, the recovery logic handles all control flow mispredictions regardless of the source of the misprediction: BTB, RAS, and so on. If the calculated BTA (`PCCorrectE`) matches what was fetched (`PCD`), then the branch predictor fetched the correct instruction.

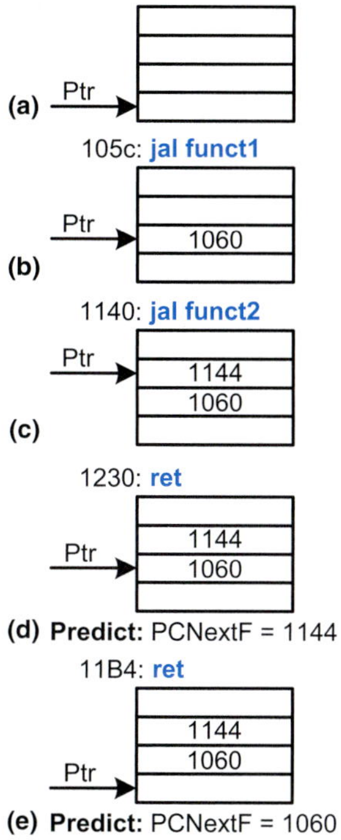

Fig. 13.10 RAS example.

13.1.6 TAGE

High-performance processors need even better prediction accuracy than Gshare can provide. A *tournament branch predictor* includes multiple subpredictors and chooses among them based on which proves most accurate for a given branch.

The *partially TAgged GEometric global history* tournament branch predictor (*TAGE*) achieves better than 99% accuracy with sufficiently large memories [Seznec06, Hennessy19]. TAGE may be too complex to complete in one cycle. Therefore, it is sometimes used as a second-level predictor to correct a faster but less accurate first-level predictor. For example, recent AMD Ryzen CPUs use TAGE as a second-level predictor [Suggs20].

TAGE uses multiple Gshare-like subpredictors, each with a different-length GHR. The lengths are typically 0, 2, 4, 8, 16, and so on, forming a geometric sequence. Shorter subpredictors warm up faster and avoid irrelevant branch history, while the longer subpredictors capture the behavior of complex branch sequences.

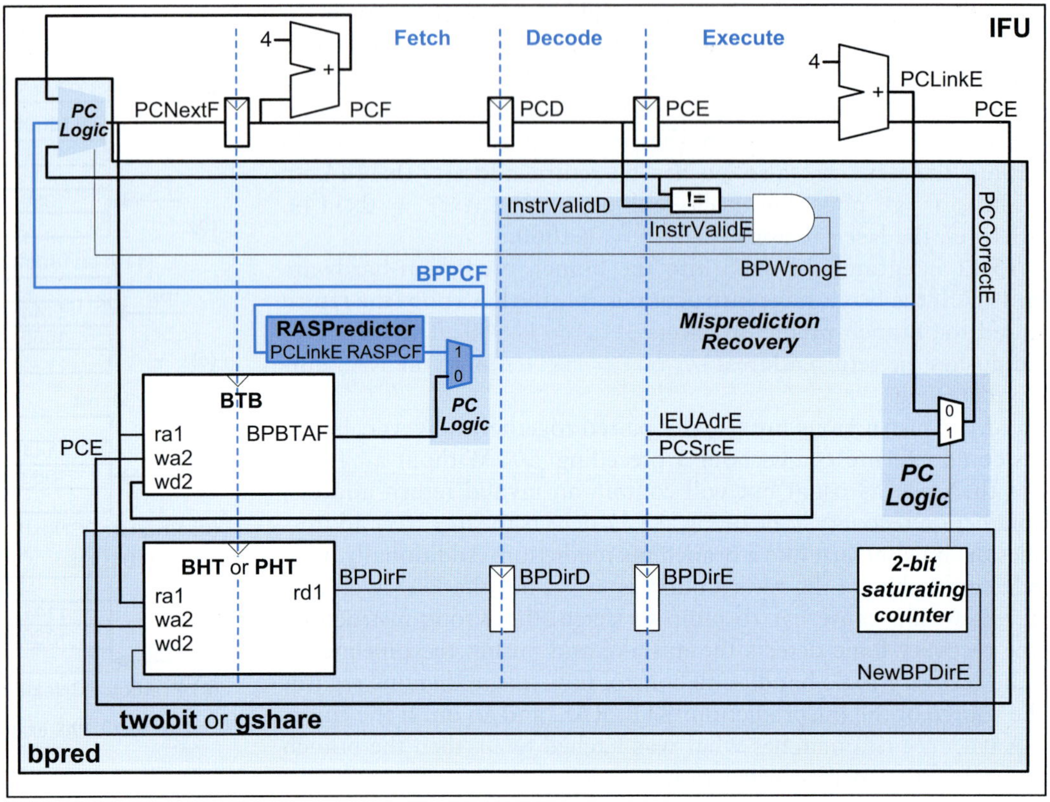

Fig. 13.11 Branch predictor with 2-bit direction predictor, BTB, and RAS.

13.2 TEST PLAN

The IFU test plan stresses the branch predictor and instruction cache to verify both correctness and performance.

13.2.1 Correctness

Wally presently uses the riscv-arch-test regression tests to verify instruction operation, including branches and jumps. When the IFU and LSU both include caches, the Zifence riscv-arch-test checks the correctness of `fence.i`, which synchronizes LSU updates to instruction memory; the test does this by first storing a new instruction to instruction memory, which will be held in the data cache, then preforming a `fence.i`, and finally fetching the new instruction.

13.2.2 Performance

The CoreMark and Embench benchmarks are small enough to fit entirely in the cache, so they do not meaningfully stress the memory system. Section 21.3.1 describes instruction cache performance verification.

Section 21.3.2 describes branch predictor performance verification. Figs. 13.6 and 13.8 show good agreement between Wally and a theoretical branch predictor.

13.3 WALLY IMPLEMENTATION

Fig. 13.12 shows the fully featured (rv64gc) Wally IFU with the PC logic and optional I$, IMMU, AHB cache interface (`ahbcacheinterface`), and branch predictor (`bpred`). The grayed-out spill logic for compressed instructions is discussed in Chapter 14. The IFU and LSU serve similar functions and thus have similar structure. Compared to the LSU in Fig. 12.2, the IFU drops the subword and endian swapping logic because instructions are always 32 bits and little-endian. The read-only I$ omits write circuity. The IFU adds PC logic, including an optional branch predictor. The IMMU relies on the LSU's HPTW to service ITLB misses. Exceptions generated during the Fetch stage are pipelined to generate `TrapM` during the Memory stage.

The IFU fetches an instruction every cycle unless `StallF` is asserted. The PMA checker decodes whether the instruction comes from cacheable memory or the IROM inside the IFU. Table A.1 summarizes typical IFU memory configurations.

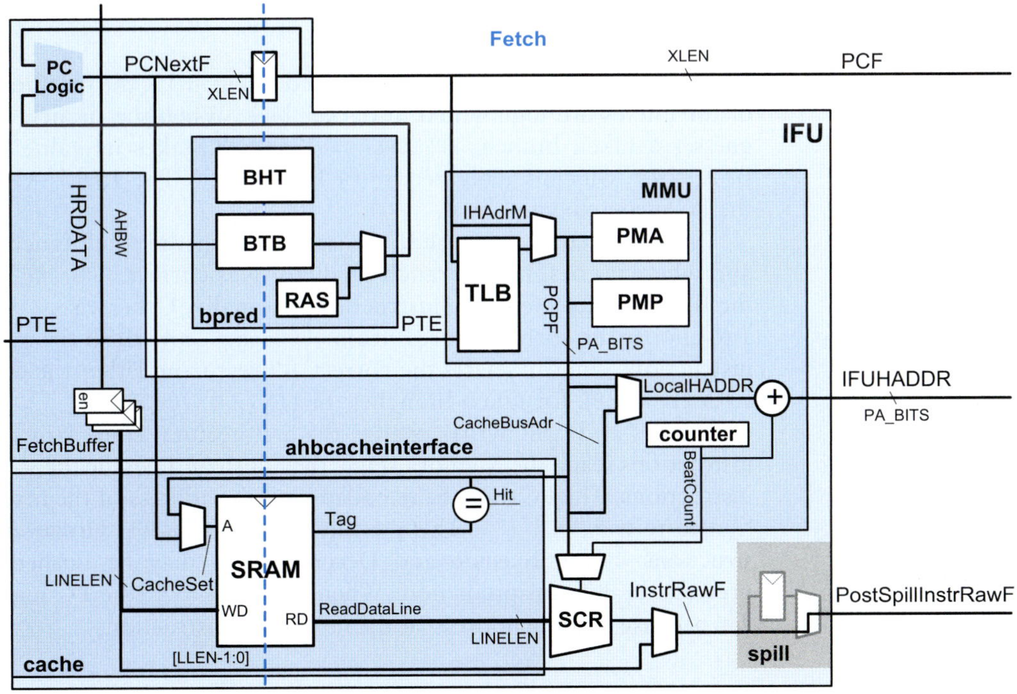

Fig. 13.12 IFU block diagram.

- **Embedded configuration (rv32e):** external memory only
- **Simple CPU configurations (rv{32/64}i):** IROM only
- **Microcontroller configuration (rv32imc):** IROM and external memory
- **Application processor (rv{32/64}gc):** I$ and external memory

13.3.1 Instruction Cache

Chapter 10 described the generic level 1 cache used by both the LSU and IFU. The IFU has a cache when `ICACHE_SUPPORTED` = 1, and it instantiates the `cache` module with the parameter `READ_ONLY_CACHE` = 1. Fig. 13.13 shows a block diagram of the I$. Compared to the generic cache of Fig. 10.24, it omits the write logic, dirty bits, and writeback and flush operations. Eliminating the writeback and flush logic also simplifies the bus and address selection logic. The cache FSM (`cachefsm`) also eliminates the Writeback, Flush, and FlushWriteback states.

13.3.1.1 Invalidation

`fence.i` was introduced in Section 10.4.2.6.

During a `fence.i` instruction, the IFU invalidates the I$ to synchronize the instruction and data caches. To do so, it asserts `InvalidateICacheM`, which drives `InvalidateCache` in the I$ to clear the valid bits in all cache lines.

13.3.2 PC Logic

Fig. 13.14 shows the PC logic. Compared to Fig. 8.17, the new branch predictor muxes are highlighted in darker blue. When the branch predictor guesses a taken branch, `BPPCSrcF` = 1 and PC takes its value from the BTA `BPPCF`, which is generated by the BTB or RAS (Fig. 13.15). Otherwise, the PC is incremented by 4.

In the Execute stage, the IEU determines whether the branch is taken and where it goes. If the branch predictor was correct, `BPWrongE` = 0 and the IFU continues fetching instructions normally. Otherwise, `BPWrongE` = 1 to choose the correct address from `PCCorrectE` and flush the misfetched instructions. `PCSrcE` selects the correct address from either the calculated BTA (`IEUAdrE`), for taken branches, or `PCE` + 4 (`PCLinkE`).

Fences and CSR writes can change the behavior of subsequent instructions, so any in-flight instructions must be restarted. For example, when `satp` or the page table changes and `sfence.vma` occurs, the subsequent instructions must be reissued using the new addressing.

Fences and CSR writes change the architectural state, so when these instructions reach the Memory stage, they flush and reissue the subsequent instructions. The PC must be reloaded with the address of the first instruction to reissue, `PCM` + 4. This next PC is normally `PCE`. However, the instructions in the Execute and Decode stages may be flushed from a previous branch misprediction. When `BPWrongM` = 1, the PC logic selects `PCF` because it is the address of the next valid instruction.

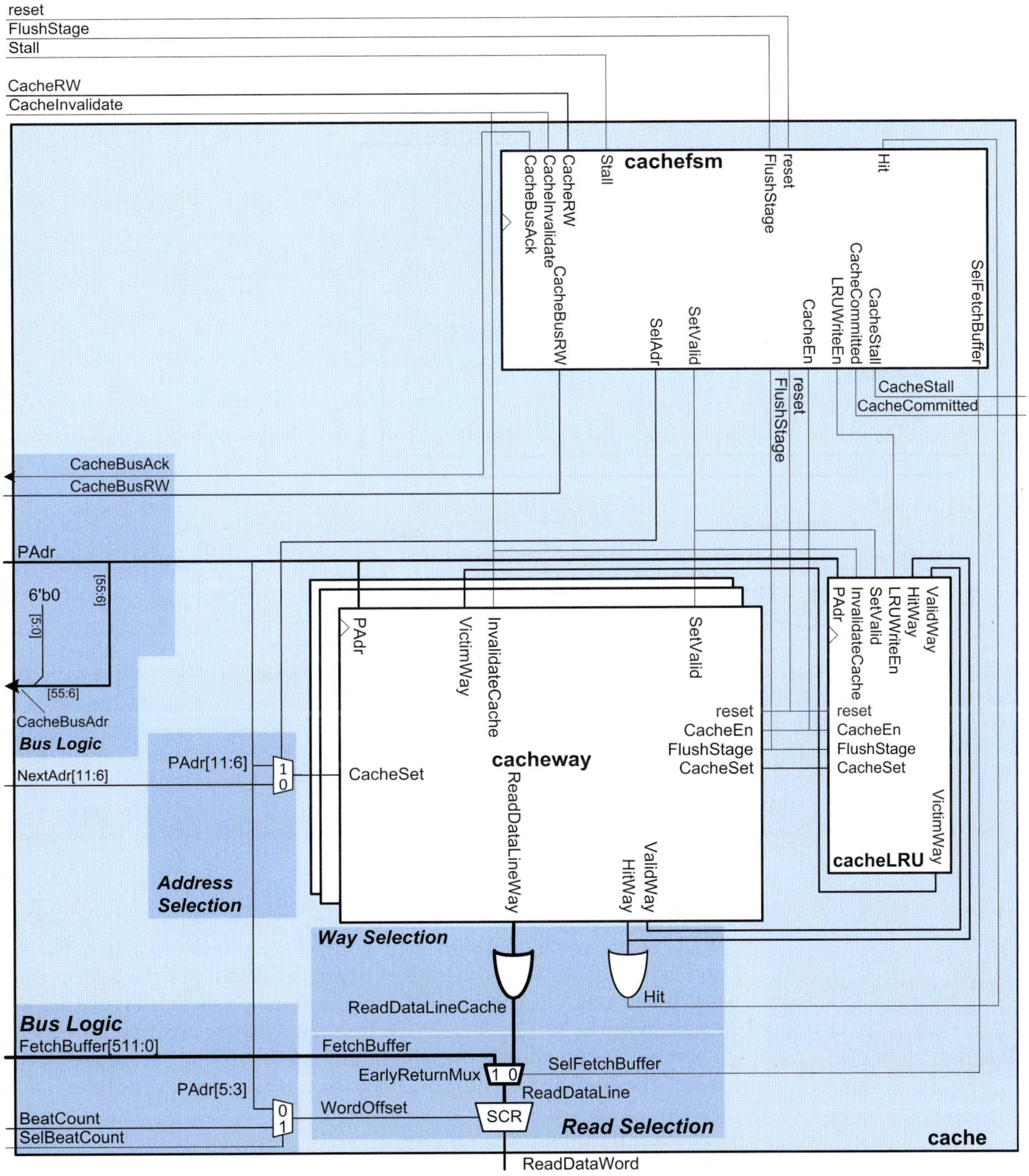

Fig. 13.13 I$ block diagram.

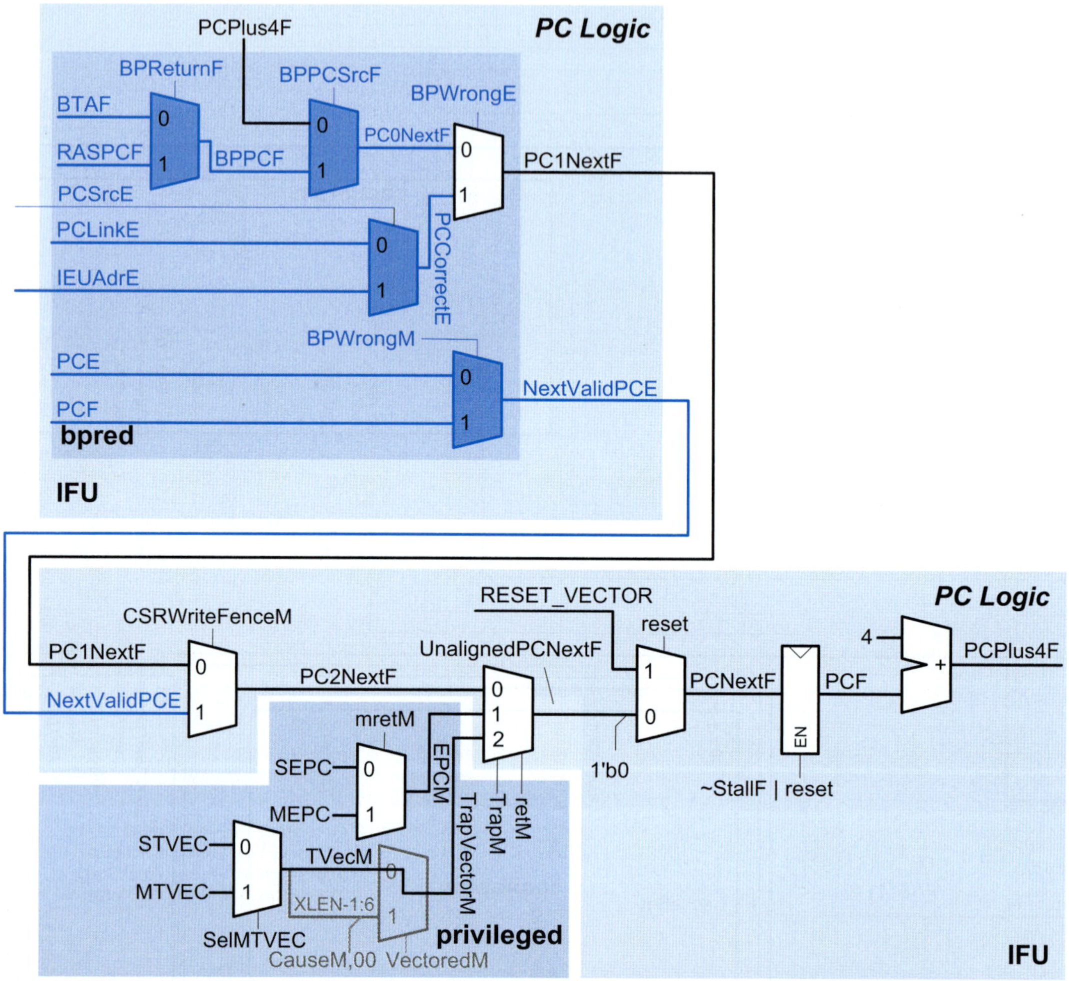

Fig. 13.14 PC logic with branch predictor.

13.3.3 Branch Predictor

At the time of writing, Wally supports additional branch predictors not discussed in the text. For more information see [Yeh91].

Wally can be configured with static or dynamic branch prediction. For dynamic prediction, Wally supports Gshare, by default, or the 2-bit predictor. When BPRED_SUPPORTED = 0, it uses a static predictor that guesses *not taken*. When BPRED_SUPPORTED = 1, it uses a dynamic predictor bpred with a branch direction predictor (2-bit saturating counter or Gshare), BTB, and RAS. The branch direction predictor is configured with the BPRED_TYPE parameter. By default, bpred uses Gshare (BPRED_TYPE = `BP_GSHARE) but can select the 2-bit predictor (BPRED_TYPE = `BP_TWOBIT). The branch direction predictor has $2^{\text{BPRED\_SIZE}}$ entries, and the BTB has $2^{\text{BTB\_SIZE}}$ entries. This section also expands the branch predictor to guess the instruction class (IC), type of instruction, being fetched.

Fig. 13.15 shows a top-level diagram of bpred. It is similar to Fig. 13.11 but includes more details of the top level while abstracting

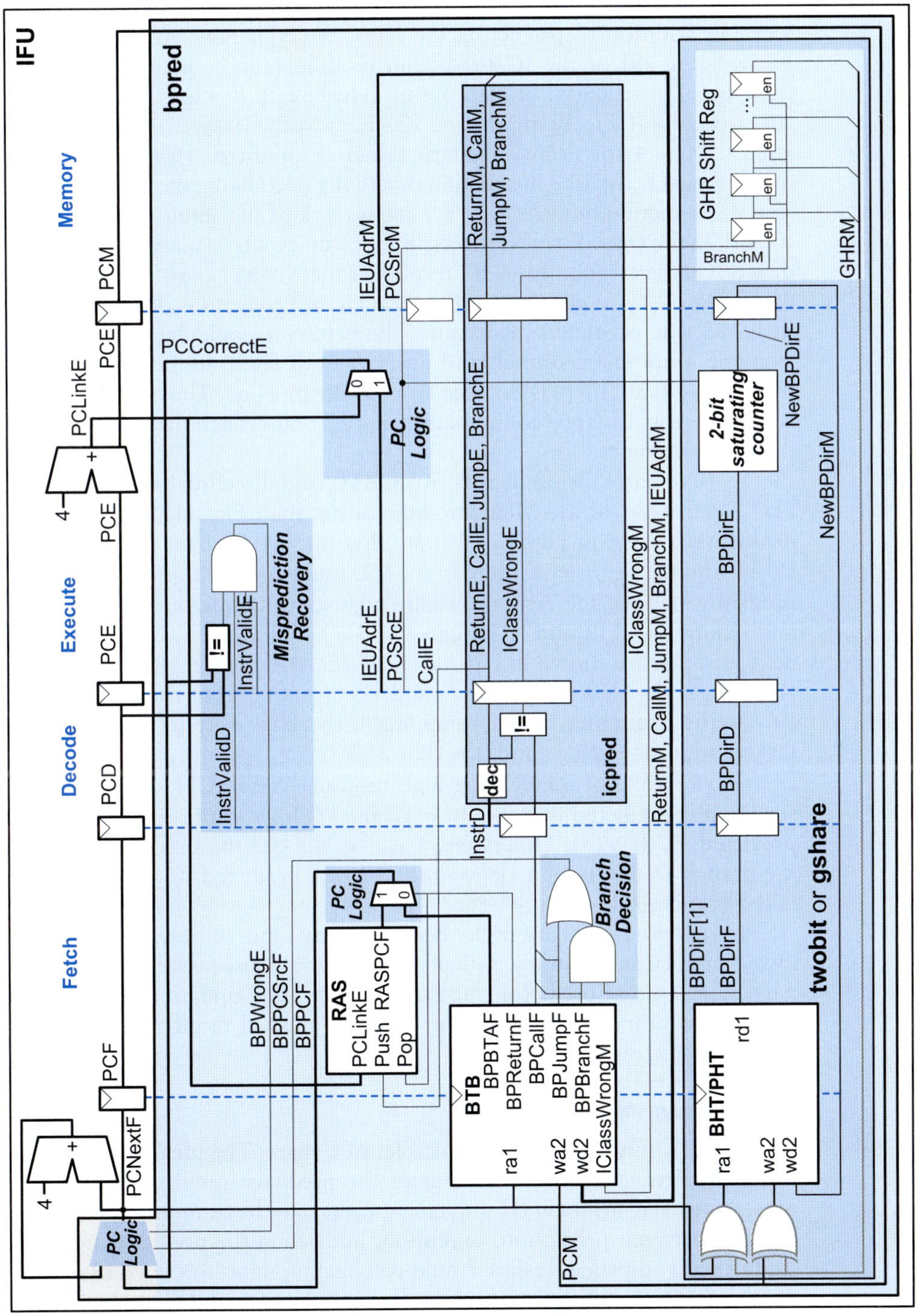

Fig. 13.15 Branch predictor top-level diagram (bpred).

some of the details of the branch direction and class predictors and PC logic. In addition to predicting the BTA, the BTB now also predicts the IC with the aid of the IC prediction module `icpred`, which determines the type of instruction: branch, jump, return, call, or noncontrol flow instruction (non-CFI). By predicting IC, the next PC (`PCNextF`) can be calculated (`PCF` + 4 for non-CFI instructions) or predicted (for branch/jump instructions) before the instruction is actually decoded in the Decode stage. The IC prediction module (`icpred`) keeps track of the actual and predicted IC and, in the Decode stage, detects if they don't match (`ICWrongD`). Mispredicted IC typically causes `bpred` to predict the wrong `PCNextF`, which leads to the same 2-cycle penalty as other branch mispredictions. However, even if the IC was predicted incorrectly, the behavior could be the same, for example a nontaken branch and an `addi` both fetch the next instruction from PC + 4. So, there is no need to flush the pipeline. The only side effect is that the performance counters tracking IC and branch direction mispredictions increment.

`bpred` updates the predictors (RAS, BTB, and direction predictor: BHT/PHT) at the end of the Memory stage rather than Execute. The PC logic details are shown in Fig. 13.14; note that only part of the PC logic is in `bpred`, and the rest is elsewhere in the IFU and privileged unit. The branch predictor may use the 2-bit or Gshare algorithm; Gshare is shown in gray and requires a GHR and XOR gates. `bpred`'s submodules, discussed in the next sections, are shown in Figs. 13.17 (Gshare PHT), 13.18 (BTB), 13.19 (`icpred`), and 13.20 (RAS). `PCSrc` and `IEUAdr` come from the IEU.

At the beginning of the Fetch stage, the BTB outputs the predicted target address `BPBTAF` and the IC: `BPReturnF`, `BPCallF`, `BPJumpF`, and `BPBranchF`. None of these being high indicates a non-CFI. If a return instruction is predicted (`BPReturnF` = 1), the PC logic chooses `RASPCF` as the predicted PC (`BPPCF`). Otherwise, `BPPCF` = `BPBTAF`, the predicted BTA. If the Branch Decision logic determines that the predicted IC is a jump or a taken branch, `PCNextF` = `BPPCF`.

Branch predictors are tricky because many types of bugs only result in lower prediction accuracy rather than incorrect processor behavior. In particular, a good predictor must take care to only update its predictions when valid branches commit. For example, flushed branches should not update predictions.

13.3.3.1 Gshare Implementation

Section 13.1.3 introduced the principles of Gshare. The ideal Gshare algorithm predicts, executes, and updates the predictor one instruction at a time. However, a pipelined implementation may have multiple branches in-flight between prediction, execution, and when the predictor updates. Additionally, pipeline flushes should roll back updates to the GHR. Wally handles these complications in a similar manner as *read after write* (*RAW*)

IEU hazards; it delays GHR and PHT updates until the Memory stage and uses forwarding to send GHR updates (from later stages) back to the Fetch stage.

A simple implementation (BPRED_TYPE = `BP_GSHARE_BASIC) avoids flushes by updating the GHR and PHT during the Memory stage but forgoes forwarding. A full implementation (BPRED_TYPE = `BP_GSHARE) adds logic to forward the GHR from a later pipeline stage to the Fetch stage. This section begins by explaining the simple implementation and then adds forwarding.

Flushed branch instructions should not update the GHR and PHT. Consider a CSR write followed by a branch. When the CSR write gets to the Memory stage, the branch will be in the Execute stage. The CSR write flushes the branch so it should not modify the GHR or PHT. By delaying GHR shifts until the Memory stage, the GHR does not require any repair.

Fig. 13.16 shows this simple implementation. It resembles Fig. 13.5 but moves the GHR and PHT updates to the end of the Memory stage.

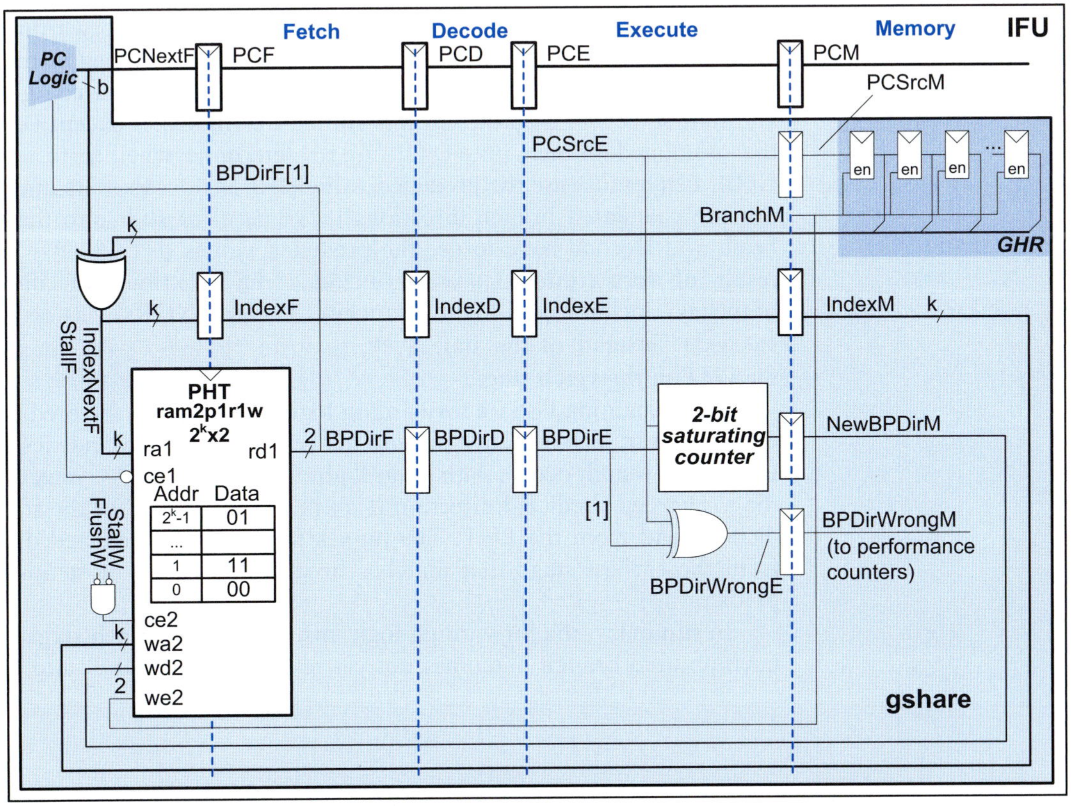

Fig. 13.16 Simple Gshare implementation (BPRED_TYPE = BP_GSHARE_SIMPLE).

Recall that the GHR is a shift register that tracks the outcomes of the last *k* branches. It is only updated on valid branches (BranchM = 1, FlushW = 0, and StallW = 0). The update shifts in PCSrcM, the outcome of the current branch.

The PHT is accessed in the Fetch stage using IndexNextF, which is an XOR hash of PCNextF and the GHR. The same entry of the PHT is subsequently updated at the end of the Memory stage using IndexM, a pipelined version of the entry. The new direction entry NewBPDirE is calculated by the 2-bit saturating counter.

The branch direction misprediction rate is measured by the performance counters from Section 8.4.3.1 as (branch prediction wrong direction)/(total branches).

A consequence of updating the GHR at the end of the Memory stage is that that the GHR lags four cycles behind the PC. So, the GHR fails to account for in-flight branches, thus degrading prediction accuracy. The full Gshare implementation of Fig. 13.17 adds forwarding logic, shown in blue, to incorporate in-flight branches into the prediction by forwarding updated versions of both the GHR and the predicted direction to the Fetch stage. We focus on GHR forwarding first, then predicted direction.

This forwarding logic is analogous to the RAW data forwarding in Section 7.2.2.3 when an instruction's source operand depends on the destination of another but the destination is not yet written to the register file.

The chain of GHR{E/D/F/NextF} muxes at the top of the forwarding logic incorporates each stage's branch outcomes into the GHR seen by the Fetch stage. If a stage has a branch instruction in it, the GHR is shifted right by 1, and that stage's branch direction is shifted in. All stages except Fetch use the calculated branch (Branch{D/E/M}) to determine whether to shift the GHR. Fetch must use the predicted value (BPBranchF). Execute and Memory use calculated branch directions (PCSrc{E/M}) to shift into the GHR; Fetch and Decode have to use the predicted values (BPDir{F/D}[1]). As a result of forwarding, GHRNextF includes the outcomes of all recent branches, including in-flight branches. Gshare XORs PCFNextF with GHRNextF, instead of the stale GHRM, to form PHTIndexNextF for reading the PHT in the Fetch stage.

The remaining Gshare forwarding logic passes along the predicted direction of a branch in a later stage to the Fetch stage if their indexes match. If an index match occurs with an in-flight branch, the BPDirF mux chooses the forwarded predicted direction (FwdNewBPDirF) instead of the stale prediction read from the PHT. The muxes forward the predicted direction from the earliest stage that matches because it has the most up-to-date prediction.

In summary, the forwarding logic ensures that the Fetch stage sees all uncommitted branch outcomes not yet reflected in the GHR and PHT.

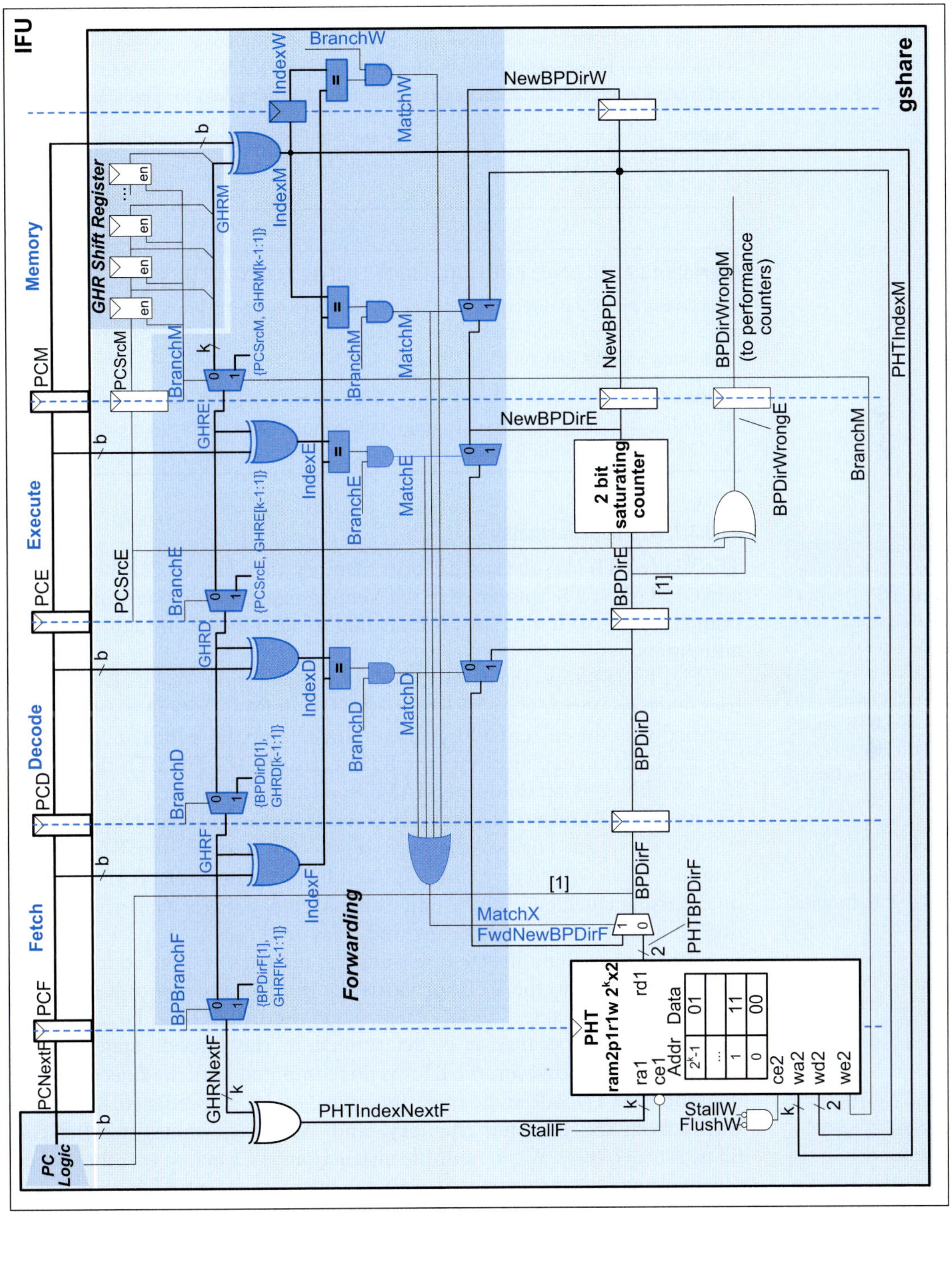

Fig. 13.17 **gshare with forwarding.**

Example 13.3 GHR CALCULATION

State the value of GHRNextF in terms of GHRM for the following three cases: when no stages contain branches, when only the Memory stage holds a branch and it is taken, and when all stages hold branches and only Execute isn't taken (or predicted taken).

Solution: If no stages hold branches, GHRNextF = GHRM, the same as without forwarding. For the other two cases, GHRNextF = {1, GHRM[k-1:1]} and GHRNextF = {1, 1, 0, 1, GHRM[k-4:4]}.

Example 13.4 BRANCH PREDICTION USING FORWARDING

What is the predicted direction (BPDirF) in the Fetch stage if a branch is in the Execute stage (BranchF = 1) and matches the Fetch stage index (IndexE = IndexF), with no other stage's index matching?

Solution: MatchE = (IndexE == IndexF) & BranchE = 1. So, MatchX = 1 and the predicted direction from the Execute stage is forwarded to the Fetch stage: BPDirF = FwdNewBPDirF = NewBPDirE.

13.3.3.2 BTB Implementation

The Wally BTB uses the same 2-port memory as in Fig. 13.7. Like the PHT and GHR, the BTB is updated in the Memory stage to avoid writing flushed branches to it, so it also uses forwarding to account for in-flight branches and jumps.

Fig. 13.18 shows the detailed BTB implementation. The BTB has 2^b entries, each comprising an XLEN-sized BTA and 4 IC bits that guess if the instruction at the current address is a branch, jump, function call, or return. The next section explains how the BTB learns the IC. The BTB thus uses a $2^b \times (\text{XLEN} + 4)$-bit dual-port RAM. Read addresses come from the lower bits of PCNextF and write addresses come from the lower bits of PCM.

On the clock edge at the start of the Fetch stage, the RAM reads TableBTBPredF, which gives the BTA and IC prediction. The RAM updates on the rising clock edge at the end of the Memory stage, when the instruction commits, and the true target and class are known.

When an in-flight instruction's address matches PCF, the address of the fetched instruction, the BTB forwards the in-flight instruction's class and BTA to the Fetch stage. In any later stage, the true IC can be forwarded (IClass{D/E/M/W}) because it is determined in the Decode stage by the icpred module. However, the BTA is not computed until the Execute stage; therefore, the Decode stage must forward BPBTAD, its predicted BTA. Forwards from the Execute, Memory, and Writeback stages use the true BTA, IEUAdr{E/M/W}. When multiple instructions' PCs match PCF, the earliest pipeline stage has priority as it contains the most recent BTA.

When compressed instructions are not supported, instructions are word-aligned so PC[1:0] = 00. Hence, the BTB is indexed by PC[b+1:2]. When compressed instructions are supported, PC might be a multiple of 2, so the BTB could be indexed by PC[b:1]. However, this trades an often useless bit (bit 1 is often 0) for a useful bit (bit b+1). Using {PC[b+1] ⊕ PC[1], PC[b:2]} to index the BTB empirically performs slightly better because PC[1] is disproportionately 0 (i.e., more uncompressed instructions exist), which enables use of PC[b+1] as a significant bit.

The BTB contains both the BTA and IC because a single large memory is denser than two smaller memories.

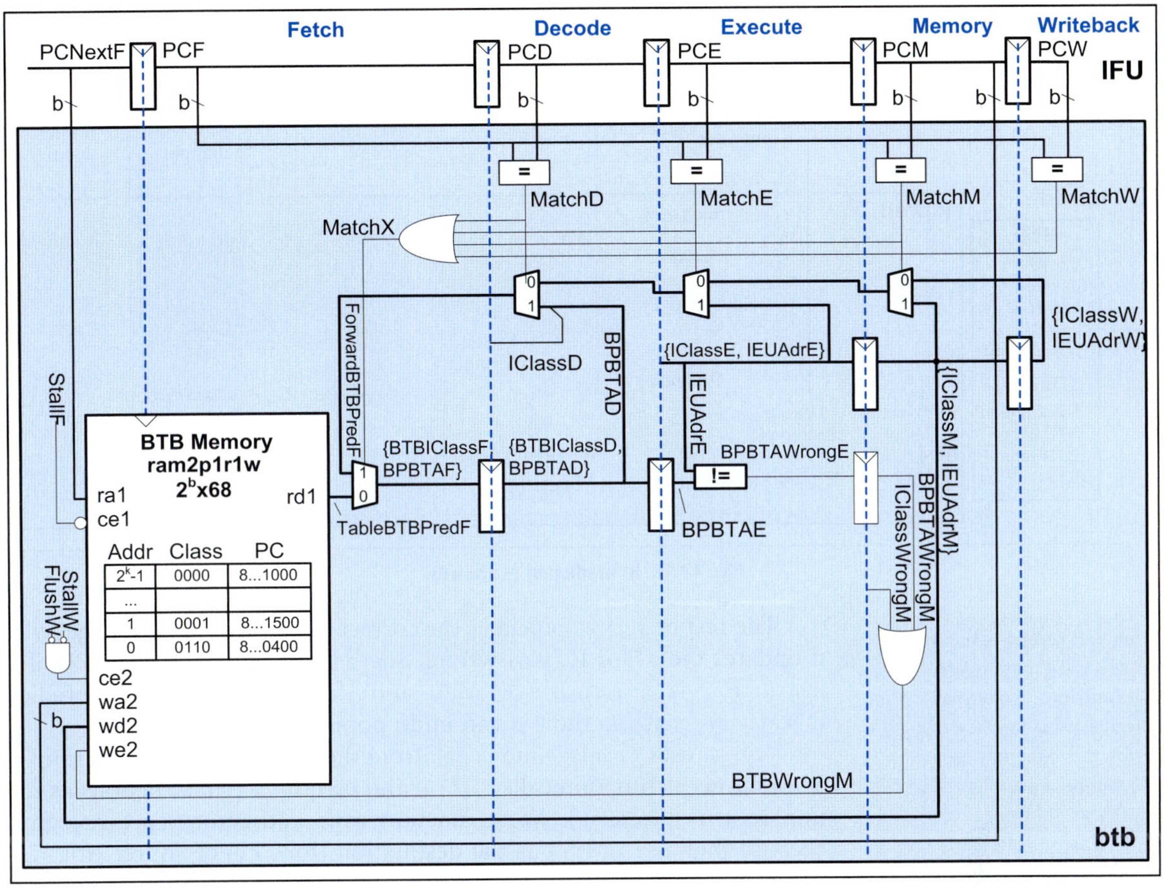

Fig. 13.18 BTB implementation with pipeline forwarding.

The BTB target misprediction rate is measured with the performance counters from Section 8.4.3.1 as the ratio of (branch/jump prediction wrong target address)/(total branches + total jumps).

The BTB may guess the wrong IC because it is smaller than the working set of the program or because multiple instructions conflict for the same location in the BTB.

13.3.3.3 Instruction Class Prediction

Fig. 13.19 shows the IC prediction module, `icpred`, which tracks the predicted and actual IC and, in the Decode stage, determines if they match. Because the actual IC is not available until the Decode stage, the BTB provides an IC prediction `BTBIClassF` in addition to the predicted BTA. `BTBIClassF` is separated into individual signals and becomes `BPBranchF`, `BPJumpF`, `BPCallF`, and `BPReturnF`. In the Decode stage, `icpred` decodes the actual IC and asserts `IClassWrongD` if it doesn't match the predicted class.

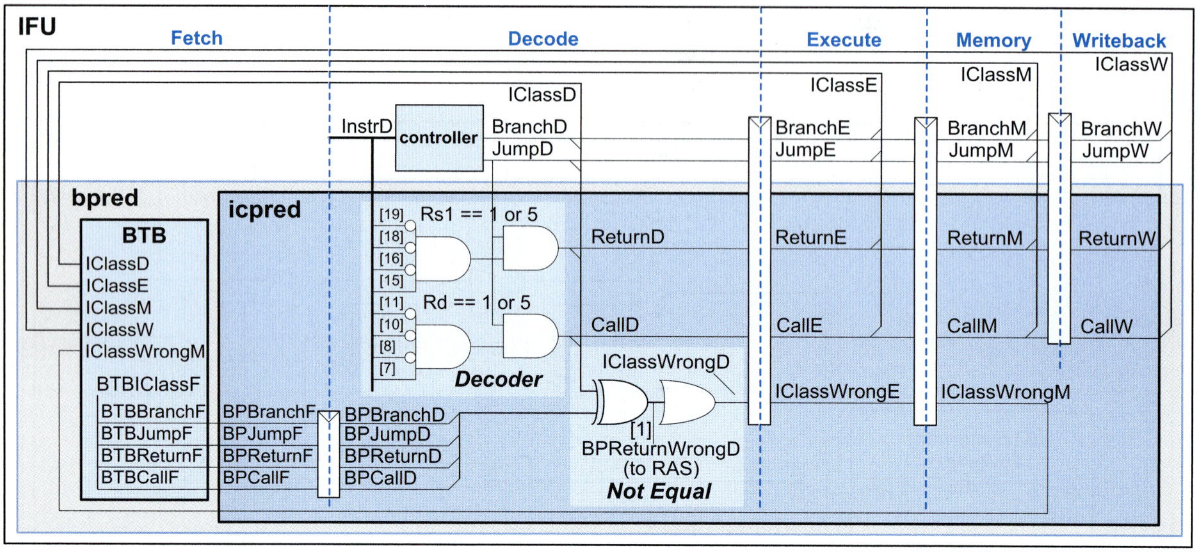

Fig. 13.19 IC prediction (icpred).

> Saving and restoring registers on the stack is common for most functions. Waterman envisions reducing code size by sharing this common register save/restore code, which he refers to as *millicode*, across function calls [Waterman16, p. 66]. He suggests invoking this shared millicode with a `jal t0, <save/restore>`, to avoid overwriting `ra`. For this reason, `CallD` asserts for both `jal ra` and `jal t0`.

> `JumpD` asserts for both `jalr` (including function returns) and `jal` instructions.

The rest of `icpred` pipelines the correct IC to the Memory stage, where it updates the `BTB` if IC was wrong. `icpred` also lets the RAS know if the instruction is a return (`ReturnD`) and if it was incorrectly predicted (`BPReturnWrongD`) so the `RAS` can undo popping its stack.

`icpred` uses `BranchD` and `JumpD` from the IEU. It recognizes certain `jal` instructions as function calls (`CallD`) and certain `jalr` instructions as function returns (`ReturnD`). Most function calls and returns are recognized because they use `x1` (`ra`) as the destination (for `jal`) or source (for `jalr`). However, `icpred` also recognizes calls and returns that use `x5` (`t0`). `icpred` checks the `rd` and `rs1` fields in the Decode stage to detect function calls and returns. `CallD` asserts for `jal {ra/t0}`. `ReturnD` asserts for `jalr x0, {ra/t0}`.

The BTB only updates when `IClassWrongM` is asserted or the BTA was wrong, saving power when the prediction is correct. The performance counters also use `IClassWrongM` to measure the number of IC mispredictions. Similarly, the performance counters use `BranchM`, `JumpM`, and `ReturnM` to track the total number of branches, jumps, and returns. As with all predictors in `bpred`, any misprediction only impacts performance by requiring a pipeline flush of the Fetch and Decode stages.

The IC misprediction rate is measured with the performance counters from Section 8.4.3.1 as: (IC predictor wrong)/(instructions retired).

13.3.3.4 RAS Implementation

Wally's RAS implementation consists of two main parts: a small asynchronous flip-flop memory to record return addresses and a `Ptr` register that indicates the top of stack. The memory is organized as a circular buffer with `RAS_SIZE` entries. Each time a call executes, the RAS pushes the return address by writing it to the memory at the top of its stack and then incrementing its stack pointer, `Ptr`. When the branch predictor predicts a return instruction, the RAS provides the return address by popping the stack (reading the memory at the top of the stack and decrementing `Ptr`). Pipeline flushes and IC mispredictions repair the RAS by undoing pushes or pops.

Fig. 13.20 shows the complete RAS implementation. The stack memory and up/down counter to push or pop `Ptr` are on the bottom right. The combinational logic converts calls, returns, pipeline flushes, and IC mispredictions into pushes and pops. When a call executes (`CallE`) and is not flushed or stalled, it pushes the return address, `PCLinkE`, onto the stack and increments `Ptr`. When a return is predicted in the Fetch stage (`BPReturnF`), it pops the return address, `RASPCF`, off the stack and decrements `Ptr`.

If the hazard unit flushes a return from either the Decode or Execute stage (`FlushedRetDE` = 1), the RAS must undo the pop by incrementing `Ptr`. If the class predictor incorrectly predicted a return (`WrongPredRetD` & $\overline{\texttt{ReturnD}}$), the RAS also increments `Ptr` to undo the faulty pop. Conversely, if the class predictor failed to predict a return (`BPReturnWrongD` & `ReturnD`), the RAS decrements `Ptr` to catch up.

The `bpred` module disables IC prediction when `INSTR_CLASS_PRED` = 0. In this configuration, the IC is determined by partially decoding `Instr` in the Fetch stage. Although this lengthens the Fetch stage, it currently doesn't affect cycle time because the Fetch stage is not in the critical path. Testing with class prediction enabled and disabled shows that the class predictor has negligible impact on branch prediction accuracy and thus does not have major performance bugs.

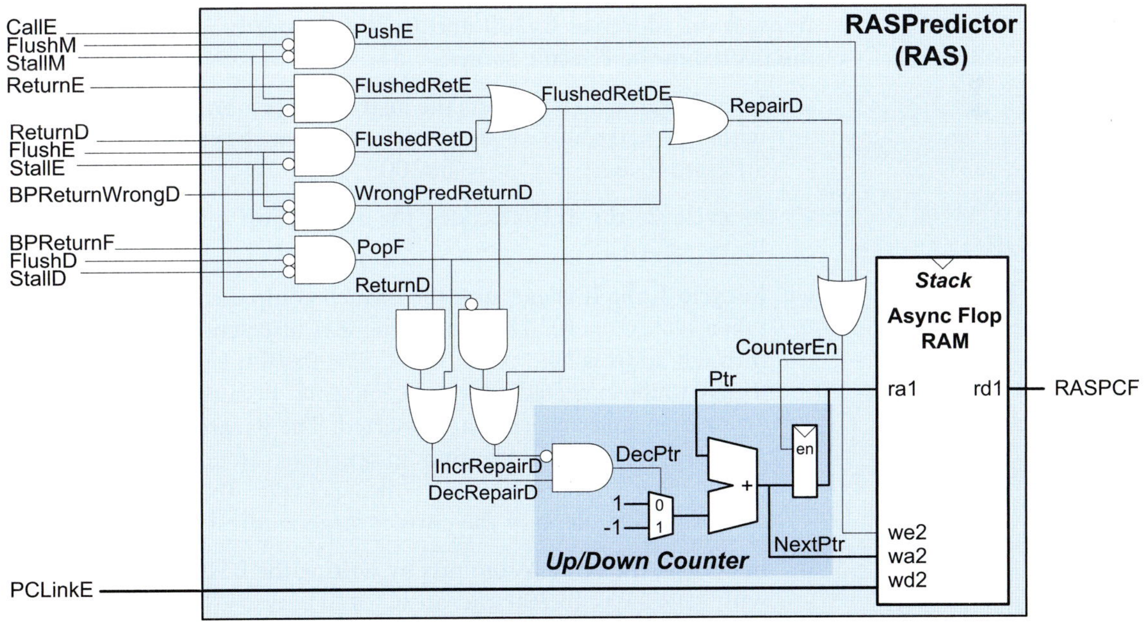

Fig. 13.20 **RAS with repair.**

Because the RAS uses asynchronous flip-flop memory, reads take place combinationally rather than requiring the address to set up before the clock edge. The memory has separate read and write ports so that the RAS can simultaneously push (write) and pop (read).

The RAS performance counter tracks the total number of return instructions committed and the number of RAS mispredictions. The "Return address stack wrong address" performance counter increments based on `RASPredPCWrongE` = (`RASPCE` != `IEUAdrE`) & `ReturnE`. The RAS misprediction rate is measured as (RAS wrong address)/(total returns) using the performance counters from Section 8.4.3.1.

13.3.3.5 Branch Misprediction Recovery

`bpred` can mispredict direction, target address, and IC, which leads to a pipeline flush if the PC of the predicted instruction is wrong, regardless of the type of misprediction. The branch predictor reports the misprediction to the hazard unit, which flushes the Decode and Execute stages. If the predicted PC is correct but the other predictors are wrong, the branch predictor still updates its state with the correct target and IC and records the misprediction in the optional performance counters.

Table 13.1 summarizes the various branch predictor outcomes. In the Execute stage, the PC logic computes `PCCorrectE`, which is `PCE` + 4 if the branch is not taken or the BTA, `IEUAdrE`, if the branch is taken. If `PCD` matches `PCCorrectE`, the correct instruction was fetched, and the program counter continues incrementing normally. Otherwise, the IFU fetched the wrong instructions. It recovers by flushing the next two instructions and setting `PCNextF` = `PCCorrectE`.

Fig. 13.21 shows a timing diagram for a branch at 0x100 incorrectly predicted as taken with a BTA of 0x200. The processor fetches two instructions, from addresses 0x200 and 0x204, before branch misprediction is discovered in the Execute stage.

- **In cycle 1**, the BTB predicts the instruction is a branch (`BPBranchF` = 1) with `BTAF` = 0x200. `gshare` predicts the branch is taken (`BPPCSrcF` = 1) and sets `PCNextF` = `BTAF` = 0x200.
- **In cycle 2**, the IEU decodes the instruction and detects a branch (`BranchD` = 1).
- **In cycle 3**, the IEU executes the branch comparison and finds it is not taken (`PCSrcE` = 0). Therefore, the next instruction should be the fall through address `PCLinkE` = `PCE` + 4 = 0x104. `bpred` compares `PCD` = 0x200 to the expected address `PCCorrectE`. Because they do not match, `BPPCWrongE` and `BPWrongE` are asserted. The hazard unit asserts `FlushD` and `FlushE` to flush the two instructions at addresses 0x200 and 0x204. `bpred` updates `PCNextF` = `PCCorrectE` = 0x104, and the processor resumes fetching at the correct target address.
- **In cycle 4**, the branch commits by writing the BTB entry: `PCM` = 0x100 with `IEUAdrM` = 0x200 and an IC of `BranchM` = 1. `gshare` updates by moving to the Weakly Taken state. The GHR shifts in a 0 for the untaken branch.

Table 13.1 Branch outcomes and recovery.

Correct Predicted PC? `PCD == PCCorrectE`	Branch Direction `PCSrcE`	Flush Pipeline?	`PCNextF`
Yes	either	No	`PCF + 4`
No	0: Not Taken	Yes	`PCCorrectE = PCE + 4`
No	1: Taken	Yes	`PCCorrectE = IEUAdrE`

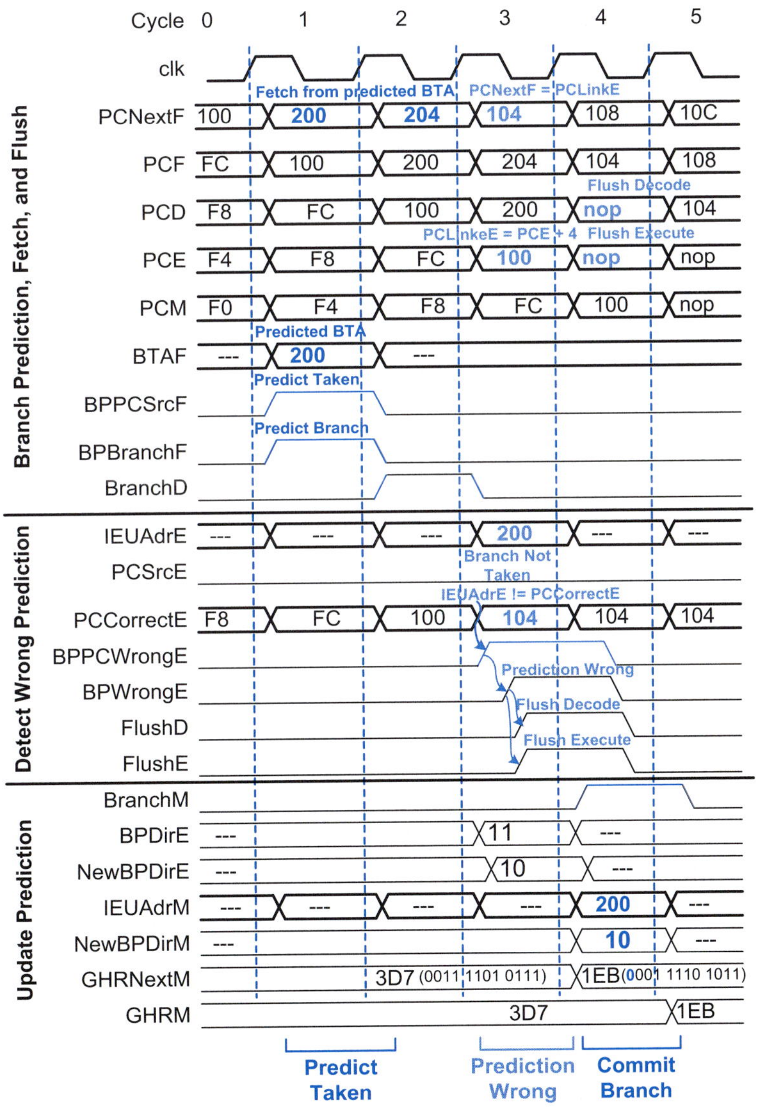

Fig. 13.21 Timing diagram of branch misprediction recovery.

13.3.4 ITLB Misses

If both the ITLB and DTLB miss in the same cycle, the HPTW gives the DTLB priority and handles the DTLB miss before handling the ITLB miss.

As introduced in Section 12.3.4, the LSU HPTW services both LSU and IFU TLB misses. When the IFU's ITLB misses, it asks the LSU's HPTW to find the missing PTE. Fig. 13.22 shows the interface between these modules, with optional signals in gray. Specifically, the ITLB asserts `ITLBMissF` and sends `PCF` to the HPTW. The page table walker handles the ITLB miss, returns `PTE` and `PageType`, and pulses `ITLBWriteF` to write `PTE` into the ITLB. If `SVADU_SUPPORTED` = 1, the HPTW atomically updates access and dirty bits in memory and the ITLB.

13.3.5 Reset, Stalls, Flushes, and Traps

On reset, the cache clears the cache line valid bits. When a line is invalid, the other bits don't matter. The valid bits are implemented with flip-flops, so they are easy to reset.

Resetting the BTB and BHT/PHT is more problematic because these prediction memories are large and, hence, far more efficient to implement with SRAM, yet SRAM lacks a reset input. Thus, the branch memories are normally not reset but power up with unknown values. As a consequence, branch prediction is nondeterministic and may vary from one hardware boot to another depending on the random values in the prediction memories. In simulation, these memories begin with 0 values because they are implemented with the System Verilog `bit` type rather than `logic`. Recall from Section 4.2.2.3 that `bit` has two values (0 and 1) in contrast to `logic`'s four: 0, 1, X, and Z.

The IFU's stalls, flushes, and traps closely resemble those of the LSU, as described in Section 12.4. The IFU asserts `IFUStallF` to stall the pipeline

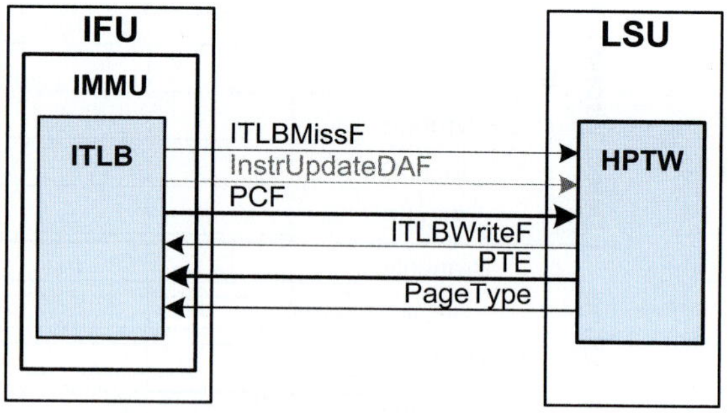

Fig. 13.22 ITLB/HPTW interface.

during multicycle cache operations (`ICacheStallF`), bus accesses, or ITLB misses. The IFU generates page, access, and misaligned instruction address faults during the Fetch stage as described in Sections 8.2.5.2 and 11.4.3. These faults are pipelined as shown in Fig. 11.28 to generate traps in the Memory stage. During a multicycle operation, such as a bus access, the IFU disables interrupts by asserting `CommittedF` to ensure bus operations complete once in flight.

The hazard unit sends `StallD` and `FlushD` to the IFU. Like the LSU, when the IFU is stalled, it completes multicycle operations and then waits

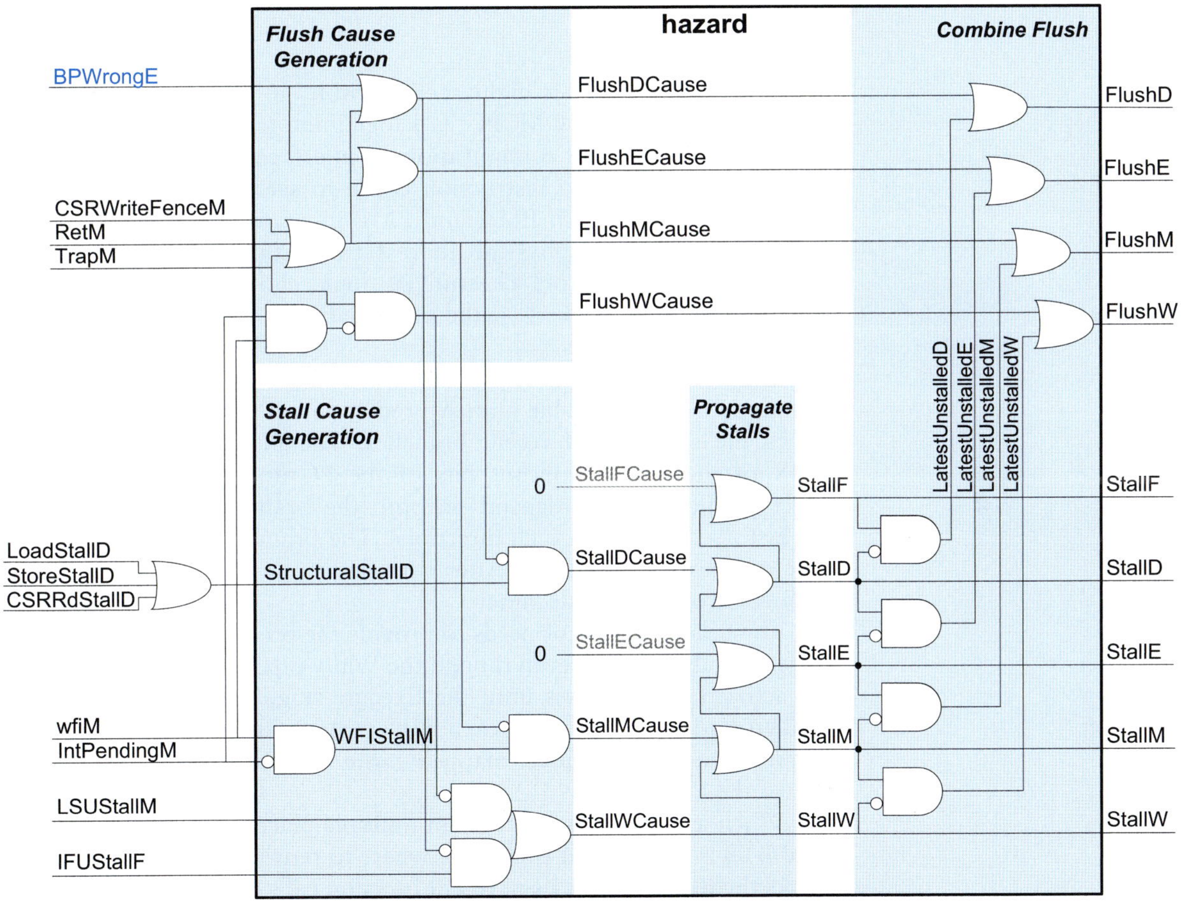

Fig. 13.23 **Hazard unit with branch predictor changes highlighted in blue.**

for the stall to be released. Flushes clear the instruction in the Decode stage to be a nop. They also cancel cache or bus operations before they begin but do not interfere with in-flight multicycle bus or cache operations.

Fig. 13.23 shows the updated hazard unit with changes to support branch prediction in blue. `BPWrongE` replaces `PCSrcE` from Fig. 9.33. In Chapter 9, the processor assumed not taken on a branch and simply fetched PC + 4. If the branch was actually taken (`PCSrcE` = 1), the processor flushed the Decode and Execute stages. Now those stages are only flushed when the branch predictor fetched the wrong instruction (`BPWrongE` = 1).

13.4 SUMMARY

Like the LSU, the IFU is highly configurable. In the simplest configuration, it has only an IROM, PC + 4 logic, and a branch multiplexer. In the most sophisticated configuration, it interfaces to the EBU for main memory access, has an instruction cache to service most accesses, and contains an MMU with PMA and PMP physical address checkers and a TLB for virtual to physical address translation. Moreover, the IFU adds a branch predictor. The full PC logic chooses the next program counter from one of PC + 4, the predicted BTA, the trap vector, or the trap return address.

Gshare is a widely-used and reasonably effective branch predictor. It maintains a PHT of 2-bit branch predictions. It also adds a GHR that tracks the direction of recent branches. The PHT is indexed with an XOR of the least significant bits of the PC and the GHR, so it can make predictions that depend on both the branch being executed and the history of recent branch directions. The branch predictor takes some time to warm up the PHT. After warming up, a PHT with 4-Ki-entries achieves around 96% accuracy.

The branch predictor needs to provide the predicted BTA in the Fetch stage to update the PC even though the Wally pipeline does not determine the actual target address until the Execute stage. Therefore, the branch predictor adds a BTB, which caches the target address of recent branches. It can be viewed as a direct-mapped cache indexed by the less significant bits of the PC.

When a function is called from multiple places in a program, a standard BTB is not good at predicting where to return. However, a RAS that tracks the return addresses of recent function calls can easily determine the return address.

The complete branch predictor includes a Gshare direction predictor, BTB, and RAS. The true branch outcome is calculated in the Execute stage. If the prediction was wrong, the pipeline must flush the incorrectly fetched instructions and restart at the true address. The predictor also updates the BHT/PHT, GHR, BTB, and RAS based on the branch outcome. Subtle logic is needed to handle stalls and flushes and to repair incorrect predictions.

Extensions: C (Compressed)

14

14.1 RISC-V Practices
14.2 Test Plan
14.3 Wally Implementation
14.4 Summary

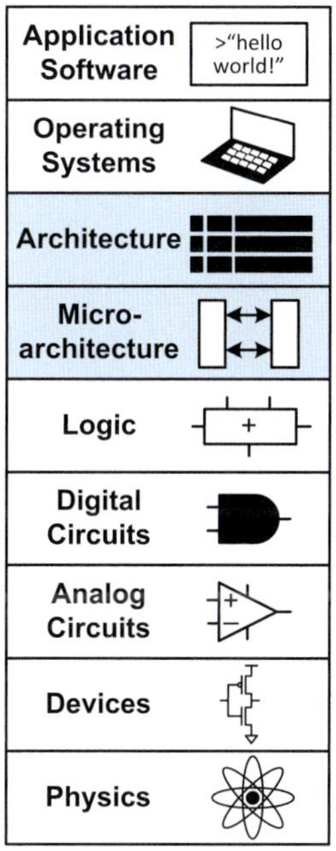

Most reduced instruction set computer (RISC) architectures were developed with 32-bit instruction encodings. As they began to compete with 16-bit microcontrollers in cost-sensitive markets, RISC was at a disadvantage because compiled programs typically required more nonvolatile memory, which is one of the costliest parts of the system. Moreover, fewer 32-bit instructions fit in a cache, resulting in higher miss rates. Therefore, many RISC architectures have introduced compressed instruction sets that provide 16-bit representations of commonly used instructions. These compressed instructions are mostly a subset of the full instruction set and can be "inflated" back to 32 bits to be executed by the normal execution units.

A processor that supports compressed instructions may execute both compressed and regular 32-bit instructions in the same program. So, a 32-bit instruction might be fetched from a 16-bit boundary, crossing cache lines or IROM words. This is called a *spill* and requires two memory accesses to fetch both halves of the instruction.

14.1 RISC-V PRACTICES

The optional RISC-V C extension defines compressed instructions. This compressed instruction set is summarized on the last pages of the book in Tables I.6 and I.22, and compressed instruction formats are in Fig. I.2. All compressed instructions are 16 bits long. To achieve this, the first source register is generally also used as the destination, and immediates are shorter than in ordinary instructions. Compressed instruction use many formats, trading regularity for density. The formats have different numbers of function (`funct`) bits. Some instructions only allocate 3 bits to specify a register, so they are restricted to using only eight registers (`x8`-`x15`).

The compressed instruction format is cleverly chosen to play gracefully with uncompressed instructions. Whereas the two least significant bits of uncompressed instructions are always 11, these bits have a different value for compressed instructions. Thus, the instruction encodings do not

Other RISC compressed instruction sets include the ARM Thumb instructions, MIPS16e, PowerPC VLE, and Hitachi SHcompact.

See the RISC-V specification for complete details about compressed instruction formats.

overlap for compressed and uncompressed instructions, and a program can freely use both. The register fields, when present, are in the same place in all compressed formats to reduce the cost of decompression, and the most significant bit of the immediate is always in instruction bit 12 to simplify sign extension.

A few compressed instructions decompress into two or more standard RISC-V instructions.

Every compressed instruction maps onto an equivalent 32-bit instruction. Therefore, instead of adding a whole new datapath, a RISC-V processor can decompress compressed instructions and then execute that decompressed instruction normally on the existing datapath.

Example 14.1 COMPRESSED INSTRUCTION ENCODING

Give the machine language encoding for `c.add x5, x6` and show how it decompresses to a 32-bit instruction.

Solution: According to Table I.6 (on the last pages of the book), `c.add` is a CR-type instruction with `op` = 10 and `funct4` = $\texttt{instr}_{15:12}$ = 1001. The `rd/rs1` field is 5 and the `rs2` field is 6. Hence, the machine language is:

```
funct4 rd/rs1 rs2   op
1001   00101  00110 10 = 0x929A
```

The instruction inflates to `add x5, x5, x6`, with machine encoding:

```
funct7  rs2   rs1   funct3 rd    op
0000000 00110 00101 000    00101 0110011 = 0x006282B3
```

According to the RISC-V Unprivileged Specification 20240411, a compiler that uses compressed instructions when possible reduces code size by 25%–30% for typical programs [RISCV-Unprivileged].

To make best use of the limited encoding space, certain compressed machine language opcodes expand into different instructions in different ISAs. For example, the instruction with `op` = 01, $\texttt{instr}_{15:13}$ = 001 is `c.jal` in RV32IC but is `c.addiw` in RV64IC. The instruction with `op` = 00, $\texttt{instr}_{15:13}$ = 011 is `c.flw` in RV32ICF but is `c.ld` in RV64IFC. Therefore, a program compiled for RV32IC will not run correctly on RV64IC.

14.1.1 Zc* Extension

The Zc* family of extensions provides additional compressed instructions to improve code density even more than the original compressed (C) extension, especially for embedded processors where the IROM is a major part of system cost.

The C extension defines various compressed floating-point loads and stores: `c.flw`, `c.flwsp`, `c.fsw`, `c.fswsp`, `c.fld`, `c.fldsp`, `c.fsd`, and `c.fsdwp`. The Zca extension is a subset of the C extension that omits these compressed floating-point load/store instructions to save hardware in embedded processors with no floating-point unit (FPU). Zcf and Zcd define these missing compressed single- and double-precision loads and stores, respectively.

Zcb defines more compressed instructions overlooked in C. When code density is important, data density is also important, so the Zcb extension adds instructions to load and store bytes and half-words. Subword arithmetic can overflow into the upper part of the word, so it is often followed by sign- or zero-extension to discard the overflow. Zcb includes these subword zero-/sign-extension instructions. Zcb also provides compressed multiply and NOT instructions, which are common in integer programs. Certain Zcb instructions rely on the Zbb basic bit manipulation extension or the M multiplication extension described in Chapters 15 and 18.

Saving and restoring registers on the stack takes many instructions in RISC-V. Other architectures have discovered dedicated PUSH and POP instructions improve code density by saving or restoring multiple registers with a single instruction. `c.popret` pops registers and also returns from a function call. These operations decompress into two or more standard RISC-V instructions and are included in the Zcmp extension. Similarly, the Zcmt extension defines table jumps, which improve code density for jumps and function calls. They use a new *jump vector table* (*JVT*) CSR that points to a table of jump addresses; table jumps do not correspond to standard RISC-V instructions. Zcmp and Zcmt redefine compressed double-precision load and store opcodes, so they are incompatible with the full C extension and Zcd.

In summary, the Zc* family of extensions are Zca, Zcd, Zcf, Zcb, Zcmp, and Zcmt:

- **Zca:** All C extension instructions except Zcf and Zcd F/D loads and stores. Excludes:

  `c.flw, c.flwsp, c.fsw, c.fswsp, c.fld, c.fldsp, c.fsd, c.fsdwp`

- **Zcf:** Compressed single-precision loads & stores (requires F extension, RV32 only):

  `c.flw, c.flwsp, c.fsw, c.fswsp`

- **Zcd:** Compressed double-precision loads and stores (requires D extension):

  `c.fld, c.fldsp, c.fsd, c.fsdsp`

- **Zcb:** Subword and other basic integer instructions:

  `c.lbu, c.lh, c.lhu, c.sb, c.sh`
  `c.zext.b, c.not`
  Zbb only: `c.sext.b, c.zext.h, c.sext.h`
  M only: `c.mul`
  RV64 only: `c.zext.w` (requires Zba extension)

- **Zcmp:** Push/pop compressed instructions (opcodes conflict with Zcd):

  `cm.push, cm.pop, cm.popret, cm.popretz, cm.mva01s, cm.mvsa01`

- **Zcmt:** Table jump instructions (opcodes conflict with Zcd):

  `cm.jt, cm.jalt`

The RV32GC/RV32IFDC extension implies Zca + Zcf + Zcd. RV64GC/RV64IFDC implies Zca + Zcd; they do not support Zcf because these instruction encodings are reused for 64-bit integer loads and stores. An RV64GC_Zcb_Zbb processor implements Zca + Zcb + Zcd, but not Zcmp or Zcmt.

Zce is shorthand for Zca + Zcb + Zcmp + Zcmt. In RV32F, Zce also includes Zcf. It excludes Zcd, whose opcodes are reused for Zcmp and Zcmt. Benchmark results show that Zce provides an average 11% further code size reduction over the C extension [codesize]. Zce is a natural replacement for the C extension in embedded processors.

Table I.22 lists the Zcb instructions and Fig. I.2 lists new formats used by these instructions (see the last pages of the book).

14.2 TEST PLAN

The C-extension instructions are tested with the riscv-arch-test suite as discussed in Section 7.4. Table 14.1 lists the number of test cases for each Zca and Zcb instruction.

14.3 WALLY IMPLEMENTATION

Wally uses the `ZCA_SUPPORTED`, `ZCB_SUPPORTED`, `ZCF_SUPPORTED`, and `ZCD_SUPPORTED` configuration parameters to enable compressed instructions. There is no `C_SUPPORTED` parameter. Instead, the C bit of the `misa` register is set if Zca is supported and either Zcd (and D) or Zcf (and F) are supported. Wally does not presently implement Zcmp or Zcmt and, therefore, does not support Zce.

Fig. 14.1 shows the Wally IFU changes required to support compressed instructions. Wally uses an optional decompression unit (`decompress`, abbreviated to `decomp` in Fig. 14.1), which receives 16-bit compressed instructions and inflates them to their equivalent 32-bit instructions. The decompression takes place at the start of the Decode stage and must complete in time to access the integer register file (`regfile`, shown as `IRF` in Fig. 14.1). The PC logic adds 2 instead of 4 for compressed instructions. The IFU takes multiple cycles to handle spills, when an instruction wraps a cache line or IROM word. The following sections describe decompression, PC logic, and spills in detail.

14.3.1 Instruction Decompression

As shown in Fig. 14.2, `decompress` is combinational logic with a 32-bit input `InstrRawD` and a 32-bit output `InstrD`. If `InstrRawD` is uncompressed (detected by bits [1:0] = 11), it is passed through to `InstrD`. Otherwise, the combinational logic examines at least bits [15:13] and [1:0] to determine the type of compressed instruction and to inflate it to a 32-bit output, `InstrD`.

Table 14.1 riscv-arch-test C cases.

Instruction	RV32Zca Tests	RV64Zca Tests	RV32Zcb Tests	RV64Zcb Tests
c.add	584	668		
c.addi	375	440		
c.addi16sp	69	132		
c.addi4spn	20	19		
c.addiw		438		
c.addw		678		
c.and	585	670		
c.andi	376	439		
c.beqz	85	149		
c.bnez	85	149		
c.ebreak	1	1		
c.j	18	18		
c.jal	18			
c.jalr	31	31		
c.jr	31	31		
c.lbu			8	8
c.ld		13		
c.lh			8	8
c.lhu			8	8
c.ldsp		31		
c.li	32	32		
c.lui	31	31		
c.lw	14	13		
c.lwsp	31	31		
c.mul			589	677
c.mv	84	148		
c.nop	14	14		
c.not			66	130
c.or	590	679		
c.sb			8	9
c.sd		132		
c.sdsp		131		
c.sext.b			65	129
c.sext.h			66	130
c.sh			8	8
c.slli	86	148		
c.srai	87	148		
c.srli	85	149		
c.sub	580	670		
c.subw		670		
c.sw	70	131		
c.swsp	69	133		
c.xor	581	662		
c.zext.b			65	129
c.zext.h			66	130
c.zext.w				130
Total	**4,632**	**7,829**	**957**	**1,496**

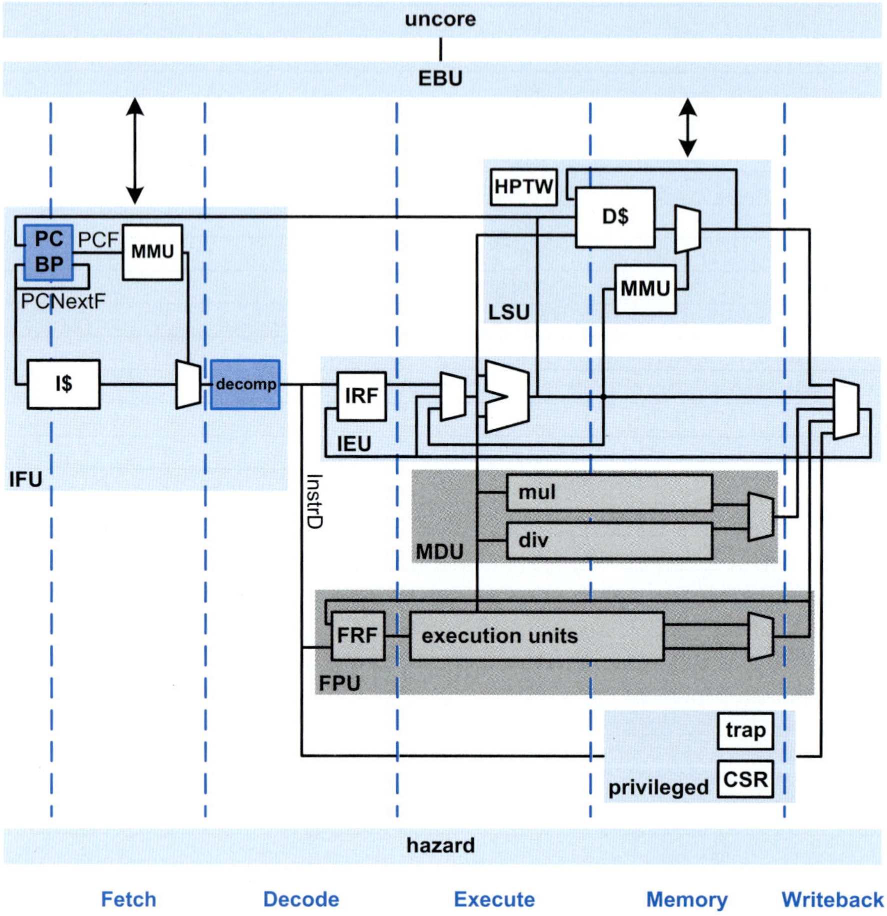

Fig. 14.1 Wally RV32IC/64IC high-level block diagram emphasizing the `decompress` unit.

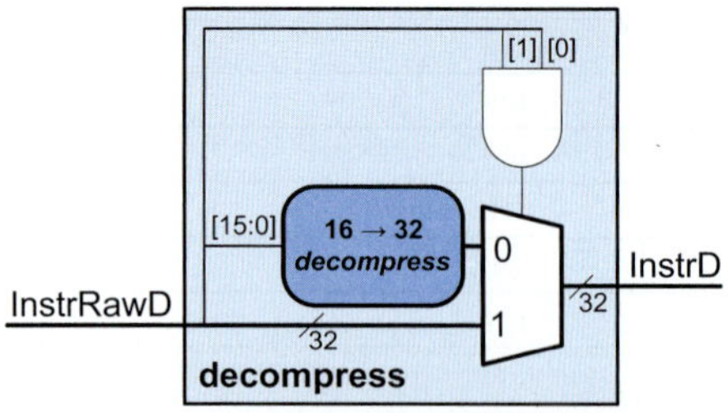

Fig. 14.2 `decompress` block diagram.

Code Example 14.1 shows a snippet of the decompress module's Verilog code. Based on InstrRawD's op field (bits [1:0]) and funct3 field (bits [15:13]), the module identifies the type of compressed instruction. It then identifies the registers and immediate from the other InstrRawD bits and cobbles them together into an uncompressed instruction. Thus, the decompressor can be viewed as a giant multiplexer that selects fields from the compressed instruction to emit an uncompressed instruction.

Code Example 14.1 EXCERPT OF **decompress** MODULE

```
assign op = instr16[1:0];
assign rs1p = {2'b01, instr16[9:7]};
assign rdp = {2'b01, instr16[4:2]};
assign rs2p = {2'b01, instr16[4:2]};
assign rds1 = instr16[11:7];

assign immCL = {5'b0, instr16[5], instr16[12:10], instr16[6], 2'b00};
assign immCS = {5'b0, instr16[5], instr16[12:10], instr16[6], 2'b00};
assign immCI = {{7{instr16[12]}}, instr16[6:2]};

always_comb
  if (op == 2'b11) InstrD = InstrRawD; // Noncompressed instruction
  else case ({op, instr16[15:13]})     // Inflate compressed instruction
    5'b00010: InstrD = {immCL, rs1p, 3'b010, rdp, 7'b0000011};                          // c.lw
    5'b00110: InstrD = {immCS[11:5], rs2p, rs1p, 3'b010, immCS[4:0], 7'b0100011}; // c.sw
    5'b01000: InstrD = {immCI, rds1, 3'b000, rds1, 7'b0010011};                         // c.addi
    ...
```

Adding Zcb support simply involves extending the decompress module with more opcodes. The ZCB_SUPPORTED configuration flag indicates that these instructions are available.

14.3.2 PC Logic

When the IFU fetches an instruction, it asserts CompressedF if the instruction is compressed. Fig. 14.3 shows how the PC logic is modified to add 2 to PCF instead of 4 for compressed instruction, with the mux that selects 2 or 4 highlighted in bright blue.

As described in Section 8.2.5.2, instruction address-misaligned faults ordinarily occur if the branch or jump target address is not a multiple of 4. When compressed instructions are supported, that fault is not permitted because target addresses may be multiples of 2.

The branch predictor of Section 13.3.3 aligned the BHT (branch history table) to a 4-byte boundary, such that a 2^k-entry table used PC[k+1:2] to index the BHT. Because compressed instructions align to 2 bytes, the branch predictor should include PC[1]. A typical program contains a

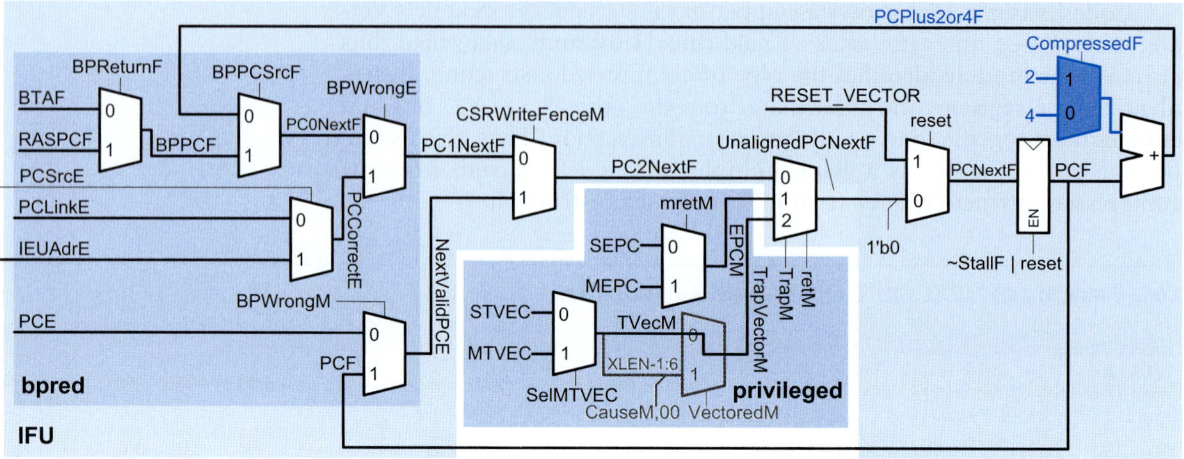

Fig. 14.3 IFU PC logic with support for compressed instructions.

mix of 32- and 16-bit instructions, so using PC[k:1] reduces the effective capacity of the BHT when most instructions are 32 bits. By indexing the BHT with {PC[$k+1$] $\oplus$ PC[1], PC[k:2]}, it maintains most of the capacity if the program has few compressed instructions.

14.3.3 Spills

Recall that a spill occurs when a 32-bit instruction spans two instruction cache lines or two IROM words. If PC is an odd multiple of 2 that maps to the end of a cache line, the cache can only return the first 16 bits of the instruction. If that instruction is 32 bits long, the I$ must read the next cache line to retrieve the remaining 16 bits. Similarly, the IROM or bus may require two accesses to fetch a 32-bit instruction spanning consecutive words.

A dual-ported cache could concurrently read two cache lines; however, dual-ported SRAMs occupy significantly more area.

For example, Fig. 14.4 shows `li s8, 0x3D` = 0x03D00C13 at address 0x31E. If cache lines are 32 bytes, the instruction spans two consecutive cache lines. The I$ first reads 0x0C13 from address 0x31E. Then the

Fig. 14.4 32-byte cache line spill highlighted in blue.

IFU adds 2 to `PCF` to read 0x03D0 from address 0x320. The spill logic (`spill`) concatenates the two halves and then outputs the full 32-bit instruction.

When an instruction fetch spills, the `spill` module pipelines the fetch over 3 cycles:

- **Step 1:** The I$/IROM sets up to read the instruction at `PCNextF`.
- **Step 2:** The I$/IROM outputs the first 2 instruction bytes, and the `spill` module saves them to a local register. The I$/IROM also sets up the next read with `PCF`+2.
- **Step 3:** The I$/IROM outputs the next instruction bytes, and the `spill` module merges the two half instructions to form the full 4-byte instruction. The PC logic also sets `PCNextF` to `PCF`+4 to set up the read for the next instruction.

Unlike the I$, the branch predictors continue to read using `PCNextF` because they do not need a separate prediction for the second instruction half.

The spill adds one stall cycle of latency when both halves hit the I$/IROM. A cache miss in either half behaves like a conventional I$ miss.

Fig. 14.5 shows the complete IFU. Compared to Fig. 13.12, the new spill circuity is highlighted in blue. The `spill` module adds a +2 adder and muxes to generate `PCSpillNextF` and `PCSpillF`. `PCSpillNextF` drives the

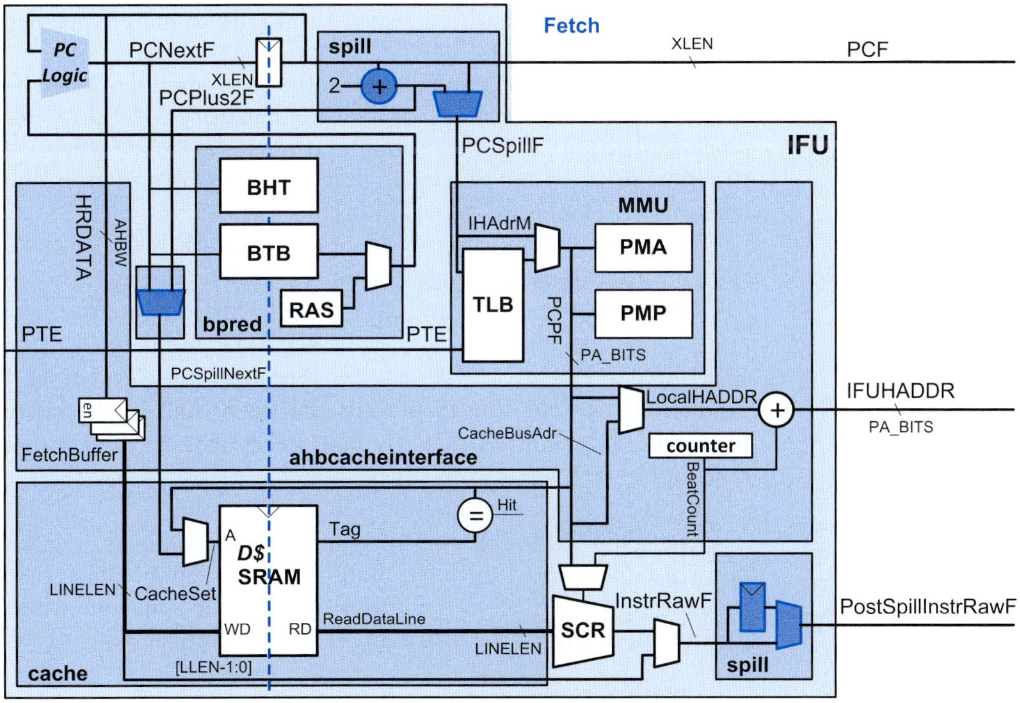

Fig. 14.5 IFU with `spill` highlighted in blue.

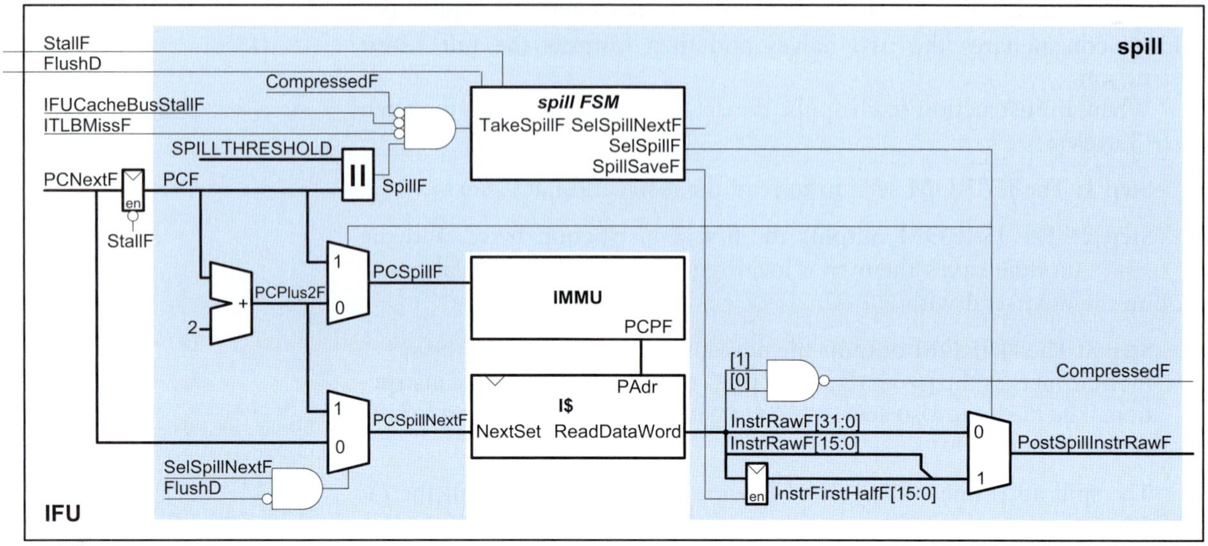

Fig. 14.6 `spill` **module.**

IROM address or the set bits of the I$. `PCSpillF` goes through the MMU for address translation to become the physical program counter `PCPF`. Finally, the spill logic adds a register and mux to merge together the two halves of the instruction.

Fig. 14.6 shows a detailed view of the `spill` module. When handling a spill, the module asserts `SelSpillNextF` that drives `IFUStallF` to stall the pipeline. The `SPILLTHRESHOLD` comparator asserts `SpillF` when a 4-byte instruction read crosses two cache lines. For example, with 32-byte cache lines, `SpillF` is asserted when `PCF[4:1]` = 1111.

Upon a spill, the first access is a usual I$ read: `PCSpillNextF` = `PCNextF`. After the clock edge, a spill is detected (`TakeSpillF` = 1 because `PCF` == `SPILLTHRESHOLD`), and all the spill FSM outputs rise. As a result, `PCSpillNextF` = `PCPlus2F` (`SelSpillNextF` = 1), and the first half of the instruction is saved in `InstrFirstHalfF` (because `SpillSaveF` = 1). Just after the next clock edge, the I$ outputs the upper half of the instruction in `InstrRaw[15:0]`, and `SelSpillF` = 1 selects the merged instruction to send to the IFU: `PostSpillInstrRawF` = {`InstrRaw[15:0]`, `InstrFirstHalf[15:0]`}. Cache misses make this process take longer.

The spill FSM in Fig. 14.7 stays in the Ready state on a regular instruction read and to read the lower half of an instruction. When the FSM detects a spill (`TakeSpillF` = 1), it transitions to the Spill state where it waits for the upper half of the instruction read to complete.

Fig. 14.6 shows an adder producing `PCPlus2F`, but the RTL actually uses a multiplexer to save hardware. When `PCF[1]` = 0, adding 2 just sets `PCF[1]`: `PCPlus2F` = {`PCF[XLEN-1:2]`, `2'b10`}. Otherwise, `PCPlus2F` = {`PCPlus4F[XLEN-1:2]`, `2'b00`}. For example, if `PCF` = 1100 = 12, `PCPlus2F` = 1110 = 14 (set `PCF[1]`). If `PCF` = 0010 = 2, `PCPlus4F` = 0110 = 6 and `PCPlus2F` = 0100 = 4 (clear `PCPlus4F[1]`).

Fig. 14.8 shows a timing diagram of an IFU fetching the load immediate (`li`) instruction (0x03D00C13) from address 0x031E (see Fig. 14.4) with cache misses on both lines.

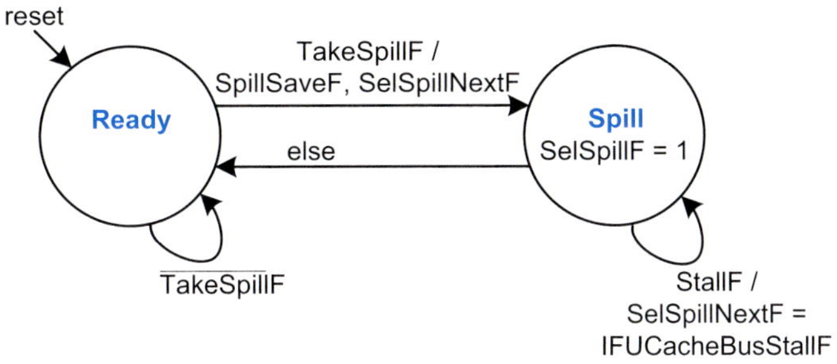

Fig. 14.7 `spillfsm`.

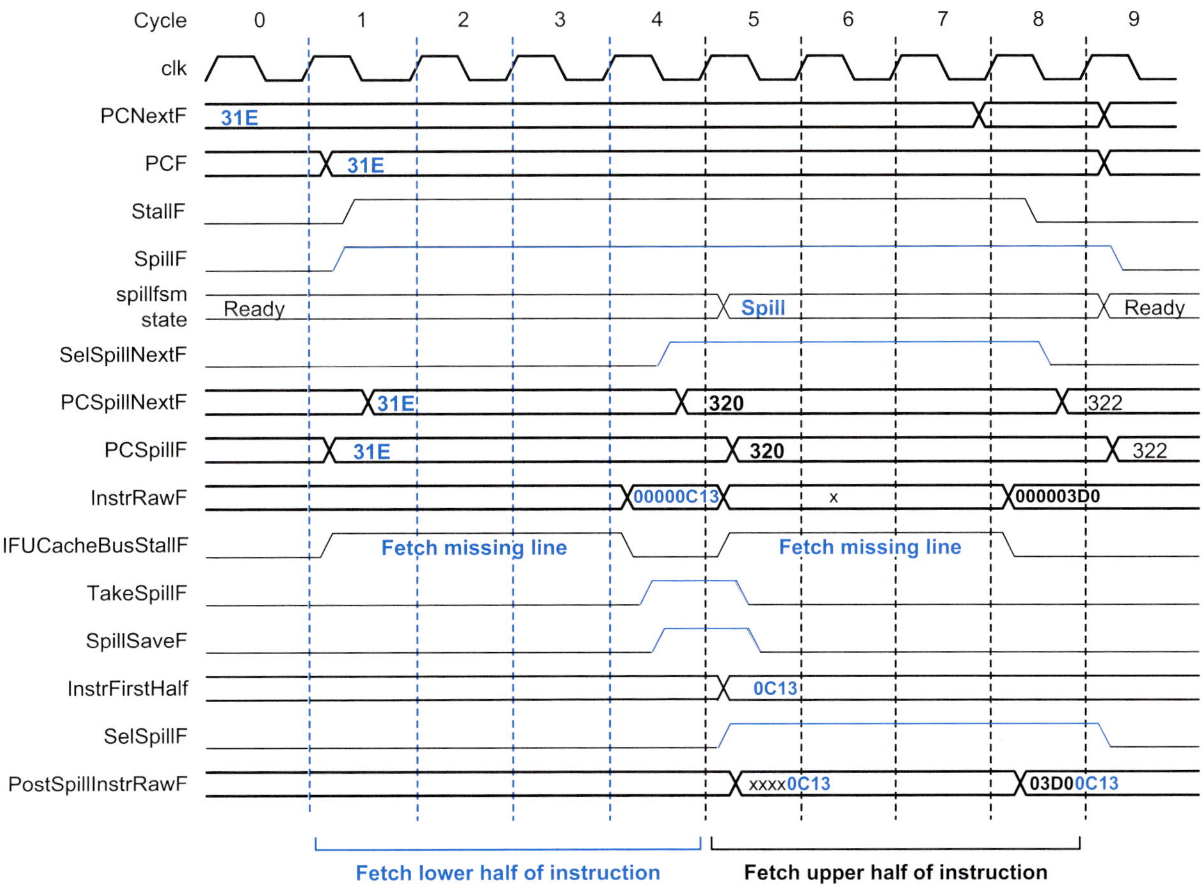

Fig. 14.8 **Spill timing diagram.**

multiplication, which are the same for all flavors. `mulh`, `mulhu`, and `mulhsu` produce the upper 32 bits of the multiplication, interpreting the operands as signed, unsigned, and signed/unsigned, respectively.

15.1.2 Carry-Save Addition

In computer arithmetic, we often need to add multiple operands—for example, when adding a multiplier's partial products. The most obvious way to add k N-bit operands is with k-1 consecutive carry-propagate adders (CPAs) as shown in Fig. 15.1(a), but this approach is slow and costly. A better technique is called *carry-save addition*. Recall that a full adder sums three 1-bit inputs of unit weight (i.e., not shifted) and produces a sum output with unit weight and a carry output with double weight (i.e., shifted left by 1 bit): $\{C, S\} = X + Y + Z$. In other words, the carry output (C) belongs in the column to the left of the sum bit, S. If we place N full adders in parallel, they can accept three N-bit operands $X_{N\text{-}1:0}$, $Y_{N\text{-}1:0}$, $Z_{N\text{-}1:0}$ and produce two N-bit outputs $S_{N\text{-}1:0}$ and $C_{N\text{-}1:0}$ that satisfy $X + Y + Z = 2C + S$, as shown in Fig. 15.1(b). This is called *carry-save redundant format* because the carry output of each column is preserved rather

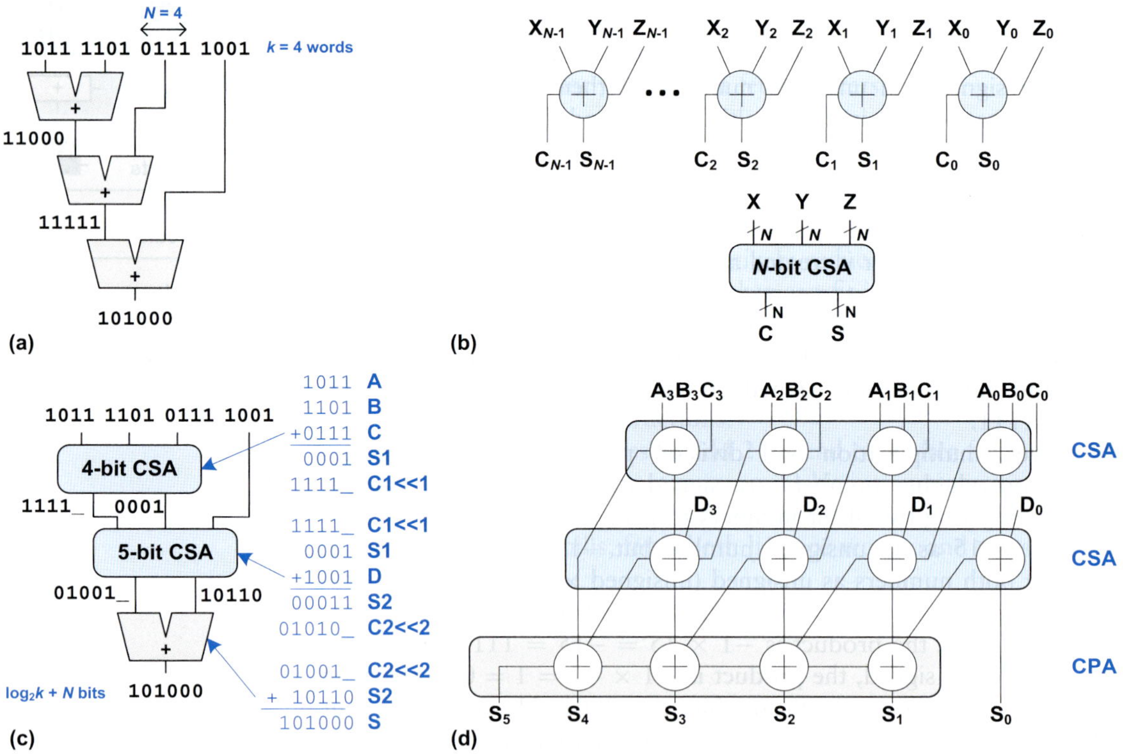

Fig. 15.1 Multiple-input adders: (a) CPAs, (b) *N*-bit CSA, (c) CSAs and 1 CPA, (d) CSAs and 1 CPA details.

than propagated along the adder. In this application, the full adders are called *3:2 carry-save adders* (*CSAs*) because they accept 3 inputs and produce two outputs in carry-save (redundant) form; CSAs are also referred to as 3-to-2 adders. To obtain the sum $X + Y + Z$ in nonredundant binary form (i.e., as a single binary number), the carry word C is shifted left by one column (because it has double weight) and added to the sum word S with an ordinary CPA. Alternatively, a fourth word could be added to the carry-save redundant result with another row of CSAs, again giving a carry-save redundant result. This result could again be summed with a CPA as shown in Fig. 15.1(c and d).

A CPA takes N full adder delays to add two N-bit numbers. In contrast, a CSA takes only 1 full adder delay to add three N-bit numbers; however, the result is in redundant format.

We use carry-save addition in Section 15.1.3 to quickly add partial products in a multiplier and in Section 16.2.3 to quickly compute the partial remainder for division and square root.

15.1.3 Multiplication

Consider multiplying an M-bit multiplicand $Y = \{y_{M\text{-}1}, y_{M\text{-}2}, \ldots, y_1, y_0\}$ by an N-bit multiplier $X = \{x_{N\text{-}1}, x_{N\text{-}2}, \ldots, x_1, x_0\}$ to produce an $M+N$-bit product P. If X and Y are both unsigned, then the product can be written as

$$P = \left(\sum_{j=0}^{M-1} y_j 2^j\right)\left(\sum_{i=0}^{N-1} x_i 2^i\right) = \sum_{i=0}^{N-1}\sum_{j=0}^{M-1} x_i y_j 2^{i+j} \tag{15.1}$$

This can be viewed as forming N appropriately shifted partial products of M bits each, then summing the columns to compute the product. 1×1-bit binary multiplication is equivalent to a logical AND operation, so generating partial products consists of taking the logical AND of the appropriate bits of the multiplier and multiplicand. Fig. 15.2(a) shows these partial products for 4×4-bit multiplication of unsigned numbers, and Fig. 15.2(b) shows the details of multiplying 1011 × 0101 = 0110111.

Large multiplication is easier to illustrate using dot diagrams. Fig. 15.3 shows a Dadda dot diagram for a 16 × 16 unsigned multiplier. The diagrams in this text are adapted from [Dadda65, Bewick94, Weste12]. Each dot represents a bit. The blue dots on the right are the multiplier bits, and horizontal boxed rows represent partial products, shifted according to their weight. Each partial product is either the multiplicand Y or 0 depending on whether the multiplier bit for that row is 1 or 0.

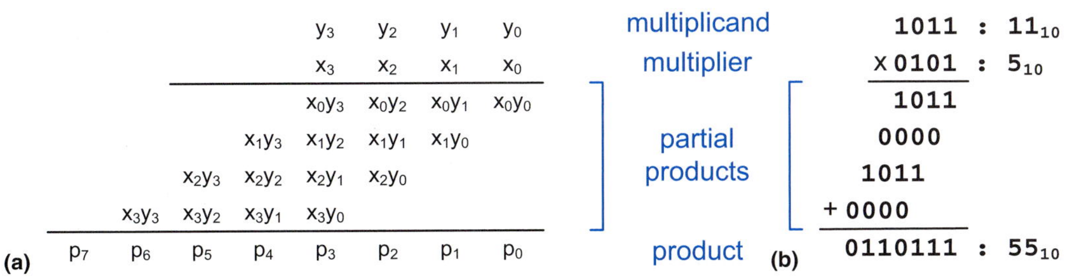

Fig. 15.2 Unsigned multiplication: (a) operations and (b) example.

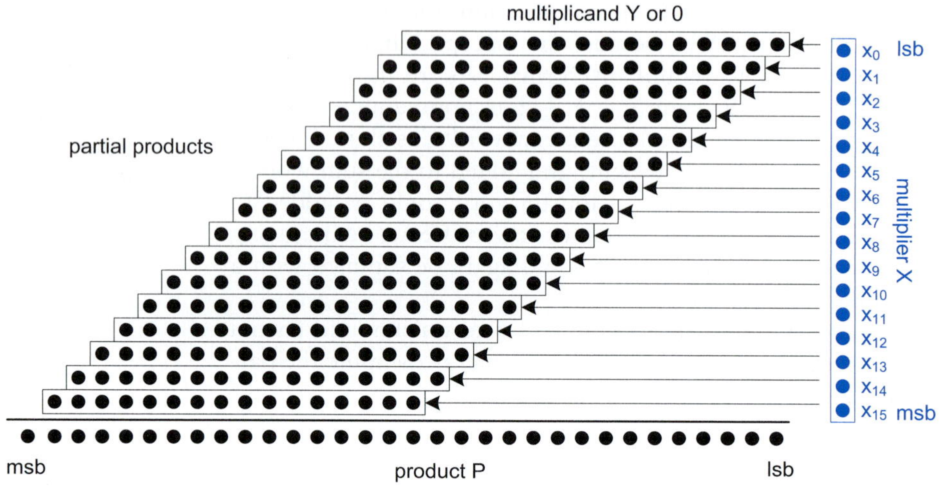

Fig. 15.3 **Multiplier dot diagram.**

The speed, cost, and power consumption of multipliers are largely determined by how the partial products are summed. A simple approach is to add the first two partial products with a CPA, then add the third partial product to the running sum, and so forth. This requires N-1 CPAs and is slow even when using a fast CPA. However, various algorithms make adding partial products faster or more efficient.

The remainder of this section addresses multiplication implementation techniques, focusing on array multipliers, Booth encoding, and tree multipliers. An array multiplier uses an array of CSAs that is simple and reasonably fast. The array can also be easily modified to perform signed multiplication. Booth encoding is an important optimization to reduce the number of partial products. Tree multipliers can sum the partial products faster than array multipliers. Although other multiplier optimizations are known, logic synthesis tools are quite mature and generally produce excellent multipliers from behavioral code that uses the * operator.

15.1.3.1 Array Multiplier

Practical multipliers use CSAs to add the partial products to reduce delay and size compared to CPAs. Fig. 15.4(a) shows a 4 × 4 *array multiplier* for unsigned numbers. Each square in the diagram contains a 2-input AND gate to compute one bit of a partial product, corresponding to a black dot in Fig. 15.3, and a full adder to add that bit to a running sum in carry-save redundant format. Note that the S_{in} and C_{in} inputs to the top row are 0 because there is no running sum yet, so the full adders could be omitted from that row. The least significant N output bits come directly from the sum outputs of the CSAs. The most significant N bits arrive in carry-save redundant format, so the multiplier needs a final CPA to add them together. In this implementation, the CPA is a ripple-carry adder made from one more row of full adders. Fig. 15.4(b) shows the same adder squashed into a rectangular form that fits a chip datapath more easily.

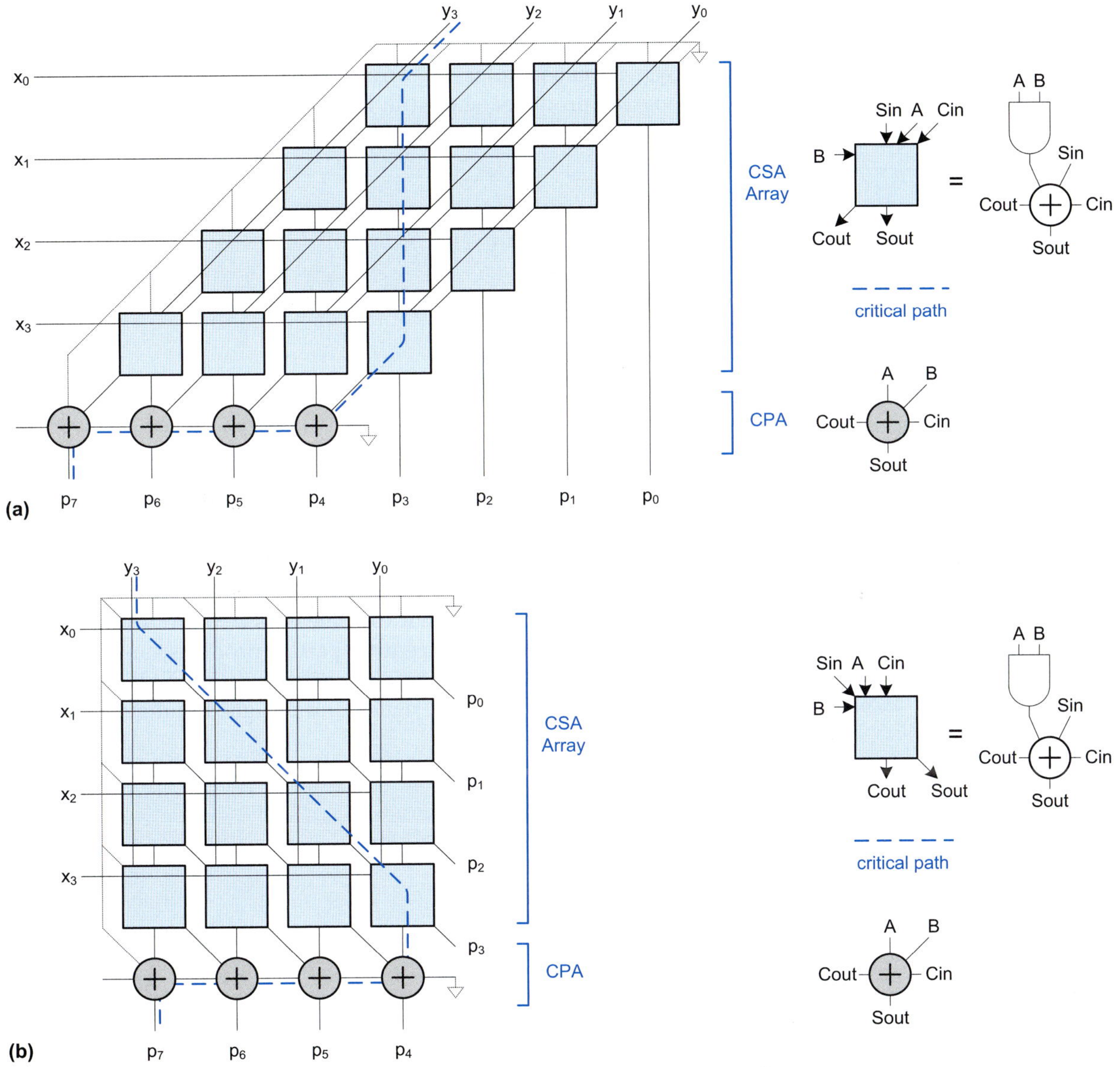

Fig. 15.4 Array multiplier: (a) concept and (b) rectangular form.

15.1.3.2 Signed Multiplication

When multiplying signed two's complement numbers, two of the partial products may be negative, as shown in Equation 15.2. Fig. 15.5 shows how these partial products, shown in blue, involve inverting the bits and adding 1. Observe how even the implicit zero (less significant) bits must be inverted.

Recall that the weight of the most significant bit of an N-bit two's complement number is -2^{N-1}.

$$\begin{aligned} P &= \left(-y_{M-1}2^{M-1} + \sum_{j=0}^{M-2} y_j 2^j\right)\left(-x_{N-1}2^{N-1} + \sum_{i=0}^{N-2} x_i 2^i\right) \\ &= \sum_{i=0}^{N-2}\sum_{j=0}^{M-2} x_i y_j 2^{i+j} + x_{N-1}y_{M-1}2^{M+N-2} \\ &\quad - \left(\sum_{i=0}^{N-2} x_i y_{M-1} 2^{i+M-1} + \sum_{j=0}^{M-2} x_{N-1} y_j 2^{j+N-1}\right) \end{aligned} \tag{15.2}$$

All of the 1s in the bottom right of Fig. 15.5 add together to create the 1 in the top row of Fig. 15.6. The two 1s in column p_6 of Fig. 15.5 add together to produce the 1 in column p_7 in Fig. 15.6. The two 1s in column p_7 of Fig. 15.5 would only produce a carry out, which is discarded, so they are not represented in Fig. 15.6.

The *modified Baugh-Wooley multiplier* [Hatamian86] reduces the hardware cost by precomputing the sum of the constant ones and pushing some of the terms upward, as shown in Fig. 15.6. Fig. 15.7 shows a rectangular array implementation, which is almost identical to the unsigned multiplier in Fig. 15.4(b) except that the blue cells contain NAND instead of AND gates to compute the negative partial products, and two cells receive carries in of 0. By conditionally inverting (using XOR gates), the same hardware can perform both signed and unsigned multiplication.

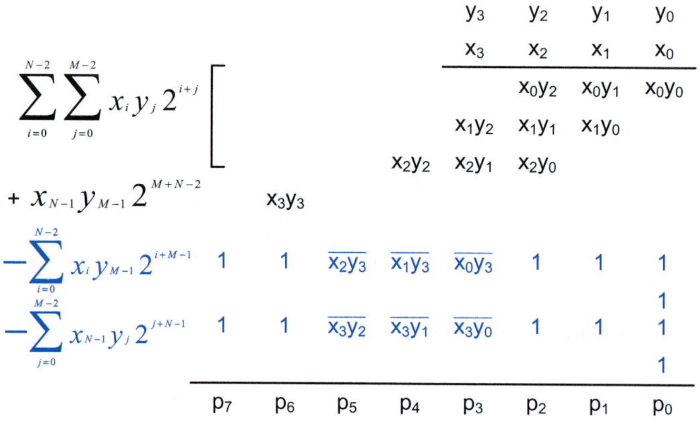

Fig. 15.5 Partial products for a signed multiplier.

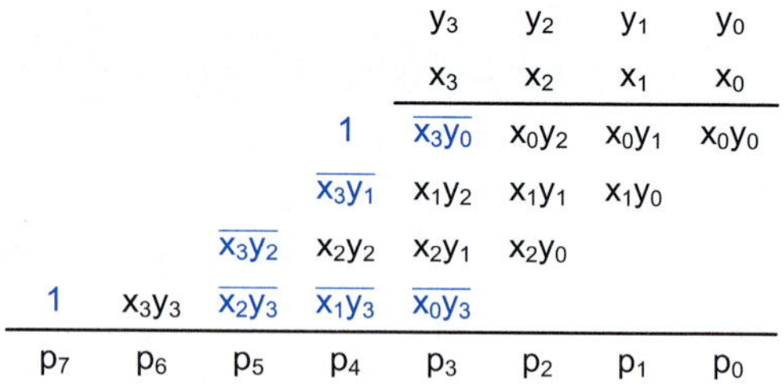

Fig. 15.6 Simplified partial products for a signed multiplier.

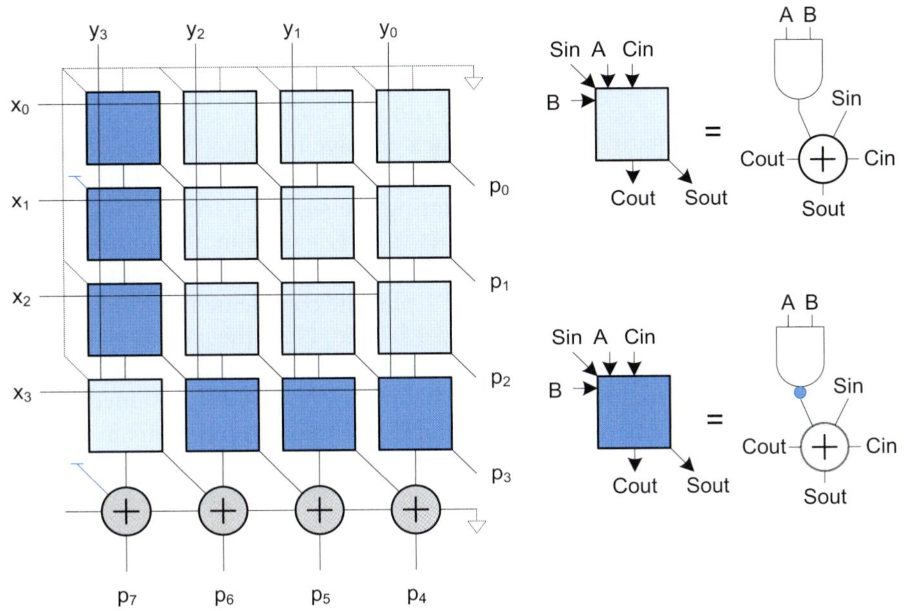

Fig. 15.7 **Modified Baugh-Wooley signed multiplier.**

15.1.3.3 Booth Encoding

The array multipliers considered so far are radix-2, producing one partial product (either multiplicand Y or 0) for each bit of the multiplier X. A radix-4 multiplier produces one partial product for every two bits of X, saving area and delay in the CSA array. The partial product is 0, Y, $2Y$, or $3Y$. $2Y$ is a simple 1-bit shift of Y, but $3Y$ involves adding $Y + 2Y$ with an expensive and slow CPA.

Modified Booth encoding [MacSorley61] provides the cost and speed benefits of higher radix multipliers without generating the hard $3Y$ multiple. Instead, radix-4 modified Booth encoding generates multiples $-2Y$, $-Y$, 0, Y, and $2Y$. It then exploits the property that $3Y = 4Y + (-Y)$ and $2Y = 4Y + (-2Y)$. Because each partial product has 4 times the weight of the previous partial product, adding $4Y$ is equivalent to incrementing the next partial product by 1. Hence, partial product i is selected by considering a pair of bits $\{x_{2i+1}, x_{2i}\}$, along with the most significant bit of the previous pair x_{2i-1}. When the most significant bit in a pair is 1, choose a negative partial product; the next partial product will need to increment to compensate. Thus, when the most significant bit of the *previous* pair is 1, add 1. For inputs with an even number of bits, one extra partial product of 0 or Y is needed to handle the increment from the most significant bit of X.

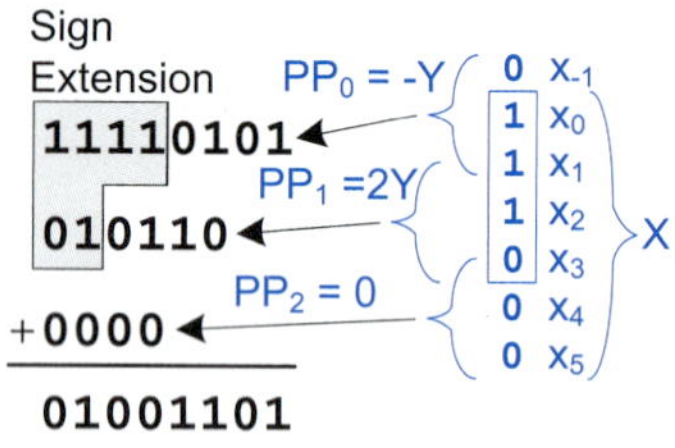

Fig. 15.8 Booth encoded example.

Table 15.1 shows the partial product logic, and Fig. 15.8 shows an example of 4×4-bit Booth-encoded multiplication with four partial products: $1011 \times 0111 = 01001101$ $(11 \times 7 = 77)$.

Fig. 15.9 shows a radix-4 Booth encoder hardware implementation. Negative partial products involve inverting Y or $2Y$ and adding 1. Again, they must be sign-extended to fill the array. Fig. 15.10 shows these partial products, with s indicating the sign of partial product i. Observe how a negative partial product ($s = 1$) has 1s in all the upper bits and also a 1 on the next row. All of the sign extension bits are either 0 or 1. They can be simplified as all 1 plus the complement of the sign bit in the first position, as shown in Fig. 15.11(a). Now the partial product array has many constant ones, whose sum can be precomputed as shown in Fig. 15.11(b). In summary, Booth-encoding an $N \times N$-bit unsigned multiplier results in $\lfloor N/2+1 \rfloor$ partial products, each up to $N + 4$ bits in length. These partial products can then be summed with CSAs as described earlier.

Booth encoding for signed two's complement multiplication is similar, but the multiplicand Y could be negative, so sign extension is performed on the sign bit of the partial product $e_i = PP_{iM}$ [Bewick94]. e_i is the exclusive-OR of the sign of Y and the sign of the multiple being selected. Only $N/2$ partial products are needed because $PP_{N/2} = 0$.

For simple radix-4 multiplication, $\{x_{2i+1}, x_{2i}\}$ of 00, 01, 10, and 11 produce partial products of 0, Y, $2Y$, and $3Y$. But with modified Booth encoding, the 10 and 11 combinations produce $-2Y$, and $-Y$ instead; then, the next partial product must add $4Y$ to produce a combined result of $4Y + (-2Y) = 2Y$ and $4Y + (-Y) = 3Y$, as desired. In other words, when the msb of a pair is 1 ($x_{2i+1} = 1$), the next partial product must add $4Y$, which is just Y in that 2-bit left-shifted position. Thus, when the previous bit is 1, $\{x_{2i+1}, x_{2i}, x_{2i-1}\} = 001, 011, 101$, and 111, the partial products are Y, $2Y$, $-Y$, and 0 (i.e., $Y + 0 = Y$, $Y + Y = 2Y$, $Y + (-2Y) = -Y$, $Y + (-Y) = 0$).

Table 15.1 Radix-4 Booth Encoding

Inputs			Partial Products	Booth Selects		
x_{2i+1}	x_{2i}	x_{2i-1}	PP_i	$Single_i$	$Double_i$	Neg_i
0	0	0	0	0	0	0
0	1	0	Y	1	0	0
1	0	0	-2Y	0	1	1
1	1	0	-Y	1	0	1
0	0	1	Y (= Y + 0)	1	0	0
0	1	1	2Y (= Y + Y)	0	1	0
1	0	1	-Y (= Y - 2Y)	1	0	1
1	1	1	0 (= Y - Y = 0)	0	0	1

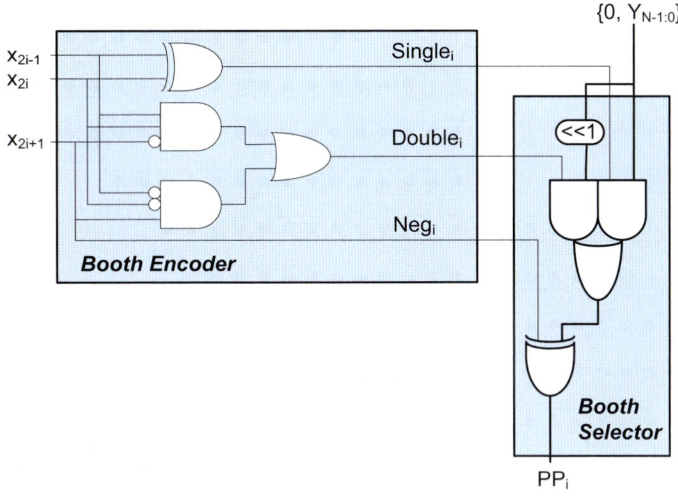

Fig. 15.9 Radix-4 Booth encoder.

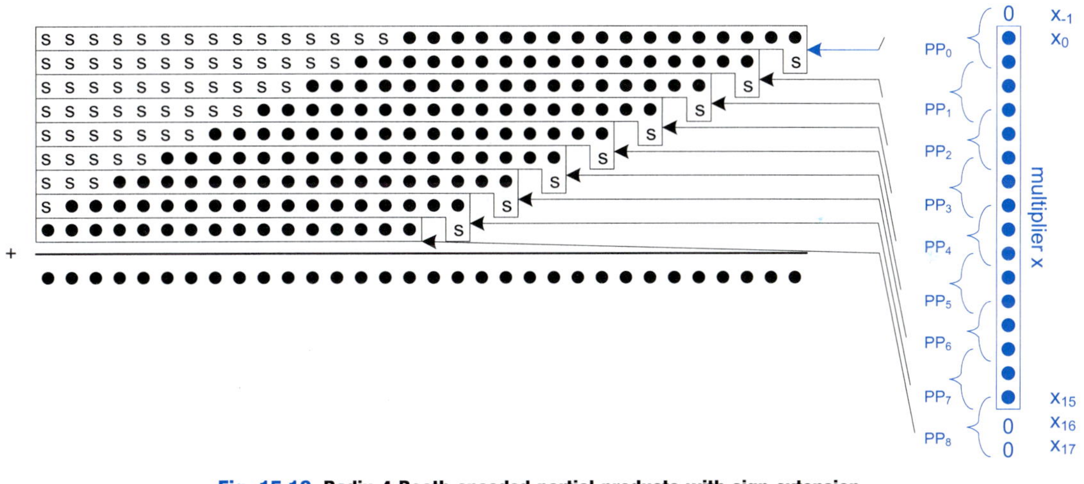

Fig. 15.10 Radix-4 Booth encoded partial products with sign extension.

15.1.3.4 Tree Multipliers

A multiplier sums the partial products in carry-save redundant format using CSAs. It produces a final pair of values that are summed by a carry-propagate adder to give the final result. The array multipliers described so far sum k partial products with a linear array of k-2 CSAs, as shown in Fig. 15.13(a). A faster approach is to use a CSA tree to sum the partial products. For example, Fig. 15.13(b) shows a Wallace tree with $\log_{3/2}(k-2)$ levels of CSAs. The Wallace tree's irregular and sometimes lengthy wiring cuts into the benefit of fewer logic levels.

> In Figs. 15.10 - 15.12, the s_i and e_i subscripts are dropped for readability. But remember that s and e are specific to a given partial product.

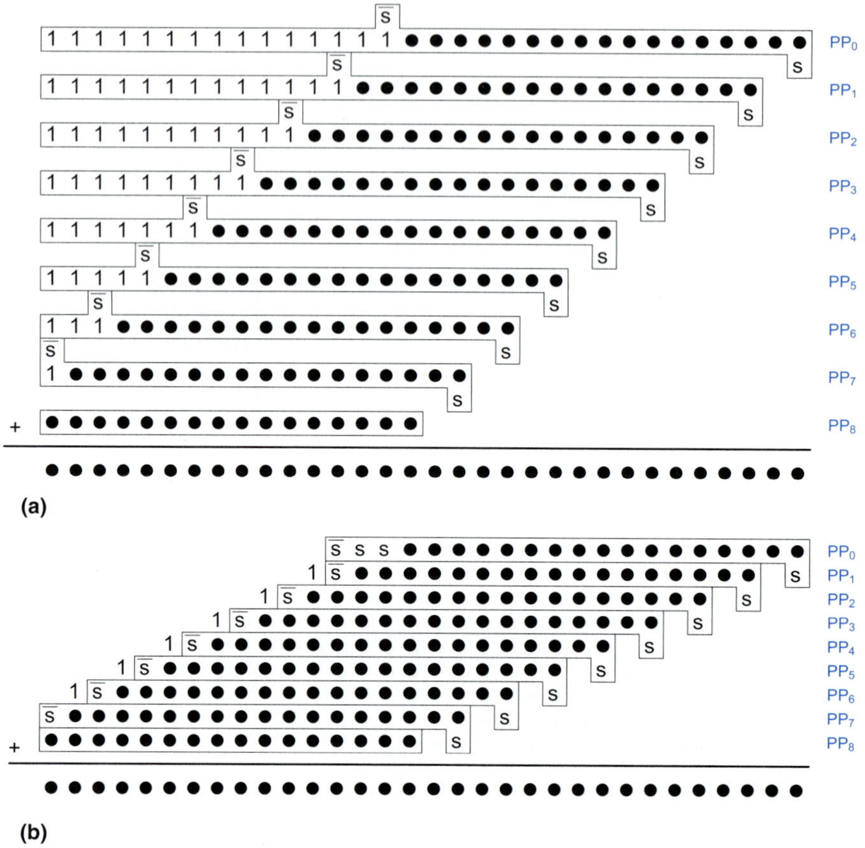

Fig. 15.11 Simplified sign extension: (a) partial products with sign extension, (b) precomputed.

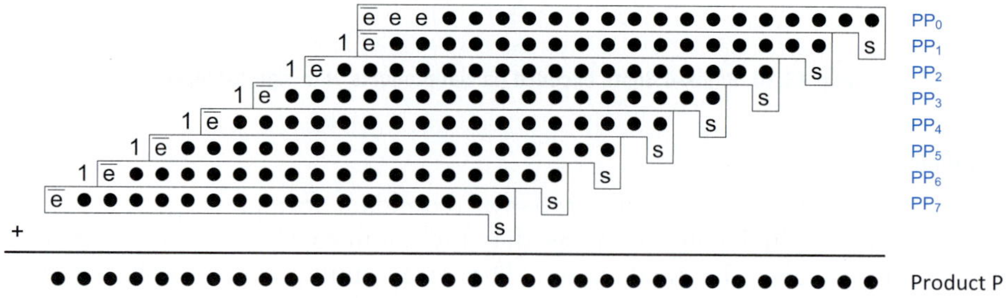

Fig. 15.12 Sign extension for Booth-encoded signed multipliers.

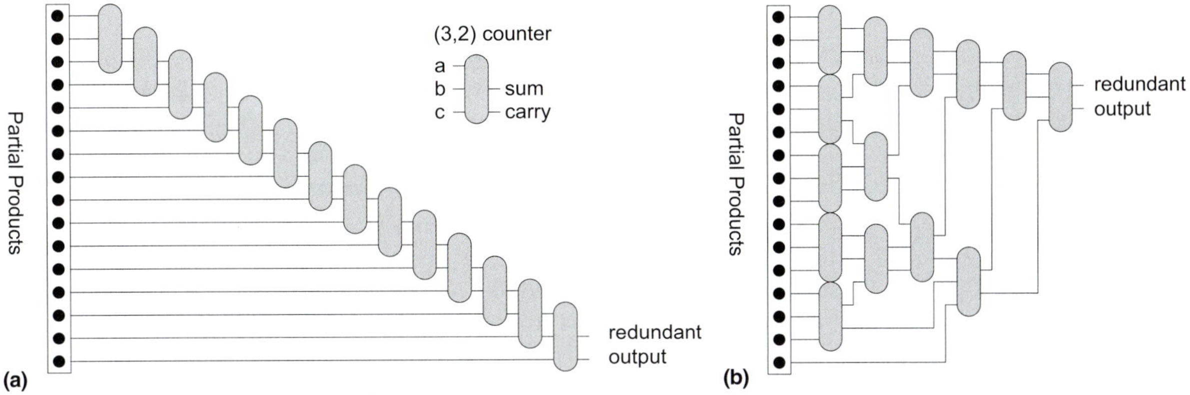

Fig. 15.13 Summing partial products: (a) linear array and (b) Wallace tree.

15.1.3.5 Other Multiplication Techniques

Radix-4 Booth encoding reduces the number of partial products to nearly half. Radix-8 Booth encoding extends this idea to reduce the partial products to nearly a third. However, it requires selecting from partial products of $\{4Y, 3Y, 2Y, Y, 0, -Y, -2Y, -3Y, -4Y\}$. The $3Y$ multiple is not trivial to obtain; it requires a CPA to compute $Y + 2Y$ before Booth selection begins. Thus Radix-8 Booth encoding only becomes beneficial for large multiplications.

Wallace tree multipliers suffer from irregular layouts with long, slow wires. Many other structures are used, including *[4:2] compressor trees* [Weinberger81, Santoro89], the 3-*Dimensional Method* [Oklobdzija96, Stelling98], and hybrids between arrays and trees such as [Dhanesha95].

All multipliers involve a CPA that can be optimized with the usual techniques summarized in [Weste12] and also by exploiting the fact that the upper and lower bits arrive earlier than the middle bits [Zimmerman96]. When a multiplication is followed by one or more additions, the additions may be folded into the CSA tree.

15.1.3.6 Multiplier Synthesis

Logic synthesis of multipliers is a mature field. Synthesis tools use the techniques described earlier and generally produce multipliers that are as good as an engineer could handcraft with extensive effort. Therefore, the most practical technique for designing multipliers is to write

```
assign p = x * y;
```

and allow synthesis to map onto your target technology, automatically taking advantage of Booth encoding and CSA trees.

15.1.4 Division

Division of N-bit integers takes a *dividend* X and a *divisor* D. It computes a *quotient* Q and *remainder* REM satisfying $X = Q \times D + REM$. If X and D are signed; then REM has the same sign as X.

In C and related programming languages, the / and % operators are defined to compute $X / D = Q$ and X % $D = REM$. Hence, for signed numbers, % may produce a negative result and is really the remainder operator, not a modulus operator. For example, $8 / 5 = 1$ and 8 % $5 = 3$. $-8 / 5 = -1$ and -8 % $3 = -2$. However, -8 modulus $3 = 1$.

The *modulus operator* finds $M \in [0, D)$ satisfying $X = Q' \times D + M$. In other word, the modulus is nonnegative. If the remainder REM is nonnegative, $M = REM$. If REM is negative, $M = REM + D$.

15.1.4.1 Radix-2 Restoring Unsigned Integer Division

In a fashion similar to elementary school long division, the quotient can be computed one bit at a time over N steps, as given in Fig. 15.14. In each step, the divisor is compared to a *partial remainder* W. If it fits, the quotient bit is 1, and the divisor is subtracted from the partial remainder; otherwise, the quotient bit is 0. Then the partial remainder is shifted left, and the process is repeated to obtain the next less significant quotient bit.

Using this simple algorithm, the quotient is computed one bit at a time starting with the most significant bit. For convenience and consistency with floating-point division covered in Section 16.2.3, we use big-endian notation, where bit 0 is the most significant bit and bit N-1 is the least significant. In other words, the quotient and dividend, Q and X, are written as follows:

$$Q = \sum_{i=0}^{N-1} q_i 2^{N-1-i} \quad X = \sum_{i=0}^{N-1} x_i 2^{N-1-i} \tag{15.3}$$

The remainder W is also referred to as the *residual* or *residue*.

At the end of this process, Q contains the N-bit quotient, and W contains the N-bit remainder. This algorithm is known as *radix-2 restoring unsigned integer division* because it produces 1 bit per cycle and discards W' when negative, restoring W to its value before the subtraction. Fig. 15.15 shows long division compared to the radix-2 restoring division algorithm for unsigned numbers using the example: 1101/0010 = 0110 R 0001 (13/2 = 6 R1).

Fig. 15.16 shows a clever hardware implementation using only a 2N-bit shift register, an N-bit subtractor, and some multiplexers. The left shift units are just wires and thus are free. The left path is initialized with a partial remainder of 0, and the right side is initialized with X at the start of a

$W = 0$
for $i = 0$ to N-1
  $W = \{W << 1, x_i\}$
  $W' = W - D$
  if $(W' \geq 0)$ $q_i = 1, W = W'$
  else $q_i = 0$

Fig. 15.14 Radix-2 restoring unsigned integer division algorithm.

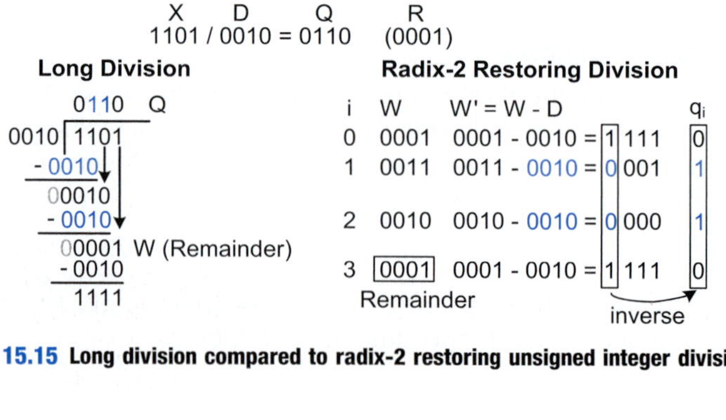

Fig. 15.15 Long division compared to radix-2 restoring unsigned integer division.

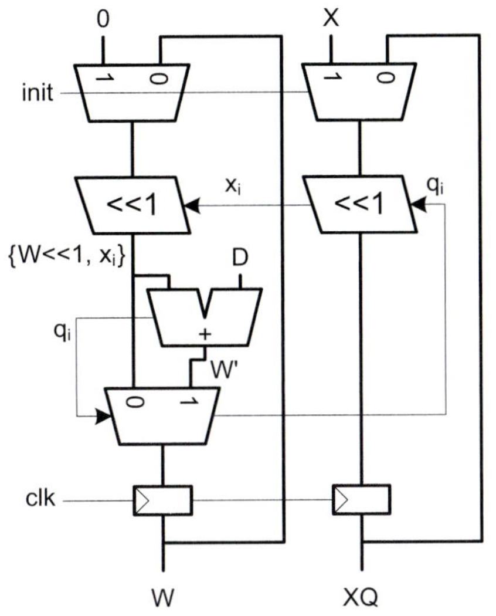

Fig. 15.16 **Radix-2 restoring integer divider.**

division. On each step, the partial remainder is shifted left by 1, and the most significant remaining bit of X is shifted into the bottom. The subtractor computes W' and its carry out indicates the comparison, which determines the quotient bit q_i. q_i is shifted into the vacancy at the least significant bit of X. After N steps, all the bits of X have been shifted out of the right register and replaced with the bits of Q. The left register holds the final remainder $REM = W$.

Arithmetic functional unit block diagrams traditionally flow from top to bottom, in contrast with our pipeline diagrams that flow from left to right.

The slowest step in Fig. 15.16 is the subtraction. Recall that subtraction involves inverting an input and adding, so the delay of the subtractor is essentially that of a CPA.

Example 15.1 RADIX-2 RESTORING DIVISION

Compute the quotient and remainder for $X/D = 1101/0010 = 13/2$ using radix-2 restoring division.

Solution: Table 15.2 shows the division steps. The black and blue XQ register shows how the black bits of X are gradually shifted out and replaced by blue bits of Q. At the end of the division, we see the W register holds the remainder 0001 and the XQ register holds the quotient 0110.

Table 15.2 **Radix-2 restoring unsigned integer division example.**

i	$W = \{W<<1, x_i\}$	$W' = W - D$	q_i	W Output	XQ Output
init				0000	1101
0	0001	1111	0	0001	1010
1	0011	0001	1	0001	0101
2	0010	0000	1	0000	1011
3	0001	1111	0	0001	0110

Table 15.3 Signed division.

X	D	Q	REM
13	2	6	1
13	-2	-6	1
-13	2	-6	-1
-13	-2	6	-1

15.1.4.2 Signed Integer Division

Recalling that the definition of Q and REM are $X = Q \times D + REM$ and that REM has the same sign as X, Table 15.3 shows examples of signed division. We see that Q is negative when X and D have different signs, and REM is negative when X is negative.

Signed integer division can be performed by taking the absolute value of the inputs and then performing unsigned integer division to compute unsigned Q and REM. Afterward, if the signs of X and D differ, negate Q. If the sign of X is negative, negate REM.

15.1.4.3 Other Integer Division Optimization

If the cycle time permits, the divider may duplicate the shifter, subtractor, and restoring mux to produce more than one bit per cycle.

The quotient of integer division is often a small integer. Some *early termination* dividers, commonly used in ARM and RISC-V microcontrollers, are optimized to detect this case and finish after a variable number of cycles [Kim24].

Recurrence techniques for floating-point dividers can also be used to speed up integer dividers and potentially produce more bits per cycle, as described in Section F.2.

15.2 RISC-V PRACTICES

As shown in Table I.5 (see the last pages of the book), RV32M has multiply, divide, and remainder instructions that operate on the full XLEN-bit words, treated as signed or unsigned values.

RV64M adds word operations `mulw`, `divw`, `divuw`, `remw`, `remuw` that act on the lower 32-bit word of each operand, which is sign-extended to fill 64-bit registers.

C and related languages prohibit mixing signed and unsigned variables in arithmetic statements; the signed input is automatically converted to unsigned. Hence, `div` and `divu` are sufficient; there is no `divsu`. However, `mulsu` is used to write multiword signed multiplication code to multiply the most significant word of one input (containing the sign bit) by the other unsigned words of the other input.

15.2.1 Signed and Unsigned Operations

As discussed in Section 15.1.3, multiplication of two N-bit numbers produces a $2N$-bit result, with the least significant N bits being the same independent of the operands' sign. These XLEN least significant bits are produced by the `mul` instruction. `mulhu`, `mulhsu`, and `mulh` compute the most significant XLEN bits for each combination of unsigned $\times$ unsigned, signed $\times$ unsigned, and signed $\times$ signed.

Similarly, division and remainder instructions have different flavors for signed and unsigned.

15.2.2 Division Special Cases

RISC-V defines results for integer division by zero and overflow. In RISC-V, division by 0 (`rs2` = 0) produces a result of `rd` = all 1s. Remainder by 0 (`rs2` = 0) produces a result of `rd` = `rs1`. For example, 0x12345678/0x00000000 produces a quotient of 0xFFFFFFFF and a remainder of 0x12345678.

Signed division of the largest negative number (e.g., 0x80000000) by −1 overflows because the result is larger than the largest positive number (e.g., 0x7FFFFFFF). RISC-V defines the result to be the largest negative number. For example, signed division of 0x80000000/0xFFFFFFFF produces a quotient of 0x80000000 and remainder 0x00000000.

RISC-V does not trigger hardware exceptions for either of these cases because no arithmetic instructions cause exceptions. The programmer or compiler can test for division by 0 with `beq rs2, zero, div0_error_handler` before performing the divide.

15.2.3 Software Arithmetic Implementations

When the M extension is not supported, RISC-V performs multiplication and division in software. For example, Code Example 15.1 shows how `a * b` for RV32I compiles into a call to the libgcc `__mulsi3` function. The function may take up to 195 instructions, so it is far slower than hardware multiplication.

Code Example 15.1 libgcc RISC-V INTEGER MULTIPLICATION LIBRARY FUNCTION

C Code

```c
int mul(int a, int b) {
  return a * b;
}

/* Equivalent C code
int mul(int a, int b) {

  int b0, result = 0;

  do {
    b0 = b & 0x1;
    if (b0) result += a;

    b = b >> 1; // Logical shift
    a = a << 1;
  } while (b);
  return result;
}
*/
```

Assembly Language

```
mul(int, int):
  addi   sp,sp,-16
  sw     ra,12(sp)   # Save return address
  call   __mulsi3    # Call mulsi3 library
  lw     ra,12(sp)   # Restore return address
  addi   sp,sp,16
  ret

__mulsi3:
  mv     a2, a0      # Save a in a2
  li     a0, 0       # Init result = 0 in a0
.L1:
  andi   a3, a1, 1   # Get lsb of b
  beqz   a3, .L2     # If lsb = 0, skip
  add    a0, a0, a2  # Else result += a
.L2:
  srli   a1, a1, 1   # b = b >> 1
  slli   a2, a2, 1   # a = a << 1
  bnez   a1, .L1     # Repeat while b != 0
  ret
```

tools from selecting their well-optimized adders or from mapping onto hardware multiplier blocks on an FPGA. Instead, our design observes that an $N \times N$ multiplier can be built from an N-1 $\times$ N-1 multiplier and three other terms to handle the most significant bits. The N-1 $\times$ N-1 multiplier can be expressed behaviorally to allow the logic synthesizer freedom to optimize.

Consider computing $P = X \times Y$, with X and Y being N-bit numbers and P being the $2N$-bit product. We can write $X = \{X_{n-1}, X'\}$, with X_{n-1} being the most significant bit and $X' = X_{n-2:0}$ being the remaining N-1 bits. If X is treated as unsigned, its value is $X' + (X_{n-1} \times 2^{N-1})$. If X is treated as a signed two's complement number, its value is $X' - (X_{n-1} \times 2^{N-1})$. The same applies to Y.

Table 15.5 defines partial products of these terms. Table 15.6 expands the multiplication in terms of these partial products for the flavors of signed and unsigned multiplications.

The signed flavors involve some negative partial products. As in Fig. 15.5, these are produced by inverting the bits and adding 1. This results in six 1s in the dot diagram, which can be simplified by pre-adding them. Fig. 15.18 shows this for a 4 $\times$ 4-bit signed $\times$ signed case. Part (a) shows the four partial products, in which the negative products are inverted, sign-extended and have a 1 added. Part (b) shows the four simplified partial products. The signed $\times$ unsigned case is similar.

In summary, each flavor of multiplication can be written as the sum of four partial products $P = P1 + P2 + P3 + P4$, as given in Table 15.7. $P1$ involves a large N-1 $\times$ N-1 bit multiplication, and the other partial

Table 15.5 Partial products.

Term	Definition	Comments
P'	$X' \times Y'$	N-1 $\times$ N-1 bit multiplication
PX	$Y_{n-1} \times X'$	N-1 bit partial product
PY	$X_{n-1} \times Y'$	N-1 bit partial product
P_{msb}	$X_{n-1} \times Y_{n-1}$	1-bit partial product

Table 15.6 Expansion of signed and unsigned multiplications.

Flavor	X	Y	P
`mulhu`	Unsigned	Unsigned	$P' + (PX + PY)2^{N-1} + P_{msb} 2^{2N-2}$
`mulh`	Signed	Signed	$P' + (-PX + -PY)2^{N-1} + P_{msb} 2^{2N-2}$
`mulhsu`	Signed	Unsigned	$P' + (PX + -PY)2^{N-1} - P_{msb} 2^{2N-2}$

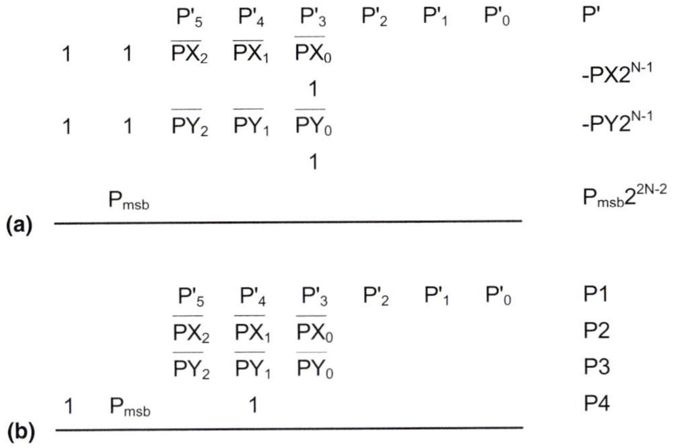

Fig. 15.18 Pre-adding 1s in dot diagram.

Table 15.7 Simplified partial products.

Flavor	*P1*	*P2*	*P3*	*P4*
`mulhu`	P'	$(PX)2^{N-1}$	$(PY)2^{N-1}$	$P_{msb}\ 2^{2N-2}$
`mulh`	P'	$(\overline{PX})2^{N-1}$	$(\overline{PY})2^{N-1}$	$P_{msb}\ 2^{2N-2} + 2^{2N-1} + 2^{N}$
`mulhsu`	P'	$(PX)2^{N-1}$	$(\overline{PY})2^{N-1}$	$\overline{P_{msb}}\ 2^{2N-2} + 2^{2N-1} + 2^{N-1}$

products are relatively small. *P*4 accounts for the 1s from taking the two's complement, as well as the product of the sign bits. The *N* least significant bits are the same for all flavors, so `mul` can be performed by doing any of these flavors and then selecting the bottom half. Logic synthesis is good at optimizing large multiplications and the additions.

Example 15.2 SIGNED MULTIPLICATION

Use the partial product method to perform signed multiplication of X × Y = 1101 × 0110.

Solution:

$X_{n-1} = 1 \qquad X' = 101$

$Y_{n-1} = 0 \qquad Y' = 110$

$P' = X' \times Y' = 101 \times 110 = 011110$

$PX = Y_{n-1} \times X' = 0 \times 101 = 000$

$PY = X_{n-1} \times Y' = 1 \times 110 = 110$

$P_{msb} = X_{n-1} \times Y_{n-1} = 1 \times 0 = 0$

$P1 = P' = 00011110$

$P2 = (PY)\ 2^{N-1} = 00111000$

$P3 = (PY)\ 2^{N-1} = 00001000$

$P4 = P_{msb} 2^{2N-2} + 2^{2N-1} + 2^{N} = 10010000$

$P = P1 + P2 + P3 + P4 = 11001000 + 00011110 = 11101110 = -18$

In decimal, $X = -3$, $Y = 6$, $P = -18$.

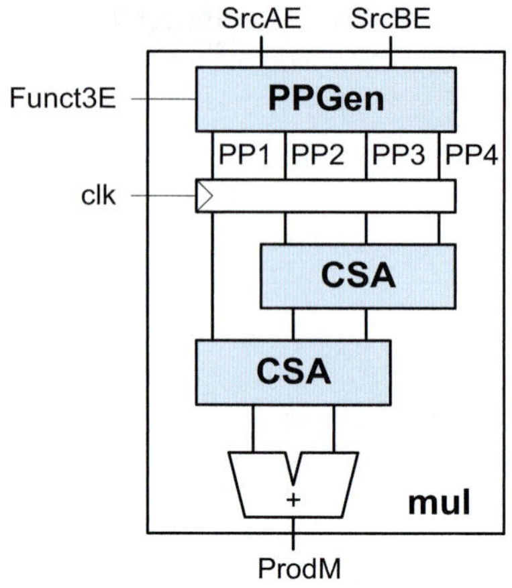

Fig. 15.19 **Wally integer multiplier unit (`mul`).**

Fig. 15.19 shows the integer multiplication unit (`mul`), which has a two-cycle latency. The Execute stage generates partial products P1–P4 according to Table 15.7, and the Memory stage adds these products. The slowest step is the N-1 × N-1 multiplier for P1.

15.4.2 Division and Remainder

Fig. 15.20 shows the Wally integer divide/remainder unit that computes $Q = X/D$ and $REM = X \% D$. It is based on the design from Fig. 15.16. It is configurable to produce k = `IDIV_BITSPERCYCLE` = 1, 2, or 4 bits per cycle by duplicating the core logic that many times (taking care that the divider doesn't become the critical path of the processor.) The latency is given in Table 15.8; division by 0 is a special case that completes in only 2 cycles.

The inputs arrive at the top (Fig. 15.20) during the Execute stage, with X in `SrcA` and D in `SrcB`. RV64 handles W-type operations in the shaded blue box by treating the inputs as 32-bit values: D is sign-extended to fill the 64-bit `DinE`, and the 32-bit value of X is placed in the top half of `XinE` so that only 32 division steps are required to process. Negators and multiplexers take the absolute value of D and X during signed division. On the first cycle, indicated by `DivStartE`, W is initialized to 0, XQ is initialized to X, and the absolute value of D is registered.

The divider then iterates, with k copies of the restoring division hardware producing k bits of result on each cycle. The pipeline stalls in the Execute stage while the iterations take place. After enough iterations, W and XQ hold the absolute value of the remainder and quotient, respectively.

Even though the Execute stage registers are stalled during division, -|D| is held in a register because the divisor comes from `SrcBE`, which comes from forwarding muxes that may change value as the Memory and Writeback stages' results change.

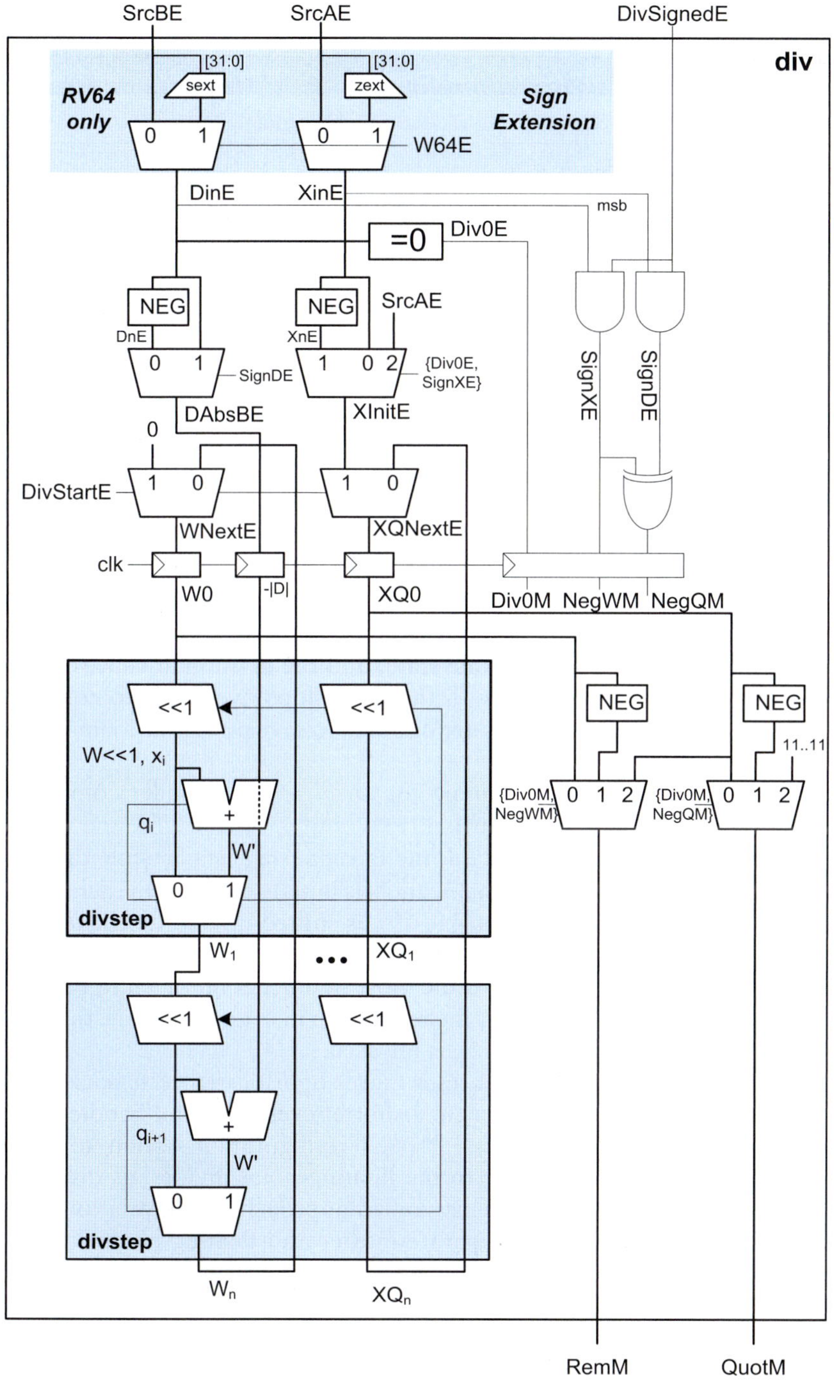

Fig. 15.20 **Wally integer divide/remainder unit (`div`).**

Table 15.8 Division and remainder latencies.

Architecture	div /divu /rem /remu	divw /divuw /remw /remuw
RV32M	$32/k + 2$	n/a
RV64M	$64/k + 2$	$32/k + 2$

Finally, the stall is released, and the divider continues to the Memory stage, in which it flips the sign of the quotient and/or remainder when necessary for signed operations. Output multiplexers also choose the special case values of X and all 1s for remainder/division by 0.

Fig. 15.21 shows the control logic and timing. A simple finite state machine in Fig. 15.21(a) manages the sequencing through Idle, Busy, and Done states to count the number of iterations needed and stall the pipeline appropriately. It begins in the Idle state. When a division operation enters the Execute stage, DivE asserts, which causes DivStartE to pulse high for a cycle, as shown in Fig. 15.21(b). Normally, the divider then enters the Busy state, which starts a step counter. DivBusyE is raised on the starting cycle and while the divider is Busy, causing a StallE Execute stage stall. After N steps, the FSM enters the Done state, and the division advances to the Memory stage. For division by 0, the FSM skips over Busy to complete in 2 cycles. The divider registers are disabled to save power (and simulation time) when idle.

Some dividers do early termination to halt when the residual is less than the divisor. Our formulation of division has a low hardware cost. Adding early termination would require an expensive shifter to extract the residual and quotient from a variable location in the W and XQ registers.

Fig. 15.21(c) shows the timing for an $N = 6$-bit divider computing $33/6 = 5$ remainder 3. In cycle 1, XInitE and DabsBE are 33 and 6, and DivStartE is pulsed. In cycles 2–8, the Execute stage stalls while the division occurs. XQ is initialized with 33, and W is initialized to 0. On each step, W is shifted left by 1, and the next bit of X is shifted in. D is subtracted and compared to find the blue quotient bit q_i and the next residual W. Meanwhile, XQ is shifted left to discard the most significant bit of X and prepend the blue quotient bit. In cycle 8, the results are in QuotM and RemM, the FSM enters the Done state, and StallE is released.

The floating-point divider uses a more sophisticated recurrence algorithm that can produce more bits in a shorter cycle time and handles early termination. Wally has an IDIV_ON_FPU configuration option to route integer division and remainder to the floating-point divider, as discussed in Sections 16.6.4 and F.2.9. This unified approach can have a lower latency and smaller area than separate integer and floating-point dividers, especially for configurations such as rv32gc where the floating-point datapath is already wider than XLEN.

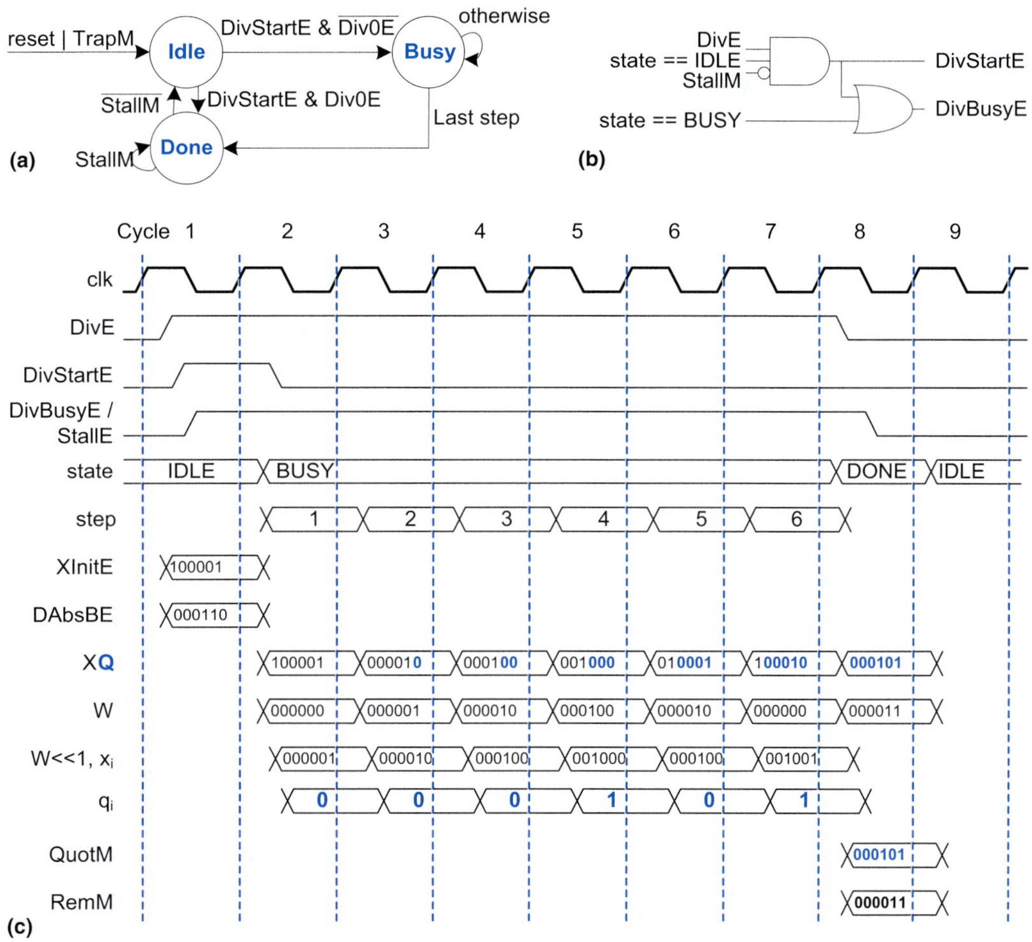

Fig. 15.21 **Integer division control: (a) FSM, (b) start and busy logic, and (c) timing.**

15.4.3 Multiply and Divide Unit

Fig. 15.22 shows a block diagram of the entire multiply and divide unit (MDU), omitting some control signals. The unit contains the multiplier and divider that operate in the Execute and Memory stages. These are followed by a result multiplexer that picks the quotient, remainder, or upper or lower half of the product based on the `funct3` field of the instruction. Finally, in the RV64 configurations, to handle W-type instructions (`mulw`, `divw`, `divuw`, `remw`, `remuw`), the bottom 32 bits of the result are sign-extended into the top 32. The MDU's result is captured in a Writeback stage flip-flop, which connects to the IEU and forwarding logic.

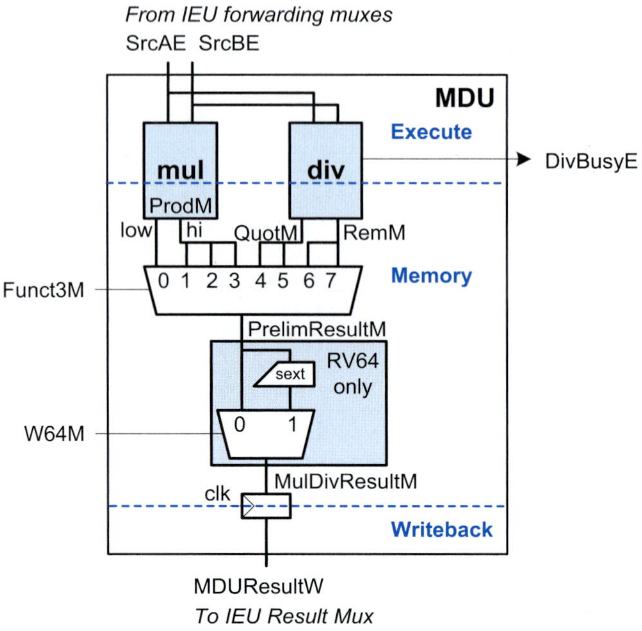

Fig. 15.22 **MDU block diagram.**

15.4.4 Hazard Unit

The multiplier is pipelined with a two-cycle latency (Execute + Memory) but a throughput of one instruction per cycle. The divider also has a minimum two-cycle latency in special cases (division by zero), but it usually stalls in the Execute stage for many cycles while completing the division. The multicycle division is not pipelined, so only one divide can operate at a time. However, the next MDU instruction can issue when the division enters the Memory stage.

Fig. 15.23 shows the revised hazard unit supporting the MDU, with MDU changes highlighted in blue. The MDU returns results in the Memory stage, so like other structural hazards, it stalls the Decode stage when the next instruction depends on the result of the multiply or divide (`MDUStallD` = 1).

The pipeline stalls at the Execute stage while the divider is busy. The divider FSM is reset to idle if the Execute stage is flushed (`FlushE`). However, when the IFU mispredicts the instruction class (`BPPredWrongE`) and needs to flush the instructions currently in the Fetch and Decode stages by asserting `FlushD` and `FlushE`, it must wait for the division to complete ($\overline{\text{DivBusyE}}$).

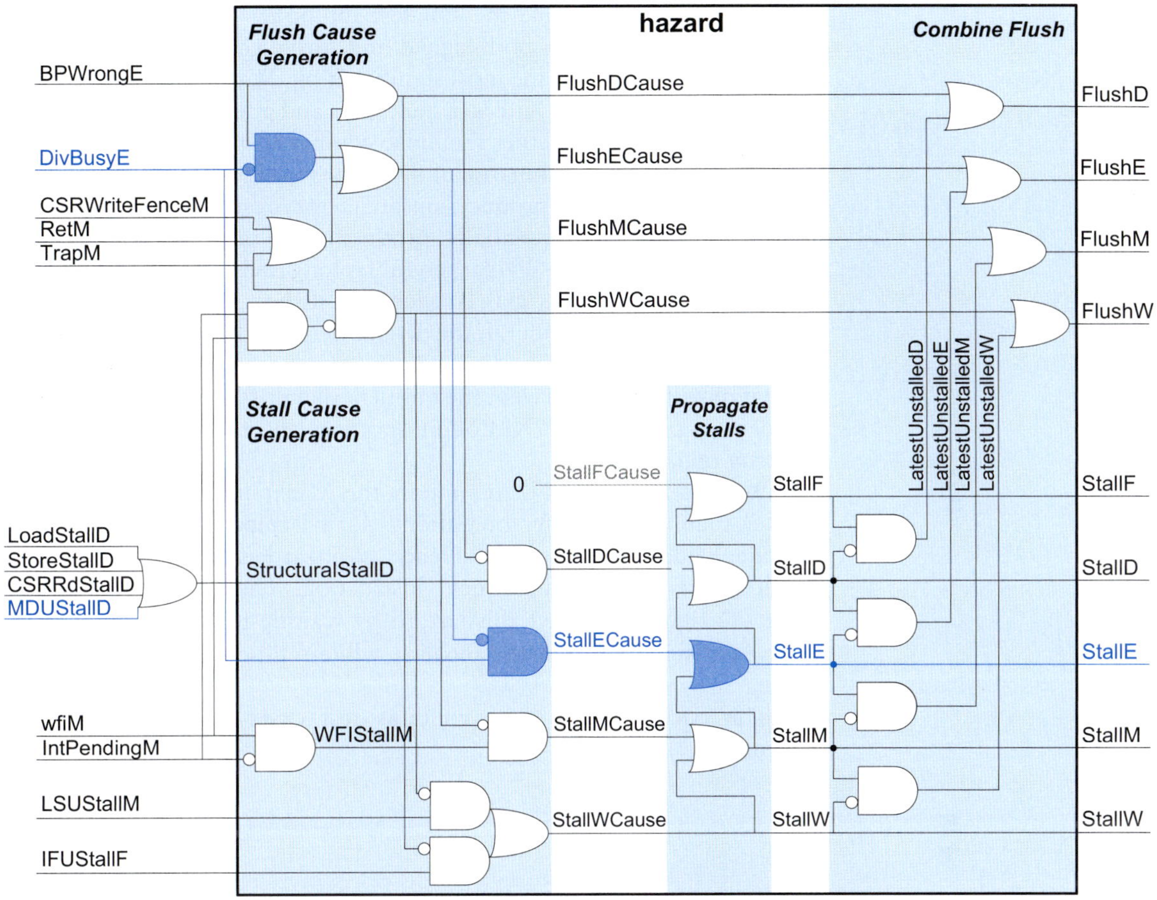

Fig. 15.23 **Hazard unit, with MDU changes highlighted in blue.**

15.5 SUMMARY

The RISC-V base instruction set omits multiply and divide instructions because they require substantially more hardware than other instructions and are not necessary in all applications. The RISC-V M extension defines multiply, divide, and remainder instructions. Different flavors are used for signed and unsigned operands and to obtain the upper or lower half of a 2XLEN-sized product produced by an XLEN × XLEN multiplication.

Multiplication involves forming partial products and adding them together. The field has been studied extensively. Booth encoding is used to reduce the number of partial products. Tree multipliers sum the partial products in parallel for greater speed. Logic synthesis understands these optimizations, so the designer can simply write `x * y` and allow the synthesis tool to produce the smallest multiplier that meets delay constraints. RISC-V defines both signed and unsigned multiplication, which differ in the handling of the most significant bit. Wally handles this by synthesizing an XLEN-1 × XLEN-1 multiplier and adding three partial products that depend on whether the most significant bit is signed.

Unfortunately, combinational division is both large and slow, so synthesizing X/D generally produces unsatisfactory results. Instead, division is usually implemented sequentially, producing one or a few bits per cycle. A radix-2 restoring integer divider resembles long division in base 2. It initializes the partial remainder to 0. At each step, it shifts 1 bit from the dividend X into the partial remainder. If the divisor D is less than or equal to the partial remainder, the quotient digit is 1, and the divisor is subtracted from the partial remainder. Otherwise, the quotient digit is 0. After XLEN steps, the divider has produced an XLEN-bit quotient, and the partial remainder is the true remainder.

Wally adds the MDU to support the M extension. Multiplication is slower than other integer operations, so it is pipelined to use both the Execute and Memory stages. Hence, a hazard exists when a multiply result is used by the subsequent instruction. The hazard unit adds an `MDUStall` signal to resolve this hazard by waiting for a cycle. It also adds a `DivBusy` signal to stall the pipeline while multicycle division is taking place.

The riscv-arch-test suite contains test cases for the M extension instructions.

16

Extensions: F/D/Q/Zfh/Zfa (Floating-Point)

16.1 **Principles**
16.2 **Floating-Point Unit Design**
16.3 **SoftFloat**
16.4 **RISC-V Practices**
16.5 **Test Plan**
16.6 **Wally Implementation**
16.7 **Summary**

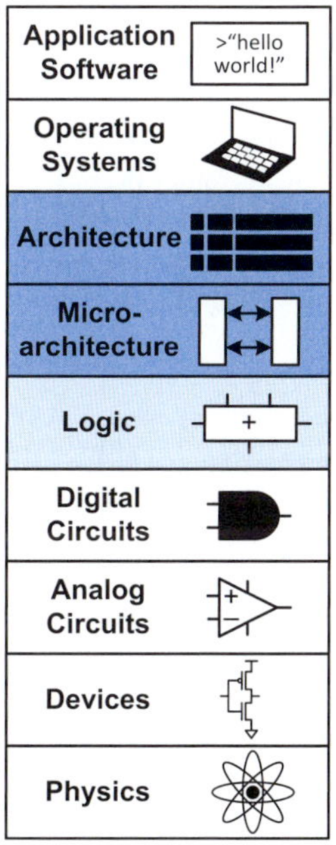

Floating-point numbers are similar to scientific notation in that they use a sign, fraction, and exponent to represent a number. They represent real numbers with a wide dynamic range but limited precision. Early mainframe and minicomputers used a chaotic soup of idiosyncratic vendor-specific floating-point formats [Beebe17]. The Institute for Electrical and Electronic Engineers (IEEE) standardized floating-point arithmetic in 1985, facilitating vendor-independent numerical programming [Severance98].

The original IEEE 754-1985 standard defines single- and double-precision floating-point number formats, including special representations for zero, infinity, not-a-number (NaN), and very small *subnormal* numbers, as well as rounding modes and result flags. Although floating-point arithmetic can be performed in software with a lengthy sequence of integer operations, most processors that run numerically intensive code support instructions that execute floating-point arithmetic in hardware.

This chapter reviews the floating-point standard, then discusses issues related to the RISC-V floating-point extensions. Although floating-point units (FPUs) have been included in commercial processors for more than 40 years, this book is the first to show a complete FPU design. Floating-point hardware has many tricky corner cases and is notoriously difficult to get right, but fortunately, there is a mature University of California, Berkeley (UCB) TestFloat suite of test cases. This chapter concludes with a detailed description of the Wally FPU, including all the computational blocks. Many implementation details are deferred to Appendix F so that this chapter flows more smoothly. Readers with deep interest in floating-point arithmetic will enjoy [Muller18] and [Ercegovac03].

16.1 PRINCIPLES

The IEEE Standard for Binary Floating-Point Arithmetic released in 1985 defines the industry standard for single- and double-precision (32- and 64-bit) floating-point representations and operations [IEEE754-1985]. The

The IEEE 754 standard is unusually difficult to read. This section summarizes the general concepts. When tracing an obscure floating-point bug, you may need to turn to the standard to sort it out.

Half-, single-, double-, quad-, and octuple-precision floating-point formats that use 16, 32, 64, 128, and 256 bits to represent a number are also called simply halves, floats, doubles, quads, and octs.

Note that some architectures, notably x86, support an 80-bit double extended precision format.

draft standard was first implemented in 1980 on the Intel 8087 math coprocessor and has been used in nearly every commercial system since the 1980s. The 2008 version of the IEEE 754 standard added fused multiply-add (FMA), which is faster and more accurate by avoiding rounding between the two steps (multiply and add), as well as half- and quad-precision (16- and 128-bit), and decimal floating-point. The 2019 version [IEEE754-2019] makes minor extensions such as octuple-precision. Most commercial processors, including the RISC-V instruction set architecture (ISA), presently support half-, single-, and double-precision IEEE 754 arithmetic. Some IBM POWER and Z-series mainframe processors support quad-precision in hardware [Lichtenau16], but quads and octs have not yet been widely needed or adopted.

16.1.1 Normalized Floating-Point Formats

Binary floating-point numbers are normally written in the following form:

$$(-1)^s \times (1.f) \times 2^{(e-bias)} \tag{16.1}$$

It may seem odd to use biased exponents rather than two's complement representation for exponents. This approach has been widespread since the 1960s because it allows programs to compare floating-point numbers using ordinary integer comparison [Beebe17]. With a biased exponent representation, larger floating-point numbers have larger values when their bit patterns are treated as signed integers.

The *sign bit s* is 0 for positive and 1 for negative numbers. The *fraction f* and *biased exponent e* have a limited number of bits of precision, N_f and N_e. The biased exponent is treated as an unsigned number, and a bias of approximately half the maximum exponent value is subtracted to give the true exponent so that positive and negative values can be represented. Because normalized numbers always have a leading 1 before the fraction, this 1 is not stored in the binary representation. This leading 1 is sometimes called the "hidden 1." The $(1.f)$ term is called the *normal significand*, or simply the *significand* and will be abbreviated m. m has 1 integer bit and N_f fraction bits. m is often called the *mantissa*, but this usage is slightly improper [Muller18, p. 17].

Table 16.1 summarizes the number of bits and the exponent bias in each IEEE floating-point format, along with the approximate range and decimal digits of precision. As shown in Fig. 16.1, the sign bit is in the most significant position, followed by the biased exponent, and then the fraction.

Table 16.1 IEEE 754 floating-point format.

Name	Common Name	Fraction bits N_f	Exponent bits N_e	*bias*	$e_{\min}$	$e_{\max}$	Range	Decimal Digits of Precision
binary16	half	10	5	15	−14	15	$\pm10^{4}$	3
binary32	float/single	23	8	127	−126	127	$\pm10^{38}$	7
binary64	double	52	11	1,023	−1,022	1,023	$\pm10^{308}$	16
binary128	quad	112	15	16,383	−16,382	16,383	$\pm10^{4932}$	34
binary256	octuple/oct	236	19	262,143	−262,142	262,143	$\pm10^{78913}$	71

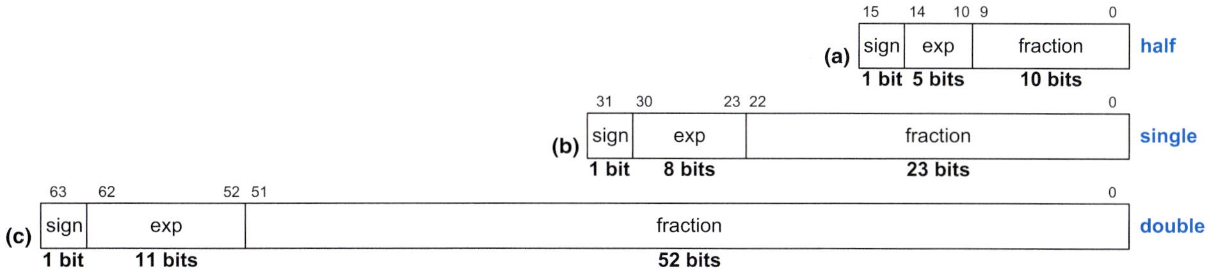

Fig. 16.1 Floating-point fields: (a) half-, (b) single-, and (c) double-precision.

Biased exponent fields of all 0s and all 1s are reserved for special cases discussed in the next section. The largest and smallest allowable true exponents for normalized numbers are called e_{max} and e_{min}. $e_{max} = bias$ and $e_{min} = 1 - bias$. For example, $e_{max} = 127$ and $e_{min} = -126$, respectively, for single-precision numbers.

Double-precision has 53 bits or $\log_{10}(2^{53}) \approx 16$ decimal digits of precision. This is sufficient to count the number of virus particles that could be lined up from the earth to the moon and back. The largest number that can be represented is approximately $2^{1024} \approx 10^{308}$.

The IEEE standard also defines 32-, 64-, and 128-bit *decimal floating-point* representations and operations that avoid rounding errors when operating on decimal values such as in financial calculations or on a pocket calculator. This chapter does not address decimal floating-point.

Example 16.1 SINGLE-PRECISION FLOATING-POINT

Write −9.625 as a single-precision floating-point number. Express your answer as an 8-digit hexadecimal value.

Solution: The sign bit is 1 because the number is negative. The magnitude, written in binary, is $1001.101 = 1.001101 \times 2^3$. Hence, the biased exponent is $3 + 127 = 130 = 10000010_2$. Dropping the implicit leading 1 yields a fraction of 001101. Fig. 16.2 shows how these fields are packed into a 32-bit number. For brevity, the number is usually expressed in hexadecimal: 0xC11A0000.

Example 16.2 DOUBLE-PRECISION FLOATING-POINT

Determine the decimal value of the double-precision floating-point number 0x3FDB000000000000.

Solution: Fig. 16.3 shows converting the number to binary and unpacking it into fields. The sign bit is 0, so the number is positive. The biased exponent is 1021. Subtracting a bias of 1023 gives the true exponent of −2. The fraction is 1011. Hence, the number is $1.1011 \times 2^{-2} = 11011 \times 2^{-6} = 27/64 = 0.421875$.

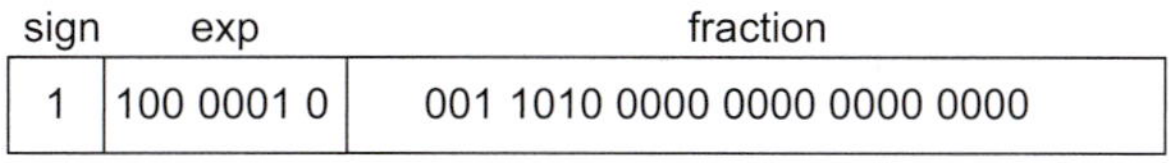

Fig. 16.2 Fields for single-precision representation of −9.625 (0xC11A0000).

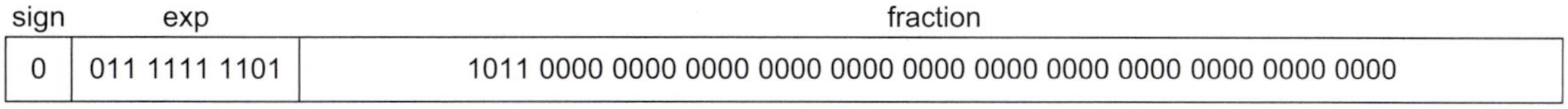

Fig. 16.3 Fields for double-precision number 0x3FDB000000000000 (1.1011×2^{-2}).

Table 16.2 IEEE 754 special cases.

Sign	Biased Exponent	Fraction	Special Case
0	All zeros	All zeros	+0
1	All zeros	All zeros	−0
any	All zeros	Nonzero	Subnormal
0	All ones	All zeros	+∞
1	All ones	All zeros	−∞
any	All ones	Nonzero	NaN

16.1.2 Special Cases

The IEEE floating-point standard defines some special cases, encoded with the largest or smallest value in the biased exponent field, as shown in Table 16.2. All the other floating-point numbers are referred to as *normalized* and have a biased exponent that is not all zeros or ones.

16.1.2.1 Zero

Because IEEE 754 floating-point numbers technically use a sign-magnitude format, representations of both positive and negative zero exist. −0 occurs when a very small negative number is rounded to zero. Comparisons ignore the sign of zero, so +0 and −0 are considered equal. Note that +0 is conveniently written with all zeros.

16.1.2.2 Infinity

Infinity (∞ *or* INF) represents a number larger than any normalized number or the limiting cases of real arithmetic with operations of arbitrarily large magnitudes. Infinities can be generated by the following operations:

- Overflow: operations on finite operands producing a large result
- Division by zero
- log(0) or the exponent of a large positive number
- Most operations with infinity as an input

Positive infinity is expressed in half-, single-, and double-precision as 0x7C00, 0x7F800000, and 0x7FF0000000000000, respectively.

16.1.2.3 NaN

Not a Number (*NaN*) indicates a result that has no other sensible answer. For example, 0/0 or $\sqrt{-1}$ produces NaN. NaNs are returned by the following types of operations:

- Indeterminate operations
  - Multiplication $0 \times \infty$
  - Division $\frac{0}{0}$ or $\frac{\infty}{\infty}$
  - Remainder $x \ \% \ y$ if x is ∞ or y is 0
  - Subtraction $\infty - \infty$
- Real operations with complex results, such as
  - Square root of a negative number
  - Logarithm of a negative number
  - Inverse sine or cosine of numbers outside the range $[-1, 1]$
- Most operations with at least one NaN input

The nonzero fraction in the NaN is called its *payload* and can carry implementation-specific information about the cause of the NaN. NaNs are also classified as *quiet* (*qNaN*) or *signaling (sNaN)*. Operations on sNaNs trigger an invalid operation exception even if the result is not NaN. Thus, a program may set uninitialized variables to a sNaN value to detect if they are accidentally used. Unfortunately, the encoding of sNaNs is also implementation-specific, making them hard to use in portable code. Operations producing a NaN result, such as those listed above, deliver a qNaN.

The most significant bit of the fraction field is 1 for quiet NaNs and 0 for signaling NaNs. RISC-V instructions that generate NaN results produce the *canonical NaN*, which has the most significant bit of the fraction set and all the other fraction bits zeroed.

16.1.2.4 Subnormal

Subnormal numbers, also called *subnorms*, are very small positive or negative numbers. They fill the *underflow gap* below the smallest positive and smallest negative normalized numbers. Their true exponent is always e_{min}, but they have no implicit leading 1 before the fraction. For example, the single-precision floating-point number 0x00300000 is subnormal because it has an exponent field of 0. The fraction field is 01100...00, so it represents the value $0.0110...00 \times 2^{-126} = 3 \times 2^{-129}$.

Numerical analysists find that subnormal numbers make it easier to write robust, stable numerical software [Coonen81]. For example, consider the following code to avoid division by zero:

Subnormal numbers were the most controversial part of the IEEE 754 specification because they are the most difficult to implement and add cost and delay to the FPU. Thus, they are sometimes implemented in software, which degrades performance when they occur.

```
if (a ≠ b) y = 1/(a – b);
```

If `(a – b)` is smaller than the smallest normalized number and subnorms don't exist, `(a – b)` will be truncated to 0 and cause a division by zero error despite the check. Subnormal numbers provide *gradual underflow*; loss of precision occurs slowly for tiny numbers. IEEE floating-point arithmetic will give a reasonable answer even when `a` and `b` are nearly equal.

Like subnormal numbers, *denormalized* numbers have a leading 0 in the significand, but unlike subnormal numbers, they have an arbitrary exponent. Subnormal numbers are denormalized numbers with a true exponent of e_{min}, so their magnitudes are always less than the smallest normalized number. The original IEEE 754 standard used denormal to refer to such numbers, but the 2008 revision of the specification switched to the more proper term *subnormal*. The literature still has a mix of the two terms.

16.1.3 Rounding

Proper rounding is costly in hardware but critical for writing well-behaved numerical software.

The result of floating-point arithmetic may not be exactly represented with the precision available and must be rounded to a representable value. IEEE 754 defines five rounding modes, as listed below:

- **RNE:** Round to nearest even
- **RNTA:** Round to nearest ties away
- **RZ:** Round toward zero
- **RN:** Round toward $-\infty$
- **RP:** Round toward $+\infty$

Round to nearest even is preferred for most numerical computation and is the default rounding mode in the IEEE 754 standard. It rounds to the closest representable value. If two values are equally close (a "*tie*"), it rounds to the even one. This is also called *unbiased rounding* because it avoids introducing systematic errors by rounding up or down on a tie. Thus, it maintains an average zero error in computation.

Round to nearest ties away rounds to the number with the larger magnitude in the event of a tie. This is required by law for certain financial calculations, such as converting Euros. The mode is also known as *round to nearest ties to maximum magnitude* (RNM).

Round toward zero truncates the fraction, rounding positive numbers down and negative numbers up. This mode is also known as truncation.

Rounding toward positive or *negative infinity* (sometimes called *directed* rounding) rounds a number up or down, respectively. They can be used to perform *interval arithmetic* [IEEE1788.1-2017], where the true result is within a bounded interval. Specifically, a computation is repeated with both rounding modes; thus, the true answer is guaranteed to lie in an interval between an upper and lower bound. If the interval is small, the program knows that the answer is close to exact and was not corrupted by the rounding error. If the interval is large, the program can detect numerical instability, and either flag the answer as unreliable or repeat the calculation with greater precision.

MAXNUM is defined as the largest normalized number ($1.1111\ldots111 \times 2^{e_{\max}}$). Results that overflow the range of finite numbers are rounded to either $\pm\infty$ or $\pm MAXNUM$ depending on the rounding mode. For example, round toward zero (RZ) rounds a huge positive number to *MAXNUM*, whereas RP rounds it to ∞.

Rounding is performed by computing the exact result and then picking the appropriate representable result. The rounded result may be one of the following:

- ***TRUNC*:** The result with the fraction truncated to the number of bits in the floating-point representation.
- ***RND*:** *TRUNC* plus 1 in the least significant bit of the representable fraction. If the fraction had been all 1s, adding 1 will produce a fraction of all 0s and increment the exponent. If the exponent was $e_{\max}$ before incrementing, the result will be ∞.
- ***MAXNUM* or ∞:** For results that overflow before rounding.

The 1 in the least significant bit is known as an *ulp* (*unit in the last place*, rhymes with *gulp*). Informally, for any real number x, ulp(x) is the gap between the two closest finite floating-point numbers straddling x. Given two consecutive finite floating-point numbers a and b such that $|x| \in (a, b]$, ulp(x) $= b - a$. ulp is helpful to reason about rounding error without getting bogged down in exponent or precision. Whereas rounding to nearest produces an error no more than ½ ulp, the other rounding modes produce an error of no more than 1 ulp.

[Muller05] explores corner cases in the definition of ulp related to infinities, NaN, and numbers very close to an exponent boundary.

Example 16.3 FINDING ulp(x)

For the half-precision floating-point number $x = 2.125$, find ulp(x).

Solution: $x = 2.125 \times 2^0 = 1.0625 \times 2^1 = 1.0001000000 \times 2^1$. x is an exactly representable floating-point number, so $b = x$. Half-precision numbers have 10 fraction bits, so the previous number is $a = 1.0000111111 \times 2^1$. Hence, ulp(2.125) $= b - a = 0.0000000001 \times 2^1 = 2^{-9}$.

Computing the exact result is troublesome, especially because division and square root can have an infinite number of bits in the exact result (e.g., for 1/3 or $\sqrt{2}$). Fortunately, rounding can be performed using only the N_f fraction bits and three additional bits: the *guard* (G), *rounding* (R), and *sticky* (T) bits. L is the least significant bit (lsb) of the N_f fraction bits of the exact result (*TRUNC*). G is the next bit of the exact result followed by the *rounding* bit R and the *sticky* bit T, which is 1 if any of the remaining bits are nonzero (i.e., T is the OR of all the remaining bits of the exact result). Together, G, R, and T are called the *rounding bits*.

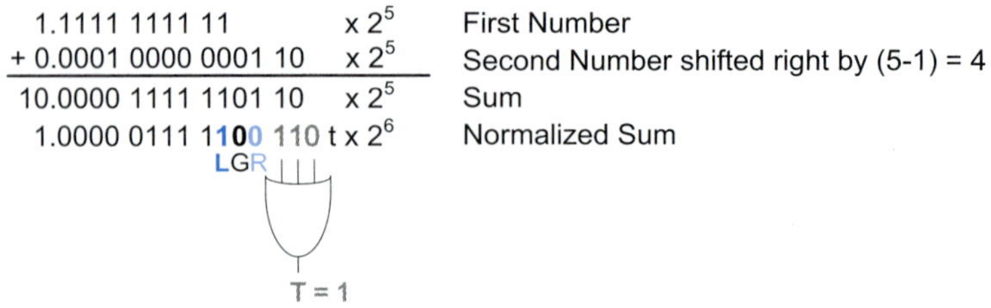

Fig. 16.4 **Addition and rounding bits.**

Example 16.4 HALF-PRECISION ADDITION AND ROUNDING

Compute the exact half-precision sum $1.1111\ 1111\ 11 \times 2^5 + 1.0000\ 0001\ 10 \times 2^1$. Find the L, G, R, and T bits.

Solution: The sum is computed by shifting the smaller (in this case, second) number right by $(5 - 1) = 4$ bits to make the exponents equal and then adding. Fig. 16.4 shows the exact sum is $1.0000\ 0111\ 1100\ 110 \times 2^6$. The half-precision result includes only the first 10 fraction bits. $L = 1$, $G = 0$, $R = 0$, and $T = \text{OR}(1, 1, 0) = 1$.

Dividing two numbers in the range [1, 2) produces a quotient in the range (1/2, 2). If this quotient is less than 1, it requires a left normalization shift by one bit to bring it back into the range [1, 2). This moves G into the least significant fraction bit. The guard bit hence guards against a loss of precision. Let L', R', and T' be the effective lsb, rounding bit, and sticky bit accounting for this possible shift. Table 16.3 summarizes their selection. R' refers to the bit used for rounding, which may or may not be R, and T' is the OR of all bits to the right of R'. If there is no shift, L' is L, and rounding (R') is done using G. The effective sticky bit T' is the OR of all the lower bits, $R \mid T$. If a shift occurs, $L' = G$ because the guard bit moves into the lsb. This leaves $R' = R$ and $T' = T$.

Floating-point subtraction also uses a guard bit to maintain precision while normalizing with a left shift of 1 after producing a significand with a 0 integer bit. However, this chapter focuses on a more general fused multiply-add, which involves a large normalization shift before L, G, R, and T are determined. The fused multiply-add (FMA) module uses the top row of Table 16.3 (the no normalization shift entry), but L, G, R, and T remain important for underflow detection as will be described in Section F.4.7.1.

The rounding decision is made based on the rounding mode, the sign, the overflow flag, and the L', R', and T' bits, according to Table 16.4. For example, RZ always truncates and rounds overflowing values down to *MAXNUM*. RP truncates negative numbers and rounds up positive numbers whenever the truncated result is not exact (R' or T' bits are 1);

Table 16.3 **Rounding bits.**

Normalization Shift	L'	R'	T'
No	L	G	$R \mid T$
Yes	G	R	T

Table 16.4 Rounding.

Sign	Overflow	L'	R'	T'	RNE	RNTA	RZ	RP	RN
0	0	x	0	0	*TRUNC*	*TRUNC*	*TRUNC*	*TRUNC*	*TRUNC*
0	0	x	0	1	*TRUNC*	*TRUNC*	*TRUNC*	*RND*	*TRUNC*
0	0	0	1	0	*TRUNC*	*RND*	*TRUNC*	*RND*	*TRUNC*
0	0	1	1	0	*RND*	*RND*	*TRUNC*	*RND*	*TRUNC*
0	0	x	1	1	*RND*	*RND*	*TRUNC*	*RND*	*TRUNC*
0	1	x	x	x	$+\infty$	$+\infty$	$+MAXNUM$	$+\infty$	$+MAXNUM$
1	0	x	0	0	*TRUNC*	*TRUNC*	*TRUNC*	*TRUNC*	*TRUNC*
1	0	x	0	1	*TRUNC*	*TRUNC*	*TRUNC*	*TRUNC*	*RND*
1	0	0	1	0	*TRUNC*	*RND*	*TRUNC*	*TRUNC*	*RND*
1	0	1	1	0	*RND*	*RND*	*TRUNC*	*TRUNC*	*RND*
1	0	x	1	1	*RND*	*RND*	*TRUNC*	*TRUNC*	*RND*
1	1	x	x	x	$-\infty$	$-\infty$	$-MAXNUM$	$-MAXNUM$	$-\infty$

RN does the opposite. RNE rounds down when the R' bit is 0 and rounds up when the R' and T' bits are 1. When only the R' bit is 1, the number is exactly halfway between two representable numbers, and RNE rounds to an even value by rounding up if $L' = 1$ and rounding down if $L' = 0$. Hence, RNE rounds up for R' & $(L' \mid T')$.

Most numerical algorithms, including matrix multiplication and filtering, involve multiplication followed by an addition: $R = X \times Y + Z$. It is faster and more accurate to compute this in one step and round at the end rather than multiply, round, add, and round again. This combined operation is called *fused multiply-add* (*FMA*) or *multiply-accumulate*.

Example 16.5 ROUNDING MODES

Given the exact sum of 1.0000 0111 1100 110 × 2^6 from Example 16.4, what is the rounded result in the RNE and RP rounding modes?

Solution: The rounding bits are $L = 1$, $G = 0$, $R = 0$, and $T = 1$. No further normalization shift is needed, so according to Table 16.3, $L' = L = 1$, $R' = G = 0$, $T' = R \mid T = 1$. As shown in Table 16.4, with RNE, the result rounds down to $TRUNC = 1.0000\ 0111\ 11$. In RP, the result rounds up to $RND = 1.0000\ 1000\ 00$.

Example 16.6 MULTIPLICATION AND ROUNDING MODES

Compute the exact product of two half-precision floating-point numbers 0x4010 × 0x4420. What is the rounded result in each of the rounding modes?

Solution: 0x4010 = 0 10000 0000010000,with a positive sign, biased exponent of 16 (unbiased exponent = 1), and fraction of .000001, making 1.000001×2^1. 0x4420 =

For trigonometric functions, logarithms, and related functions, it is difficult to find the exact result required for rounding. For example, if the exact result is very close to halfway between two representable numbers, an unbounded amount of computation might be needed to determine whether it is slightly higher or lower than the boundary. This is called the *Table Maker's Dilemma* [Muller18]. If exact rounding is impractical, the function may arbitrarily pick *TRUNC* or *RND*, one of the two closest representable numbers. This is called a *faithful result* and differs from the correct result by no more than one ulp (unit in the last place).

0 10001 0000100000, with a positive sign, biased exponent of 17, and fraction of .00001, making 1.00001 $\times$ 2^2. The exact product is $(1{+}2^{-6})(1{+}2^{-5})(2^1)(2^2) = (1 + 2^{-5} + 2^{-6} + 2^{-11})2^3 = 1.0000\ 1100\ 001 \times 2^3$. $L = 0$, $G = \mathbf{1}$, $R = 0$, $T = 0$. There is no further shift, so $L' = L = 0$, $R' = G = \mathbf{1}$, $T' = R \mid T = 0$. Half-precision numbers have only 10 fraction bits, so the right-most bit is not representable, and we must round down to $TRUNC = 1.0000\ 1100\ 00 \times 2^3$ = 0x4830 or round up to $RND = 1.0000\ 1100\ 01 \times 2^3$ = 0x4831. The exact result is a tie half-way between *TRUNC* and *RND*. RNE rounds to the even answer, *TRUNC*. RNTA rounds the tie up to *RND*. RZ and RN round down to *TRUNC*. RP rounds up to *RND*.

16.1.4 Flags

The IEEE 754 standard defines five types of exceptions that must be signaled by setting a status flag or taking a trap whenever they arise: Invalid, Division by Zero, Overflow, Inexact, and Underflow. Inexact, Overflow, and Underflow are never raised when Invalid or Division by Zero are raised or when any of the inputs are *NaN* or *INF*. The Overflow and Underflow flags depend on the *intermediate result* obtained by rounding the exact answer to the specified precision but with an unbounded exponent. These flags are raised in the circumstances described next, with I_e being the exponent of the intermediate result.

Invalid

- Operations on an *sNaN*
- $0 \times \infty$ or $\infty \times 0$
  - IEEE specifically mentions a case where $\infty \times 0 + qNaN$ may or may not set the Invalid flag. But for RISC-V, this operation causes the Invalid flag to be set.
- $\frac{0}{0}$ or $\frac{\infty}{\infty}$
- Remainder (i.e., $rem(x, y)$) where y is 0 or x is ∞ but neither is a *NaN*
- Square root of a negative input operand.
- Effective subtraction of two infinities (e.g., $\infty - \infty$)
- Conversion of a floating-point number to an integer format where one of the input operands is a *NaN*, ∞, or a value that would convert to an integer outside the range of the result format under the applicable rounding attribute.
- Comparison ($<$, $\leq$) of *NaN*.
- The operation $\log(x)$ where x is *NaN*, ∞, or 0.

- Inverse sine or cosine of a value outside the domain of $[-1, 1]$.

Division by zero

- Dividing a nonzero finite number by zero
- Computing log(0)

Overflow

- If the intermediate result is larger than the largest finite number

$$Overflow = I_e > e_{max} \tag{16.2}$$

Note that Overflow can be set even when the answer is not infinite. For example, in half-precision, 0x7800 + 0x7800 = $1.0 \times 2^{15} + 1.0 \times 2^{15}$, which has an intermediate result of 1.0×2^{16}. This is rounded to either ∞ or *MAXNUM*, depending on the rounding mode, but in either case, the Overflow flag is set.

Inexact

- If the rounded result differs from what would have occurred with unbounded precision. This may occur during overflow and underflow as well.
- $R' \mid T'$ indicates whether there are any nonzero results beyond the truncated result.

$$Inexact = R'|T'|Overflow \tag{16.3}$$

Underflow

- If the result is *tiny* and a loss of accuracy occurs. A result is considered tiny when it is subnormal ($|result| < 2^{e_{min}}$). A loss of accuracy occurs when the Inexact flag is set, which indicates that the answer differs from what would have occurred with unbounded precision. The IEEE 754 specification has an ambiguity that tininess may be detected either before rounding or after rounding to an intermediate result; not all implementations set this flag in the same way. RISC-V detects tininess after rounding the exact answer to the *intermediate result* with bounded precision but unlimited exponent range. Note that underflow is signaled when a very small result is rounded to zero but not when the exact result is zero.

$$Underflow = (I_e < e_{min})\&Inexact \tag{16.4}$$

Example 16.7 INTERMEDIATE AND FINAL RESULTS

Consider an exact result of $1.1111\ 1111\ 110 \times 2^{-15}$ being expressed in half-precision format with 10 fraction bits and $e_{min} = -14$, using RP rounding mode. Find the intermediate and final results.

Solution: The intermediate result is $1.1111\ 1111\ 11 \times 2^{-15}$, rounded to 10 fraction bits but with an exponent less than e_{min}. The rounding is down because the blue R' bit is 0, and there is no sticky bit, T'. On the other hand, to obtain the final result, the exact result is first shifted into subnormal form with the exponent bounded to $-e_{min}$: 0.1111 1111

1110×2^{-14}. It is then rounded up because the blue R' bit is 1, giving $1.0000\ 0000\ 00 \times 2^{-14}$. Thus, the intermediate and final result differ.

Some algorithms aim to avoid under/overflow. For example, [Beebe17, p. 113] shows a function that computes the length of a vector $\sqrt{x^2 + y^2}$. A direct implementation will overflow or underflow if either input is very large or very small. The function checks if either flag is set and, if so, recomputes the calculation as $x\sqrt{1 + \left(\frac{y}{x}\right)^2}$ for $x > y$. This alternate computation avoids unnecessary under/overflow but is more time consuming and requires additional code.

Underflow is the most subtle of the flags. Whereas underflow is intuitively associated with subnormal results, the Underflow flag is not raised if the subnormal result is exact. Moreover, it is raised even when the final answer is zero or the smallest normalized number $2^{e_{min}}$ if the intermediate result is tiny. Underflow warns the programmer of risk of inaccuracy related to tiny, inexact results. Some examples of underflow include:

- Some half-precision examples with $e_{min} = -14$ illustrate corner cases of tininess and underflow.
- $0x0000 + 0x0400 = 0.0 + 0.1 \times 2^{-14} = 0.1 \times 2^{-14}$. This result is tiny but exact, so the Underflow flag is not set.
- $0x0800 \times 0x0800 = 1.0 \times 2^{-13} \times 1.0 \times 2^{-13} = 1.0 \times 2^{-26}$. If the exponent range were unbounded, this would be the properly rounded result, which is tiny but nonzero. Define *MINNUM* to be the smallest subnormal number, $0.0000\ 0000\ 01 \times 2^{e_{min}} = 2^{-24}$. Because the exponent range is bounded and the result is less than *MINNUM*, the result must be rounded to either 0.0 or *MINNUM* depending on the rounding mode. Hence, the result is inexact. The Underflow flag is set even when the result is rounded to 0.0 because the result with an unbounded exponent is tiny and nonzero.
- $0x07FE \times 0x3400 + 0x0200 = 1.1111\ 1111\ 10 \times 2^{-14} \times 1.0 \times 2^{-2} + 0.1 \times 2^{-14} = 1.1111\ 1111\ 110 \times 2^{-15}$ (see Example 16.7). The intermediate result can be exactly represented with 10 fraction bits because the exponent range is unbounded. However, because the exponent range is actually bounded, the number cannot be represented and (in RNE) is rounded up to 1.0×2^{-14}. This result is inexact and tiny so the Underflow flag is set even though the final result is normalized.
- $0x07FF \times 0x3400 + 0x0200 = 1.1111\ 1111\ 11 \times 2^{-14} \times 1.0 \times 2^{-2} + 0.1 \times 2^{-14} = 1.1111\ 1111\ 111 \times 2^{-15}$, which cannot be exactly represented with 10 fraction bits even if the exponent range is unbounded. In RNE, the number is rounded up to 1.0×2^{-14}, so it is not tiny. This result is inexact but not tiny, so the underflow flag is not set. In RZ, the number is rounded down to $0.1111\ 1111\ 11 \times 2^{-14}$, which is tiny and inexact, so the Underflow flag is set.

16.1.5 Fixed-Point Notation

Floating-point hardware uses fixed-point arithmetic internally to operate on the significand. To keep track of the number of integer and fraction bits and whether the number is signed or unsigned, we use the Q and U formats. U*a*.*b* indicates an unsigned fixed-point number with *a* integer bits and *b* fraction bits. Q*a*.*b* indicates a two's complement signed number with *a* integer and *b* fraction bits. In this chapter, we also use U*a*, Q*a*, and $B^b a$ to indicate unsigned, two's complement signed, and biased integers with *a* bits and a bias of *b*.

Fixed-point numbers are commonly used in application-specific computing such as signal processing and machine learning where low power consumption and inexpensive massively parallel hardware are more important than precision. For example, Q1.15 format is used to represent normalized numbers between ±1, which is convenient because such numbers remain in the same range when multiplied. However, the product of two Q1.15 numbers is often represented as a Q1.31 number to preserve accuracy.

Example 16.8 FIXED-POINT REPRESENTATIONS

Interpret the bit pattern 10111 as U3.2, U2.3, Q2.3, U5, Q5, and $B^{15}5$.

Solution: Table 16.5 interprets the bit patterns. Remember that the most significant bit of a two's complement number has a negative weight (e.g., for an *n*-bit two's complement number *a*):

$$-a_{n-1}\cdot 2^{n-1} + \sum_{i=0}^{n-2} a_i \cdot 2^i \qquad (16.5)$$

A U*a*.*b* number represents a value bounded by $[0, 2^a)$, and a Q*a*.*b* number represents a value bounded by $[-2^{a-1}, 2^{a-1})$. Performing operations increases the range and the number of bits required. When adding two fixed-point numbers with *a* and *c* integer bits, the result may require max(*a*, *c*) + 1 integer bits. Exactly multiplying U*a*.*b* × U*c*.*d* numbers produces a U(*a*+*c*).(*b*+*d*) result. Exactly multiplying Q*a*.*b* × Q*c*.*d* numbers produces a Q(*a*+*c*−1).(*b*+*d*) result. Observe that signed products have one fewer integer bit than unsigned products because the range of signed numbers is smaller, although the signed product may be sign-extended to *a* + *c* integer bits for consistency.

Table 16.5 Fixed-point interpretation examples.

Format	Interpretation	Value
U3.2	$2^2 + 2^0 + 2^{-1} + 2^{-2}$	5.750
U2.3	$2^1 + 2^{-1} + 2^{-2} + 2^{-3}$	2.875
Q2.3	$-2^1 + 2^{-1} + 2^{-2} + 2^{-3}$	−1.125
U5	$2^4 + 2^2 + 2^1 + 2^0$	23
Q5	$-2^4 + 2^2 + 2^1 + 2^0$	−9
$B^{15}5$	$2^4 + 2^2 + 2^1 + 2^0 - 15$	8

16.2 FLOATING-POINT UNIT DESIGN

A floating-point unit (FPU), as shown in Fig. 16.5, typically contains several execution units:

- **`fma:`** Fused multiply-add (FMA) unit for add, subtract, multiply, and FMA
- **`fdivsqrt:`** Divide and square root
- **`fcvt:`** Conversions between integer and floating-point formats
- **`fcmp:`** Comparisons and min/max

Signal names in this chapter and in Appendix F are italicized instead of in code font to be consistent with the theoretical development. Unit/module names are still in code font.

The FMA unit computes $W = X \times Y + Z$, where W is the result, X and Y are the *multiplicands*, and Z is the *addend*. The FMA can perform floating-point multiplies $X \times Y$ or adds $X + Z$ by tying Z to 0.0 or Y to 1.0, respectively. It can also invert input signs to compute $X \times Y - Z$, $-(X \times Y) + Z$, $-(X \times Y) - Z$, or $X - Z$.

The `fdivsqrt` unit computes $W = X \div Y$ or $W = \sqrt{X}$.

All of these execution units need to extract the significand, exponent, and sign, and identify the special cases. Thus, that logic is shared in an *unpacking* unit (`unpack`). The logic for normalization, rounding, special-case handling, and flag generation is similar across the `fma`, `fdivsqrt`, and `fcvt` modules, so it is also shared in a *postprocessing* unit (`postprocess`). The FMA requires three inputs (X, Y, and Z), but other units need only one or two inputs. The FPU produces the floating-point result and flags. It may also output an integer result (when converting from `float` to `int`) and some comparison bits.

The unpacking unit is also referred to as the *unpack unit* or *unpacker*. Likewise, the postprocessing unit is also referred to as the *postprocess unit* or *postprocessor*.

Handling subnormal numbers in hardware was once considered difficult, so some FPUs would trap on subnormal values and handle them in software instead. Designers have discovered hardware subnormal support is not too expensive [Schwartz05], and RISC-V doesn't support such traps, so this section focuses on FPUs that handle subnormal numbers and all special cases in hardware.

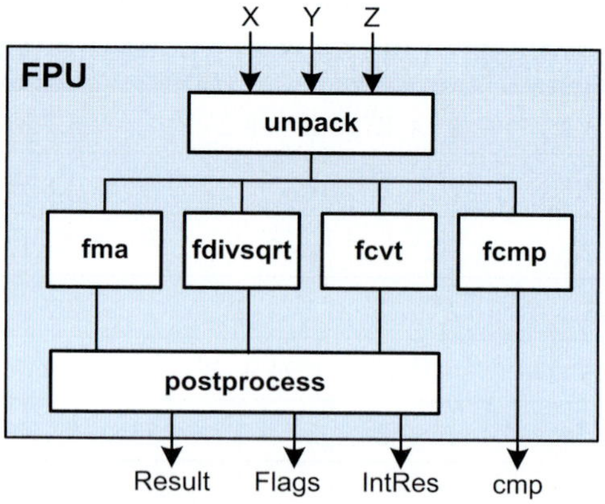

Fig. 16.5 Generic FPU block diagram.

This section describes the unpacker, postprocessing unit, and various floating-point execution units. All of these use parameterized signals to support arbitrary precision (FLEN) operands. The units have many corner cases, especially related to rounding and special case handling. Section 16.4 discusses the RISC-V floating-point instruction set, and Section 16.6 describes the Wally implementations in more detail, including configurable units capable of supporting multiple precisions on shared hardware.

The Wally FPU is located in the Execute stage and uses a two-stage pipeline to unpack inputs, execute the operations, and postprocess, including normalization, rounding, and special cases. The FPU then writes the result to the register file in the Memory stage. The FPU can limit the Wally cycle time. Most commercial FPUs use a 3-7 stage floating-point pipeline to improve cycle time [Muller18].

16.2.1 Unpack Unit

The unpack unit unpacks lower-precision inputs, determines special cases, and extracts the sign, exponent, and significand fields of each input so that this bookkeeping work can be shared across the other functional units. Fig. 16.6(a) shows a schematic of the unpack unit comprising three channels to unpack each of the three inputs. Fig. 16.6(b) shows how the unit unpacks a single channel.

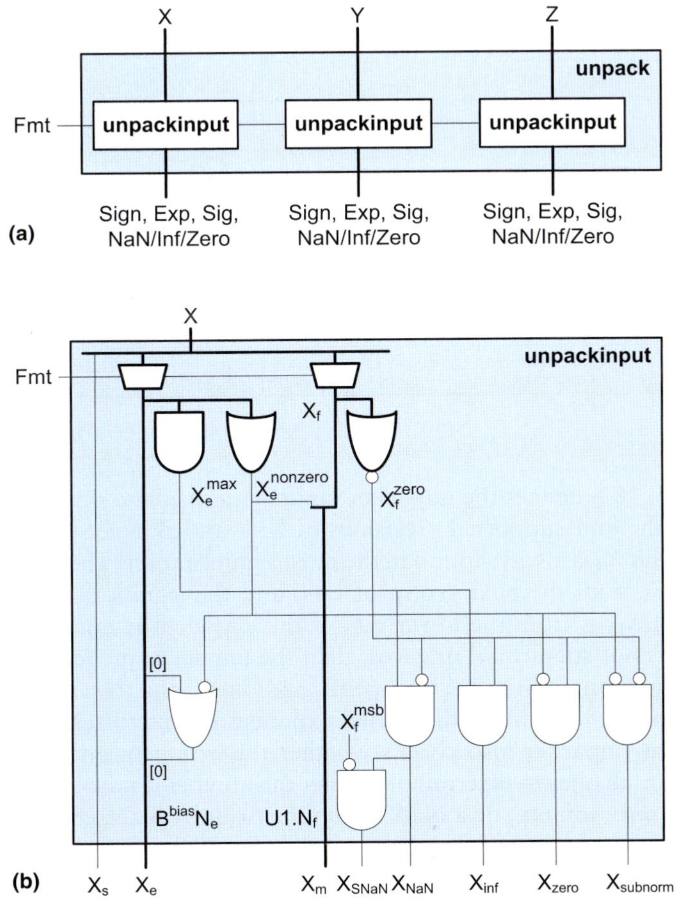

Fig. 16.6 **The unpack unit: (a) top level and (b) unpack hardware for a single channel.**

Table 16.6 Unpacking signals.

Signal	Format	Quad	Double	Single	Half	Comments
Inputs						
X	FLEN	$X_{127:0}$	$X_{63:0}$	$X_{31:0}$	$X_{15:0}$	Input
Intermediate Signals						
X_e^{max}	Boolean	AND(X_e)				Exponent is max(INF, NaN)
X_e^{nonzero}	Boolean	OR(X_e)				Exponent is nonzero
X_f	U0.N_f	$X_{112:0}$	$X_{51:0}$	$X_{22:0}$	$X_{9:0}$	Fraction part of X
X_f^{zero}	Boolean	$X_f == 0$				Fraction part is zero
Outputs						
X_s	Q1	X_{127}	X_{63}	X_{31}	X_{15}	Sign
X_e	$B^{bias}N_e$	$X_{126:112}$	$X_{62:52}$	$X_{30:23}$	$X_{14:10}$	Biased Exponent
		$+\ \overline{X_e^{\text{nonzero}}}$				
X_m	U1.N_f	$X_e^{\text{nonzero}}, X_f$				Significand
$X_{subnorm}$	Boolean	$\overline{X_e^{\text{nonzero}}}\ \overline{X_f^{\text{zero}}}$				Subnormal
X_{zero}	Boolean	$\overline{X_e^{\text{nonzero}}} X_f^{\text{zero}}$				Zero
X_{inf}	Boolean	$X_e^{\text{max}} X_f^{\text{zero}}$				Infinity
X_{NaN}	Boolean	$X_e^{\text{max}} \overline{X_f^{\text{zero}}}$				NaN
X_{SNaN}	Boolean	$X_{NaN} \overline{X_{f_{\text{msb}}}}$				Signaling NaN

Recall that sNaNs are recognized by a 0 in the most significant bit of the fraction.

Table 16.6 defines the unpacker inputs, intermediate signals, and outputs for the four supported precisions of X. Y and Z have similar signals. The sign bit X_s is always found in the most significant bit. The biased exponent X_e is from the next group of bits, and the significand (also called mantissa) X_m is from the lower bits. If the exponent is nonzero (i.e., the number is not subnormal or zero), then the unpacker prepends a leading 1 to the fraction. Otherwise, it prepends a leading 0 and then either handles it as a zero or increments the biased exponent to one to represent a subnorm. The unpacker also checks whether the exponent and fraction are all zeros or all ones to determine whether the number is normalized, denormalized, zero, infinity, or a NaN (signaling or quiet NaN). Section 16.6.2 discusses unpacking lower-precision inputs for Wally.

Example 16.9 UNPACKING

Unpack the half-precision input $X = 0xC900$.

Solution: The number is 1 10010 0100000000 in binary. The sign $X_s = X_{15} = 1$. The biased exponent $X_e = X_{14:10} = 10010 = 18$, so $X_e^{max} = 0$, and $X_e^{nonzero} = 1$. $X_f = 0100000000$, so $X_f^{zero} = 0$. The significand $X_m = 1.01$. All the other conditions are 0. Thus, we have the normalized floating-point number -1.01×2^3.

16.2.2 Fused Multiply-Add

Many numerical algorithms such as matrix multiplication, finite impulse response filtering, or the fast Fourier transform (FFT) involve multiplications followed by adds, such as multiplying data by coefficients and adding the result to a running total. Designers discovered that fusing the multiply and add into a single instruction runs nearly as fast as either one alone and doubles the number of FLOPs (*FLoating-point OPerations per second*) that a system can perform. A fused multiply-add (FMA) unit computes $W = X \times Y + Z$, where W is the result, X and Y are the *multiplicands*, and Z is the *addend*. It can perform floating-point multiplies or adds by tying Z to 0.0 or Y to 1.0 and can also invert input signs to compute $X \times Y - Z$, $-(X \times Y) + Z$, $-(X \times Y) - Z$, or $X - Z$. Moreover, FMA is more precise than separate multiplies and adds because it avoids rounding the intermediate result. Thus, the FMA is the workhorse of an FPU.

The original IEEE 754-1985 specification did not define FMA. By the mid-1990s, high-performance processors widely supported the nonstandard operation for speed and accuracy. The IEEE standard was finally revised in 2008 [IEEE 754-2008] to add FMA. Graphics processing units (GPUs) also began widely adopting IEEE 754 fused FMA operations starting in the early 2010s. The adoption of FMA operations in GPUs marked a significant advancement in their computational capabilities, aligning them more closely with the needs of scientific computing and other fields requiring high precision.

16.2.2.1 FMA Algorithm

FMA is easiest to understand by first considering floating-point multiply (fmul) and addition (fadd). To multiply two floating-point numbers $W = X \times Y$, multiply their significands and add their exponents. The product P of two normalized floating-point significands in the range [1, 2) is in the range [1, 4), so it may be necessary to right-shift the product by 1 and increment the exponent to get a normalized result M. If the result cannot be exactly expressed as a floating-point number, it is rounded according to the specified rounding mode. Finally, special cases of zero, infinity, and NaNs are handled, and the flags are set. Fig. 16.7 summarizes these steps, neglecting subnormal number handling. The slowest step is the large significand multiplication.

1. Multiply the significands of X and Y:	$P_m = X_m \times Y_m$
2. Add the exponents of X and Y:	$P_e = X_e + Y_e - bias$
3. Find the leading 1 for normalization shift:	if ($P_m \geq 2.0$) $M_{cnt} = -1$ else $M_{cnt} = 0$
4. Shift the result to renormalize:	$M_m = P_m << M_{cnt}$; $M_e = P_e - M_{cnt}$
5. Round the result:	$R = \text{round}(M)$
6. Handle flags and special cases:	$W = \text{specialcase}(R, X, Y)$

Fig. 16.7 Simplified FMUL algorithm.

Adding two floating-point numbers $W = X + Z$ is trickier because their binary points will not line up if their exponents differ. To align the binary points, shift the significand of Z by the exponent difference, which might be positive or negative. Then add to get the sum S. The result could be huge if Z is much bigger than X. It could also be nearly 0 if X and Z are nearly equal but had opposite signs. Normalize by shifting the sum into the range [1, 2) and adjusting the exponent by the shift amount. Finally, round and handle special cases and flags. Fig. 16.8 summarizes these steps. The slowest steps are the alignment and normalization shifts.

1. Determine the alignment shift count:	$A_{cnt} = (X_e - Z_e)$
2. Shift the significand of Z into alignment:	$A_m = Z_m >> A_{\text{cnt}}$
3. Add the aligned significands:	$S_m = X_m + A_m$
4. Find the leading 1 for normalization shift:	M_{cnt} = # of bits to shift
5. Shift the result to renormalize:	$M_m = S_m << M_{\text{cnt}}$; $M_e = X_e - M_{\text{cnt}}$
6. Round the result:	$R = \text{round}(M)$
7. Handle flags and special cases:	$W = \text{specialcase}(R, X, Z)$

Fig. 16.8 Simplified fadd algorithm.

The FMA combines the multiply and add steps, as given in Fig. 16.9. The slow significand multiplication P and alignment shift A steps take place concurrently. S is the sum $P + A$. M is the normalized result after S is shifted to have a single 1 to the left of the binary point. R is the result after rounding, and W is the final result, accounting for special cases. This section describes computing P, A, and S. M, R and W are computed by the postprocessing unit described in Section 16.2.6.

1. Multiply the significands of X and Y:	$P_m = X_m \times Y_m$
2. Add the exponents of X and Y:	$P_e = X_e + Y_e - bias$
3. Determine the alignment shift count:	$A_{cnt} = (P_e - Z_e)$
4. Shift the significand of Z into alignment:	$A_m = Z_m >> A_{\text{cnt}}$
5. Add the aligned significands:	$S_m = P_m + A_m$
6. Find the leading 1 for normalization shift:	M_{cnt} = # of bits to shift
7. Shift the result to renormalize:	$M_m = S_m << M_{\text{cnt}}$; $M_e = P_e - M_{\text{cnt}}$
8. Round the result:	$R = \text{round}(M)$
9. Handle flags and special cases:	$W = \text{specialcase}(R, X, Y, Z)$

Fig. 16.9 Simplified FMA algorithm.

FMA is faster than multiply followed by add because the hard step of multiplying the significands of A and B is performed in parallel with the slow step of aligning the significand of Z. Wally's FMA uses the same hardware for multiplication alone or addition alone. Some FMAs are optimized to skip steps, delivering multiplication or addition results in fewer cycles than a full multiply-add.

The next three examples illustrate these steps, including the possibility of massive cancellation. All significands are written in binary.

Example 16.10 FMA: FUSED MULTIPLY ADD

Compute $(48)(5/16) + 28 = (1.1 \times 2^5)(1.01 \times 2^{-2}) + 1.11 \times 2^4$.

Solution:

- First multiply the significands $P_m = X_m \times Y_m = 1.1 \times 1.01 = 1.111$.
- Add the exponents $P_e = 5 + (-2) = 3$.
- Now add this product to Z: $1.111 \times 2^3 + 1.11 \times 2^4$. To add the significands, the exponents must be equal. Thus, align Z to be 11.1×2^3. The alignment shift is by $A_{cnt} = (P_e - Z_e) = 3 - 4 = -1$. The aligned significand is $A_m = Z_m >> A_{cnt} = 1.11 >> -1 = 11.1$. Add these aligned significands: $S_m = P_m + A_m = 1.111 + 11.1 = 101.011$.
- So, the sum is $S = 101.011 \times 2^3$. Floating-point numbers are written in normalized form (M) with a single 1 to the left of the binary point, so normalize by shifting S right by -$M_{cnt} = 2$ and increasing its exponent by 2: $M = 1.01011 \times 2^5$. The result is exact and does not require rounding. In decimal, this is $W = 43$, as expected.

Example 16.11 FMA WITH MASSIVE CANCELLATION

Compute $(48)(5/16) - 13 = (1.1 \times 2^5)(1.01 \times 2^{-2}) - 1.101 \times 2^3$.

Solution:

- X and Y are the same as in the last example, so $P_m = 1.111$, and $P_e = 3$.
- The exponents are already equal, so no alignment shift is required.
- Add these aligned significands: $S_m = 1.111 + (-1.101) = 0.010$. So, the sum is $S = 0.010 \times 2^3$.
- Normalize by shifting S left by 2 and decreasing the exponent by 2: $M = 1.0 \times 2^1$. Again, the result is exact and does not require rounding. In decimal, this is $W = 2$, as expected. This is an example of *massive cancellation* where many leading digits become 0 when nearly equal values are subtracted.

Example 16.12 FMA WITH ROUNDING

Compute $(48)(5/16) + 2^{35} = (1.1 \times 2^5)(1.01 \times 2^{-2}) + 1.0 \times 2^{35}$ and round the result toward positive infinity to produce a single-precision floating-point result with 23 fraction bits. Express your answer in hexadecimal.

Solution:

- X and Y are the same as in the last example, so $P_m = 1.111$ and $P_e = 3$. $Z_e = 35$.
- The alignment shift brings Z left by $A_{cnt} = (3 - 35) = -32$ bits. This difference is greater than the 23 bits of precision of the result, so P is insignificant compared to Z and can be discarded except that the sticky bit (T') becomes 1 for rounding.
- The normalized result $M = S = Z$ is inexact and requires rounding. Because the number is positive and the sticky bit is set, we round up by 1 ulp, as given in

Table 16.4. The rounded answer is 1.00000000000000000000001 $\times 2^{35}$. Thus, the sign is 0, the biased exponent is $127 + 35 = 162 = 10100010$, and the fraction is 00000000000000000000001. Fig. 16.10 shows this number in floating-point format as 0x51000001.

sign	exp	fraction
0	101 0001 0	000 0000 0000 0000 0000 0001

Fig. 16.10 **Fields for single-precision representation of 1.00000000000000000000001** $\times$ **2^{35}.**

16.2.2.2 FMA Implementation

Fig. 16.11 shows a hardware implementation of the FMA algorithm from Fig. 16.9. This section provides a mathematical description of each block and explains the fixed-point formats used in each signal. Whereas this

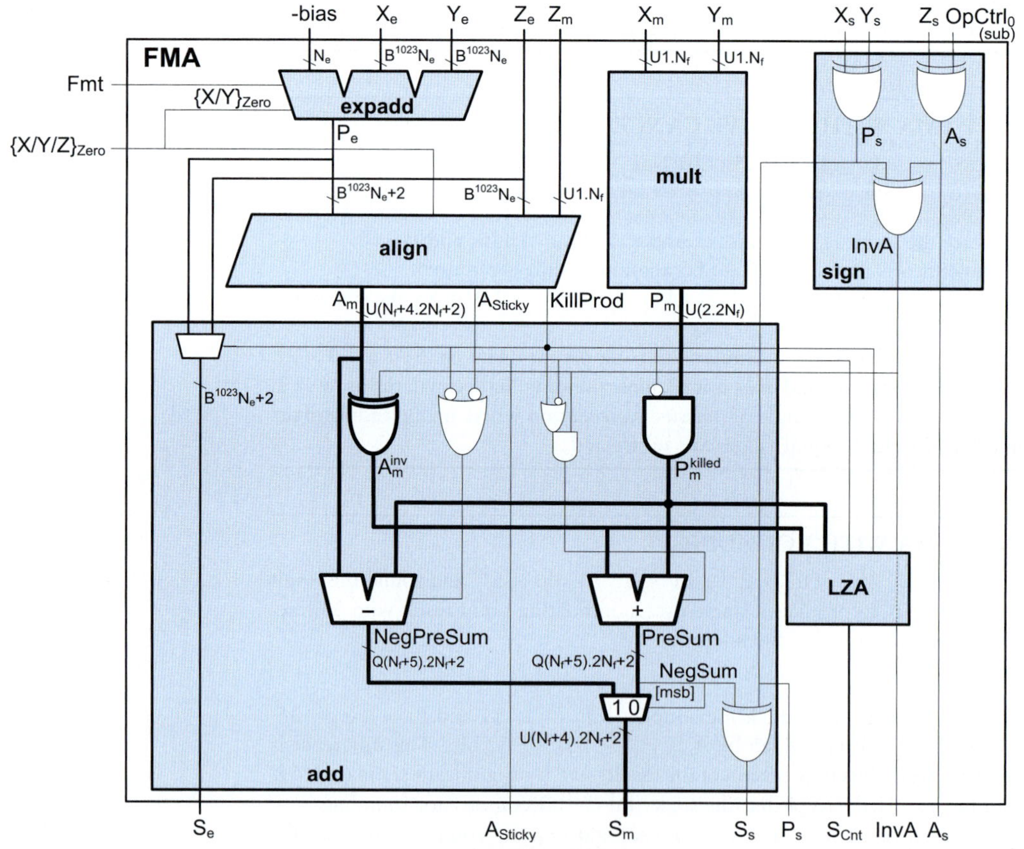

Fig. 16.11 **FMA datapath.**

design focuses on a simple single-path approach like the classic IBM RS/6000 design from [Montoye90], Section F.1.7 discusses multipath optimizations. See `src/fpu/fma.sv` for Wally implementation details.

The details of each block are given in Appendix F.1.

Wally's FMA operates in the Execute stage. The stage suffixes are dropped from signal names within the FMA for brevity.

16.2.3 Division and Square Root

Recall from Section 15.1.4 that division of N-bit numbers takes a *dividend* X and a *divisor* D. It computes a *quotient* Q and *remainder* REM satisfying $X = Q \times D + REM$. Or stated another way, $X/D = Q + REM/D$. If X and D are signed, then REM has the same sign as X.

A full hardware implementation of RV64G requires computing integer division and remainder operations on 32- and 64-bit operands and computing properly rounded floating-point division and square root on 24- and 52-bit single- and double-precision operands, including subnormal numbers.

The simple division algorithm from Section 15.1.4 involves a slow carry propagate addition in each step to compute the partial remainder. A faster *digit recurrence* algorithm keeps the partial remainder in *carry-save redundant form* (see Section 15.1.2) so that each step's addition reduces to one full adder delay. However, it is impossible to exactly determine the correct quotient digit from a redundant partial remainder without converting back to nonredundant form, which eliminates the performance benefit. Instead, we guess the quotient digit based on the most significant bits of the partial remainder. The guess might be incorrect, but close. We then introduce a *redundant digit set* including negative quotient digits. If the quotient digit proves to be too high, we later correct by selecting a negative quotient digit. Radix-2 division uses a redundant digit set of $\{1, 0, -1\}$ to compute 1-bit digits. Radix-4 division is faster, generating 2-bit digits on each step, typically using a *minimally redundant digit set* of $\{2, 1, 0, -1, -2\}$. Some designers have explored higher radices, but this section focuses on radix-2 and radix-4. The redundant quotient digits are converted back to nonredundant form in an *on-the-fly* fashion.

Digit recurrence gets its name from a *recurrence relation*, which is an equation that defines the next term of a sequence as a function of the previous term. In digit recurrence division, each digit depends on the partial remainder, which depends on the previous digits.

Remarkably, square root can be computed digit by digit in much the same way as division, also keeping the intermediate results in carry-save redundant format and selecting digits of the answer from a redundant digit set. The similarities are so strong that the two operations can be performed on a shared hardware unit.

This idea for fast division was invented almost concurrently by [Sweeney57], [Robertson58], and [Tocher58] and thus became known as *SRT division*. Ercegovac and Lang have studied the topic extensively [Ercegovac94], [Ercegovac03], and their more generalized techniques are now known as *digit recurrence* algorithms. Readers interested in a proof of the technique should consult their books.

The recurrence algorithm assumes a normalized divisor in the range $[1, 2)$. To support subnormal floating-point numbers, the inputs must be preshifted into normalized form. Recurrence division is sometimes avoided for integers because of this normalization issue, but because the preshifters are needed for floating-point subnorms, they can also be used for integer division. Moreover, a variable latency design supports different operand sizes as well as early termination detection. Shorter operands are kept in the most significant bits so that the least significant register bits can be clock-gated for power savings. Hence, RISC-V is well-suited to a configurable unified divide/square root unit (`fdivsqrt`) capable of supporting all of these floating-point and integer operations. The system has the following parameters:

XLEN: Integer register length (32, 64, or 128)

N_s: Floating-point significand length $= N_f + 1$ (11, 24, 53, or 113, or 0 for no FP)

$N = \max(\text{XLEN}, N_s)$: Size of largest divider operands

Fig. 16.12 outlines the basic radix-2^r (radix-R) recurrence division algorithm. The algorithm accepts a dividend X and a divisor D. It produces the quotient Q one digit q_j at a time. It keeps the partial remainder W in carry-save redundant form $W = WS + WC$ so that addition can be performed quickly. Comparing the full divisor to the full partial remainder would be slow, so the quotient selection logic (QSLC) estimates the quotient digit based on the most significant bits of the partial remainder and divisor. It uses enough bits so that the estimate will either be correct or just slightly wrong and correctable with the digit calculation in the next step. It then chooses $-q_jD$ from a set of precomputed divisor multiples, adds it to the partial remainder with a carry-save adder (CSA), and shifts left by one digit. It repeats until all the quotient digits are known (or terminates early when the partial remainder becomes 0 and the result is exact). Fig. 16.13 shows a block diagram of the divider.

$WC_{-1} = 0,\ WS_{-1} = X$
for $j = 0$ to n
  $q_j = \text{QSLC}(WC_{\text{msbs}\ j-1},\ WS_{\text{msbs}\ j-1},\ D_{\text{msbs}})$
  $\{WC_j,\ WS_j\} = \{WC_{j-1} + WS_{j-1} - q_jD\} \ll \text{r}$

Fig. 16.12 Recurrence division algorithm.

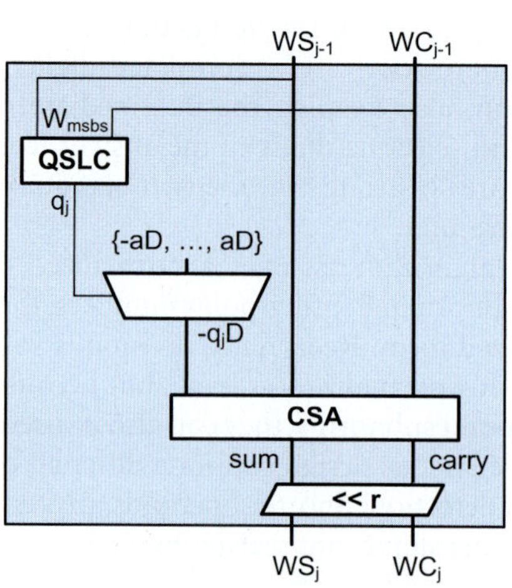

Fig. 16.13 Recurrence division hardware.

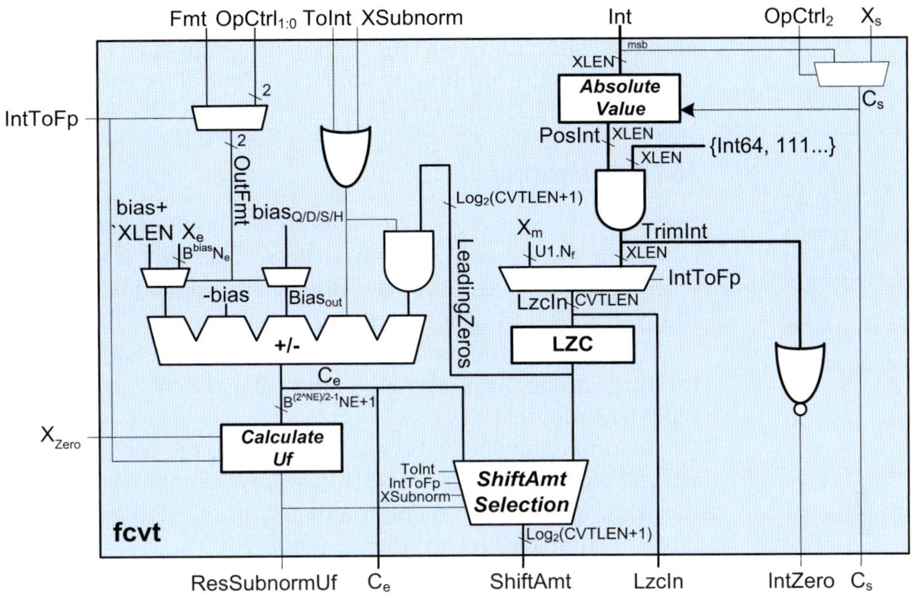

Fig. 16.14 `fcvt` **unit.**

Appendix F.2 describes floating-point division and square root in further detail.

16.2.4 Conversion

The floating-point convert unit (`fcvt`) handles conversion from int to floating-point, floating-point to int, and floating-point to different-precision floating-point.

Conversions from shorter to longer floating-point format involve adding the difference in biases to the exponent and placing zeros in the least significant bits of the fraction. Conversions from longer to shorter format involve subtracting the difference in biases from the exponent and rounding the fraction to the new precision. The result may overflow or underflow to create an infinite or zero answer. Both types may shift the significand to handle subnorms.

Conversions from floating-point to int involve an alignment shift, rounding, and taking the two's complement. Conversion from int to floating-point involves taking the absolute value, counting the leading zeros, normalizing the magnitude and adjusting the exponent, rounding, and setting the sign bit.

Fig. 16.14 shows the main components, which are mostly shared across types of conversions. These components include taking an absolute value to handle negative integers, counting leading zeros for integers and subnormal floating-point numbers, calculating the normalization shift amount, and adjusting the exponent accordingly. As with the FMA, normalization, rounding, special cases, and flags are handled by the post-processing unit. Appendix F.3 describes each block in detail.

16.2.5 Comparison

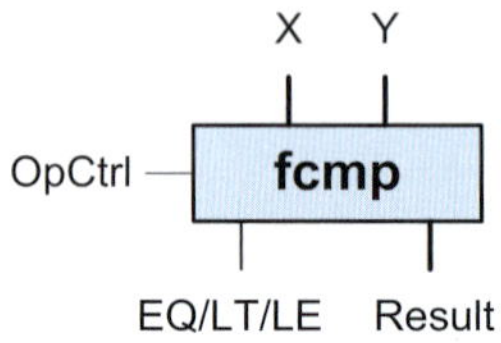

Fig. 16.15 `fcmp` unit.

Fig. 16.15 shows a floating-point comparison unit (`fcmp`) that computes =, <, and ≤ comparisons. The unit also performs `fmin` and `fmax` operations. It produces Boolean outputs for comparisons and an FLEN *Result* output for min/max.

The unit computes *EQ* and *LT* (equal and less than) flags, then selects the appropriate output based on these flags, *Zero* and *NaN* signals, and the type of comparison.

Comparisons of NaNs are always 0 except for the inequality case (≠) which returns 1. For non-NaN cases, we compare the absolute values of the floating-point numbers by discarding the sign bit and doing unsigned integer comparison on the remaining bits. An interesting feature of floating-point numbers with biased exponents is that larger floating-point numbers are also larger when treated as integers. This even works for subnormal numbers and zeros. The exponent is in the upper bits, and numbers with larger biased exponents are strictly larger. If the exponents are the same, numbers with larger fractions are also larger.

Hence, <, ≤, and = comparison instructions (`flt`, `fle`, `feq`) first compute flags *EQ*, and LT_{abs} according to Equations 16.6 and 16.7). *EQ* is true if all of the bits are equal. LT_{abs} is true if the absolute value of *X* is less than the absolute value of *Y*. Next, compute the less-than flag *LT* according to Table 16.7 to correct for signs. Select the comparison result according to Table 16.8. Note that +0 and −0 are considered equal. Finally, for NaNs, the comparison result is always 0.

Table 16.7 Floating-point < (*LT*).

X_s	Y_s	LT ($X < Y$)
0	0	LT_{abs}
0	1	0
1	0	1
1	1	$\sim(LT_{abs} \mid EQ)$

Table 16.8 Floating-point comparisons.

Type	X_{zero} & Y_{zero}	*Result*
`feq`	0	*EQ*
`feq`	1	1
`flt`	0	*LT*
`flt`	1	0
`fle`	0	$LT \mid EQ$
`fle`	1	1

Table 16.9 Floating-point min/max.

Type	*LT*	*Result*
`fmin`	0	Y
`fmin`	1	X
`fmax`	0	X
`fmax`	1	Y

$$EQ = \{X_s, X_e, X_f\} == \{Y_s, Y_e, Y_f\} \tag{16.6}$$

$$LT_{abs} = \{X_e, X_f\} < \{Y_e, Y_f\} \tag{16.7}$$

`fmin` and *fmax* instructions are handled similarly based on the *LT* flag as shown in Table 16.9. Note that −0 is considered less than +0 for these operations. When only one input is NaN, the output is the non-NaN input. When both inputs are NaN, the output is a quiet NaN.

16.2.6 Postprocessing Unit

The postprocessing unit (`postprocess`) performs normalization, rounding, special-case handling, and flag generation common to all floating-point execution units including `fma`, `fdivsqrt`, and `fcvt`.

Fig. 16.16 shows a block diagram of the `postprocess` unit. The heavy lines show the primary flow of wide buses. The unit receives raw results from the execution units that need to be normalized. Like the alignment shifter, the normalization shifter only shifts in one direction for simplicity. Thus, the inputs are prealigned to the right and then left-shifted for normalization. For each execution unit, the postprocessor performs a shift calculation involving prealigning the input for the normalization shifter and computing how much to shift. The postprocessor selects the input from the active execution unit and performs a left shift to normalize, then adjusts the exponent. The normalization shifter also produces subnormal results when the exponent is small. Example 16.10 shows normalization for an FMA. An additional 1-bit left shift may be needed to compensate for an error in the LZA (leading zero anticipator) and for quotients less than 1. The rounding unit uses the normalized result, the round and sticky bits, the signs, and the rounding mode to produce a correctly rounded result. Finally, the postprocessing unit handles the floating-point special cases and flag generation. The postprocessing unit relies on various control signals, input attributes (e.g., *Inf*, *NaN*, and *Zero*), and minor intermediate results. Appendix F.4 describes each block.

16.3 SOFTFLOAT

Checking floating-point arithmetic is notoriously difficult, so we need a reliable reference model. John Hauser has developed the open-source UCB SoftFloat library [Hauser18] that defines float_16t, float_32t, float_64t, and float_128t and operations on these types, including add, sub, mul, div, sqrt, mulAdd (FMA), remainder, comparisons, and type conversions. For example, f64_mulAdd takes three float64_t arguments and produces a float64_t result. The softfloat_roundingMode global variable specifies the rounding mode. RISC-V uses SoftFloat as a reference model, and we use the associated TestFloat tool to generate corner cases for testing.

SoftFloat is also in the `cvw/addins/berkeley-softfloat-3` directory. See Section 16.5.2 about compiling TestFloat, which uses the SoftFloat library.

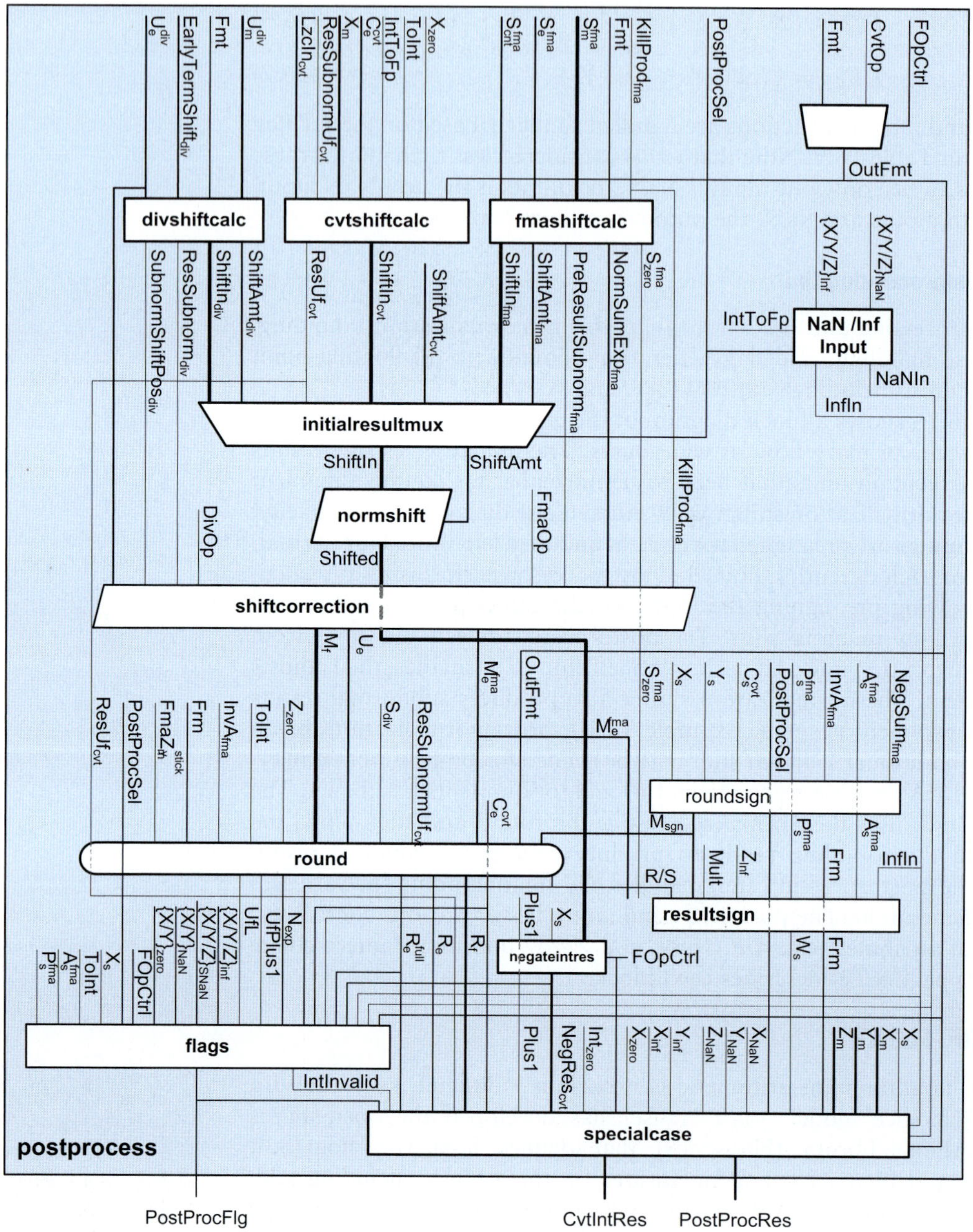

Fig. 16.16 Postprocessing unit (`postprocess`).

In the next section, we describe `softfloat_demo`, which shows how to call SoftFloat to convert between hexadecimal and floating-point representations, perform a floating-point operation with a specified rounding mode, and print the result and flags. In the following section, we describe fpcalc, a simple utility to perform floating-point calculations and visualize the numbers.

16.3.1 SoftFloat Demo

Code Example 16.1 shows the `softfloat_demo.c` program from `examples/fp/softfloat_demo`. The program uses SoftFloat to compute 1.5 × 1.5 + the smallest subnormal number (1.4013e−45). It prints the results (rounded to 2.25) and the flags (Inexact is set because of the rounding). Compile and run it using:

```
$ make
$ ./softfloat_demo
X: 0x3fc0_0000 = 1.5
Y: 0x3fc0_0000 = 1.5
Z: 0x0000_0001 = 1.4013e-45
result = X*Y+Z: 0x4010_0000 = 2.25
   Flags: Inexact 1 Underflow 0 Overflow 0 DivideZero 0 Invalid 0
```

The program initializes SoftFloat by setting the rounding mode to RNE, clearing the flags, and setting the underflow logic to detect tininess before rounding (the default for RISC-V). It then assigns values to single-precision `float32_t` variables `x`, `y`, `z`. SoftFloat declares `float32_t` to be a 32-bit number with a value field (or member) `.v` that can be written with an ordinary unsigned number, so `x.v` = 0x3fc00000 gives that hexadecimal pattern to `x`, equivalent to 1.5. The program performs the FMA operation, then prints the result and the flags. It uses a union of type single-precision (`sp`) to convert the 32-bit unsigned value into a floating-point value for printing. It also uses the `softfloat_exceptionFlags` global variable to report the flags.

16.3.2 fpcalc

`examples/fp/fpcalc` contains a simple floating-point calculator that performs computations on half-, single-, or double-precision floating-point numbers in hexadecimal form and reports the results in hexadecimal, decimal, and scientific notation. This is a handy way to visualize an operation when debugging floating-point test cases. See the following examples of using the program.

This calculator is also conveniently available online as WebFloat: s-y-wu.github.io/webfloat/

Code Example 16.1 **softfloat_demo.c**

```
// softfloat_demo.c
// David_Harris@hmc.edu 27 February 2022
//
// Demonstrate using SoftFloat for a floating-point computation, then print results.

#include <stdio.h>
#include <stdint.h>
#include "softfloat.h"
#include "softfloat_types.h"
typedef union sp {
  uint32_t v;
  float f;
} sp;

void printF32 (char *msg, float32_t f) {
  sp conv;
  int i, j;
  conv.v = f.v; // Use union to convert between hexadecimal and floating-point views
  printf("%s: ", msg);  // Print out nicely
  printf("0x%04x_%04x = %g\n", (conv.v >> 16),(conv.v & 0xFFFF), conv.f);
}

void printFlags(void) {
  int NX = softfloat_exceptionFlags % 2;
  int UF = (softfloat_exceptionFlags >> 1) % 2;
  int OF = (softfloat_exceptionFlags >> 2) % 2;
  int DZ = (softfloat_exceptionFlags >> 3) % 2;
  int NV = (softfloat_exceptionFlags >> 4) % 2;
  printf ("Flags: Inexact %d Underflow %d Overflow %d DivideZero %d Invalid %d\n",
          NX, UF, OF, DZ, NV);
}

void softfloatInit(void) {
  // Rounding modes: RNE: softfloat_round_near_even
  //                 RZ:  softfloat_round_minMag
  //                 RP:  softfloat_round_max
  //                 RM:  softfloat_round_min
  softfloat_roundingMode = softfloat_round_near_even;
  softfloat_exceptionFlags = 0; // Clear exceptions
  softfloat_detectTininess = softfloat_tininess_afterRounding;
}

int main() {
  float32_t x, y, z, r;
  x.v = 0x3fc00000;
  y.v = 0x3fc00000;
  z.v = 0x00000001;

  softfloatInit();
  r = f32_mulAdd(x, y, z);
  printF32("X", x); printF32("Y", y); printF32("Z", z);
  printF32("result = X*Y+Z", r); printFlags();
}
```

Example 16.13 ADDITION USING FPCALC

Add simple floats: $2 + 1 = 3$.

Solution:

```
$ ./fpcalc 0x40000000 + 0x3f800000
X: 0x4000_0000 = 2 = +1.0 × 2^1: Biased Exp 128 Fract 0x0
Y: 0x3f80_0000 = 1 = +1.0 × 2^0: Biased Exp 127 Fract 0x0
0x40000000 + 0x3f800000: 0x4040_0000 = 3 = +1.1 × 2^1: Biased Exp 128 Fract 0x400000
exceptions: Inexact 0 Underflow 0 Overflow 0 DivideZero 0 Invalid 0
```

Example 16.14 DEFAULT ROUNDING (RNE) USING FPCALC

Add floats with rounding to nearest (RNE), the default rounding mode, which sets the Inexact flag: $1.0 + 1.4 \times 10^{-45} = 1.0$.

Solution:

```
$ ./fpcalc 0x3f800000 + 0x00000001
X: 0x3f80_0000 = 1 = +1.0 × 2^0: Biased Exp 127 Fract 0x0
Y: 0x0000_0001 = 1.4013e-45 = Subnorm: +0.00000000000000000000001 × 2^-126: Biased Exp 0 Fract 0x1
0x3f800000 + 0x00000001: 0x3f80_0000 = 1 = +1.0 × 2^0: Biased Exp 127 Fract 0x0
exceptions: Inexact 1 Underflow 0 Overflow 0 DivideZero 0 Invalid 0
```

Example 16.15 ROUNDING TO +∞ (RP) USING FPCALC

Add floats with rounding to positive infinity (RP): $1.0 + 1.4 \times 10^{-45}$ is slightly more than 1.0.

Solution:

```
$ ./fpcalc 0x3f800000 + 0x00000001 RP
X: 0x3f80_0000 = 1 = +1.0 × 2^0: Biased Exp 127 Fract 0x0
Y: 0x0000_0001 = 1.4013e-45 = Subnorm: +0.00000000000000000000001 × 2^-126: Biased Exp 0 Fract 0x1
0x3f800000 + 0x00000001: 0x3f80_0001 = 1 = +1.00000000000000000000001 × 2^0: Biased Exp 127 Fract 0x1
exceptions: Inexact 1 Underflow 0 Overflow 0 DivideZero 0 Invalid 0
```

Example 16.16 MULTIPLICATION USING FPCALC

Multiply half-precision numbers that overflow to ∞ and set the Overflow flag.

Solution:

```
$ ./fpcalc 0x7200 × 0x4600
X: 0x7200 = 12288 = +1.1 × 2^13: Biased Exp 28 Fract 0x200
Y: 0x4600 = 6 = +1.1 × 2^2: Biased Exp 17 Fract 0x200
0x7200 × 0x4600: 0x7c00 = inf = +inf: Biased Exp 31 Fract 0x0
exceptions: Inexact 1 Underflow 0 Overflow 1 DivideZero 0 Invalid 0
```

flags are set at the end of the calculation. RISC-V does not have any floating-point traps.

Programs often need to access the rounding mode or flags separately, so RISC-V also has `frm` and `fflag` CSRs that allow accessing these bits of the `fcsr` directly without needing to mask the other fields, as shown in Fig. 16.17(b) and (c).

For the underflow flag, RISC-V detects tininess after rounding (see Section 16.1.4).

16.4.2 Registers and NaN Boxing

RISC-V has 32 floating-point registers. If the D extension is supported, the registers are all 64 bits wide, and single-precision values are held in the least significant half of the register with all 1s in the upper half. This is known as *NaN boxing* because the double-precision value is a NaN with a payload of the single-precision value in the bottom half. NaN boxing generalizes to any supported precision; when lower-precision values are represented in a register capable of higher precision, the higher-precision values are NaNs.

Example 16.19 NAN BOXING

Show how to store the half-precision number 1.0 = 0x3C00 as a NaN-boxed quad-precision number.

Solution: The half-precision number goes in the bottom 16 bits and the upper 112 bits are filled with 1s: 0xFFFFFFFFFFFFFFFFFFFFFFFFFFFF3C00.

16.4.3 NaNs in RISC-V

The IEEE 754 specification and default SoftFloat implementation recommend propagating the payload from one of the NaN inputs to a NaN output. In contrast, RISC-V produces a *canonical quiet NaN* (or, simply, *canonical NaN*) as the output of any floating-point operation resulting in NaN. This canonical NaN has a positive sign and all fraction bits of 0 except the most significant bit. Table 16.11 lists the canonical NaNs for various precisions.

The RISC-V designers concluded that propagating NaN payloads is not used in portable code and thus was not worth the hardware cost of a mux to choose the payload.

Table 16.11 Canonical NaNs.

Precision	Canonical NaN
Half	7E00
Single	7FC00000
Double	7FF8000000000000
Quad	7FFF8000000000000000000000000000

Table 16.12 RISC-V floating-point moves.

Instruction	Operation
fmv.x.w	Move from single-precision floating-point to integer register
fmv.w.x	Move from integer to single-precision floating-point register
fmv.x.d	Move from double-precision floating-point to integer register (RV64D only)
fmv.d.x	Move from integer to double-precision floating-point register (RV64D only)

Floating-point minimum and maximum instructions FMIN and FMAX select the non-NaN operand if one operand is a NaN.

16.4.4 Floating-Point Move Instructions

The `fmv` family of instructions moves raw bits between integer and floating-point registers without conversion, as summarized in Table 16.12. If the integer register is wider than the floating-point format, moves to floating-point take the lower bits of the integer register, and moves from floating-point sign-extend the bits to the width of the integer register.

16.4.5 Zfa Additional Floating-Point Instructions

As the RISC-V ecosystem of compilers and libraries matured, developers discovered a variety of handy floating-point instructions were absent from the initial F, D, Q, and Zfh specifications. The Zfa extension instructions are summarized in Table I.26 at the back of the book. This section describes each group of instructions.

16.4.5.1 `fli`

Loading a floating-point constant is surprisingly expensive. Loading it from memory may take many cycles to fetch from DRAM. It can take many instructions to assemble an immediate integer with adds and shifts, then convert to floating-point. The `fli` instruction loads one of 32 constants (commonly used in numerical libraries) into a floating-point register, as given in Table 16.13.

16.4.5.2 `fround` and `froundnx`

`fround` rounds a floating-point number to an integer according to the rounding mode and represents the result in floating-point. If the input is a signaling NaN, it sets the Invalid flag. `froundnx` is similar but also sets the Inexact flag if the rounded value differs from the input value. These instructions correspond to the IEEE754-2019 standard *roundToIntegeral*

Table 16.13 Floating-point load-immediate values and hexadecimal patterns.

rs1	Value	Decimal	fli.h	fli.s	fli.d	fli.q
0	−1.0	−1.0	BC00	BF800000	BFF00...0	BFFF00...0
1	$2^{e_{min}}$	Minimum positive normal	0400	00800000	00100...0	000100...0
2	2^{-16}	1 / 65,536	0100	37800000	3EF00...0	3FEF00...0
3	2^{-15}	1 / 32,768	0200	38000000	3F000...0	3FF000...0
4	2^{-8}	1 / 256	1C00	3B800000	3F700...0	3FF700...0
5	2^{-7}	1 / 128	2000	3C000000	3F800...0	3FF800...0
6	2^{-4}	0.0625	2C00	3D800000	3FB00...0	3FFB00...0
7	2^{-3}	0.125	3000	3E000000	3FC00...0	3FFC00...0
8	1.00×2^{-2}	0.25	3400	3E800000	3FD00...0	3FFD00...0
9	1.01×2^{-2}	0.3125	3500	3EA00000	3FD40...0	3FFD40...0
10	1.10×2^{-2}	0.375	3600	3EC00000	3FD80...0	3FFD80...0
11	1.11×2^{-2}	0.4375	3700	3EE00000	3FDC0...0	3FFDC0...0
12	1.00×2^{-1}	0.5	3800	3F000000	3FE00...0	3FFE00...0
13	1.01×2^{-1}	0.625	3900	3F200000	3FE40...0	3FFE40...0
14	1.10×2^{-1}	0.75	3A00	3F400000	3FE80...0	3FFE80...0
15	1.11×2^{-1}	0.875	3B00	3F600000	3FEC0...0	3FEFC0...0
16	1.00×2^{0}	1.0	3C00	3F800000	3FF00...0	3FFF00...0
17	1.01×2^{0}	1.25	3D00	3FA00000	3FF40...0	3FFF40...0
18	1.10×2^{0}	1.5	3E00	3FC00000	3FF80...0	3FFF80...0
19	1.11×2^{0}	1.75	3F00	3FE00000	3FFC0...0	3FFFC0...0
20	1.00×2^{1}	2.0	4000	40000000	40000...0	400000...0
21	1.01×2^{1}	2.5	4100	40200000	40040...0	400040...0
22	1.10×2^{1}	3	4200	40400000	40080...0	400080...0
23	2^{2}	4	4400	40800000	40100...0	400100...0
24	2^{3}	8	4800	41000000	40200...0	400200...0
25	2^{4}	16	4C00	41800000	40300...0	400300...0
26	2^{7}	128	5800	43000000	40600...0	400600...0
27	2^{8}	256	5C00	43800000	40700...0	400700...0
28	2^{15}	32,768	7800	47000000	40E00...0	400E00...0
29	2^{16}	65,536	7C00	47800000-	40F00...0-	400F00...0-
30	∞		7C00	7F800000	7FF00...0	7FFF00...0
31	Canonical NaN		7E00	7FC00000	7FF80...0	7FFF80...0

and *roundToIntegralExact* operations. With suitable selection of rounding mode, it can perform the standard C library functions `trunc()`, `ceil()`, `floor()`, and `round()`.

16.4.5.3 `fmvh` and `fmvp`

These instructions are used in RV32D and RV64Q to transfer information between a floating-point register and a pair of integer registers. `fmvh` moves the upper half of a 2XLEN-sized floating-point register into an XLEN-sized integer register. It is used in conjunction with `fmv`, which moves the lower half. `fmvp` moves a pair of XLEN-sized integer registers into the lower and upper halves of a 2XLEN-sized floating-point register.

16.4.5.4 `fminm` and `fmaxm`

The `fminm` / `fmaxm` instructions are similar to `fmin` / `fmax` but produce a canonical NaN if either input is NaN (rather than if both are NaN). They correspond to the IEEE754-2019 standard *minimum/maximum* operations, while `fmin` / `fmax` corresponds to *minimumNumber/maximumNumber*.

16.4.5.5 `fltq` and `fleq`

`fltq` / `fleq` are similar to `flt` / `fle` comparison instructions but do not set the invalid flag when an input is a quiet NaN. They correspond to the IEEE754-2019 *compareQuiet* operations, and `flt` / `fle` correspond to *compareSignaling* operations.

16.4.5.6 `fcvtmod.w.d`

`fcvtmod.w.d` is an unusual instruction that accelerates handling of JavaScript `Numbers`. `Number` is a double-precision floating-point data type, but it is sometimes treated as a signed integer modulo 2^{32}.

`fcvtmod.w.d` is similar to `fcvt.w.d` but always rounds to zero. The bottom 32 bits are taken from the unbounded result and are sign-extended to XLEN and written to an integer register. Infinities and NaNs are converted to 0. The instruction sets the same flags as `fcvt.w.d`.

16.5 TEST PLAN

Floating-point verification is notoriously tricky. Wally tests the FPU using the riscv-arch-test suite to prove basic functionality of the pipeline and the units and also tests each unit using the robust UCB TestFloat suite to verify corner cases.

16.5.1 riscv-arch-test

The riscv-arch-test suite contains rv32i_m/F, rv32i_m/D, rv64i_m/F, and rv64i_m/D test suites for single- and double-precision floating-point instructions. It is derived from the IBM Floating-Point Test-Suite for IEEE [Aharoni08]. The IBM test suite defines categories of tests identified in Table 16.14. IBM uses the FPgen test generator to create inputs covering all the situations. The number of cases is approximate and sometimes varies across instructions or FLEN.

Approximately 91% of the tests use the RNE rounding mode; the remainder test the other rounding modes.

The floating-point tests are the slowest part of the regression, and the FMA tests are the biggest part of the floating-point tests. To speed up regression, the FMA tests are separated out to run in parallel with the other tests. Thus, four FMA and other floating-point test suites cover rv{32/64} and f/d, such as rv32gc_arch32d_fma and rv64gc_arch64f.

Note that TestFloat does not produce vectors for RISC-V instructions `fclass`, `fsgnj`, `fsgnj{n/x}`, `fmax` or `fmin`.

16.5.2 TestFloat

As mentioned in Section 16.3, TestFloat produces test cases to check compliance of floating-point implementations using the UCB SoftFloat library to compute expected results. Wally has a testbench to run the TestFloat tests directly on Verilog FPUs rather than by creating RISC-V programs.

TestFloat has `-level 1` and the more comprehensive `-level 2` options of test generation. Level 2 generates tens or hundreds of millions of tests for 2-input operations, so we use level 2 for 1-input square root and convert operations and level 1 for all others.

TestFloat covers the instructions in Table 16.15. It generates separate test vectors files for every rounding mode rm ∈{rd, rne, ru, rz, rnm} and precision fp ∈ {f16, f32, f64, f128}. Integer conversions also support integer precisions ip ∈ {i32, u32, i64, u64}.

TestFloat and SoftFloat are submodules in `$WALLY/addins`. Softfloat has a compiler flag called SPECIALIZE_TYPE=RISCV to support RISC-V behavior. Wally's Makefile uses this flag for RISC-V compatibility. A Makefile builds SoftFloat and TestFloat and then uses TestFloat to create floating-point testvectors. Invoke with:

```
$ cd $WALLY/tests/fp
$ make -jobs
```

It runs TestFloat for each operation, precision, and rounding mode, then converts the vectors into a format that Verilog simulators can read using `$readmemh`. It places the vectors in `$WALLY/tests/fp/vectors`.

The full set of TestFloat testvectors is 8 GB, which is why it is generated rather than kept in the repository.

For example, `f32_mulAdd_rne.tv` is shown in Fig. 16.18. The first three 32-bit hexadecimal numbers are the three source operands. The fourth 32-bit number is the result. The next two digits are a 5-bit hexadecimal number representing flags: {Invalid, Divide by Zero, Overflow, Underflow, Inexact}.

`make` also creates test vectors for the combined integer divide/remainder and floating-point divide/square root unit in `tests/fp/combined_IF_vectors/IF_vectors`, as shown in Fig. 16.19. It requires the

Table 16.14 **IBM test categories.**

Category	Description	Applicable Instructions	Number of Cases
B1	Floating-point basic types: all combinations of all signs of: 0, 1, min subnorm, subnorm, max subnorm, min norm, norm, max norm, INF, NaNs	fadd/fsub/fmul/fdiv/fmadd/fsqrt fclass; fsgnj/fsgnjn/fsgnjx feq/fle/flt; fmin/fmax CvtFP2Int; CvtInt2FP; fmv	33 for 1 input 580 for 2 input 13,835 for 3 input
B2	Near FP basic values	fadd/fsub/fmul/fdiv/fmadd/fsqrt	280
B3	Rounding: sign, LSB, guard, sticky	fadd/fsub/fmul/fdiv/fmadd/fsqrt	1124
B4	Overflow and near overflow	fadd/fsub/fmul/fdiv/fmadd/fsqrt	144
B5	Underflow and near underflow	fadd/fsub/fmul/fdiv/fmadd/fsqrt	224
B6	Tiny rounding intermediate tasks	fmul/fdiv/fmadd	144
B7	Sticky bit calculation	fadd/fsub/fmul/fdiv/fmadd/fsqrt	164
B8	Vicinity of rounding boundaries	fadd/fsub/fmul/fdiv/fmadd/fsqrt	1,684
B9	Special significands on inputs	fmul/fdiv/fsqrt	1,324
B10	Add: shift	fadd/fsub	78
B11	Add: shift combined with special significands	fadd/fsub	7,016
B12	Add: cancellation	fadd/fsub	54
B13	Add: cancellation and subnormal result	fadd/fsub	204
B14	FMA: shift	fma	84
B15	FMA: shift combined with special significands	fma	10,096 for half 38,129 for float 102,005 for double
B16	FMA: cancellation	fma	211
B17	FMA: cancellation and subnormal result	fma	211
B18	FMA: special events	fma	281
B19	Compare: different input field relations	feq/fle/flt; fmin/fmax	1,072
B20	Divide: trailing zeros	fdiv/fsqrt	106
B21	Divide: divide by zero exception	fdiv	680
B22	CvtFP2Int: special input exponents	CvtFP2Int	41
B23	CvtFP2Int: overflow	CvtFP2Int	46
B24	CvtFP2Int: Underflow	CvtFP2Int	106
B25	CvtInt2FP: special inputs	CvtInt2FP; fmv	33
B26	CvtInt2FP: number of significant bits	CvtInt2FP; fmv	33
B27	CvtInt2FP: NaN	CvtFP2Int; fmv	33
B28	Convert: Special Values	CvtFP2Int; fmv	33
B29	Convert: Rounding	CvtFP2Int; fmv	81

Source: riscv-isac.readthedocs.io/en/stable/code.html

Table 16.15 **TestFloat test cases.**

RISC-V Instruction	TestFloat Case Name	Number of Tests/Rounding Mode/Precision
`fadd`	<fp>_add_<rm>.tv	46,464
`fsub`	<fp>_sub_<rm>.tv	46,464
`fmul`	<fp>_mul_<rm>.tv	46,464
`fdiv`	<fp>_div_<rm>.tv	46,464
`fsqrt`	<fp>_sqrt_<rm>.tv	2,448/8,800/26,112/69,108
`fmadd`[a]	<fp>_mulAdd_<rm>.tv	6,133,248
`feq/flt/fle`[b]	<fp>_eq/le/lt_<rm>.tv	46,464
convert float to float `fcvt.h/s/d/q.h/s/d/q`	<fp>_to_<fp>_<rm>.tv	2,448/8,800/26,112/69,108
convert float to int `fcvt.w/wu/l/lu.h/s/d/q`	<fp>_to_<ip>_<rm>.tv	2,448/8,800/26,112/69,108
convert int to float `fcvt.h/s/d/q.w/wu/l/lu`	<ip>_to_<fp>_<rm>.tv	15,500/63,756

[a] `fmsub`, `fnmadd`, and `fnmsub` are the same as `fmadd` except sign handling is not tested via TestFloat.
[b] Comparisons do not round, so the results are the same in all rounding modes. RISC-V deviates from the default IEEE behavior of SoftFloat in four ways:
1. RISC-V NaN boxes lower-precision floating-point numbers stored in wider registers.
2. Floating-point to unsigned integer conversions of negative numbers produce all 1s in IEEE but all 0s in RISC-V.
3. Floating-point to integer conversions that overflow the positive range produce the most negative integer (0x8000...) in IEEE, but the largest positive integer (0x7FFFF... signed or 0xFFFF... unsigned) in RISC-V.
4. Operations with a NaN input preferably propagate the payload to the output in IEEE but produce a canonical NaN in RISC-V.

```
8683F7FF_C07F3FFF_00000000_07839504_01
00000000_00000000_3C072C85_3C072C85_00
9EDE38F7_3E7F7F7F_DF7EFFFF_DF7EFFFF_01
...
```

Fig. 16.18 `f32_mulAdd_rne.tv.`

```
0_00000000000000000000000000000008_00000000000000000000000000000008_0000000000000000000000000
0000001_XX_X
0_00000000000000000000001000000000_00000000000000000000000000000000_0000000000000000fffffffff
fffffff_XX_X
...
```

Fig. 16.19 `cvw_64_div-01.tv.`

compiled RISCOF test cases from `tests/riscof/work`. The format of these vectors is `operation_A_B_Y_flags_roundingMode`, with the operation code defined in Table 16.16.

Table 16.16 Divider test vector operations.

Code	Instruction	Operation
0	`div`	Integer divide
1	`fdiv`	Floating-point divide
2	`fsqrt`	Floating-point square root
3	`rem`	Integer remainder
4	`divu`	Integer unsigned division
5	`remu`	Integer unsigned remainder
6	`divw`	Integer 32-bit division
7	`divuw`	Integer unsigned 32-bit division
8	`remw`	Integer 32-bit remainder
9	`remuw`	Integer unsigned 32-bit remainder

To run the tests on an RV64 configuration that supports all four floating-point precisions, use one of the following commands, with choices of tests being `add`, `sub`, `mul`, `div`, `sqrt`, `fma`, `cmp`, `cvtint`, or `cvtfp`:

```
$ wsim fdqh_rv64gc fma -t testbench_fp
$ wsim fdqh_rv64gc div -t testbench_fp -gui
$ wsim fdqh_rv64gc cvtint -t testbench_fp -sim verilator
```

These commands invoke Questa or Verilator on `$WALLY/ testbench/ testbench-fp.sv`, optionally with the Questa GUI. This testbench reads the TestFloat vectors from `$WALLY/tests/fp/vectors` and runs them through the unpacker and each of the units. The testbench then chooses the result and flags of the unit being tested and compares them to the expected values.

16.6 WALLY IMPLEMENTATION

Fig. 16.20 shows the changes to the `wallypipelinedcore` to support floating-point. The FPU operates in the Execute and Memory stages with

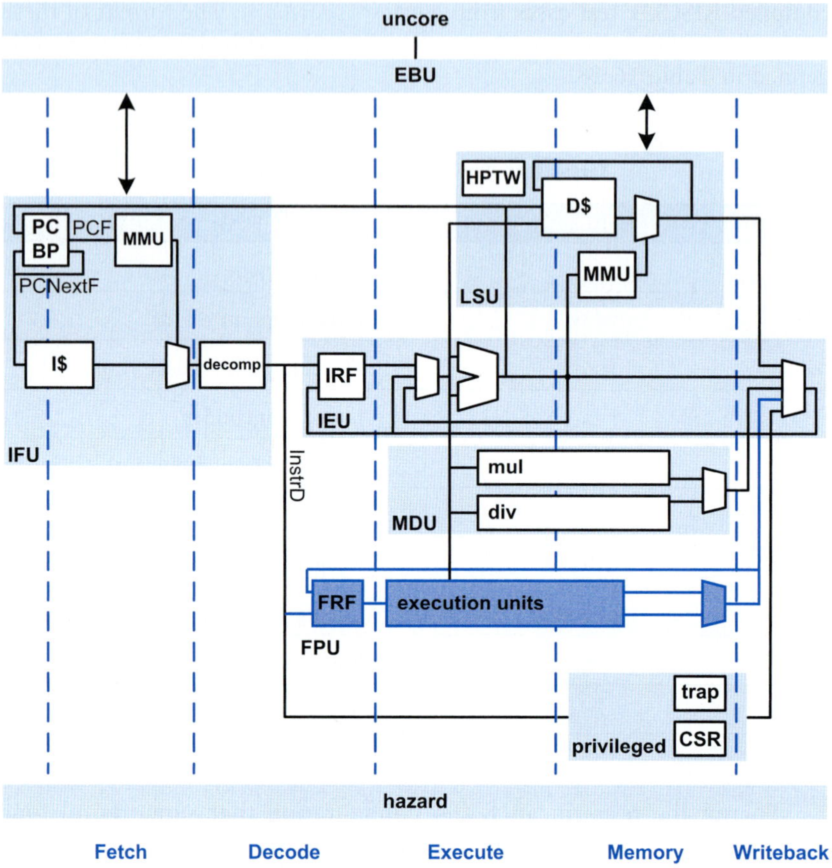

Fig. 16.20 **Wally R{32/64}IFDMC high-level block diagram, emphasizing the FPU.**

a two-cycle latency and single-cycle throughput, except that division and square root have longer latencies and stall the pipeline. The privileged unit includes a module to support the `fcsr` and keep track of whether the floating-point registers are dirty. The FPU connects to the LSU for loads and stores, to the IEU for moves and converts, to the privileged unit for the `fcsr`, and to the hazard unit for stalls.

16.6.1 Floating-Point Unit

Fig. 16.21 shows a block diagram of the Wally FPU. The FPU is parameterized to support half-, single-, double-, and quad-precision or any subset thereof. The highest precision supported by the FPU is called FLEN. For example, FLEN = 64 for an RV32IFD configuration because it supports single- and double-precision.

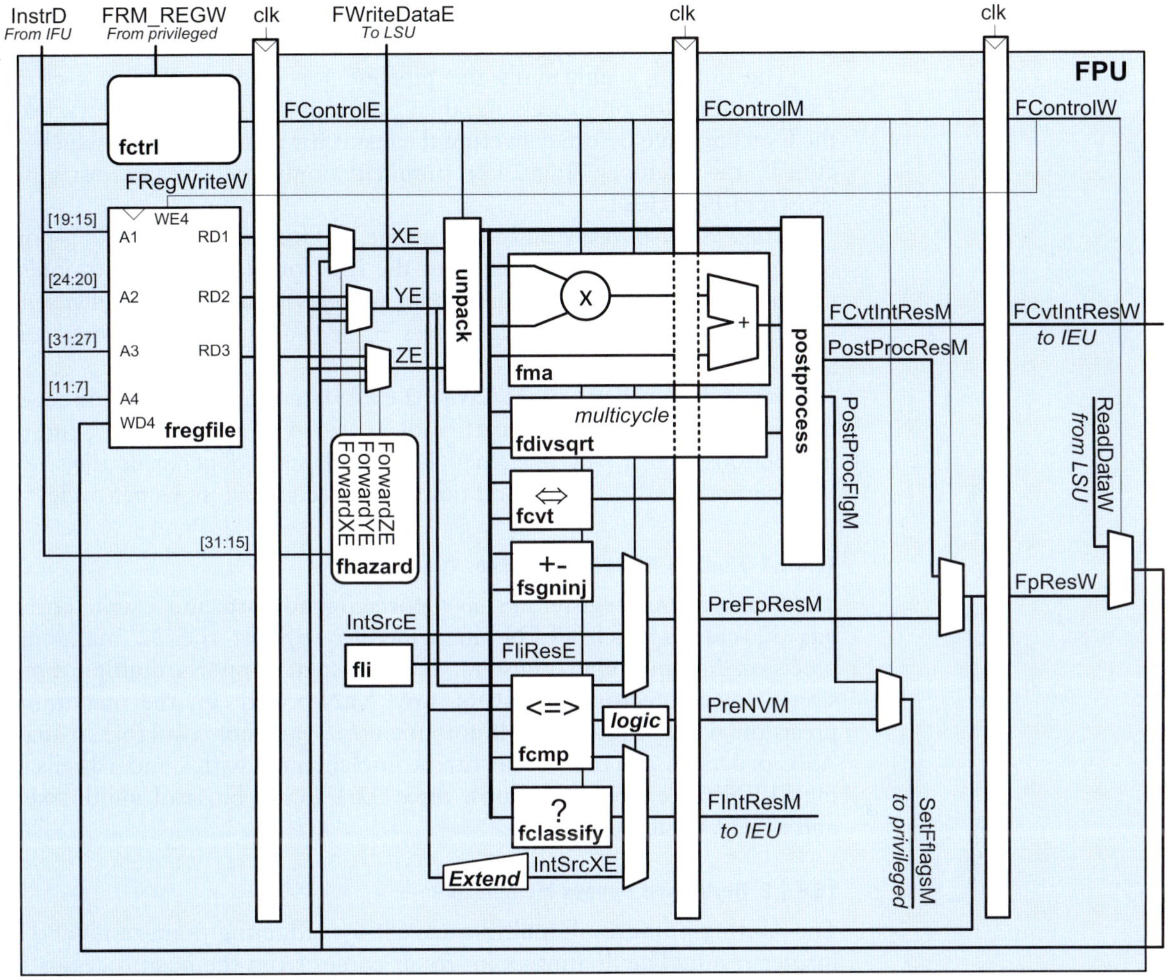

Fig. 16.21 Wally FPU.

The supported precisions depend on the F (32-bit), D (64-bit), and Q (128-bit) extensions in the `misa` (M-mode ISA register), as well as the `ZFH_SUPPORTED` (16-bit) configuration. Legal combinations are: F, FD, FDQ, HF, HFD, and HFDQ.

The `fregfile` register file holds the highest precision FLEN, with lower precisions NaN-boxed. The execution units, such as the FMA, are also sized for the highest supported precision but can handle lower precisions. The unpacking unit (`unpack`) extends lower-precision operands to FLEN. The postprocessing unit (`postprocess`) rounds to the target precision.

The optional RISC-V Zfhmin extension provides only the following half-precision instructions: `flh`, `fsh`, `fmv.x.h`, `fmv.h.x`, `fcvt.*.h`, and `fcvt.h.*`. It is intended for systems that use half-precision as a storage format and perform computations at higher precision. Wally presently optionally supports Zfh but not Zfhmin.

The FMA handles `fadd`, `fsub`, and `fmul` as well as `fmadd`, `fmsub`, `fnmadd`, and `fnmsub`.

The `fdivsqrt` unit stalls the pipeline for many cycles. It uses the Execute stage for preprocessing, then performs many recurrence cycles in the Execute stage before delivering the result for postprocessing, which occurs in the Memory stage. For simplicity, only the Execute portion is shown in Fig. 16.21.

The FPU also receives `SrcAE` from the IEU for integer-to-floating-point conversion and delivers `FIntResM` to the IEU for floating-point-to-integer moves and conversions, comparisons, and classifications. The FPU sends `FWriteDataE` to the LSU and receives back `ReadDataW` for floating-point stores and loads.

`fma` and `fcvt` instructions have a 2-cycle latency, so subsequent dependent instructions incur a one-cycle stall. `fcmp`, `fsigninj`, `fclassify`, and `fmv` instructions have a 1-cycle latency. Division latency depends on the size of the operands and the radix and number of recurrence steps per cycle.

16.6.1.1 Floating-Point Register File

The floating-point register file has 4 ports: 3 read ports and 1 write port. It has 32 entries, each FLEN-bits wide, to support the 32 maximum-precision floating-point registers. If the system supports multiple precisions, lower-precision numbers are NaN-boxed in the maximum-precision word. Because multiport memories are not available in many ASIC processes, the register file can be implemented with standard cells using FLEN*32 enabled flip-flops, three 32:1 FLEN-bit read multiplexers, and a 5:32 write decoder.

16.6.1.2 Result and Bypass Multiplexers

The floating-point result multiplexers choose a floating-point result and an integer result. The floating-point result comes from the postprocessor for `fma`, `fdivsqrt`, or `fcvt` instructions or from `fcmp`, `fsgninj`, or the IEU (for int-to-float moves). The postprocessor result arrives late, so the other results are premultiplexed in the execute stage to save time and registers. The floating-point result is written to the floating-point register file and may be selected by the bypass (also called forwarding) muxes for a dependent operation.

The integer result comes from the postprocessor (for float-to-int conversions), `fcmp`, `fclassify`, or `SrcX` (for float-to-int moves). The IEU `Result` multiplexer sends that integer result to the IEU register file, as described in Section 16.6.13. The multiplexing occurs in the Execute stage and the results are sent to the IEU at the start of the Memory stage in time to forward to the next instruction.

The three bypass multiplexers at the start of the Execute stage select `SrcX`, `SrcY`, and `SrcZ` from the floating-point register file or from previous instructions in the Memory or Writeback stages.

16.6.2 Unpacking Unit

The unpacking unit (`unpack`) extracts fields from the inputs as described in Section 16.2.1. Because Wally supports multiple precisions on the same configurable hardware, it also extends lower precisions to FLEN by appending trailing 0s to the fractions and by adding a suitable offset to the exponents. The `fmt` field of the instruction represents the desired precision according to Table 16.17. PLEN ($\leq$ FLEN) is the length of the operand. For example, in RV32IFD, XLEN = 32 because the base architecture is 32 bits and FLEN = 64 because the register file and datapath accommodate double precision. In an `fadd.s f1, f2, f3` instruction, the precision is single, so PLEN = 32.

Table 16.17 Floating-point format codes.

fmt	Precision	PLEN
00	Single	32
01	Double	64
10	Half	16
11	Quad	128

Unpacking also verifies that lower-precision numbers are properly NaN-boxed by having all 1s in the leading bits. It asserts the NaN output if they are not boxed correctly. Thus, if a program inadvertently attempts to perform a double-precision operation on a single-precision number in a double-precision register, the number is treated as NaN to warn about the mistake.

The unpacker then pulls out the sign X_s from the most significant bit of the specified precision ($X_{\text{PLEN}-1}$). The raw fraction comes from the bottom bits and the raw exponent from the middle bits. The unpacker checks for special cases of all 0s and all 1s in the exponents. The fraction is extended to the full length by appending 0s to the least significant bits.

Extending the exponent to the desired length and bias initially seems difficult but simplifies gracefully. For example, consider extending an 8-bit single-precision exponent with a bias of 127 to an 11-bit double-precision exponent with a bias of 1,023. We add 1,023 and subtract 127 to get the new exponent. In binary, we add 01111111111 − 01111111 = 01110000000 to the raw 8-bit exponent. If the raw exponent were $e_7e_6e_5e_4e_3e_2$, adding 01110000000 gives $e_7\bar{e}_7\bar{e}_7\bar{e}_7e_6e_5e_4e_3e_2e_1e_0$. In other words, the lower 7 bits remain the same, and if e_7 is 0, the upper bits are 0111; otherwise, they are 1000. Thus, extending an exponent reduces to inserting complemented versions of the raw exponent's most significant bit. If the exponent is subnormal (all original exponent bits are 0), replace e_0 with 1 to bump up the exponent.

Example 16.20 UNPACK A HALF-PRECISION NUMBER

Consider an FPU that supports single- and half-precision operands. The half-precision number 0x8300 $= -0.11 \times 2^{-14} = -1.1 \times 2^{-15}$ is stored in NaN-boxed form as 0xFFFF0300. Unpack it to the fields of a single-precision number.

Solution: 0x8300 = 1 00000 1100000000, so the sign is $X_s = 1$, the raw fraction is $X_f =$ 1100000000, and the raw exponent is $X_e =$ 00000. Extend this exponent from 5 to 8 bits by adding 127 − 15 = 112 (01111111 − 01111 = 01110000), then add 1 (because the exponent was originally subnormal) to get: 01110001 = 113, or an unbiased exponent of −14. Because the original exponent was zero, the number was subnormal and we don't prepend a 1 to the fraction; so, the significand is 0.110…0. This is normalized to a significand of 1.10…0 (giving a fraction of 10…0), and the exponent is decremented by 1 to 01110000 to account for normalization. Hence, the single-precision number is 1 01110000 100…0 = 0xB8400000.

The remainder of the special case detection is the same as in the fixed-size unpacker from Table 16.6.

16.6.3 Fused Multiply-Add

Compared to the basic FMA from Section 16.2.2, the RISC-V FMA provides the option of inverting the addend, the result, or both: $W = \pm (X \times Y \pm Z)$. Thus, it supports all four flavors of RISC-V instructions: `fmadd`, `fmsub`, `fnmadd`, `fnmsub`. The FMA is also used to compute `fadd`, `fsub`, and `fmul` by setting one of the sources to 1.0 or 0.0.

16.6.4 Divide and Square Root

k is `FDIVSQRT_STEPS_PER_CYCLE` in the configuration files.

The Wally divide and square root unit (`fdivsqrt`) uses minimally-redundant radix-2 or radix-4 steps with hardware for each step replicated by a configurable amount k = 1 to 4 to perform k steps each cycle.

Division and square root perform a variable number of steps to produce $b = N_f + 1$ fraction bits, with N_f depending on the target precision. Radix-2^r operations produces r-bit digits so requires b/r steps. Division needs an additional step to produce the integer digit. The unit contains k copies of the recurrence hardware to perform k steps per cycle. Preprocessing and postprocessing each involve an extra cycle. Hence, the number of cycles is:

$$Cycles = \left\lceil \frac{1 + \frac{b}{r}}{k} \right\rceil + 2 \tag{16.8}$$

This latency is given in Table 16.18.

Wally optionally uses the floating-point divider for integer division. When `IDIV_ON_FPU` is 1 in the configuration files, the integer division logic

Table 16.18 **Wally's divide and square root unit (`fdivsqrt`) latency.**

	Latency				
	Radix-2			Radix-4	
Precision	$k = 1$	$k = 2$	$k = 4$	$k = 1$	$k = 2$
Half	15	8	5	9	5
Single	28	15	8	15	8
Double	57	29	15	30	16
Quad	117	59	30	60	31

from Section 15.4.3 is removed, and the FPU divider is enhanced to support integer division. These enhancements include the integer preprocessing and postprocessing from Section F.2.9, an FSM modification to select the number of cycles for integer division, and a multiplexer in the IEU datapath to choose the division result from the FPU rather than the MDU. `DivBusyE` is disabled; instead, `FDivBusyE` is used to stall both integer and floating-point division.

rv64gc is configured with Radix-4, $k = 4$, and `IDIV_ON_FPU` = 1. rv32gc is configured with Radix-4, $k = 2$, `IDIV_ON_FPU` = 0, `IDIV_BITSPERCYCLE` = 2. rv32gc has smaller operands, so it can get away with less division hardware and still have reasonable latency. Also, having different configurations makes the regression suite test both floating-point and integer division hardware and different k.

16.6.5 Convert

Wally's floating-point convert unit (`fcvt`) handles the RISC-V conversion instructions defined in Table 16.19, depending on which floating-point sizes are supported. Note that the first letter specifies the destination format (half/single/double/quad) and the second specifies the source format, so `fcvt.s.d` converts a double to a single. `w` and `wu` are signed and unsigned 32-bit integers, and `l` and `lu` are signed and unsigned 64-bit integers. For example, `fcvt.s.wu` converts an unsigned 32-bit number to a 32-bit single-precision floating-point number.

RV64 doubles the number of supported integer conversions by adding `l` and `lu` everywhere `w` and `wu` appear for RV32.

In RISC-V, float-to-int conversions set the Inexact flag when the result is inexact. In the IEEE standard, this operation is called *ConvertToIntegerExact*.

Table 16.19 Floating-point conversions supported in RV32.

Modes Supported	Float-Float Conversions	RV32 Float-Int Conversions	RV32 Int-Float Conversions
F	None	`fcvt.w.s`, `fcvt.wu.s`	`fcvt.s.w`, `fcvt.s.wu`
FD	`fcvt.s.d`, `fcvt.d.s`	F + `fcvt.w.d`, `fcvt.wu.d`	F + `fcvt.d.w`, `fcvt.d.wu`
FDQ	FD + `fcvt.s.q`, `fcvt.d.q`, `fcvt.q.s`, `fcvt.q.d`	FD + `fcvt.w.q`, `fcvt.wu.q`	FD + `fcvt.q.w`, `fcvt.q.wu`
HF	`fcvt.h.s`, `fcvt.s.h`	F + `fcvt.w.h`, `fcvt.wu.h`	F + `fcvt.h.w`, `fcvt.h.wu`
HFD	HF + FD + `fcvt.h.d`, `fcvt.d.h`	HF + FD	HF + FD
HFDQ	HFD + FDQ + `fcvt.h.q`, `fcvt.q.h`	HF + FDQ	HF + FDQ

16.6.6 Move

The FPU module handles moves from floating-point to integer registers by selecting `FIntResE` as a sign-extended version of the floating-point bits. Similarly, moves from integer to floating-point registers place the integer bits in the lsbs of the floating-point register and 1s in the upper bits for the NaN box, creating the move result `AlignedSrcAE`.

16.6.7 Compare

In RISC-V, when both inputs are NaN, `fmin` and `fmax` produce a canonical NaN result.

16.6.8 Sign Injection

The `fsgn` sign injection block handles the `fsgnj`, `fsgnjn`, and `fsgnjx` instructions that modify the sign of X based on the sign of Y according to Table 16.20. The hardware involves only a few 1-bit multiplexers and gates.

Sign injection supports pseudoinstructions for more common operations of floating-point moves, negation, and absolute value.

```
fmv     ft1, ft0          fsgnj     ft1, ft0, ft0
fneg    ft1, ft0          fsgnjn    ft1, ft0, ft0
fabs    ft1, ft0          fsgnjx    ft1, ft0, ft0
```

Table 16.20 Sign injection.

Instruction	Funct3	Result
`fsgnj`	000	$\{Y_s, X_e, X_f\}$
`fsgnjn`	001	$\{\overline{Y_s}, X_e, X_f\}$
`fsgnjx`	010	$\{X_s \oplus Y_s, X_e, X_f\}$

16.6.9 Classify

The `fclassify` classification module handles the `fclass` instruction that writes a 10-bit code to the integer register file based on the characteristics of X according to Table 16.21. All of the characteristics are determined during unpacking, and the module contains only a handful of gates.

Table 16.21 Classification.

Bit	Meaning
0	$-\infty$
1	Negative normalized
2	Negative subnormal
3	-0
4	$+0$
5	Positive subnormal
6	Positive normalized
7	$+\infty$
8	Signaling NaN
9	Quiet NaN

16.6.10 Zfa Additional Floating-Point Instructions

When `ZFA_SUPPORTED` = 1, the FPU is enhanced to support the additional instructions from Section 16.4.5. The new and modified modules include:

- **fli:** New module with tables of immediates, indexed with `rs1` and the destination size
- **fround:** New rounding module, described next
- **fpu:** Two new move multiplexers to select a pair of registers in `fmvp` (`IntSrcE`) and the upper half of the floating-point register for `fmvh` (`IntSrcXE`)
- **fcmp:** Modifications supporting the NaN handling and flags for `fminm`/`fmaxm`/`fltq`/`fleq`
- **specialcase:** `fcvtmod.w.d` result and flag multiplexing

16.6.10.1 `fround`

`fround` resembles `fcvt` in that it rounds a floating-point number to an integer but differs in that it expresses the result in floating-point form. Reusing `fcvt` logic and adding the conversion back to floating-point would significantly increase the FPU postprocessing critical path or add an extra cycle that the pipeline does not presently support. Therefore, Wally adds a dedicated `fround` module that is surprisingly simple.

The `fround` module receives the exponent X_e, significand X_m, and sign bit X_s from the unpacker, expanded to the full datapath width FLEN. The basic idea is to examine the unbiased exponent. If it is negative, the input is less than 1 and will round to either 0 or 1 based on the significand bits. Otherwise, the unit will round to an integer by discarding the least significant fraction bits. The unit computes L', R', and T' to round according to Table 16.4. A few special cases exist: If the input exponent is greater than or equal to N_f, then the input is a large integer value that does not require rounding; it is passed to the output. If the input is NaN or 0, the output will also be the same.

Fig. 16.22 summarizes the `fround` algorithm, and Fig. 16.23 shows a block diagram of the unit. The unit generates a thermometer-encoded *TMask* signal with 1s in the least significant columns indicating which bits contribute to the sticky bit T'. If the input is 1.0 or larger and hence has an integer portion to round, it also creates an *IMask* signal with 1s in the most significant columns indicating which bits contribute to the truncated integer result *TRUNC*. Finally, it generates one hot *HotE* and *HotEP1* signals that indicate the positions of the L' and R' bits (the least

W = fround(X)
  $E = X_e - bias$ // Unbiased exponent
  // Calculate L', R', and T'
  if ($E < 0$) // Negative exponents round to 0 or 1.
    $L' = 0$ // Lsb $L' = 0$
    if ($E = -1$) $R' = 1$; $TMask = 0.1111...111$ // If ($E = -1$) $0.5 \leq X < 1$. Round bit is 1
    else $R' = 0$; $TMask = 1.1111...111$ // If ($E < -1$), $X < 0.5$. Round bit is 0
  else // Positive exponents truncate fraction and may add 1
    $IMask = 1.0000\ldots000 >>> E$ // (in U1.N_f form); implies thermometer code (1.1…100…0)
    $TMask = \overline{IMask} >>> 1$ // $\overline{1}.\overline{1} \ldots \overline{1}\overline{0}\overline{0} \ldots \overline{0} >>> 1 = 0.0\ldots001\ldots1$
    $HotE = IMask \; \& \; \overline{(IMask << 1)}$ // A 1 in column E, where bit 0 is the integer bit,
    // bit 1 is the most significant fraction bit, etc.
    $HotEP1 = HotE >> 1$ // A 1 in column $E + 1$
    L' = OR(X_m & HotE) // $X_m[E]$, where $X_m[0]$ is the integer bit,
    // $X_m[1]$ is the most significant fraction bit, etc.
    R' = OR(X_m & HotEP1) // $X_m[E+1]$
    $TRUNC = X_m \; \& \; IMask$ // Truncated fraction, corresponds to truncated integer value
    $RND = TRUNC + HotE$ // $TRUNC + (1 >> E)$, corresponds to next integer
  T' = OR(X_m & $TMask$) // T' = OR($X_m[E+2{:}N_f]$) if $E >= 0$, OR(X_f) if $E = -1$, 1 if $E < -1$
  // Round and set Flags
  Flags = 0 // Initialize Flags (might be updated in a later step)
  if (X is NaN)
    W = Canonical NaN
    Invalid = (X is signaling NaN)
  else if ($E \geq N_f$)
    $W = X$ // Is exact; this also handles infinity
  else
    $RoundUp$ = RoundingLogic(X_s, L', R', T', rm) // Table 16.4
    if ($E < 0$) // $0 \leq X < 1$ rounds to 0 or 1
      if ($RoundUp$) $\{W_s, W_e, W_f\} = \{X_s, \text{bias}, 0\}$ // ± 1.0
      else $\{W_s, W_e, W_f\} = \{X_s, 0, 0\}$ // ± 0
    else // $X \geq 1$ rounds to an integer or overflows to infinity
      if ($RoundUp$) $R_m = RND$ else $R_m = TRUNC$ // Round up to RND or down to TRUNC
      if ($R_m = 2.0$) $\{W_s, W_e, W_f\} = \{X_s, X_e{+}1, 0\}$ // Round up to $1.0 \times 2^{E+1}$ incrementing exponent
      else $\{W_s, W_e, W_f\} = \{X_s, X_e, R_f\}$ // Rounded fraction, retain sign and exponent
    if (is an `froundnx` instruction) Inexact = $R' \mid T'$
  // Finally, pack W into specified precision (including NaN boxing when needed)

Fig. 16.22 `fround` algorithm.

significant bit of the integer result and the first fraction bit afterward). It extracts L', R', and T' with AND-OR masking and reduction logic. Rounding logic determines whether to round up, which is performed by adding 1 to the significand in the $HotE$ column. If it rounds up to 2.0, the exponent is incremented, and the fraction is 0. Otherwise, the fraction R_f is the significand R_m without the leading 1. If the input is 0 or has an

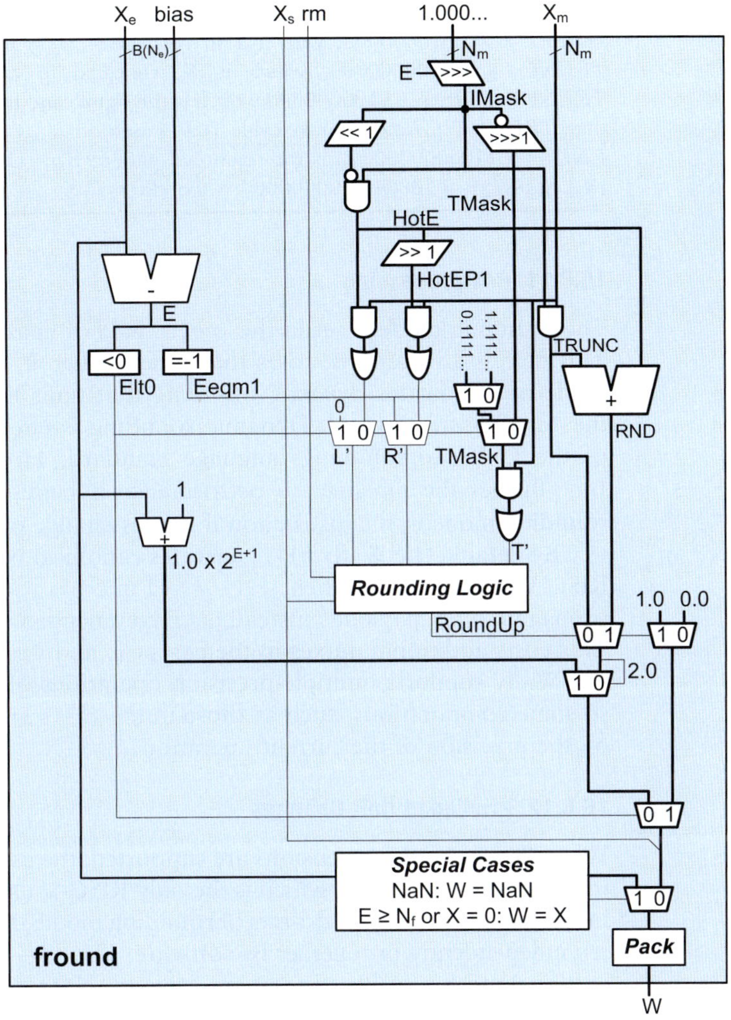

Fig. 16.23 `fround` **unit.**

exponent of at least N_f, it already represents an exact integer and does not require rounding. The unit also handles flag generation and special case of NaN. Finally, the output is packed back into the desired format.

Example 16.21 `fround` ALGORITHM

Express 6.75 as a half-precision floating-point number. Use the algorithm of Fig. 16.22 to round to an integer value (RNE mode) in half-precision floating-point format.

Solution: $X = 6.75 = 1.1011000000 \times 2^2$, with $X_s = 0$, $X_e = 15 + 2 = 17 = 10001_2$, and $X_f = 1011000000$. This exponent $E = 2$ is between 0 and 10, so we need to round. $IMask = 1.0000000000 >>> 2 = 1.1100000000$. $TMask = \overline{IMask} >>> 1 =$

0.0001111111. $HotE$ = 0.0100000000, with a 1 in column 2 of the fraction, and $HotEP1$ = 0.0010000000, with a 1 in column 3. L' = 0, R' = 1, and T' = 1. $TRUNC$ = 1.1000000000, which is X truncated to an integer (6). RND = $TRUNC + HotE$ = 1.1100000000, which is the significand for the next integer greater than $TRUNC$ (7). According to Table 16.4, based on L', R', and T', round to R_m = RND = 1.1100000000. R_f is the fractional part of R_m, .1100000000. Hence, W = $\{X_s, X_e, R_f\}$ = 0 10001 1100000000 = 0x4700 = 7.0.

16.6.11 Postprocessing

The rounding mode is set in the `funct3` field of floating-point instruction. This field is also referred to as the `rm` field when it indicates rounding. A mode of 7 indicates *dynamic rounding* that finds the rounding mode in the `fcsr` register instead. Dynamic rounding is effectively a requirement of the C99 programming language standard. However, changing the `fcsr` flushes the pipeline, so performance is better when specifying the rounding mode in the instruction if modes change often.

By default, the Wally FPU generates canonical NaNs according to the RISC-V specification. When `IEEE754` = 1 in `config_wally.vh`, the FPU will propagate NaN payloads instead, as recommended in the IEEE standard. This costs additional muxes in the `postproc` and `fcmp` units.

Wally supports multiple-precision operations on the same hardware, so some computations, such as those in the `fma's shiftcalc` unit, depend on the precision of the current operation.

16.6.12 Privileged Unit Changes

When floating-point extensions are supported, the user-mode CSR module (`csru`) contains the `fcsr`, which is the only RISC-V user-mode CSR state. It includes 5 bits of flags and 3 bits of rounding mode; these fields can be written independently or together by software. The flags can also be set by operations and are sticky, remaining set until cleared by writing the `fcsr`. Fig. 16.24 shows a schematic of the `csru`.

The FPU also updates the `mstatus.FS` field, as described in Section 8.2.3.1. Specifically, `mstatus.FS` is set to 11 (dirty) whenever `FloatRegWriteW` asserts.

16.6.13 Integer Execution Unit Changes

When floating-point is supported, the IEU receives `FIntResM` and `FCvtIntResW` from the FPU. The FPU produces `FIntResM` for `fclass`, `fcmp`, and `fmv` instructions, which take a single cycle, and `FCvtIntResW` for `fcvt` (to integer) instructions, which take two cycles. The IEU datapath adds two muxes to choose these results, as shown in Fig. 16.25.

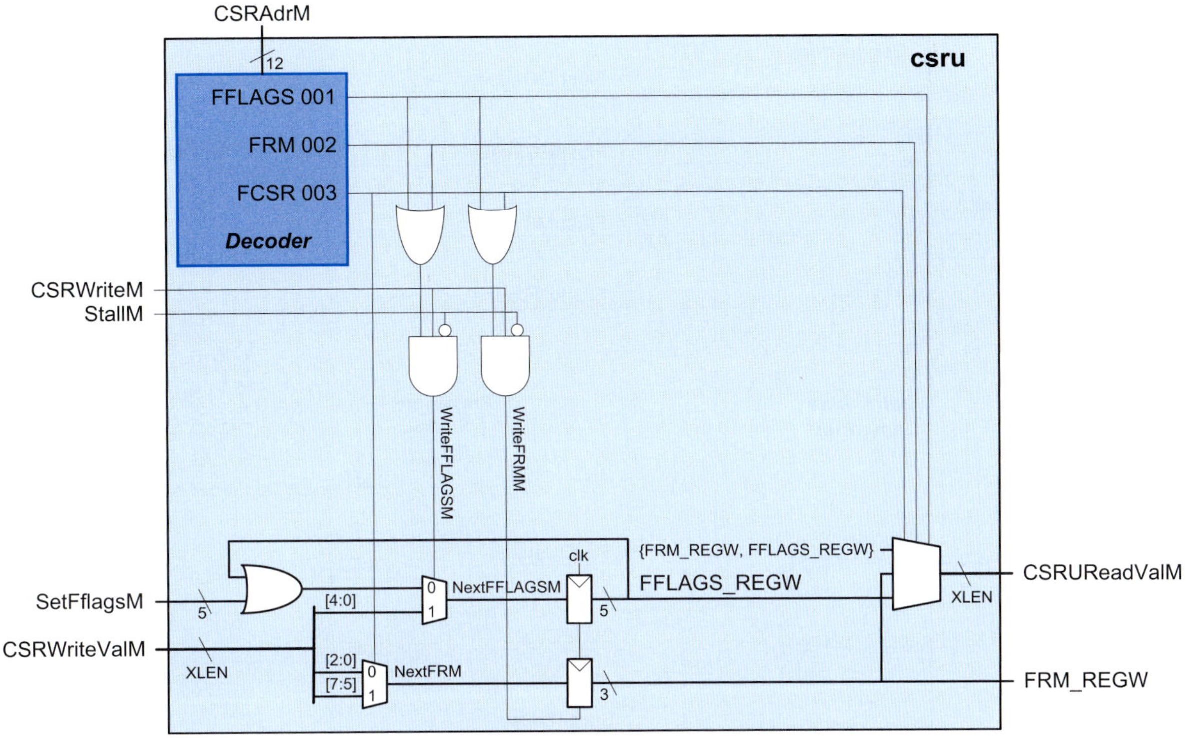

Fig. 16.24 User-mode CSR module (`csru`) with `fcsr`.

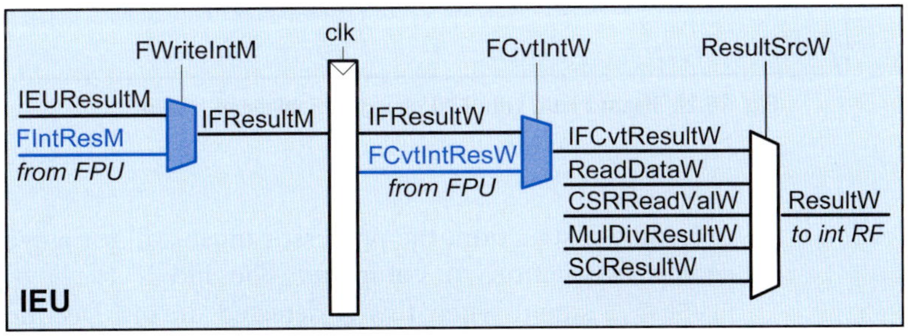

Fig. 16.25 IEU, highlighting the FPU signals.

16.6.14 Hazard Unit Changes

The hazard unit adds stalls for instructions with two-cycle latency and for divides with long and variable latency. Fig. 16.26 shows the complete Wally hazard unit, with FPU changes highlighted in blue.

Like the integer multiplier and the LSU, the FPU's `fma` and `fcvt` modules have a two-cycle latency, introducing a structural hazard. If the sources of these floating-point instructions are dependent on the result of the

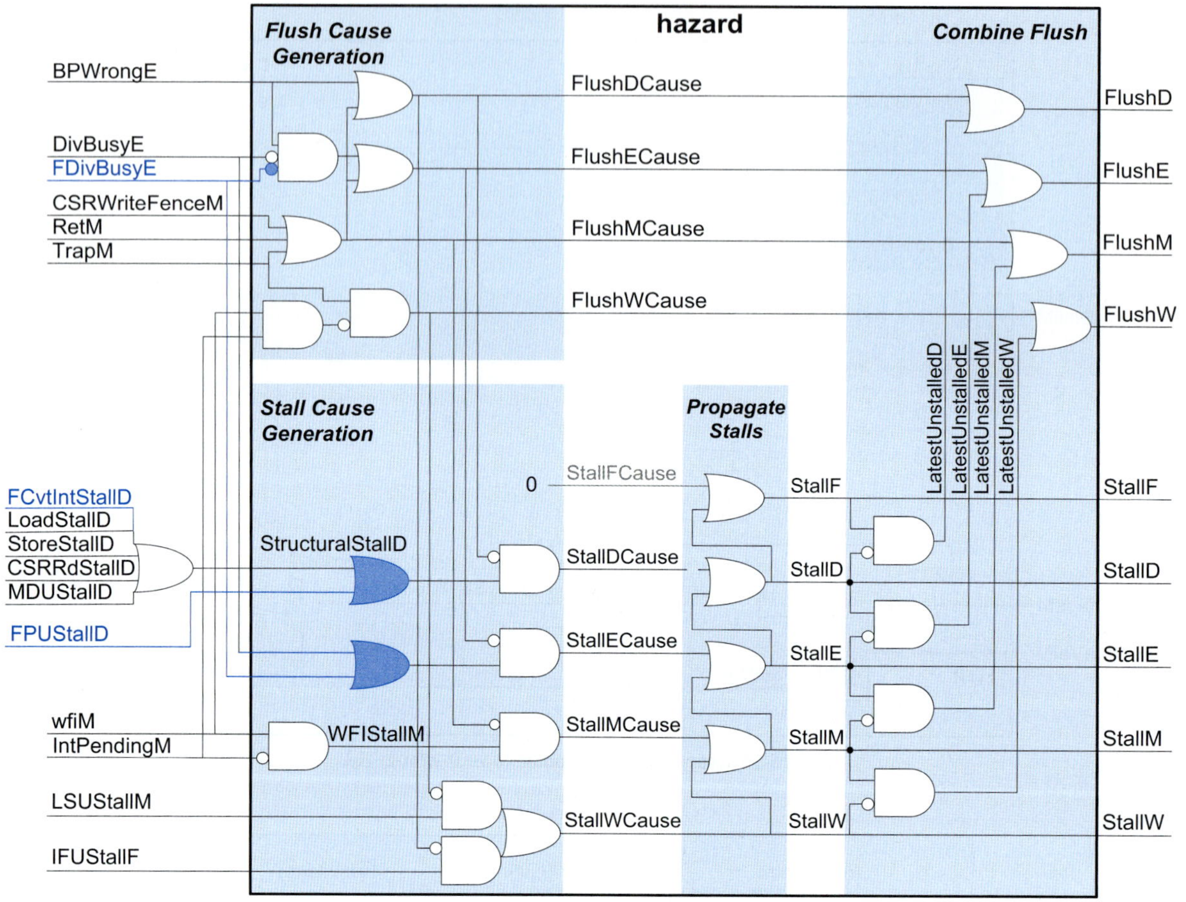

Fig. 16.26 **Hazard unit, with FPU changes highlighted in blue.**

preceding instruction, the processor must stall for a cycle to allow the preceding instruction to complete. The FPU's `fhazard` unit generates an `FPUStallD` signal when this is detected during Decode. Floating-point to integer conversion must stall when an IEU instruction is dependent on the result of `fcvt`. The IEU's controller generates `FCvtIntStallD` when this is detected during Decode. The hazard unit's `StallDCause` logic ORs `FPUStallD` and `FCvtIntStallD` with the other structural stall causes.

The `fdivsqrt` unit asserts `FDivBusyE` while it is busy, and the hazard unit stalls the pipeline until it finishes. As with `DivBusyE` from Section 15.4.4, a branch instruction class misprediction waits for the division to complete before flushing the Execute stage.

FPU to IEU operations such as `fcmp`, `fclass`, and `fmv` have single-cycle latency, so no stalls are needed. Integer to floating-point conversion doesn't require new stalls either; most integer instructions have single-cycle latency, and load/store/MDU stalls already exist.

16.6.15 LSU Changes

As shown in gray in Fig. 12.2, the LSU (load/store unit) adds an `FWriteDataM` port containing data from the FPU for floating-point stores. It also adds a multiplexer to select between IEU and FPU write data to implement floating-point stores. The mux width is LLEN = max(FLEN, XLEN). For example, on RV32GC, FLEN = 64 and XLEN = 32, so LLEN = 64 to accommodate the doubles.

The external bus remains AHBW (typically XLEN), so if FLEN > AHBW, the system uses a data cache to service the wide accesses rather than going out to external memory. Moreover, the `ReadDataWordAHB` mux is needed for cache writebacks to reduce the FLEN-sized `ReadDataWordM` to an AHBW-sized `LSUHWDATA` word. Presently, an LSU without a cache is unable to support the wide accesses needed for RV32D.

16.7 SUMMARY

The IEEE 754 Floating-Point Standard revolutionized numerical computing by allowing programmers to write code that behaves identically on different computers and performs gracefully in exceptional circumstances. Floating-point numbers are represented in 16-, 32-, 64-, or 128-bit half, single, double, and quad formats with a sign bit, a biased exponent, and a fraction with an implicit leading 1. Normalized numbers are of the form $(-1)^s \times (1.f) \times (2^{e\text{-}bias})$, where s is the sign, f is the fraction bits, and e is the exponent. The largest exponent is reserved for infinity and NaN. The smallest exponent is used for zero and subnormal numbers, which provide gradual underflow for more stable numerical computing. By default, results are rounded to the nearest representable value but can also be rounded toward positive or negative infinity or toward zero. The standard also provides sticky flags to indicate if a result has overflowed, underflowed, or was inexact, invalid, or involved division by zero.

An FPU typically includes an FMA (fused multiply-add) unit to perform addition, subtraction, multiplication, and FMA operations; a fdivsqrt unit for division and square roots; a convert unit for int-float, float-int, and float-float conversions; and a compare unit. FPUs often support multiple-precision operands on shared hardware. This involves the following:

- Unpacking the specified-precision operands into the longest supported format

- Detecting special cases
- Performing the operation in the longest format
- Normalization shift
- Rounding to the specified precision
- Handling special cases
- Updating flags

The FMA block computes $W = X \times Y \pm Z$. Thus, it can also be used to perform addition, subtraction, and multiplication by setting Y to 1 or Z to 0. Wally uses the FMA for all four operations to save chip area. The FMA first multiplies the significands and adds the exponents of X and Y. It aligns the significand of Z by right-shifting by the difference between the exponents of the product and of Z. It then adds or subtracts the significands of the product and aligned Z. If the result significand is negative, take the absolute value and flip the sign of the result. To normalize the result significand, it finds the leading 1 and shifts it into the integer position, adjusting the exponent accordingly. Then it rounds the results and handles special cases and flags. High-performance FMAs have several logic optimizations, such as a dual path adder to remove absolute value logic from the critical path, and a leading zero anticipator to speed up normalization.

The divsqrt block (`fdivsqrt`) uses a recurrence algorithm to produce one digit of the quotient or square root on each step. The division algorithm requires normalized significand inputs in the range [1, 2), so subnormal significands are left shifted into this range. Square root requires inputs in the range [1/4, 1) with an even exponent, so they are right-shifted by one or two bits. The key to recurrence divsqrt is to avoid slow carry-propagate addition in each step. To do this, the residual is maintained in carry-save redundant form, representing a weighted version of the partial remainder or error between the approximate and true square root. Furthermore, the recurrence algorithm uses a symmetric redundant digit set for the quotient or square root digits: {−1, 0, 1} for radix 2, or {-2, -1, 0, 1, 2} for radix 4. At each step, the algorithm selects a result digit based on the most significant bits of the residual and the divisor or square root approximation, then updates the residual with a carry-save adder. The algorithm uses an on-the-fly converter to form the nonredundant binary result without a CPA. If the cycle time is long enough, the unit can perform more than one step per cycle to reduce latency. If the result is exact, the recurrence can terminate early. Again, the result may need normalization, rounding, special case handling, and flags. Integer division can share the divsqrt recurrence unit by normalizing the integer inputs to have a leading 1.

The convert block handles int-to-float, float-to-int, and float-to-float conversions. The algorithm is easiest to understand by considering each case separately, but the hardware has many similarities. A shared unit for all three types of conversions involves an exponent calculation, an LZC (leading zero counter), and a significand shifter, as well as rounding, special cases, and flags.

The compare block takes advantage of the fact that positive floating-point numbers can be ordered with simple integer comparison. Therefore, it involves an integer comparator and some logic to handle sign correction and special cases.

Floating-point has many tricky corner cases and is notoriously difficult to get right. We check it against the UCB SoftFloat reference model using TestFloat and riscv-arch-test (IBM) test vectors.

RISC-V supports IEEE floating-point using the F, D, Q, and Zh extensions. The `FS` bits of the `mstatus` register must be turned on to enable floating-point. The floating-point control and status register (`fcsr`) is used to set rounding modes and check flags. Lower-precision operands are NaN-boxed with 1s in the upper bits when stored in the full-precision floating-point registers. Operations that produce a NaN always generate a canonical NaN with a 1 in the most significant bit of the fraction and 0s in the other bits, rather than propagating a NaN from one of the inputs.

Wally implements a simple FPU operating in the Execute and Memory stages with a critical path comparable to the cache access paths in the rv{32/64}gc configurations. The `fdivsqrt` unit supports both floating-point and integer division and uses multiple copies of recurrence hardware to produce many result bits per cycle.

17

Extensions: A (Atomic)

17.1 **Principles**
17.2 **RISC-V Practices**
17.3 **Test Plan**
17.4 **Wally Implementation**
17.5 **Summary**

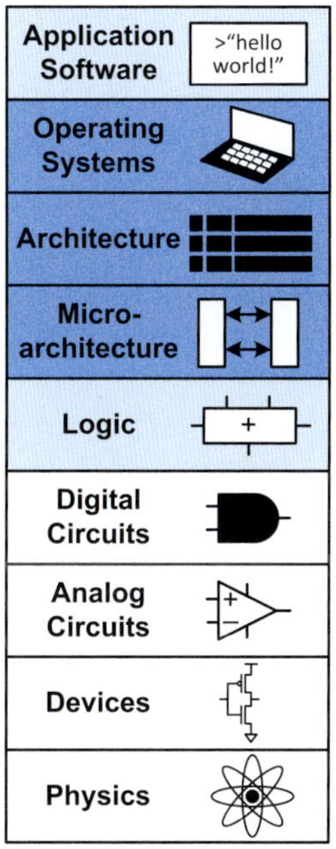

This chapter introduces principles of concurrency, consistency, coherence, synchronization, and atomic instructions. It describes the RISC-V atomic instructions and their implementation in Wally.

Recall that a *process* is a running program that comprises one or more *threads*. Sophisticated computer systems run multiple threads at the same time. This is known as *concurrency*. Users want concurrency because they want to do multiple things at once (e.g., listening to music while printing a term paper while simulating Verilog code while compressing a video file while monitoring for incoming emails and Slack messages). Each thread runs on a virtual user-level hart. The operating system (OS) may time-multiplex many virtual user-level harts onto a single physical core or may spread them across multiple cores to boost performance. For example, video compression runs faster when using multiple cores.

When concurrent threads access memory, the loads and stores might be interleaved in different ways. *Consistency* describes the order in which memory accesses are seen across the machine. *Coherence* is when all threads and cores eventually see the same values stored in memory. Often a specific order of memory accesses is necessary for correct operation; ensuring a specified order is called *synchronization*. Processors provide *atomic* instructions to facilitate synchronization. Such instructions reserve a memory address and then write to it or read, modify, and write it without interference from another thread. Readers interested in delving further into these topics may enjoy [Nagarajan20], [Scott13], and [Arpaci-Dusseau21].

Atomic instructions derive their name from the ancient Greek word *atomos*, meaning indivisible.

17.1 PRINCIPLES

Concurrent systems have subtle and potentially nondeterministic behavior when multiple threads attempt to access the same data. This is clearly an issue in systems with multiple cores, but it may also occur when a single core context switches among multiple threads. Multicore systems face further challenges maintaining coherence among separate caches.

Furthermore, concurrent systems raise questions of memory consistency: the allowable ordering of reads and writes to different memory locations. The most intuitive model of sequential consistency requires each thread to read and write in program order but allows different threads to interleave accesses in arbitrary ways. However, this model is overly restrictive for high-performance processors. The more relaxed *release consistency* model allows most memory accesses to be in any order and asks the programmer to explicitly acquire and release access to a shared variable when order matters. Acquiring and releasing access to a shared variable is called *synchronization*.

Threads must synchronize to avoid improper access to shared variables. Two common synchronization techniques are *locks* and *condition variables*. A thread places a lock on a shared variable, which guarantees that only that thread can access it until the thread releases the lock. Condition variables interact with the OS scheduler so that one thread can sleep until notified that another thread has produced or consumed data. Incorrect synchronization can also cause threads to *deadlock*, *livelock*, or *starve*, as discussed in Section 17.1.5.

17.1.1 Concurrency

> *"In computer science, as in real life, concurrency makes it much more difficult to reason about events."*
>
> **– Michael L. Scott [Scott13]**

Concurrency is essential to a good user experience and to high performance but introduces subtle issues that are easily mishandled. Concurrency arises for a single core running multiple threads, for multiple cores sharing memory, and for distributed systems that communicate through message passing. Although this book focuses on a single-core processor, we also briefly outline the issues faced by shared-memory and distributed systems.

In a single-core system running a multitasking OS such as Linux, the S-mode OS time-multiplexes among many user-level harts, each running a different thread. The SoC periodically produces a timer interrupt, which causes the OS scheduler to *context switch* to the next active thread. On an interrupt, the core enters S-mode, saves the PC, and jumps to an interrupt handler. The interrupt handler saves the state of the current thread's user-level hart, loads the state for the next thread's user-level hart, and returns to U-mode to run the next thread. When two threads are accessing a shared resource such as a memory location, file, or peripheral, the order of accesses among the threads may be important.

In a shared-memory multiprocessor, multiple cores have access to the same memory. For example, a typical phone, laptop, or server is a shared-memory multiprocessor. In 2025, a low-end processor might have only 4-8 cores, but a high-performance AMD or Intel laptop processor has

16–24 cores, and a high-performance server may have 2–8 sockets for chips, each containing up to 60–288 cores. In any of these configurations, each core typically has small (16–32 KiB) L1 instruction and data caches that can be accessed quickly and a medium (256 KiB–8 MiB) L2 cache. The chip also has a large (32–1,152 MiB) L3 cache that serves all the cores.

Fig. 17.1 shows an AMD Epyc 9004 server motherboard, which is a good example of a shared-memory multiprocessor. The Gigabyte MZ73-LM0 motherboard contains two sockets and a 12-channel DDR5 (double data rate 5) memory interface with 24 DIMM (dual inline memory module) sockets. Fully populated with 24 32-GiB modules, the motherboard holds 768 GiB of memory. Each 72 × 75-mm socket contains an AMD processor. The AMD 9684X processor has 13 dies on a multichip module, including the I/O controller and a dozen CPU dies built on TSMC's 5-nm process. The processor can consume up to 400 watts and has 1,152 MB of L3$ using cache chips stacked vertically on top of the CPU die. Each CPU die has 8 dual-threaded cores, for a total of 96 cores or 192 threads. Each Zen 4 core operates at a base frequency of 2.55 GHz and a maximum

Fig. 17.1 AMD Epyc 9004 motherboard. (Figure courtesy of GigaByte. Reprinted with permission.)

frequency of 3.7 GHz. The cores each have 32 KiB L1 instruction and data caches and a 1-MiB 8-way set associative L2$ cache. Their branch predictors have 1,536- and 7,168-entry L1 and L2 branch target buffers (BTBs) and a 32-entry return address stack (RAS). For instruction and data translation lookaside buffers (ITLBs and DTLBs), the L1 and L2 ITLBs have 64 and 512 entries, and the DTLBs have 72 and 3,072 entries. The core can decode 4 instructions per cycle, so the server has a peak throughput of 2 sockets $\times$ 96 cores $\times$ 3.7 GHz $\times$ 4 instructions/cycle = 2.8 trillion instructions per second.

17.1.2 Coherence

Shared-memory multiprocessors face the challenge of *cache coherence*. The goal of cache coherence is to ensure that the memory system behaves as if the caches were not there. In particular, every store to memory must eventually be seen by subsequent loads from the same memory address. For example, imagine that a value at address 1000 is in the L1 cache of two different cores. One core writes a new value to address 1000, changing the value in its L1 cache. After sufficient time for the write to propagate through the memory system, the second core must get the new value when it reads from address 1000. If the second core simply kept the old value in its L1 cache indefinitely and never wrote it back to main memory, the memory system is called *incoherent*. Early shared-memory multiprocessors used *snooping* to maintain coherence. In such systems, when one core writes to its cache, the write address is also broadcast over a shared bus. Other cores snoop the bus and invalidate any cache entries at this address. Snooping does not scale gracefully because of contention for the shared bus. Larger multiprocessors use *directory*-based coherence, maintaining a shared or distributed directory of cache lines. They often use the *MESI* protocol, in which each line is marked as Modified, Exclusive, Shared, or Invalid [Hennessy19].

Although cache coherence is primarily a consideration for multicore processors, coherence between the instruction and data caches is also important for single-core processors. If the processor writes to a certain memory address and then executes an instruction from that same address, this *self-modifying code* should see the new instruction. As discussed in Sections 10.4.2.6 and 10.4.5.4, the processor must perform a fence operation to flush any dirty lines from the data cache and invalidate them in the instruction cache to keep the caches coherent.

Very large systems would pay an unacceptable overhead keeping memory coherent across all the cores. Instead, such *clusters* or warehouse-scale computers are built from distributed processors with independent memories tied together on a high-speed network. Processors that wish to communicate pass *messages* over the network rather than using shared memories.

17.1.3 Consistency

Coherence describes the behavior of reads and writes to the same memory location, and *consistency* defines the relative ordering of reads and writes to different memory locations.

Suppose two concurrent threads T1 and T2 run Code Example 17.1 with shared variables a, b, c, and d, which are all initialized to 0 in memory. What values could c and d have after the threads finish?

Code Example 17.1 CONCURRENT THREADS ILLUSTRATING CONSISTENCY

T1	T2
a = 1	b = 1
c = b	d = a

If T1 runs first, c = 0, and d = 1. If T2 runs first, c = 1, and d = 0. If T1 and T2 run concurrently, it is possible that c = 1 and d = 1. This could occur even on a single core. For example, T1 could execute a = 1 and then context switch to T2. T2 could complete and then context switch back to T1.

Surprisingly to many programmers, c and d could both be 0. For example, Intel and AMD processors have *write buffers* to improve performance. If T1 and T2 executed concurrently on different cores, the writes to a and b could both be placed into buffers on each core. The other thread on the other core would read the old value of these variables (0) because the writes are not yet visible to a cache coherence system. Moreover, most high-performance processors use out-of-order execution. T1 could execute in reverse order (c = b, a = 1) because that would not violate any data dependencies within T1. Similarly, T2 could execute in reverse order. In this case, c and d would both be 0 again, even if writes were instantly visible to all threads.

Code Example 17.1 demonstrates that even very simple concurrent threads can have nondeterministic and counterintuitive behavior related to memory consistency. Programmers need to understand what behaviors are possible on a particular computer. Thus, architects define *memory consistency models* that specify the legal behaviors of concurrent programs. This section explores various models trading off simplicity and performance. Section 17.1.4 describes synchronization required to guarantee correct behavior considering these consistency issues. Consistency is a subtle subject, and [Nagarajan20] is presently the best source if you wish to delve deeper.

Leslie Lamport [1941–] is a computer scientist and recipient of the 2013 Turing Award for contributions to distributed and concurrent systems, including the notion of sequential consistency. He is also the developer of the open-source LaTeX system for document preparation. Lamport received his bachelor's degree and PhD in mathematics from MIT and Brandeis and worked at SRI, DEC, Compaq, and Microsoft Research. (Image courtesy of Leslie Lamport. Reprinted with permission.)

17.1.3.1 Sequential Consistency

The most straightforward consistency model is called *sequential consistency* [Lamport79, Hennessy19]:

- Memory accesses by a particular thread occur in program order
- Memory accesses by multiple threads are arbitrarily interleaved

In the case of Code Example 17.1, sequentially consistent orderings of memory accesses are given in Table 17.1. Systems with sequential consistency do not reorder instructions and cannot use write buffers; thus, they never produce `c` = `d` = 0. An ordering such as `c = b → d = a → a = 1 → b = 1` also never occurs because sequential consistency requires that each thread's accesses occur in program order.

Table 17.1 Sequentially consistent orders of access.

Ordering	c	d
a = 1 → c = b → b = 1 → d = a	0	1
a = 1 → b = 1 → c = b → d = a	1	1
a = 1 → b = 1 → d = a → c = b	1	1
b = 1 → d = a → a = 1 → c = b	1	0
b = 1 → a = 1 → d = a → c = b	1	1
b = 1 → a = 1 → c = b → d = a	1	1

Just as coherence makes caches invisible, sequential consistency makes shared memory look like a single memory module [Nagarajan20]. Sequential consistency makes it easier to reason about concurrent programs and avoids some counterintuitive behaviors. However, it comes at a high cost to performance. It prohibits out-of-order memory accesses that are legal in a nonconcurrent program. More significantly, it makes shared-memory multiprocessors painfully slow at writing because a write to one cache cannot complete until it invalidates any lines containing the same address in different caches. These issues led architects to consider alternate consistency models that support write buffers and out-of-order issue of memory operations at the expense of a more subtle programming model.

17.1.3.2 Total Store Ordering

The SPARC architecture formalized the *total store ordering* (*TSO*) consistency model to support multicore processors with write buffers, and Intel/AMD x86 processors adopted TSO as well. It is similar to sequential consistency except that it allows memory accesses by a particular thread to perform loads before a previous store completes, as long as the accesses are to distinct memory addresses. Stores, however, must occur in program order. TSO allows the counterintuitive result of c = d = 0 in the earlier example. However, most practical programs that properly synchronize access to shared variables see no difference between sequential consistency and TSO.

In situations where sequential consistency matters, the processor can use a *fence* operation. A fence ensures that all memory operations before the fence are seen by all threads to complete before any memory operation after the fence occurs. The fence can be implemented by stalling the core(s) until the write buffers finish updating the caches. A fence can be an expensive (time-consuming) operation, especially in large systems with long memory latencies.

Fences are sometimes called *memory barriers* (*membars*) or *syncs*.

Code Example 17.2 adds fences to Code Example 17.1. Now, even on a TSO machine, T1 finishes writing a = 1 before it may read b. Similarly, T2 finishes writing b = 1 before it may read a. The threads continue to interleave in arbitrary ways but never produce a sequentially inconsistent result of c = d = 0.

Code Example 17.2 CONCURRENT THREADS WITH FENCES

T1	T2
a = 1	b = 1
FENCE	FENCE
c = b	d = a

17.1.3.3 Release Consistency

TSO facilitates multicore processors with write buffers, but TSO is not flexible enough to accommodate other important optimizations, including instruction reordering. Advanced compilers and speculative out-of-order processors seek to arbitrarily reorder loads and stores to distinct addresses. A *coalescing write buffer* can merge several writes to the same cache line into a single operation. A more relaxed consistency model permits these performance optimizations.

Define memory access ordering rules X → Y to mean operation X must complete before operation Y completes. Four orderings of reads and writes are possible: R → R, R → W, W → R, and W → W. Recall that sequential consistency requires all four to occur in program order for a particular thread. Total store order removes the W → R constraint. *Partial store order* removes both W → R and W → W to allow for reordering writes with coalescing write buffers. *Relaxed ordering* allows threads to reorder independent memory accesses in any fashion [Adve95]. Of course, within a thread, accesses to a particular address take place in program order.

Relaxed ordering comes in several subtly different flavors, but the *release consistency* model, which requires the programmer to perform explicit synchronization on shared variables when order matters, is widely used; it is the model for RISC-V, ARMv8, and MIPS. The C/C++ language standard is based on release consistency or stronger. Power, Alpha, and SPARC architectures use other relaxed ordering variants. Release consistency avoids introducing unnecessary ordering in unshared variables. Release consistency uses fences or acquire/release operations (see Section 17.1.3.4) to synchronize between threads only when required for correct program operation. Release consistency gives maximum flexibility for hardware optimization but also requires a highly sophisticated programmer to write correct synchronization routines.

Under a release consistency model without fences, Code Example 17.1 could produce the sequentially inconsistent result `c` = `d` = 0. Adding fences in Code Example 17.2 is sufficient to prevent this result.

Code Example 17.3 gives another example where, in the absence of fences, release consistency differs from sequential consistency and TSO. Suppose `a`, `b`, and `flag` are initially 0. The producer thread `T1` writes a nonzero value to `a` and `b`, then sets the `flag`. The consumer thread `T2` waits until the `flag` is set, then reads `a` and `b` into `c` and `d`. Consider behavior with and without the `FENCE`. Under sequential consistency, `c` = `d` = 1. Under TSO, the program also produces `c` = `d` = 1 even when there are no fences, because `T2` will never see `flag` = 1 until after `T1` stores 1 into `a` and `b`. However, under release consistency, without fences, the stores could be reordered arbitrarily. `T2` might observe `flag` = 1 before `a` or `b` have become 1 and, hence, might set `c` and/or `d` to 0. Less intuitively, `T2` could reorder the loads such that it reads `a` or `b` = `0` (and thus sets `c` or `d` to 0) before finding `flag` = 1. When the fences are added, the `T1` fence ensures all other threads see `a` = `b` = 1 before they see `flag` = 1. Similarly, the `T2` fence ensures that `T2` reads `flag` = 1 before reading `a` or `b`.

Code Example 17.3 PRODUCER-CONSUMER EXAMPLE SHOWING RELEASE CONSISTENCY ORDERING

T1: Producer	T2: Consumer
`a = 1`	`While (!flag); // Wait for flag`
`b = 1`	`FENCE`
`FENCE`	`c = a;`
`flag = 1`	`d = b`

Even with the fences, release consistency does not impose strict ordering. For example, `T2` might observe `b` becoming 1 before `a` becomes 1. This is harmless because `T2` doesn't care about `a` and `b` until after `flag` is raised, and it gives the hardware flexibility to reorder operations for performance.

Notice that the fence is conservative. For example, if `T1` performed other subsequent loads or stores to independent variables, they could not be reordered before the fence even if that wasn't important to program operation.

17.1.3.4 Acquire and Release Constraints for Release Consistency

Programs in a release consistency model must specify ordering more often than in TSO or sequentially consistent models because release consistency lacks the implicit ordering of the other models. Fences are sufficient but can be overkill because they order all operations, not just the essential ones. Moreover, each fence is another instruction to fetch and execute, slowing program performance. To address these issues, release consistency models provide *acquire* and *release* constraints to specify ordering of certain memory operations. These constraints are typically specified by bits in the memory operation, so no separate fence instruction is needed.

A memory operation is called an *acquire operation* if it has an acquire constraint, a *release operation* if it has a release constraint, and a *sequentially consistent operation* if it has both acquire and release constraints. Acquire and release operations in some thread T enforce how other threads observe ordering of other memory operations from thread T. All threads will observe an acquire operation in T before any later memory operations in T. All threads will observe a release operation in T after all earlier memory operations in T. All threads will observe a sequentially consistent operation in T after all earlier memory operations in T and before any later operations in T.

Code Example 17.4 adapts Code Example 17.3 to use acquire/release constraints instead of fences. In `T1`, the store to `flag` has a release constraint, ensuring that `T2` observes `flag` becoming 1 after `a` and `b` became 1. In `T2`, the load of flag has an acquire constraint, ensuring that `T2` reads `a` and `b` after `flag` becomes 1.

Code Example 17.4 PRODUCER-CONSUMER EXAMPLE WITH ACQUIRE/RELEASE CONSTRAINTS

```
T1: Producer
a = 1
b = 1
flag = 1; <release>
```

```
T2: Consumer
while (!flag); <acquire>
c = a;
d = b
```

17.1.3.5 Consistency Summary

A memory consistency model describes how reads and writes to different memory locations are ordered. The programmer must understand the consistency model to understand the possible behaviors of a multithreaded program; they must also understand fences or other synchronization required to achieve the desired behavior. Microarchitects design the out-of-order issue logic and memory system to obtain the desired level of consistency. Three common models are sequential consistency, TSO, and release consistency.

Sequential consistency is the most natural to a programmer. In this model, each thread issues memory accesses in program order, but multiple threads may interleave in arbitrary order. Hence, shared memory behaves to the programmer as one single large memory, even if it is physically implemented with per-core caches. Unfortunately, sequential consistency comes at a high performance cost. It prevents out-of-order issue of memory accesses that would be legal ways to increase performance in a single-threaded system. It also costs power and performance because a write to an address in one core's cache must invalidate any line in another core's cache that contains the same address. It prevents the use of store buffers that hide the latency of stores and coalesce multiple stores to consecutive addresses into a single cache line update.

TSO is similar to sequential consistency except that a thread can reorder a load to occur before a previous store to a different address completes. This allows microarchitects to use store buffers and out-of-order issue. Most properly synchronized programs behave the same under TSO as under sequential consistency. When stronger ordering is necessary, the programmer inserts fence operations that ensure all memory accesses before the fence are seen by all threads to have completed before any memory access after the fence occurs.

Release consistency is an even more relaxed ordering that allows maximum flexibility for hardware optimization, including out-of-order execution. It comes at the cost of a thread having to acquire a shared variable before using it and then release the variable before another thread can use it. Fences or acquire/release constraints are inserted to ensure ordering where it matters. Section 17.2 shows how RISC-V supports these operations for release consistency.

Getting these constraints right takes careful thought by a sophisticated developer. Programmers normally rely on carefully designed synchronization libraries to avoid introducing subtle consistency bugs.

17.1.4 Synchronization

Concurrent threads can misbehave when they interleave in undesirable ways. *Synchronization* is the art of avoiding incorrect interleaving. Most synchronization problems can be classified as *mutual exclusion* or *producer-consumer*. These problems are usually solved with *locks* and *condition variables*, respectively. This section describes the problems, along with synchronization mechanisms and hardware support to solve the problems.

17.1.4.1 Mutual Exclusion: Locks

Imagine that you and your mother share access to your college bank account. You have $500 in the account and need to withdraw $400 to buy textbooks at the campus bookstore. Your mother promises to deposit $300 in your account to help out.

You and your mother each visit an ATM at the same time. Your ATM invokes the Withdrawal function with thread T1, and your mother's invokes the Deposit function with thread T2, as shown in Code Example 17.5. The functions run concurrently and could interleave in arbitrary ways even if the system is sequentially consistent. Suppose they run in the order shown in Table 17.2. At the end of the day, your mother's deposit vanished, you have only $100 in your account, and you are poor and angry.

Code Example 17.5 MUTUAL EXCLUSION BANKING PROBLEM

T1: Withdrawal	T2: Deposit
b1 = getBalance()	b2 = getBalance()
b1 = b1 − withdrawalAmount	b2 = b2 + depositAmount
setBalance(b1)	setBalance(b2)

Table 17.2 Mutual exclusion ordering example.

Thread	Statement	Notes
T1	b1 = getBalance()	b1 = 500
T2	b2 = getBalance()	b2 = 500
T2	b2 = b2 + depositAmount	b2 = 800
T2	setBalance(b2)	balance = 800
T1	b1 = b1 − withdrawalAmount	b1 = 100
T1	setBalance(b1)	balance = 100

Mutual exclusion locks are also referred to as *mutex locks* or simply *mutexes*.

This is an example of a *race condition* in which the result depends on the order of operations in the interleaved threads. To solve this problem, the program must use *mutual exclusion*, where code within a *critical section* cannot interleave. `T1` and `T2` each are executing *critical sections* (from `getBalance` to `setBalance`) that should not be allowed to interleave. They should be synchronized so that either the withdrawal or deposit occurs first and runs to completion before the other thread runs at all. If `T1` ran before `T2`, your balance would drop to $100 and then rise to $400. If `T2` ran before `T1`, your balance would rise to $800 and then drop to $400. Although the exact sequence is nondeterministic, the final answer is correct.

Race conditions can occur even on systems with a single core because the OS may context switch between threads at an inopportune time. For example, even if `T1` and `T2` both executed on a single computer core, the computer could switch from `T1` to `T2` and back to `T1` as in Table 17.2.

Locks

Threads use locks around critical sections to achieve mutual exclusion. A *lock* is a shared variable used to ensure only one thread can be in its critical section at a time. Code Example 17.6 fixes the banking program using locks. The `lock(L)` function uses a lock variable `L`. It waits until `L` is 0, then "locks" it: sets `L` to 1. `unlock(L)` clears `L` to 0. Only one of the threads can hold the lock at a time. Therefore, only one thread at a time can change the balance, and the transactions occur properly, although their order is unpredictable.

Code Example 17.6 MUTUAL EXCLUSION BANKING PROBLEM WITH LOCKS

T1: Withdrawal	T2: Deposit

`lock(L)   // Critical section start`
`b1 = getBalance()`
`b1 = b1 - withdrawalAmount`
`setBalance(b1)`
`unlock(L) // Critical section done`
	`lock(L)   // Critical section start`
	`b2 = getBalance()`
	`b2 = b2 + depositAmount`
	`setBalance(b2)`
	`unlock(L) // Critical section done`

Code Example 17.7 shows three methods of implementing lock and unlock. The lock variable `L` is a pointer to a memory location that holds 0 when the lock is available or 1 when the lock is claimed.

Code Example 17.7 LOCK FUNCTION IMPLEMENTATIONS

(a) Incorrect software

```
void lock(int *L) {
  while (*L); // Wait
  *L = 1;
}

void unlock(int *L) {
  *L = 0;
}
```

(b) Interrupt-based

```
void lock() {
  DisableInterrupts();
}

void unlock() {
  EnableInterrupts();
}
```

(c) Test and set

```
void lock(int *L) {
  while (TestAndSet(L));
  FENCE
}

void unlock(int *L) {
  FENCE
  *L = 0;
}
```

In method (a), the `lock` function waits until `L` is 0, then sets `L` to 1. This is called *spinning*. `unlock` restores `L` to 0. This approach has a subtle bug. Imagine that `T1` and `T2` execute `lock()` concurrently. Both detect that `L` is 0 and exit the while loop. Both advance to `*L` = 1 and think they have claimed the lock and can run their critical section. This may occur even on a single-core processor if a context switch occurs from `T1` and `T2` between when `T1` detected `L` = 0 and when it sets `L` = 1.

Method (b) works on a single-core processor. The lock function simply disables interrupts. In particular, it turns off the timer interrupt that could allow the OS scheduler to switch threads. Because interrupts are disabled, when a thread has the lock, it is guaranteed to continue executing until it releases the lock. This method has many drawbacks. If it holds the lock for a long time, other threads will be starved from executing, so the system may feel unresponsive. A buggy or malicious thread may never release the lock, hanging the system. For these reasons, an OS typically does not let user code disable interrupts. Moreover, on a multicore system, the threads could be executing on different cores so the race condition could occur even if interrupts are disabled on both cores.

Method (c) relies on the hardware support of an *atomic* `TestAndSet` instruction. The instruction reads the value at an address, writes a 1 to that address, and returns the value read. *Atomic instructions* cannot be interrupted in the middle; the read and write happen back to back such that no other thread may access the memory system in between the read and write. Either `T1` or `T2` will call `TestAndSet` first, detecting `L` was 0 and setting it to 1. That thread now has the lock and can execute the critical section. When the other thread calls `TestAndSet`, it reads `L` = 1 and waits in the while loop until the lock is released. Even if both `T1` and `T2` attempt `TestAndSet` at the same time, the memory system will arbitrate between the two requests and only issue one at a time. In a system with release consistency, fences are needed after acquiring the lock and before releasing

The `TestAndSet` instruction dates back to the earliest multiprocessors, such as the Burroughs B5000 in the 1960s [Mayer82].

it to ensure the critical section runs while the lock is held. Sequential consistency or TSO guarantees such ordering so the fence could be omitted.

Theoreticians have devised clever algorithms to reliably lock without atomic instructions [Dijkstra65]. These algorithms are tricky, and they are even trickier when they must support more than two threads. Moreover, these algorithms require sequential consistency, which most computers don't provide. Atomic support for locks has become a standard instruction set feature for concurrent systems.

Better Locking

A good lock should not just protect the critical section but also be *fair* and *efficient*. When multiple threads are waiting on the same lock, each thread should have a fair chance of getting the lock when it becomes available, and a thread should never *starve* waiting indefinitely while other threads always get the lock. An efficient lock doesn't waste large amounts of CPU time waiting.

The lock from Code Example 17.7(c) is called a *spin lock* and is simple but neither efficient nor fair. It spins (waits) in the while loop until the lock is available. Spinning for a long time is a waste of processor resources. Moreover, when the lock becomes free, there are no guarantees which thread will get it next, so an unlucky thread might starve.

A spin lock is analogous to trying to buy hard-to-get concert passes from a website. Users keep refreshing the website hoping a pass becomes available, then trying to purchase the pass before somebody else gets it. This is a waste of time, and you might never get your pass. In contrast, a *ticket lock* resembles taking a ticket at the deli counter. The butcher calls customers in the order of tickets. You still have to wait, but you are guaranteed eventual service. Waiting is inefficient, so it is even better to put the thread to sleep if the lock is unavailable and to have it wake up when its ticket is ready. However, context switching a sleeping process out and in is also expensive, so it is best to spin for a little while in the hopes of acquiring the lock quickly. This *sleep lock* is analogous to a customer service phone line, where the agent will help you if they are available soon or take your phone number and call you back in the order your call was received. Ticket and sleep locks take advantage of more capable atomic instructions and support from the OS, as discussed next.

Atomic Instructions

Typical atomic instructions include `TestAndSet`, `Swap`, `FetchAndIncrement`, `CompareAndSwap`, and the `LoadReserved-StoreConditional` pair. The atomic `TestAndSet` instruction is sufficient to build a spin lock, which in turn can be used to build other synchronization operations. The `Swap` instruction is similar; it atomically reads a memory location and writes a new value back to that location; it differs from `TestAndSet` in that the value written is configurable rather than always 1. Either of these instructions is sufficient for building synchronization mechanisms, but more powerful atomic instructions make synchronization more efficient.

The atomic `FetchAndIncrement` instruction reads a value from memory, adds one, and writes it back. `FetchAndIncrement` can be used in place of `TestAndSet`, but it is more powerful because it facilitates ticket locks and semaphores, as will be discussed soon.

The even more powerful atomic `CompareAndSwap` instruction takes three inputs: a memory address, a comparison value, and a swap value. It reads a value from memory. If the read value matches the comparison value, it then writes the swap value back to the same memory location. In either case, it returns the read value. `TestAndSet`, `FetchAndIncrement`, and other lock-free and wait-free synchronization algorithms can be implemented with `CompareAndSwap`, as shown in Code Example 17.8 for `TestAndSet`. `CompareAndSwap` has been part of the IBM 370 architecture since 1970 and is supported by x86, ARM, and most but not all other architectures. `CompareAndSwap` does not play as gracefully with RISC architectures because it requires three source operands.

`CompareAndSwap` is susceptible to the *ABA problem*, in which the value in memory changes from A to B and back to A without being detected in a comparison. To avoid this problem, `CompareAndSwap` often uses double-wide operands, storing the value in one word and a counter in the second word so that the counter can be atomically incremented at the same time the value is swapped.

Code Example 17.8 IMPLEMENTING **TestAndSet** WITH **CompareAndSwap**

```
int TestAndSet(int *adr) {
  while (CompareAndSwap(adr, 0, 1) == 1); // Spin until adr contains 0, then set *adr = 1
}
```

The `LoadReserved-StoreConditional` pair of instructions (abbreviated as `lr-sc`) supports the same synchronization algorithms as `CompareAndSwap`; avoids the implementation difficulties of reading more than two registers; and avoids the need to read, modify, and write memory in a single complex instruction. `LoadReserved` (`lr`) acts like a normal load, but it also flags that the memory address is reserved. `StoreConditional` (`sc`) attempts to write a value to a memory address. If the address is still reserved, the write succeeds, and `sc` returns a success code (e.g., 0). If the reservation is no longer valid, the store is rejected, and `sc` returns a failure code (e.g., nonzero). The reservation is invalidated by another `lr` to a different address, any write to the reserved address, or an `sc` to any address. In other words, the `lr-sc` pair only completes the store and returns success if the pair operates in an atomic-like fashion, but across two separate instructions. If the pair fails, it can be repeated until it succeeds. This raises a potential of livelock, addressed in Section 17.1.5.2. Code Example 17.9 shows how to implement other atomic operations and a spin lock using `lr-sc`.

RISC-V defines the *Zacas* (*atomic compare and swap*) extension that defines a `CompareAndSwap` instruction. `amocas.w rd, rs1, rs2` performs the following:

```
temp = mem[rs1]
if (mem[rs1] == rd)
  mem[rs1] = rs2
rd = temp
```

This instruction requires pipeline modifications because it reads `rd` as well as `rs1` and `rs2`. There is also an `amocas.d` double-wide version that operates on consecutive pairs of registers for RV32.

MIPS introduced `lr-sc` instructions. RISC-V, ARM, and other RISC architectures provide similar instructions.

Code Example 17.9 IMPLEMENTING ATOMIC OPERATIONS WITH `lr-sc`

```
int TestAndSet(int *adr) {
  int val;
  while (1) {        // Repeat until lr-sc succeeds atomically
    val = LoadReserved(adr);
    if (StoreConditional(adr, 1) == 0) return val;          // Return if successful
  }
}

int Swap(int *adr, int swap) {
  int val;
  while (1) {        // Repeat until lr-sc succeeds atomically
    val = LoadReserved(adr);
    if (StoreConditional(adr, swap) == 0) return val;       // Return if successful
  }
}


int FetchAndIncrement(int *adr) {
  int val;
  while (1) {        // Repeat until lr-sc succeeds atomically
    val = LoadReserved(adr);
    if (StoreConditional(adr, val+1) == 0) return val;      // Return if successful
  }
}

int CompareAndSwap(int *adr, int comp, int swap) {
  int val;
  while (1) {        // Repeat until lr-sc succeeds atomically
    val = LoadReserved(adr);
    // Return value read from memory if it doesn't compare or if swap succeeds
    f (val != comp | StoreConditional(adr, swap) == 0) return val;
  }
}

void lock(int *L) { // Spin lock using lr-sc
  int val;
  while (1) {        // Repeat until lr-sc succeeds atomically and finds lock was zero
    while (LoadReserved(L)); // Spin until lock is free
    if (StoreConditional(adr, 1) == 0) return;              // Return if successful
  }
}
```

Ticket Locks

A *ticket lock* aims for fairness by ensuring each thread gets the lock in the order it was requested. The lock consists of two parts: turn and ticket. The ticket counts up, like a ticket at a deli counter. Each thread that requests the lock takes the next value of the ticket. The turn also counts up, indicating which ticket is presently being served, like the "Now Serving" sign at the deli counter. Each thread that releases the lock increments the turn to make the lock available to the next ticketholder. Code Example 17.10

shows an implementation of a ticket lock using `FetchAndIncrement` or `lr-sc` [Arpaci-Dusseau19]. As with the spin lock, taking a lock requires an atomic operation, but releasing it does not.

Code Example 17.10 TICKET LOCK

```
typedef struct lock_t {
  int ticket;
  int turn;
} lock_t;

void lock_init(lock_t *L) {
  L -> ticket = 0; L->turn = 0;
}

void lockWithFetchAndIncrement(lock_t *L) {
  int myTurn = FetchAndIncrement(&L->ticket);
  while (L->turn != myTurn); // Spin until it is my turn
}

void lockWithLRSC(lock_t *L)
  int myTurn;
  do { // Repeat until lr-sc succeeds atomically
    myTurn = LoadReserved(&L->ticket);
  } while (StoreConditional(&L->ticket, myTurn+1));
  while (L->turn != myTurn); // Spin until it is my turn

}

void unlock(lock_t *L) {
  L->turn++;
}
```

Sleep Locks

Ticket locks are fair but still inefficient because threads spin while waiting. A better strategy would be for a thread to go to sleep rather than spin when the lock is busy, then wake up when its turn has arrived. This requires coordination with the OS, which manages thread scheduling. It also involves maintaining a queue of sleeping threads to reawaken in the order they requested the busy lock. Sections 17.1.4.3 and 17.1.4.4 examine implementing a *sleep lock* with a semaphore, which keeps track of the number of resources available, or a *futex*, which is similar to a semaphore but with reduced overhead. Both are described further in this section.

17.1.4.2 Producer-Consumer: Condition Variables

Locks solve the problem of mutual exclusion, allowing only one thread at a time to enter a critical section of code and manipulate a shared variable. The other common synchronization problem is known as *producer-consumer* or *bounded buffer*, in which *producer* threads *put* results into

If the producer finds that the buffer is full, rather than spinning, it calls `cond_wait(&getEntry, &L)`. `cond_wait` is an OS call whose arguments are the addresses of a condition variable and a lock. It places the condition variable and lock addresses in a wait queue, releases the lock, and schedules the thread to sleep. When the OS receives a `cond_signal` call for a condition variable, it reacquires the lock and wakes up the next thread in the wait queue.

When the producer wakes back up, it checks the `while` loop condition once more to ensure the buffer really has space. (Another thread might have swooped in and refilled the buffer before this producer woke up.) Upon success, it puts the next element into the buffer and signals the `putEntry` condition variable to wake up any consumer that might be sleeping. Finally, it releases the lock.

17.1.4.3 Semaphores

A *semaphore* is another synchronization mechanism that can implement both locks and condition variables [Dijkstra68]. A semaphore is an integer that is initialized to the number of shared resources available. The semaphore is atomically decremented or incremented using the OS calls `sem_wait` and `sem_post` to take or release resources, respectively. If the semaphore becomes negative because no resource is available, `sem_wait` puts the thread to sleep. If any threads are sleeping, `sem_post` wakes one.

Semaphore-Based Mutex

Code Example 17.13 shows how to build a semaphore-based mutual exclusion lock (mutex), also called a sleep lock with a semaphore. The semaphore `S` is initialized to 1 because a single thread can hold the lock. The lock is acquired by calling `sem_wait`, which decrements `S` to 0. If another thread tries to acquire the lock that is already held, it will decrement `S` to a negative value and go to sleep until the lock becomes available. Unlocking with `sem_post` increments `S` and wakes up a sleeping thread that was waiting for the lock.

Code Example 17.13 SLEEP LOCK USING A SEMAPHORE

```
int S = 1; // Lock is initially available

void lock(int *S) {
  sem_wait(S);
}

void unlock(int *S) {
  sem_post(S);
}
```

Semaphore-Based Producer-Consumer

Code Example 17.14 modifies Code Example 17.12 to solve the producer-consumer synchronization problem with semaphores instead of condition variables [Arpaci-Dusseau19]. It replaces the `entries` variable, the lock `L`, and the `putEntry` and `getEntry` condition variables with two semaphores: `empty` and `full`. It initializes `empty` to `SIZE` and `full` to 0 because all the buffer entries are empty, and none are full. The producer `waits` on `empty` to ensure an empty entry is available. Initially, `empty` is positive, so the producer can proceed. When the buffer becomes full, `empty` becomes negative and the producer sleeps until a consumer `posts`. `put` and `get` are critical sections of code, so they must be surrounded by `lock` and `unlock` so only one thread can manipulate the shared buffer variables at a time.

Code Example 17.14 SEMAPHORE-BASED PRODUCER-CONSUMER

Shared Buffer

```
int buf[SIZE];
int head = 0, tail = 0;
int empty = SIZE, full = 0; // Initialize semaphores
int L = 0; // Lock
```

T1: Producer

```
void put(int val) {
  buf[head] = val;
  head = (head + 1) % SIZE;
}

void producer(int n) {
  for (int i = 0; i < n; i++) {
    // Wait until buffer has space
    sem_wait(&empty);
    lock(&L);
    put(i);
    unlock(&L);
    sem_post(&full);
  }
}
```

T2: Consumer

```
void get(void) {
  int val = buf[tail];
  tail = (tail+1) % SIZE;
  return val;
}

void consumer(int n) {
  for (int i = 0; i < n; i++) {
    // Wait until buffer is nonempty
    sem_wait(&full);
    lock(&L);
    printf("%d\n", get());
    unlock(&L);
    sem_post(&empty);
  }
}
```

17.1.4.4 Futex

Semaphores use OS calls, which have a high overhead of context switching and saving/restoring registers. For applications such as multithreaded databases that need extensive locks but rarely face contention, the OS call overhead is enormous. A *futex* [Franke02] solves this problem by attempting to take a lock in U-mode and only calling the OS if the lock is contended. Futex initially stood for *fast user mutex* but can also be used to address producer-consumer synchronization like a condition variable or semaphore.

Linux and Windows provide futex capabilities, which are currently the most efficient synchronization primitives.

A futex library provides `futex_wait` and `futex_wake` system calls, similar to `sem_wait` and `sem_post`. `futex_wait` puts a thread to sleep until it is notified. `futex_wake` notifies a sleeping thread to wake up. Futex operations attempt synchronization with atomic operations in U-mode and only call the OS if the synchronization spins too long. This is much faster than an OS call in the uncontended case; it also avoids inefficient lengthy spinning. In comparison, a semaphore always involves an OS call even in the uncontended case.

`futex_wait(int *F, int expected)` takes the address `F` of an integer futex variable and an expected value. If the futex variable does not match the expected value, it returns immediately. Otherwise, the calling thread goes to sleep until woken by `futex_wake`.

`futex_wake(int *F, int num)` wakes `num` threads waiting for the futex `F`. `num` is usually 1 to wake up one waiter but sometimes is set to `INT_MAX` to wake up all waiting threads.

Code Example 17.15 shows how to implement sleep locks with futex [Lockless23]. A lock is still an integer at some address `L` in memory. It is initialized to 0 and then set to 1 when a single thread has the lock or 2 when one thread has the lock and another wants it. The `lock()` function uses `CompareAndSwap` to atomically set the lock to 1 if it was 0. If it doesn't succeed immediately, it spins for a little while in the hopes that the lock becomes available soon. If that fails, it sets the lock to 2 to indicate contention, then calls `futex_wait` to sleep until the lock is available. When `futex_wait` returns, the `lock()` function checks that the lock really was available (`c` = 0) and waits again if not. The `unlock()` function checks if the lock was contended. If not contended, it tries to change the lock from 1 to 0 and returns if successful. Otherwise, the code calls `futex_wake` to wake up one of the sleeping threads that is waiting for the lock.

Code Example 17.15 LOCKS USING FUTEX

```
void lock_init(int *L) {
        *L = 0;
}

void lock(int *L) {
  int c;

  // Spin a little while and try to take lock
  for (int i = 0; i < 100; i++)
    if ((c = CompareAndSwap(L, 0, 1)) == 0) return;

  // The lock is now contended
  if (c == 1) c = Swap(L, 2);

  while (c) { // Wait in the kernel while futex is 2
    futex_wait(L, 2);
    c = Swap(L, 2);
  }
}

void unlock(int *L) {
  // Unlock, and if not contended then exit
  if (*L == 2) *L = 0;                  // Contended
  else if (Swap(L, 0) == 1) return;     // Uncontended
  futex_wake(L, 1);                     // Wake up 1 waiting thread
}
```

17.1.4.5 Synchronization Libraries

Although this chapter has explored building synchronization operations from low-level atomic instructions and fences, real programs should use standard synchronization libraries. Standard libraries have many advantages. It is easy to identify the synchronization taking place. The code is portable across machines. They include the necessary fences for a particular CPU's memory consistency model. Most importantly, standard libraries have been thoroughly debugged. A study of 12 major programs such as Apache Web Server, Mozilla Firefox, and MySQL discovered that every one of the programs had at least some ad-hoc synchronization and that 22%–67% of the ad-hoc synchronization contained undetected bugs [Xiong10].

The POSIX thread library is a widely-used standard for creating and synchronizing threads in C/C++. It defines locks, condition variables, and semaphores:

- `pthread_mutex_lock`
- `pthread_mutex_unlock`
- `pthread_cond_wait`

- `pthread_cond_signal`
- `sem_wait`
- `sem_post`

Linux defines the `SYS_futex` call to perform the scheduling portions of futex operations.

There is growing interest in lock-free synchronization algorithms, but these are outside the scope of this chapter.

17.1.5 Deadlock, Livelock, and Starvation

Concurrent threads can get stuck for several reasons: deadlock, livelock, and starvation. *Deadlock* occurs when two or more threads are waiting for each other and are unable to proceed. *Livelock* is closely related to deadlock; the threads continue taking action but never make forward progress. *Starvation* occurs when a lower priority thread never gets resources because it is outcompeted by higher priority threads. These problems lead to the familiar annoyance of computers locking up and requiring a reboot.

17.1.5.1 Deadlock

In a later retelling, the philosophers are eating rice with pairs of chopsticks. In a post-COVID retelling, no sane philosopher would agree to share utensils.

Deadlock is illustrated by the classic "dining philosophers problem" [Dijkstra71]. Five philosophers sit around a circular table. Each has a plate containing a difficult kind of spaghetti that must be eaten with two forks. There is a fork between each plate, as shown in Fig. 17.2. When a philosopher gets hungry, she picks up the fork on the left, then the fork on the right, then eats. Once satisfied, she puts down the fork on the left and then the fork on the right. She faces a dire problem that if all her colleagues become hungry simultaneously, they will all pick up their left forks, leaving no right fork. They are now deadlocked, unable to get a right fork and unwilling to put down the left fork.

Deadlock requires four conditions, referred to as *Coffman Conditions*, to occur [Coffman71]:

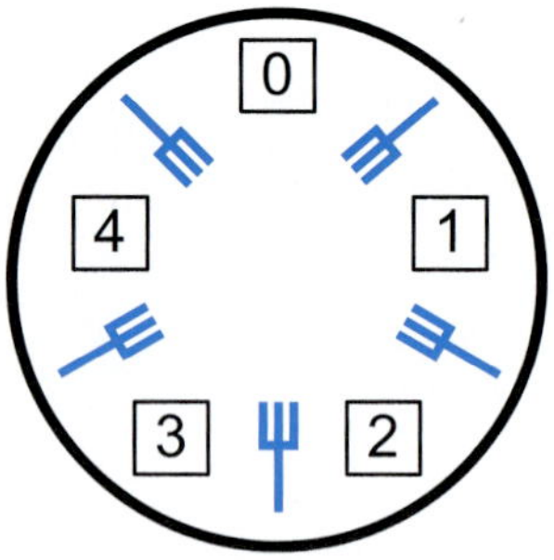

Fig. 17.2 Dijkstra dining philosophers problem. [Dijkstra71]

- **Mutual exclusion:** Threads take exclusive control of resources they require.
- **Hold and wait:** Threads hold allocated resources while waiting for additional resources.
- **No preemption:** Resources cannot be forcibly removed from a waiting thread.
- **Circular wait:** A circular chain of threads exists such that each holds a resource requested by the next thread.

In the dining philosophers problem, philosophers are threads, and forks are resources. The problem has mutual exclusion because only one philosopher can hold a fork at a time. It has hold and wait because a philosopher won't set down a fork until done eating. It has no preemption because a hungry philosopher won't wrestle a fork out of her colleague's hand. And it has circular wait because each philosopher could be holding a left fork and waiting for a right fork held by the next philosopher.

Concurrent programs face the same issue. Locks are a common example of mutual exclusion. Suppose two threads in a simple system both want to print documents. They each want a lock on the printer and on a spooling file on the disk. If one thread gets the lock on the printer and the other gets the lock on the spool file and neither can preempt the other, they will deadlock. The threads become unresponsive until the user kills them or reboots the computer.

A better system can break deadlocks by preventing one of the Coffman Conditions using the following strategies:

- **Mutual exclusion:** Threads use lock-free data structures, when possible, to minimize the need for exclusive control.
- **Hold and wait:** Each thread requests all its resources at once and does not proceed until all resources have been granted.
- **No preemption:** If a thread is denied access to a resource, it releases all the resources it already holds (so that another thread can proceed), then tries again later to obtain all resources.
- **Circular wait:** Threads are required to acquire resources in a certain order so that circular chains can't exist.

For example, the dining philosophers could set two forks at each plate to avoid the need for mutual exclusion caused by sharing forks. Or the philosophers could pick up both forks at a time, rather than the left and then right. If they pick up a left fork and find the right already in use, they could set down the left fork and try again after pondering deep thoughts for a random amount of time. Or the forks could be labeled 1–5, and a philosopher could pick up the lower-numbered fork before taking the higher-numbered one, avoiding the circular wait.

Other approaches involve:

- **Avoidance:** When a thread requests a resource, the OS denies the request if granting the resource would put the system in an unsafe state.
- **Detection and recovery:** The OS detects deadlock and removes resources until the deadlock is solved.
- **Ignorance:** Assume deadlock will not occur. Let the user reboot the computer when it occurs. This may be acceptable if deadlocks are less frequent than other system failures. The Windows OS is notorious for using this strategy, and even Linux is susceptible to this for certain deadlocks.

No, you can't always get what you want

You can't always get what you want

You can't always get what you want

But if you try sometime you'll find

You get what you need.

—**The Rolling Stones**

17.1.5.2 Livelock

Livelock occurs when the threads repeat the same sequence of operations without making real progress. Suppose the dining philosophers break the no preemption problem by setting down their left fork if the right fork is already in use. However, they are ravenous and immediately pick up the left fork again before their neighbor can take it. They enter livelock, repeatedly picking up and setting down the left fork without ever getting both forks.

Section 17.2.3.2 defines RISC-V constrained `lr-sc` loops that are guaranteed to eventually succeed and hence avoid livelock.

A disadvantage of `lr-sc` over `CompareAndSwap` is that the former can enter livelock. Two threads might spin to acquire the same lock. If they exactly interleave, each one might invalidate the other thread's reservation such that neither ever successfully completes the `StoreConditional`.

17.1.5.3 Starvation

Starvation occurs when one or more threads never get the resources they need. For example, if threads have priority and a condition variable or semaphore wakes up the highest priority waiting thread, a lower priority thread might never be woken when the system is busy.

Starvation is avoided by fair synchronization algorithms such as ticket locks (see Section 17.1.4.1) or by raising the priority of a thread that has been waiting too long.

17.2 RISC-V PRACTICES

RISC-V uses the release consistency model to allow maximum flexibility in hardware and software optimizations while ensuring that synchronized programs behave correctly. The base integer instruction set includes fence instructions to synchronize memory operations when needed. The *atomic* (A) extension provides atomic instructions to support synchronization across threads; this extension is necessary for single-core processors with multiple threads as well as for multicore processors.

17.2.1 Consistency Models

RISC-V uses a release consistency model called *RISC-V Weak Memory Ordering* (*RVWMO*) [RISCV-Unprivileged]. Each thread runs on its own hart. The hart may be a physical core or may be a virtual hart time-multiplexed by the OS onto the physical core. The thread appears to execute in order from the perspective of other memory instructions in the same hart, but memory instructions from another hart may observe the memory instructions from the first hart being executed in a different order. For example, stores can be sent to a write buffer rather than completing immediately. A speculative out-of-order hart may reorder instructions as long as the result is indistinguishable by that hart from executing in program order.

If a system needs a critical section of a given hart's memory accesses to appear in order from another hart's perspective, the given hart must add explicit synchronization through fences or acquire and release. To do so, a hart must use an *acquire* operation before a critical section of code that manipulates shared variables and follow that section with a *release* operation. The acquire operation acquires the lock and ensures that all memory operations in the subsequent critical section are seen by other threads as happening after the lock is acquired. The release operation releases the lock and ensures that all memory operations in the critical section are seen by other threads as happening before the lock is released. Therefore, all threads have a consistent view of the shared variables. The acquire and release operations can be enforced by either a fence instruction or by a more selective acquire or release annotation on an atomic operation, as described in the following subsections.

Recall from Section 11.2.2.1 that a RISC-V core can use physical memory attributes (PMAs) to designate certain address ranges as main memory and others as peripheral I/O. Peripheral reads are called *input*, and peripheral writes are called *output*. The RVWMO can enforce consistency separately for memory and I/O accesses; for example, using a lock to control access to a shared data structure in main memory does not have to wait for an access to a peripheral to complete. Section 17.2.2 defines fences that order memory or I/O accesses relative to one another.

RISC-V optionally supports the *RISC-V Total Store Ordering* (*RVTSO*) consistency model, which is called the Ztso extension. RVTSO makes it easier to port programs written for other TSO architectures such as x86 or SPARC. It may come for free on simple in-order processors because they never access memory out of program order. The compiled Executable and Linkable Format (ELF) file sets bit 4 of the `e_flags` field in the ELF header if a program requires Ztso to run correctly.

17.2.2 Fences

The `fence` instruction is an I-type instruction listed in Table I.8 (on the last pages of the book). It is part of the RV{32/64}I base instruction set and is supported on all RISC-V processors. Fences can enforce ordering for memory reads (r), memory writes (w), peripheral input (i), and peripheral output (o). A fence takes two arguments that specify which types of accesses before the fence must complete before which types of accesses after the fence may occur.

The 12-bit immediate field specifies which types of accesses are ordered, as shown in Fig. 17.3. The `fm` field is normally 0000. The `PI`, `PO`, `PR`, and `PW` bits define the predecessor set, and the `SI`, `SO`, `SR`, and `SW` bits define the successor set. The fence orders accesses so other peripherals or RISC-V harts cannot see any access in the successor set before it sees all the accesses in the predecessor set. For example, `fence rw, rw` has an

fm	PI	PO	PR	PW	SI	SO	SR	SW
31:28	27	26	25	24	23	22	21	20

Fig. 17.3 `fence` **immediate field format.**

immediate field of 0000_0011_0011 (`PR`, `PW`, `SR`, and `SW` are 1), and ensures that all reads and writes to memory before the fence complete before any read or write to memory after the fence completes. For example, placing a `fence rw, rw` between each memory operation ensures that they take place in a sequentially consistent fashion.

> Simple in-order pipelined cores with no store buffers, such as Wally, are inherently sequentially consistent and thus don't require fences between memory accesses. `fence` can be treated as a nop on these cores.

The `fence` pseudoinstruction is a shorthand for `fence iorw, iorw`, which enforces sequential consistency across both memory and I/O peripheral accesses. Weaker fences are also available, as described in Table 17.3. The `fence.tso` instruction is encoded as a `fence rw, rw` with `fm` = 1000. It implies total store order, which is weaker than sequential consistency. Other fences with nonzero `fm` are reserved. Code Example 17.18 gives an example of using a fence in a lock. Similarly, in Code Example 17.3, the producer thread `T1` can use a `fence w, w` to ensure other cores see the variables written before `flag` is written to 1, and `T2` can use a `fence r, r` to ensure that the flag is read as 1 before the variables are read.

However, peripherals with direct memory access (DMA) capability that can read or write main memory independently of the core introduce memory consistency issues and can require synchronization. Section 9.1.4 describes such synchronization.

Recall that RISC-V also supports `fence.i` (synchronize instruction memory) and `sfence.vma` (synchronize page table). `fence.i` requires the Zifencei extension. It synchronizes the instruction and data caches, as discussed in Sections 10.4.2.4 and 13.3.1.1. `sfence.vma` requires virtual memory support and synchronizes the TLB as described in Section 11.2.4.

> The `menvcfg.FIOM` bit stands for Fence of I/O implies Memory.

When the `menvcfg.FIOM` bit is set, a `fence` involving the `PI`, `PO`, `SI`, or `SO` bits at a privilege level below M-mode behaves as if the corresponding `PR`, `PW`, `SR`, or `SW` bits are also set. Similarly, `senvcfg.FIOM` modifies I/O fences to also imply memory fences in U-mode. These bits help the OS to emulate virtual I/O devices in software, as described in the RISC-V Privileged Specification Release 20211203, Section 4.1.10.

17.2.3 Atomic Instructions

The RISC-V atomic (A) extension defines two forms of atomic instructions: atomic memory operations (AMOs) and `LoadReserved-StoreConditional` (`lr-sc`), as listed in Table I.24 in the last pages of the book. AMO instructions fetch data, operate on it, and store it back into the same memory address from which it was fetched.

Both forms of atomic instructions, AMOs and `lr-sc`, have acquire and release bits (`aq` and `rl`) that can optionally be set to order the atomic operations relative to other memory accesses within the same hart, without

Table 17.3 Common RISC-V fences.

Fence	imm	Meaning
`fence iorw, iorw`	0xFF	Other cores see all input, output, reads, and writes before the fence complete before any input, output, read, or write after the fence completes. This gives sequential consistency across both memory and I/O accesses.
`fence rw, rw`	0x33	Other cores see all reads and writes before the fence complete before any read or write after the fence completes. This gives sequential consistency for memory accesses.
`fence r, rw`	0x23	Other cores see all reads before the fence complete before any read or write after the fence. This is an *acquire* operation.
`fence rw, w`	0x31	Other cores see all reads and writes before the fence complete before any write after the fence. This is a *release* operation.
`fence r, r`	0x22	Other cores see all reads before the fence complete before any read after the fence. Linux calls this a *read memory barrier*: `smp_rmb()`.
`fence w, w`	0x11	Other cores see all writes before the fence complete before any write after the fence. Linux calls this a *write memory barrier*: `smp_wmb()`.
`fence.tso`	0x833	Other cores see all reads before the fence complete before any read or write after the fence completes and all writes before the fence complete before any write after the fence completes. This gives total store order. A hart can optionally implement `fence.tso` as `fence rw, rw` for stronger sequential consistency.

the need for an additional fence. When `aq` is set, no later memory operation in the hart can be observed to take place before this atomic access. When `rl` is set, all previous memory operations in the hart will be observed to take place before this atomic access. When both `aq` and `rl` are set, the atomic access is sequentially consistent and will be observed to occur after all previous accesses and before any subsequent accesses. If neither are set, the atomic operation may be seen in any order relative to other memory operations in the hart. The acquire and release bits do not enforce ordering of memory relative to I/O operations; `fence` should be used if this is necessary.

17.2.3.1 Atomic Memory Operations

The atomic extension provides a family of AMOs that read a value from memory, modify it, and write it back to memory in one uninterruptable step. AMOs are also referred to as `FetchAndOp` because they fetch data from memory and perform an operation. They are not as flexible as `lr-sc` but are simpler, do not livelock, and perform better on highly parallel systems because they involve a single instruction and always succeed.

AMOs are of the form `amo<op>.w rd, rs2, (rs1)`. The operation `<op>` is one of `swap`, `add`, `and`, `or`, `xor`, `max`, `maxu`, `min`, or `minu`. These instructions read a value from memory address `rs1` into `rd`. They also compute

RV64 also supports `.d` doubleword AMOs.

(rs1) op rs2 and put the value back into the same address. For example, if a0 = 1000, t0 = 2 and memory address 1000 contains 5, then amoadd.w s0, t0, (a0) reads 5 from address 1000, puts 5 into s0, and writes the value 2 + 5 = 7 back to memory address 1000. amoswap writes the value from rs2 to memory and puts the original memory value in rd.

AMOs can implement the TestAndSet, Swap, and FetchAndIncrement atomic operations from Section 17.1.4.1. Assuming t1 = 1, a0 = adr, a1 = swap, and val is in s0:

- val = TestAndSet(adr)   amoswap.w s0, t1, 0(a0)
- val = Swap(adr, swap)   amoswap.w s0, a1, 0(a0)
- val = FetchAndIncrement(adr)   amoadd.w s0, t1, 0(a0)

amoor, amoand, and amoxor can be used to atomically set, clear, or toggle bits in memory-mapped device registers.

Code Example 17.16 shows a spin lock using amoswap to perform TestAndSet. The lock is at address a0. It first spins using lw until it detects that the lock is free (0). It uses amoswap to test if the lock is still free and set it busy. Spinning on lw before amoswap, called TestAndTestAndSet, is more efficient than spinning on amoswap because lw doesn't stress the shared memory system as much as an atomic operation. The code then enters a critical section and finally releases the lock with another amoswap storing 0. The amoswap that acquires the lock is marked aq to ensure no later memory operations in the critical section are seen to take place before the lock is acquired. Similarly, the amoswap releasing the lock is marked rl to ensure that all memory operations in the critical section are seen to take place before the lock is released. If the aq and rl were not specified, accesses to shared variables in the critical section might appear to happen when the lock is not held, resulting in improper synchronization.

Code Example 17.16 SPIN LOCK USING **amoswap**

```
  # Acquire lock
  li t0, 1                      # Initialize swap
lock:
  lw t1, (a0)                   # Check if lock is held
  bnez t1, lock                 # Retry if lock is held
  amoswap.w.aq t1, t0, (a0)     # Attempt to acquire lock
                                # No critical section memory accesses appear before here
  bnez t1, lock                 # Retry, if held

  #
  # Critical Section Code: do stuff with shared variable. . .
  #

  # Release lock
  amoswap.w.rl x0, x0, (a0)     # Release lock by storing 0
                                # No critical section memory accesses appear after here
```

When an AMO instruction causes an exception, it is reported as a store instead of a load. Thus, the exceptions are formally called store/AMO. The possible exceptions are store/AMO access fault, store/AMO address misaligned, and store/AMO page fault.

17.2.3.2 LoadReserved-StoreConditional

As described in Section 17.1.4.1, an `lr-sc` pair is used to perform complex AMOs. `lr.w rd, (rs1)` loads a word from the memory address specified by `rs1` into `rd` and marks the address as reserved. `sc.w rd, rs2, (rs1)` attempts to store the value in `rs2` into the memory address specified by `rs1`. If the address is still reserved, it succeeds and clears `rd` to 0. If the address is not reserved, `sc` fails (doesn't change memory) and sets `rd` to a nonzero failure code. The reservation is canceled by a write to the reserved address from any hart or by the reserving hart's `sc` to any address.

RV64 also supports the `lr.d` and `sc.d` double-word forms.

Strictly speaking, `lr` reserves a block of contiguous addresses rather than a single word. This block is called the *reservation set*. A write to any address within the reservation set cancels the reservation. The reservation set size is often one full cache line, so the reservation is tracked with an extra status bit on the cache line. Cache lines are often 512 bits (64 bytes), so this is a common reservation set size.

During a context switch or change of relevant page table entries, the OS invalidates any pending reservations by performing a `StoreConditional` to a scratch address.

Code Example 17.17 shows a RISC-V assembly language implementation of `CompareAndSwap` using `lr` and `sc` based on the C version from Code Example 17.9.

Code Example 17.17 COMPAREANDSWAP USING **lr-sc**

```
# int CompareAndSwap(int *adr, int comp, int swap);
# adr is in a0, comp in a1, swap in a2, val in t0, return value in a0
CompareAndSwap:
  lr.w t0, (a0)               # val = LoadReserved(adr)
  bne t0, a1, done            # if (val != comp) return val
  sc.w t1, a2, (a0)           # t1 = StoreConditional(adr, swap)
  bnez t1, CompareAndSwap     # Repeat if StoreConditional fails
done:
  mv a0, t0                   # Return val
  ret
```

As described in Section 17.1.5.2, `lr–sc` is at risk of livelock if two concurrent threads constantly cancel each other's reservations. RISC-V prevents livelock with an *eventual success guarantee* for *constrained lr-sc loops* [RISCV-Unprivileged]. In simple terms, the execution environment guarantees that one of the threads will eventually successfully store. This is straightforward for single-core processors because the OS switches threads relatively infrequently, so even if one thread has the misfortune of being switched out before its sc succeeds, the other will succeed. Multicore processors need a monitor to detect livelock and allow one thread to succeed.

Using lr.w.aq and sc.w.aqrl makes the lr-sc sequence sequentially consistent so it appears in program order relative to other loads and stores from the same hart. This is the desired behavior when lr-sc is used to take a lock.

Code Example 17.18 shows a spin lock using lr-sc to perform TestAndTestAndSet. The lock is at address a0. As in Code Example 17.16, it first spins using lw until it detects that the lock is free (0). It then uses lr and sc to try to acquire the lock. If another thread had grabbed it, the lr-sc spins until the lock becomes available again. Mostly spinning on lw instead of lr-sc is more efficient because lw doesn't stress a shared-memory system as much as atomic operations. lr.w.aq and sc.w.aqrl ensure sequential consistency around the lock. The code then enters a critical section and finally unlocks. Unlocking involves a fence rw, w to ensure all accesses to shared variables complete first, then a store of 0 to the lock variable. Nobody else can be holding the lock, so a simple store word (sw) suffices to release the lock. This example is more complex than the preferred amoswap lock method of Code Example 17.16, but illustrates lr-sc and a fence.

In Code Example 17.18, a sw.rl x0, 0(a0) could be used in place of the fence and sw, but the RISC-V atomic specification does not provide acquire/release for ordinary stores or loads.

Code Example 17.18 SPIN LOCK USING **lr-sc**

```
  # Acquire lock
  li t0, 1                       # Initialize swap
test:
  lw t1, (a0)                    # Check if lock is held
  bnez t1, test                  # Retry if lock is held
lock:
  lr.w.aq t1, (a0)               # Attempt to acquire lock
  bnez t1, lock                  # Retry if somebody else go the lock first
  sc.w.aqrl t1, t0, (a0)         # t1 = 0 and lock = 1 if successful
  bnez t1, lock                  # Retry, if held

  #
  # Critical Section Code: do stuff with shared variable...
  #

  # Release lock
  fence rw, w                    # Ensure critical section completes before here
  sw x0, 0(a0)                   # Unlock the lock by storing 0
```

17.2.4 Atomic Subsets

Although the RISC-V atomic (A) extension supports both AMOs and `lr-sc` instructions, AMOs may be preferred because they scale better to highly parallel systems and also are easy to implement on cacheless microcontrollers. The Zaamo extension supports only the AMO instructions of the A extension. The Zalrsc extension only supports the `lr-sc` instructions of the A extension. The A extension implies Zaamo + Zalrsc.

17.2.5 Profile Extensions

The atomic extension imposes few restrictions on the implementation, but operating systems prefer to rely on some sensible expectations. The following extensions are expected in the RVA20U64 and later profiles, as given in Table A.6.

17.2.5.1 Ziccif

An OS normally expects that main memory is suitable for holding programs and that interrupts cannot occur midway through fetching an instruction. The RVA20U64 and later profiles formalize this understanding with the Ziccif extension, which requires that main memory (i.e., memory with cacheable and coherence PMAs) must support instruction fetch. Moreover, the instruction fetch must be atomic (i.e., an interrupt never takes place while an instruction is only partially fetched).

Wally's IFU (instruction fetch unit) can fetch instructions from any memory address. The fetch is atomic; even if a spill occurs, the IFU stalls the rest of the pipeline until the full instruction is fetched. Therefore, Wally inherently supports the Ziccif extension.

17.2.5.2 Ziccrse

As discussed in Section 17.2.3.2, `lr-sc` could livelock in certain multicore systems unless precautions are taken. The *RsrvEventual* property indicates that the memory system is designed such that `lr-sc` will eventually succeed. The Ziccrse extension formalizes this by requiring that all main memory has the *RsrvEventual* property.

Wally has a single core and meets this property with a reasonable OS scheduler. Therefore, it inherently supports Ziccrse.

17.2.5.3 Ziccamoa

As with Ziccrse, the PMA checker can prohibit some or all AMO instructions from accessing a region of memory. The *AMOArithmetic* property indicates that a memory region supports all AMO instructions. The Ziccamoa extension requires that all main memory has the *AMOArith* property. Wally's PMA checker defaults to supporting all atomic instructions in main memory, so it supports Ziccamoa.

17.2.5.4 Za128rs and Za64rs

RISC-V allows a reservation set of arbitrary size, such that a store to any address within the reservation set cancels a `LoadReserved` to any other address in the set. When the OS allocates memory for a lock, it prefers to have no other variables within the same reservation set so other accesses do not interfere with acquiring the lock. If the reservation set is large, this wastes memory, and if the OS doesn't know the size of the reservation set, it cannot allocate memory effectively.

The Za128rs extension tells the OS that the reservation set is no larger than 128 bytes. Similarly, the Za64rs extension indicates that the reservation set is no larger than 64 bytes. The RVA22 profiles expect Za64rs to limit wasted memory. 64 bytes is a natural implementation choice because it is a typical cache line size.

17.3 TEST PLAN

At the time of this writing, riscv-arch-test has AMO tests listed in Table 17.4, but lacks `lr-sc` tests.

The `tests/wally-riscv-arch-test` directory contains simple `WALLY-lrsc-01` tests. They are located in the `privileged` subdirectory because the test infrastructure does not yet look for an `A` subdirectory. `WALLY-lrsc` performs a successful `LoadReserved` and `StoreConditional` word and double-word operation followed by an unsuccessful operation to check basic functionality.

17.4 WALLY IMPLEMENTATION

Wally is an in-order single-core processor, so it is inherently sequentially consistent. It also has a single instruction cache and single data cache, so caches are inherently coherent except that store operations are not reflected in the instruction cache. Wally needs the `fence.i` instruction for coherence between the instruction and data caches (see Sections 10.4.2.6 and 10.4.5.4) and `sfence.vma` for TLB coherence (see Section 11.2.4). In Wally, all other fences are treated as nops, and the acquire/release bits of atomic instructions are ignored.

When Atomic extensions are supported (`ZAAMO_SUPPORTED` = 1 and/or `ZALRSC_SUPPORTED` = 1), the LSU (load/store unit) adds an `atomic` module, as shown in Fig. 17.4. This module handles `lr-sc` instructions and AMOs. It is relatively simple because it does not need to handle interactions with other cores. The LSU does not have an FSM to sequence atomic read-modify-write operations over multiple cycles, so the atomic module relies on the LSU having a data cache. The Wally PMA checker only allows atomic operations on cachable main memory. The cache read (`ReadDataWordM`), modify (`IMAWriteDataM`), and write (`LSUWriteDataM`) take place in the same cycle.

Table 17.4 riscv-arch-test for A extension

Instruction	RV32IA Tests	RV64IA Tests
amoadd.w	69	133
amoand.w	69	133
amomax.w	69	133
amomax.w	69	133
amomin.w	69	133
amominu.w	69	133
amoor.w	69	133
amoswap.w	69	133
amoxor.w		133
amoadd.d		133
amoand.d		133
amomax.d		133
amomax.d		133
amomin.d		133
amominu.d		133
amoor.d		133
amoswap.d		133
amoxor.d		133
Total	552	2,394

The `atomic` module takes advantage of the existing LSU hardware to add atomic support with minimal change. `LoadReserved` resembles an ordinary load but adds state to keep track of which address is reserved. `StoreConditional` resembles an ordinary store but is squashed if the address is no longer reserved. The LSU already can perform a read at the start of the Memory stage and a write at the end of the Memory stage. AMO operations take the read result, perform an ALU operation on it in the `atomic` module, and mux this ALU result back into the write path. Specifically, the LSU reads `ReadDataM` at the start of the Memory stage.

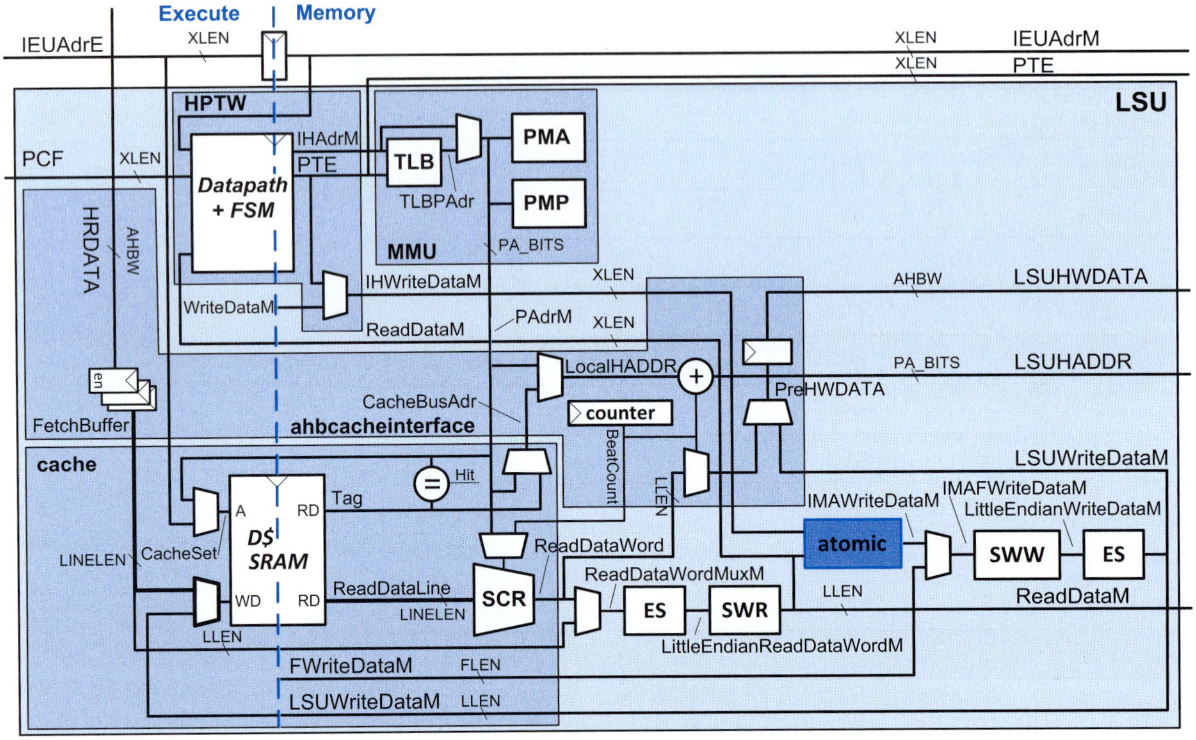

Fig. 17.4 LSU with atomic module.

The atomic module performs an ALU operation to combine this data with IHWriteDataM from a register and calls the result IMAWriteDataM. After subword muxing and endian swap, the result becomes LSUWriteDataM to write back to the cache at the end of the Memory stage.

Wally has a single core, so lr-sc will not livelock and requires no monitor to ensure eventual success. The OS is responsible for invalidating a reservation during a context switch by performing a StoreConditional to a scratch memory address.

17.4.1 Atomic Module

The atomic module is shown in Fig. 17.5. It comprises an AMO ALU and an LRSC module (amolru and lrsc). During AMO operations, it selects IMAWriteDataM from amoalu to write to memory. The LRSC module maintains the reservation address and squashes StoreConditional operations that do not match the reservation.

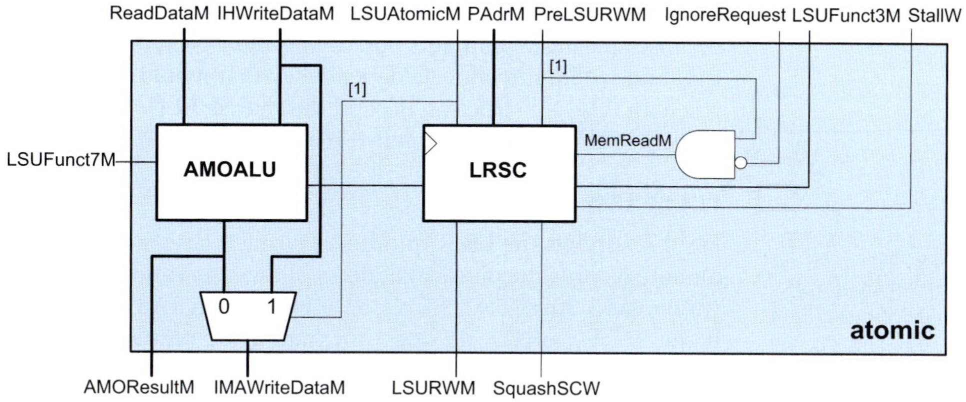

Fig. 17.5 `atomic` **module.**

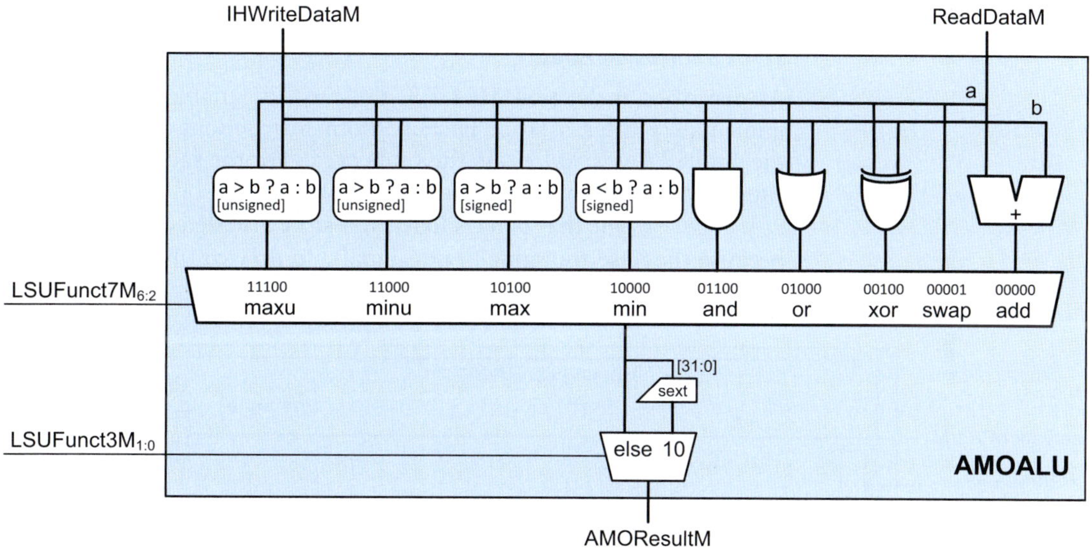

Fig. 17.6 `amoalu` **module.**

17.4.1.1 AMO ALU

Fig. 17.6 shows the AMO ALU (`amoalu`). It performs an operation on `ReadDataM` and `IHWriteDataM` to produce `AMOResultM`. RV64 provides sign-extension multiplexers for 32-bit word operations. Although `amoalu` has many similarities with the integer ALU, it operates in the Memory stage rather than the Execute stage, because it needs the read memory data as an operand. The AMO ALU also provides max/min operations. Note that the four comparators are closely related and can use a shared carry chain. The

AMO ALU uses dedicated hardware rather than sharing the integer ALU because sharing would add busses and multiplexers between the IEU and LSU and would introduce a dependency that would stall the Execute stage when an AMO instruction needed the shared ALU. The extra hardware is modest in the context of a full-featured application core.

17.4.1.2 LRSC module

Observe that this implementation tracks the reservation with an address register rather than a bit on each cache line. Therefore, the size of the reservation set is the native word size (4 bytes on RV32 and 8 bytes on RV64).

Fig. 17.7 shows the LRSC module. During a `LoadReserved` instruction, the module records the physical address of the reservation and marks the reservation valid. Any `StoreConditional` instruction invalidates the reservation. If the `StoreConditional` does not match the physical address of a valid reservation, the `LRSC` module asserts `SquashSCM` to squash the store. This squash turns off the `LSURWM` write enable and writes a 1 to the destination register to indicate failure. Fig. 17.8 shows how the IEU `Result` multiplexer (from Fig. 7.13) takes an additional input for the `StoreConditional` result, highlighted in blue.

17.4.1.3 Atomic Hazards

As described in Section 10.4.2.3, the cache is unable to perform a load immediately after a store because both instructions would need to use the data SRAM's address in the same cycle. For the same reason, the cache cannot perform an AMO after a store or AMO or before a load. The `StoreStallD` logic that detects load after store also detects these AMO cases because they use the same `MemRW` signals, so no extra hardware is needed.

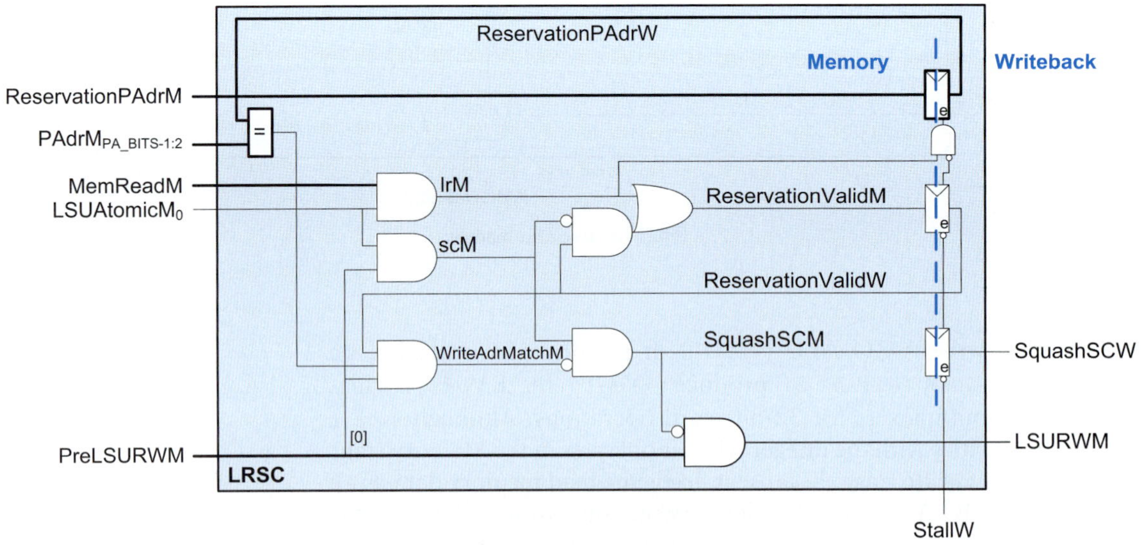

Fig. 17.7 **`lrsc` module.**

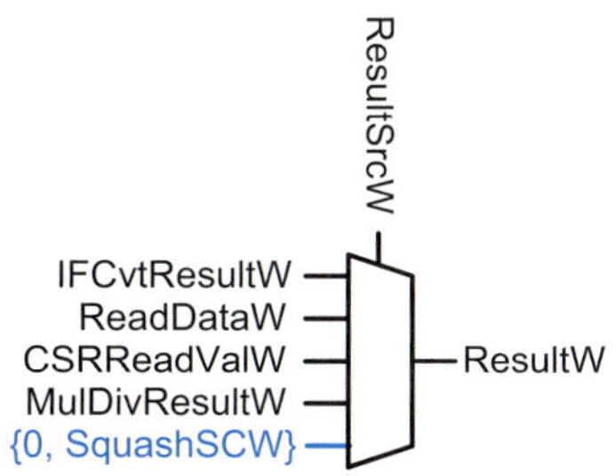

Fig. 17.8 IEU `Result` multiplexer supporting `StoreConditional`.

17.4.2 Timing Diagrams

The timing for atomic operations is similar to other LSU operations. They rely on a data cache in the LSU. Loads, including `LoadReserved`, read the cache at the start of the Memory stage. Stores, including `StoreConditional`, write the cache at the end of the Memory stage. AMOs read memory at the start of the Memory stage, perform an operation, and write the result back to the cache at the end of the Memory stage.

Code Example 17.19 exercises `LoadReserved`, successful and unsuccessful `StoreConditional`, and an AMO. Assume the program runs on an RV32IA machine in little-endian mode with virtual memory disabled, so the virtual and physical addresses are the same, and the subword and endian swap blocks do not modify the data. Assume memory location 0x80001000 initially contains 42.

The program performs a `LoadReserved` to address 0x80001000, reading the value 42 and reserving the address. It then performs a `StoreConditional` back to the same address, writing the value 43 and clearing the reservation. The second `StoreConditional` fails, returning a success code of 0. It attempts to repeat the `StoreConditional`, now squashing the store and getting a failure code of 1 because the reservation is no longer valid. Finally, it performs an `amoadd` at the same address, reading 43 and writing 45.

Code Example 17.19 ATOMIC PROGRAM

```
li a0, 0x80001000           # a0 = 80001000
li a1, 2                    # a1 = 2
lr.w t0, (a0)               # t0 = mem[80001000] = 42; reserve 80001000
addi t0, t0, 1              # t0 = 43
sc.w t1, t0, (a0)           # Successful StoreConditional to 80001000; t1 = 0
sc.w t2, t0, (a0)           # Unsuccessful StoreConditional to 80001000; t2 = 1
amoadd.w s0, a1, 0(a0)      # s0 = 43; mem[80001000] = 45
```

Fig. 17.9 shows the waveforms while running this example. `MemRWM` is 10 for loads, 01 for stores, and 11 for AMO. `LSUAtomicM` is 00 for

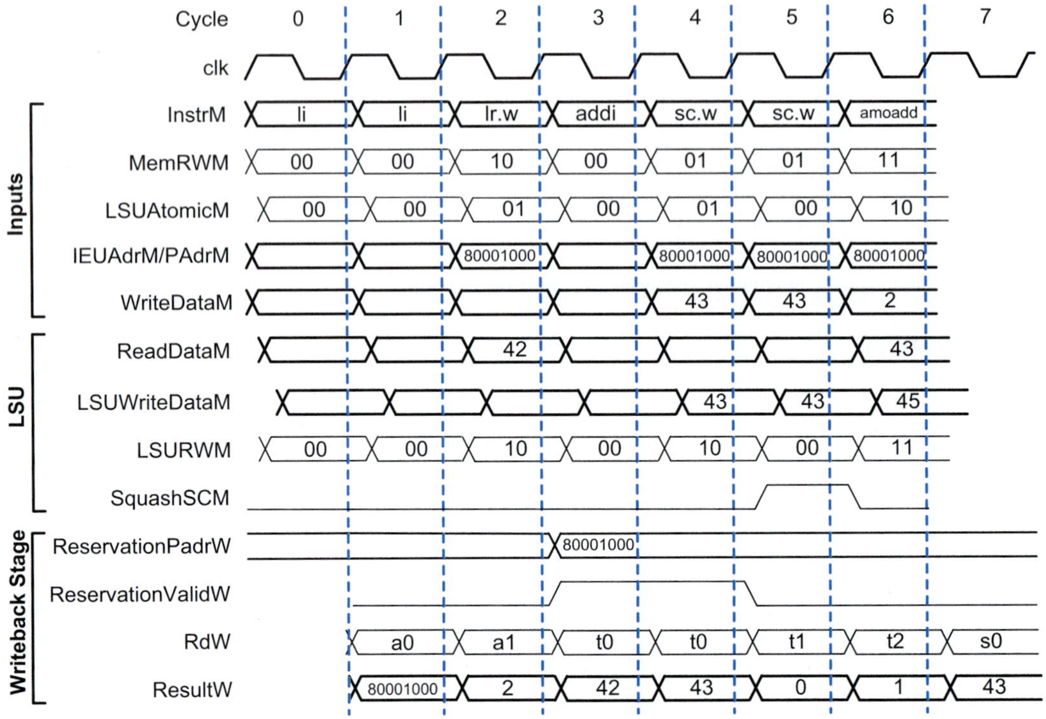

Fig. 17.9 **Atomic waveforms.**

nonatomic operations, 01 for `lr-sc`, and 10 for AMOs. The first `StoreConditional` succeeds in cycle 4, writing 0 to `t0` and clearing `ReservationValidW` in cycle 5. The second `StoreConditional` in cycle 5 finds the reservation is no longer valid and fails. It raises `SquashSCM`, clears `LSURWM` to prevent storing, then writes 1 to `t0` in cycle 6. The `amoadd` in cycle 6 reads 43 from memory and 2 from `a1`. It adds these operands and writes 45 back to memory, then returns 43 to `s0`.

17.5 SUMMARY

Modern computers run processes concurrently by time-multiplexing processes and threads on a single core and/or using multiple cores. Programmers want coherence, consistency, and synchronization in concurrent systems. Coherence means that all processes eventually see the same value stored to memory. Consistency describes the ordering of reads and writes with respect to different locations. Programmers intuitively expect a sequential consistency model, which means that each thread performs reads and writes in program order but that different threads may interleave in arbitrary ways. Unfortunately, guaranteeing sequential consistency precludes important hardware optimizations such as write buffers, so high-performance systems use more relaxed total store order or release consistency models. Under release consistency, the programmer must explicitly insert fences or acquire and release operations when a certain ordering is necessary to synchronize multiple threads. Common synchronization requirements and techniques are as follows:

- **Mutual exclusion (locks):** Only one thread can access a shared variable at a time.
- **Producer-consumer (condition variables):** Threads communicate by putting and getting information in a bounded buffer and waiting when the buffer is empty or full.

Processors provide atomic instructions to facilitate synchronization. The RISC-V atomic (A) extension defines `LoadReserved-StoreConditional` (`lr-sc`) and atomic memory operations (AMOs). RISC-V has a release consistency model, so it also defines a set of fences to enforce memory consistency. Wally is an in-order single-core processor, so it is inherently sequentially consistent. However, it still needs atomic instructions to synchronize multiple threads. It also uses `fence.i` to maintain coherence between the instruction and data caches and `sfence.vma` for TLB coherence.

18 Extensions: Zb* & Zk* (Bit Manipulation and Cryptography)

18.1 Cryptography Principles
18.2 RISC-V Practices
18.3 Test Plan
18.4 Wally Implementation
18.5 Summary

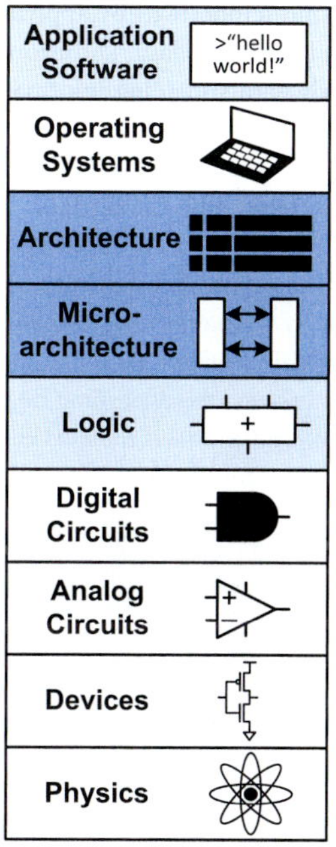

RISC-V defines bit manipulation (Zb*) and cryptography (Zk*) extensions. Bit manipulation instructions are an extension to ordinary integer instructions. They include rotation, setting or clearing bits, and counting the number of 1s in a register. Although programs can be written without bit manipulation instructions, competing computer architectures provide them to improve code speed and size at the expense of a modest amount of new hardware. For example, Section 21.2.3 shows these extensions make Embench 5% faster and 4% denser than an rv32imc baseline. In Section 21.1.3, CoreMark is 8% faster with bit manipulation instructions.

Cryptography is critical for maintaining security. Consider Alicia (A) and Ben (B) attempting to communicate. The key tasks of cryptography are:

- **Secrecy:** Only A and B can see what is sent.
- **Integrity:** What A sends is what B receives.
- **Authenticity:** Nobody can impersonate A or B.

The most widely used cryptographic operations are *encryption* and *decryption* for secrecy, *hashing* for integrity, and *signatures* for authenticity. Pure software implementations are slow, power hungry, and vulnerable to timing attacks that can be exploited remotely. Modern instruction sets provide dedicated cryptography instructions that are faster, simpler, and provide better performance than pure software implementations. Moreover, having cryptographic instructions promotes standardized software and reduced code size, which helps reduce the risk of inadvertent security flaws. RISC-V provides the Zk* family of cryptographic extensions to support existing standardized cryptographic algorithms and practices for encryption, decryption, and hashing. Cryptography extensions are discussed in the same chapter as bit manipulation because they share the same ALU and have some instructions in common.

Cryptography is also referred to simply as crypto, not to be confused with cryptocurrency.

This chapter discusses cryptography principles, including finite field arithmetic and its applications for Advanced Encryption Standard (AES)

Now we consider multiplication by x. Again, consider a polynomial $a = x^2 + 1$ in GF(2^3) with the irreducible polynomial $m(x) = x^3 + x + 1$. $f(x) = m(x) - x^n = x + 1$. Compute the product ax as:

$$
\begin{aligned}
ax &= (x^2+1)x \bmod (x^3+x+1) && \text{Definition of multiplication} \\
&= (x^3+x) \bmod (x^3+x+1) && \text{Polynomial multiplication} \\
&= x^3 \bmod (x^3+x+1) + x \bmod (x^3+x+1) && \text{Distributive property} \\
&= (x+1) + x \bmod (x^3+x+1) && x^n \bmod m(x) = f(x);\ x^n = x^3,\ f(x) = x+1 \\
&= (x+1) + (x) && p(x) \bmod m(x) = p(x);\ p(x) = x \\
&= (x+\text{x}) + (1) && \text{Associativity} \\
&= 1 && x + x = 0 \text{ in a binary extension field}
\end{aligned}
$$

Multiplication by higher powers of x involves repeated multiplication by x. ax^2 is computed from ax above as:

$$
\begin{aligned}
ax^2 &= (ax)x \\
&= (1)x \\
&= x
\end{aligned}
$$

Multiplication by an arbitrary polynomial can be expressed as the sum of multiplications by powers of x. For example, with $a = x^2 + 1$ and $b = x + 1$, the product ab is computed as:

$$
\begin{aligned}
ab &= a(x+1) && \text{Substitute } b \\
&= ax + a && \text{Distributive property} \\
&= (1) + (x^2+1) && \text{Substitute precomputed } ax \\
&= x^2 + (1+1) && \text{Commutativity} \\
&= x^2 + (0) && 1 + 1 = 1 \oplus 1 = 0 \\
&= x^2 && \text{Addition}
\end{aligned}
$$

In binary representation, multiplying a by x gives $ax = a \times 10 = (a << 1) \bmod m$. Thus, multiplication by x is a left shift by 1. If the result has a 0 shifted into column n, it is less than m, and the modulus does nothing. Otherwise, as with polynomials, the modulus reduces the result by dropping the 1 from column n and XORing the other bits with f, the binary representation of $f(x)$. Formally, for an arbitrary element a of GF(2^n) written as an n-bit binary number $a_{n-1:0}$

$$
ax = \begin{cases} (a_{n-2:0} << 1) & \text{when } a_{n-1} = 0 \\ (a_{n-2:0} << 1) \oplus f & \text{when } a_{n-1} = 1 \end{cases} \tag{18.1}
$$

For example, in GF(2^3), we can write a as a 3-bit binary number $a = a_2 a_1 a_0$. Thus, $(a << 1) \bmod m = (a_2 a_1 a_0 0) \bmod m$. When

$a_2 = 0$, this is just $(0a_1a_00) \bmod m = a_1a_00$. When $a_2 = 1$, this is $(1a_1a_00) \bmod m = (a_1a_00 \bmod m) + (1000 \bmod m) = a_1a_00 + f = a_1a_00 \oplus f$.

Let us repeat the multiplication examples above using binary representation. In binary, $a_{2:0} = 101$. $m = 1011$, so $f = 011$.

$$ax = (101 << 1) = 1010, \text{which reduces to } 010 \oplus 011 = 001$$
$$ax^2 = (001 << 1) = 0010, \text{which needs no reduction} = 010$$

In binary, multiplication by higher powers of x involves repeated shifts and XOR. Multiplying arbitrary polynomials in binary representation is the addition of selected multiplications by powers of x. For example, $a(x^2 + x) = ax^2 + ax = 001 + 010 = 001 \oplus 010 = 011$.

Example 18.1 AES MULTIPLICATION

AES treats bytes as elements of GF(2^8) with $m(x) = x^8 + x^4 + x^3 + x + 1$. Thus, $f(x) = x^4 + x^3 + x + 1$; $f = 00011011$. Let $a = 11000101 = x^7 + x^6 + x^2 + 1$ and $b = 10100001 = x^7 + x^5 + 1$. Compute the product ab.

Solution: First, find the products ax^k for $k \in [0, 7]$, as shown in Table 18.3. For each k, use $ax^k = ax^{k-1} \times x$, left-shifting the previous result and XORing to reduce if necessary. The products are represented in binary for easy and compact computation.

Table 18.3 ax^k for $k \in [0,7]$

Product	Expression	Binary Computation	Result
ax^0	A	11000101	11000101
ax^1	$(ax^0 \times x) \bmod m(x)$	1000101 << 1 ⊕ 00011011	10010001
ax^2	$(ax^1 \times x) \bmod m(x)$	0010001 << 1 ⊕ 00011011	00111001
ax^3	$(ax^2 \times x) \bmod m(x)$	0111001 << 1	01110010
ax^4	$(ax^3 \times x) \bmod m(x)$	1110010 << 1	11100100
ax^5	$(ax^4 \times x) \bmod m(x)$	1100100 << 1 ⊕ 00011011	11010011
ax^6	$(ax^5 \times x) \bmod m(x)$	1010011 << 1 ⊕ 00011011	10111101
ax^7	$(ax^6 \times x) \bmod m(x)$	0111101 << 1 ⊕ 00011011	01100001

Next, sum the terms based on b:

$$\begin{aligned} ab &= a(x^7 + x^5 + 1) \\ &= ax^7 + ax^5 + ax^0 \\ &= 01100001 \oplus 11010011 \oplus 11000101 \\ &= 01110111 \\ &= x^6 + x^5 + x^4 + x^2 + x^1 + 1 \end{aligned}$$

Table 18.6 Key expansion example.

i	$w_{i\text{-}1}$	After RotWord	After SubWord	$Rcon_{i/4}$	t	$w_{i\text{-}4}$	w_i = key$_i$ $i < 4$ $= t \oplus w_{i\text{-}4}$ $i \geq 4$
0							2b7e1516
1							28aed2a6
2							abf71588
3							09cf4f3c
4	09cf4f3c	cf4f3c09	8a84eb01	01000000	8b84eb01	2b7e1516	a0fafe17
5	a0fafe17				a0fafe17	28aed2z6	88542cb1
6	88542cb1				88542cb1	abf71588	23a33939
7	23a33939				23a33939	09cf4f3c	2a6c7605
8	2a6c7605	6c76052a	50386be5	02000000	52386be5	a0fafe17	f2c295f2
9	f2c295f2				f2c295f2	88542cb1	7a96b943
10	7a96b943				7a96b943	23a33939	5935807a
11	5935807a				5935807a	2a6c7605	7359f67f
12	7359f67f	59f67f73	cb42d28f	04000000	cf42d28f	f2c295f2	3d80477d
13	3d80477d				3d80477d	7a96b943	4716fe3e
14	4716fe3e				4716fe3e	5935807a	1e237e44
15	1e237e44				1e237e44	7359f67f	6d7a883b
…							
40	575c006e	5c006e57	4a639f5b	36000000	7c639f5b	ac7766f3	d014f9a8
41	d014f9a8				d014f9a8	19fadc21	c9ee2589
42	c9ee2589				c9ee2589	28d12941	e13f0cc8
43	e13f0cc8				e13f0cc8	575c006e	b6630ca6

Example 18.3 PLAINTEXT ENCRYPTION USING ROUND KEYS

Encrypt the plaintext 0x3243f6a8_885a308d_313198a2_e0370734 with the 128-bit round keys from Example 18.2.

Solution: Fig. 18.11 (on page 734) shows the encryption steps. The input is first XORed with the round key. In each round, AddRoundKey, SubBytes, ShiftRows, and MixColumns are applied. On the last round, MixColumns is omitted. Specifically, consider the upper left entry $s_{0,0}$ in Round 1. The entry at the start of the round comes from the initial AddRoundKey: 32 $\oplus$ 2b = 19. Referring to Fig. 18.7, SubBytes(19) = d4. ShiftRows does not modify the top row, so the element remains 19. MixColumns computes [02 03 01 01] [d4 bf 5d 30]^T = (d4)x $\oplus$ (bf)(x + 1) $\oplus$ (5d)1 $\oplus$ (30)1 = ((d4 << 1) $\oplus$ 11b) $\oplus$ ((bf << 1) $\oplus$ 11b) $\oplus$ bf $\oplus$ 5d $\oplus$ 30 = b3 $\oplus$ 65 $\oplus$ bf $\oplus$ 5d $\oplus$ 30 = 04. The upper left entry of the Round Key is a0 from the row 4 w_i entry of Table 18.6.

18.1.2.6 Decryption

Because the AES algorithm is a symmetric cryptographic algorithm, decryption (also called the *inverse cipher*) consists of the same steps as encryption but with operations replaced by their inverses, as shown in Fig. 18.12. The key schedule also changes. AddRoundKey is its own inverse because $a = -a$ in GF(2^n). Therefore, the hardware XORs with the

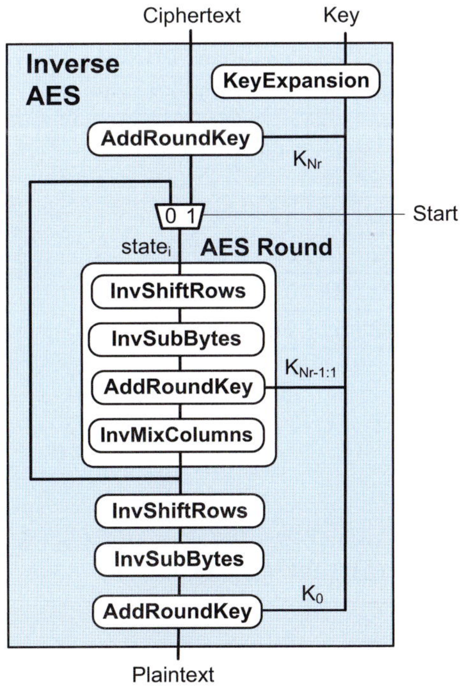

Fig. 18.12 Overall AES decryption architecture.

Round Number	Start of Round	After SubBytes	After ShiftRows	After MixColumns	Round Key Value
input	32 88 31 e0				2b 28 ab 09
	43 5a 31 37				7e ae f7 cf
	f6 30 98 07				15 d2 15 4f
	a8 8d a2 34				16 a6 88 3c
1	19 a0 9a e9	d4 e0 b8 1e	d4 e0 b8 1e	04 e0 48 28	a0 88 23 2a
	3d f4 c6 f8	27 bf b4 41	bf b4 41 27	66 cb f8 06	fa 54 a3 6c
	e3 e2 8d 48	11 98 5d 52	5d 52 11 98	81 19 d3 26	fe 2c 39 76
	be 2b 2a 08	ae f1 e5 30	30 ae f1 e5	e5 9a 7a 4c	17 b1 39 05
2	a4 68 6b 02	49 45 7f 77	49 45 7f 77	58 1b db 1b	f2 7a 59 73
	9c 9f 5b 6a	de db 39 02	db 39 02 de	4d 4b e7 6b	c2 96 35 59
	7f 35 ea 50	d2 96 87 53	87 53 d2 96	ca 5a ca b0	95 b9 80 f6
	f2 2b 43 49	89 f1 1a 3b	3b 89 f1 1a	f1 ac a8 e5	f2 43 7a 7f
3	aa 61 82 68	ac ef 13 45	ac ef 13 45	75 20 53 bb	3d 47 1e 6d
	8f dd d2 32	73 c1 b5 23	c1 b5 23 73	ec 0b c0 25	80 16 23 7a
	5f e3 4a 46	cf 11 d6 5a	d6 5a cf 11	09 63 cf d0	47 fe 7e 88
	03 ef d2 9a	7b df b5 b8	b8 7b df b5	93 33 7c dc	7d 3e 44 3b
10	eb 59 8b 1b	e9 cb 3d af	e9 cb 3d af		d0 c9 e1 b6
	40 2e a1 c3	09 31 32 2e	31 32 2e 09		14 ee 3f 63
	f2 38 13 42	89 07 7d 2c	7d 2c 89 07		f9 25 0c 0c
	1e 84 e7 d2	72 5f 94 b5	b5 72 5f 94		a8 89 c8 a6
output	39 02 dc 19				
	25 dc 11 6a				
	84 09 85 0b				
	1d fb 97 32				

Fig. 18.11 Encryption example. [NIST-FIPS-197-upd1, Appendix B]

same round keys for both encryption and decryption. InvSubBytes uses an inverse S-box that flips inputs and outputs. InvShiftRows shifts right instead of left. InvMixColumns uses the transformation matrix from Fig. 18.13, which involves a more complicated sequence of shifts and XORs than the MixColumns matrix.

$$\begin{bmatrix} 0e & 0b & 0d & 09 \\ 09 & 0e & 0b & 0d \\ 0d & 09 & 0e & 0b \\ 0b & 0d & 09 & 0e \end{bmatrix}$$

Fig. 18.13 InvMixColumns matrix.

18.1.3 Hash Functions

Cryptographic hash functions are important elements of cryptography. They transform a variable-length message into a short numerical fingerprint called a *message digest*. Hashes are used to verify data integrity and as a building block for digital signatures. Any alteration to a message will corrupt the message digest. Passwords are usually stored in hashed form instead of plaintext, so stealing the password file will not reveal the password itself. A good hash function has several properties:

- **Avalanche effect:** Any change in the message will, with very high probability, change many bits of the digest.
- **Pre-image resistant:** Given a message digest, it is computationally infeasible to find the original message.
- **Collision resistant:** Given a message digest, it is computationally infeasible to find another message that produces the same digest.
- **Fast and easy to compute**

SHA-2 is the hash function used in most internet protocols, such as TLS, SSL, and PGP, well as to verify transactions in Bitcoin and other cryptocurrencies. It was designed by the US National Security Agency (NSA) and was first published in 2001. It is now a standard maintained by the National Institute of Standards and Technology [NIST-FIPS-180-4]. SHA-2 generates 224-, 256-, 384-, or 512-bit message digests, replacing the SHA-1, MD4, and MD5 algorithms that produced shorter digests that are no longer considered secure. This section describes the SHA-256 and SHA-512 flavors of SHA-2 that produce 256- or 512-bit digests. Other flavors of SHA-2 are similar but truncate the digest to fewer bits after it is computed, trading compactness for reduced security.

This chapter refers to cryptographic hash functions as hash functions for brevity. Other kinds of noncryptographic hash functions exist that are optimized for different goals.

Table 18.7 summarizes the structure of SHA-256 and SHA-512. The hash operates on a message M comprising N m-bit blocks; it is padded if necessary to be an integral number of blocks. Each block is formed from 16 w-bit words. Word j of block i is denoted $M_j^{(i)}$. The message digest (also called the hash) H is formed from 8 w-bit words. Each block goes through r rounds of hashing, which involve applying some left shifts, right rotates, and XOR operations. The hashing steps involve sigma (Σ/σ) functions that are expressed in terms of right rotations (ror) and right shifts

Table 18.7 SHA-2 structure and sigma operations.

SHA-2 Algorithm	Variable	SHA-256	SHA-512
Message digest size (bits)	d	256	512
Block size (bits)	m	512	1,024
Word size (bits)	w	32	64
Rounds	r	64	80
$\Sigma_0^{(d)}(x)$		$(x \text{ ror } 2) \oplus (x \text{ ror } 13) \oplus (x \text{ ror } 22)$	$(x \text{ ror } 28) \oplus (x \text{ ror } 34) \oplus (x \text{ ror } 39)$
$\Sigma_1^{(d)}(x)$		$(x \text{ ror } 6) \oplus (x \text{ ror } 11) \oplus (x \text{ ror } 25)$	$(x \text{ ror } 14) \oplus (x \text{ ror } 18) \oplus (x \text{ ror } 41)$
$\sigma_0^{(d)}(x)$		$(x \text{ ror } 7) \oplus (x \text{ ror } 18) \oplus (x >> 3)$	$(x \text{ ror } 1) \oplus (x \text{ ror } 8) \oplus (x >> 7)$
$\sigma_1^{(d)}(x)$		$(x \text{ ror } 17) \oplus (x \text{ ror } 19) \oplus (x >> 10)$	$(x \text{ ror } 19) \oplus (x \text{ ror } 61) \oplus (x >> 6)$

The random-looking initialization and round constants in Table 18.7 correspond to the most significant fractional parts of the cube and square roots of the first 8 and r prime numbers, respectively. They are called *nothing-up-my-sleeve numbers* and give confidence that they were not specially chosen by the NSA to provide a backdoor.

(>>) of the words. The hash is initialized with 8 w-bit constants $H_j^{(0)}$ and uses r w-bit round constants K_t tabulated in the SHA-2 specification.

Fig. 18.14 describes the SHA-2 algorithm, which is the same for SHA-256 and SHA-512 except for the parameter values, constants, and definition of sigma functions, as given in Table 18.7. Addition is defined to be modulo 2^w, so overflow is discarded. The algorithm also uses *Choose* (Ch) and *Majority* (Maj) functions which, respectively, choose y or z based on x and output 1 if a majority (at least 2 of the 3) inputs is TRUE. These functions are defined as:

$$\text{Ch}(x, y, z) = xy \oplus \bar{x}z = xy \mid \bar{x}z = x\ ?\ y : z \text{ (bitwise)}$$
$$\text{Maj}(x, y, z) = xy \oplus xz \oplus yz = xy \mid xz \mid yz = 1$$

SHA-2 initializes the 8-word hash $H^{(0)}$ to known constants. It then processes each of the N blocks. It begins by expanding the block into an R-word *message schedule* $W_{S\text{-}1:0}$. The first 16 words of the message schedule are the 16 words of the block, and the remaining words of the schedule depend on rotates and shifts of some of the earlier words. Next, the algorithm initializes 8 *working variables* a_j based on the hash of the previous block. It then introduces diffusion and confusion into the hash by applying r rounds. In each round, the working variables are shifted by one word, and a hashing function is applied to certain working variables. At round t, the nonlinear hashing function depends on the round constant K_t and message schedule word W_t. The arbitrary-seeming choices of the hashing function draw from the *Merkle-Damgård method* of building

for $j = 0$ to 7
    $H_j^{(0)} = initial\ value(d, j)$ // Initialize hash
for $i = 1$ to N // Process each block in order
    for $j = 0$ to 7 // Initialize working variables
        $a_j = H_j^{(i-1)}$
    for $t = 0$ to $r - 1$ // Hashing rounds
        if $(t < 16)$ $W_t = M_t^{(i)}$ // Message schedule
        else $W_t = \sigma_1^{\{d\}}(W_{t-2}) + W_{t-7} + \sigma_0^{\{d\}}(W_{t-15}) + W_{t-16}$
        $T = a_7 + \Sigma_1^{(d)}(a_4) + \mathrm{Ch}(a_0, a_5, a_6) + K_t + W_t$
        for $j = 7$ downto 1 // Shift words
            $a_j = a_j - 1$
        $a_0 = \Sigma_0^{(d)}(a_0) + \mathrm{Maj}(a_0, a_1, a_2) + T$ // Hash word 0
        $a_4 = a_4 + T$ // Hash word 4
    for $j = 0$ to 7
        $H_j^{(i)} = H_j^{(i-1)} + a_j$ // Update hash after block i
$H = \left\{H_0^{(N)}\ H_1^{(N)}\ H_2^{(N)}\ H_3^{(N)}\ H_4^{(N)}\ H_5^{(N)}\ H_6^{(N)}\ H_7^{(N)}\right\}$ // Final D-bit hash

Fig. 18.14 SHA-2 algorithm.

For the sake of brevity and to emphasize the similarities between SHA-256 and SHA-512, the SHA-2 algorithm in Fig. 18.14 is described with notation that is simpler than but equivalent to that of [NIST-FIPS-180-4]. The working variables are called $a_{7:0}$ rather than $\{a, \ldots, h\}$. T_1 is called T and T_2 is folded into the a_4 equation. The rounds are also called iterations.

collision-resistant hashes. After all the rounds are complete, the 8 words of the hash for block i are computed as the sum of the working variables and the words of the hash for block i–1. After all blocks are processed, the message digest is the hash for the final block. This diffusion and compression block for a_j is sometimes called a compression function and is visualized in Fig. 18.15.

Most Linux distributions come with programs to compute the SHA-256 hash function to verify data integrity. For example, the command at the top of page 738 produces the hash for "Hello World!" using SHA-256. If somebody changed the exclamation point to a question mark, `sha256sum` would give a different hash, revealing that the message had been corrupted.

You can interactively watch SHA-256 run at sha256algorithm.com.

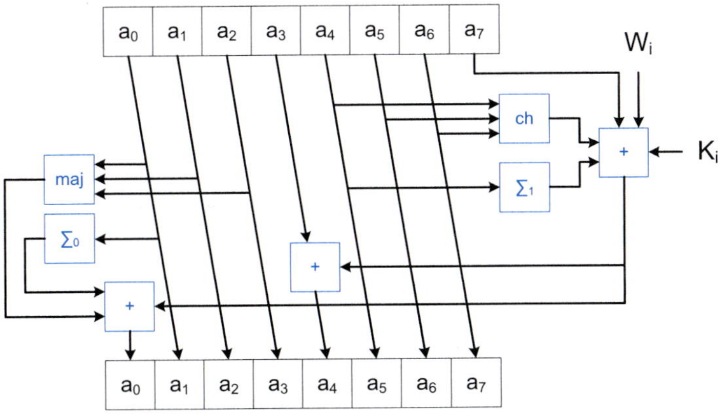

Fig. 18.15 SHA256 compression function.

```
$ echo -n "Hello World!" | sha256sum
7f83b1657ff1fc53b92dc18148a1d65dfc2d4b1fa3d677284
addd200126d9069 -
```

Scalar cryptography instructions operate on the ordinary register file. They provide lightweight cryptography acceleration practical even for low-cost microcontrollers. Vector cryptography instructions are faster because they can process multiple elements concurrently from the wider vector register file. The RISC-V vector extensions are outside the scope of this book, but the crypto instructions are similar.

RISC-V also defines the Zksed and Zksh extensions for the Shāngmì SM4 block cipher and SM3 hash functions. These are primarily of interest to markets where the Chinese government requires these algorithms; they are not described in this chapter. The Chinese Cybersecurity Law established a legal obligation for internet operators to turn over data to the government [Laskai21].

18.2 RISC-V PRACTICES

The RISC-V base integer instruction set (RV{32/64}I) has a minimal set of instructions sufficient to write any program, but bit manipulation instructions make programs faster and smaller. RISC-V has ratified the four separate bit manipulation extensions listed in Table 18.8. A configuration may support any or all of these extensions. For example, a microcontroller might support Zba, Zbb, and Zbs but omit Zbc to save cost if it rarely performs cryptographic operations. The B extension is shorthand for Zba + Zbb + Zbs. RV64 adds W64-type operations for Zba and Zbb, which act on only the lower 32 bits of each operand. The instructions are listed in Table I.23 (on the last pages of the book).

RISC-V has also ratified various scalar cryptography extensions listed in Table 18.9. The Zbk* family of instructions performs bit manipulations for cryptography and partially overlaps with the Zb* extensions. The Zkn* family of instructions performs key steps in AES decryption and encryption and SHA-2 hashing. The Zkn extension is shorthand for all these instructions, collectively called NIST cryptography. If instructions have data-dependent delay, hackers can use *timing attacks* to infer information about a cryptographic key. Therefore, the Zkt extension specifies that a list of simple and common instructions execute in data-independent time to prevent such attacks; this list excludes division and remainder and is automatically satisfied by most microarchitectures, including Wally. The Zkr extension defines a `seed` CSR that provides entropy bits to initialize a cryptographic random number generator. The Zk extension is shorthand for Zkn + Zkt + Zkr, which provides all the features that cryptographic library code would need.

18.2.1 Zba: Address Generation

To access an array, a program calculates an element's address by adding the array's base address to the element index multiplied by the element size. The Zba extension makes accessing halfwords, words, and double words more efficient by incorporating multiplications by 2, 4, and 8

Table 18.8 Bit manipulation extensions.

Extension	Name	Summary
Zba	Address generation	Shifts and add to access 2/4/8-byte elements
Zbb	Basic bit manipulation	Min/max, rotations, counting, sign/zero extension, etc.
Zbc	Carry-less multiplication	For Galois-field operations such as cryptography
Zbs	Single-bit instructions	Set, clear, invert, or extract single bits
B	Bit manipulation	Zba + Zbb + Zbs

Table 18.9 Scalar cryptography extensions.

Extension	Name	Summary
Zbkb	Bit manipulation	Bit manipulation instructions for cryptography
Zbkc	Carry-less multiplication	Subset of Zbc
Zbkx	Crossbar permutations	Useful for implementing S-boxes
Zknd	Decryption/key scheduling	Functions of the AES block cipher for decryption
Zkne	Encryption/key scheduling	Functions of the AES block cipher for encryption
Zknh	Hash functions	For SHA-2 family of cryptographic hash
Zkn	NIST crypto instructions	Zbkb + Zbkc + Zbkx + Zknd + Zkne + Zknh
Zkt	Data-independent latency	Instructions execute in data-independent time
Zkr	Entropy source	For random number generation
Zk	Standard scalar crypto	Zkn + Zkt + Zkr

into an add instruction. Code Example 18.2 shows how the `sh2add` (shift left by 2 and add) instruction saves an instruction in the common case of accessing an array element.

Code Example 18.2 ADDRESS GENERATION

C Program

```
int i, b;   # s0, s1
int a[10]; # Base address in s2

b = a[i];
```

Assembly Without Zba

```
slli t0, s0, 2  # i * 4
addi t0, s2, t0 # &a[i]
lw s1, 0(t0)    # b = a[i]
```

Assembly With Zba

```
sh2add t0, s0, s2 # &a[i]
                  # t0=s2+s0<<2
lw s1, 0(t0)      # b = a[i]
```

The `.uw` instructions for RV64 are helpful for generating code for the 32-bit `unsigned int` data type. In Code Example 18.2, if `i` is an `unsigned int`, then the RV64 assembly language code becomes `sh2add.uw t0, s0, s2`. Code Example 18.3 illustrates how `add.uw` simplifies zero extension of a 32-bit `unsigned int`.

Code Example 18.3 ZERO EXTENSION OF UNSIGNED INT

C Program

```
long         a, y; // s0, s2
unsigned int b;    // s1

y = a + b;
```

Assembly Without Zba

```
slli s1, s1, 32
srli s1, s1, 32 # Zero ext.
add  s2, s0, s1
```

Assembly With Zba

```
add.uw s2, s1, s0
```

18.2.2 Zbb: Basic Bit Manipulation

Zbb provides a miscellaneous set of widely used bit manipulation operations.

18.2.2.1 Rotates

roli is unnecessary because it is equivalent to rori with a complementary version of the immediate. For example, roli t0, s0, 1 can be performed as rori t0, s0, 31.

Rotate operations (rol, ror, rori) circularly shift an input left or right and fill the empty positions with the bits shifted off the end. For example:

```
rori 0000_1111_0000_0000_0000_0000_1000_0100, 4
  = 0100_0000_1111_0000_0000_0000_0000_1000
```

C does not have a rotate operator, but rotations can be performed by merging a pair of logical shifts. Code Example 18.4 shows C code and its assembly equivalent without and with the rotate instruction.

Code Example 18.4 LEFT ROTATE (RV32)

C Program

```
# Rotate left by n
int rol(unsigned int a, int n) {
  return (a << n) | (a >> (32-n));
}
```

Assembly Without Zbb

```
sll t0, a0, a1   # a << n
sub t1, zero, t1 # -n
srl t1, a0, t1   # a >> 32-n
or  a0, t0, t1   # rol
ret
```

Assembly With Zbb

```
rol a0, a0, a1



ret
```

The assembly language code in Code Example 18.4 optimizes the shift of $32-n$ to a shift of just $-n$ because RV32 shift amounts are interpreted modulo 32.

Rotations show up in many algorithms from discrete mathematics, including hashing, cryptography, and cyclic redundancy checks (CRCs) used in error correction.

18.2.2.2 Inverted Operations

Inverted operations (andn, orn, xnor) perform the base operation (and, or, xor) with the second operand inverted. This is essentially free in hardware because subtraction already requires the ability to invert the second operand.

For example, andn *masks* off the asserted bits, clearing them to zero.

```
andn 1010_1111_1101_0000, 0000_0000_1110_0000 = 1010_1111_0001_0000
```

18.2.2.3 Max/Min

Maximum and minimum operations have signed and unsigned flavors. For example,

```
max  0xFE000000, 0x12345678 = 0x12345678 # Signed
maxu 0xFE000000, 0x12345678 = 0xFE000000 # Unsigned
```

Code Example 18.4 shows C code and its assembly equivalent without and with the max instruction.

Code Example 18.5 MAX

C Program	Assembly Without Zbb	Assembly With Zbb

`int max(int m, int n) {`
`  return m > n ? m : n;`
`}`
	`    blt a1, a0, skip`
	`    mv  a0, a1`
	`skip:`
	`    ret`
		`max a0, a0, a1`
		
		`ret`

18.2.2.4 Extension

The extension operations (`zext.h`, `sext.h`, and `sext.b`) zero- or sign-extend half-words or bytes (16- or 8-bit values). For example, the following code sign-extends the least significant byte.

```
sext.b  0000_1111_0000_0000_0000_0000_1000_0100
       =1111_1111_1111_1111_1111_1111_1000_0100
```

Code Example 18.6 shows C code to add two signed 16-bit short integers. In RV32, the upper 16 bits of the 32-bit register are copies of the sign bit. Ordinary 32-bit addition can overflow into the 17th bit, producing the wrong answer. Hence, the addition is followed by a sign extension operation. 16-bit sign extension can be done with a pair of shifts or the `sext.h` instruction.

Other sign or zero extensions can be performed using pseudoinstructions:

```
# Zero-extend the LSB
zext.b rd, rs1 ->
andi   rd, rs1, 255
# Zero-extend the lower
#32 bits
zext.w rd, rs1 ->
add.uw rd, rs1, 0
# Sign-extend the lower
# 32 bits
sext.w rd, rs1 ->
addiw  rd, rs1, 0
```

Code Example 18.6 EXTENSION

C Program	Assembly Without Zbb	Assembly With Zbb

`short sum(short m, short n) {`
`  return m + n;`
`}`
	`add  a0, a0, a1`
	`slli a0, a0, 16`
	`srai a0, a0, 16`
	`ret`
		`add    a0, a0, a1`
		`sext.h a0, a0`
		
		`ret`

18.2.2.5 Counting, Combining, Reversing

`cpop` (population count) counts the number of ones in a register. For example,

```
cpop 0000_1111_0000_0000_0000_0000_1000_0100 = 6
```

`clz` (count leading zeros) and `ctz` (count trailing zeros) count the number of zeros at the most significant or least significant end of a register. For example, cpop has many applications, such as counting 1s in a bitstream, computing Hamming distance, and implementing associative arrays with an efficient data structure called *hashed array mapped trie (HAMT)* [Bagwell00]. Computing Hamming distance is important for cryptanalysis, and the National Security Agency (NSA) reportedly asked for the cpop instruction in early supercomputers.

```
clz 0000_1111_0000_0000_0000_0000_1000_0100 = 4
ctz 0000_1111_0000_0000_0000_0000_1000_0100 = 2
```

orc.b (OR combine bytes) checks each byte in a register. If the byte is nonzero, it sets the corresponding result byte to all ones. Otherwise, the result byte remains all zeros. For example,

```
orc.b 00001111_00000000_00000000_10000100
    = 11111111_00000000_00000000_11111111
```

rev8 (reverse bytes) reverses the bytes in a register. For example,

```
rev8  00001111_00000000_00000000_10000100
    = 10000100_00000000_00000000_00001111
```

Cryptography uses the rev8 byte reversal instruction described in Section 18.2.2.5 to convert between big and little-endian formats for input and output data.

orc.b is helpful to find a NULL (0) byte for string manipulation functions. rev8 is helpful to reverse the endianness. See the RISC-V unprivileged specification for example code using these instructions to find the length of a string and to compare two strings (using functions strlen and strcmp).

C does not have any of these operators. GCC currently does not infer them from C code either, but it does produce them using built-in counting functions:

```
int __builtin_popcount(unsigned int x);
int __builtin_clz(unsigned int x);
int __builtin_ctz(unsigned int x);
```

18.2.3 Zbc: Carry-less Multiplication

```
       0111
     x 1010
   --------
       0000
      0111
     0000
  + 0111
   --------
    0110110
```

Fig. 18.16 Carry-less multiplication example.

Carry-less multiplication produces partial products like regular multiplication but sums each column using XOR, discarding carries out of each column. Fig. 18.16 shows an example of 4 × 4-bit carry-less multiplication. In the blue column, the answer is $0 \oplus 1 \oplus 0 \oplus 1 = 0$, with no carry out to the next column.

Carry-less multiplication is a building block for multiplication in the Galois field GF(2^n). It is used in a variety of algorithms, including CRC, Galois/Counter Mode symmetric key encryption, and fast universal hash functions.

The @ symbol indicates carry-less multiplication in Table I.23. clmul and clmulh produce the low and high half of the 2XLEN-bit carry-less product. clmulr (carry-less multiply reversed) is used in CRC calculations.

18.2.4 Zbs: Single-Bit Operations

Zbs provides instructions to set, clear, invert, and extract a bit within a register. `bset` sets the bit to 1. `bclr` clears the bit to 0. `binv` inverts the bit. `bext` produces a result that is either 0 or 1, matching the specified bit in the source register. These instructions are often used to modify bitfields within a word. In memory-mapped I/O, they are used to turn on and off control bits or to check the value of a status bit. For example:

```
bset 0000_1111_0000_0000_0000_0000_1000_0100, 5
   = 0000_1111_0000_0000_0000_0000_1010_0100
bclr 0000_1111_0000_0000_0000_0000_1000_0100, 2
   = 0000_1111_0000_0000_0000_0000_1000_0000
binv 0000_1111_0000_0000_0000_0000_1000_0100, 1
   = 0000_1111_0000_0000_0000_0000_1000_0110
bext 0000_1111_0000_0000_0000_0000_1000_0100, 3
   = 0
```

Code Examples 18.7 and 18.8 show that the assembly language code for bit manipulation is shorter with Zbs.

Code Example 18.7 SINGLE-BIT OPERATIONS

C Program

```
// Clear, invert, set, extract
// a, n in a0, a1
// b, c, d, e in s0, s1, s2, s3
  b = a & ~(1 << n); // Clear

  c = a ^  (1 << n); // Invert
  d = a |  (1 << n); // Set
  e = (a >> n) & 1;  // Extract
```

Assembly Without Zbs

```
li   t0, 1        # t0 = 1
sll  t0, t0, a1   # t0 = 1 << n

not  t1, t0       # t1 = ~(1 << n)
and  s0, a0, t1   # b = a & ~(1 << n)
xor  s1, a0, t0   # c = a ^  (1 << n)
or   s2, a0, t0   # d = a |  (1 << n
srl  s3, a0, a1   # a >> n
andi s3, a3, 1    # (a >> n) & 1
```

Assembly With Zbs

```



bclr s0, a0, a1

binv s1, a0, a1
bset s2, a0, a1
bext s3, a0, a1
```

Code Example 18.8 BIT CLEAR OPERATION

C Program

```
int bitclear(int a, int n) {
  return a & ~(1 << n);
}
```

Assembly Without Zbs

```
li  t0, 1         # t0 = 1
sll t0, t0, a1    # t0 = 1 << n
not t0, t0        # t0 = ~(1 << n)
and a0, a0, t0    # b = a & ~(1 << n)
ret
```

Assembly With Zbs

```
bclr a0, a0, a1




ret
```

18.2.5 Zbkb: Bit Manipulation for Cryptography

Zbkb provides bit manipulation instructions specifically for cryptography. In addition to crypto-related instructions also available in Zbb, it provides `pack`, `packh`, `packw`, `brev8`, `zip`, and `unzip`.

18.2.5.1 Packing

Packing instructions place the bottom portions of two source register into a destination register. This is useful for rearranging small data elements within wider registers. For example, AES uses packing when loading keys that are not word aligned. Code Example 18.9 shows the assembly language code for the `pack` function that stores low halves of registers into a destination. The code is shorter with Zbkb.

- `pack` places the bottom half of `rs1` and `rs2` into the bottom and top half of `rd`, respectively.
- `packh` places the bottom byte of `rs1` and `rs2` into a halfword in `rd`.
- `packw` (RV64Zbkb only) places the bottom halfword of `rs1` and `rs2` into the lower word of `rd`.

```
pack  0x01234567, 0x89ABCDEF = 0xCDEF4567 (RV32)
packh 0x01234567, 0x89ABCDEF = 0x0000EF67
```

Code Example 18.9 PACK OPERATION

C Program

```
unsigned int pack(unsigned int a,
                  unsigned int b) {
  return ((a << 16) >> 16) | (b << 16);
}
```

Assembly Without Zbkb

```
slli  a0, a0, 16 # a  << 16
srli  a0, a0, 16 # () >> 16
slli  a1, a1, 16 # b  << 16
or    a0, a1, a0 # lo | hi
ret
```

Assembly With Zbkb

```
pack a0, a0, a1

ret
```

18.2.5.2 Bit Reversal

The `brev8` instruction reverses the bits within each byte of a register. This is a step in the GHASH (Galois Hash) authentication function for AES encryption and, without `brev8`, this step is painful to implement in software.

```
brev8 00001111_10101010_00110011_01111111
    = 11110000_01010101_11001100_11111110
```

18.2.5.3 Bit Interleaving

64-bit rotations are difficult to implement with 32-bit arithmetic. Various algorithms including SHA-3 require such 64-bit rotations. Hence, RV32Zbkb provides `zip` and `unzip` instructions to facilitate longer rotations. `zip` interleaves bits (like a zipper interleaves its two sides) by placing the bits in the low half of a word into the even bits of the destination and

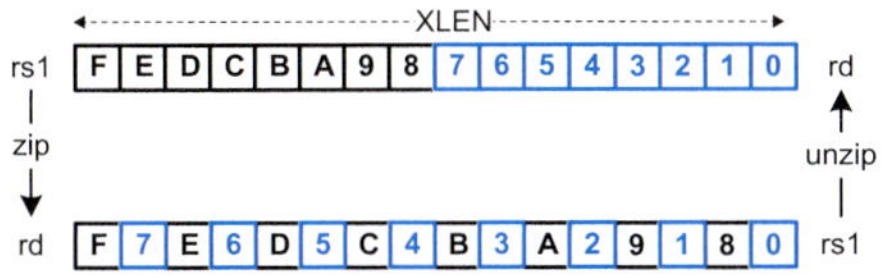

Fig. 18.17 zip and unzip instructions showing interleaving of bit positions for XLEN = 16.

places the bits in the high half into the odd bits. unzip does the inverse, as shown in Fig. 18.17.

```
zip   00001111_10101010_00110011_01111111
    = 00000101_10101111_10011101_11011101
unzip 00001111_10101010_00110011_01111111
    = 00111111_01010111_00110000_01011111
```

For example, an RV32 SHA-3 hash implementation uses unzip and zip to transform 64-bit rotations into pairs of 32-bit rotations on the even and odd halves [github.com/rvkrypto/rvkrypto-fips/blob/main/sha3/sha3_f1600_rvb32.c].

18.2.6 Zbkx: Crossbar Permutations

Crossbar permutations are used to select nibbles or bytes from one register based on indices stored in another register. For example, xperm4 uses each nibble of rs2 as an index to select a nibble from rs1. The xperm4 example below shows rs2 selecting the 5th nibble of rs1 (2) as the least significant nibble in rd, as highlighted in blue. If the index is out of bounds, the permutation returns 0. For example, the A index in the xperm4 example below is out of bounds, so the corresponding result nibble is 0, as highlighted in gray. xperm8 is similar but selects bytes.

```
xperm4 0xE3217DB3, 0x71023A45 = 0xEB3D7012
xperm8 0xAA_BB_CC_DD, 0x01_00_72_03 = 0xCC_DD_00_AA
```

Crossbar permutations are useful for implementing S-boxes that create nonlinearities in many symmetric key cryptosystems, including AES. The substitution could be done with a table lookup from memory, but cache systems introduce data-dependent delays that can leak information about the key.

Code Example 18.10 shows an example using xperm8 to implement an arbitrary S-box in software with a data-independent delay. It processes eight byte-wide inputs in parallel packed into a 64-bit word and produces eight outputs in the same format. It uses sbox, a 32-element array of 64-bit words that contains the 256-entry × 1-byte S-box lookup table. Whereas the mask is used to step through the upper 5 bits of each byte, xperm8 uses the lower 3 bits if the upper 5 bits match the mask. None of the memory accesses depend on the input, so the delay is independent of the input.

Define an *Sbox* operation that takes an 8-byte input and produces an 8-byte output that contains 8 AES substitution-boxes (S-boxes) that operate on the individual bytes.

Define a *MixColumns* operation that takes a 4-byte input column and produces the corresponding 4-byte output column by applying the MixColumns transformation matrix. Recall that MixColumns multiplies elements by 01, 02, or 03 using GF(2^8). Recall from Section 18.1.1.2 that multiplying by 01 has no effect. Multiplying by 02 (0b10) is a multiply by x (which is a left shift, followed by an XOR for overflow). Multiplying by 03 (0b11) is a multiply by $(x + 1)$, which is the XOR of the original value and the value multiplied by x. Addition is XOR in this field. If the column vector is $z = [z_3\ z_2\ z_1\ z_0]^T$, MixColumns produces:

$$\text{MixColumns}(z) = \begin{bmatrix} 02 & 03 & 01 & 01 \\ 01 & 02 & 03 & 01 \\ 01 & 01 & 02 & 03 \\ 03 & 01 & 01 & 02 \end{bmatrix} \begin{bmatrix} z_0 \\ z_1 \\ z_2 \\ z_3 \end{bmatrix}$$

$$= \begin{bmatrix} 02z_0 + 03z_1 + 01z_2 + 01z_3 \\ 01z_0 + 02z_1 + 03z_2 + 01z_3 \\ 01z_0 + 01z_1 + 02z_2 + 03z_3 \\ 03z_0 + 01z_1 + 01z_2 + 02z_3 \end{bmatrix} = \begin{bmatrix} xz_0 \oplus xz_1 \oplus z_1 \oplus z_2 \oplus z_3 \\ z_0 \oplus xz_1 \oplus xz_2 \oplus z_2 \oplus z_3 \\ z_0 \oplus z_1 \oplus xz_2 \oplus xz_3 \oplus z_3 \\ xz_0 \oplus z_0 \oplus z_1 \oplus z_2 \oplus xz_3 \end{bmatrix} \tag{18.3}$$

Finally define the `aes64esm` instruction, which takes two 64-bit inputs and produces a 64-bit output:

`aes64esm` (x, y):

w = ShiftRows(x, y) // Wires permute inputs

s = Sbox(w) // S-box operations on all 8 bytes in parallel

// GF(2^8) multiplications and XORs on two columns in parallel

return $t = \{\text{MixColumns}(s_{63:32}), \text{MixColumns}(s_{31:0})\}$

If x and y contain the first and second pair of columns, respectively, w contains the first pair of columns after ShiftRows. s contains this pair of columns after SubBytes. t contains the pair of columns after MixColumns is applied to each.

Code Example 18.11 shows code for an AES encryption round using `aes64esm`. The state matrix s is stored in {`a1`, `a0`} in column-major form. The RoundKey is in {`a3`, `a2`}. AddRoundKey uses two XOR instructions to update the state registers into {`t1`, `t0`}. Then two `aes64esm` instructions perform the ShiftRows, SubBytes, and MixColumns operations, putting the new state back into {`a1`, `a0`}. Observe how the `aes64esm` source registers, `t0` and `t1`, are simply reversed to select the pair of columns.

Code Example 18.11 RV64Zkne AES ENCRYPTION ROUND

```
AES_ENCRYPT_ROUND:
    xor         t0, a0, a2    // AddRoundKey columns 0, 1
    xor         t1, a1, a3    // AddRoundKey columns 2, 3
    aes64esm    a0, t0, t1    // Rest of round for columns 0, 1
    aes64esm    a1, t1, t0    // Rest of round for columns 2, 3
```

Code Example 18.11 is adapted from a RISC-V AES encryption library at github.com/riscv/riscv-crypto/blob/main/benchmarks/aes/zscrypto_rv64/aes_enc.S.

18.2.7.2 RV32Zkne: 32-bit AES Encryption

RV32 holds the 128-bit state matrix s in four 32-bit registers, as shown in Fig. 18.19 and as expressed in vector form next. Encryption is performed byte by byte because RV32Zkne instructions with two source registers are not able to provide the entire state matrix at once as was done in RV64Zkne.

$a = [s_{3,0}\ \ s_{2,0}\ \ s_{1,0}\ \ s_{0,0}]$ // a holds column 0 of s
$b = [s_{3,0}\ \ s_{2,0}\ \ s_{1,0}\ \ s_{0,0}]$ // b holds column 1 of s
$c = [s_{3,2}\ \ s_{2,2}\ \ s_{1,2}\ \ s_{0,2}]$ // c holds column 2 of s
$d = [s_{3,3}\ \ s_{2,3}\ \ s_{1,3}\ \ s_{0,3}]$ // d holds column 3 of s

Define the `aes32esmi` instruction, which takes two 32-bit inputs and a 2-bit byte-select bs. The first input x is initialized as the key and transformed into a new column of the state matrix 1 byte at a time using four `aes32esmi` instructions. The second input y is an old column of the state

a_0 $s_{0,0}$	b_0 $s_{0,1}$	c_0 $s_{0,2}$	d_0 $s_{0,3}$
a_1 $s_{1,0}$	b_1 $s_{1,1}$	c_1 $s_{1,2}$	d_1 $s_{1,3}$
a_2 $s_{2,0}$	b_2 $s_{2,1}$	c_2 $s_{2,2}$	d_2 $s_{2,3}$
a_3 $s_{3,0}$	b_3 $s_{3,1}$	c_3 $s_{3,2}$	d_3 $s_{3,3}$

Fig. 18.19 Packing the state matrix into four 32-bit registers.

matrix. The instruction selects one of the four byte-wide elements of the column and applies the S-box for SubBytes. It then multiplies by [3 1 1 2] in GF(2^8), and rotates the result left by *bs* to implement the selected row of the MixColumns matrix. Finally, it XORs with *x* to accumulate the rows of the matrix and complete the AddRoundKey step.

`aes64esm`(*x*, *y*, bs):

$si = y[bs]$	// Select one of the four elements of *y*
$s = sbox(si)$	// S-box on one element
$col = \{xs \oplus s, s, s, xs\}$	// One row of MixColumns matrix
mixed = *col* rol *bs*	// Rotate *col* into proper place
$return\ t = x \oplus mixed$	

Code Example 18.12 shows code for an AES encryption round using `aes32esmi`. The state matrix is stored in {`a3`, `a2`, `a1`, `a0`}. The RoundKey is in {`t3`, `t2`, `t1`, `t0`}. Then 16 `aes32esmi` instructions perform the AES round AddRoundKey, ShiftRows, SubBytes and MixColumns steps, putting the results back into {`t3`, `t2`, `t1`, `t0`}. Observe how the second argument varies to perform ShiftRows, and the byte select iterates from 0 to 3 to handle each byte in a column. The final round, not shown, uses the `aes32esi` instruction instead, which omits the MixColumns step.

Code Example 18.12 RV32Zkne AES ENCRYPTION ROUND

```
AES_ENCRYPT_ROUND:
    aes32esmi   t0, t0, a0, 0
    aes32esmi   t0, t0, a1, 1
    aes32esmi   t0, t0, a2, 2
    aes32esmi   t0, t0, a3, 3
    aes32esmi   t1, t1, a1, 0
    aes32esmi   t1, t1, a2, 1
    aes32esmi   t1, t1, a3, 2
    aes32esmi   t1, t1, a0, 3
    aes32esmi   t2, t2, a2, 0
    aes32esmi   t2, t2, a3, 1
    aes32esmi   t2, t2, a0, 2
    aes32esmi   t2, t2, a1, 3
    aes32esmi   t3, t3, a3, 0
    aes32esmi   t3, t3, a0, 1
    aes32esmi   t3, t3, a1, 2
    aes32esmi   t3, t3, a2, 3
```

18.2.7.3 RV{32/64}Zknd: AES Decryption

The inverse of AES encryption is decryption, which uses `aes32dsmi`, `aes32dsi`, `aes64dsm`, and `aes64ds` instructions. These instructions apply inverse operations, in the same order as encryption. They use an inverse S-box and the InvMixColumns matrix from Fig. 18.13 (or the [b d 9 e] row for aes32dsmi). ShiftRows is inverted by shifting in the opposite direction.

18.2.7.4 RV64Zkn{e/d} AES Key Scheduling

Recall from Fig. 18.10 that RoundKeys are generated using a sequence of rotations, S-box substitutions, and XORs. RV32 does this in software using RISC-V rori and xor instructions along with aes32esi for an S-box.

RV64Zkn{e/d} introduce aes64ks1i, aesks2, and aes64im instructions to accelerate key scheduling by processing two 32-bit words at a time in a 64-bit register and applying four S-boxes in parallel. See the ISA specification and [github.com/riscv/riscv-crypto/blob/main/benchmarks/aes/zscrypto_rv64/aes_128_ks.S] for details.

18.2.8 Zknh: SHA-2 Instructions

SHA-2 is another common and computationally intensive operation. Recall from Section 18.1.3 that each round applies the σ_0 and σ_1 functions to create the message schedule and the Σ_0 and Σ_1 functions to perform the hash. These functions are defined using rotations, shifts, and XOR in Table 18.7. Recall that whereas SHA-256 uses 32-bit words, SHA-512 uses 64-bit words, and that the four functions also differ accordingly. The RISC-V Zknh extension provides instructions to accelerate these functions.

For SHA-256, the instructions are called sha256sig0, sha256sig1, sha256sum0, and sha256sum1, with sig corresponding to σ and sum corresponding to Σ. Each instruction takes a register containing a 32-bit value and applies the rotations, shifts, and XORs to produce a 32-bit output. RV64Zknh sign-extends the results to fill the upper 32 bits, but these bits are ignored during hashing.

sha256sig0: rd = SignExt($\sigma_0^{(256)}(rs1_{31:0})$) = (rs1 ror 7) ⊕(rs1 ror 18) ⊕(rs1 >> 3)
sha256sig1: rd = SignExt($\sigma_1^{(256)}(rs1_{31:0})$) = (rs1 ror 17) ⊕(rs1 ror 19) ⊕(rs1 >> 10)
sha256sum0: rd = SignExt($\Sigma_0^{(256)}(rs1_{31:0})$) = (rs1 ror 2) ⊕(rs1 ror 13) ⊕(rs1 ror 22)
sha256sum1: rd = SignExt($\Sigma_1^{(256)}(rs1_{31:0})$) = (rs1 ror 6) ⊕(rs1 ror 11) ⊕(rs1 ror 25)

In RV64Zknh, SHA-512 uses a similar set of instructions: sha512sig0, sha512sig1, sha512sum0, and sha512sum1. They operate on 64-bit words and use the SHA-512 rotates and shifts.

sha512sig0: rd = $\sigma_0^{(512)}(rs1)$ = (rs1 ror 1) ⊕(rs1 ror 8) ⊕(rs1 >> 7)
sha512sig1: rd = $\sigma_1^{(512)}(rs1)$ = (rs1 ror 19) ⊕(rs1 ror 61) ⊕(rs1 >> 6)
sha512sum0: rd = $\Sigma_0^{(512)}(rs1)$ = (rs1 ror 28) ⊕(rs1 ror 34) ⊕(rs1 ror 39)
sha512sum1: rd = $\Sigma_1^{(512)}(rs1)$ = (rs1 ror 14) ⊕(rs1 ror 18) ⊕(rs1 ror 41)

Code Example 18.13 is adapted from: github.com/riscv/riscv-crypto/blob/main/benchmarks/sha512/zscrypto_rv64/sha512.c.

In RV32Zknh, SHA-512 is somewhat more difficult because the 64-bit words do not fit in a single register. Instead, the words are kept in a pair of registers, and a pair of instructions are needed to compute the upper and lower half of each function.

Code Example 18.13 shows how a SHA-512 library implements the round and schedule operations with the RV64Zknh instructions. Note how C invokes the assembly language instructions.

Code Example 18.13 SHA-512 ROUND AND SCHEDULE OPERATIONS

```
static inline uint_xlen_t _sha512sig0 (uint_xlen_t rs1) {uint_xlen_t rd; __asm__
("sha512sig0  %0, %1" :"=r"(rd) : "r"(rs1)); return rd;}
...

#define ROR64(X,Y) ((X>>Y) | (X << (64-Y)))
#define SHR64(X,Y) ((X>>Y)               )
#define CH(X,Y,Z)  ((X&Y)^(~X&Z))
#define MAJ(X,Y,Z) ((X&Y)^(X&Z)^(Y&Z))
#define SUM_0(X)   (_sha512sum0(X))
#define SUM_1(X)   (_sha512sum1(X))
#define SIGMA_0(X) (_sha512sig0(X))
#define SIGMA_1(X) (_sha512sig1(X))

#define ROUND(A, B, C, D, E, F, G, H, K, W) { \
     H  = H + SUM_1(E) + CH(E,F,G) + K + W    ; \
     D  = D + H                               ; \
     H  = H + SUM_0(A) + MAJ(A, B, C)           ; \
}

#define SCHEDULE(M0, M1, M9, ME) { \
     M0 = SIGMA_1(ME) + M9 + SIGMA_0(M1) + M0; \
}
```

18.3 TEST PLAN

The bit manipulation and cryptographic instructions are tested with the riscv-arch-test suite. The configuration files to generate these random tests are similar to the one described in Section 7.4. Tables 18.10 and 18.11 list the number of test cases for each instruction.

18.4 WALLY IMPLEMENTATION

Wally extends the integer ALU to support cryptography and bit manipulation operations, as highlighted in Fig. 18.20.

18.4.1 ALU Modifications for Bit Manipulation

The `ZBA_SUPPORTED`, `ZBB_SUPPORTED`, `ZBC_SUPPORTED`, and `ZBS_SUPPORTED` configuration options determine which bit manipulation instructions are

Table 18.10 riscv-arch-test for Zb* extensions.

Instruction	RV32B Tests	RV64B Tests
add.uw	-	701
andn	158	705
bclr	292	421
bclri	291	158
bext	291	419
bexti	158	158
binv	291	421
binvi	158	158
bset	291	419
bseti	158	158
clmul	162	288
clmulh	162	288
clmulr	162	288
clz	68	132
clzw	-	132
cpop	68	132
cpopw	-	132
ctz	68	132
ctzw	-	132
max	572	700
maxu	572	700
min	572	700
minu	572	700
orc.b	72	136
orn	592	720
rev8	154	155
rol	269	525
rolw	-	525
ror	269	525
rori	158	158
roriw	-	158
rorw	-	525
sext.b	87	151
sext.h	88	152
sh1add	572	700
sh1add.uw	-	700
sh2add	572	700
sh2add.uw	-	700
sh3add	572	700
sh3add.uw	-	700
slli.uw	-	161
xnor	592	720
zext.h	88	152
Total	**9,151**	**17,137**

Table 18.11 riscv-arch-test for Zk* extensions.

Instruction	RV32Zk Tests	RV64Zk Tests
aes32dsi	281	-
aes32dsmi	281	-
aes32esi	281	
aes64ds	-	57
aes64dsm	-	57
aes64es	-	57
aes64esm	-	57
aes64im	-	150
aes64ks1i	-	86
aes64ks2	-	280
brev8	154	136
pack	269	525
packh	269	525
packw	-	525
sha256sig0	86	150
sha256sig1	86	150
sha256sum0	86	150
sha256sum1	86	150
sha512sig0	-	150
sha512sig0h	153	-
sha512sig0l	153	-
sha512sig1	-	150
sha512sig1h	153	-
sha512sig1l	153	-
sha512sum0	-	150
sha512sum0r	153	-
sha512sum1	-	150
sha512sum1r	153	-
unzip	132	-
xperm4	289	545
xperm8	289	395
zip	132	-
Total	**3,639**	**4,595**

supported, if any. The Wally rv32gc and rv64gc configurations support all four flavors, and the other configurations support none.

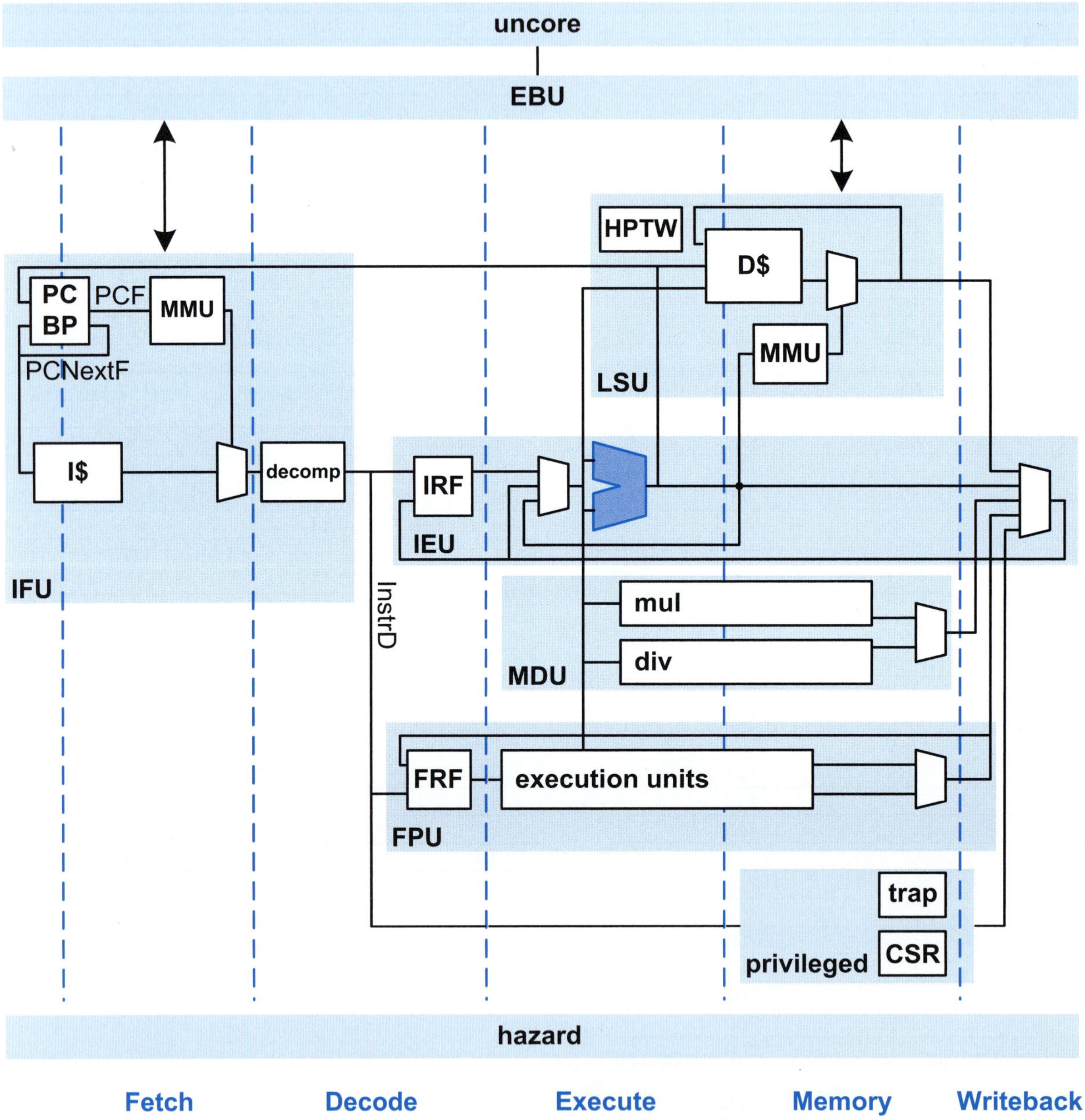

Fig. 18.20 RV{32/64}GC high-level block diagram emphasizing the ALU with bit manipulation extensions.

Fig. 18.21 shows a diagram of the ALU with optional enhancements for bit manipulation in the `bitmanipalu` module, highlighted in blue. The major changes are:

- **Zba:** Preshifter to shift `A` left by 0, 1, 2, or 3 bits. For RV64 (cross-hatched), conditionally zero-extend `A` for `.uw` instructions.
- **Zbb:** Block for min/max, counting, extension, combining, and byte reversal. The shifter is also enhanced to support rotations.
- **Zbc:** Block for carry-less multiplication.
- **Zbs:** Decoder to produce one-hot mask from `B`. Reduction OR block for bit extract operations.

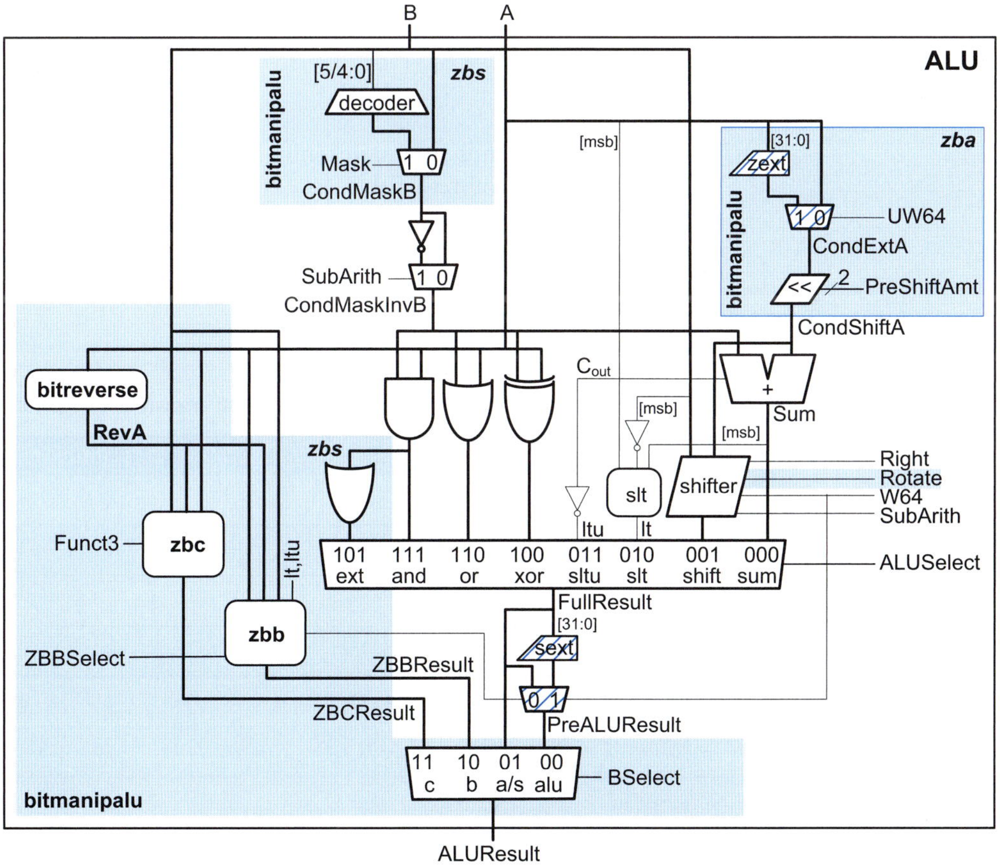

Fig. 18.21 ALU, with optional enhancements for bit manipulation in blue boxes.

Verilog lacks a graceful idiom to reverse bits in a bus. The least bad way is with a generate loop.

```
for (i=0; i<WIDTH; i++)
  assign RevA[WIDTH-i-1]
  = A[i];
```

When any of the bit manipulation extensions are supported, the ALU uses a `BSelect` multiplexer to choose `ALUResult` from the regular ALU (`PreALUResult`) or from the bit manipulation modules. The `Rotate` input to the shifter (`shift`) is only used by the `zbb` unit. The `bitreverse` module reverses the bits in `A` to produce `RevA` by using wires only, not gates. Bit reversal is used for counting trailing zeros in Zbb and for carry-less multiplication in Zbc.

18.4.1.1 Zba

The Zba extension adds a preshifter and zero-extender on `A` to support shift-and-add and `.uw` instructions.

18.4.1.2 Zbb

The Zbb extension adds a `zbb` block to perform min/max (`min`, `minu`, `max`, `maxu`), extension (`sext.b`, `sext.h`, `zext.h`), counting, (`cpop`, `clz`, `ctz`), combining (`orc.b`), and byte reversal (`rev8`). It also enhances the shifter to support rotates (`rol`, `ror`) and uses the existing inverter on the `B` input to support inverted logical operations (`orn`, `andn`, `xnor`).

Rotates

The `zbb` module extends the funnel shifter source generator from Section 7.1.3.3 to support rotates. Table 18.12 describes the source generator for RV64, which produces the source to shift, `Z`. In the table, $\texttt{RotA} = \{\texttt{A}_{31:0}, \texttt{A}_{31:0}\}$ for W64 rotates (`rolw`, `rorw`), and $\texttt{RotA} = \texttt{A}_{63:0}$ for ordinary rotates (`rol`, `ror`). Table 18.13 describes the source generator for RV32, which is simpler without the W64 operations.

Table 18.12 64-bit shifter source generator and shift amount that supports rotates.

Shift Type	z	Shift Amount
`srl`	$\{63'b0, A_{63:0}\}$	k
`sra`	$\{\{63\{A_{63}\}\}, A_{63:0}\}$	k
`sll`	$\{A_{63:0}, 63'b0\}$	$\overline{k}$
`srlw`	$\{95'b0, A_{31:0}\}$	k
`sraw`	$\{64'b0, \{31\{A_{31}\}\}, A_{31:0}\}$	k
`sllw`	$\{32'b0, A_{31:0}, 63'b0\}$	$\overline{k}$
`ror` / `rorw`	$\{RotA_{62:0}, RotA_{63:0}\}$	k
`rol` / `rolw`	$\{RotA_{63:0}, RotA_{63:1}\}$	$\overline{k}$

Table 18.13 **32-bit shifter source generator and shift amount that supports rotates.**

Shift Type	z	Shift Amount
`srl`	{31'b0, $A_{31:0}$}	k
`sra`	{{31{A_{31}}}, $A_{31:0}$}	k
`sll`	{$A_{31:0}$, 31'b0}	$\overline{k}$
`ror`	{$A_{30:0}$, $A_{31:0}$}	k
`rol`	{$A_{31:0}$, $A_{31:1}$}	$\overline{k}$

Example 18.4 FUNNEL SHIFTER

Show how the 64-bit funnel shifter performs `rorw 0x01234567_C9ABCDEF, 8`.

Solution: `RotA` = {$A_{31:0}$, $A_{31:0}$} = C9ABCDEF_C9ABCDEF. Z = {$RotA_{62:0}$, $RotA_{63:0}$} = 49ABCDEF_C9ABCDEF_C9ABCDEF_C9ABCDEF; note that Z is 127 rather than 128 bits. $k = 8$. The funnel shifter shifts right by 8 bits and selects the bottom 64 bits of the result `FullResult` = EFC9ABCD_EFC9ABCD. Finally, the W64 mux truncates to 32 bits and sign extends: `ALUResult` = `FFFFFFFF_EFC9ABCD`. Thus, the shifter performed an 8-bit right rotation on the 32-bit word.

Inverted Operations

For inverted operations (`orn`, `andn`, `xnor`), the ALU performs `A op` $\overline{B}$ using the inverter already available from subtraction, so no new datapath hardware is required. The ALU controller drives `SubArith` = 1 to select $\overline{B}$ as the second ALU operand.

`zbb` Module

Fig. 18.22 shows the `zbb` module. Each of the min/max, extension, counting, and byte operations are described below. For I-type instructions, `B` contains the immediate, and certain lsbs of `B` are used for control.

The min/max (`min`, `minu`, `max`, `maxu`) operations select either `A` or `B` based on either the *less than* (`lt`) or *less than unsigned* (`ltu`) flag from the ALU.

Sign and zero extension select from among the three possible values for `sext.b`, `sext.h`, and `zext.h` using the `ext` module.

Fig. 18.23 shows the `cnt` module that handles `cpop`, `clz`, and `ctz`. It contains `popcnt` and `lzc` modules for counting either population (total number of ones) or leading zeros. It counts trailing zeros by reversing the input bits and counting leading zeros. Recall that `RevA` is the bit-reversed version of A. Fig. 18.24 shows a simplified 8-bit version of the

The Verilog idiom for a configurable-width leading zero counter and priority encoder involves an unusual loop syntax.

```
int i;
always_comb begin
  i = 0;
  while ((i < WIDTH) &
         ~num[WIDTH-1-i])
    i = i + 1;
  ZeroCnt =
    i[$clog2(WIDTH+1)-1:0];
end
```

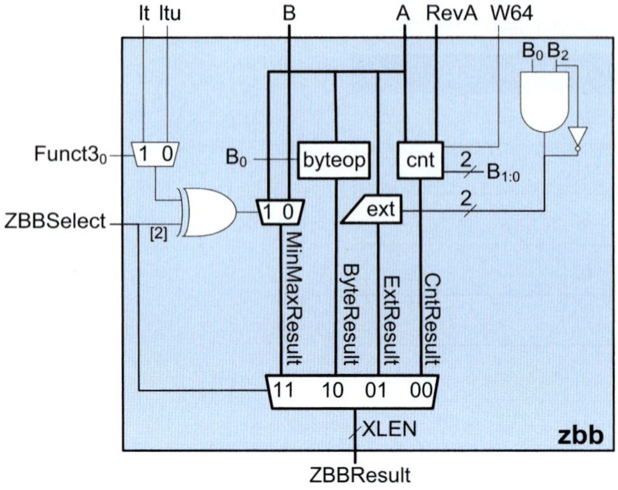

Fig. 18.22 **zbb module.**

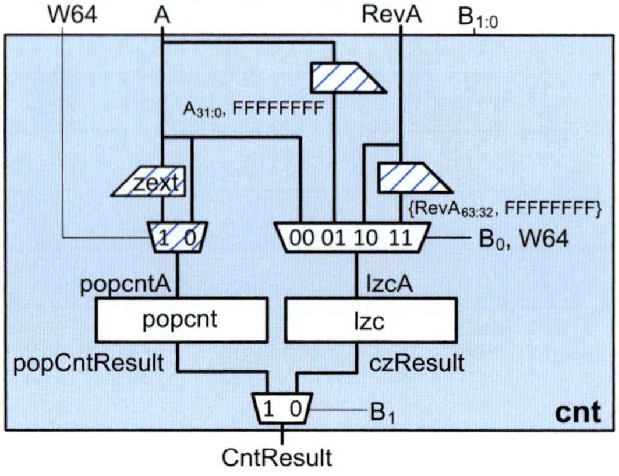

Fig. 18.23 **cnt module.**

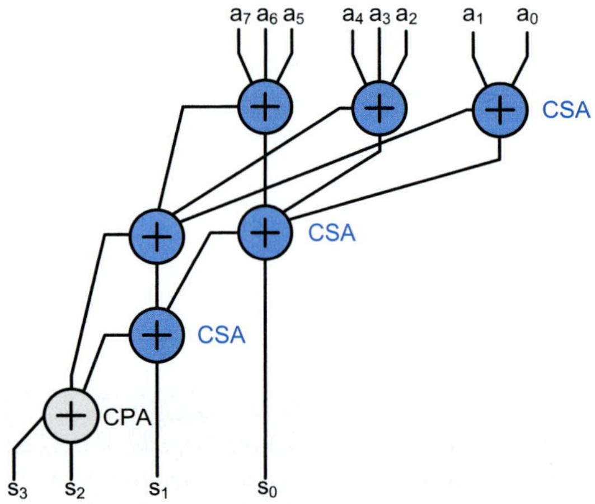

Fig. 18.24 **8-bit popcnt module.**

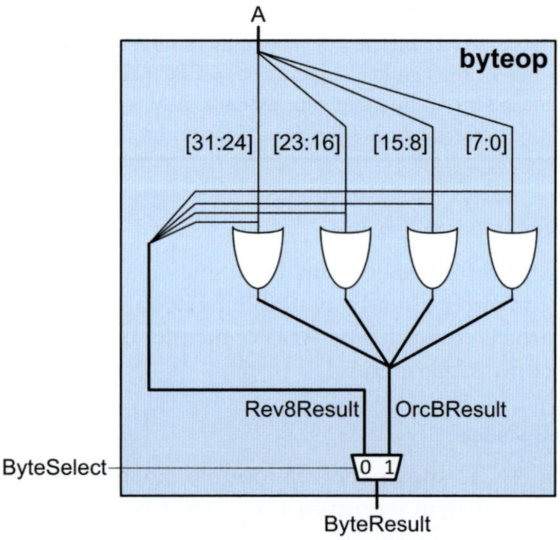

Fig. 18.25 **32-bit byteop module.**

popcnt module needed for the cpop instruction. In general, the module uses an XLEN:2 tree of carry-save adders (CSAs), followed by a carry propagate adder (CPA). The leading zero counter is simply a priority encoder with a binary output.

Fig. 18.25 shows the byteop module that handles orc.b and rev8. It contains one 8:1 OR gate for each byte to do combining and a network of wires to reverse bytes.

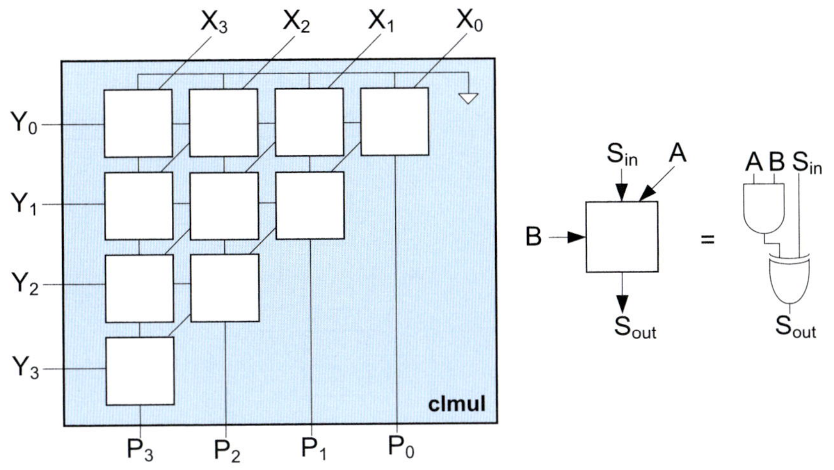

Fig. 18.26 **4-bit `clmul` module.**

18.4.1.3 Zbc

The `zbc` unit performs three flavors of carry-less multiplication: `clmul`, `clmulh`, and `clmulr`. Multiplying two XLEN-bit inputs X and Y gives a 2XLEN-bit product P. `clmul` returns the lower half and `clmulh` returns the upper half. `clmulr` is similar to `clmulh` and is discussed later in this section.

Carry-less multiplication is similar to ordinary multiplication from Fig. 15.4(a) but does not propagate carries between columns. Hence, the full adders reduce to simple XOR gates to sum the carries within a column modulo 2. Only the least significant half of the multiplier is necessary to obtain an XLEN-sized result. Fig. 18.26 shows a 4-bit version of a the `clmul` (carry-less multiplication) module. In this section, the @ operator refers to this carry-less multiplication.

The `clmulh` instruction computes the upper XLEN bits of a carry-less multiplication. Fig. 18.27(a) shows a complete 2XLEN-sized carry-less multiplier, where XLEN = 4, with the upper half highlighted in blue. If the blue bits are flipped horizontally and vertically as shown in Fig. 18.27(b), we see they have the same structure as Fig. 18.26 except that the inputs are reversed, `Y` is shifted left by 1, and then the outputs are reversed: `clmulh = rev(rev(X) @ (rev(Y) << 1))`. Hence, `clmulh` is executed on the same `clmul` module by bit-reversing and shifting the inputs.

`clmulr` is similar to `clmulh` but omits the shift on `Y`: `clmulr = rev(rev(X) @ rev(Y))`. Hence, `clmulh` = `clmulr` >> 1.

Fig. 18.28 shows a `zbc` module that performs all three flavors of carry-less multiplication. It uses a single `clmul` carry-less multiplication array and

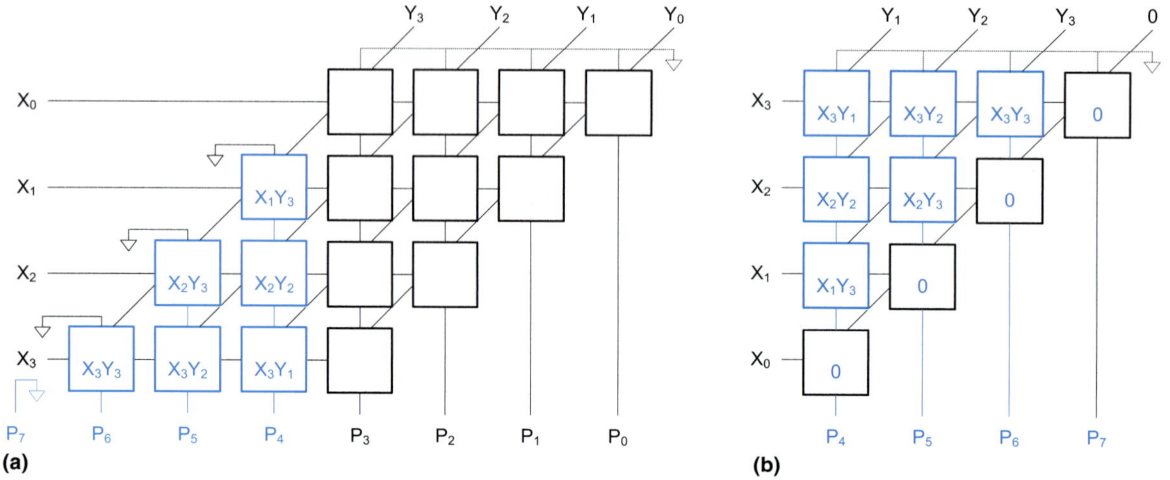

Fig. 18.27 Carry-less multiplication: (a) 8-bit and (b) 4-bit `clmulh` using a 4-bit `clmul` array.

conditionally reverses the inputs and outputs using two `bitreverse` modules and the already bit-reversed input `RevA` from the `bitmanipalu` module.

Code Example 18.14 shows the Verilog for the carry-less multiplier. SystemVerilog lacks a built-in idiom, so the module uses two nested loops to create the $N^2/2$ AND and XOR gates. Unfortunately, for $N = 64$, the inner loop is evaluated nearly 2,048 times. Zbc slows down rv64gc simulations by at least 20% because of these loops.

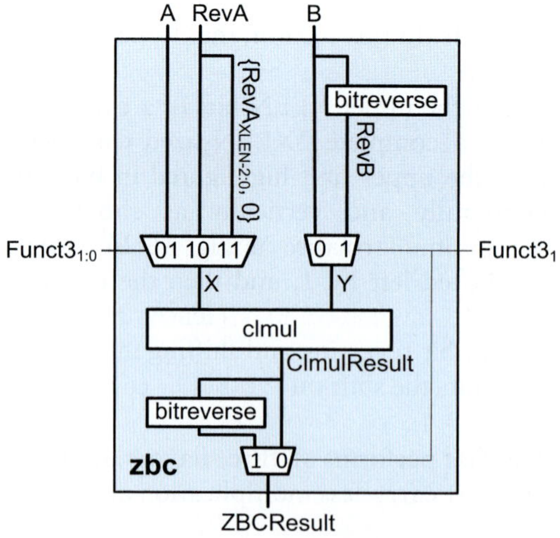

Fig. 18.28 `zbc` module.

Code Example 18.14 CARRY-LESS MULTIPLY MODULE (`clmul`)

```
module clmul #(parameter N=32) (
  input  logic [N-1:0] X, Y,          // Operands
  output logic [N-1:0] ClmulResult); // Zbc result

  logic [(N*N)-1:0] S;               // Intermediate signals for carry-less multiply

  integer i, j;

  always_comb begin
    for (i=0; i<N; i++) begin: outer
      S[N*i] = X[0] & Y[i];
      for (j = 1; j <= i; j++) begin: inner
        S[N*i+j] = (X[j] & Y[i-j]) ^ S[N*i+j-1];
      end
      ClmulResult[i] = S[N*i+j-1];
    end
  end
endmodule
```

18.4.1.4 Zbs

As shown in Fig. 18.21, the Zbs extension uses a decoder to create a mask with a single one in the column specified by `B`: `CondMaskB = 1 << B`. The ALU then performs logical operations:

- `bset:` `rd = A | CondMaskB`
- `bclr:` `rd = A & ~CondMaskB`
- `binv:` `rd = A ^ CondMaskB`
- `bext:` `rd = |(A & CondMaskB)`

The ALU also adds an XLEN:1 OR gate for `bext` to determine if the masked bit is true.

18.4.2 ALU Modifications for Cryptography

Fig. 18.29 shows a diagram of the ALU with optional enhancements for cryptography highlighted in different shades of blue and grey. The major changes are:

- **Zbkb:** Extension to bit manipulations for cryptography beyond Zbb.
- **Zbkx:** Block for permutations instructions.
- **Zknde:** Block for AES decryption and/or encryption.
- **Zknh:** Block for SHA-2 operations.

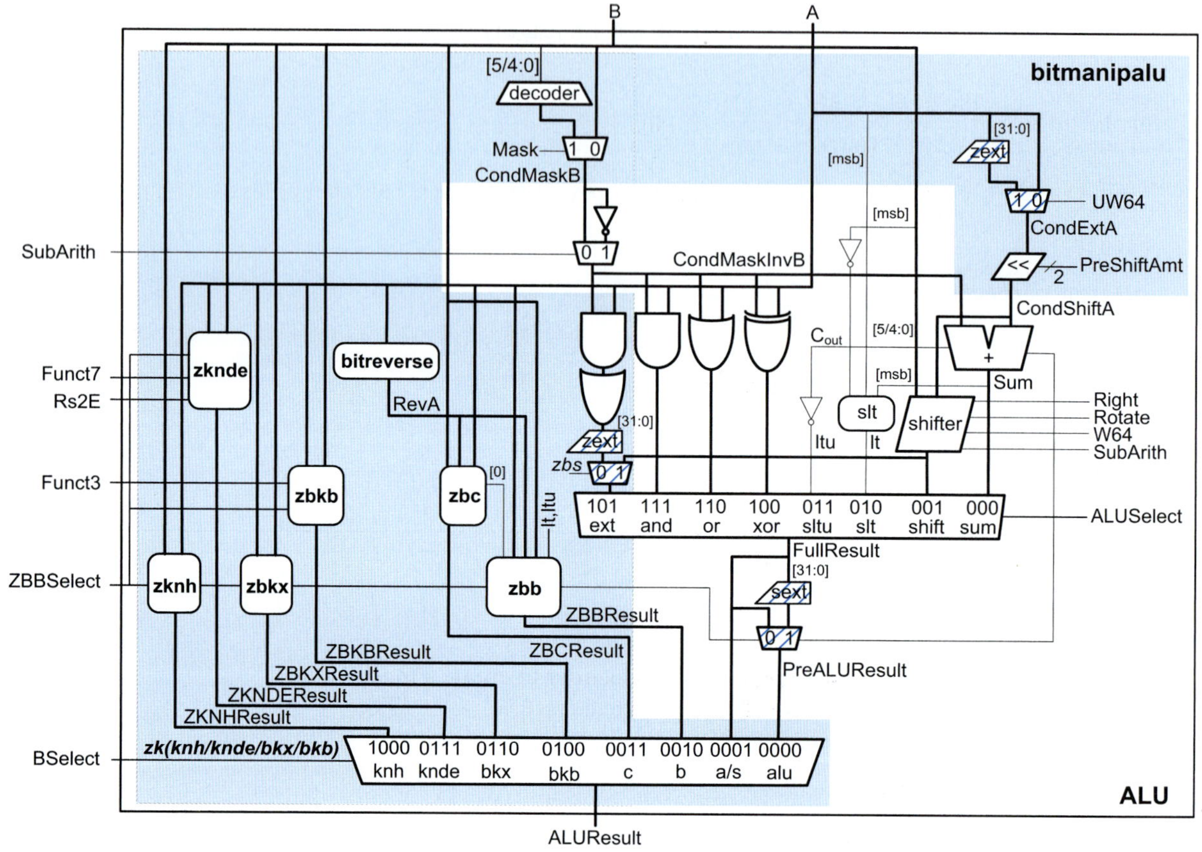

Fig. 18.29 `alu` **enhancements for cryptography.**

When any of cryptography extensions are supported, the ALU adds the supported extensions to the `bitmanipalu` and extends the `ALUResult` multiplexer to choose the new results.

18.4.2.1 Zbkb

The Zbkb extension includes a subset of the Zbb instructions, along with new `zip`/`unzip`, `pack`/`packh`/`packw`, and `brev8` instructions. The Zbkb unit, shown in Fig. 18.30, executes these new crypto bit manipulation instructions. Note that the `packw` instruction is only implemented in RV64, and `zip`/`unzip` are only implemented on RV32.

The `zipper` module contains a rat's nest of wiring to interleave bits and a 2-input mux to choose `zip` or `unzip`. The `packer` also contains wiring to rearrange the bits and a 3-input mux to choose among the flavors of `pack`. `brev8` simply contains wiring to reverse bits within the bytes.

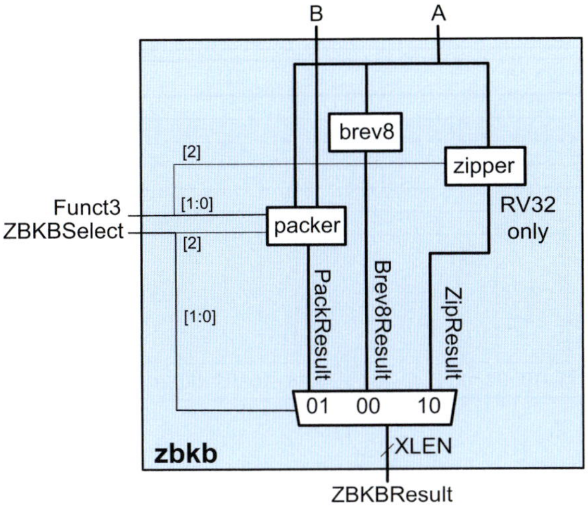

Fig. 18.30 `zbkb` **module.**

18.4.2.2 Zbkx

The `zbkx` module performs `xperm4` and `xperm8` permutations, as shown in Fig. 18.31.

A `xperm`k module of size k (4 or 8) produces n = XLEN/k chunks of k-bit outputs by selecting from one of n k-bit chunks of the input `A` based on an index in the corresponding k-bit chunk of the input `B`. If the index exceeds n (i.e., its upper bits are nonzero), the index is illegal and the chunk is set to 0. This is done with n k-bit n:1 multiplexers, and some OR-AND logic for illegal indices. Fig. 18.32 shows an `xperm8` module for rv32zbkx ($k = 8$, $n = 4$).

The logic for an `xperm`k block can be described succinctly with SystemVerilog's part select notation (see Section 4.2.3.3.3) choosing k-bit chunks `B[i+:k]`, as shown in Code Example 18.15 for k=8. The right shift by a

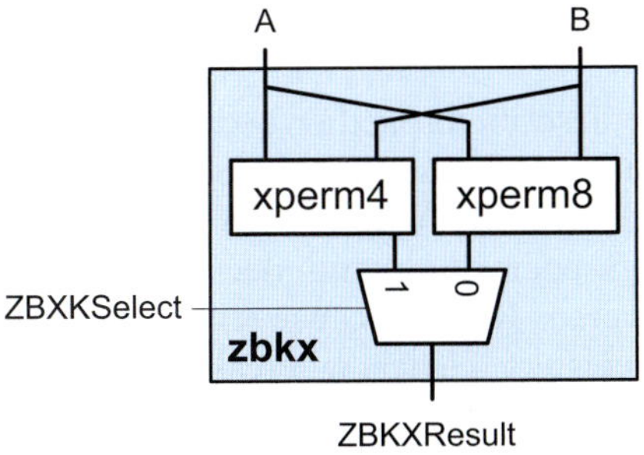

Fig. 18.31 `zbkx` **module.**

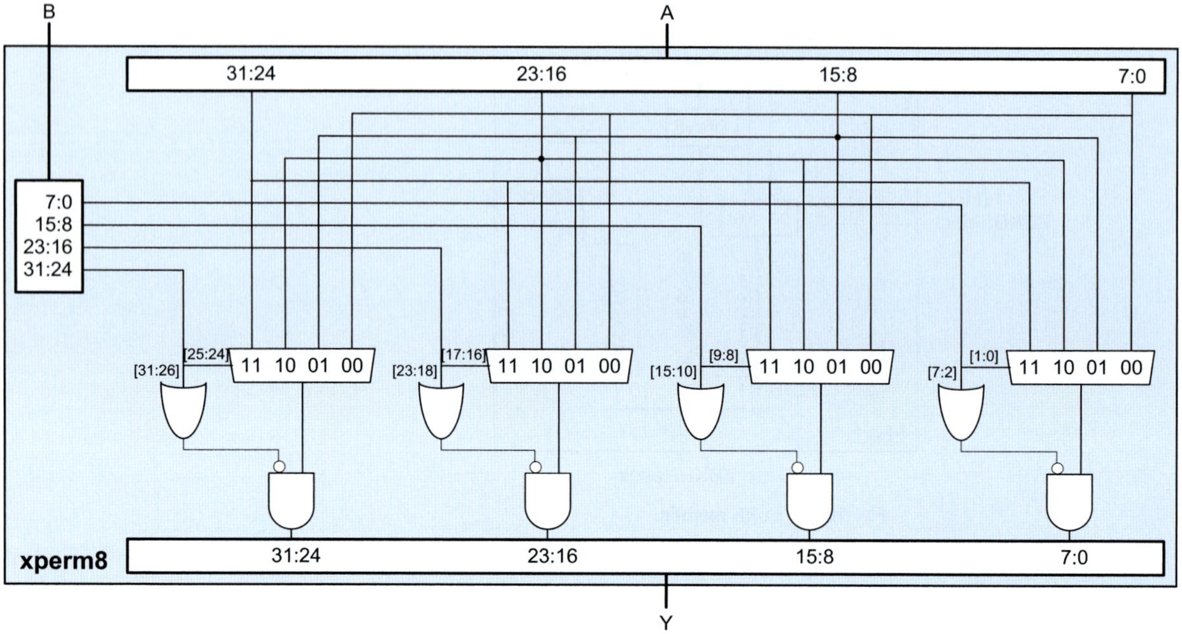

Fig. 18.32 `xperm8` **for rv32zbkx.**

multiple of 8 implies the n:1 mux and the OR-AND logic for handling out-of-range indices.

Code Example 18.15 `xperm8` USING PART-SELECT NOTATION

```
for (i = 0; i < WIDTH; i = i + 8) begin: xperm8calc
  xperm8lookup = A >> {B[i+:8], 3'b0};
  xperm8[i+:8] = xperm8lookup[7:0];
end
```

18.4.2.3 Zknd and Zkne

Zknd and Zkne are described together because they share common hardware. Recall that they accelerate the AES SubBytes, ShiftRows, and MixColumns operations with a single instruction. However, as described in Section 18.2.7, AES encryption and decryption instructions are fairly different for RV32 versus RV64 because RV32 cannot hold the entire state matrix in a pair of registers. Hence, whereas RV32 only processes one byte at a time, RV64 processes 8 bytes at a time. The largest portion of the modules is the S-boxes, which are 256 entry × 8-bit lookup tables. RV64 uses 8

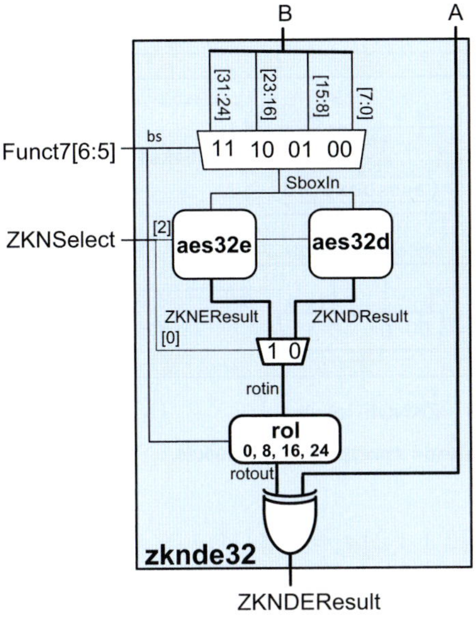

Fig. 18.33 **zknde32 encrypt/decrypt module.**

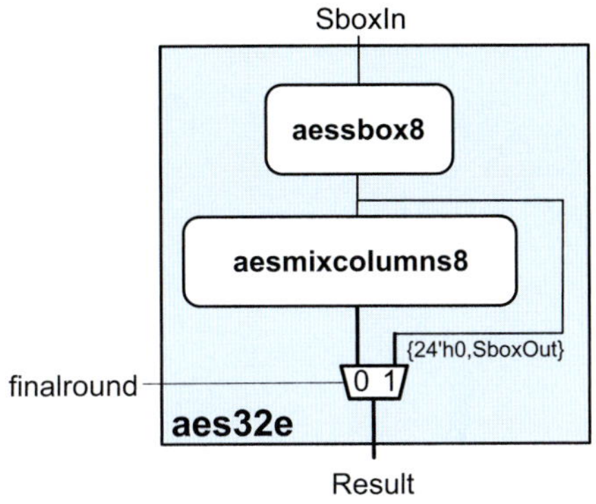

Fig. 18.34 **32-bit aes32e encrypt module.**

S-box and 8 inverse S-box tables, while RV32 only uses one of each and only delivers 1/8th of the throughput. This section begins with the 32-bit implementation, then describes the 64-bit implementation.

zknde32

RV32 instructions `aes32esm`, `aes32es`, `aes32dsmi`, and `aes32dsi` accelerate AES one byte at a time selected from a 32-bit register containing 4 bytes of the state matrix. The `zknde32` module shown in Fig. 18.33 selects the byte from the 32-bit input. The `aes32e` and `aes32d` modules handle encryption and decryption steps, respectively. Either could be omitted if only Zkne or Zknd is supported. `zknde32` then selects the desired operation, rotates the result into place, and XORs it into a running result.

Fig. 18.34 shows the `aes32e` module. It applies `aessbox8` to the selected entry, then `aesmixcolumns8` to the output word to perform one step of the SubBytes and MixColumns steps, respectively. `aessbox8` performs one step of SubBytes according to the lookup table from Fig. 18.7. `aesmixcolumns8` computes 32 bits of MixColumns based on the 1-byte input a according to Equation 18.4, which involves one GF(2^8) multiplication by x and one XOR. A multiplexer optionally skips the MixColumns step for the final round (`aes32es`). The `aes32d` module is similar except that it uses the `aesinvsbox8` and `aesinvmixcolumns8` modules.

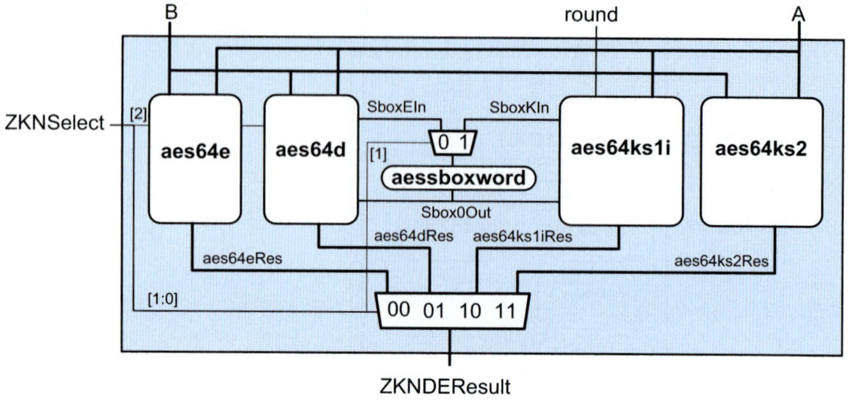

Fig. 18.35 `zknde64` **encrypt/decrypt module.**

$$\begin{aligned}\texttt{aesmixcolumns8} &= \{3\ 1\ 1\ 2\}a = \{x+1,1,1,x\}a \\ &= \{xa \oplus a, a, a, xa\}\end{aligned} \tag{18.4}$$

zknde64

RV64 instructions accelerate AES by processing eight bytes at a time. The zknde64 module shown in Fig. 18.35 supports key scheduling instructions `aes64ks1i`, `aes64ks2`, and `aes64im`; encryption instructions `aes64esm` and `aes64es`; and decryption instructions `aes64dsm` and `aes64ds`. Like `zknde32`, it contains `aes64e` and/or `aes64d` hardware for encryption and/or decryption depending on whether Zkne and Zknd are supported. However, the hardware works on 8 bytes at a time rather than 1 byte, achieving eight times the throughput but requiring 8 `sbox` and `invsbox` modules rather than one of each. Because it works on the entire word, it does not need the logic from Fig. 18.33 to select a byte or to roll the output into place and XOR it with a running total. However, `aes64ks1i` requires 4 S-boxes in `aessboxword`, which are shared with `aes64e` if Zkne is enabled.

Fig. 18.36 shows the `aes64e` module. The `aesshifrow64` module simply contains wires to rearrange bits according to Equation 18.2. The `aes64es`/`aes64esm` instructions apply 8 S-boxes in parallel; 4 are shared with `aes64ks1i` in `zknde64`, and the other 4 are in `aessboxword` here. `aesmixcolumns32` computes one column of the output using four GF(2^8) multiplications according to Equation 18.3. This MixColumns step is skipped on the last round by `aes64es`.

`aes64d` is similar to `aes64e`, but it uses inverse operations. It contains all 8 inverse S-boxes because none are shared with key scheduling. InverseMixColumns contains 12 Galois field multipliers because the inverse matrix coefficients from Fig. 18.13 are larger. The block also performs

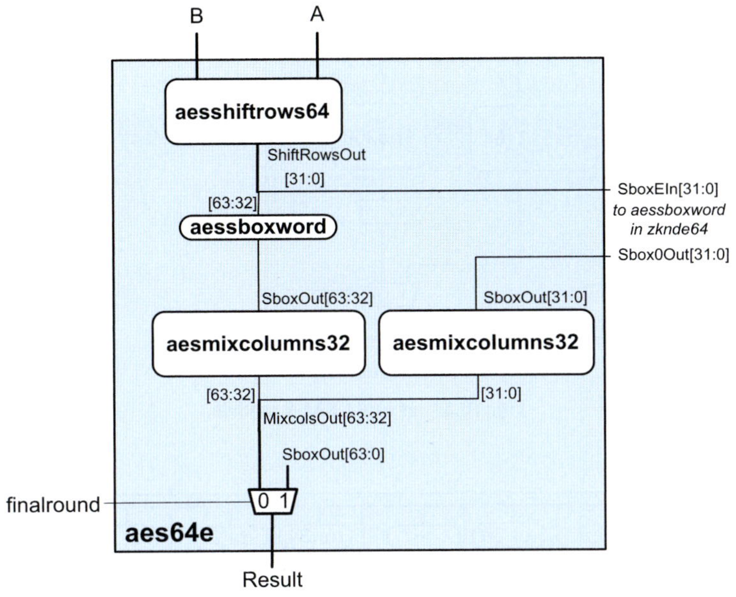

Fig. 18.36 `aes64e` **encrypt module.**

`aes64im` by skipping the first steps and only applying InverseMixColumns to two 32-bit words.

`aes64ks1i` looks up a round constant based on the round. It rotates 32 bits of the incoming key, sends them back to `aessboxword` to apply an S-box to each byte, and XORs the results with the round constant. `aes64ks2` performs two 32-bit XORs. Both modules are relatively small.

18.4.2.4 Zknh

Fig. 18.37 shows a block diagram of the RV64 `zknh64` module, which provides separate modules performing the various SHA-2 σ and Σ instructions for each of SHA-256 and SHA-512. The RV32 `zknh32` module is similar but sign-extends SHA-256 outputs to 64 bits and has a different `sha512_32` module to operate on half of a 64-bit word at a time.

Fig. 18.38 shows the `sha512_64` module. It supports four instructions, each of which need three 64-bit rotations or shifts. Thus, it performs all 12 rotations and shifts, chooses the three needed for the given instruction, and XORs them together. `sha256` is similar, but it performs different 32-bit shifts and rotations. `sha512_32` is also similar but assembles the upper or lower 32 bits of a 64-bit result from shifts operating on two 32-bit registers.

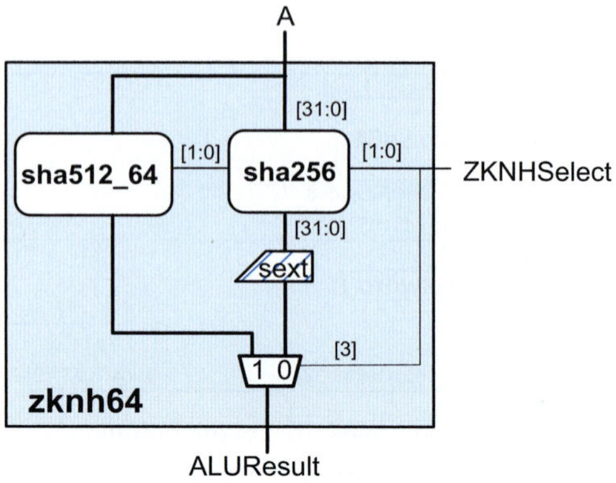

Fig. 18.37 `zknh64` SHA-2 module.

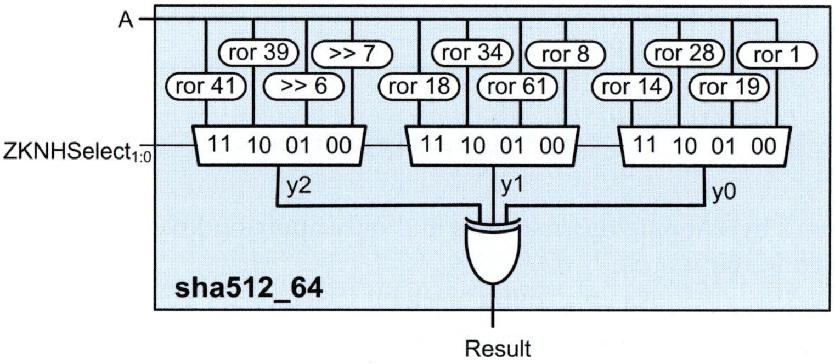

Fig. 18.38 `sha512_64` module.

18.4.2.5 Zkt

Wally implements all relevant instructions with constant latency, so it inherently satisfies the Zkt extension. Division and remainder have variable latency, but they are not part of the Zkt specification.

18.4.3 Controller Modifications

When any of the bit manipulation or cryptography extensions are supported, the IEU (integer execution unit) controller instantiates the BMU control (`bmuctrl`) module that decodes the supported instructions. It produces control signals used by the bit manipulation modules in Section 18.4.1, along with an `IllegalBitmanipInstrD` signal. If `IllegalBitmanipInstrD` = 0, then the controller has found a legal instruction.

18.5 SUMMARY

In summary, the Zb* bit manipulation extensions are a family of four optional extensions that provide modest speedups for certain common code sequences:

- **Zba:** Shifts and adds to facilitate address generation.
- **Zbb:** Basic bit operations: min/max, rotate, zero/sign extension, count, combine, reverse.
- **Zbc:** Carry-less multiplication for cryptography and error checking codes.
- **Zbs:** Single-bit set/clear/invert/extract.
- **B:** Zba + Zbb + Zbs

Cryptographic extensions are added for Zb* and Zk* that provide instructions to accelerate cryptography:

- **Zbkb:** Bit manipulations generally useful for cryptographic algorithms.
- **Zbkc:** Subset of Zbc needed for cryptography.
- **Zbkx:** Permutations useful for implementing S-boxes in data-independent time.
- **Zknd:** Instructions for accelerating the decryption and key schedule functions of the AES block cipher.
- **Zkne:** Instructions for accelerating the encryption and key schedule functions of the AES block cipher.
- **Zknh:** Instructions for accelerating the σ and Σ functions for the SHA-2 family of hash functions.
- **Zkn:** NIST cryptography: Zbkb + Zbkc + Zbkx + Zkne + Zknd + Zknh
- **Zkt:** Common data processing instructions must have data-independent latency.
- **Zkr:** Seed CSR to provide entropy for a random number generator.
- **Zk:** Zkn + Zkt + Zkr

Wally modifies the integer IEU to support the Zb* and Zkn extensions but lacks Zkr. The additional hardware is modest except that Zbc implies a large array of $XLEN^2/2$ AND and XOR gates for the carry-less multiplier array, and RV64Zknd/Zkne implies sixteen 256 × 8-bit lookup tables for AES S-boxes.

Other Extensions

19

19.1 Cache Management: Zicbom, Zicboz, Zicbop
19.2 Misaligned Loads and Stores: Zicclsm
19.3 TLB Management: Svinval
19.4 Pause Hint: Zihintpause
19.5 Conditional Instructions: Zicond
19.6 Custom Extensions
19.7 Summary

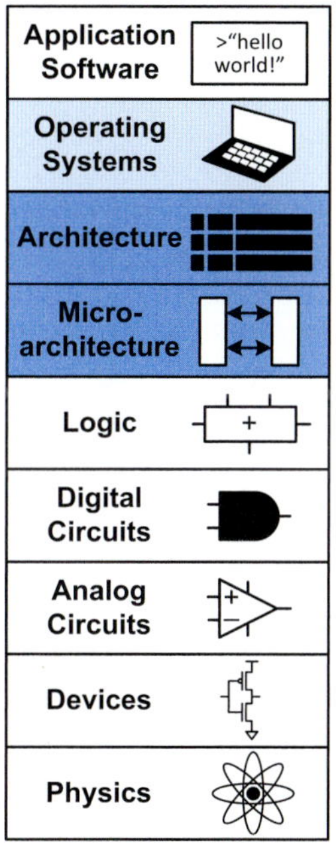

RISC-V is evolving rapidly. Table A.2 lists various extensions that had been ratified by the time of this writing, the profiles that use them, and the chapters that describe them. Most of the extensions are described in earlier chapters with their required hardware. This chapter describes extensions (particularly those needed for the RVA22S64 profile) that do not naturally fit elsewhere in the book:

- **Zicbom, Zicboz, Zicbop:** Cache management operations (CMOs)
- **Zicclsm:** Misaligned loads and stores
- **Svinval:** TLB management
- **Zihintpause:** Pause hint
- **Zicond:** Conditional instructions

Some extensions define *hints*. Hints are encoded as ordinary instructions that have no effect on architectural state but that suggest the core change nonarchitectural state to improve performance or power. For example, the `prefetch.r` instruction suggests that the cache prefetch a cache line that the program anticipates will be needed soon.

19.1 CACHE MANAGEMENT: Zicbom, Zicboz, Zicbop

The cache block management instructions give a program fine-grained control over cache lines to improve performance and maintain coherence with certain I/O devices. These extensions provide instructions to zero, clean, flush, invalidate, and prefetch cache lines.

19.1.1 RISC-V CMO Instructions

The cache management operation (CMO) extensions are defined in Table I.21 at the back of the book. They are divided into three independent extensions (Zicbom, Zicboz, and Zicbop), but most cores implement all or none.

cbo.clean, cbo.inval, cbo.flush, and cbo.zero are encoded as I-type instructions except that the immediate field is used to distinguish the various operations, and is not added to the base address in rs1. Prefetch hints are encoded as ori x0, rs1, imm, which is a nop because writes to x0 do nothing.

19.1.1.1 Zicbom Cache Block Management

The Zicbom *cache block management* instructions, cbo.clean, cbo.inval, and cbo.flush, are used to maintain coherence between the data cache and main memory when an *incoherent agent* (e.g., a peripheral with direct memory access) might write data to the main memory without updating the cache. The device driver may clean the cache to write back dirty lines to main memory before the incoherent agent reads this memory and may invalidate stale cache lines after the incoherent agent writes to memory.

These instructions all operate on a virtual address specified by rs1. cbo.clean writes the contents of a dirty cache block (line) to main memory. cbo.inval invalidates a cache block. cbo.flush performs a clean followed by invalidate as one atomic operation.

19.1.1.2 Zicboz Cache Block Zero

The Zicboz *cache block zero* instruction cbo.zero writes zeros to an entire cache block (line) or to the underlying physical memory if the addresses are not cacheable. Programs sometimes need to initialize regions of memory to zero; for example, the operating system (OS) clears the .bss section of memory when starting a process. Using cbo.zero to clear one cache line at a time is faster than storing zeros one word at a time and also avoids the need to fetch a cache line from main memory that is about to be zeroed anyway.

19.1.1.3 Zicbop Cache Block Prefetch

When a program knows it is likely to access a memory address soon, it can prefetch the corresponding cache line to hide the latency of waiting for the cache line. Zicbop defines three prefetch hint instructions (prefetch.i/r/w), which suggest that the processor prefetch instructions, read data, or write data, if beneficial. Prefetching for data reads and writes is often identical. The hints are encoded as ori x0, rs1, imm, which is technically a nop because x0 is hardwired to 0. However, the hint indicates that the line containing a certain address should be prefetched into the cache. The bottom 5 bits of imm are treated as 0s, so the virtual address is computed as Address = [rs1] + {imm[11:5], 00000}.

19.1.1.4 Environment Configuration Registers

As described in Section 8.2.3.6, the menvcfg and senvcfg registers enable the use of CMO instructions outside of M-mode. Each of these registers has CBZE, CBCFE, and CBIE[1:0] enable fields. menvcfg's enable must be set to use a CMO instruction in S- or U-mode. If S-mode is supported, senvcfg's enable must also be set to use the instruction in U-mode.

When CBZE = 0 and the hart is in a lower privilege mode, the cbo.zero instruction causes an illegal-instruction fault. Similarly, when CBCFE = 0 and the hart is in a lower privilege mode, cbo.clean and cbo.flush are illegal. CBIE = 00 disables cbo.inval and CBIE = 11 enables it. CBIE = 01 causes cbo.inval to operate as a flush, which invalidates the line but also ensures dirty data is first written back to the memory hierarchy, making the instruction safer to use.

19.1.2 Test Plan

The CBO instructions are tested with tests/wally-riscv-arch-test/riscv-test-suite/rv{32/64}i_m/privilege/src/WALY-cbo{m/z}-01.S. These tests assume a 64-byte cache line.

The WALLY-cbom-01 test copies a reference region of memory to another region aligned to the size of a cache way. Then it invalidates, cleans, or flushes the copied region and checks main memory for the correct data.

The WALLY-cboz-01 test copies a region of memory with known data, then overwrites it with cbo.zero, and then verifies that the region contains zeros. riscv-arch-test/riscv-test-suite/rv{32/64}i_m/CMO/src/cbo.zero-01.S also recently added cbo.zero tests.

19.1.3 Wally Implementation

The Wally ZICBOM_SUPPORTED and ZICBOZ_SUPPORTED configuration parameters indicate whether Wally should add hardware support for CBO instructions. The CBO extension modifies the controller, privileged, lsu, pmachecker, and cache units. The IEU controller detects CBO instructions, and gates them with the appropriate environment configuration registers, envcfg. It sends CBOpM[3:0], which specifies {zero, clean, flush, inval}, to the LSU. The MMU then checks for exceptions, and the LSU sends the request to either the cache or the AHB bus.

19.1.3.1 CMO LSU Modifications

Fig. 19.1 updates the LSU with the required CMO changes shown in blue. CBO instructions add additional access types to the PMA checker, mux zeros into WriteDataM for cbo.zero, and update the cache and AHB cache interface.

This section describes the cache modifications for the cbo.* instructions. cbo.zero operates on both cacheable and uncacheable memory, so it is more efficient to zero IHWriteDataM in the LSU rather than in both the cache and the AHB cache interface. The LSU uses the MMU's CacheableM attribute to select if the cache directly zeros the data or if the operation should be transmitted to the bus as an AHB burst request, as described in Section 19.1.3.3.

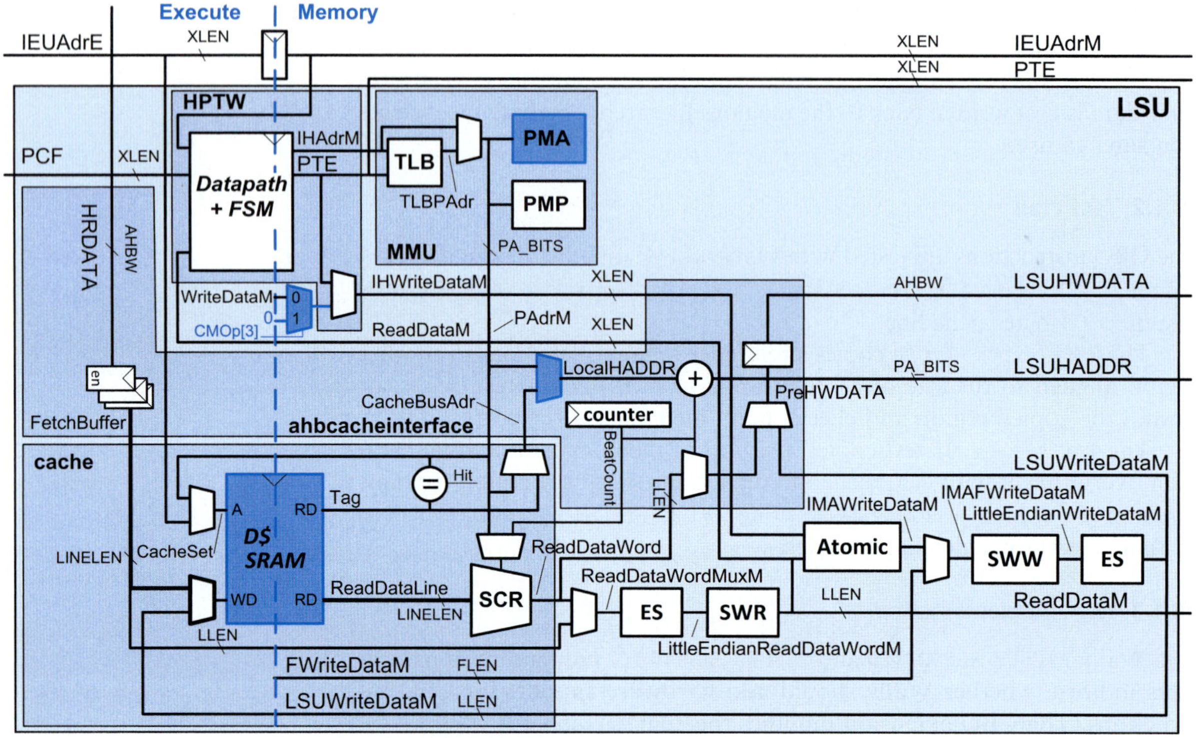

Fig. 19.1 **LSU with CMO support additions in darker blue.**

As with a regular load or store, during a `cbo.*` operation, the LSU receives the virtual address for the CMO instruction. If virtual memory is enabled, the TLB translates it to a physical address, throws a store page fault if the page is not accessible, and sets the access/dirty bits if necessary. The PMA and PMP check that the physical address is accessible and throw a store access fault if not. The PMA adds CMO to the types of valid accesses for the RAMs, as shown in Fig. 19.2.

19.1.3.2 CMO Cache Modifications

Fig. 19.3 shows the top-level cache with the changes highlighted in blue. `CMOp` is sent to the cache FSM, which controls the various operations. The FSM produces a new `ClearValid` output to clear the valid bit for CMO invalidate and flush operations. For `cbo.zero`, the cache always selects `LineWriteData` = 0.

The `cacheway` tag and valid logic determines if this address matches a line in the way. `cbo.inval` clears the valid bit for the matching line and updates the LRU (least recently used) bits, such that the invalidated way

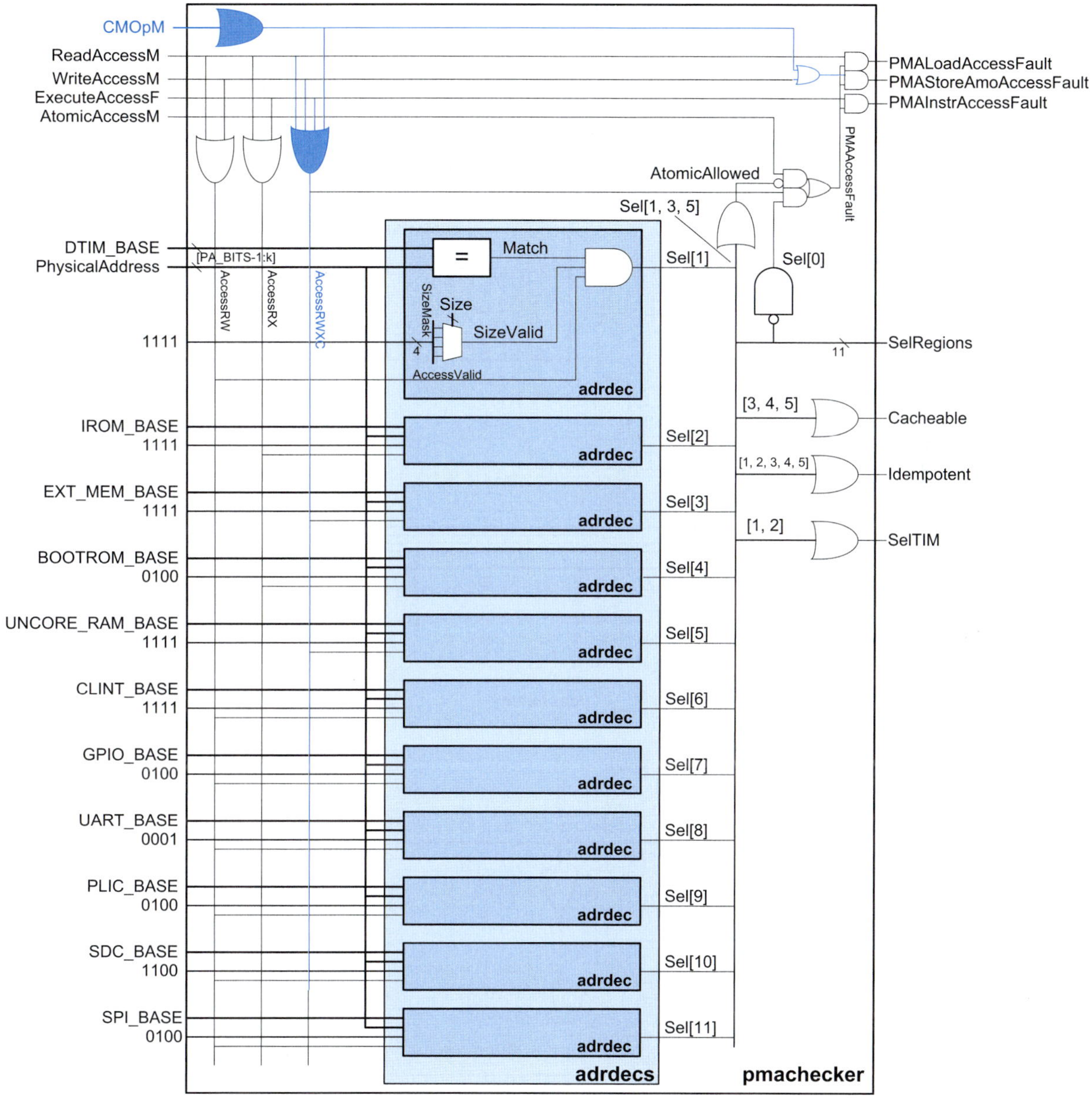

Fig. 19.2 PMA checker with CMO support additions in darker blue.

becomes the next victim. `cbo.clean` writes the matching line back to main memory if it is dirty, and clears the dirty bit. `cbo.flush` cleans and invalidates the line. `cbo.zero` writes all zeros to the matching line and marks it dirty and valid.

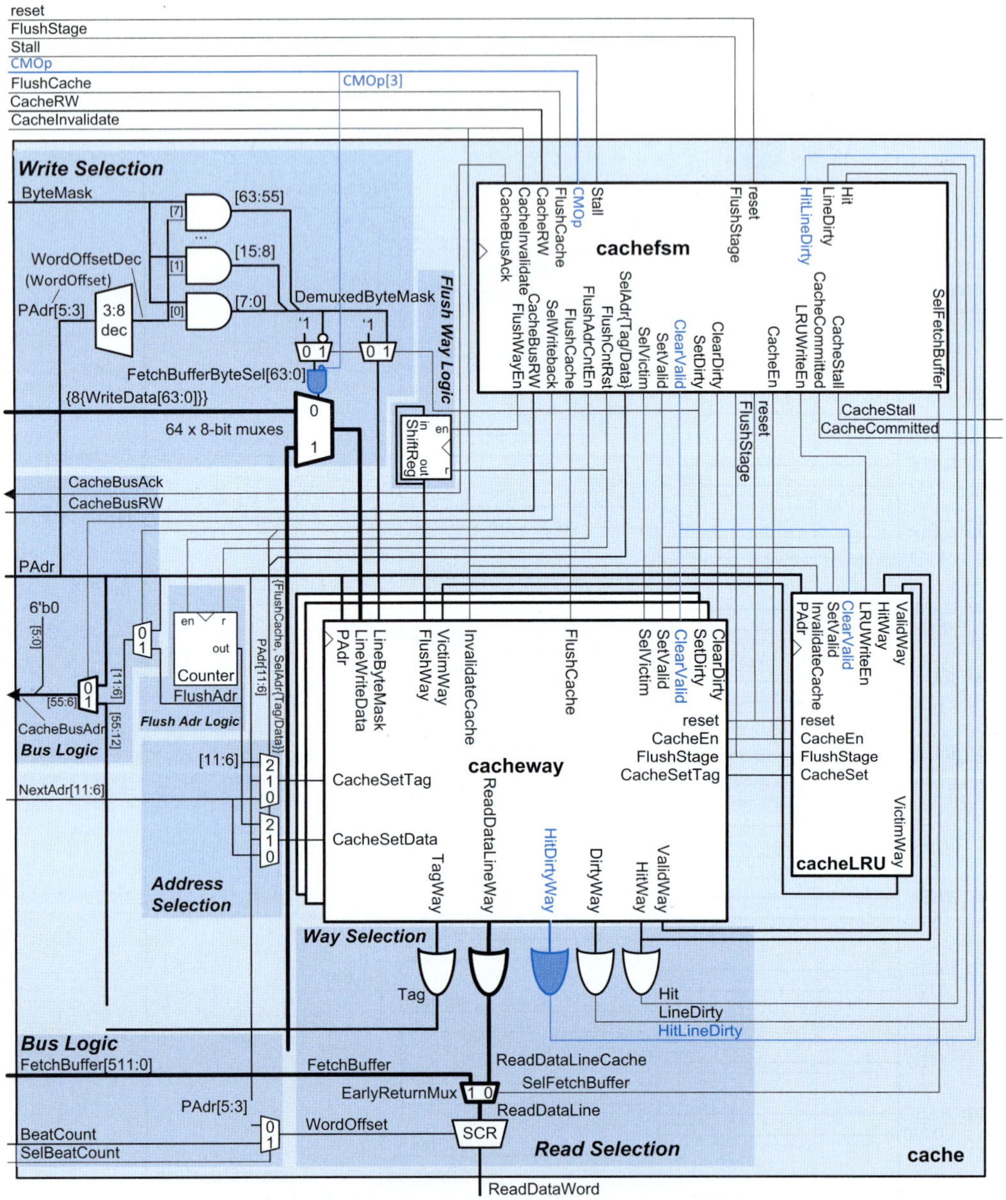

Fig. 19.3 Data cache with CMO support additions in darker blue.

Fig. 19.4 shows the cache FSM updates, highlighted in blue. Table 19.1 shows the FSM inputs and outputs that have been modified from Table 10.2.

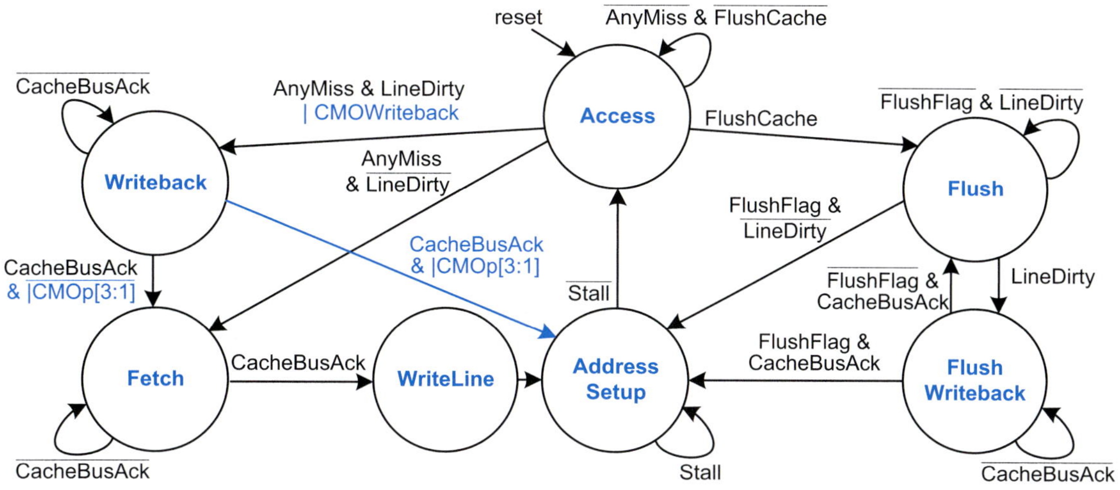

Fig. 19.4 `cachefsm` **with CMO support additions in blue.**

Table 19.1 `cachefsm` **inputs and outputs with CMO support additions in blue.**

<table>
<tr><th>FSM Input</th><th>Criteria to Assert</th></tr>
<tr><td>CMOp[3:0]</td><td>cmo.{zero/clean/flush/inval}</td></tr>
<tr><td>HitLineDirty</td><td>|HitDirtyWay</td></tr>
<tr><th>FSM Internal</th><th>Criteria to Assert</th></tr>
<tr><td>CMOZeroNoEviction</td><td>CMOp[3] & ~LineDirty</td></tr>
<tr><td>CMOWriteback</td><td>((CMOp[1] | CMOp[2]) & Hit & HitLineDirty) | (CMOp[3] & LineDirty)</td></tr>
<tr><th>FSM Output</th><th>Criteria to Assert</th></tr>
<tr><td>SetValid</td><td>WriteLine | Access & CMOZeroNoEviction | Writeback & CacheBusAck & CMOp[3]</td></tr>
<tr><td>ClearValid</td><td>Access & (CMOp[0] | CMOp[2] & ~HitLineDirty) |
Writeback & CMOp[2] & CacheBusAck</td></tr>
<tr><td>SetDirty</td><td>(Access & (AnyUpdateHit | CMOZeroNoEviction)) | (WriteLine & CacheRW[0]) |
(WriteBack & CMOp[3] & CacheBusAck)</td></tr>
<tr><td>ClearDirty</td><td>(WriteLine & ~CacheRW[0] | (Flush & LineDirty)) | (Writeback & ((CMOp[1] |
CMOp[2]) & CacheBusAck))</td></tr>
<tr><td>SelVictim</td><td>Access & (AnyMiss & LineDirty | ~Hit & CMOZeroNoEviction) | WriteLine |
Writeback & ((~(CMOp[1] | CMOp[2]) & ~CacheBusAck) |
(CMOp[3] & CacheBusAck))</td></tr>
<tr><td>LRUWriteEn</td><td>Access & (AnyHit | CMOZeroNoEviction) | WriteLine | Writeback & CMOp[3] &
CacheBusAck</td></tr>
<tr><td>SelAdrTag</td><td>Access & (AnyMiss | (|CMOp)) | Fetch | Writeback | WriteLine</td></tr>
<tr><td>SelAdrData</td><td>Access & (AnyMiss | (|CMOp) | CacheRW[0]) | Fetch | Writeback | WriteLine</td></tr>
<tr><td>CacheBusRW[1:0]</td><td>[1]: Access & AnyMiss & ~LineDirty | Writeback & CacheBusAck & ~|CMOp | Fetch
& ~CacheBusAck

[0]: Access & LoadMiss & LineDirty | (Writeback | FlushWriteback) &
~CacheBusAck</td></tr>
</table>

An alternate implementation for cbo.* instructions could transition from Writeback to Access but would require lowering CacheStall during the transition, thus increasing the critical path. By first transitioning to AddressSetup, CacheStall only depends on the current state.

cbo.inval remains in the Access state. It asserts ClearValid to invalidate the line if there is a hit.

cbo.clean and cbo.flush both transition to the Writeback state if the line hits and is dirty. Once in Writeback, the cache FSM waits for CacheBusAck as it did for bus operations in Section 10.4.3. However, for cbo.* instructions, it continues directly to AddressSetup rather than Fetch because it does not need to fetch a missing line. In the last cycle of Writeback, the cache FSM also asserts ClearValid for cbo.flush to invalidate the line now that it has been cleaned. During Writeback, CacheBusRW[0] is asserted to initiate writing the dirty cache line back to main memory via the AHB bus. Both cbo.clean and cbo.flush clear the dirty bit in the last cycle of Writeback.

cbo.zero writes zeros to a cache line. If cbo.zero hits the cache or misses but the victim line is clean, it writes zeros to the entire line and sets the valid and dirty bits; it also updates the tag on a miss. This takes a single cycle from the Access state. If cbo.zero misses the cache and the victim line is dirty, it transitions to the Writeback state to evict the dirty victim. It writes zeros to the line, sets valid and dirty, and updates the tag on the last cycle of Writeback. Like the other cbo.* instructions, it transitions to the AddressSetup state to finish.

Fig. 19.5 shows how cacheway writes a 0 to the valid bit of the selected cache line when ClearValid = 1. Similarly, but not diagrammed, cacheLRU updates the LRUMemory for that line when ClearValid is asserted.

SelVictim controls whether cacheway outputs the cache line corresponding to a hit or ignores the hit logic to select the VictimWay cache line for writeback or FlushWay cache line for an entire cache flush resulting from fence.i. SelVictim cannot control DirtyWay because the cachefsm drives SelVictim based on the value of DirtyWay. Unlike other cache operations, cbo.clean and cbo.flush need to know if the hit cache line is dirty, indicated by the new HitDirtyWay output.

19.1.3.3 Uncached cbo.zero

The MMU can mark certain main memory pages as uncacheable using page-based memory types (PBMT), as described in Section 11.2.2.1.1. cbo.zero must still write a cache line worth of zeros to such memory, but it must directly do so through the AHB bus. The BusCMOZero = CMOp[3] & ~Cacheable signal indicates this situation.

The LSU leverages the ahbcacheinterface's existing CacheWriteback state (see Fig. 12.5) to iterate over the beats of a burst request with only two changes to the hardware. The transition from Adr to CacheWriteback becomes HREADY & (CacheRW[0] | BusCMOZero). The CacheBusAck output is not asserted when BusCMOZero is high because the cachefsm did not make this memory request and should not receive this acknowledgment.

19.1.3.4 Prefetch

Wally is an in-order processor that stalls during cache fills. Therefore, prefetch would stall the processor and not improve performance. Hence, the prefetch.* hints are ignored.

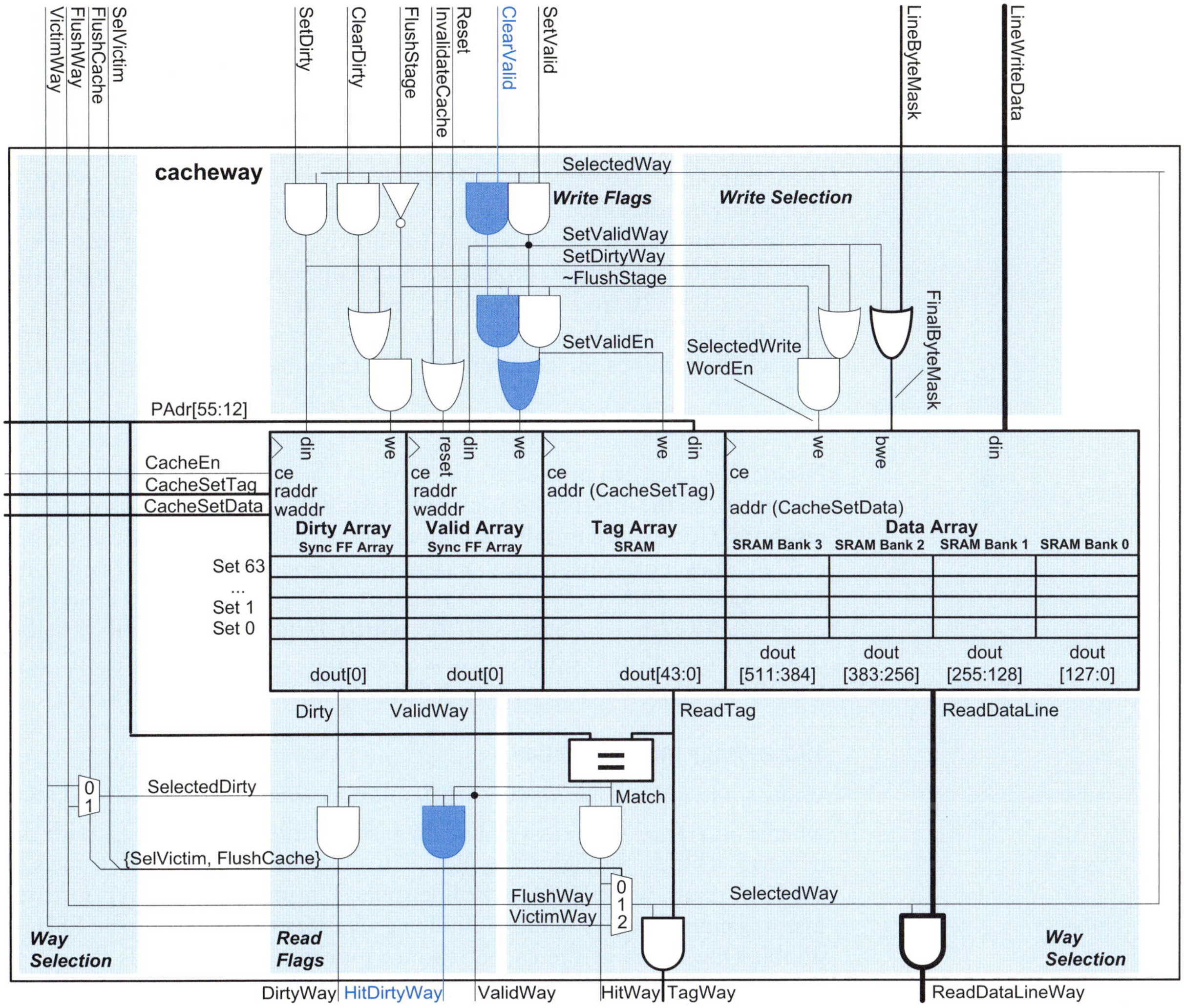

Fig. 19.5 `cacheway` **with CMO support additions in darker blue.**

19.2 MISALIGNED LOADS AND STORES: Zicclsm

By default, RISC-V produces a misaligned-load or misaligned-store exception when the address is not a multiple of the size of the element being accessed. For example, `lw t0, 6(x0)` produces a misaligned-load exception because address 6 is not a multiple of the 4-byte word size. The trap handler may emulate the misaligned access through a sequence of smaller aligned accesses, such as:

```
lh t0, 6(x0)          # Load lower half word
lh t1, 8(x0)          # Load upper half word
slli t1, t1, 16       # Move upper half word into upper bytes
or t0, t0, t1         # Combine lower and upper half words
```

RISC-V allows misaligned accesses to be handled by hardware rather than trapping. The Zicclsm *(load/store misaligned)* extension indicates that a hart handles misaligned loads and stores to cacheable coherent memory in hardware. They may be handled slowly, but they will not trap. Even with Zicclsm, misaligned atomic memory operations (AMOs) might still trap. The RVA20 and later profiles require support for Zicclsm. This is valuable for porting legacy code, which may contain misaligned structures, and which becomes unacceptably slow if each misaligned access traps.

19.2.1 Test Plan

No load or store can produce {Load/Store}MisalignedFaultM when Zicclsm is enabled, so the misaligned riscv-arch-tests will not function correctly. To test both Zicclsm and misaligned faults, Wally rv64gc enables Zicclsm, but rv32gc does not. Misaligned access hardware handling is tested with tests/wally-riscv-arch-test/riscv-test-suite/rv64i_m/privilege/src/WALLY-misaligned-access-01.S, which sequences over all possible combinations of byte offsets, access sizes, load and store operations, and enough memory accesses to cross 64-byte cache line boundaries.

19.2.2 Wally Implementation

Wally adds support for misaligned loads and stores to cacheable memory when ZICCLSM_SUPPORTED = 1. Fig. 19.6 shows the modified LSU, with the new align block highlighted in blue. The align block affects three parts of the LSU: reads, writes, and cache line spills. The implementation prioritizes fast misaligned access while minimizing changes to the rest of the LSU. It doubles the width of the cache read and write buses and adds the align module to shift misaligned data into the correct position. If the address spills across two cache lines, it issues requests to both lines and merges the data for loads. Misaligned atomic operations still throw a store/AMO-misaligned fault.

The data cache from Section 10.4 reads and writes a native word size WORDLEN = LLEN = max(FLEN, XLEN). To support Zicclsm, Wally extends WORDLEN to LLEN*2, which allows the cache to output the entire access in one request for misaligned addresses within a single cache line.

Fig. 19.7 shows the details of the align module, with the read, write, and spill logic all together. It computes ByteOffsetM as IEUAdrM modulo the access size. If this offset is nonzero for a read or write, it asserts MisalignedM and shifts read or write data by ByteOffsetM bytes to align the data.

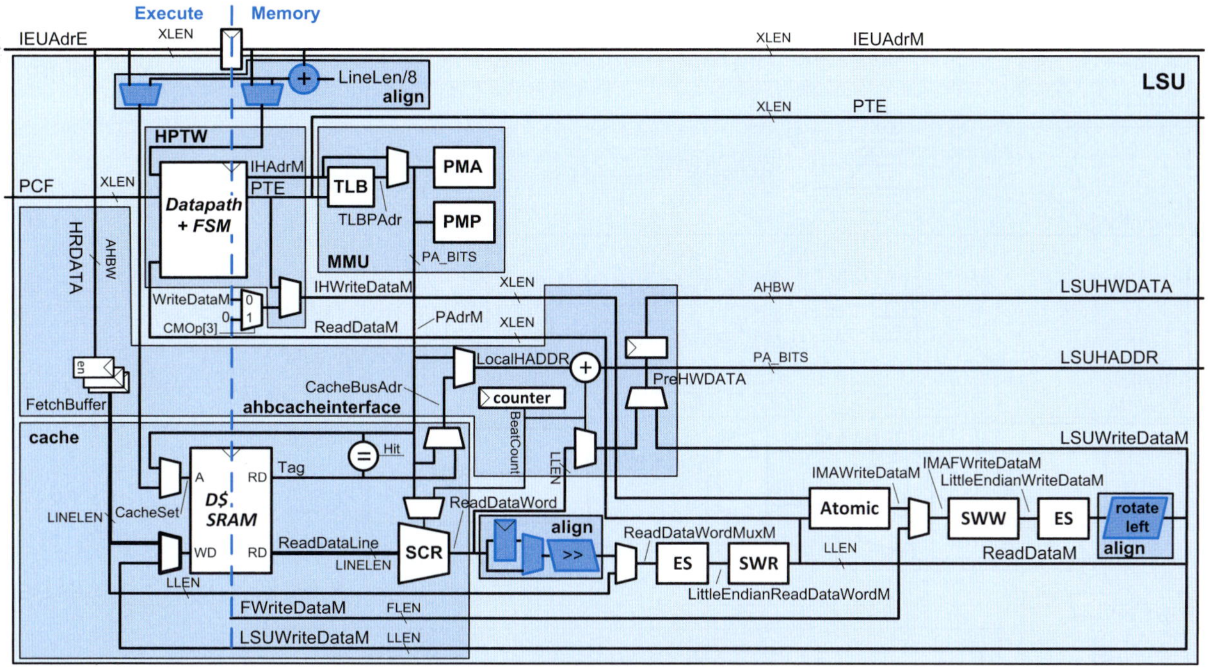

Fig. 19.6 **LSU with Zicclsm support.**

A spill (`ValidSpillM`) occurs when a misaligned address (`MisalignedM`) is in the last native-sized access of a cache line (`PotentialSpillM`). During a load spill, the align FSM asserts `SpillSaveM` to capture the `DCacheDataWordM` into register `ReadDataWordFirstHalfM`. `SelSpillE` is also asserted to increment `IEUAdrSpillE` by the size of the cache line in bytes. If the request is a store, the data cache writes the lower part of the word with the lower bits of `LSUWriteDataM`.

Fig. 19.8 shows a simplified diagram of the `align` FSM. See the code for full details. During a load spill, the FSM transitions to Spill and asserts `SelSpillM`, which increments `IEUAdrSpillM` by the size of the cache line in bytes. The MMU translates the address if address translation is enabled to the physical address `PAdrM`, and the cache outputs the second half of the load. `SelSpillM` also selects the `DCacheReadDataWordM` concatenated with `ReadDataWordFirstHalfM` which is then shifted by `ShiftAmount`.

19.3 TLB MANAGEMENT: Svinval

The Svinval TLB management extension defines `sinval.vma`, `sfence.w.inval`, and `sfence.inval.ir` instructions. They are closely

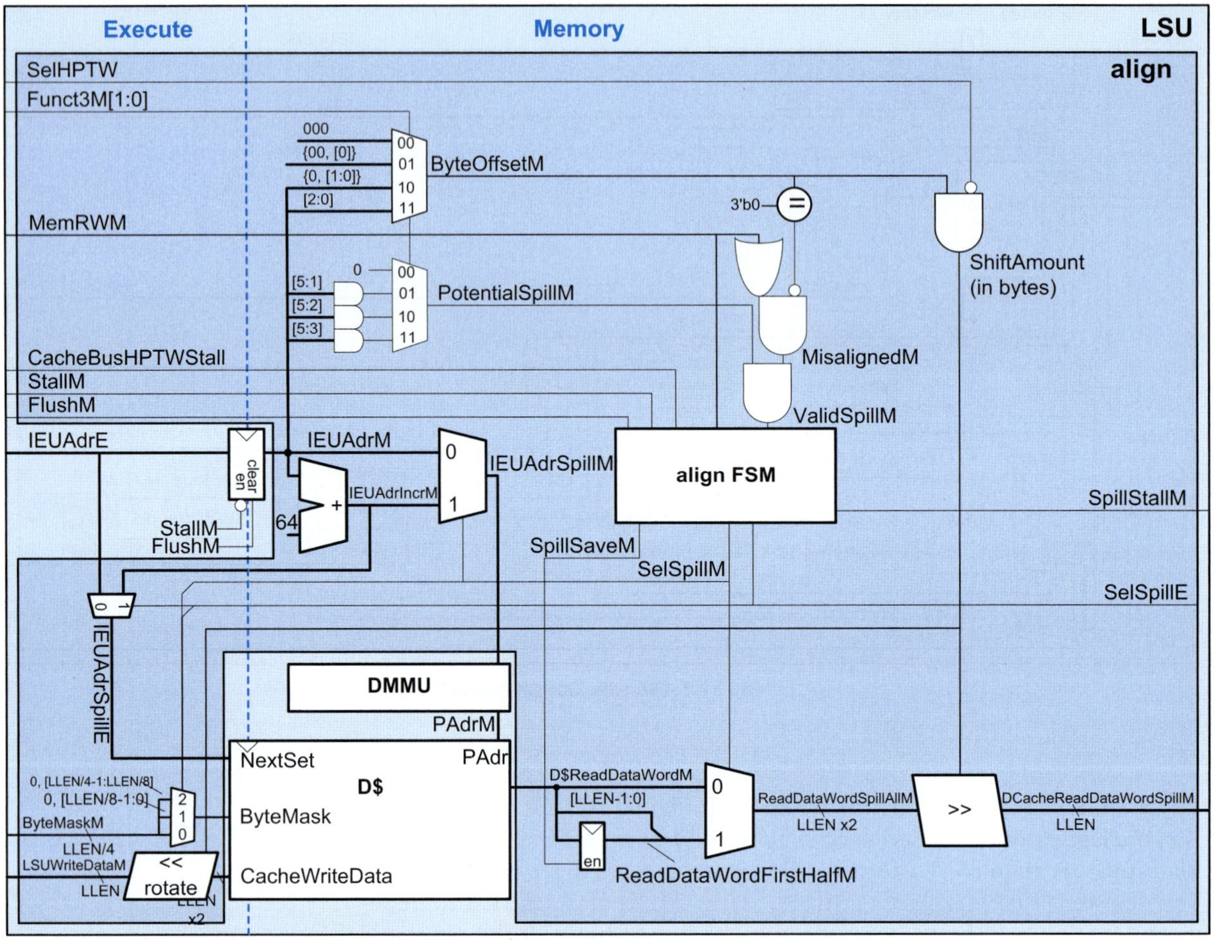

Fig. 19.7 **`align` module (assumes 64-byte cache line).**

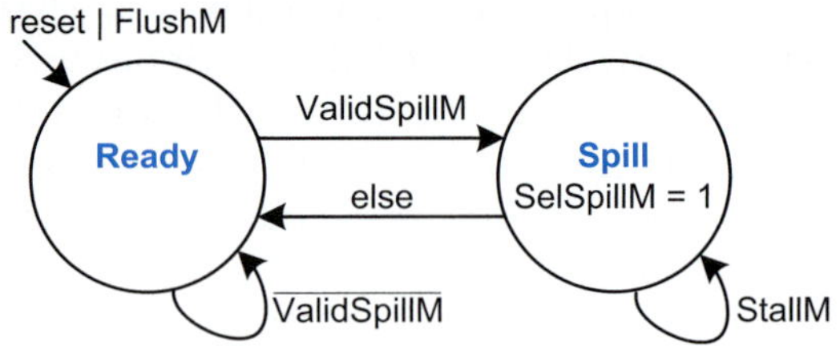

Fig. 19.8 **`align` FSM.**

related to `sfence.vma` but relax the ordering relative to other memory instructions, potentially improving performance on out-of-order or multicore systems. Their encoding matches `sfence.vma` except in bits 29:27.

Wally is an in-order single-core processor with inherent total store order (TSO) behavior, so these fine-grained TLB management instructions are unnecessary. Thus, the Svinval instructions are conservatively implemented as `sfence.vma` when `SVINVAL_SUPPORTED` = 1.

19.4 PAUSE HINT: Zihintpause

Section 17.1.4.1 described spin locks, which can be used to wait until a resource is available. The processor wastes power while spinning and, in a multithreaded system, it wastes execution resources potentially available to another thread. The Zihintpause extension introduces a `pause` instruction, which is a hint suggesting that the hart temporarily reduce the instruction issue rate while spinning. A spin lock loop should contain exactly one `pause` instruction. Code Example 19.1 illustrates a spin lock, based on Code Example 17.7, that waits until the lock is free (0), then takes the lock. If not immediately successful, it uses a `pause` hint before trying again.

`pause` is a hint encoded as a `fence w` (machine code: 0x0100000F), which does nothing.

Code Example 19.1 SPIN LOCK WITH PAUSE

```
spinlock:                      # Address of lock is in a0
        lr.w t0, (a0)          # Read the lock
        bnez t0, retry         # Spin until free
        li   t1, 1
        sc.w t0, t1, (a0)      # Try to write a 1 to take lock
        bnez t0, retry         # Spin until successful
        ret                    # Got the lock!
retry:                         # No lock yet
        pause                  # Pause hint to reduce spin power
        j spinlock             # Try again
```

Wally presently has no power management features to control instruction issue rate. It accepts the `pause` hint but ignores it.

19.5 CONDITIONAL INSTRUCTIONS: Zicond

Branches are expensive. When they are mispredicted, they flush the pipeline and reduce performance. Even when branches are predicted correctly, they take up space in the branch predictor memories. They also impact extracting instruction-level parallelism for superscalar and out-of-order microarchitectures.

Many architectures define *conditional* instructions to avoid branches by performing an operation only if a condition is true. For example, MIPS, Alpha, PowerPC, SPARC, x86, and ARM all have conditional move instructions to copy one register to another depending on the state of a third register or condition code. ARM and IA-64 provide more extensive *predication* in which any instruction can be conditionally executed based on condition codes. However, RISC-V lacks space in the instruction encodings to specify a third source register or condition code for full predication. Furthermore, conditional move instructions are difficult to implement in microarchitectures that use register renaming because they require reading three registers: the condition, the value to move, and the old value of the destination register that must be written to the renamed register if the condition is not satisfied.

RISC-V introduces a Zicond extension for conditional operations. This extension defines `czero.eqz` and `czero.nez`, listed in Table I.25 (on the last pages of the book). `czero.eqz rd, rs1, rs2` sets `rd` = 0 if `rs2` = 0, or `rd` = `rs1` otherwise. `czero.nez` is similar but checks if `rs2` ≠ 0.

Conditional zero instructions, `czero.*`, are easy to implement, even on a microarchitecture with register renaming and two register read ports per instruction issued. A programmer can use these instructions to write short sequences of code that operate conditionally. Table 19.2 lists some sequences. `cond` is `eq` or `ne` (equal or not equal zero). A conditional operation `add.eq rd, rs1, rs2, rc` computes `rd = (rc == 0) ? rs1 + rs2 : rs1`. In other words, it conditionally makes `rd` equal to `rs1` + `rs2` if `rc` equals 0, and just `rs1` otherwise. `add.ne` adds if `rc` does not equal 0. The conditional

Table 19.2 Conditional operation sequences.

<table>
<tr><th>Conditional Operation</th><th>Cond: eq</th><th>Cond: ne</th></tr>
<tr><td>add.cond rd, rs1, rs2, rc
rd = (rc cond 0) ? rs1+rs2 : rs1</td><td>czero.nez rd, rs2, rc
add rd, rs1, rd</td><td>czero.eqz rd, rs2, rc
add rd, rs1, rd</td></tr>
<tr><td>sub.cond rd, rs1, rs2, rc
rd = (rc cond 0) ? rs1-rs2 : rs1</td><td>czero.nez rd, rs2, rc
sub rd, rs1, rd</td><td>czero.eqz rd, rs2, rc
sub rd, rs1, rd</td></tr>
<tr><td>or.cond rd, rs1, rs2, rc
rd = (rc cond 0) ? rs1|rs2 : rs1</td><td>czero.nez rd, rs2, rc
or rd, rs1, rd</td><td>czero.eqz rd, rs2, rc
or rd, rs1, rd</td></tr>
<tr><td>xor.cond rd, rs1, rs2, rc
rd = (rc cond 0) ? rs1^rs2 : rs1</td><td>czero.nez rd, rs2, rc
xor rd, rs1, rd</td><td>czero.eqz rd, rs2, rc
xor rd, rs1, rd</td></tr>
<tr><td>and.cond rd, rs1, rs2, rc
rd = (rc cond 0) ? rs1&rs2 : rs1</td><td>and rd, rs1, rs2
czero.eqz rtmp, rs1, rc
or rd, rd, rtmp</td><td>and rd, rs1, rs2
czero.nez rtmp, rs1, rc
or rd, rd, rtmp</td></tr>
<tr><td>sel.cond rd, rs1, rs2, rc
rd = (rc cond 0) ? rs1 : rs2</td><td>czero.nez rd, rs1, rc
czero.eqz rtmp, rs2, rc
or rd, rd, rtmp</td><td>czero.eqz rd, rs1, rc
czero.nez rtmp, rs2, rc
or rd, rd, rtmp</td></tr>
<tr><td>mov.cond rd, rs1, rc
rd = (rc cond 0) ? rs1 : rd</td><td>czero.nez rd, rd, rc
czero.eqz rtmp, rs1, rc
or rd, rd, rtmp</td><td>czero.eqz rd, rd, rc
czero.nez rtmp, rs1, rc
or rd, rd, rtmp</td></tr>
</table>

Table 19.3 Code examples with branches, conditional operations, and conditional move.

Code	Branches	Conditional Zero	Conditional Move

`if (cond)`
`  x = 0;`
	`  beq s0, zero, skip`
	`  mv  s1, zero`
	`skip:`
		`czero.nez s1, s1, s0`	`cmov.nez s1, zero, s0`

`if (cond)`
`  x = 1;`
	`  beq s0, zero, skip`
	`  li  s1, 1`
	`skip:`
		`li        t0, 1`
		`czero.nez t0, t0, s0`
		`czero.eqz s1, s1, s0`
		`or        s1, s1, t0`
			`li       t0, 1`
			`cmov.nez s1, t1,   s0`

`if (cond == 0)`
`  x = x + y;`
	`  bne s0, zero, skip`
	`  add s1, s1,   s2`
	`skip:`
		`czero.eqz t0, s2, s0`
		`add       s1, t0, s1`
			`add      t0, s1,   s2`
			`cmov.eqz s1, t0,   s0`

`if (y >= 7)`
`  x = 2;`
	`  li  t0, 7`
	`  blt t1, t0,   skip`
	`  li  s1, 2`
	`skip:`
		`li        t0, 7`
		`slt       t0, s2, t0`
		`czero.eqz s1, s1, t0`
		`li        t1, 2`
		`czero.nez t1, t1, t0`
		`or        s1, s1, t1`
			`li       t0, 7`
			`slt      t0, s2,   t0`
			`li       t1, 2`
			`cmov.eqz s0, t1,   t0`

operations in the left column of Table 19.2 are purely explanatory; they are not pseudoinstructions supported by RISC-V assemblers and must be replaced by the actual sequence of instructions in the columns on the right. Note that some of the conditional operations involve a three-instruction sequence and a temporary register.

Table 19.3 compares how some common code sequences can be written with branches, conditional zero, or conditional moves. Assume `cond`, `x`, and `y` are in `s0`, `s1`, and `s2`, respectively. Conditional zero can sometimes replace a branch without increasing program length, but other times they take more instructions than a branch solution. Conditional moves (`cmov`) are more powerful than `czero.*` but are not part of the RISC-V instruction set because of the implementation difficulty.

19.5.1 Test Plan

The Zicond-extension instructions are tested with the riscv-arch-test suite. Table 19.4 lists the number of tests for each instruction.

Table 19.4 riscv-arch-test for the Zicond extension.

Instruction	RV32Zicond Tests	RV64Zicond Tests
`czero.eqz`	613	741
`czero.nez`	613	741
Total	**1,226**	**1,482**

19.5.2 Wally Implementation

The `czero.*` instructions are executed on the ALU using the existing AND gate. The AND gate's second source is set to all 1s or all 0s to either select the first source `rs1` or 0, respectively, based on `rs2`. Fig. 19.9 shows how the ALU from Fig. 18.21 is enhanced with Zicond support, highlighted in darker blue. A mux produces the `ZeroCondMaskInvB` of all 0 or all 1 for `czero` instructions based on `rs2 == 0` (`BZero`) The controller is also modified to drive `ALUSelect` = 111 to choose the AND output when it decodes `czero` instructions.

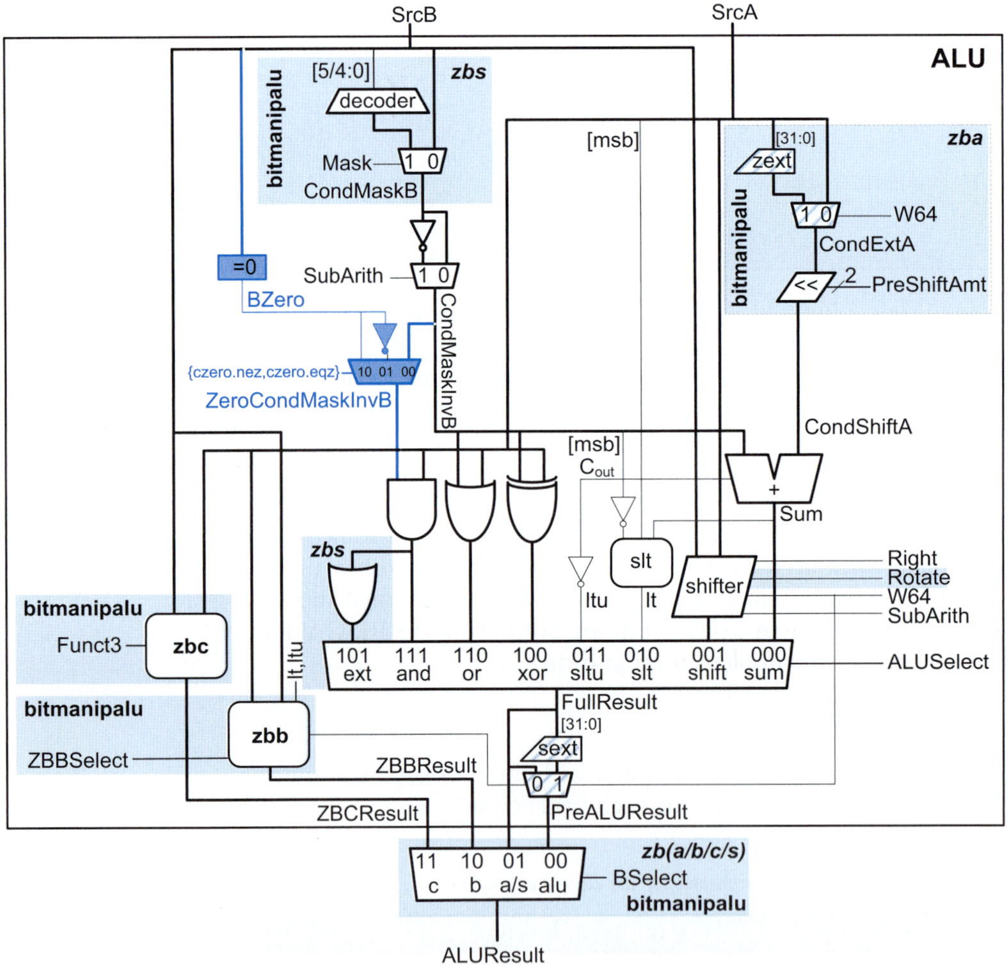

Fig. 19.9 ALU with Zicond support (crypto support not shown).

19.6 CUSTOM EXTENSIONS

RISC-V facilitates custom extensions for specific applications. The general process of creating a custom extension is to start with a precisely written specification, critically reviewed by an experienced team. In a test-driven design methodology, create a test plan and a set of tests to fully exercise the custom extension. Then update the HDL implementation. If the extension adds new instructions, the controller must decode the instructions and send appropriate commands to functional units. If the extension adds new CSRs, the privileged unit must add support. A custom extension usually requires a simulator and patches to the compiler and possibly the OS to run the extension.

Some companies create proprietary extensions to differentiate their products in the market. However, when a custom extension proves useful, there are many incentives to publish and standardize it:

- More software and hardware developers will use the extension.
- The extension become accessible on the main branch of the compiler, simulator, and OS.
- The ecosystem support burden is shared among many developers.
- The extension will not conflict with subsequent standard extension encodings and CSRs.
- The inventor gains credibility in the open-source RISC-V community and in the marketplace.

Relatively simple or noncontroversial extensions are standardized through a RISC-V International fast track process:

- A proposer documents the extension's name, motivation, use cases, and high-level description.
- The proposer requests that the RISC-V Architecture Review Committee open a fast-track review.
- Appropriate tech chairs review the plan milestones.
- The proposer provides a draft specification for internal review by an appropriate RISC-V committee.
- The proposer provides a final draft specification and checklist of deliverables.
- The final draft specification goes for a 2-week public review period.
- If no objections are raised, the RISC-V Technical Steering Committee votes to approve the extension.

More complex extensions, such as vector, matrix, CHERI, and crypto, are usually developed by a task group over a longer period of time with collaboration from many stakeholders.

19.7 SUMMARY

RISC-V is rapidly ratifying more extensions. This chapter covers extensions required for the RVA22S64 profile that don't fit naturally elsewhere in the book. They include cache management, misaligned loads and stores, TLB management, a pause hint, and conditional instructions.

Bibliography

Descriptor	Reference
6SN7	"6H8C (6SN7 double triode) Vostok 1980, 2022 [6SN7]," 2022. [Online]. Available: https://commons.wikimedia.org/wiki/File:6%D0%9D8%D0%A1_(6SN7_double_triode)_Vostok_1980_03.jpg.
Adve95	S. V. Adve and K. Gharachorloo, "Shared memory consistency models: a tutorial," *WRL Research Report 95/7*, 1995.
Aharoni08	M. Aharoni, "Floating-Point Test-Suite for IEEE," IBM Labs in Haifa, Jan. 2008. [Online]. Available: https://research.ibm.com/haifa/projects/verification/fpgen/papers/ieee-test-suite-v2.pdf.
AHB21	AMBA AHB Protocol Specification IHI0033C, ARM Ltd., 2021.
Aiken46	H. H. Aiken and G. M. Hopper, "The Automatic Sequence Controlled Calculator – I," *Electrical Engineering*, vol.65 no.8-9, pp. 384–391, Aug. 1946.
AMBA99	*AMBA Specification* Rev 2.0 IHI0011A, ARM Ltd., 1999.
APB21	*AMBA APB Protocol Specification* IHI0024D, ARM Ltd., 2021.
AppleIIe	Pratyeka, "Front view of an Apple IIe system, including computer chassis, monitor, and external 5¼" floppy disk drive," 2016. [Online]. Available: https://en.wikipedia.org/wiki/Apple_IIe#/media/File:Iie-system.jpg.
ARM11	*ARM11 Processor Technical Reference Manual*. Texas Instruments, 2005.
Arpaci-Dusseau19	R. H. Arpaci-Dusseau and A. C. Arpaci-Dusseau, *Operating Systems: Three Easy Pieces*, Arpaci-Dusseau Books, 2019.
Atkins68	D. Atkins, "Higher radix division using estimates of the divisor and partial remainders," *IEEE Trans. Computers*, vol. C-17, no. 10, pp. 925–934, Oct. 1968.
AXI21	*AMBA AXI and ACE Protocol Specification* IHI0022H.c, ARM Ltd., 2021.
[Babbage]	R. Roffe, Image of Charles Babbage, Wellcome Collection, 1833. [Online]. Available: https://wellcomecollection.org/works/krsuk329/images?id=d9hxrch7.

Bagwell00 P. Bagwell, "Ideal hash trees," 2001. [Online]. Available: https://lampwww.epfl.ch/papers/idealhashtrees.pdf.

Beck03 K. Beck, *Test-Driven Development by Example*, Pearson, 2003.

Beebe17 N. Beebe, *The Mathematical-Function Computation Handbook: Programming Using the MathCW Portable Software Library*, Springer, 2017.

Bell75 G. Bell and W. Strecker, "Computer structures: What have we learned from the PDP-11?" *IEEE Computer Conference*, Florida, Nov. 1975.

Bell76 G. Bell and W. Strecker, "Computer structures: what have we learned from the PDP-11?" *ACM SIGARCH Computer Architect. News*, pp. 1–14, Jan. 1976.

Bennett19 D. Patterson, J. Bennett, P. Dabbelt, C. Garlati, G. S. Madhusudan, and T. Mudge, "Embench™: An evolving benchmark suite for embedded IoT computers from an academic-industrial cooperative: towards the long overdue and deserved demise of Dhrystone," *RISC-V Workshop*, Zurich, Switzerland, June 2019. [Online]. Available: https://github.com/embench/embench-iot.

Bewick94 G. W. Bewick, "Fast Multiplication: Algorithms and Implementation," Ph.D. dissertation, Stanford University, 1994. [Online]. Available: http://i.stanford.edu/pub/cstr/reports/csl/tr/94/617/CSL-TR-94-617.pdf.

Binkley09 D. Binkley, M. Davis, D. Lawrie, and C. Morrell, "To CamelCase or Under_score," *Intl. Conf. on Program Comprehension*, Vancouver, British Columbia, Canada, pp. 158–167, 2009.

Bohr07 M. Bohr, "A 30 year retrospective on Dennard's MOSFET scaling paper," *IEEE Solid-State Circuits Society Newslett.*, vol. 12, no. 1, pp. 11–13, 2007.

bug The Naval Surface Warfare Center, Dahlgren, VA., U.S. Naval History and Heritage Command Photograph, "The First 'Computer Bug'," 1988. [Online]. Available: https://www.history.navy.mil/content/history/nhhc/our-collections/photography/numerical-list-of-images/nhhc-series/nh-series/NH-96000/NH-96566-KN.html.

C17 "C programming," *Wikimedia*. [Online]. Available: upload.wikimedia.org/wikipedia/commons/0/07/C_Programming.pdf.

Cekleov97 M. Cekleov and M. Dubois, "Virtual-address caches part 1: problems and solutions in uniprocessors," *IEEE Micro*, pp. 64–77, 1997.

Chacon14 S. Chacon and B. Straub, *Pro Git*, Apress, 2014. [Online]. Available: https://git-scm.com/book/en/v2.

codesize RISC-V Code Size Reduction Group, Archived 2024. [Online] Available: https://github.com/riscvarchive/riscv-code-size-reduction?tab=readme-ov-file.

Coffman71 E. Coffman, M. Elphick, and A. Shoshani, "System deadlocks," *ACM Computing Surveys*, vol. 3, no. 2, pp. 67–78, June 1971.

Colossus The National Archives (United Kingdom), catalogued under document record FO850/234, "A Colossus Mark 2 codebreaking computer being operated by Dorothy Du Boisson (left) and Elsie Booker (right)," 1943.

[Online]. Available: https://en.wikipedia.org/wiki/Colossus_computer#/media/File:Colossus.jpg.

Colwell02 B. Colwell, "If you didn't test it, it doesn't work," *Computer*, vol. 35, no. 5, pp. 11–13, May 2002.

Colwell05 R. Colwell, *The Pentium Chronicles: The People, Passion, and Politics Behind Intel's Landmark Chips*, Wiley, 2005.

Coonen81 J. Coonen, "Underflow and the denormalized numbers," *IEEE Computer*, vol. 4, no. 3, pp. 75–87, Mar. 1981.

Cooper13 V. Cooper, *Getting Started with UVM: A Beginner's Guide*, Verilab Publishing, 2013.

Corbato62 F. Corbato, M. Merwin-Daggett, and R. Daley, "An experimental time-sharing system," *Proc. Spring Joint Computer Conference*, vol. 21, pp. 335–344, 1962.

corei9 F. Fritz and JmsDoug, "Intel Core i9-13900K die," 2023. [Online]. Available: https://commons.wikimedia.org/wiki/File:Intel_Core_i9-13900K_Labelled.

Cornea04 M. Cornea, "Software implementations of division and square root operations for Intel Itanium processors," *Proc. Workshop on Computer Architecture Education*, ACM, 2004.

Cortadella92 J. Cortadella and J. Llaberia, "Evaluation of A + B = K conditions without carry propagation," *IEEE Trans. Computers*, vol. 41, no. 11, pp. 1484–1488, Nov. 1992.

crtg "Sample constrained random test generator file," *Github*. [Online]. Available: https://github.com/riscv-software-src/riscv-ctg/blob/master/sample_cgfs/rv32i.cgf.

Dadda65 L. Dadda, "Some schemes for Parallel Multipliers," Alta Frequenza, vol. 34, pp. 349–356, 1965.

Dally98 W. Dally and J. Poulton, *Digital Systems Engineering*, Cambridge Univ. Press, 1998.

Davis64 E. M. Davis, W. E. Harding, R. S. Schwartz, and J. J. Corning, "Solid logic technology: Versatile, high-performance microelectronics," *IBM Journal of Research and Development*, vol. 8, no. 2, pp. 102–114, Apr. 1964.

DeHon99 A. DeHon, "Balancing interconnect and computation in a reconfigurable computing array (or, why you don't really want 100% LUT utilization)," *Proceedings of the 1999 ACM/SIGDA seventh international symposium on Field programmable gate arrays (FPGA '99)*, ACM, New York, NY, USA, pp. 69–78, 1999.

Dennard74 R. H. Dennard, F. H. Gaensslen, H. -N. Yu, V. L. Rideout, E. Bassous, and A. R. LeBlanc, "Design of ion-implanted MOSFET's with very small physical dimensions," *IEEE Journal of Solid-State Circuits*, vol. 9, no. 5, pp. 256–268, Oct. 1974.

Dhanesha95 H. Dhanesha, K. Falakshahi, and M. Horowitz, "Array-of-arrays architecture for parallel floating point multiplication," *Proceedings Sixteenth Conf. on Advanced Research in VLSI*, Chapel Hill, NC, USA, pp. 150–157, 1995.

Dijkstra65 E. W. Dijkstra, "Cooperating sequential processes," *Technical Report EWD-123*, 1965.

Dijkstra68 E. W. Dijkstra, "The structure of the 'THE'-multiprogramming system," *Proceedings of the first ACM symposium on Operating System Principles (SOSP '67)*, ACM, pp. 10.1–10.6, 1968.

Dijkstra71 E. W. Dijkstra, "Hierarchical ordering of sequential processes," *Acta Informatica* 1, pp. 115–138, 1971.

Dobberpuhl92 D. W. Dobberpuhl et al., "A 200-MHz 64-b dual-issue CMOS microprocessor," *IEEE Journal of Solid-State Circuits,* vol. 27, no. 11, pp. 1555–1567, Nov. 1992.

Ellis21 L. Ellis, "It's time for IEEE to retire 'master/slave'," *EE Times*, 2021. [Online]. Available: https://www.eetimes.com/its-time-for-ieee-to-retire-master-slave/.

Ercegovac03 M. Ercegovac and T. Lang, *Digital Arithmetic*, Elsevier, 2003.

Ercegovac94 M. Ercegovac and T. Lang, *Division and Square Root: Digit-Recurrence Algorithms and Implementations*, Springer, 1994.

Eves69 H. Eves, *In Mathematical Circles*, Brookes/Cole Publishers, 1969.

FE310 *SiFive FE310-G003 Manual v1p1*, SiFive, 2020. [Online]. Available: https://sifive.cdn.prismic.io/sifive/3af39c59-6498-471e-9dab-5355a0d539eb_fe310-g003-manual.pdf.

Ferguson10 N. Ferguson, B. Schneier, and T. Kohno, *Cryptography Engineering: Design Principles and Practical Applications*, Wiley Publishing, 2010.

Fleming86 P. Fleming and J. Wallace, "How not to lie with statistics: The correct way to summarize benchmark results," *Communications of the ACM*, vol. 29, no. 2, pp. 218–221, Mar. 1986.

Flynn97 D. Flynn, "AMBA: Enabling reusable on-chip designs," *IEEE Micro,* pp. 20–27, July/Aug. 1997.

Franke02 H. Franke, R. Russell, and M. Kirkwood, "Fuss, futexes and furwocks: Fast user-level locking in Linux," *AUUG Conference Proceedings*, vol. 85. AUUG, Inc, 2002.

FU540 *SiFive FU540-C000 Manual v1p0*, SiFive, 2018. [Online]. Available: https://static.dev.sifive.com/FU540-C000-v1.0.pdf.

Gallmann21 N. Gallmann, P. Vogel, P. Schiavone, and L. Benini, "From swift to mighty: A cost-benefit analysis of Ibex and CV32E40P regarding application performance, power, and area," *Workshop Computer Architecture Research with RISC-V*, 2021. [Online]. Available: https://carrv.github.io/2021/papers/CARRV2021_paper_8_Gallmann.pdf.

Gelsinger01 P. P. Gelsinger, "Microprocessors for the new millennium: Challenges, opportunities, and new frontiers," *2001 IEEE International Solid-State Circuits Conference Digest of Technical Papers ISSCC*, San Francisco, CA, USA, pp. 22–25, 2001.

Gelsinger06 P. Gelsinger, "Into the Core . . . ," 2006. [Online]. Available: https://web.stanford.edu/class/ee380/Abstracts/060607-EE380-Gelsinger.pdf.

Golson16 S. Golson and L. Clark, "Language wars in the 21st century: Verilog versus VHDL—revisited," *Synopsys Users Group Conference* (SNUG), 2016.

Greenblatt74 R. Greenblatt, "The LISP machine," *Working Paper 79*, MIT Artificial Intelligence Laboratory, Cambridge, MA, USA, Nov. 1974.

Halfhill95 T. Halfhill, "The truth behind the Pentium bug," *BYTE.com*, 1995. [Online]. Available: https://web.archive.org/web/20060209005434/http://www.byte.com/art/9503/sec13/art1.htm.

Harris21 S. Harris and D. Harris, *Digital Design and Computer Architecture: RISC-V Edition*, Elsevier, 2021.

Harris23 D. Harris, J. Stine, M. Ercegovac, A. Nannarelli, K. Parry, and C. Turek, "Unified digit selection for radix-4 recurrence division and square root," *IEEE Trans. Computers*, vol. 73, no. 1, pp. 292–300, Jan. 2024.

Hatamian86 M. Hatamian and G. L. Cash, "A 70-MHz 8-bit/spl times/8-bit parallel pipelined multiplier in 2.5-/spl mu/m CMOS," *IEEE Journal of Solid-State Circuits*, vol. 21, no. 4, pp. 505–513, Aug. 1986.

Hauser18 J. Hauser, "Berkeley TestFloat release 3e general documentation," 2018. [Online]. Available: http://www.jhauser.us/arithmetic/TestFloat-3/doc/TestFloat-general.html.

Hennessy19 J. Hennessy and D. Patterson, *Computer Architecture: A Quantitative Approach*, 6th ed., Morgan Kaufmann, 2019.

Hennessy20 J. Hennessy and D. Patterson, *Computer organization and design RISC-V edition: The hardware software interface*, 2nd Edition, Morgan Kaufmann, 2020.

Hennessy82 J. Hennessy, N. Jouppi, S. Przybylski, C. Rowen, T. Gross, F. Baskett, and J. Gill, "MIPS: A microprocessor architecture," *Proceedings of the 15th Annual Workshop on Microprogramming (MICRO 15)*, IEEE Press, 17–22, 1982.

Hennessy90 J. L. Hennessy and D. A. Patterson, *Computer Architecture: A Quantitative Approach*, Ap Professional Publishers Inc., 1990.

Hinton01 Hinton, G., Upton, M., Sager, D. J., Boggs, D., Carmean, D. M., Roussel, P., ... & Murray, R., "A 0.18-/spl mu/m CMOS IA-32 processor with a 4-GHz integer execution unit," *IEEE Journal of Solid-State Circuits*, vol. 36, no. 11, pp. 1617–1627, Nov. 30, 2001.

Hopper J. S. Davis, "Commodore Grace M. Hopper, USN (covered)," 1984. [Online] Available: https://commons.wikimedia.org/wiki/File:Commodore_Grace_M._Hopper,_USN_(covered).jpg.

Horel99 T. Horel and G. Lauterbach, "UltraSPARC-III: Designing third-generation 64-bit performance," *IEEE Micro*, vol. 19, no. 3, pp. 73–85, May-June 1999.

IBM64 IBM, "The IBM System/360." [Online]. Available: https://www.ibm.com/history/system-360.

IBMPC IBM, "The IBM PC." [Online]. Available: https://www.ibm.com/history/personal-computer.

IC3-21	FBI, *Internet Crime Report*, 2021. [Online]. Available: https://www.ic3.gov/Media/PDF/AnnualReport/2021_IC3Report.pdf.
IDC23	IDC. [Online]. Available: https://www.idc.com/getdoc.jsp?containerId=prAP50994023.
IEEE 1800.2-2017	IEEE, *Universal Verification Methodology*, IEEE Standard 1800.2-2017, 2017.
IEEE1076-2019	IEEE, *VHDL Language Reference Manual*, IEEE Standard 1076-2019, 2019.
IEEE1666-2011	IEEE, *Standard SystemC Language Reference Manual*, IEEE Standard 1666-2011, 2012.
IEEE1788.1-2017	IEEE, *Interval Arithmetic*, IEEE Standard 1788.1-2017, 2017.
IEEE1800-2017	IEEE, *SystemVerilog—Unified Hardware Design, Specification, and Verification Language*, IEEE Standard 1800-201, 2018.
IEEE1800-2023	IEEE, *SystemVerilog—Unified Hardware Design, Specification, and Verification Language*, IEEE Standard 1800-2023, 2024.
IEEE754-1985	IEEE, *Binary Floating-Point Arithmetic*, ANSI/IEEE Standard *754-1985*, 1985.
IEEE754-2008	IEEE, *Floating-Point Arithmetic*, IEEE Standard 754-2008, 2008.
IEEE754-2019	IEEE, *Floating-Point Arithmetic*, IEEE Standard 754-2019, 2019.
Intel4004	IEEE Spectrum, "The surprising story of the first microprocessors." [Online]. Available: https://spectrum.ieee.org/the-surprising-story-of-the-first-microprocessors.
Intel69	Intel, "Intel's first product—the 3101," [Online]. Available: https://www.intel.com/content/www/us/en/history/virtual-vault/articles/intels-first-product-3101.html.
Intel71	Intel, "Announcing a new era of integrated electronics: The Intel 4004." [Online]. Available: https://www.intel.com/content/www/us/en/history/virtual-vault/articles/the-intel-4004.html.
Intel89	"80486—Intel," *WikiChip*. Intel Corporation, i486 Microprocessor, 1989. [Online]. Available: https://en.wikichip.org/wiki/intel/80486#Datasheet.
iphone13	Tech Insights, "What's inside the Apple iPhone 13 Pro teardown?" [Online]. Available: http://www.techinsights.com/blog/teardown/apple-iphone-13-pro-teardown.
Isen09	C. Isen, L. K. John, and E. John, "A tale of two processors: Revisiting the RISC-CISC debate," *2009 SPEC Benchmark Workshop*, Springer, vol. 5419, pp. 57–75, 2009.
Keating02	M. Keating and P. Bricaud, *Reuse Methodology Manual for System-on-a-Chip Designs*, Springer, 2002.
Keating07	M. Keating, D. Flynn, R. Aitken, A. Gibbons, and K. Shi, *Low Power Methodology Manual*, Springer, 2007.
Kidder81	T. Kidder, *The Soul of a New Machine*, Little, Brown, & Co., 1981.
Kilburn62	T. Kilburn, D. Edwards, M. Lanigan, and F. Sumner, "One-level storage system," *IRE Trans. Electronic Computers*, vol. EC-11, no. 2, pp. 223–235, Apr. 1962.

Kim25	K. Kim, K. Parry, D. Harris, C. Turek, A. Maiuolo, R. Thompson, and J. Stine, "Shared Recurrence Floating-Point Divide/Sqrt and Integer Divide/Remainder With Early Termination" IEEE Transactions on Computers, 2024.
Lamport79	L. Lamport, "How to make a multiprocessor computer that correctly executes multiprocess programs," *IEEE Trans. Computers*, vol. C-28, no. 9, pp. 690–691, Sept. 1979.
Larabel19	M. Larabel, "The Linux kernel enters 2020 at 27.8 million lines in Git but with less developers for 2019." [Online]. Available: https://www.phoronix.com/scan.php?page=news_item&px=Linux-Git-Stats-EOY2019.
Laskai21	L. Laskai and A. Segal, "The Encryption Debate in China: 2021 Update," 2021. [Online]. Available: https://carnegieendowment.org/posts/2021/03/the-encryption-debate-in-china-2021-update.
Lichtenau16	C. Lichtenau, S. Carlough, and S. M. Mueller, "Quad precision floating point on the IBM z13," *IEEE Symp. Computer Arithmetic*, pp. 87–94, 2016.
Liu12	W. Liu and A. Nannarelli, "Power efficient division and square root unit," *IEEE Trans. Computers*, vol. 61, no. 8, pp. 1059–1070, 2012.
Lockless23	Lockless Inc., "Mutexes and condition variables using futexes," 2023. [Online]. Available: http://locklessinc.com/articles/mutex_cv_futex.
MacSorley61	O. L. Macsorley, "High-speed arithmetic in binary computers," *Proc. IRE*, vol. 49, no. 1, pp. 67–91, Jan. 1961.
Magnussen97	P. Magnussen, "Understanding stacks and registers in the Sparc architecture(s)," 1997. [Online]. Available: https://cseweb.ucsd.edu/~gbournou/CSE131/sparcstack.html.
market25	T. Alsop, "Semiconductor market size worldwide from 1987 to 2025," 2025. [Online]. Available: http://www.statista.com/statistics/266973/global-semiconductor-sales-since-1988.
MarkI	J. Kopplin, "An illustrated history of computers: Part 3," 2002. [Online]. Available: http://www.cs.kent.edu/~rothstei/10051/HistoryPt3.htm.
Mayer82	A. J. W. Mayer, "The architecture of the Burroughs B5000: 20 years later and still ahead of the times?" *SIGARCH Comput. Archit. News 10*, pp.3–10, 1982.
McFarling93	S. McFarling, "Combining branch predictors," *WRL Technical Note TN-36*, Digital Equipment Corporation, 1993.
Mecklenburg04	R. Mecklenburg, *Managing Projects with GNU Make*, 3rd. ed., O'Reilly, 2004.
Mehta20	A. Mehta, *SystemVerilog Assertions and Functional Coverage*, 3rd. ed., Springer, 2020.
memory	*Ken Shirriff's blog*. [Online]. Available: http://static.righto.com/images/ibm-360-mem/core-closeup2.jpg.
Met Museum of Art/Intel	Intel, "October 17, 1986: Raising the bar with the 386." [Online]. Available: https://timeline.intel.com/1985/raising-the-bar-with-the-386.

Montoye90 R. Montoye, E. Hokonek, and S. Runyan, "Design of the IBM RISC System/6000 floating-point execution unit," *IBM J. Research & Development*, vol. 34, no. 1, pp. 59–70, 1990.

Moody01 G. Moody, "The Greatest OS that (N)ever Was," *Wired*, Aug. 1, 1997.

Moore03 G. E. Moore, "No exponential is forever: But "forever" can be delayed! [semiconductor industry]," *2003 IEEE International Solid-State Circuits Conference, 2003 Digest of Technical Papers ISSCC.*, San Francisco, CA, USA, vol. 1, pp. 20–23, 2003.

Moore65 G. Moore, "Cramming more components onto integrated circuits," *Electronics*, vol. 38, no. 8, pp. 114–117, Apr. 1965.

Muller05 J.-M. Muller, "On the definition of ulp," *Research Report 2005-09*, École Normale Supérieure de Lyon, 2005.

Muller18 J. Muller et al., *Handbook of Floating-Point Arithmetic*, 2nd ed., Springer, 2018.

MultilayerAHB04 Multi-layer AHB V2.0 Technical Overview DVI0045B, ARM Ltd., 2004.

Nagarajan20 V. Nagarajan, D. J. Sorin, M. D. Hill, and D. A. Wood, "A primer on memory consistency and cache coherence," 2nd ed., *Synthesis Lectures on Computer Architecture*, vol. 15, no. 1, pp. 1–294, 2020.

NameConv "Naming convention (programming)," *Wikipedia*. [Online]. Available: https://en.wikipedia.org/wiki/Naming_convention_(programming).

NIST-FIPS-180-4 National Institute of Standards and Technology, *Secure Hash Standard*, Pub 180-4, Aug. 2015. [Online]. Available: https://nvlpubs.nist.gov/nistpubs/FIPS/NIST.FIPS.180-4.pdf.

NIST-FIPS-197-upd1 National Institute of Standards and Technology, *The Advanced Encryption Standard (AES)*, Pub 197. [Online]. Available: https://nvlpubs.nist.gov/nistpubs/FIPS/NIST.FIPS.197-upd1.pdf.

OBI22 Open Bus Interface 1.4, Silicon Labs, 2022. [Online]. https://github.com/openhwgroup/core-v-docs/blob/master/cores/obi/OBI-v1.2.pdf https://github.com/openhwgroup/programs/blob/master/TGs/cores-task-group/obi/OBI-v1.4.pdf.

Oklobdzija96 V. G. Oklobdzija, D. Villeger, and S. S. Liu, "A method for speed optimized partial product reduction and generation of fast parallel multipliers using an algorithmic approach," *IEEE Transactions on Computers*, vol. 45, no. 3, pp. 294–306, Mar. 1996.

onsemi61 Fairchild/ON Semiconductor Corporation via The Chip History Center, "First Monolithic Silicon IC Chip. Invented by Robert Noyce, Fairchild Semiconductors," 1961. [Online]. Available: https://www.chiphistory.org/chc_upload/content/jpg/7/1499079245/1499079245.jpg.

opensource.org The Open-Source Initiative. [Online]. Available: https://opensource.org.

OSHA22 Open Source Hardware Association, "A resolution to redefine SPI signal names," 2022. [Online]. Available: https://www.oshwa.org/a-resolution-to-redefine-spi-signal-names/.

oshwa.org Open Source Hardware Association. [Online]. Available: https://www.oshwa.org.

Ousterhout09 J. Ousterhout and K. Jones, *Tcl and the Tk Toolkit,* Upper Saddle River, NJ, USA: Addison-Wesley, 2009.

Pan92 S.-T. Pan, I. So, and J. T. Rahmeh, "Improving the accuracy of dynamic branch prediction using branch correlation," *Proc. Architecture Support for Programming Languages and Operating Systems*, pp. 76–84, 1992.

Patterson25 D. Patterson et al., "Embench IOT 2.0 and DSP 1.0: Modern Embedded Computing Benchmarks," IEEE Computer, vol. 58, no 5, May 2025, pp. 37–47.

Patterson80 D. R. Ditzel and D. A. Patterson, "Retrospective on high-level language computer architecture," *Proceedings of the 7th Symposium Computer Architecture (ISCA '80)*, ACM, pp. 97–104, 1980.

PC16550 Texas Instruments, "PC16550D universal asynchronous receiver/transmitter with FIFOs," 1995.

PDP-11 D. McGuire, "Later PDP–11/70 with disks and tape," 2009. [Online]. Available: https://en.wikipedia.org/wiki/PDP-11#/media/File:PDP-11-70-DDS570.jpg.

PLIC22 RISC-V International, "RISC-V Platform-Level Interrupt Controller Specification," 2022. [Online]. Available: http://github.com/riscv/riscv-plic-spec/blob/master/riscv-plic.adoc.

Prabhu95 J. A. Prabhu and G. B. Zyner, "167 MHz radix-8 divide and square root using overlapped radix-2 stages," *IEEE Symposium on Computer Arithmetic*, pp. 155–162, 1995.

Quinnell07 E. Quinnell, E. E. Ewartzlander, Jr., and C. Lemonds, "Floating-point fused multiply-add architectures," *Asilomar Conf. Signals, Systems, and Computers*, pp. 331–337, 2007.

RISCV23 "RISC-V Celebrates Momentum of 2022 and Accelerating Adoption in 2023," RISC-V International, 2023. [Online]. Available: https://riscv.org/blog/2023/01/risc-v-celebrates-momentum-of-2022-and-accelerating-adoption-in-2023/.

RISCV-ABI RISC-V International, "RISC-V ELF Processor-Specific ABI," [Online]. Available: http://github.com/riscv-non-isa/riscv-elf-psabi-doc.

riscv-arch-test RISC-V International, "RISC-V Foundation Architecture Test Special Interest Group (SIG)," [Online]. Available: http:github.com/riscv-non-isa/riscv-arch-test.

RISCV-ASM RISC-V International, "RISC-V assembly programmer's manual," [Online]. Available: http://github.com/riscv-non-isa/riscv-asm-manual.

RISCV-Privileged RISC-V International, "The RISC-V Instruction Set Manual, Volume II: Privileged Architecture." [Online]. Available: https://riscv.org/specifications/ratified/.

RISCV-Profiles RISC-V International, "RISC-V Profiles." [Online]. Available: https://riscv.org/specifications/ratified/.

RISCV-SAIL RISC-V International, "RISC-V Sail Model." [Online]. Available: https://github.com/riscv/sail-riscv.

RISCV-SBI RISC-V International, "RISC-V SBI Specification." [Online]. Available: https://github.com/riscv-non-isa/riscv-sbi-doc.

RISCV-Unprivileged RISC-V International, "The RISC-V Instruction Set Manual, Volume I: Unprivileged Architecture." [Online]. Available: https://riscv.org/specifications/ratified/.

Ritchie20 H. Ritchie and M. Roser, "Moore's law," 2020. [Online]. Available: https://commons.wikimedia.org/wiki/File:Moore%27s_Law_Transistor_Count_1970-2020.png.

Ritchie84 D. Ritchie, "The evolution of the Unix Time-sharing System," *Communications of the ACM*, 1984, pp. 365–375.

Robertson58 J. Robertson, "A new class of digital division methods," *IRE Trans. Electronic Computers*, vol. EC-7, pp. 218–222, Sept. 1958.

RV22 M. Himelstein, "RISC-V state of the union," 2022. [Online]. Available: https://open-src-soc.org/2022-05/media/slides/RISC-V-International-Day-2022-05-05-09h00-Mark-Himelstein.pdf.

RVVI-TRACE "RISC-V verification interface," *Github*. [Online]. Available: https://github.com/riscv-verification/RVVI.

RyzenVi9 B. Justice, "Intel Core i9-12900K vs AMD Ryzen 9 5900X Performance Review," Feb. 8, 2022. [Online]. Available: https://www.thefpsreview.com/2022/02/08/intel-core-i9-12900k-vs-amd-ryzen-9-5900x-performance-review/6/.

Saarinen20 M.-J. O. Saarinen, "A lightweight ISA extension for AES and SM4," *International Workshop on Secure RISC-V*, Apr. 2020.

SAIL "The Sail ISA specification language," *Github*. [Online]. Available: https://github.com/rems-project/sail.

Santoro89 M. R. Santoro and M. A. Horowitz, "SPIM: A pipelined 64*64-bit iterative multiplier," *IEEE Journal of Solid-State Circuits*, vol. 24, no. 2, pp. 487–493, Apr. 1989.

Schmookler01 M. S. Schmookler and K. J. Nowka, "Leading zero anticipation and detection - a comparison of methods," *IEEE Symp. Computer Arithmetic*, pp. 7–12, 2001.

Schotts19 W. Schotts, "The Linux command line," 2024. [Online]. Available: http://linuxcommand.org/tlcl.php.

Schwartz05 E. M. Schwartz, M. Schmookler, and S. D. Trong, "FPU implementations with denormalized numbers," *IEEE Trans. Computers*, vol. 54, no. 7, pp. 825–836, July 2005.

Scott13 M. Scott, *Shared-Memory Synchronization*, Morgan & Claypool, 2013.

Seidel03 P. Seidel, "Multiple path IEEE floating-point fused multiply-add," *International Midwest Symp. Circuits and Systems*, pp. 1359–1362, 2003.

Seligman23 E. Seligman, T. Schubert, and M. V. A. K. Kumar, *Formal verification: An essential toolkit for modern VLSI Design*, 2nd ed., Elsevier, 2023.

Sequin82 C. H. Sequin and D. A. Patterson, "Design and Implementation of RISC I," *UC Berkeley Technical Report*, 1982.

Severance16 C. Severance, *Python for Everybody*, 2016. [Online]. Available: http://do1.dr-chuck.com/pythonlearn/EN_us/pythonlearn.pdf.

Severance98 C. Severance, "IEEE 754: An interview with William Kahan," *Computer*, vol. 31, no. 3, pp. 114–115, Mar. 1998.

Seznec06 A. Seznec and P. Michaud, "A case for (partially) TAgged GEometric history length branch prediction," *J. Instruction-Level Parallelism*, vol. 8, pp. 23, 2006.

Shannon45 C. E. Shannon, "A mathematical theory of cryptography," *Bell System Technical Memo*, MM 45-110-02, Sept. 1945.

Shannon49 C. E. Shannon, "Communication theory of secrecy systems," *Bell System Technical Journal*, vol. 28-4, pp. 656–715, 1949.

SHD24 SHD Group, "RISC-V market report: Application forecasts in a heterogeneous world," *SHD101A.b-24*, Jan. 2024.

Shilov20 A. Shilov, "The future of leading-edge chips according to TSMC: 5nm, 4nm, 3nm and beyond," TechRadar, 2020. [Online]. Available: https://www.techradar.com/news/the-future-of-leading-edge-chips-according-to-tsmc-5nm-4nm-3nm-and-beyond.

SLT IEEE Spectrum, "Building the system/360 mainframe nearly destroyed IBM." [Online]. Available: https://spectrum.ieee.org/building-the-system360-mainframe-nearly-destroyed-ibm#toggle-gdpr.

SPEC2000 Standard Performance Evaluation Corporation, "SPEC CPU2000 V1.3 Benchmark," 2000. [Online]. Available: https://www.spec.org/cpu2000/.

SPEC92 Standard Performance Evaluation Corporation, "SPEC CPU92 Benchmark," 1992. [Online]. Available: https://www.spec.org/cpu92/.

Spike RISC-V International, "Spike RISC-V ISA simulator." [Online]. Available: http://github.com/riscv-software-src/riscv-isa-sim.

Stallman22 R. Stallman, R. McGrath, and P. Smith, "GNU make," 2022. [Online]. Available: http://www.gnu.org/software/make/manual/make.pdf.

Stallman23 R. Stallman, R. Pesch, S. Shebs, et al., "Debugging with GDB," 2023. [Online]. Available: http://sourceware.org/gdb/current/onlinedocs/gdb.pdf.

Stallman86 R. M. Stallman, "What is the Free Software Foundation?" *GNU's Bulletin*, vol. 1, no. 1, Feb. 1986.

Statista25 "Arm Holdings chip unit shipments worldwide from fiscal year 2017 to 2024, by quarter (in billions)," Statista, 2025. [Online]. Available: https://www.statista.com/statistics/1131983/arm-based-chip-unit-shipments-as-reported-by-licensees-worldwide/.

StDenis07 T. St. Denis, *Cryptography for Developers*, Syngress Publishing, 2007.

Stelling98 P. F. Stelling, C. U. Martel, V. G. Oklobdzija, and R. Ravi, "Optimal circuits for parallel multipliers," *IEEE Transactions on Computers*, vol. 47, no. 3, pp. 273–285, Mar. 1998.

Suggs20 D. Suggs, M. Subramony, and D. Bouvier, "The AMD 'Zen 2' processor," *IEEE Micro*, vol. 40, no. 2, pp. 45–52, 2020.

Sutherland06 S. Sutherland and D. Mills, "Standard gotchas subtleties in the Verilog and SystemVerilog standards that every engineer should know," *Synopsys User Group Conference Proc. (SNUG)*, 2006/7.

Sutherland13 S. Sutherland, "I'm still in love with my X," *Design and Verification Conference*, 2013.

Sutherland99 I. Sutherland, R. Sproull, and D. Harris, *Logical Effort: Designing Fast CMOS Circuits*, Morgan Kaufmann, 1999.

Swann23 R. Swann and J. E. Stine, "Evaluation of a modular approach to AES hardware architecture and optimization," *J Sign Process Syst*, vol. 95, no. 7, pp. 797–813, 2023.

Sweeny57 J. Cocke and D. Sweeney, "High speed arithmetic in a parallel device," *IBM Technical Report*, 1957. [Online]. Available: http://archive.computerhistory.org/resources/text/IBM/Stretch/pdfs/06-08/102632302.pdf.

Taylor12 M. B. Taylor, "Is dark silicon useful? Harnessing the four horsemen of the coming dark silicon apocalypse," *DAC Design Automation Conference 2012*, San Francisco, CA, USA, pp. 1131–1136, 2012.

Thompson23 R. Thompson, A. Sutton, D. Harris, L. Moore, and J. Stine, "Challenges in CPU DV: Why 'It boots Linux' is not verification," *RISC-V Summit*, Nov. 2023. [Online]. Available: https://www.youtube.com/watch?v=MM9tUC7rR88.

ThompsonRitchie "Ken Thompson (left) and Dennis Ritchie (right), creators of the Unix operating system," 1973. [Online] Available: https://en.wikipedia.org/wiki/History_of_Linux#/media/File:Ken_Thompson_and_Dennis_Ritchie–1973.jpg.

TileLink20 TileLink Specification 1.8.1, SiFive, 2020.

Tocher58 K. Tocher, "Techniques of multiplication and division for automatic binary computers," *Quart. J. Mech. Appl. Math.*, vol. 11, pt. 3, pp. 364–384, 1958.

Trimberger15 S. M. Trimberger, "Three ages of FPGAs: a retrospective on the first thirty years of FPGA technology," *Proceedings of the IEEE*, vol. 103, no. 3, pp. 318–331, Mar. 2015.

Turing "File:Alan Turing Aged 16.jpg," *Wikimedia.* [Online]. Available: https://commons.wikimedia.org/wiki/File:Alan_Turing_Aged_16.jpg.

Usage22 "Usage share of operating systems," *Wikipedia.* [Online]. Available: https://en.wikipedia.org/wiki/Usage_share_of_operating_systems.

Vasquez21 F. Vasquez and C. Simmonds, *Mastering Embedded Linux Programming: Create Fast and Reliable Embedded Solutions With Linux 5.4 and the Yocto Project 3.1 (Dunfell),* Packt Publishing, 2021.

vonNeumann John von Neumann, Los Alamos National Laboratory, 2008. [Online]. Available: https://commons.wikimedia.org/wiki/File:JohnvonNeumann-LosAlamos.gif.

vonNeumann45 J. von Neumann, "First draft of a report on the EDVAC," June 30, 1945.

Waterman16 A. Waterman, "Design of the RISC-V instruction set architecture," *UC Berkeley Technical Report UCB/EECS-2016-1*, Jan. 3, 2016. [Online]. Available: https://www2.eecs.berkeley.edu/Pubs/TechRpts/2016/EECS-2016-1.html.

Weinberger81 A. Weinberger, "4:2 Carry-Save Adder Module," *IBM Technical Disclosure Bulletin*, vol. 23, Jan. 1981.

Weste12	N. Weste and D. Harris, *CMOS VLSI Design*, 4th. ed., Pearson, 2012.
WikiChipProcTech	"Process Technology History—Intel," *WikiChip*. [Online]. Available: https://en.wikichip.org/wiki/intel/process.
Wilson10	The 2010 Wilson Research Group Functional Verification Study. [Online]. Available: https://blogs.sw.siemens.com/verificationhorizons/2011/03/30/part-1-the-2010-wilson-research-group-functional-verification-study/.
Wishbone10	OpenCores, Wishbone B4 Specification, 2010. [Online]. Available: https://cdn.opencores.org/downloads/wbspec_b4.pdf.
Xiong10	W. Xiong, S. Park, J. Zhang, Y. Zhou, and Z. Ma, "Ad hoc synchronization considered harmful," *9th USENIX Symp. Operating Systems Design and Implementation*, pp. 163–176, Oct. 2010.
Yeh91	T.-Y. Yeh and Y. Patt, "Two-level adaptive training branch prediction," *Proceedings of the 24th Annual International Symposium on Microarchitecture (MICRO 24)*, pp. 51–61, 1991.
Zafar19	R. Zafar, "Apple A13 & beyond: How transistor count and costs will go up," 2019. [Online]. Available: https://wccftech.com/apple-5nm-3nm-cost-transistors.
Zimmerman96	R. Zimmermann, "Non-heuristic optimization and synthesis of parallel-prefix adders," *Proc.Intl. Workshop on Logic and Architecture Synthesis*, pp. 123–132, Dec. 1996.

Index

Note: Page numbers followed by f indicate figures, t indicate tables, and b indicate boxes.

A

Access faults, 353
Actel, 784.e79
Activity factor, 256
Addend generation, 784.e248
Address decoder and multiplexer, AHB, 397f
Address misaligned exception, 353
Address translation, 470
AddRoundKey, 723–724
Ad-hoc testing, 196
Advanced encryption standard (AES), 715–716, 722–731
Advanced microcontroller bus architecture (AMBA) buses
  advanced eXtensible interface, 404–410
  advanced high-performance bus, 395–404
    additional signals, 397–400
    burst transfers, 402–403
    multilayer, 404
    peripherals, 400
    SRAM interface, 400–401
    -to-APB bridge, 401–402
  advanced peripheral bus, 392–394
AES decryption architecture, 729f
Agile project management, 143–144
Alignment shift, 784.e195f, 784.e196t
align module, 778f
Altera, 784.e79
AMD
  Epyc 9004 motherboard, 675f
  DRAM controller, 784.e82
  IP Modules, 784.e82t
AMOALU atomic module, 709–710
Android, 784
Application binary interface (ABI), 50, 95–99, 96t, 97f
Application processors, 32
Application-specific integrated circuit (ASIC), 141
Architectural test pool, 134
Area and delay *vs.* configurations, 291–294, 292f, 294t–295t
Area and delay *vs.* target clock period, 287–291, 290f
Arithmetic circuitry, 303–311, 304t
Arithmetic/logic unit (ALU), 33
  decoder, 313, 314t
  operation, 58t
Array multiplier, 594, 595f
Arrays of modules, 176
Arty A7 board, 784.e81, 784.e81
  AHB burst performance, 784.e93
  components, 784.e83
  FPGA block diagram, 784.e82f
  FPGA program, 784.e87
  in-circuit logic analyzer, 784.e93–784.e96
  PmodSD DRAM expansion card, 784.e83f
  SD Flash Card, 784.e84–784.e86
  setup, 784.e84–784.e96
  UART, 784.e88–784.e92
  Wally Vivado, 784.e88f
Assembler directives, 103–104, 104t
Assertions, 173, 195, 241
Associative caches, 436–438
Atomic instructions, 673
  atomic memory operations, 701–703
  LoadReservede-StoreConditional, 703–704
Atomic memory operations, 784.e135f
Atomic module
  amoalu, 709–710
  atomic hazards, 710
  LRSC module, 710
Atomic waveforms, 712f
AXI channel signals, 406t–407t

B

Bad synchronizer, 143–144
Basic bit manipulation, 784.e141f
Behavioral/register transfer level (RTL), 141
Benchmarking, 76–77, 77t
  CoreMark, 784.e24–784.e27
  Embence, 784.e27–784.e38
  goal, 784.e24
  performance verification
    for CoreMark, 784.e39
    for Embench, 784.e39
    Wally's branch predictor, 784.e40–784.e44
    Wally's instruction and data caches, 784.e39
Better locking, 686
Bidirectional I/O, 154f
Bidirectional *vs.* unidirectional buses, 385–386

Binary files, 784
Bit manipulation (Zb*)
  riscv-arch-test, 749t
  Wally implementation, 748–764
    ALU Modifications
      inverted operations, 753
      rotates, 752–753
      Zba extension, 752
      Zbb extension, 752
      zbb module, 753–754
      Zbc, 755–757
      Zbs, 757
    controller modifications, 764
Bitstream, 784.e78
Bit swizzling, 169
Bit widths, 784.e252–784.e254
Boolean equations, 141
Booth encoding, 594, 597–598
  array multipliers, 597
  example, 598f
  modified, 597
  radix-4, 598t, 599f
  signed 2's complement multiplication, 598
Booting Linux, 196
  modifying buildroot
    configuration files, 784
    rebuilding the image, 784.e15
    root file system overlay, 784
  on QEMU, 784.e12–784
  on Wally in Questa simulation, 784
Bootloader, 784.e3
Boot trace, 784.e3–784
Branch direction predictor, 784.e40
Branches, 784f
Branch logic, 312–313, 313f
Branch misprediction, 547
Branch offset, 35
Branch prediction principles
  branch target buffer, 554–556
  Gshare, 551–552
  misprediction penalty, 545–547
  return address stack, 556–557
  simple dynamic branch prediction, 548–551
  static branch prediction, 548
  TAGE, 557
Branch predictor
  branch misprediction recovery, 572
  BTB implementation, 568–569
  Gshare implementation, 564–568
  instruction class prediction, 569–570
  RAS implementation, 570–572
Branch predictor performance verification
  CoreMark branch accuracy, 784.e44
  direction and target reference model, 784.e40–784.e42
  embench branch accuracy sweeps, 784.e42–784.e44
Branch target address (BTA), 545–547, 554
Branch target buffer (BTB), 547, 554–556
BTA. *See* Branch target address (BTA)
BTB. *See* Branch target buffer (BTB)
B-type instructions, 35
Buildroot, 784
Buildroot Linux, 784
  platforms, 784.e48
  principles
    Bootloader, 784.e3
    device tree, 784.e50–784
    embedded Linux, 784.e3
    init, 784
    kernel, 784
    root filesystem, 784.e51–784
    shell, 784
  on Wally, 784.e52–784
    boot overview, 784.e6
    building Linux, 784
    device tree, 784.e12
    file locations, 784t
    init process, 784.e12
    kernel, 784.e63–784
    memory map, 784
    openSBI Firmware. *See* OpenSBI Firmware
    root file system, 784.e12
    trap handlers, 784.e69–784
    zero stage bootloader, 784.e6
Build systems, 784
Burst EBU, 525
Burst transfers, 402–403
Bus cache FSM, 522f
Bus hazards, 423–426
Bus interface
  other SOC buses, 410
  principles
    bidirectional *vs.* unidirectional buses, 385–386
    handshaking, 386
    high-speed off-chip links, 390–391
    multiple controllers, 387–389
    multiple peripherals, 386–387
    other bus features, 389–390
  test plan, 410–411
Bypassing, 321–322
Byte write enables, 164–165, 164f

C

Cache coherence, 676
Cache design tradeoffs, 438–439
Cache flush and invalidation, 441–442
Cache FSM, 454–456
Cache line
  fetches, 520–523
  writeback, 523–525
Cache load hit, 534
Cache management operation (CMO) extension, 389, 767
  RISC-V instructions, 767–769
    environment configuration registers, 768–769
    Zicbom cache block management, 768
    Zicbop cache block prefetch, 768
    Zicboz cache block zero, 768
  test plan, 769
  Wally implementation
    cachefsm Inputs and Outputs, 773t
    cache modifications, 770–774, 771f, 773f
    LSU modifications, 769–770, 770f
    prefetch, 774
    uncached cbo.zero, 774
Caches
  modules, 241–242
  performance, 430–431
  RISC-V practices, 442
  store hit, 534–536
  test plan, 442
  way module, 446–449
  way read hit, 448–449
  way write hit, 450–453
  writes, 434–436
Capitalization conventions, 182t
Card detect (CD), 784.e84
Carry-less multiplication, 784.e141f
Carry-propagate adders (CPAs), 592–593

Carry-save addition, 592–593
C extension, 580
Chip enable, 165–166
Clock/data recovery (CDR), 391
Clock gating, 257
Coalescing write buffer, 679
Code coverage, 195, 217–221
Code examples, 108b–114b
Coherence, 676
Combinational logic
using always_comb statements, 149–151, 150f
using assign statements, 148–149, 149t, 150f
Xs and Zs, Oh My, 151–153, 152f–153f
Combinational read, 161
Comparator, 308–311
Comparisons operations, 34
Compatible time sharing system (CTSS), 336
Compiler explorer, 123, 123f
Complex instruction set computer (CISC) architecture, 55
Computer-aided design (CAD), 250
Concurrency, 673–676
Concurrent assertions, 216–217
Condition variables, 682–683
Cone of influence, 227
Configurable hardware, 173–181, 189
arrays of modules, 176
conditional generates, 174–175
disabling signals, 173
generate loops, 175
with parameter structure, 178–181
values of configurable width, 174
Consistency, 673, 677–682
acquire and release constraints, 681
release consistency, 679–681
sequential, 678
summary, 682
total store ordering, 678–679
Constrained random tests, 226
Content-addressable memories (CAMs), 168
Control and status registers (CSRs), 31, 46, 212–213, 273, 299, 526–527
counters, 346–348
environment configuration registers, 348–349
interaction with peripherals, 348
machine information registers, 346, 371
misa and XLEN, 371
misa register, 346
performance counters, 371
status register, 343–346
Controller design, 59–60, 60f, 60t, 311–312, 312f
ALU decoder, 313, 314t
branch logic, 312–313, 313f
IFU, 314
load/store unit, 314, 315f
main decoder, 312, 313t
sample program, 316, 317f
Convert float-to-float special case handling, 784.e271t
Convert float-to-int special case handling, 784.e270t
Convert logic, 784.e258t
Core local interruptor (CLINT) peripheral, 99
memory-mapped registers, 784, 784.e1t
testing plan, 784.e1
top-level interface, 784, 784f
Wally implementation, 784
CoreMark
branch accuracy, 784.e44
clock cycles, 784.e26
compile, 784.e25
compiler flags, 784t
Dhrystone, 784.e24
instruction mix, 328t
instruction set impace, 784.e27
performance *vs.* ISA, 784.e28t, 784.e29t
porting, 784.e26–784.e27
run output, 784.e25f
scores, 77t
source code, 784.e24
steps, 784.e24
Verilog simulation, 784.e25–784.e26
Core running process, 47f
Cores, 47, 48f
CORE-V-Wally, 199
Cortex-M4 reference processor, 784.e28–784.e29
Counters, 346–348
Coverage-driven testing, 197
Coverage points, 197
C programming, 784.e167
Cryptographic hash functions, 731–734
Cryptography (Zk*), 715
principle
advanced encryption standard, 722–731
AddRoundKey, 723–724
architecture, 724f
array structure, 723f
decryption, 729–731
hash functions, 731–734
KeyExpansion, 726–729
MixColumns, 725–726
ShiftRows, 724–725
SubBytes, 724
finite fields, 716–722
GF(2_n), 718–722
GF(p), 717–718
riscv-arch-test, 749t
Wally implementation, 748–764
ALU Modifications, 757–764
Zbkb extension, 758
zbkx module, 759–760
Zknd and Zkne, 760–763
zknde32, 761–762
zknde64, 762–763
Zknh, 763–764
Zkt, 764
controller modifications, 764
Custom extensions, 783–784
Cycles per instruction (CPI), 784.e26

D

Data cache, 429
integration
burst EBU, 525
bus cache FSM, 522f
cache line fetches, 520–523
cache line writeback, 523–525
simplified LSU, 518f
Data tightly integrated memory (DTIM), 299
Deadlock, 696–697
Debugging, 203
with GDB, 128–132, 129t
hints, 235–239
Decryption, 729–731
Default subordinate, 396

Design Compiler, 281–282
Design strategies, 142–144
Design verification (DV) plan, 194, 202–203, 223t, 227–228
Device tree, 784.e50–784
D flip-flop, 154
Dhrystone, 784.e24
Digital signal processing (DSP), 784.e79–784.e80
Digit selection, 784.e248
Directed tests, 196, 222–226
Direct-mapped cache, 432–434
Direct memory access (DMA), 389
Directory-based coherence, 676
Direct ripple implementation, 176f
Disassembled files, 784.e3
Divider control schematic, 161f
Divider FSM, 160f
Divide special case handling, 784.e269t
Divide/Square root shift calculation, 784.e264t
Division
  N-bit integers, 602
  other integer division optimization, 604
  radix-2 restoring unsigned integer, 602–603
  signed integer, 604
Divisor latch access bit (DLAB), 784–
Double-data-rate (DDR), 144–145
Double-precision, 616.e3f, 616.e3b, 616.e4b
dRead port, 161
Dynamic RAM (DRAM), 168, 429
Dynamic random access memory (DRAM), 784.e81
Dynamic rounding, 616.e48

E

Edge-triggered read, 161
Embedded processors, 32
Embence
  benchmark, 784.e27–784.e28
  code size, 784.e36t
  Cortex-M4 processor, 784.e31
  flags, 784.e35t
  instruction set impact, 784.e35–784.e38
  JSON files, 784.e31
  porting, 784.e33–784.e34
  1.0 programs, 784.e30t
  Python lief library, 784
  repository, 784.e29
  size calculation, 784.e33
  speed, 784.e37t
  troubleshooting, 784.e38
  Wally benchmarking, 784.e30
  Wally Embench 1.0 rv32imc, 784.e32t
Embench branch accuracy sweeps, 784.e43f
Embench scores, 78t
EmbToolkit, 784
Emulation, 198–199
Emulator, 51
Enabled flip-flops, 156–157, 156f
Encryption example, 730f
Endianness, 37, 37f
Endian swapper, 378–379
Environment configuration registers, 348–349
Equality comparator, 150f
Exceptions, 49, 466–467
  access faults, 353
  address misaligned, 353
  ecall and ebreak, 354
  generating instructions, 341–342
  illegal instructions, 352–353
  page faults, 354
Executable and Linkable Format (ELF), 89
Execution environment (EE), 46, 49–51, 50f
External bus unit (EBU), 411, 415–417, 784.e136f
External interrupts, 784

F

Factorial function, 42b
  optimized, 42b
  recursive, 43b
Faster tree implementation, 176f
fclassify classification, 662–663
fcmp unit, 640
Field programmable gate array (FPGA), 78, 141, 163, 194, 250
  Actel, 784.e79
  vs. ASICs, 784.e78
  in emulation system, 784.e78–784.e79
  generic architecture, 784.e80f
  interesting bugs, 784.e100–784.e104
    AHB's burst, 784.e101–784.e102
    concurrent exceptions, 784.e102
    critical path analysis, 784.e103f, 784.e104f
  lookup table, 784.e79
  Moore's law, 784.e79
  principles, 784.e79–784.e80
  results, 784.e99–784.e100
  Wally
    implementation. *See* Wally implementation. *See* FPGA
    specifications, 784.e99t
    utilization, 784.e100t
Finite fields, 716–722
Finite state machines (FSMs), 157, 158f
First C program, 115–120, 120t
First-in, first-out (FIFO), 214
Flash memory controllers, 144–145
Flattened device tree (FDT), 784.e51
Flip-flops, 154–157, 154f–155f
Floating-point, 617, 784.e193
  compare unit, 784.e148f
  comparisons, 640
  conversion. *See* Floating-point conversion unit
  conversions supported in RV32, 662t
  divider operand sizes, 784t
  fields, 619f
  format codes, 659
  fused multiply-add unit, 784.e146f
  less than flag logic, 640
  Min/Max, 640t
  Postprocessing. *See* Floating-point postprocessing unit
Floating-point conversion unit, 784.e193
  examples, 784.e261–784.e262
  result exponent, 784.e257
    float-to-float, 784.e257
    float-to-int, 784.e257–784.e259
    int-to-float, 784.e259–784.e260
    ResSubnormUf, 784.e260
  shift amount selection, 784.e260
    float-to-float, 784.e260
    float-to-int, 784.e260
    int-to-float, 784.e260

Floating-point division and square root, 784.e147f
  algorithm for unsigned integer division, 784.e201–784.e202
  CPA delay, 784.e202, 784.e202
  divider latency, 784.e218–784.e219
  early termination, 784.e221–784.e222
  integer division, 784.e224–784.e225
    examples, 784.e230–784.e233
    integer latency, 784.e226–784.e227
    integer postprocessing, 784.e227–784.e228
    integer preprocessing, 784.e225–784.e226
  minimally redundant radix-4, 784.e202
  postprocessing, 784.e223
    exponent, 784.e223–784.e224
    normalization, 784.e224
    rounding, 784.e224
    sign, 784.e223
    special cases and flags, 784.e224
  quotient conversion, 784.e219–784.e220
  radix-2 division
    examples, 784.e208–784.e209
    implementation, 784.e211f
    quotient digit set, 784.e206
    quotient selection optimization, 784.e209–784.e212
    WC and WS, 784.e208
  radix-4 division
    example, 784.e216
    implementation, 784.e216
    minimally redundant divisor, 784.e212
    P-D diagram, 784.e213, 784.e213
    QSLC constants, 784.e215t
    quotient digit, 784.e214–784.e215
  recurrence division algorithm, 784.e202–784.e204
    containment of partial remainder, 784.e205–784.e206
    examples, 784.e204
    quotient digit set, 784.e204–784.e205
    quotient representation, 784.e206
    redundant partial remainder, 784.e204–784.e205
  square root. *See* Floating-point square root
  subnormal floating-point preprocessing, 784.e222–784.e223
Floating-point multiply-accumulate
  add aligned significands, 784.e195–784.e198
  addend signs, 784.e193
  add exponents, 784.e193–784.e194
  alignment shift, 784.e194–784.e195
  leading zero anticipator, 784.e198–784.e200
  multipath FMA optimization, 784.e200
  multiply significands, 784.e193
  product signs, 784.e193
Floating-point postprocessing, 784.e149f
  conversion, 784.e265t
  correction shift, 784.e266t
  initial result mux, 784.e265t
  normalization shift, 784.e265t
  rounding, 784.e267
  shift calculation, 784.e262
    conversion, 784.e264
    FMA, 784.e263t
    unified divide/square root, 784.e264t
  sign logic, 784.e267t
  special cases and flags, 784.e268–784.e271
    other flags, 784.e271–784.e272
    result selection, 784.e272
Floating-point square root
  addend input selection, 784.e238–784.e241
  ASEL table, 784.e238t
  containment analysis, 784.e235
  containment of residual, 784.e235–784.e236
  early termination, 784.e242–784.e243
  on-the-fly conversion, 784.e241–784.e242
  operation, 784.e233, 784.e233
  postprocessing, 784.e247–784.e248
  radix-2 example, 784.e243
  radix-2 implementation, 784.e245
  radix-2 square root digit selection, 784.e237
  radix-4 example, 784.e245
  radix-4 implementation, 784.e247
  radix-4 square root digit selection, 784.e237–784.e238
  recurrence algorithm, 784.e234–784.e235
Floating-point units (FPUs), 198, 239–240, 580
  design
    comparison, 640–641
    conversion, 639–640
    division and square root, 637–639
    fdivsqrt unit, 630
    fused multiply-add, 633
    postprocessing unit, 641
    unpack unit, 631–633
  generic block diagram, 630
  IEEE 754
    infinity, 620
    Not a Number, 621
    rounding, 625
    special cases, 620t
    subnormal numbers, 621–622
    zero, 620
  principles
    fixed-point notation, 628–629
    flags, 616.e10–616.e12
    normalized floating-point formats, 618–619
  RISC-V practices. *See* RISC-V practices. *See* FPUs
FMA special case handling, 784.e269t
Formal verification (FV), 195
  algorithmic improvements, 227
  cone of influence, 227
  definition, 226
  OpenHW Group, 227
Forwarding logic, 324–325, 325f
fpcalc, 643
FPGA Hardware, 784.e81t
Free and Open Source Silicon Foundation (FOSSi), 784.e27–784.e28
FreeRTOS, 784
fround algorithm, 664
fround unit, 665
fsgn sign injection, 662
Fulladder simulation waveforms, 208f, 212f
Fully associative cache, 437–438
Functional coverage, 195, 245–246
  architectural, 221
  constrained random tests, 226

covergroup and coverpoint statements, 222
directed tests, 222–226
DV plan, 221
measurement, 222
riscvISACOV, 214t, 222, 239f
Futex, 693–694

G

Galois fields, 717
General-purpose input/output (GPIO) peripheral, 383–384
controls, 784
loopback testing, 784.e1
memory map, 784t
pins connected to switches and LED, 784f
test plan, 784.e1
top-level interface, 784–, 784.e1f
Wally implementation
peripherals highlighting, 784, 784f
RV32 configuration, 784.e1–
set/clear logic, 784, 784f
General-purpose operating systems, 784
GF(2_n), 718–722, 719t
GF(*p*), 717
Git
basic flow, 784.e172–784.e175
Ben's workspace, 784.e181, 784.e183
branching and merging, 784.e176–784.e177
capabilities, 784.e181
features, 784.e170
HEADS, 784.e175
layers of data storage, 784.e182f
merge conflicts, 784.e175–784.e176
repository, 784.e170–784.e171
repository setting, 784.e172
snapshot, 784.e170–784.e171, 784.e175
staging and undo, 784.e178–784.e179
starting account, 784.e171–784.e172
submodules, 784.e179–784.e180
tags, 784.e177
version control tool, 784.e181
Git file states, 784.e179f
GitHub account, 784.e171
Git layers, 784.e200f
Global history shift register (GHR), 552
Globally unique identifier partition table (GPT), 784.e84
GNU Compiler Collection (GCC)
ABI, 95–99, 96t, 97f
CLINT, 99
compiler explorer, 123, 123f
definition, 91
driving, 99
first C program, 115–120, 120t
inline assembly language, 120
linking and ELF, 93–94
mixed C and assembly language, 121
quick tutorial, 91–93, 92b
SBI, 99
standard library and startup code, 122–123
Good synchronizer, 144
Gshare, 551–552, 564–568
GTKWave, 212f

H

Handshaking, 386, 690
Hardware description language (HDL), 141, 201. *See also* Sequential logic
combinational logic. *See* Combinational logic
design strategies, 142–144
reuse, 144–145
with systemverilog. *See* Systemverilog
version control, 144
Hardware emulation, 198–199
Hardware page table walkers (HPTWs), 475, 515
integration, 531–532
Hardware thread, 45
Harts, 47–49, 48f
Hazard unit (hazard), 273f, 325–327, 326f–327f, 784.e129f
High-speed off-chip links, 390–391
Hitchhiker's guide, Linux commands
bundling, 784.e161
cast of characters, 784.e157
commonly used, 784.e158
compressing, 784.e161
standard input and output, 784.e157
transferring files, 784.e161
working with process, 784.e159–784.e160
productivity
document viewing and printing, 784.e165
environment variables, 784.e164
installing packages, 784.e165
Linux file system, 784.e163–784.e164
multiple terminal windows, 784.e161–784.e165
text editors, 784.e162–784.e163
web browsing, 784.e164–784.e165
scripting and programming
c programming, 784.e167
makefiles, 784.e167–784.e169
python, 784.e166–784.e167
shell scripts, 784.e165–784.e166
server connection, 784.e153–784.e154
password change, 784.e154
X2Go, 784.e153–784.e154
working with files, 784.e154
Hold time analysis, 254
Hold time constraint, 251
Host target interface (HTIF), 106
HPTW critical path
after optimization, 784.e104f
before optimization, 784.e103f
HPTWs. *See* Hardware page table walkers (HPTWs)
HTRANS transfer types, 395t
Huge pages, 486–488

I

IEEE 754 Floating-Point Format, 617
IFU. *See* Instruction fetch unit (IFU)
Illegal-instruction exception, 352–353
Immediate assertions test, 215
ImperasDV, 213

lock-step-compare system, 213f
In-circuit logic analyzer, 784.e93–784.e96, 784.e86
Init process, 784
Inline assembly language, 120
Input/output elements (IOEs), 784.e79
Input/output (I/O) pins, 784.e78
Institute for Electrical and Electronic Engineers (IEEE), 617
Instruction address-misaligned exception, 353
Instruction cache, 429
Instruction class prediction, 784.e40
Instruction fetch unit (IFU), 56–57, 198, 241–242, 314, 377–378, 784f
block diagram, 559f
branch prediction. *See* Branch prediction principles
signal declarations, 184b
test plan, 558–559
correctness, 558
performance, 558–559
Wally high-level block diagram, 546f
Wally implementation
branch predictor. *See* Branch predictor
instruction cache, 560
ITLB misses, 574
PC logic, 560
reset, stalls, flushes, and traps, 574–576
Instruction Read-Only Memory (IROM), 52, 56–57, 299
Integer divider, 784.e143f
Integer execution unit (IEU), 56–57, 198, 784f
Integer and floating-point registers, 212–213
Integer instructions, 300t
Integer multiplier, 784.e142f
Integer multiply/divide unit, 784.e142f
Integer recurrence division preprocessing, 784.e226f
Integrated circuit fabrication technologies
chip manufacturers, 258
Cypress Semiconductor, 259
inverter characteristics, 259–260, 259t
NDA, 258–259
28-nm node, 258–259
planar processes, 257–258, 258f
SKY130, 259
Intellectual property (IP), 784.e79–784.e80
Interfaces, 445–446
Interrupt enable register (IER), 784.e16–784.e17
Interrupts, 49, 354–355, 466–467
triggering and generation, 355–357
vectoring, 357
Inverse cipher, 729–731
InvMixColumns matrix, 731f
I-type instructions, 34

J

J-type instructions, 35
Jump and link register, 34
Jump vector table (JVT), 581

K

Kernel, 784
KeyExpansion
example, 728t
key schedule, 727f
unit variables, 727t

L

Least recently used (LRU) replacement policy, 439
Library considerations, 167–168
Line control register (LCR), 784.e16–784.e17
Line fetch, 452
Line status register (LSR), 784.e16–784.e17, 784.e26
Line writeback, 452
Linking, 34
Lint, 203–204
Linters, 203
Linux-based execution environments, 50f
Linux boot messages, 784.e3f
Linux operating system (OS), 784.e48
file
binary, 784
boot trace, 784.e3–784
disassembled, 784.e3
locations, 784t
memory, 784.e3
Hitchhiker's guide. *See* Hitchhiker's guide, Linux
Livelock, 698
Load/store unit (LSU), 56–57, 198, 241–242, 314, 315f, 784.e130f
block diagram, 517f
cache, AHB cache interface, and EBU connections, 519f
cache load hit, 534, 536–537
cache store hit, 534–537
data cache integration. *See* Data cache integration
disabling interrupts, 533
full MMU and TLB address translation, 530f
memory operations, 529t
response to stalls and flushes, 533–534
stall generation, 532–533
test plan, 544
TLB access fault, 543–544
trap generation, 533
uncached access, 537–540
Wally LSU overview, 515–518
Locality, 432
Local repository\, 784.e170–784.e171
Locks, 684–686
Lock-step-compare, 240–241
Log files
area reports, 284, 285f
mapped output files, 270–273, 271f
other reports, 284
performance reports, 277–282, 277f–278f
power reports, 282–284, 283f
quality of results (QOR) reports, 275, 276f
synth.out, 269, 271f
Logic design engineer, 254

Logic elements (LEs), 784.e78
Logic simulators, 200t, 207–212
Logic synthesis, 249
Logic verification process
  definition, 193
  device under test (DUT), 193
Lookup table (LUT), 784.e79
Loopback testing, 784.e1
Loop grade inflation, 41b
LSU. *See* Load/store unit (LSU)

M

Machine architecture, 99–100, 100t
Machine information registers, 346
Machine language, 43–45, 44f
MacOS, 784
Main decoder, 312, 313t
Market, RISC-V, 80, 81t–84t
Maximum and minimum integer values, 784.e270t
Memories, 160–168
  block characteristics, 266–267, 267t
  consistency models, 677
  exceptions, 494–495
  files, 784.e3
  hierarchy, 429
Memory management and HPTW integration
  HPTW integration, 531–532
  PMA checker, 527–529
  PMP checker, 529
  TLB and cache integration, 529–531
Memory management unit (MMU), 241–242, 784.e132f
  definition, 469
  principles
    physical memory checking, 477–478
    virtual memory, 469–477
  RISC-V. *See* RISC-V practices
  test plan, 495
  Wally implementation. *See* Wally implementation
Memory management units (MMUs), 784
Memory map, 51–52, 51f, 386
M extension
  principles
    carry-save addition, 592–593
    division. *See* Division
    multiplication. *See* Multiplication
    signed arithmetic, 591–592
    unsigned arithmetic, 591–592
  riscv-arch-test, 606, 606t
  Wally implementation
    division and remainder, 610–612, 611f
    hazard unit, 614, 615f
    multiplication, 607–610
    multiply and divide Unit, 613, 614f
Microarchitecture
  CPU and CPI, 56
  HDL code, 56
  single-cycle, 56–61, 58f
Min/max special case handling, 784.e270t
Misprediction penalty, 545–547
Miss penalty, 430–431
MixColumns operation, 725–726, 726f
Mixed C and assembly language, 121
Mixed-mode clock manager, 784.e81
Modem control register (MCR), 784.e16–784.e17
Modem status register (MSR), 784.e16–784.e17
Modified Baugh-Wooley signed multiplier, 597f
Modified booth encoding, 597
Modules, 188–189
Moore's law, 784.e79
Multiple controllers, 387–389
Multiple-input adders, 592f
Multiple peripherals, 386–387
Multiplication, 593–601
  array multiplier, 594
  booth encoding, 594, 597–598
  dot diagram, 594f
  logical AND, 593
  multiplier synthesis, 601
  other techniques, 601
  signed multiplication, 596
  tree multipliers, 594, 599
  unsigned, 593
Multiplicative inverse table, 723f
Multiplier dot diagram, 594f
Multiplier synthesis, 601
Multiported RAMs, 166
Multithreaded/multicore processor, 45

N

National Aeronautical and Space Administration (NASA), 194
Naturally aligned power-of-two (NAPOT), 784.e63
Network on Chip (NoC), 390
New User Setup and Toolchain Installation, 91
Nightly build, 234–235
Nightly regression, 234
Nonrecurring engineering costs (NREs), 784.e78
Nonvolatile program memory, 53
Normalized floating-point formats, 618
N-way set associative cache, 436–437

O

On-chip peripheral bus (OPB), 410
One-bit with many-bit gating, 170
On-the-fly conversion (OTFC), 784.e219f
OpenEmbedded system, 784
Open Hardware Manager, 784.e89f
OpenHW Group, 200t, 227
OpenSBI Firmware, 784.e60–784.e61
  PLIC initialization, 784.e62–784.e63
  PMP configuration, 784.e63
  UART initialization, 784.e60–784.e62
Open-source build systems, 784
Open Target, 784.e89f
Operands, 299
Operating system (OS), 784.e84
OpernWrt, 784

P

Page-based memory types (Svpbmt), 490–491
Page coloring, 476–477
Page table entry (PTE), 471–472
Partial store order, 679–680
Pattern history table (PHT), 552

PC16550, 784.e16–784.e17
Performance measurement, 107
Performance, power, and area (PPA), 249–250
  activity factor, 256
  clock period and frequency, 250–251
  datapath units, 260–266, 262f, 262t, 263f, 265t
  dynamic and leakage power, 255
  dynamic energy, 256
  setup time constraint, 251
  static timing analysis, 253–255, 255f
  Synopsys Design Compiler, 257
  time constraints, 251–253, 251f–253f
Peripherals
  core local interruptor, 784
  general-purpose input/output, 784
  memory-mapped I/O, 784.e23
  platform-level interrupt controller, 784
  serial peripheral interface, 784.e1
  testing, 784.e1
  universal asynchronous receiver/ transmitter, 784.e1
Personal computers (PCs), 784
Phase-locked loop (PLL), 391
Physical addressing, 469–470
Physical design engineer, 254
Physical harts, 45
Physical memory attributes checker, 784.e134f
Physical memory checking, 477–478
Physical memory protection (PMP), 784.e60
Physical memory protection checker, 784.e135f
Physical page number (PPN), 471–472
Pipelined microarchitecture, 65–67, 66f
Pipelined processor performance, 327–328
Pipeline hazards
  CSR reads, 375–376
  CSR writes, 376
  description, 321–322
  fences, 376
  hazard example, 322–323
  mret and sret, 375
  precise traps, 374
  WFI, 376–377
Pipeline registers, 323, 323f
Pipeline stalls, 466
Planar processes, 257–258
Platform-level interrupt controller (PLIC) peripheral, 784
  hart context, 784
  interrupt pending, 784
  memory-mapped registers, 784.e1t
  M- mode interrupt, 784
  S- mode interrupt, 784
  test plan, 784
  Wally implementation, 784.e1
pLRU implementation, 456–459
PMA checker, 491–494, 501–502
  memory management and HPTW integration, 527–529
  page-based memory types (Svpbmt), 490–491
  PMP checker, 490–491
  regions, 493t
  registers, 784.e63
Port connections, 171–172, 184–185
Positive edge-triggered flip-flop, 154
Power gating, 257
PPN. *See* Physical page number (PPN)
Precise traps, 352
Prefix comparators, 311f
Preprocessor, 172–173
Priority circuit, 151f
Privileged instructions
  CSR access instructions, 341
  exception generating instructions, 341–342
  trap return instructions, 342
  wait for interrupt, 342
  wally implementation, 370
Privileged operations
  instruction fetch unit and privileged changes, 377–378
  principles
    control and status registers (CSRs), 338
    security and protection, 336–337
    traps, 337–338
Privileged specification, 32
Privileged test plan, 360–362
  generating exceptions, 362
  generating interrupts, 363
Privileged unit, 784.e137f
Privilege modes, 46
Processes and threads, 46–47
Process ID (PID), 476
Process variation, 260, 261t
Producerconsumer/bounded buffer, 689–690
Product's and sum's exponents, 784.e194t
Product significand outputs, 784.e193t
Profile extensions
  Za128rs and Za64rs, 706
  Ziccamoa, 705
  Ziccif, 705
  Ziccrse, 705
Program counter (PC), 212–213
Programmable logic array (PLA), 196–197
Pseudoinstructions, 37–38
Pseudo least recently used (pLRU), 456–459
PTE. *See* Page table entry (PTE)
PTXdist, 784
Pull request, 784.e171
Python, 784.e166–784.e167

Q

QEMU simulation
  commands, 128t
  open-source emulator, 127
  run options, 96t
  simulation of, 127–128
Quality of results (QOR) reports, 275, 276f
Questa GUI, 207–209
Quick Emulator (QEMU), 784.e48
Quotient selection constants (QSLC)
  preferred radix-4 division, 784.e215t
  viable radix-4 division, 784.e215t

R

Radix-2 adder input generation, 784.e239t
Radix-4 Booth encoder, 599f

Radix-2 division example, 784.e210f
Radix-2 division P-D diagram, 784.e207f
Radix-4 OTFC implementation, 784.e221f
Radix-2 restoring unsigned integer division, 602–603, 603f
Random access memory (RAM), 163–164, 163f
RAS. *See* Return address stack (RAS)
Read after write (RAW) data hazard, 322–323
Read-only memory (ROM), 162–163, 162f
Read/write port, 161
Real-time operating system (RTOS), 784
Receiver buffer register (RBR), 784.e1
Recurrence division, 784.e205t
  algorithm, 784.e205f
  hardware, 784f
Reduced instruction set computer (RISC)
  architectures, 579
  C extension
    decompress block diagram, 584f
    floating-point loads, 580
    instruction decompression, 582–585
    PC logic, 585–586, 586f
    riscv-arch-test suite, 582, 583t
    spills, 586–590, 586f, 588f–589f
    Wally implementation, 582–590
    Wally RV32IC/64IC, 584f
    Zc* extension, 580–582
  compressed instruction format, 579–580
  compressed instruction maps, 580
  compressed instruction sets, 579b
  spill, 579
Reference models, 198, 229
Register, 784.e79
Register transfer level (RTL), 249
Relaxed ordering, 679–680
Release consistency model, 674
Relocation, 469–470
Replacement policy, 439–441
Reset behavior, 349
Resettable flops, 155–156, 156f
Reset vector, 52
Residual sign correction and shift, 784.e228t
Result multiplexers, 658
Result selection logic, 784.e273t
Return address stack (RAS), 547, 556–557
RISCOF
  definition, 137
  Wally RISCOF test suites, 137, 138t
  wally-riscv-arch-test, 138–140, 139f
RISC-V
  bit manipulation
    test plan, 748
  cryptography (Zk*), 715
  M extension. *See* M extension
  practices
    division special cases, 604–605
    signed operations, 604
    software arithmetic implementations, 605–606
    unsigned operations, 604
    Zmmul, 606
riscv-arch-test, 133–136, 134f, 331t–332t
RISC-V atomic instructions
  practices
    atomic instructions, 700–704
    atomic subsets, 705
    consistency models, 698–699
    fences, 699–700
    profile extensions, 705–706
  principles, 673–698
    coherence, 676
    concurrency, 674–676
    concurrent systems, 673
    consistency, 677–682
    deadlock, 696–697
    livelock, 698
    release consistency model, 674
    starvation, 698
    synchronization, 674, 682–696
  test plan, 706
  Wally implementation
    atomic module, 708–710
    timing diagrams, 711–712
RISC-V International (RISC-V) Foundation, 1–2, 26
  architecture, 32
  assembly language
    B-type instructions, 35
    C to, examples, 38–43
    endianness, 37, 37f
    instruction types, 32, 33t
    I-type instructions, 34
    J-type instructions, 35
    pseudoinstructions, 37–38
    register conventions, 37
    R-type instructions, 32–34
    RV64I additional instructions, 36
    S-type instructions, 34
    U-type instructions, 35–36
  CSRs, 31
  extensions and profiles
    common extensions, 52, 52t
    RISC-V profiles, 53–54
  load/store ISA, 31
  machine language, 43–45, 44f
  microarchitecture. *See* Microarchitecture
  microprocessors, 32
  vs. other architectures, 54–55
  processor design, 31
  width XLEN, 31
riscvISACOV, 214t, 222, 239f
  RV32I add covergroup, 224f
RISC-V practices
  carry-less multiplication, 738
  crossbar permutations, 741
  fences, 495
  FPUs
    control and status register, 647f
    move instructions, 649
    NaNs, 621
    registers and NaN boxing, 648
    Zfa additional instruction
     fcvtmod.w.d, 651
     fli, 651
     fltq and fleq, 651
     fminm/fmaxm, 651
     fmvh and fmvp, 651
     fround, 651
    froundnx, 649–651
  hardware page table walker (HPTW), 495
  memory exceptions, 494–495
  memory protection, 489–494
    PMA checker, 489–491
    PMP checker, 491–494
  privileged operations
    CSRs. *See* Control and status registers (CSRs)
    privileged instructions, 340–342
    privilege modes, 340
    reset behavior, 349
    trap handler, 357
    traps, 350–357
  single-bit operations, 739

systems
core local interruptor, 784.e23
platform-level interrupt controller, 784.e23
virtual memory, 478–489
dirty and accessed bits, 488
mstatus/sstatus register, 488–489
satp (supervisor address translation and protection) register, 478–479
variable page sizes, 483–488
virtual address and page table structure, 479–482
Zba, 734–735
Zbb, 736–738
counting, combining, reversing, 737–738
extension, 737
inverted operations, 736
Max/Min, 736–737
rotates, 736
Zbkb, 739
bit interleaving, 740–741
bit reversal, 740
packing, 740
Zknd and Zkne, 742–747
RVZknd: AES Decryption, 746–747
RV32Zkne: 32-bit AES Encryption, 745–746
RV64Zkne: 64-Bit AES Encryption, 743–745
RV64Zkn AES Key Scheduling, 747
Zknh, 747–748
Root file system, 784.e51–784
compressed, 784.e12
overlay, 784
using BusyBox, 784.e3f
Rounding logic, 784.e268t
Round to nearest ties away, 622
Round to nearest ties to maximum magnitude (RNM), 622
Round toward zero, 622
R-type instructions, 32–34
Running synthesis, 268–269
Run Spike, 107
RV64gc, area and delay *vs.* features, 295–297, 296f, 297t
RV32GC/RV32IFDC extension, 582
RVVI-TRACE Interface, 206t, 214
RVVI-TRACE standard interface (RVVI-TRACE), 213

S

Sail specification
definition, 124
RISC-V simulators, 124
Sample program, 57–59
S-box contents, 725f
Scratch register (SCR), 784–
Second-level cache, 429
Security and protection, 336–337
Self-checking testbench, 204f
Self-modifying code, 676
Semaphore-based mutex, 692
Semaphore-based producer-consumer, 693
Sequential consistency, 678
Sequential logic
bidirectional I/O, 154f
definition, 153–154
flip-flops, 154–157, 154f–155f
Sequentially consistent operation, 681
Serial clock (SCK), 784.e1
Serial communication, 784.e14
Serial data in (SDI), 784.e1
Serial data out (SDO), 784.e1
Serial peripheral interface (SPI), 784.e1
A/D converter interfaced, 784.e22f
controller-peripheral interface, 784f
loopback testing, 784
memory-mapped registers, 784–, 784.e1t
PC16550 block diagram, 784.e1f
physical interface, 784.e1
SiFive FE540, 784
signaling, 784f
test plan, 784.e23
top-level interface, 784f
Wally implementation, 784.e23
Setup time constraint, 251
SHA-2 algorithm, 733f
Shell, 784
Shell scripts, 784.e165–784.e166
Shifter, 306–307, 306f
Shift operations, 33
ShiftRows operation, 724–725, 726f
Signature checking, 212–214
Signed arithmetic, 591–592
Signed division, 604t
Signed multiplication, 596
Signed multiplier
modified Baugh-Wooley, 597f
partial products, 596f
simplified partial products, 596f
Sign injection, 616.e44t
Sign outputs, 784.e194t
Simulated harts, 45
Simulator, 51
Single-cycle core, 301–317, 302f
add/subtract, 305
architectural state, 302–303
arithmetic circuitry, 303–311, 304t
comparator, 308–311
immediate extension, 303
set less than, 308
shifter, 306–307, 306f
W-type instructions, 308
Single-cycle microarchitecture, 56–61, 58f
Single-precision
Single-threaded processor, 45
Sleep locks, 689
Snapshot, 784.e170–784.e171
SoC, 78–79, 79f
SoftFloat, 641–646
fpcalc, 643–646
softFloat demo, 643
Software interrupts, 784
Software threads, 46–47
Software tool flow
requirements and tool overview, 90
RISC-V toolchain, 89f
Spacing and indentation, 185–187
SPARC, 55
Spike simulation, 105–107, 106t
Spill
32-byte cache line, 586f
IFU, 587f
instruction fetch, 587
module, 588f
spillfsm, 589f
timing diagram, 589f
Spin lock, 686
Split transfer, 389
Square root special case handling, 784.e269t

SRAM interface, AHB, 400–401
Standard cell library, 249
Standard library and startup code, 122–123
Starvation, 698
Static branch prediction, 548
Static random access memory (SRAM), 266
Static timing analysis, 253–255, 255f
Status bits, 435
Status register, 343–346
Store address-misaligned exception, 353
Strongly Not Taken (SNT), 158
Strongly Taken (ST), 158
Structural hazards, 451–452
S-type instructions, 34
SubBytes operation, 724, 725f
Sub-cacheline read (SCR) module, 534
Subnormal numbers, 621–622
Subword loads, 34
Supervisor address translation and protection (satp) register, 478–479
Supervisor binary interface (SBI), 50, 99
Svinval TLB management extension, 777–779
Symmetric-key encryption, 716f
Synchronization, 673, 682–696
  futex, 693–694
  mutual exclusion
    atomic instructions, 686–688
    better locking, 686
    locks, 684–686
    sleep locks, 689
    ticket locks, 688–689
  producer-consumer: condition variables, 689–692
  semaphores, 692
    semaphore-based mutex, 692
    semaphore-based producer-consumer, 693
  synchronization Libraries, 695–696
Synchronous design
  memories and flip-flops, 187
  sequential, 187
Synchronous memories, 319–321, 320f, 320t
Synonym problem, 476
Synopsys Design Compiler, 141, 257
Synopsys VCS, 212
Synthesis configurations, 287, 288t–289t
SystemC, 142
System-on-chip (SoC), 784.e48
  address decoding and device Mux, 418–420
  designs, 784.e185
SystemVerilog, 61, 61b–65b, 141–142, 328–329
  behavioral HDL (RTL) code, 145
  code, 784.e78
  definition, 145
  hierarchically/structurally, 145
  language subsets, 194t
  modules, 145–148, 146f
SystemVerilog Assertions (SVAs), 214

T

Tcl
  advanced commmands
    Aliases, 784.e192
    control flow, 784.e189–784.e191
    display data on screen, 784.e189
    exiting Tcl, 784.e192
    file, 784.e189
    file-related, 784.e189
    Glob, 784.e189
    lists, 784.e188
    procedures, 784.e191
    running Tcl scripts, 784.e191–784.e192
  embeddable scripting language, 784.e185
  system-on-chip designs, 784.e185
  use, 784.e185–784.e187
    arguments and variables, 784.e186–784.e187
    command syntax, 784.e187
    nested commands, 784.e187
Technology readiness levels (TRLs), 194, 199, 200t
Testbenches, 204–206, 208f
Test-driven development, 202
TestFloat test cases, 654
Test plan, 329–333
  riscv-arch-test, 651–652
  TestFloat, 652–653
Test suites
  architectural test pool, 134
  defnition, 133
  riscv-arch-test, 133–136, 134f
  test file structure, 135–136
Test vector generation, 196–197
Thrashing, 436
32-Bit shifter source generator, 308f
3-Bit register, 154f
Ticket locks, 688–689
Tightly integrated memory (TIM), 545
TileLink, 410
TIM. *See* Tightly integrated memory (TIM)
Time division multiple access (TDMA), 388
Timer interrupts, 784–
Tizen, 784
Toggle rates, 256
Tool Control Language (TCL), 250
Top-level cache logic
  address, way, and cache line read, 460
  cache flush logic, 463–466
  read selection and bus interface logic, 462–463
  write selection logic, 462
Top-level testbench and testvectors, 229–230
Total negative slack (TNS), 254
Total store ordering (TSO), 678–679
Tournament branch predictor, 557
Translation lookaside buffer (TLB), 474–475, 784.e133f
  access fault, 543–544
  and cache integration, 529–531
  miss, 540–542
Transmit holding register (THR), 784.e26
Transparent/D latch, 157, 157f
Trap handler, 49, 357
  external interrupts, 784
  M-mode interrupts, 784.e12
  S-mode interrupts, 784.e69–
  software interrupts, 784
  timer interrupts, 784.e69–784.e70
Trap return instructions, 342
Traps, privileged operations, 371–373
  exceptions, 352–354
  interruption, 354–357
  precise traps, 352
Trap unit, 784.e138f
Tree multipliers, 594, 599

Tree pseudo-LRU, 440
Troubleshooting, 784.e38
Trunc multiplexer, 308f
Trusted execution environments (TEEs), 784.e60
TSMC28 ROM generators, 266
Two-bit branch predictor state transition diagram, 158f–159f, 159b

U

UART initialization, 784.e62, 784.e62t
Unbiased rounding, 622
Uncached access, 537–540
Unified Coverage Database (UCDB), 217–218
Unified divide and square root
  addend generation, 784.e248
  bit widths, 784.e252–784.e254
  implementation, 784.e254–784.e256
  multiplicitive algorithms, 784.e256–784.e257
  on-the-fly conversion, 784.e251
  postprocessing, 784.e251–784.e252
  preprocessing, 784.e248
  unified digit selection, 784.e248
Unit in the last place, 623
Unit *vs.* system testing, 198
Universal asynchronous receiver/transmitter (UART)
  asynchronous, 784
  PC16550 register map, 784–, 784.e1f
  receiver buffer register, 784.e1
  RISC-V Linux, 784.e1
  SPI and I2C serial links, 784.e1
  system clock frequency, 784.e1, 784.e1
  test plan, 784
  timing, 784f
  top-level interface, 784, 784f
  transmit holding register, 784.e1
  VT100 text, 784f
  Wally implementation, 784.e1
  waveforms, 784
Universal Asynchronous Receiver/Transmitter (UART), 784.e1, 784.e83
Unpacking signals, 632
Unpack unit, 784.e145f
Unprivileged Specification, 32
Unreachable logic exclusion, 222–226
Unsigned arithmetic, 591–592
Unsigned multiplication, 593f
Upstream repository, 784.e171
U-type instructions, 35–36

V

Vectored interrupts, 357
Vendor-specific electronic design automation tools, 784.e79–784.e80
Verification strategies, 67–69, 70f, 76, 76t
  definition, 195–196
  design and verification approaches, 200–202
  emulation, 198–199
  reference models, 198
  test vector generation, 196–197
  TRL, 199, 204f
  unit *vs.* system testing, 198
Verilator, 210–211
Verilog Change Dump, 206
Verilog syntax
  bit swizzling, 169
  delays, 170–171
  numbers, 168, 169t
  one-bit with many-bit gating, 170
  port connections, 171–172
  preprocessor
    define, 172
    include, 172
Version control, 144
Virtual harts, 45
Virtual memory
  interaction of processes, virtual memory, and caches, 476–477
  multilevel page tables, 473–474
  page faults, 472–473
  pages, 471–472
  physical addressing, 469–470
  process, 469
  protection, 473
  translation lookaside buffer, 474–475
  variable page size, 475–476
Virtual peripherals, 784.e48
Vivado, 784.e86
VxWorks, 784

W

Wallace tree multipliers, 601
Wally
  additional test, 784.e126t
  AHB and APB interfaces, 784.e136f
  configurations, 784.e108t
    additional parameters, 784.e113t, 784.e115t
    derivative, 784.e115t, 784.e118t
    key parameters, 784.e110t, 784.e112t
  default rvgc memory map, 784.e108t
  directory structure, 784.e107f
  key Wally commands, 784.e109, 784.e109
  PC logic, 784.e128f
  regression and test suites, 784.e109
  regression tests, 784.e125t
  repository, 784.e106
  RISC-V extensions and profiles, 784.e118t, 784.e124t
  test suites, 784.e125t, 784.e126t
Wally block diagram, 73, 74f
Wally code base, 74
Wally configurations, 73, 190–191
Wally diagrams
  AHB and APB interfaces, 784.e136f
  ALU with bit manipulation and cryptograph enhancements, 784.e140f
  atomic memory operations, 784.e135f
  basic bit manipulation, 784.e141f
  branch prediction unit, 784.e128f
  cache, 784.e131f
  carry-less multiplication, 784.e141f
  external bus unit, 784.e136f
  floating-point compare unit, 784.e148f
  floating-point convert unit, 784.e148f
  floating-point divide and square root unit, 784.e147f

floating-point fused multiply-add unit, 784.e146f
floating-point postprocessing unit, 784.e149f
floating-point unit, 784.e144f
hazard unit, 784.e129f
high-level, 784.e109–
instruction fetch unit, 784f
integer divider, 784.e143f
integer executiom unit, 784f
integer multiplier, 784.e142f
integer multiply/divide unit, 784.e142f
load/store unit, 784.e130f
memory management unit, 784.e132f
PC logic, 784.e128f
physical memory attributes checker, 784.e134f
physical memory protection checker, 784.e135f
privileged unit, 784.e137f
top-level schematic, 784.e127f
translation lookaside buffer, 784.e133f
trap unit, 784.e138f
uncore unit, 784.e139f
unpack unit, 784.e145f

Wally device tree, 784.e3f

Wally implementation
block diagram, 497–507
bus interface
bus hazards, 423–426
combined bus and tightly integrated memories, 420–421
external bus unit, 415–417
LSU and IFU AHB interface, 412–415
system-on-chip, 417–420
classify, 662–663
compare, 662
convert, 661
core local interruptor, 784
divide and square root, 660–661
division and remainder, 610–612
fence instruction, 511–512
floating-point unit
register file, 658
result and bypass multiplexers, 658–659
FPGA
Arty A7. *See* Arty A7 board
external bus connection, 784.e83
hardware required, 784.e80
mixed-mode clock manager, 784.e81
SD card, 784.e83–784.e84
SPI peripheral, 784.e83
supporting other boards, 784.e96–784.e99
uses IP, 784.e81
Xilinx SC7A100TCSG324-1, 784.e80
zero stage bootloader, 784.e99
general-purpose input/output, 784
hazard unit, 614
hazard unit changes, 667
HPTW
block diagram, 508
datapath, 508f
PTE status bit updates, 510–511
stalls, 510
state machine, 509f
instruction fetch unit (IFU)
branch predictor. *See* Branch predictor
instruction cache, 560
ITLB misses, 574
PC logic, 560
reset, stalls, flushes, and traps, 574–576
integer execution unit changes, 666
load/store unit changes, 669
MMU faults, 511
move, 662
multiplication, 607–610
multiply-add, 660
multiply and divide unit, 607, 613
platform-level interrupt controller, 784.e1
PMA checker, 498–500
PMP checker, 501–502
postprocessing, 666
privileged unit changes, 666
serial peripheral interface, 784.e23
sign injection, 662
TLB, 502–507
svnapot extension, 507
tlbcam, 502–503
tlbcontrol, 505–506
tlblru, 505
tlbmixer, 504
tlbram, 504
top-level block diagram, 495–496, 496f
universal asynchronous receiver/transmitter, 784.e1
unpacking unit, 659
Zfa additional floating-point instructions, 649–651
fround, 649–652

Wally LSU
application processor configuration, 517
high-level block diagram, 516f
LSU block diagram, 517f

Wally memory map, 784.e1

Wally pipelined processor, 318f

Wally privileged unit, 364–370

Wally-riscv-arch-test, 138–140, 139f

Wally's branch predictor
branch direction predictor, 784.e40
branch target buffer, 784.e40
instruction class prediction, 784.e40
return address stack, 784.e40

Wally's divide and square root unit, 661

Wally's floating-point convert unit, 661

Wally style guidelines
capitalization, 182, 182t
declaring signals, 183–184
lowRISC and Freescale, 181
port connections, 184–185
signal naming conventions, 181–185
spacing and indentation, 185–187
suffixes and prefixes, 183

Wally system-on-chip, 70–79, 71t–72t

Wally verification
assertions, 241
code coverage, 241–245
debugging hints, 235–239
design verification plan, 223t, 227–228
FPU, 239–240
functional coverage, 245–246
lint, 230
lock-step-compare, 240–241
reference model, 229
regression, 233–235
simulation, 231–233
top-level testbench and testvectors, 229–230

Wally zero stage bootloader, 784.e99

Waterfall project management, 143–144

Way invalidation, 453

Weakly Not Taken (WNT), 158

Weakly Taken (WT), 158
Windows, 784
Worst negative slack (WNS), 254
Write buffers, 677
Write port, 161
Write-through and writeback, 436
W-type instructions, 308

X

Xilinx, 784.e79

Z

Za64rs, 706
Za128rs, 706
Zca, 581
Zcb, 581
Zcd, 581
Zce shorthand, 582
Zcf, 581
Zc* family, 581
Zcmp, 581
Zcmt, 581
Zero stage bootloader (ZSBL), 784
Zfa additional floating-point instructions, 649–651
Ziccamoa, 705
Ziccif, 705
ZICCLSM
  misaligned loads and stores, 775–777
  test plan, 776
  Wally implementation, 776–777
Ziccrse, 705
Zicond conditional instructions
  alu with Zicond support, 782f
  operation sequences, 780t
  riscv-arch-test, 781t
  test plan, 781
  Wally implementation, 782
Zicsr extension, 53

RISC-V Instruction Set Summary

31:25		24:20	19:15	14:12	11:7	6:0	
funct7		rs2	rs1	funct3	rd	op	R-Type
imm$_{11:0}$			rs1	funct3	rd	op	I-Type
imm$_{11:5}$		rs2	rs1	funct3	imm$_{4:0}$	op	S-Type
imm$_{12,10:5}$		rs2	rs1	funct3	imm$_{4:1,11}$	op	B-Type
imm$_{31:12}$					rd	op	U-Type
imm$_{20,10:1,11,19:12}$					rd	op	J-Type
fs3	fmt	fs2	fs1	funct3	fd	op	R4-Type
5 bits	2 bits	5 bits	5 bits	3 bits	5 bits	7 bits	

Fig. I.1 RISC-V 32-bit instruction formats

- `XLEN`: 32 for RV32 / 64 for RV64
- `imm`: signed immediate in imm$_{11:0}$
- `uimm`: 5/6-bit unsigned immediate in imm$_{5/4:0}$
- `upimm`: 20 upper bits of a 32-bit immediate, in imm$_{31:12}$
- `Address`: memory address: rs1 + SignExt(imm$_{11:0}$)
- `[Address]`: data at memory location Address
- `BTA`: branch target address: PC + SignExt({imm$_{12:1}$, 1'b0})
- `JTA`: jump target address: PC + SignExt({imm$_{20:1}$, 1'b0})
- `label`: text indicating instruction address
- `SignExt`: value sign-extended to `XLEN`
- `ZeroExt`: value zero-extended to `XLEN`
- `csr`: control and status register

Table I.1 RV32/64I: RISC-V integer instructions

op	funct3/ imm$_{14:12}$	funct7/ instr$_{31:25}$	Type	Instruction	Description	Operation
0000011	000	imm$_{11:5}$	I	lb rd, imm(rs1)	load byte	rd = SignExt([Address]$_{7:0}$)
0000011	001	imm$_{11:5}$	I	lh rd, imm(rs1)	load half	rd = SignExt([Address]$_{15:0}$)
0000011	010	imm$_{11:5}$	I	lw rd, imm(rs1)	load word	rd = SignExt([Address]$_{31:0}$)
0000011	100	imm$_{11:5}$	I	lbu rd, imm(rs1)	load byte unsigned	rd = ZeroExt([Address]$_{7:0}$)
0000011	101	imm$_{11:5}$	I	lhu rd, imm(rs1)	load half unsigned	rd = ZeroExt([Address]$_{15:0}$)
0010011	000	imm$_{11:5}$	I	addi rd, rs1, imm	add immediate	rd = rs1 + SignExt(imm)
0010011	001	000000*	I	slli rd, rs1, uimm	shift left logical immediate	rd = rs1 << uimm
0010011	010	imm$_{11:5}$	I	slti rd, rs1, imm	set less than immediate	rd = (rs1 < SignExt(imm))
0010011	011	imm$_{11:5}$	I	sltiu rd, rs1, imm	set less than imm. unsigned	rd = (rs1 < SignExt(imm))
0010011	100	imm$_{11:5}$	I	xori rd, rs1, imm	xor immediate	rd = rs1 ^ SignExt(imm)
0010011	101	000000*	I	srli rd, rs1, uimm	shift right logical immediate	rd = rs1 >> uimm
0010011	101	010000*	I	srai rd, rs1, uimm	shift right arithmetic imm.	rd = rs1 >>> uimm
0010011	110	imm$_{11:5}$	I	ori rd, rs1, imm	or immediate	rd = rs1 | SignExt(imm)
0010011	111	imm$_{11:5}$	I	andi rd, rs1, imm	and immediate	rd = rs1 & SignExt(imm)
0010111	imm$_{14:12}$	imm$_{31:25}$	U	auipc rd, upimm	add upper immediate to PC	rd = {upimm, 12'b0} + PC
0100011	000	imm$_{11:5}$	S	sb rs2, imm(rs1)	store byte	[Address]$_{7:0}$ = rs2$_{7:0}$
0100011	001	imm$_{11:5}$	S	sh rs2, imm(rs1)	store half	[Address]$_{15:0}$ = rs2$_{15:0}$
0100011	010	imm$_{11:5}$	S	sw rs2, imm(rs1)	store word	[Address]$_{31:0}$ = rs2$_{31:0}$
0110011	000	0000000	R	add rd, rs1, rs2	add	rd = rs1 + rs2
0110011	000	0100000	R	sub rd, rs1, rs2	sub	rd = rs1 − rs2
0110011	001	0000000	R	sll rd, rs1, rs2	shift left logical	rd = rs1 << rs2$_{5/4:0}$ (RV64/32)
0110011	010	0000000	R	slt rd, rs1, rs2	set less than	rd = (rs1 < rs2)
0110011	011	0000000	R	sltu rd, rs1, rs2	set less than unsigned	rd = (rs1 < rs2)
0110011	100	0000000	R	xor rd, rs1, rs2	xor	rd = rs1 ^ rs2
0110011	101	0000000	R	srl rd, rs1, rs2	shift right logical	rd = rs1 >> rs2$_{5/4:0}$ (RV64/32)
0110011	101	0100000	R	sra rd, rs1, rs2	shift right arithmetic	rd = rs1 >>> rs2$_{5/4:0}$ (RV64/32)
0110011	110	0000000	R	or rd, rs1, rs2	or	rd = rs1 | rs2
0110011	111	0000000	R	and rd, rs1, rs2	and	rd = rs1 & rs2
0110111	imm$_{14:12}$	imm$_{31:25}$	U	lui rd, upimm	load upper immediate	rd = {upimm, 12'b0}
1100011	000	imm$_{12,10:5}$	B	beq rs1, rs2, label	branch if =	if (rs1 == rs2) PC = BTA
1100011	001	imm$_{12,10:5}$	B	bne rs1, rs2, label	branch if ≠	if (rs1 ≠ rs2) PC = BTA
1100011	100	imm$_{12,10:5}$	B	blt rs1, rs2, label	branch if <	if (rs1 < rs2) PC = BTA
1100011	101	imm$_{12,10:5}$	B	bge rs1, rs2, label	branch if ≥	if (rs1 ≥ rs2) PC = BTA
1100011	110	imm$_{12,10:5}$	B	bltu rs1, rs2, label	branch if < unsigned	if (rs1 < rs2) PC = BTA
1100011	111	imm$_{12,10:5}$	B	bgeu rs1, rs2, label	branch if ≥ unsigned	if (rs1 ≥ rs2) PC = BTA
1100111	000	imm$_{11:5}$	I	jalr rd, imm(rs1)	jump and link register	PC = [Address], rd = PC + 4
1101111	imm$_{14:12}$	imm$_{20,10:5}$	J	jal rd, label	jump and link	PC = JTA, rd = PC + 4

* = 0 for RV32, uimm$_5$ for RV64.

Table I.2 RV64I: Extra integer instructions

op	funct3	funct7/ $imm_{11:5}$	Type	Instruction	Description	Operation
0000011	011	$imm_{11:5}$	I	ld rd, imm(rs1)	load double word	rd = [Address]$_{63:0}$
0000011	110	$imm_{11:5}$	I	lwu rd, imm(rs1)	load word unsigned	rd = ZeroExt([Address]$_{31:0}$)
0011011	000	$imm_{11:5}$	I	addiw rd, rs1, imm	add immediate word	rd = SignExt((rs1 + SignExt(imm))$_{31:0}$)
0011011	001	0000000	I	slliw rd, rs1, uimm	shift left logical immediate word	rd = SignExt((rs1$_{31:0}$ << uimm)$_{31:0}$)
0011011	101	0000000	I	srliw rd, rs1, uimm	shift right logical immediate word	rd = SignExt((rs1$_{31:0}$ >> uimm)$_{31:0}$)
0011011	101	0100000	I	sraiw rd, rs1, uimm	shift right arith. immediate word	rd = SignExt((rs1$_{31:0}$ >>> uimm)$_{31:0}$)
0100011	011	$imm_{11:5}$	S	sd rs2, imm(rs1)	store double word	[Address]$_{63:0}$ = rs2
0111011	000	0000000	R	addw rd, rs1, rs2	add word	rd = SignExt((rs1 + rs2)$_{31:0}$)
0111011	000	0100000	R	subw rd, rs1, rs2	subtract word	rd = SignExt((rs1 − rs2)$_{31:0}$)
0111011	001	0000000	R	sllw rd, rs1, rs2	shift left logical word	rd = SignExt((rs1$_{31:0}$ << rs2$_{4:0}$)$_{31:0}$)
0111011	101	0000000	R	srlw rd, rs1, rs2	shift right logical word	rd = SignExt((rs1$_{31:0}$ >> rs2$_{4:0}$)$_{31:0}$)
0111011	101	0100000	R	sraw rd, rs1, rs2	shift right arithmetic word	rd = SignExt((rs1$_{31:0}$ >>> rs2$_{4:0}$)$_{31:0}$)

In RV64I, registers are 64 bits, but instructions are still 32 bits. The term "word" generally refers to a 32-bit value. In RV64I, immediate shift instructions use 6-bit immediates: $uimm_{5:0}$; but for word shifts, the most significant bit of the shift amount ($uimm_5$) must be 0. Instructions ending in "w" (for "word") operate on the lower half of the 64-bit registers. Sign- or zero-extension produces a 64-bit result.

Table I.3 RVF/D/Q/Zfh: RISC-V single-, double-, quad-, and half-precision floating-point instructions

op	funct3	funct7/ $instr_{31:25}$	rs2/ $imm_{4:0}$	Type	Flags NV DZ OF UF NX	Instruction	Description	Operation
1000011	rm	fs3, fmt	fs2	R4	x – x x x	fmadd.* fd,fs1,fs2,fs3	multiply-add	fd = fs1 * fs2 + fs3
1000111	rm	fs3, fmt	fs2	R4	x – x x x	fmsub.* fd,fs1,fs2,fs3	multiply-subtract	fd = fs1 * fs2 − fs3
1001011	rm	fs3, fmt	fs2	R4	x – x x x	fnmsub.* fd,fs1,fs2,fs3	negate multiply-sub	fd = −(fs1 * fs2 − fs3)
1001111	rm	fs3, fmt	fs2	R4	x – x x x	fnmadd.* fd,fs1,fs2,fs3	negate multiply-add	fd = −(fs1 * fs2 + fs3)
1010011	rm	00000, fmt	fs2	R	x – x – x	fadd.* fd,fs1,fs2	add	fd = fs1 + fs2
1010011	rm	00001, fmt	fs2	R	x – x – x	fsub.* fd,fs1,fs2	subtract	fd = fs1 − fs2
1010011	rm	00010, fmt	fs2	R	x – x x x	fmul.* fd,fs1,fs2	multiply	fd = fs1 * fs2
1010011	rm	00011, fmt	fs2	R	x x x x x	fdiv.* fd,fs1,fs2	divide	fd = fs1 / fs2
1010011	rm	01011, fmt	00000	I	x – – – x	fsqrt.* fd,fs1	square root	fd = sqrt(fs1)
1010011	000	00100, fmt	fs2	R	– – – – –	fsgnj.* fd,fs1,fs2	sign injection	fd = fs1, sign = sign(fs2)
1010011	001	00100, fmt	fs2	R	– – – – –	fsgnjn.* fd,fs1,fs2	negate sign injection	fd = fs1, sign = −sign(fs2)
1010011	010	00100, fmt	fs2	R	– – – – –	fsgnjx.* fd,fs1,fs2	xor sign injection	fd = fs1, sign = sign(fs2)^sign(fs1)
1010011	000	00101, fmt	fs2	R	x – – – –	fmin.* fd,fs1,fs2	min	fd = min(fs1, fs2)
1010011	001	00101, fmt	fs2	R	x – – – –	fmax.* fd,fs1,fs2	max	fd = max(fs1, fs2)
1010011	010	10100, fmt	fs2	R	x – – – –	feq.* rd,fs1,fs2	compare =	rd = (fs1 == fs2)
1010011	001	10100, fmt	fs2	R	x – – – –	flt.* rd,fs1,fs2	compare <	rd = (fs1 < fs2)
1010011	000	10100, fmt	fs2	R	x – – – –	fle.* rd,fs1,fs2	compare ≤	rd = (fs1 ≤ fs2)
1010011	001	11100, fmt	00000	I	– – – – –	fclass.* rd,fs1	classify	rd = classification of fs1
0000111	010	$imm_{11:5}$	$imm_{4:0}$	I	– – – – –	flw fd, imm(rs1)	load float	fd = [Address]$_{31:0}$
0100111	010	$imm_{11:5}$	fs2	S	– – – – –	fsw fs2,imm(rs1)	store float	[Address]$_{31:0}$ = fd
1010011	rm	11000, fmt	000,lu	I	x – x – x	fcvt.dst.src rd, fs1	convert fp to int	rd = dst(fs1)
1010011	rm	11010, fmt	000,lu	I	– – x – x	fcvt.dst.src fd, rs1	convert int to fp	fd = dst(rs1)
1010011	rm	01000, dfmt	000,sfmt	I	x – x x x	fcvt.dst.src fd, fs1	convert fp to fp	fd = dst(fs1)
1010011	000	11100, fmt	00000	I	– – – – –	fmv.x.{h/w/d} rd, fs1	move fp to int	rd = fs1
1010011	000	11110, fmt	00000	I	– – – – –	fmv.{h/w/d}.x fd, rs1	move int to fp	fd = rs1
						RVD only		
0000111	011	$imm_{11:5}$	$imm_{4:0}$	I	– – – – –	fld fd, imm(rs1)	load double	fd = [Address]$_{63:0}$
0100111	011	$imm_{11:5}$	fs2	S	– – – – –	fsd fs2,imm(rs1)	store double	[Address]$_{63:0}$ = fs2
						RVQ only		
0000111	100	$imm_{11:5}$	$imm_{4:0}$	I	– – – – –	flq fd, imm(rs1)	load quad	fd = [Address]$_{127:0}$
0100111	100	$imm_{11:5}$	fs2	S	– – – – –	fsq fs2,imm(rs1)	store quad	[Address]$_{127:0}$ = fs2
						RVH only		
0000111	001	$imm_{11:5}$	$imm_{4:0}$	I	– – – – –	flh fd, imm(rs1)	load half	fd = [Address]$_{15:0}$
0100111	001	$imm_{11:5}$	fs2	S	– – – – –	fsh fs2,imm(rs1)	store half	[Address]$_{15:0}$ = fs2

* = {h/s/d/q}. fs1, fs2, fs3, fd: floating-point registers. fs1, fs2, and fd are encoded in fields rs1, rs2, and rd; only R4-type also encodes fs3. fmt: precision of computational instruction (half=10_2, single=00_2, double=01_2, quad=11_2). sfmt: source format. rm: rounding mode (0=to nearest, 1=toward zero, 2=down, 3=up, 4=to nearest (max magnitude), 7=dynamic). sign(fs1): the sign of fs1. dst or src = {h/s/d/q} for fp, {w/wu/l/lu} for 32/64 bit signed/unsigned ints. {w/wu} = 32-bit signed/unsigned int. {l/lu} = 64-bit (long) signed/unsigned int. For fp move instructions (fmv), {h/w/d} denotes a half-/single-/double-precision fp number held in a fp register, and x is a fp number (held in an integer register), with the same-precision as the fp source/destination. Floating-point exception flags are: NV: invalid operation, DZ: divide by zero, OF: overflow, UF: underflow, and NX: inexact.

Table I.4 Register names and numbers

Name	Register Number	Use
zero	x0	Constant value 0
ra	x1	Return address
sp	x2	Stack pointer
gp	x3	Global pointer
tp	x4	Thread pointer
t0–2	x5–7	Temporary registers
s0/fp	x8	Saved register / Frame pointer
s1	x9	Saved register
a0–1	x10–11	Function arguments / Return values
a2–7	x12–17	Function arguments
s2–11	x18–27	Saved registers
t3-6	x28–31	Temporary registers

15 14 13	12	11 10	9 8 7	6 5	4 3 2	1 0	
funct4		rd/rs1		rs2		op	CR
funct3	imm	rd/rs1		imm		op	CI
funct3	imm		rs1'	imm	rs2'	op	CS
funct6			rd', rs1'	funct2	rs2'	op	CA
funct6			rs1'	imm	rd' / rs2'	op	CLB / CSB
funct6			rs1'	f1 imm	rd' / rs2'	op	CLH / CSH
funct6			rd', rs1'	funct5		op	CU
funct3	imm		rs1'	imm		op	CB
funct3	imm	funct	rd', rs1'	imm		op	CB'
funct3	imm					op	CJ
funct3	imm			rs2		op	CSS
funct3	imm				rd'	op	CIW
funct3	imm		rs1'	imm	rd'	op	CL
3 bits	3 bits		3 bits	2 bits	3 bits	2 bits	

Fig. I.2. RISC-V compressed (16-bit) instruction formats

Table I.5 RVM: RISC-V multiply and divide instructions

op	funct3	funct7	Type	Instruction	Description	Operation
0110011	000	0000001	R	mul rd, rs1, rs2	multiply	rd = (rs1 * rs2)$_{XLEN-1:0}$
0110011	001	0000001	R	mulh rd, rs1, rs2	multiply high signed signed	rd = (rs1 * rs2)$_{2*XLEN-1:XLEN}$
0110011	010	0000001	R	mulhsu rd, rs1, rs2	multiply high signed unsigned	rd = (rs1 * rs2)$_{2*XLEN-1:XLEN}$
0110011	011	0000001	R	mulhu rd, rs1, rs2	multiply high unsigned unsigned	rd = (rs1 * rs2)$_{2*XLEN-1:XLEN}$
0110011	100	0000001	R	div rd, rs1, rs2	divide (signed)	rd = rs1 / rs2
0110011	101	0000001	R	divu rd, rs1, rs2	divide unsigned	rd = rs1 / rs2
0110011	110	0000001	R	rem rd, rs1, rs2	remainder (signed)	rd = rs1 % rs2
0110011	111	0000001	R	remu rd, rs1, rs2	remainder unsigned	rd = rs1 % rs2

RV64M adds mulw, divw, remw, divuw, remuw with op = 59. These operate on only the lower 32 bits of a register.

Table I.6 RVC/Zca: RISC-V compressed (16-bit) integer instructions

op	instr$_{15:10}$	funct2	Registers	Type	Compressed Instruction	32-Bit Equivalent
C/Zca: Compressed instructions excluding floating-point loads and stores						
00	000---	–	imm ≠ 0	CIW	c.addi4spn rd', sp, imm	addi rd', sp, ZeroExt(imm)*4
00	010---	–		CL	c.lw rd', imm(rs1')	lw rd', (ZeroExt(imm)*4)(rs1')
00	110---	–		CS	c.sw rs2', imm(rs1')	sw rs2', (ZeroExt(imm)*4)(rs1')
01	000000	–	rs1 = 0, imm = 0	CI	c.nop	addi x0, x0, 0
01	000---	–	rd ≠ 0, imm ≠ 0	CI	c.addi rd, imm	addi rd, rd, SignExt(imm)
01	010---	–	rd ≠ 0	CI	c.li rd, imm	addi rd, x0, SignExt(imm)
01	011---	–	rd ≠ {0,2}, imm ≠ 0	CI	c.lui rd, imm	lui rd, {14{imm_5}, imm}
01	011---	–	rd = 2, imm ≠ 0	CI	c.addi16sp sp, imm	addi sp, sp, SignExt(imm)*16
01	100-00	–	imm ≠ 0	CB'	c.srli rd', imm	srli rd', rd', imm
01	100-01	–	imm ≠ 0	CB'	c.srai rd', imm	srai rd', rd', imm
01	100-10	–		CB'	c.andi rd', imm	andi rd', rd', SignExt(imm)
01	100011	00		CA	c.sub rd', rs2'	sub rd', rd', rs2'
01	100011	01		CA	c.xor rd', rs2'	xor rd', rd', rs2'
01	100011	10		CA	c.or rd', rs2'	or rd', rd', rs2'
01	100011	11		CA	c.and rd', rs2'	and rd', rd', rs2'
01	101---	–		CJ	c.j label	jal x0, label
01	110---	–		CB	c.beqz rs1', label	beq rs1', x0, label
01	111---	–		CB	c.bnez rs1', label	bne rs1', x0, label
10	000---	–	rd ≠ 0, imm ≠ 0	CI	c.slli rd, imm	slli rd, rd, imm
10	010---	–	rd ≠ 0	CI	c.lwsp rd, imm	lw rd, (ZeroExt(imm)*4)(sp)
10	1000--	–	rs1 ≠ 0, rs2 = 0	CR	c.jr rs1	jalr x0, 0(rs1)
10	1000--	–	rd ≠ 0, rs2 ≠ 0	CR	c.mv rd, rs2	add rd, x0, rs2
10	1001--	–	rs1 = 0, rs2 = 0	CR	c.ebreak	ebreak
10	1001--	–	rs1 ≠ 0, rs2 = 0	CR	c.jalr rs1	jalr ra, 0(rs1)
10	1001--	–	rd ≠ 0, rs2 ≠ 0	CR	c.add rd, rs2	add rd, rd, rs2
10	110---	–		CSS	c.swsp rs2, imm	sw rs2, (ZeroExt(imm)*4)(sp)
C/Zca (RV32 only)						
01	001---	–		CJ	c.jal label	jal ra, label
C/Zca (RV64 only, reuses encodings for c.jal above and c.flw, c.fsw, c.flwsp, and c.fswsp from Table I.22)						
00	011---	–		CL	c.ld rd', imm(rs1') (replaces c.flw)	ld rd', (ZeroExt(imm)*8)(rs1')
00	111---	–		CS	c.sd rs2', imm(rs1') (replaces c.fsw)	sd rs2, (ZeroExt(imm)*8)(rs1')
01	001---	–	rd ≠ 0	CI	c.addiw rd, imm (replaces c.jal)	addiw rd, rd, (SignExt(imm))
01	100111	00		CA	c.subw rd', rs2'	subw rd', rd', rs2'
01	100111	01		CA	c.addw rd', rs2'	addw rd', rd', rs2'
10	011---	–	rd ≠ 0	CI	c.ldsp rd, imm (replaces c.flwsp)	ld rd, (ZeroExt(imm)*8)(sp)
10	111---	–		CSS	c.sdsp rs2, imm (replaces c.fswsp)	sd rs2, (ZeroExt(imm)*8)(sp)

rs1', rs2', rd': 3-bit register designator for registers 8–15: 000_2 = x8 or f8 , ... , 111_2 = x15 or f15. Table I.9 lists additional RISC-V compressed instructions.

Table I.7 Common RISC-V pseudoinstructions

Pseudoinstruction	RISC-V Instructions	Description	Operation
nop	addi x0, x0, 0	no operation	
li rd, imm$_{11:0}$	addi rd, x0, imm$_{11:0}$	load 12-bit immediate	rd = SignExtend(imm$_{11:0}$)
li rd, imm$_{31:0}$	lui rd, imm$_{31:12}$* addi rd, rd, imm$_{11:0}$	load 32-bit immediate	rd = imm$_{31:0}$
mv rd, rs1	addi rd, rs1, 0	move (also called "register copy")	rd = rs1
not rd, rs1	xori rd, rs1, -1	one's complement	rd = ~rs1
neg rd, rs2	sub rd, x0, rs2	two's complement	rd = −rs1
seqz rd, rs1	sltiu rd, rs1, 1	set if = 0	rd = (rs1 == 0)
snez rd, rs2	sltu rd, x0, rs2	set if ≠ 0	rd = (rs1 ≠ 0)
sltz rd, rs1	slt rd, rs1, x0	set if < 0	rd = (rs1 < 0)
sgtz rd, rs2	slt rd, x0, rs2	set if > 0	rd = (rs1 > 0)
beqz rs1, label	beq rs1, x0, label	branch if = 0	if (rs1 == 0) PC = label
bnez rs1, label	bne rs1, x0, label	branch if ≠ 0	if (rs1 ≠ 0) PC = label
blez rs2, label	bge x0, rs2, label	branch if ≤ 0	if (rs1 ≤ 0) PC = label
bgez rs1, label	bge rs1, x0, label	branch if ≥ 0	if (rs1 ≥ 0) PC = label
bltz rs1, label	blt rs1, x0, label	branch if < 0	if (rs1 < 0) PC = label
bgtz rs2, label	blt x0, rs2, label	branch if > 0	if (rs1 > 0) PC = label
ble rs1, rs2, label	bge rs2, rs1, label	branch if ≤	if (rs1 ≤ rs2) PC = label
bgt rs1, rs2, label	blt rs2, rs1, label	branch if >	if (rs1 > rs2) PC = label
bleu rs1, rs2, label	bgeu rs2, rs1, label	branch if ≤ (unsigned)	if (rs1 ≤ rs2) PC = label
bgtu rs1, rs2, label	bltu rs2, rs1, offset	branch if > (unsigned)	if (rs1 > rs2) PC = label
j label	jal x0, label	jump	PC = label
jal label	jal ra, label	jump and link	PC = label, ra = PC + 4
jr rs1	jalr x0, 0(rs1)	jump register	PC = rs1
jalr rs1	jalr ra, 0(rs1)	jump and link register	PC = rs1, ra = PC + 4
ret	jalr x0, 0(ra)	return from function	PC = ra
call label	jal ra, label	call nearby function	PC = label, ra = PC + 4
call label	auipc ra, offset$_{31:12}$* jalr ra, offset$_{11:0}$(ra)	call far away function	PC = PC + offset, ra = PC + 4
la rd, symbol	auipc rd, symbol$_{31:12}$* addi rd, rd, symbol$_{11:0}$	load address of global variable	rd = PC + symbol
l{b|h|w} rd, symbol	auipc rd, symbol$_{31:12}$* l{b|h|w} rd, symbol$_{11:0}$(rd)	load global variable	rd = [PC + symbol]
s{b|h|w} rs2, symbol, rs1	auipc rs1, symbol$_{31:12}$* s{b|h|w} rs2, symbol$_{11:0}$(rs1)	store global variable	[PC + symbol] = rs2
csrr rd, csr	csrrs rd, csr, x0	read CSR	rd = csr
csrw csr, rs1	csrrw x0, csr, rs1	write CSR	csr = rs1
csrs/csrc csr, rs1	csrrs/csrrc x0, csr, rs1	set/clear bits in CSR	csr = csr | rs1 / csr = csr & ~rs1

*If bit 11 of the immediate/offset/symbol is 1, the upper immediate is incremented by 1. offset/symbol are the 32-bit PC-relative addresses of a label/global variable.

Table I.8 Privileged / CSR / fence instructions

op	funct3	imm$_{11:0}$/funct7	Registers	Type	Instruction	Description	Operation
1110011	000	000000000000	rs1,rd=0	I	ecall	transfer control to OS	
1110011	000	000000000001	rs1,rd=0	I	ebreak	transfer control to debugger	
1110011	000	000100000010	rs1,rd=0	I	sret	return from supervisor exception	PC = sepc
1110011	000	001100000010	rs1,rd=0	I	mret	return from machine exception	PC = mepc
1110011	000	000100000101	rs1,rd=0	I	wfi	wait for interrupt	
0001111	000	000000110011	rs1,rd=0	I	fence	synchronize loads and stores	
0001111	001	000000000000	rs1,rd=0	I	fence.i	synchronize instruction memory	
1110011	000	0001001	rd=0	R	sfence.vma rs1,rs2	synchronize page table	Note: rs1 = vaddr,rs2 = asid
1110011	001	csr		I	csrrw rd,csr,rs1	CSR read/write	rd = csr,csr = rs1
1110011	010	csr		I	csrrs rd,csr,rs1	CSR read/set	rd = csr,csr |= rs1
1110011	011	csr		I	csrrc rd,csr,rs1	CSR read/clear	rd = csr,csr &= ~rs1
1110011	101	csr	rs1=uimm	I	csrrwi rd,csr,uimm	CSR read/write immediate	rd = csr,csr = ZeroExt(uimm)
1110011	110	csr	rs1=uimm	I	csrrsi rd,csr,uimm	CSR read/set immediate	rd = csr,csr |= ZeroExt(uimm)
1110011	111	csr	rs1=uimm	I	csrrci rd,csr,uimm	CSR read/clear immediate	rd = csr,csr &= ~ZeroExt(uimm)

RISC-V Privileged Architecture (Version 1.12) Summary

(Excluding Hypervisor and Debug Features)

Table I.9 Control and Status Registers (CSRs)

Name	Address	h Address (RV32)	Size	Type	Description
			Traps		
mstatus/sstatus	300/100	310	64/XLEN	MRW/SRW	Status
medeleg	302	312	XLEN	MRW	Exception delegation
mideleg	303		XLEN (12)	MRW	Interrupt delegation
mie/sie	304/104		XLEN (12)	MRW/SRW	Interrupt enable
mtvec/stvec	305/105		XLEN	MRW/SRW	Trap vector
mscratch/sscratch	340/140		XLEN	MRW/SRW	Scratch (for trap stack pointer)
mepc/sepc	341/141		XLEN	MRW/SRW	Exception program counter
mcause/scause	342/142		XLEN	MRW/SRW	Trap cause
mtval/stval	343/143		XLEN	MRW/SRW	Bad address or instruction
mip/sip	344/144		XLEN (12)	MRW/SRW	Interrupt pending
			Virtual Memory		
satp	180		XLEN	SRW	Address translation & protection
			Physical Memory Protection		
pmpcfg0/2/.../14	3A0/3A2/.../3AE	3A1/3A3/.../3AF	64 (8 x 8)	MRW	PMP configuration
pmpaddr0/1/.../63	3B0/3B1/.../3EF		XLEN	MRW	PMP address
			Floating-Point		
fcsr/fflags/frm	003/001/002			URW	Floating-point CSR/flags/rounding
			Counters / Performance Monitoring		
mcounteren/scounteren	306/106		32	MRW/SRW	Counter enable
mcountinhibit	320		32	MRW	Counter inhibit
mhpmevent3/4/.../31	323/324/.../33F	723/724/.../73F	XLEN	MRW	Counter event selectors
mcycle/cycle	B00/C00	B80/C80	64	MRW/URO	Cycle counter
time	C01	C81	64	URO	Time (from CLINT)
minstret/instret	B02/C02	B82/C82	64	MRW/URO	Instructions retired
mhpmcounter3/4/.../31 hpmcounter3/4/.../31	B03/B04/.../B1F C03/C04/.../C1F	B83/B84/... C83/C84/...	64	MRW/ URO	Performance-monitoring counters
			Machine Information Registers & Configuration		
misa	301		XLEN	MRW	Instruction set
menvcfg/senvcfg	30A/10A	31A/-	64/XLEN	MRW/SRW	Environment configuration
mseccfg	747	757	64	MRW	Security configuration
mvendorid	F11		32	MRO	JEDEC vendor ID
marchid	F12		XLEN	MRO	Vendor-specific architecture ID
mimpid	F13		XLEN	MRO	Vendor-specific implementation ID
mhartid	F14		XLEN	MRO	Hart ID
mconfigptr	F15		XLEN	MRO	Configuration pointer
			Sstc Extension		
stimecmp	14D	15D	64	SRW	Supervisor timer compare
			Zkr Extension		
seed	015		32	MRW	Seed to provide entropy for random number generator

{M/S/U}{RW/RO} = {Machine/Supervisor/User}{Read & Write/Read Only}. Gray registers are read only. For 64-bit registers in RV32, the bottom 32 are in the base register and the upper 32 are in a corresponding h register (e.g., mcycleh).

Table I.10 CSR bitfields

Register	XLEN-1	XLEN-2	XLEN-3	...	37	36	35	34	33	32	31:23	22	21	20	19	18	17	16	15	14	13	12	11	10	9	8	7	6	5	4	3	2	1	0
mstatus	SD				MBE	SBE	**SXL**		**UXL**			TSR	TW	TVM	MXR	SUM	MPRV	XS		FS		MPP		VS		SPP	MPIE	UBE	SPIE		MIE		SIE	
sstatus	SD								**UXL**						MXR	SUM		XS		FS				VS		SPP		UBE	SPIE				SIE	
misa	MXL												V	U		S		Q				M				I	H		F	E	D	C		A
menvcfg	STCE	PBMTE	ADUE																								CBZE	CBCFE	CBIE					FIOM
senvcfg																											CBZE	CBCFE	CBIE					FIOM
fcsr																											frm			NV	DZ	OF	UF	NX
{m/s}cause	INT																																	cause
medeleg																		See Table I.13																
mideleg																							MEI		SEI		MTI		STI		MSI		SSI	
mie																							MEIE		SEIE		MTIE		STIE		MSIE		SSIE	
mip																							MEIP		SEIP		MTIP		STIP		MSIP		SSIP	
sie																									SEIE				STIE				SSIE	
sip																									SEIP				STIP				SSIP	

Gray register bits are read only. For 64-bit registers in RV32, the bottom 32 are in the base register and the upper 32 are in a corresponding h register (e.g., mcycleh). RV32 mstatus and sstatus place SD in bit 31. The bolded fields are in RV64 only: mstatush drops the SXL and UXL fields, leaving only MBE and SBE as bits 5:4; sstatus has no UXL field, so no sstatush is needed.

Table I.11 CSR bitfield legend

Bit Field	Meaning	Description
SD	**S**tate **D**irty	1 = any of XS, FS, VS dirty
{M/S/U}BE	{**M**achine/**S**upervisor/**U**ser} **B**ig **E**ndian	1 = Big-endian data
{M/S/U}XL	{**M**achine/**S**upervisor/**U**ser} Base ISA **XL**EN	01= 32; 10 = 64; 11 = 128
TSR	**T**rap **SR**ET	For virtualization
TW	**T**imeout **W**ait	1 = Trap on WFI after timeout
TVM	**T**rap **V**irtual **M**emory	For virtualization
MXR	**M**ake e**X**ecutable **R**eadable	1 = Allow loads from executable pages
SUM	**S**upervisor **U**ser **M**emory	1 = Supervisor can access user memory
MPRV	**M**odify **PR**i**V**ilege	1 = Apply memory privileges from MPP
{X/F/V}S	{Custom(**X**)/**F**loating-point/**V**ector} **S**tate dirty	00 = off; 01 = initial; 10 = clean; 11 = dirty
{M/S}PP	{**M**achine/**S**upervisor} **P**revious **P**rivilege	00 = User; 01 = Supervisor; 11 = Machine
{M/S}PIE	{**M**achine/**S**upervisor} **P**revious **I**nterrupt **E**nable	Restore IE = PIE after MRET/SRET
{M/S}IE	{**M**achine/**S**upervisor} global **I**nterrupt **E**nable	0 = disable all interrupts
STCE	**S**upervisor **T**imer **C**ompare **E**nable	1 = enable Sstc extension (see Table I.9)
PBMTE	**P**age-**B**ased **M**emory **T**ypes **E**nable	1 = enable Svpbmt extension (see Section 11.2.2.1.1XREF)
ADUE	**A**ccess/**D**irty **U**pdate **E**nable	1 = enable Svadu extension (see Section 8.2.3.5XREF)
CB{Z/CF}E	**C**ache **B**lock {**Z**ero/**C**lean&**F**lush} **E**nable	1 = enable CBO instructions (see Table I.21)
CBIE	**C**ache **B**lock **I**nvalidate **E**nable	00 = disable; 01 = flush; 11 = invalidate
FIOM	**F**ence of **I/O** implies **M**emory	For virtualization
frm	**F**loating-point **R**ounding **M**ode	0 = RNE; 1 = RTZ; 2 = RDN; 3 = RUP; 4 = RMM; 7 = DYN
fflags {NV, DZ, OF, UF, NX}	{i**NV**alid, **D**ivide by **Z**ero, **O**ver**F**low, **U**nder**F**low, i**N**e**X**act}	Floating-point flags
INT	Trap is an **INT**errupt	1 = trap is an interrupt
cause	**Cause** of trap	See Table I.13
{M/S}{E/T/S}I{E/P}	{**M**achine/**S**upervisor}{**E**xternal/**T**imer/**S**oftware} **I**nterrupt{**E**nable/**P**ending}	MEIP set by PLIC MTIP set by CLINT when mtime > mtimecmp MSIP set by CLINT SEIP set by PLIC or mip.SEIP STIP set by mip.STIP or when mtime > stimecmp SSIP set by mip.SSIP

Table I.12 misa bitfields

Bit	Character	Description
0	A	Atomic
1	B	Bit manipulation (Zba+Zbb+Zbs)
2	C	Compressed
3	D	Double-precision floating-point
4	E	RV32E Embedded ISA
5	F	Single-precision floating-point
6	G	At least IMAFD supported
7	H	Hypervisor
8	I	RV32/64/128I integer ISA
12	M	Integer multiply/divide
16	Q	Quad-precision floating-point
18	S	Supervisor mode
20	U	User mode
21	V	Vector
XLEN-1:XLEN-2	MXL	XLEN: 01 = 32, 10 = 64, 11 = 128

Table I.13 cause interrupt and exception codes

Cause	Interrupt	Exception
0		Instruction address misaligned
1	Supervisor Software	Instruction access fault
2		Illegal instruction
3	Machine Software	Breakpoint
4		Load address misaligned
5	Supervisor Timer	Load access fault
6		Store/AMO address misaligned
7	Machine Timer	Store/AMO access fault
8		Environment call from U-mode
9	Supervisor External	Environment call from S-mode
10		
11	Machine External	Environment call from M-mode
12		Instruction page fault
13		Load page fault
14		
15		Store/AMO page fault

RISC-V Virtual Memory Summary

Table I.14 satp bitfields

Field	Meaning	RV64	RV32
Mode	RV64: 0 = bare; 8 = SV39; 9 = SV48; 10 = SV57 RV32: 0 = bare; 1 = RV32	63:60	31
ASID	Address space ID	59:44	30:22
PPN	Physical page number of root page table	43:0	21:0

Table I.15 Virtual address bitfields

Format	63:57	56:48	47:39	38:32	31:30	29:22	21	20:12	11:0
SV32	n/a				VPN[1]		VPN[0]		Page Offset
SV39	extend			VPN[2]		VPN[1]		VPN[0]	
SV48	extend		VPN[3]						
SV57	extend	VPN[4]							

Table I.16 XWR encodings

X	W	R	Meaning
0	0	0	Pointer to next level of page table
0	0	1	Read-only page
0	1	0	Reserved
0	1	1	Read-write page
1	0	0	Execute-only page
1	0	1	Read-execute page
1	1	0	Reserved
1	1	1	Read-write-execute page

Table I.17 Physical address bitfields

Format	55:48	47:39	38:34	33:30	29:22	21	20:12	11:0
SV32	n/a			PPN[1]		PPN[0]		Page Offset
SV39	PPN[2]				PPN[1]		PPN[0]	
SV48	PPN[3]		PPN[2]					
SV57	PPN[4]	PPN[3]						

Virtual addresses are 32 bits for RV32 and 64 bits for RV64.
Physical addresses are 34 bits for RV32 and 56 bits for RV64.

Table I.18 Page table entry bitfields

ISA	63	62:61	60:54	53:32	31:10	9:8	7	6	5	4	3	2	1	0
RV32	n/a				PPN	**RSW**: **R**eserved for **S**oft**w**are	**D**irty	**A**ccessed	**G**lobal	**U**ser	e**X**ecutable	**W**ritable	**R**eadable	**V**alid
RV64	**NAPOT:** 0 = normal 1 = 64 KiB huge page	**PBMT**: 0 = Use PMA 1 = Uncacheable Memory 2 = Uncacheable I/O	reserved	PPN										

RISC-V Physical Memory Protection (PMP) Summary

Table I.19 pmpaddr0-63 bitfields

ISA	63:54	53:32	31:0
RV32	n/a	n/a	33:2
RV64	0	55:34	33:2

Table I.20 pmpcfg0-15 bitfields

7	6:5	4:3	2	1	0
Locked	00	**A**lignment: 0 = off 1 = **TOR**: **T**op **O**f **R**ange 2 = **NA4**: **N**aturally **A**ligned **4**-byte 3 = **NAPOT**: **N**aturally **A**ligned **P**ower **o**f **T**wo	e**X**ecutable	**W**ritable	**R**eadable

Each pmpcfg register contains 8 (RV64) or 4 (RV32) copies of the 8-bit configuration fields shown in Table I.20, so each pmpcfg register supports up to 8 (RV64) or 4 (RV32) pmpaddr registers. To support all 64 pmpaddr registers, RV64 uses 64-bit pmpcfg0, 2, 8,... 14 and RV32 uses 32-bit pmpcfg0, 1, 2,...15.

RISC-V Cache Management Operation (CMO) Instructions

Table I.21 Zicbo*: CMO instructions

op	funct3	$imm_{11:0}$	Registers	Type	Instruction	Description	Operation
				Zicbom: Cache Block Management			
0001111	010	000000000000	rd = 0	I	`cbo.inval (rs1)`	invalidate block	invalidate block holding [rs1]
0001111	010	000000000001	rd = 0	I	`cbo.clean (rs1)`	clean (write back) block	write back block holding [rs1]
0001111	010	000000000010	rd = 0	I	`cbo.flush (rs1)`	flush block	flush block holding [rs1]
				Zicboz: Cache Block Zero			
0001111	010	000000000100	rd = 0	I	`cbo.zero (rs1)`	zero block	zero block holding [rs1]
				Zicbop: Cache Block Prefetch			
0010011	110	imm[11:5], 00000	rd = 0	I	`prefetch.i imm(rs1)`	instruction prefetch	I$ Prefetch [Address]
0010011	110	imm[11:5], 00001	rd = 0	I	`prefetch.r imm(rs1)`	data read prefetch	D$ Read Prefetch [Address]
0010011	110	imm[11:5], 00011	rd = 0	I	`prefetch.w imm(rs1)`	data write prefetch	D$ Write Prefetch [Address]

RISC-V Additional Compressed Instructions

Table I.22 RVC/Zcb/Zcf/Zcd: Additional compressed (16-bit) instructions (continued from Table I.6)

op	$instr_{15:10}$	funct	Type	Compressed Instruction	32-Bit Equivalent
		Zcb: Simple code-saving compressed instructions (some require Zba, Zbb, M, or RV64)			
00	100000	–	CLB	`c.lbu rd', imm(rs1')`	`lbu rd', (ZeroExt(imm)) (rs1')`
00	100001	1	CLH	`c.lh rd', imm(rs1')`	`lh rd', (ZeroExt(imm)*2)(rs1')`
00	100001	0	CLH	`c.lhu rd', imm(rs1')`	`lhu rd', (ZeroExt(imm)*2)(rs1')`
00	100010	–	CSB	`c.sb rs2', imm(rs1')`	`sb rs2', (ZeroExt(imm)) (rs1')`
00	100011	0	CSH	`c.sh rs2', imm(rs1')`	`sh rs2', (ZeroExt(imm)*2)(rs1')`
01	100111	11000	CU	`c.zext.b rd'`	`andi rd', rd', 0xFF`
01	100111	11001	CU	`c.sext.b rd'`	`sext.b rd', rd'`
01	100111	11010	CU	`c.zext.h rd'`	`zext.h rd', rd'`
01	100111	11011	CU	`c.sext.h rd'`	`sext.h rd', rd'`
01	100111	11101	CU	`c.not rd'`	`xori rd', rd', -1`
01	100111	10	CA	`c.mul rd', rs2'`	`mul rd', rd', rs2'`
01	100111	11100	CU	`c.zext.w rd'`	`add.uw rd', rd', zero`
		{C/Zcf}F (RV32 only): Compressed single-precision floating-point loads and stores			
00	011---	–	CL	`c.flw fd', imm(rs1')`	`flw fd', (ZeroExt(imm)*4)(rs1')`
00	111---	–	CS	`c.fsw fs2', imm(rs1')`	`fsw fs2', (ZeroExt(imm)*4)(rs1')`
10	011---	–	CI	`c.flwsp fd, imm`	`flw fd, (ZeroExt(imm)*4)(sp)`
10	111---	–	CSS	`c.fswsp fs2, imm`	`fsw fs2, (ZeroExt(imm)*4)(sp)`
		{C/Zcd}D: Compressed double-precision floating-point loads and stores			
00	001---	–	CL	`c.fld fd', imm(rs1')`	`fld fd', (ZeroExt(imm)*8)(rs1')`
00	101---	–	CS	`c.fsd fs2', imm(rs1')`	`fsd fs2', (ZeroExt(imm)*8)(rs1')`
10	001---	–	CI	`c.fldsp fd, imm`	`fld fd, (ZeroExt(imm)*8)(sp)`
10	101---	–	CSS	`c.fsdsp fs2, imm`	`fsd fs2, (ZeroExt(imm)*8)(sp)`

RISC-V Bit Manipulation Instructions

Table I.23 RVB/Zb*: Bit manipulation instructions

op	funct3	funct7/ imm$_{11:5}$	rs2/ imm$_{4:0}$	Type	Instruction	Description	Operation
Zba: Address Generation							
0110011	010	0010000	rs2	R	sh1add rd,rs1,rs2	shift left by 1 and add	rd = rs2 + (rs1 << 1)
0110011	100	0010000	rs2	R	sh2add rd,rs1,rs2	shift left by 2 and add	rd = rs2 + (rs1 << 2)
0110011	110	0010000	rs2	R	sh3add rd,rs1,rs2	shift left by 3 and add	rd = rs2 + (rs1 << 3)
Zba (RV64 only): Shift and Add 32-bit Unsigned Word (U.W.)							
0111011	010	0010000	rs2	R	sh1add.uw rd,rs1,rs2	shift u.w. left 1, add	rd = rs2 + ZeroExt(rs1$_{31:0}$) << 1
0111011	100	0010000	rs2	R	sh2add.uw rd,rs1,rs2	shift u.w. left 2, add	rd = rs2 + ZeroExt(rs1$_{31:0}$) << 2
0111011	110	0010000	rs2	R	sh3add.uw rd,rs1,rs2	shift u.w. left 3, add	rd = rs2 + ZeroExt(rs1$_{31:0}$) << 3
0111011	000	0000100	rs2	R	add.uw rd,rs1,rs2	add u.w.	rd = rs2 + ZeroExt(rs1$_{31:0}$)
0011011	001	000010*	uimm$_{4:0}$	I	slli.uw rd,rs1,uimm	shift left logical imm. u.w.	rd = ZeroExt(rs1$_{31:0}$) << uimm
Zbb: Basic Bit Manipulation							
0110011	100	0000101	rs2	R	min rd,rs1,rs2	minimum	rd = min(rs1, rs2)
0110011	101	0000101	rs2	R	minu rd,rs1,rs2	unsigned minimum	rd = minu(rs1, rs2)
0110011	110	0000101	rs2	R	max rd,rs1,rs2	maximum	rd = max(rs1, rs2)
0110011	111	0000101	rs2	R	maxu rd,rs1,rs2	unsigned maximum	rd = maxu(rs1, rs2)
0110011	100	0100000	rs2	R	orn rd,rs1,rs2	or inverted operand	rd = rs1 | ~rs2
0110011	111	0100000	rs2	R	andn rd,rs1,rs2	and inverted operand	rd = rs1 & ~rs2
0110011	110	0100000	rs2	R	xnor rd,rs1,rs2	xor inverted operand	rd = rs1 ^ ~rs2
0110011	001	0110000	rs2	R	rol rd,rs1,rs2	rotate left	rd = rol(rs1, rs2$_{4/5:0}$)
0110011	101	0110000	rs2	R	ror rd,rs1,rs2	rotate right	rd = ror(rs1, rs2$_{4/5:0}$)
0010011	101	0010100	00111	I	orc.b rd,rs1	bitwise or, combine on bytes	rd = orc.b(rs1)
011•011	100	0000100	00000	I	zext.h rd,rs1	zero-extend halfword	rd = ZeroExt(rs1$_{15:0}$)
0010011	001	0110000	00000	I	clz rd,rs1	count leading zeros	rd = clz(rs1)
0010011	001	0110000	00010	I	cpop rd,rs1	population count	rd = popcnt(rs1)
0010011	001	0110000	00001	I	ctz rd,rs1	count trailing zeros	rd = clz(rev(rs1))
0010011	001	0110000	00100	I	sext.b rd,rs1	sign-extend byte	rd = SignExt(rs1$_{7:0}$)
0010011	001	0110000	00101	I	sext.h rd,rs1	sign-extend halfword	rd = SignExt(rs1$_{15:0}$)
0010011	101	011010•	11000	I	rev8 rd,rs1	byte-wise reverse	rd = revbyte(rs1)
0010011	101	011000*	uimm$_{4:0}$	I	rori rd,rs1,uimm	rotate right immediate	rd = ror(rs1, uimm)
Zbb (RV64 only): Basic Bit Manipulation on 32-bit Words							
0111011	001	0110000	rs2	R	rolw rd,rs1,rs2	rotate left word	rd = SignExt(rol32(rs1$_{31:0}$, rs2$_{4:0}$))
0111011	101	0110000	rs2	R	rorw rd,rs1,rs2	rotate right word	rd = SignExt(ror32(rs1$_{31:0}$, rs2$_{4:0}$))
0011011	101	0110000	uimm$_{4:0}$	I	roriw rd,rs1,uimm	rotate right immediate word	rd = SignExt(ror32(rs1$_{31:0}$, uimm))
0011011	001	0110000	00000	I	clzw rd,rs1	count leading zeros word	rd = clz(rs1$_{31:0}$)
0011011	001	0110000	00010	I	cpopw rd,rs1	population count word	rd = popcnt(rs1$_{31:0}$)
0011011	001	0110000	00001	I	ctzw rd,rs1	count trailing zeros word	rd = clz(rev(rs1$_{31:0}$))
Zbc: Carry-Less (CL) Multiplication							
0110011	001	0000101	rs2	R	clmul rd,rs1,rs2	cl multiply (lower half)	rd = (rs1 @ rs2)$_{XLEN-1:0}$
0110011	011	0000101	rs2	R	clmulh rd,rs1,rs2	cl multiply (upper half)	rd = (rs1 @ rs2)$_{2*XLEN-1:XLEN}$
0110011	010	0000101	rs2	R	clmulr rd,rs1,rs2	cl multiply reversed	rd = rev((rev(rs1) @ rev(rs2))$_{XLEN-1:0}$)
Zbs: Single-Bit Operations							
0110011	001	0100100	rs2	R	bclr rd,rs1,rs2	bit clear	rd = rs1 & ~(1 << rs2$_{4/5:0}$)
0110011	001	0110100	rs2	R	binv rd,rs1,rs2	bit invert	rd = rs1 ^ (1 << rs2$_{4/5:0}$)
0110011	001	0010100	rs2	R	bset rd,rs1,rs2	bit set	rd = rs1 | (1 << rs2$_{4/5:0}$)
0110011	101	0100100	rs2	R	bext rd,rs1,rs2	bit extract	rd = |(rs1 & (1 << rs2$_{4/5:0}$))
0010011	001	010010*	uimm$_{4:0}$	I	bclri rd,rs1,uimm	bit clear immediate	rd = (rs1 & ~(1 << uimm)
0010011	001	011010*	uimm$_{4:0}$	I	binvi rd,rs1,uimm	bit invert immediate	rd = (rs1 ^ (1 << uimm
0010011	001	001010*	uimm$_{4:0}$	I	bseti rd,rs1,uimm	bit set immediate	rd = (rs1 | (1 << uimm)
0010011	101	010010*	uimm$_{4:0}$	I	bexti rd,rs1,uimm	bit extract immediate	rd = |(rs1 & (1 << uimm))

* = 0 for RV32, uimm$_5$ for RV64.

• = 0 for rv32, 1 for rv64. rev8 and zext.h are versions of more general GREV, PACK, and PACKW instructions that may be provided in a future enhancement of the bit manipulation specification and that use the op and funct7 fields in more general ways.

RISC-V Atomic Memory Operation (AMO) Instructions

Table I.24 RVA: RISC-V AMO instructions

op	funct3	funct7/ $imm_{11:5}$	rs2	Type	Instruction	Description	Operation
0101111	010	00010, aq, rl	00000	I	lr.w rd, (rs1)	load reserved	rd=SignExt([rs1]), reserve [rs1]
0101111	010	00011, aq, rl	rs2	R	sc.w rd, rs2, (rs1)	store conditional	if ([rs1] still reserved) $[rs1]_{31:0}$ = $rs2_{31:0}$, rd = 0 #success else rd = 1 #fail
0101111	010	00001, aq, rl	rs2	R	amoswap.w rd, rs2, (rs1)	amo swap	rd=[rs1], $[rs1]_{31:0}$ = rs2
0101111	010	00000, aq, rl	rs2	R	amoadd.w rd, rs2, (rs1)	amo add	rd=[rs1], $[rs1]_{31:0}$ = [rs1] + rs2
0101111	010	00100, aq, rl	rs2	R	amoxor.w rd, rs2, (rs1)	amo xor	rd=[rs1], $[rs1]_{31:0}$ = [rs1] ^ rs2
0101111	010	01100, aq, rl	rs2	R	amoand.w rd, rs2, (rs1)	amo and	rd=[rs1], $[rs1]_{31:0}$ = [rs1] & rs2
0101111	010	01000, aq, rl	rs2	R	amoor.w rd, rs2, (rs1)	amo or	rd=[rs1], $[rs1]_{31:0}$ = [rs1] | rs2
0101111	010	10000, aq, rl	rs2	R	amomin.w rd, rs2, (rs1)	amo min	rd=[rs1], $[rs1]_{31:0}$ = min ([rs1], rs2)
0101111	010	10100, aq, rl	rs2	R	amomax.w rd, rs2, (rs1)	amo max	rd=[rs1], $[rs1]_{31:0}$ = max ([rs1], rs2)
0101111	010	11000, aq, rl	rs2	R	amominu.w rd, rs2, (rs1)	amo min unsign.	rd=[rs1], $[rs1]_{31:0}$ = minu([rs1], rs2)
0101111	010	11100, aq, rl	rs2	R	amomaxu.w rd, rs2, (rs1)	amo max unsign.	rd=[rs1], $[rs1]_{31:0}$ = maxu([rs1], rs2)

In this table, [rs1] = SignExt($[rs1]_{31:0}$), the sign-extended contents of memory address rs1. RV64A adds .d versions of all these instructions (i.e., lr.d, amoswap.d, etc.), with funct3 = 011; these operate on a 64-bit operand (double word). The aq (acquire) and rl (release) bits indicate additional ordering. When aq = 1, later memory operations in this hart are only seen after AMO. When rl = 1, other RISC-V harts will see all previous memory operations in this hart before AMO. If aq = rl = 1, all later/previous memory operations in this hart must occur after/before AMO.

RISC-V Integer Conditional Instructions

Table I.25 Zicond: Integer conditional instructions

op	funct3	funct7	rs2	Type	Instruction	Description	Operation
0110011	101	0000111	rs2	R	czero.eqz rd, rs1, rs2	conditional zero on equal zero	rd=(rs2 == 0) ? 0 : rs1
0110011	111	0000111	rs2	R	czero.nez rd, rs1, rs2	conditional zero on not equal zero	rd=(rs2 != 0) ? 0 : rs1

RISC-V Additional Floating-Point Instructions

Table I.26 Zfa: Additional RISC-V floating-point instructions

op	funct3	funct7/ $imm_{11:5}$	rs2/ $imm_{4:0}$	Type	Flags NV DZ OF UF NX	Instruction	Description	Operation
1010011	000	11110, fmt	00001	I	– – – – –	fli.* fd, rs1	load immediate	fd = imm[rs1]
1010011	010	00101, fmt	fs2	R	x – – – –	fminm.* fd, fs1, fs2	min	fd = min(fs1, fs2)
1010011	011	00101, fmt	fs2	R	x – – – –	fmaxm.* fd1, fs1, fs2	max	fd = max(fs1, fs2)
1010011	rm	01000, fmt	00100	I	x – – – –	fround.* fd, fs1	round to integer	fd = round(fs)
1010011	rm	01000, fmt	00101	I	x – – – x	froundnx.* fd, fs1	round to integer	fd = round(fs)
1010011	101	10100, fmt	fs2	R	x – – – –	fltq.* rd, fs1, fs2	compare <	rd = (fs1 < fs2)
1010011	100	10100, fmt	fs2	R	x – – – –	fleq.* rd, fs1, fs2	compare ≤	rd = (fs1 ≤ fs2)
						Only RV32D		
1010011	000	1110001	00001	I	– – – – –	fmvh.x.d rd, fs1	move high	rd = fs1[63:31]
1010011	000	1011001	rs2	R	– – – – –	fmvp.d.x fd, rs1, rs2	move pair	fd = {rs2, rs1}
						Only RV64Q		
1010011	000	1110011	00001	I	– – – – –	fmvh.x.q rd, fs1	move high	rd = fs1[127:64]
1010011	000	1011011	rs2	R	– – – – –	fmvp.q.x fd, rs1, rs2	move pair	fd = {rs2, rs1}
						Only D Extension		
1010011	001	1100001	01000	I	x – – – x	fcvtmod.w.d rd, fs1, rtz	modular convert	rd = SignExt(trunc(fs1) mod 2^{32})

imm[rs1] is a floating-point immediate specified by rs1, as shown in Table 16.52. In fcvtmod.w.d, rtz is the rounding mode.

RISC-V Cryptography Instructions

Table I.27 : Zbk*/Zkn*: RISC-V scalar cryptography instructions

op	funct3	funct7/ instr$_{31:25}$	rs2/ instr$_{24:20}$	Type	Instruction	Description	Operation
Zbkb: Bit Manipulation for Cryptography (also includes rol, ror, rori, andn, orn, xnor, rev8 from Zbb, Table I.23)							
0110011	100	0000100	rs2	R	pack rd,rs1,rs2	pack XLEN/2 in XLEN	rd = {$rs2_{XLEN/2-1:0}$, $rs1_{XLEN/2-1:0}$}
0110011	111	0000100	rs2	R	packh rd,rs1,rs2	pack 8 in 16	rd = ZeroExt{$rs2_{7:0}$, $rs1_{7:0}$}
0010011	101	0110100	00111	I	brev8 rd,rs1	bit reverse	rd = reversebitsinbytes(rs1)
Zbkb (RV32 only): Bit Manipulation for Cryptography							
0010011	001	0000100	01111	I	zip rd,rs1	hi/low -> odd/even	rd_{even} = $rs1_{15:0}$; rd_{odd} = $rs1_{31:16}$
0010011	101	0000100	01111	I	unzip rd,rs1	odd/even -> hi/low	$rd_{15:0}$ = $rs1_{odd}$; $rd_{31:16}$ = $rs1_{even}$
Zbkb (RV64 only): Bit Manipulation for Cryptography on 32-bit Words (also includes rolw, rorw, rorrwi, Table I.23)							
0111011	100	0000100	rs2	R	packw rd,rs1,rs2	pack 16 in 32	rd = ZeroExt({$rs2_{15:0}$, $rs1_{15:0}$})
Zbkc: Carry-less Multiply (see clmul and clmulh from Zbc, Table I.23)							
Zbkx: Bit Manipulation for Cryptography							
0110011	010	0010100	rs2	R	xperm4 rd,rs1,rs2	permute nibbles	rd=lookup4(rs1,rs2)
0110011	100	0010100	rs2	R	xperm8 rd,rs1,rs2	permute bytes	rd=lookup8(rs1,rs2)
Zknd (RV32 only): AES Decryption							
0110011	000	{bs, 10101}	rs2	R	aes32dsi rd,rs1,rs2,bs	final round	rd = rs1 ^ (isbox(rs2[bs]) << bs*8)
0110011	000	{bs, 10111}	rs2	R	aes32dsmi rd,rs1,rs2,bs	middle round	rd = rs1 ^ (imixcols(isbox(rs2[bs])) << bs*8)
Zknd (RV64 only): AES Decryption & Inverse Key Schedule							
0110011	000	0011101	rs2	R	aes64ds rd,rs1,rs2	final round	rd = isbox(ishiftrows(rs2,rs1))
0110011	000	0011111	rs2	R	aes64dsm rd,rs1,rs2	middle round	rd = imixcols(isbox(shiftrows(rs2,rs1)))
0010011	001	0011000	00000	I	aes64im rd,rs1	InvMixCols	rd = imixcols(rs1)
Zkne (RV32 only): AES Encryption							
0110011	000	{bs, 10001}	rs2	R	aes32esi rd,rs1,rs2,bs	final round	rd = rs1 ^ (sbox(rs2[bs]) << bs*8)
0110011	000	{bs, 10011}	rs2	R	aes32esmi rd,rs1,rs2,bs	middle round	rd = rs1 ^ (mixcols(sbox(rs2[bs])) << bs*8)
Zkne (RV64 only): AES Encryption							
0110011	000	0011001	rs2	R	aes64es rd,rs1,rs2	final round	rd = sbox(shiftrows(rs2,rs1))
0110011	000	0011011	rs2	R	aes64esm rd,rs1,rs2	middle round	rd = mixcols(sbox(shiftrows(rs2,rs1)))
Zknd and Zkne (RV64 only): AES Key Schedule							
0010011	001	0011000	{1, rnum}	I	aes64ks1i rd,rs1,rnum	key schedule	rd = {2{rcon[rnum] ^ sbox(rot(rs1))}}
0110011	000	0111111	rs2	R	aes64ks2 rd,rs1,rs2	key schedule	w0 = $rs1_{63:32}$ ^ $rs2_{31:0}$; rd = {w0 ^ $rs2_{63:32}$, w0}
Zknh: SHA2-256 Hash*							
0010011	001	0001000	00010	I	sha256sig0 rd,rs1	$\sigma_0^{(256)}$ function	rd = (rs1 ror 7) ^ (rs1 ror 18) ^ (rs1 >> 3)
0010011	001	0001000	00011	I	sha256sig1 rd,rs1	$\sigma_1^{(256)}$ function	rd = (rs1 ror 17) ^ (rs1 ror 19) ^ (rs1 >> 10)
0010011	001	0001000	00000	I	sha256sum0 rd,rs1	$\Sigma_0^{(256)}$ function	rd = (rs1 ror 2) ^ (rs1 ror 13) ^ (rs1 ror 22)
0010011	001	0001000	00001	I	sha256sum1 rd,rs1	$\Sigma_1^{(256)}$ function	rd = (rs1 ror 6) ^ (rs1 ror 11) ^ (rs1 ror 25)
Zknh (RV32 only): SHA2-512 Hash							
0110011	000	0101110	rs2	R	sha512sig0h rd,rs1,rs2	$\sigma_0^{(512)}$ high word	rd = ({rs2,rs1} >> 1)$_{31:0}$ ^ ({rs2,rs1} >> 8)$_{31:0}$ ^ (rs1 >> 7)
0110011	000	0101010	rs2	R	sha512sig0l rd,rs1,rs2	$\sigma_0^{(512)}$ low word	rd = ({rs2,rs1} >> 1)$_{31:0}$ ^ ({rs2,rs1} >> 8)$_{31:0}$ ^ ({rs2,rs1} >> 7)$_{31:0}$
0110011	000	0101111	rs2	R	sha512sig1h rd,rs1,rs2	$\sigma_1^{(512)}$ high word	rd = ({rs2,rs1} >> 29)$_{31:0}$ ^ ({rs2,rs1} >> 19)$_{31:0}$ ^ (rs1 >> 6)
0110011	000	0101011	rs2	R	sha512sig1l rd,rs1,rs2	$\sigma_1^{(512)}$ low word	rd = ({rs2,rs1} >> 29)$_{31:0}$ ^ ({rs2,rs1} >> 19)$_{31:0}$ ^ ({rs2,rs1} >> 6)$_{31:0}$
0110011	000	0101000	rs2	R	sha512sum0r rd,rs1,rs2	$\Sigma_0^{(512)}$ half	rd = ({rs2,rs1} >> 7)$_{31:0}$ ^ ({rs2,rs1} >> 2)$_{31:0}$ ^ ({rs2,rs1} >> 28)$_{31:0}$
0110011	000	0101001	rs2	R	sha512sum1r rd,rs1,rs2	$\Sigma_1^{(512)}$ half	rd = ({rs2,rs1} >> 9)$_{31:0}$ ^ ({rs2,rs1} >> 14)$_{31:0}$ ^ ({rs2,rs1} >> 18)$_{31:0}$
Zknh (RV64 only): SHA2-512 Hash							
0010011	001	0001000	00110	I	sha512sig0 rd,rs1	$\sigma_0^{(512)}$ function	rd = (rs1 ror 1) ^ (rs1 ror 8) ^ (rs1 >> 7)
0010011	001	0001000	00111	I	sha512sig1 rd,rs1	$\sigma_1^{(512)}$ function	rd = (rs1 ror 19) ^ (rs1 ror 61) ^ (rs1 >> 6)
0010011	001	0001000	00100	I	sha512sum0 rd,rs1	$\Sigma_0^{(512)}$ function	rd = (rs1 ror 28) ^ (rs1 ror 34) ^ (rs1 ror 39)
0010011	001	0001000	00101	I	sha512sum1 rd,rs1	$\Sigma_1^{(512)}$ function	rd = (rs1 ror 14) ^ (rs1 ror 18) ^ (rs1 ror 41)

bs = 2-bit byte select. rnum = 4-bit round number. *RV64's SHA2-256 instructions operate on rs1's low 32 bits and sign-extend the final result to 64 bits.